D1382871

Concepts of Classical Optics

Concepts of Classical Optics

....

JOHN STRONG *The Johns Hopkins University*

Illustrations by ROGER HAYWARD

W. H. FREEMAN AND COMPANY

San Francisco 1958

Dedicated to my parents:

FRED STRONG *and*
ANNA LAURA (BENNETT) STRONG

Preface

THIS TEXTBOOK is intended for an intermediate course in optics. It can be used for a course running through either one or two terms. Some familiarity with electricity and magnetism, at the intermediate level, and some knowledge of calculus, vectors, and complex numbers are prerequisite.

In the collection and organization of material for this book certain topics that have been popular in other textbooks have been omitted—mainly topics that the student will encounter in subsequent courses in modern physics and in spectroscopy. And only the very beginning of crystal optics is included.

On the other hand, various short topics of lively research interest appear later as appendices. These are intended to give the student the flavor of current activities and interests in our field. The appendices may be used informally, simply as reference material, or may be incorporated formally by the instructor.

The problems have been devised to serve as work incentives for the serious student. The student who works them out will assure himself of a full understanding of the material covered in the text. And furthermore, many items of optics that are not presented in the main body of the text are revealed in the problems.

The mathematical treatments are largely restricted to the limiting cases where qualitative validity can be retained with little or no quantitative compromise. This restriction facilitates the exposition of the concepts of classical optics with the minimum of concern for mathematical details. The use of complex numbers, and the vector representation of them, further simplify the mathematical expositions.

Many of the problems are designed to be solved by such methods as are used in the text. By these methods, which we may characterize as *the mathematics of modest rigor*, transcendental and irrational functions, when their arguments are much less than unity, are reduced to simple algebraic form by series expansions. Any fluency that the student can acquire in the application of these methods will advance his power in physics generally.

Optics has been a mother of concepts to both experimental and theoretical science. On the side of appreciation, there are many beautiful patterns of color to be seen in optical experiments. On the other side, physical optics has moving mysteries in its theoretical structure. And a practical knowledge of geometrical optics is necessary for effectiveness in any applied science. In short, optics is the most important background science—even if, currently, it appears to have yielded to particle physics in philosophic popularity.

I wish to acknowledge Mrs. Elsa Clark's indispensable help in preparation of the manuscript. I am indebted to Henry Fisk Carlton for teaching me much about writing. And finally, I need especially to acknowledge the generous understanding and vigorous cooperation of my publisher during all the ordeal of getting this manuscript ready and in press.

October 1957   JOHN STRONG

Table of Contents

INTRODUCTION   xvii

CHAPTER I. Light as Wave Motion   1

    1-1. Huygens' Ideas   2
    1-2. Newton's Ideas   8
    1-3. Complex Numbers   10
    1-4. Simple Harmonic Motion   11
    1-5. Polarized Waves in a Stretched String   16
    1-6. Velocities of Mechanical Waves   20

CHAPTER II. Superposition of Wave Motions   22

    2-1. Sum of Two Cosine Functions of the Same Frequency   23
    2-2. Standing Waves   24
    2-3. Graphical Summation of Wave Motions   25
    2-4. Superposition of Polarized Wave Motions   27
    2-5. General Procedures of Superposition   28
    2-6. Single-slit Diffraction   29
    2-7. Superposition of Waves with Random Phases   31
    2-8. Superposition of Waves of Different Frequencies   34
    2-9. Transients   34
    2-10. An Apparent Failure of the Principle of Superposition 35

CHAPTER III. Electromagnetic Waves   38

    3-1. History of Electromagnetic Waves   40
    3-2. Maxwell's Equations   42
    3-3. Maxwell's Displacement Current   44
    3-4. Derivation of the Differential Wave Equation   45
    3-5. Nature of Electromagnetic Waves   46

3-6. Light Waves in Glass   48

3-7. Boundary Conditions   48

3-8. Fresnel Reflection Coefficients   50

3-9. Conventions of Sign   52

3-10. Photometric Magnitudes   54

3-11. Interrelation of Photometric Units   57

3-12. Energy Density of Radiation   58

CHAPTER **IV. Interaction of Light and Matter**   61

4-1. Harmonic Oscillator   63

4-2. Electrical Polarization and N 64

4-3. Optical Properties of Sodium Vapor   66

4-4. Optical Properties of Glass   67

4-5. Formula of Clausius and Mossotti   68

4-6. Cauchy's Formula   68

4-7. Optical Properties of Metals at Low Frequencies   69

4-8. Optical Properties of Metals for Light   71

4-9. Dipole Emission   73

4-10. Radiation Pressure   75

4-11. Blackbody Radiation Laws   76

4-12. Total Radiation   78

CHAPTER **V. Velocities and Scattering of Light**   80

5-1. Spectral Lines in Absorption and Emission   83

5-2. Natural Line Width   85

5-3. Doppler Line Width   86

5-4. Pressure Broadening   88

5-5. Modulation Broadening   90

5-6. Observed Group Velocity   92

5-7. Explanation of Group Velocity   93

5-8. Phase Velocity in Sodium Vapor   96

5-9. Scattering   100

5-10. Scattering by Sodium Vapor   102

5-11. Sky Light   104

CHAPTER **VI. Polarized Light and Dielectric Boundaries**   105

6-1. Polarization in Nature   107

6-2. Polarizers and Analyzers   110

6-3. Law of Malus   113
6-4. Reflection of Polarized Light   114
6-5. Total Reflection   117
6-6. Circularly and Elliptically Polarized Light   119
6-7. Transmission of Polarized Light   120
6-8. Transmission of a Pile of Plates   122
6-9. Frustrated Total Reflection   124

CHAPTER **VII. Double Refraction—Calcite and Quartz**   127

7-1. Calcite Double Refraction   128
7-2. Spheroidal Huygens Wavelets   129
7-3. Calculation of r_{ext} for a Calcite Rhomb   130
7-4. Refraction by Two Calcite Rhombs in Series   131
7-5. Explanation of the e-Wave Surface   133
7-6. Various Refractions of the e-Ray by Calcite   135
7-7. Uses of Calcite   139
7-8. Analysis of Polarized Light   144
7-9. Uses of the Half-wave Plate   146
7-10. Calcite and Quartz Contrasted   149
7-11. Optical Activity   152
7-12. Induced Double Refraction   154

CHAPTER **VIII. Interference of Two Sources Laterally Separated**   157

8-1. Coherence of Light   158
8-2. Young's Experiment   160
8-3. Analysis of Two-beam Interference   161
8-4. Other Two-beam Interference Experiments   165
8-5. Visibility of Interference Bands   168
8-6. White Light Interference Bands   171
8-7. Spectral Analysis with Two-beam Interference   172
8-8. Stellar Interferometers   173
8-9. Interference of Polarized Light   178

CHAPTER **IX. Fresnel Diffraction**   180

9-1. Kirchhoff's Differential and the Case of No Diffraction   181
9-2. Circular Obstructions and Apertures   186
9-3. Zone Plates   187

9-4. Diffraction Beyond a Rectangular Aperture . 190

9-5. Cornu's Spiral 193

9-6. No Diffraction, by Cornu's Spiral 196

9-7. The Straight Edge 197

9-8. Fresnel Diffraction by Slit and Strip 198

9-9. Babinet's Principle 200

CHAPTER **X. Fraunhofer Diffraction** 201

10-1. The Rectangular Aperture 201

10-2. Square and Circular Apertures Contrasted . 203

10-3. Interference and Diffraction Combined 207

10-4. Diffraction-limited Resolving Power 210

10-5. Spectroscopic Resolving Power 212

10-6. Diffraction Gratings 216

10-7. Images of Coherent Sources 219

CHAPTER **XI. Coherent Sources Separated in Depth** 222

11-1. Thin Dielectric Plate 223

11-2. Fizeau Bands 227

11-3. Haidinger Bands 231

11-4. FECO Bands and the Channeled Spectrum 232

11-5. Michelson Interferometer for Monochromatic Light 235

11-6. Michelson Interferometer for White Light 237

11-7. Michelson's Measurement of the Meter 238

11-8. Multiple Sources Separated in Depth 242

11-9. Dielectric Films on Glass 244

11-10. Metallic Films 245

CHAPTER **XII. Applications of Physical Optics** 247

12-1. Reflection-reducing Overcoats 248

12-2. Multiple Overcoats 252

12-3. Interference Filters 253

12-4. Low-order Multiple-beam Interferometry 255

12-5. High-order Multiple-beam Spectroscopy 257

12-6. Free Spectral Range 259

12-7. Fabry-Perot Resolving Power 260

12-8. The Transmission and Reflection Echelons 264

12-9. The Lummer-Gehrcke Plate   268
12-10. Energy-limited Resolving Power   270

CHAPTER **XIII. Images of Points by Single Surfaces**   273

13-1. General Considerations   274
13-2. Optical Surfaces—Polish   278
13-3. Optical Surfaces—Geometry   280
13-4. Longitudinal Equations for a Single Surface   282
13-5. The Gauss Equation   287
13-6. Newton's Equation   288
13-7. Nodal Equation   289
13-8. Mirrors   290
13-9. Parabolic Telescope Mirror   291
13-10. Foucault's Knife-edge Test   294
13-11. Gaviola's Caustic Test   298

CHAPTER **XIV. Images of Points by Systems of Surfaces**   301

14-1. Plane Parallel Plate, and Prisms   301
14-2. Longitudinal Gaussian Equations   306
14-3. Longitudinal Newtonian Equations   311
14-4. Immersion Lenses   312
14-5. Freedom from Spherical Aberration   314
14-6. Spherical Aberration of a Lens   315
14-7. Longitudinal Chromatic Aberration   319
14-8. Optical Materials   323

CHAPTER **XV. Magnification, Aperture, and Field**   326

15-1. Magnification by a Single Surface   328
15-2. Extended Sine Relationship   329
15-3. Magnification by a System of Surfaces   330
15-4. Aplanatic Points   332
15-5. Magnifiers and Microscopes   336
15-6. Telescopes   338
15-7. Aperture   341
15-8. Depth of Field   343
15-9. Field   344
15-10. Telecentric Systems   348

15-11. Field Lens   349

CHAPTER **XVI. Image Defects**   351

16-1. Third-order Theory   352
16-2. Image Testing   353
16-3. Spherical Aberration, Coma, and Achromatism   355
16-4. Coma of a Parabolic Mirror   357
16-5. Astigmatism of a Single Surface   362
16-6. Coddington's Equations for a Thin Lens   365
16-7. Curvature of Field   366
16-8. Distortion   368
16-9. Optical Systems   368

APPENDIX **A**

Applications of Interferometry   373
by W. Ewart Williams

APPENDIX **B**

Interferometers   377
by J. Dyson

APPENDIX **C**

The Kösters Double-image Prism   393
by J. B. Saunders

APPENDIX **D**

Interferometry with Savart's Plate   400
by A. C. S. van Heel

APPENDIX **E**

Apodization   410
by Pierre Jacquinot

APPENDIX **F**

Application of Fourier Transformations in Optics: Interferometric Spectroscopy   419
by George A. Vanasse and John Strong

APPENDIX G

Some Modern Concepts of Light   435
by L. Witten

APPENDIX H

The Speed of Light   450
by C. Harvey Palmer, Jr.

APPENDIX I

Radiation Detectors and Measuring Devices   468
by Harold W. Yates

APPENDIX J

Microwave Experiments and Their Optical Analogues   507
by Gordon Ferrie Hull, Jr.

APPENDIX K

The Wave Theory of Microscopic Image Formation   525
by F. Zernike

APPENDIX L

Modern Trends in Methods of Lens Design   537
by M. Herzberger

APPENDIX M

Graphical Ray Tracing   544
by E. W. Silvertooth

APPENDIX N

Fiber Optics   553
by Narinder S. Kapany

APPENDIX O

Optical Filters   580
by Robert G. Greenler

APPENDIX **P**

Diffraction Gratings   597
by R. P. Madden and John Strong

APPENDIX **Q**

Mathematical Review   616
by Trevor Williams

PROBLEMS   657

INDEX   683

Introduction

"THE MORE I dive into this matter of [optics] and push my researches up to the very spring head of it, so much the more am I impressed with its great honorableness and antiquity; and especially when I find so many great demi-gods and heroes, prophets of all sorts, who one way or another have shed distinction upon it, I am transported with the reflection that I myself belong, though but subordinately, to so emblazoned a fraternity." (From Chap. LXXXII of Melville's *Moby-Dick*.)

Thus goes Ishmael's testimony of enthusiasm for whaling, paraphrased to apply to our science.

A short history of optics appears below to afford us at least a minimum setting for subsequent expositions. Here the names of our heroes, and the incidents by which they gave our science its honorableness, can only briefly be mentioned.

1. Optics in Ancient Times

If the ancients failed to reach the heights in optics, the most sophisticated of their sciences, that they reached in art, in literature, and in law, it was not because of any lack of talent of the type that is effective in science today; rather it was because then man did not have the scientific method. The scientific method did not exist until much later, notably after Bacon and Galileo had lived. By virtue of our inheritance of it and of the experimental revelations that that method accredits, we now enjoy transcendent appreciations of light and color. Because of the scientific method we ourselves may be able to embellish the present structure of optical science with contributions that can last throughout all time to come, even though we may not have the intellectual power of those great Greeks, and even if we live in an age too late to participate in the establishment of basic concepts.

In its ancient beginnings, optics as a science consisted mainly of the law of reflection, enunciated by Euclid (300 B.C.), and a first approximation to the law of refraction, as expressed by Ptolemy of Alexandria (70–147 A.D.).

Ptolemy measured the angles of incidence and refraction of light rays at air-water, air-glass, and water-glass boundaries. From his measurements (unusual at that time of experimental sterility) Ptolemy stated that, for a given boundary, the ratio of his measured angles was a constant. We know now that his statement is only an approximation, and that it is rather the ratio of the sines of his angles which is constant.

Although ancient science was weak, from the lack of experiments, the ancients knew empirical optical properties of crystal spheres, such as were used as magnifiers by engravers to facilitate their art, as well as properties of plane and curved metal mirrors. Aristotle (384–322 B.C.) recognized that rainbow colors were due to droplets of water. Also, Seneca of Rome (from 4 B.C. to 65 A.D.) recorded the colors that are produced from white light when it penetrates a glass prism, leaving it for Newton to show that a second prism can be used to recombine those colors and thus reconstruct white light from them.

In ancient times the Platonists (Plato, 427–347 B.C.) speculated that vision was due to a "divine fire," a stream of particles emitted by the eye, and that the particles, after combination with solar rays at the object seen, returned to the eye to give it its perceptions. There is, indeed, in a very real intuitive sense, something projected out and away from the eye; that something, however, is not a stream of particles but a projection of geometry. No doubt the ancient "divine fire" or tactile theory of vision satisfied a compulsive need to believe that in some way the eye must touch the objects around us that it helps the mind to perceive so nicely. This feeler-emission explanation retained some repute through many centuries—until the Arabian physicist Alhazen (965–1020) finally effectively banished it.

Aristotle's contrasting proposal was that light was an activity in a medium.

Various forms of these opposing views have oscillated. The pure or modified idea of Platonic *particles* dominated over Aristotle's proposed *activity* until Hooke (1635–1703) as well as Huygens (1629–1695) brought forth *activity* ideas anew. However, Newton's (1642–1727) influence brought the particle theory to dominance soon thereafter. It was sustained until Young (1773–1829) and Fresnel (1788–1827) established the wave theory, confirmed by Foucault (1819–1868). Now, finally, in our present century, it has become necessary again to ascribe to light emanations some of the properties of particles as well as waves.

2. From the Dark Ages to the Renaissance

In spite of the fact that the first organization of optics in ancient times was an effective start, all scientific progress vanished in Europe with the beginnings of the Dark Ages. This period extended roughly from the sack of Alexandria (389 A.D.) to the thirteenth century—after Arabian political power declined, the European cultural heritage was recovered from the Arabs. The Dark Ages began to end in Europe as men began to give priority to the direct knowledge revealed by experiment, as contrasted to the sophistry of pedants, based on established writing. Precursors of the coming rebirth of science appeared in the writings of Friar Bacon (1214–1294). This was also the time of St. Thomas Aquinas (1227–1274).

3. *The Seventeenth Century*

Although a full history of the revitalization of our science would mention the writings of Leonardo da Vinci (1452–1519) and Giambattista della Porta (1538–1615), and historical incidents of early times, the events of the rebirth of optics are largely a recapitulation of the works and discoveries of but a few personalities: Kepler (1571–1630), Snell (1591–1626), Grimaldi (1618–1663), Bartholinus (1625–1698), Huygens (1629–1695), Roemer (1644–1710), and, above all, Galileo (1564–1642). Their works and discoveries aroused the wide interest that never thereafter waned. The discoveries in the beginning years of the seventeenth century in astronomy were particularly significant in this regard: the rotation of the sun as revealed by sunspots, the mountains on the moon, the phases of Venus, the satellites of Jupiter, the stars in the Milky Way, the rings of Saturn and satellites of Saturn. Except for Cassini's (1625–1712) discovery of the satellites of Saturn, and Huygens' participation in the clarification of observations of Saturn's rings, all these discoveries were made by Galileo.

4. *The Eighteenth Century*

The eighteenth century was introduced by Newton's publication of his *Optics* (in 1704). Nothing can be said about this book that is both brief and adequate. The student should read Einstein's *Foreword*, Whittaker's *Introduction*, Cohen's *Preface*, and Roller's *Analytical Table of Contents*, as well as Newton's own advertisements of the first and subsequent editions of this imagination-moving masterpiece; these writings are available, all together, in a Dover publication.

The eighteenth century is notable also for an optical discovery born of effort to terminate controversy that had been started by Galileo—the controversy between proponents of the Copernican and of the Ptolemaic concepts of the planetary system. Although the discovery of Jupiter's satellites and of the gibbous phase of Venus of the previous century had played a large role to effect acceptance of the heliocentric system of Copernicus, an important expectation as a consequence of the theory of Copernicus (1473–1543), not required by the older system of Ptolemy (*loc. cit.*), and not yet observed, was an annual parallax of the near stars as seen against the "backdrop" of those farthest away. If the Copernican idea was the true one, it was argued, astronomers should observe an annual parallax produced by the earth's motion around the sun. Indeed, Bradley (1693–1762) was searching for just this required annual parallax when he made the century's most fruitful optical discovery. Instead of this parallax, he discovered (in 1728) a parallax-like annular motion, but it was arrayed in the wrong direction. He explained this manifest annular motion (the same for all stars) as the aberration of light. Bradley was thus able to confirm Roemer's as yet unaccepted velocity, which had been derived from the irregularity in eclipses of Jupiter's moons, measured a half century earlier. Bradley never found the annular stellar parallax he had sought because of its extreme smallness. That parallax was not a reality until Bessel (1784–1846) found it in 1838. Even the nearest star, Proxima, at 4.3 light-years distance, has an apparent annual movement among the fixed stars of less than one second of arc.

The most important optical invention of the eighteenth century was the achromatic lens; Hall's (1703–1771) unpublished realization of it was in 1730, and Dolland's (1706–1761) patent was in 1757.

5. *The Nineteenth Century*

The unequivocal establishment of the wave nature of light came with the beginning of the next century. The experiments that established it, by Young in Scotland and Fresnel in France, accelerated the development of our science to its liveliest pace yet. And soon afterward the new light waves were established as being transverse in character.

Before the century was half expended, manifold new experimental results had been consolidated in beautiful theories; and, for example, the science of crystal optics had been established.

At mid-century Newton's persistently accepted and long established concept of light as corpuscles was finally contravened by Foucault's measurements of the velocity of light in both air and water. (It was his doctor's thesis.) Foucault's measurements gave final support to the critics of the corpuscular concept—critics that included the mathematician Euler (1707–1783) and our own philosopher-statesman, Benjamin Franklin (1706–1790).

It was during these mid-century times that Faraday's (1791–1867) discovery of the connection between light and magnetism became, as Sommerfeld (1868–1951) has put it, "an impressively strong hint of the electromagnetic nature of light." And, indeed, Faraday himself sensed the meaning of that hint.

The nineteenth century saw a continuing development of theory, generally following, but often transcending, the requirements of experiment. Those developments had their culmination in Maxwell's (1831–1879) equations for electromagnetic radiation and in the Lorentz (1853–1928) electron theory of matter.

Experimentally, the nineteenth century saw Hertz (1857–1894) generate electromagnetic waves, and discover the photoelectric effect.

6. Our Own Times

Hertz's experimental discoveries of the waves predicted by Maxwell tightened the coherence of our science; however, a by-product of this discovery led to disruptive perplexities—the perplexities that came from further study of the photoelectric effect, which Hertz also discovered.

In addition, other perplexities arose: for example, the disparity between the predictions of thermodynamics and the measurements of heat radiation. New spectroscopic observations created other mysteries. And, finally, the negative results of the experiments of Michelson (1852–1931) and Morley (1838–1923) created still further theoretical frustrations.

But with the coming of our own century, these difficulties and frustrations were resolved finally by means of the quantum ideas of Planck (1858–1947), Bohr (1885–   ), and Einstein (1879–1955), and by Einstein's concept of relativity.

7. The Future

What, then, shall we, or our children, expect with the beginning of the next century? Do we now see all the fundamental structure of our science fully erected, or will Ishmael's maritime philosophizing, quoted below, again apply?

"Hardly have we mortals by long toilings extracted from the world's vast bulk its small but valuable sperm; and then, with weary patience, cleansed ourselves of its defilements, and learned to live here in clean tabernacles of the soul; hardly is this done, when—*There she blows!*—the ghost is spouted up, and away we sail to fight some other world, and go through young life's old routine again." (From Chap. XCVIII of Melville's *Moby-Dick*.)

Chapter I

Light as Wave Motion

SOME OF THE philosophers of the seventeenth century, in thinking on what must come to the eye from a luminous body to excite the sensation of light and color, pictured light as a wave motion.

Robert Hooke was a notable proponent of this picture. But it was one thing for him to have the idea that light was a wave motion, and quite another to give that nascent idea sufficient substance to make it a significant contribution to physics.

It remained for Christian Huygens to come forth with sufficiently specific ideas to establish the wave concept of light as an hypothesis subject to validification. Huygens' ideas were generalizations of his notions about a mechanism by which a wave motion would propagate or advance on the surface of water; a notion which was inspired, no doubt, as he watched and contemplated expanding ripples as they propagated across the surface of a Dutch canal. Huygens generalized his notions of the explanation of the progress of a two-dimensional wave front on a water surface, to the three-dimensional medium in which he imagined light propagated, which he called *ether*.

Huygens pictured a mechanism by which a later form of a wave front could come out of an earlier or preceding form; and with his mechanism he could explain the rectilinear propagation of light. In particular, he advanced an argument to show how light could be a wave motion and at the same time not travel around and behind obstacles, as water waves and sound waves do. And, in addition, Huygens' mechanism explained the law of reflection of light, as at a mirror surface, and the law of refraction, as at the surface of a lens.

After these early successes of Huygens, and even after later interference experiments of Young and Fresnel in the beginning 1800's, a strong case indeed was made for explaining light as wave motion; but it was not clear that the wave motions implied were transverse, not longitudinal. From

1

Huygens' time until the middle of the nineteenth century, and even after the experiments of Young and Fresnel, the wave concept was not generally accepted. Sir Isaac Newton was the most notable among those who greatly influenced the sustained opposition to it. Newton believed that light was some kind of corpuscular emanation.

We shall see below how both Newton's notions of light as corpuscles, as well as Huygens' concept of light as a wave motion, equally explain the law of reflection, as at a mirror surface, and the law of refraction, as at the surface of a lens. But between these equal explanations there was a great difference. Newton's notions required the corpuscles to propagate faster in glass or water than in air. In contrast Huygens' ideas required just the reverse, that the waves propagate less quickly through glass or water than through an equal path in air. It remained in 1850 for Foucault to measure the relative velocities of light both through air and through water; and the result of his experiment was decisively in favor of Huygens' wave ideas—he got an inferior velocity in water.

But opinions do not always quickly accommodate themselves to new facts. Such was the case here. This philosophical inertia is apparent in a report on the opinion of 1853 that appeared in the October issue of *The Scientific American* of that year:

"There are in vogue two theories by which the phenomena of light are explained, the one that of Descartes, Huygens and Euler, commonly called the undulatory theory, the other that of Newton and Brewster, known as the theory of emanations. Both are unsatisfactory in certain respects. The advocates of the undulatory theory maintain that light is in all respects similar to sound, and the colors are compared to the notes of an octave. But, carrying out the parallel with sound, what would be the result if an immense multitude assembled together were each at the same time to shout with a different cry? Would a listener be able to hear distinctly the voice of any one? Most certainly not; yet gazing among the myriad orbs which spangle the starry vault, the eye can readily single out the smallest, whose light is sufficient to affect it, and contemplate it, untroubled by the light of the more powerful luminaries shining in other parts of the heavens."

1-1. *Huygens' Ideas*

Mach sets forth Huygens' ideas in an unimprovable manner in his classic, *The Principles of Physical Optics*. We can do no better than quote him:†

---

† From Ernst Mach, *The Principles of Physical Optics*. Reprinted with the permission of Dover Publications, Inc., New York. Mach gives an excellent account of the history of the concepts of optics. As he puts it in his Preface, "I hope that I have laid bare, not without success, the origin of the general concepts of optics and the historical threads in their development, extricated from metaphysical ballast."

"The medium in which light is propagated cannot be the air, since light passes through a Torricellian vacuum, which is impermeable to sound. It was more likely to be a medium which can readily permeate all matter, consisting also of elastic particles able to impart their impulse one to another. Their elasticity, according to Huygens, was afforded by their being composed of still smaller elastic particles, and so on, a conception which appeared quite plausible to Huygens. He illustrated the transference of impulse by a row of contiguous elastic spheres, for which a velocity imparted to the first is transferred in a short, but finite, time to the last, all the intermediate spheres remaining at rest. Such impulses may traverse the series in opposite directions simultaneously, and cross one another.

"The conception of a regular sequence of equidistant waves sent out by the source of luminosity was alien to Huygens; in fact he explicitly refuted the idea of a periodicity analogous to that of sound waves. His conception of light waves was rather of an irregular succession of isolated pulses, whose effects only become noticeable when several of the weak individual impulses coming from different centres add up or unite to form a stronger wave. Impulses such as these are imparted not only in one direction along a straight line, but to all the particles in contact with the pulsating one. Thus in general the spreading out occurs simultaneously on all sides spherically.

"These conceptions give a method of derivation of the *optically effective* waves from the 'elementary waves'; this has been designated the *Huygens principle*."

We can best understand Huygens' principle by applying it to a spherical wave front expanding about a source point as it propagates through the ether. At some prescribed instant of time we consider that this parent spherical wave front instantly disappears, and in its stead a myriad of daughter wavelets appear, one from each elementary surface area of the parent wave front surface. These daughter disturbances themselves expand as spherical waves in the ether. According to Huygens' principle the disturbances that the adjacent daughter wavelets produce are only manifest as a resultant disturbance on their common forward envelope. Elsewhere, not on that envelope, the disturbances produced by isolated daughter wavelets are to be ignored. Thus Huygens imagined two transformations as continuously taking place: A fission transformation of a parent wave front into a myriad of daughter wavelets; and, inversely, a fusion transformation of these wavelet motions along the common envelope, at a later time, back again into a subsequent single wave front.

The bow of a small boat passing along a canal or pond produces a succession of disturbances, or Huygens' wavelets, which thus combined produce its V-shaped wake. In the case of this boat the legs of the V on the water surface

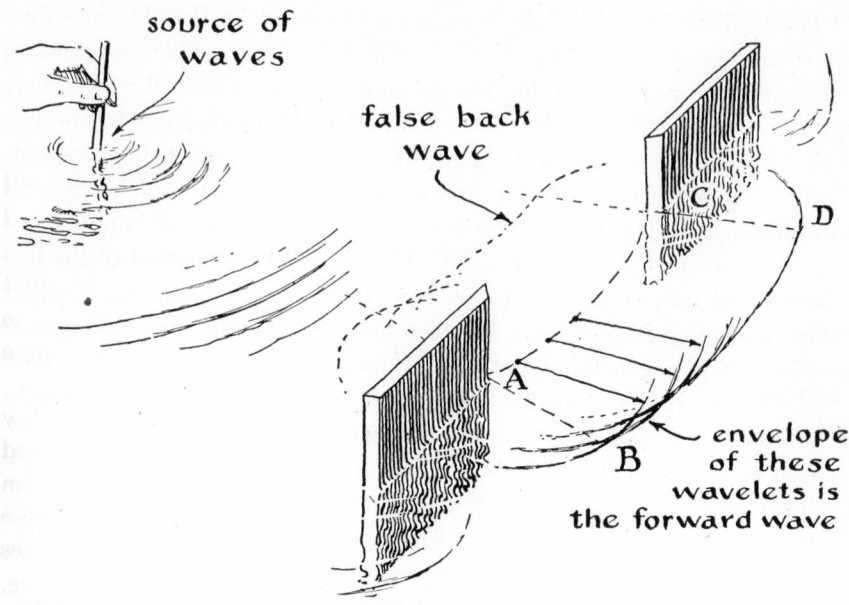

FIG. 1-1 Huygens wavelet envelope, illustrated by waves on a surface of water.

represent the common surface envelope of the disturbances recently produced. In the case of light Huygens conceived that the individual wavelets which do not lie on such a common envelope, and are not thus reinforced by neighbor wavelets, were *too feeble to be seen*.

Fig. 1-1 illustrates an application of Huygens' common-tangent construction. Here a surface wave front has struck an aperture in an opaque obstruction, and the figure illustrates the conceived daughter wavelets that were then formed. Beyond the aperture their common tangent extends only from D to B. Similarly, in the case of light, an observing screen placed beyond an aperture is predicted to show full light within the silhouette of that aperture, with a sudden transition to darkness at its border. However, when such a silhouette is examined for its fine details it is found to be bordered by Fresnel's diffraction fringes. Therefore these predictions, based on Huygens' construction, cannot be closely relied upon. They are correct, however, in first approximation, as everyone knows from seeing the projected pattern of a window when the sunlight falls on an opposite wall through it. Huygens' common-tangent constructions were matured by the physicist Fresnel; and his theoretical method of summing Huygens' wavelets gives a more successful description of such a silhouette border, including its fringes. As we shall see, Fresnel's mathematical methods were further refined by Kirchhoff. In Kirchhoff's modification we have a theory which gives enormously more detail than Huygens' simple graphical construction of the common-tangent, or Fresnel's more sophisticated method of wavelet summation. But even with these refinements the theory is still not completely true, as a determined comparison of its predictions with careful observations will show. We shall concern ourselves later extensively with both the Fresnel and Kirchhoff refinements over Huygens' common-tangent construction; but for now, let us see how his primitive graphical construction gave substance to the wave concept of light, as first proposed by Hooke.

Fig. 1-1 illustrated Huygens' demonstration of straight-line propagation on a water surface by means of a common-line envelope. Fig. 1-2 shows the construction generalized to the case of light, by means of a common surface envelope. The envelope, in each case, extends only so far as adjacent wavelets intersect on a common line or surface. In so far as we take it that the wavelets which are not tangent to this common envelope are "too feeble to produce light," this construction predicts darkness beyond the straight-line projections of the aperture; in other words, Huygens' construction predicts the

---

FIG. 1-2   Huygens wavelet envelope as applied to light waves in an extended medium.

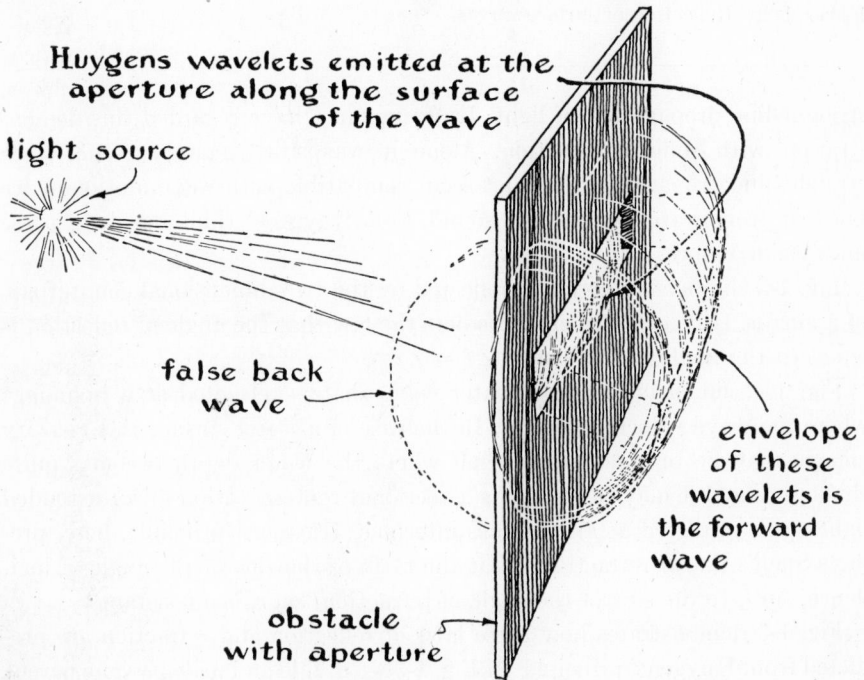

Huygens wavelets emitted at the
aperture along the surface
of the wave

light source

false back
wave

envelope
of these
wavelets is
the forward
wave

obstacle
with aperture

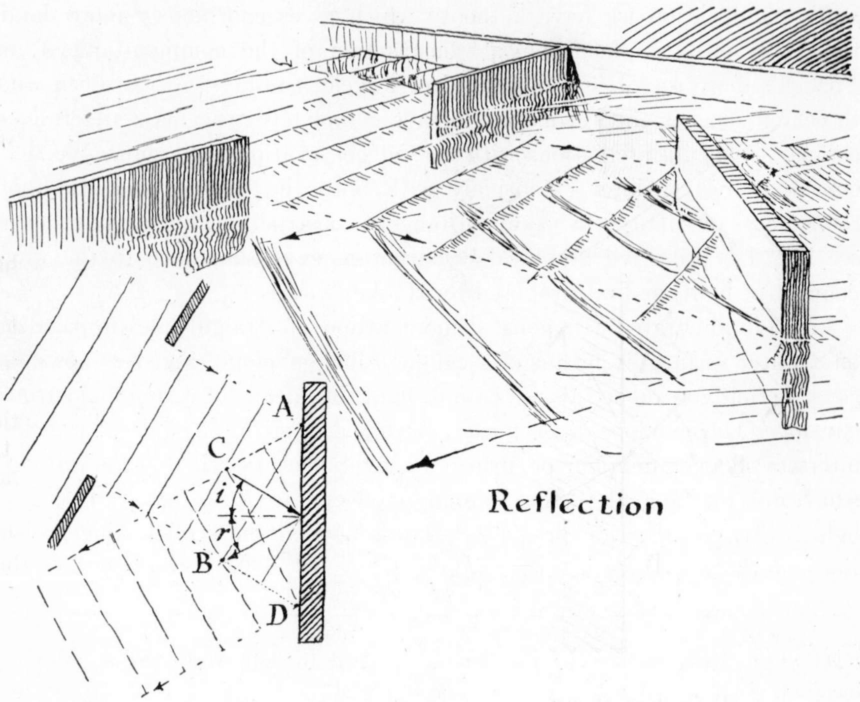

FIG. 1-3 Reflection of surface waves.

straight-line propagation of light. Huygens must have regarded this demonstration with some satisfaction. Alone it was a satisfactory result since straight-line propagation does not seem compatible with wave motion as we know it from water waves and sound. But Huygens' thinking along these lines yielded more.

Fig. 1-3 shows a water wave, reflected by the two-dimensional counterpart of a mirror. Huygens' principle predicts the law that the angle of reflection is equal to the angle of incidence: $\angle i = \angle r$.

Fig. 1-4, similarly, shows a water wave that is refracted at a boundary where the wave velocity changes. In the case of a water surface this velocity change may be produced by a shelf where the water depth becomes quite shallow. Fig. 1-4 may be taken as a sectional representation of an extended light wave refracted at an air-glass interface. Huygens' principle, here, predicts Snell's law of refraction—that the ratio of the sine of the angle of incidence, $\sin i$, to the sine of the angle of refraction, $\sin r$, is a constant.

Fig. 1-5 demonstrates how these laws of reflection and refraction are predicted from Huygens' principle. In Fig. 1-5 we divide an incident plane parent

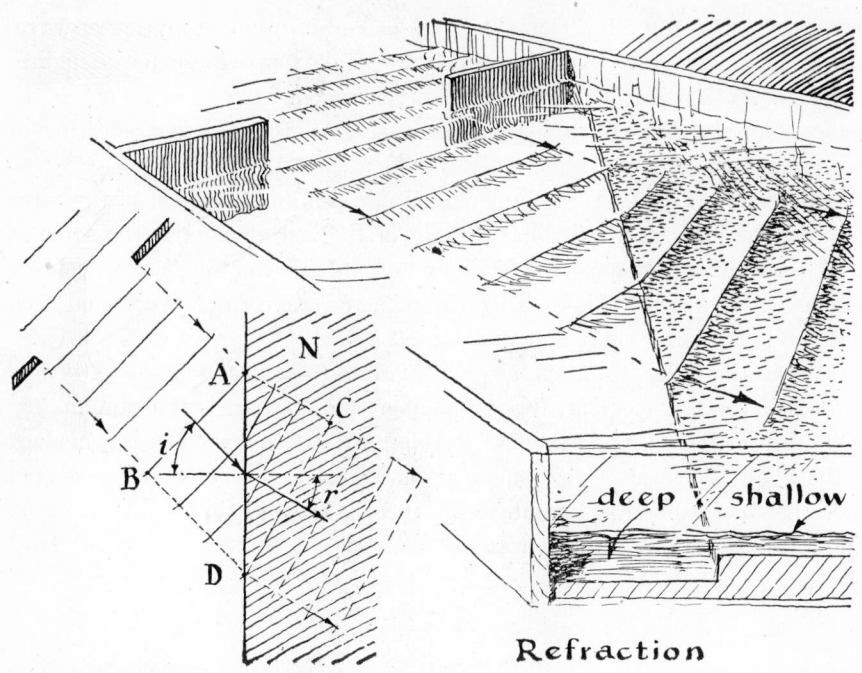

FIG. 1-4   Refraction of surface waves.

FIG. 1-5   Huygens' construction applied to reflection (a) and refraction (b).

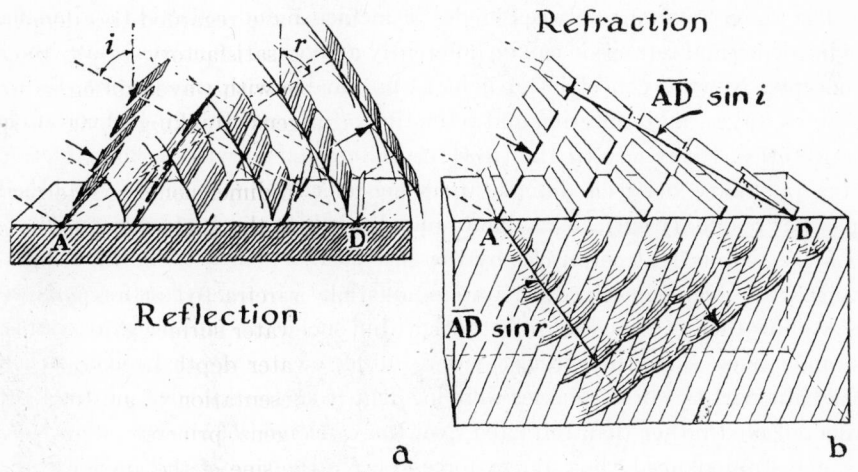

wave front into only a few elements, and instead of taking a myriad of wave-lets as being generated simultaneously, we take only these few daughter wavelets. We consider these daughter wavelets generated successively as the elements of the parent wave front successively arrive at the reflecting or refracting surface.

In reflection, Fig. 1-5a, the interval of time from first contact of the wave front on the glass until the final contact of the last element is the same as the interval from generation of the first wavelet until the final wavelet is generated. Because of this equality the reflected wave front is oriented such that $AD \sin i = AD \sin r$, and $\angle r = \angle i$.

In refraction, Fig. 1-5b, the interval of time from first contact until final contact of the last element, for the incident wave front, is the same as the interval from generation of the first wavelet until the last wavelet is generated in the glass. Because of this equality the refracted wave front is oriented such that corresponding distances in the figure, $AD \sin i$, and $AD \sin r$, after they are divided by appropriate velocities, v_0 and v_g, are equal. Thus

$$\frac{AD \sin i}{v_0} = \frac{AD \sin r}{v_g} \qquad \text{or} \qquad \frac{\sin i}{\sin r} = \frac{v_0}{v_g} = N$$

Huygens' construction requires $v_0 > v_g$ since experience shows $i > r$. The ratio of velocities is the index of refraction, N.

1-2. *Newton's Ideas*

The above relations between angles of incidence and reflection and angles of incidence and refraction can be differently arrived at by means of Newton's concepts. Newton conceived of light as an emanation of corpuscles, rather than as waves, and reflection and refraction as altered straight-line corpuscle trajectories. He conceived the trajectory alterations occurring on refraction at a boundary between different homogeneous media, such as a polished glass-air boundary surface, as due to unbalanced short-range forces of *attraction* exerted on light corpuscles by the glass near its surface. As these short-range forces were conceived, they were much like van der Waals forces as we now know them. Corpuscle trajectories within such a medium were, of course, straight lines, since the corpuscles were subjected to balanced forces. When the light particles or corpuscles were incident from the outside, and the light was reflected rather than refracted, Newton conceived of an opposite action, namely, unbalanced short-range forces of *repulsion*. Newton avoided the difficulty inherent in thus explaining simultaneous reflection and refraction, at external incidence, by his proposal that the glass periodically assumed "fits of repulsion, and fits of attraction."

In either case the resultant force acting on one of Newton's corpuscles was always directed normally to the air-glass surface because, in an isotropic material like glass, there would be no asymmetry to give this force a component parallel to the glass surface. Since a force directed normally to the reflecting or refracting glass surface could not change the parallel component of corpuscle velocity, Newton's unbalanced forces only altered the perpendicular velocity component.

In reflection, the perpendicular component of corpuscular velocity is simply reversed. On external incidence, the reversal at reflection required the repulsive forces between glass and corpuscle. On internal incidence this reversal required the attractive forces. Fig. 1-6a shows the y-component of velocity reversed by surface repulsive forces, with the x-component unchanged. Thus, on reflection, the angle of reflection is predicted to be equal to the angle of incidence.

As for refraction, the energy of an externally incident particle falling under the influence of the short-range attractive force near the glass surface would be expected to increase by the amount $\int \mathfrak{F}_y \, dy$, where $\mathfrak{F}_y$ is the perpendicular component of unbalanced force. This integral is, of course, independent of the angle of incidence. Fig. 1-6b shows the perpendicular y-component of corpuscle velocity increased by Newton's postulated short-range vertical forces, the x-component of velocity being unchanged. Expressed quanti-

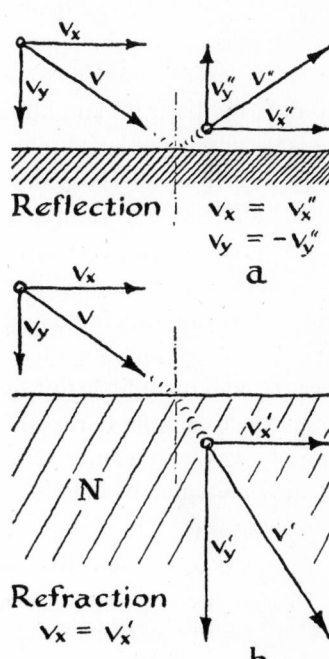

Reflection
$$v_x = v_x''$$
$$v_y = -v_y''$$
a

Refraction
$$v_x = v_x'$$
b

FIG. 1-6   Newton's corpuscular concept applied to reflection (a) and refraction (b).

tatively, $v_0 \sin i = v_g \sin r$. In contrast to the Huygens result Newton's ideas give $\dfrac{v_g}{v_0} = N$;

$$\frac{\sin i}{\sin r} = \frac{v_g}{v_0} = N$$

Thus the attraction of the glass surface for Newton's corpuscles leads also to Snell's law of refraction. But whereas Huygens' deduction of Snell's law requires $v_g < v_0$, Newton's deduction requires $v_g > v_0$. As we pointed out before, Foucault's measurements of the relative velocity in air and water showed $v_g < v_0$. Accordingly, here we shall follow Huygens' rather than Newton's ideas, and treat light as a transverse wave motion.

1-3. Complex Numbers†

In our development of the concept of light as a transverse wave motion, our discussions will be greatly facilitated if we use complex numbers. This use of "complex" mathematics is not intended to make the subject arbitrarily difficult—we use complex numbers because optics is inherently complicated and their use makes substantial simplifications in the mathematical expositions. Once understood, the use of complex variables frees one to think with more facility about the physical concepts of the science, without a distracting concern with mathematical details.

The reason why the use of complex variables simplifies mathematical exposition in optics is not far to seek. In order to describe a wave motion we have to describe both its amplitude and the phase or "timing" of its oscillations. It is because complex numbers are double numbers that they are appropriate here. When we write a complex number in exponential form, its scalar coefficient can express the amplitude of wave motion, while its imaginary exponent can express its phase. And in addition to this appropriateness complex numbers obey the laws of ordinary algebra, and in their exponential expressions they are easily integrated and differentiated. This appropriateness and facility combine to endow our quantitative expositions in optics with an awesome beauty, and not a little mystery. We shall first apply the exposition of complex numbers to simple harmonic motion; then to mechanical waves in a string; next to waves in a surface; and finally to mechanical wave motions in an extended medium. In Chapter III we shall recapitulate the theory of electromagnetic waves, applying our complex number exposition.

Ultimately we shall use complex notation in the addition of the myriads

---

† For an excellent intuitive account of the use of complex numbers see E. A. Guillemin's *Communication Networks* (John Wiley & Sons, New York), Vol. I, Chaps. I–III.

of Huygens wavelets to determine their effect in aggregate; or to determine when the segments of wavelets, in the aggregate, are indeed *"too feeble to produce light."* Such additions take account of both the strength and phase of the added wavelets, as Fresnel did; and furthermore our addition of wavelets will incorporate all the finesse that was introduced by Kirchhoff.

1-4. *Simple Harmonic Motion*

Fig. 1-7 shows a damped oscillating mass. Here the mass is mounted on a spring and constrained so that it can move only in a vertical direction (x of a coordinate system). A dashpot introduces a frictional restraint to its motion, which restraining force is proportional to the velocity of motion. And in addition, we suppose that the mass is driven by a periodic force of $\mathcal{F} \cos \omega t$ newtons. Although later we shall use the symbol m for mass, for the present we represent mass, in kilograms, by the symbol λ. The spring restraint is described by a compliance, γ, expressed in meters of displacement of the mass per newton of applied force. The damping coefficient ρ is expressed in newtons per meter per second of velocity. We write down the equation for the equi-

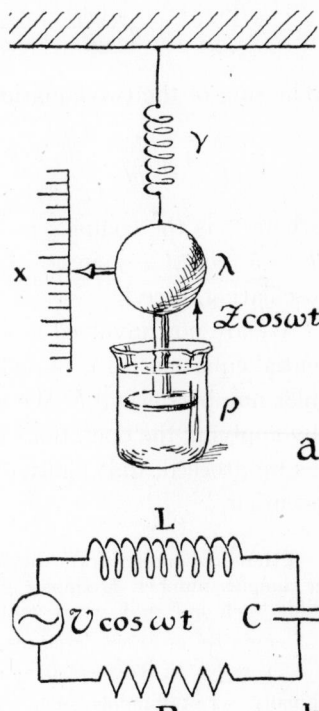

FIG. 1-7 Model of a damped harmonic oscillator (a), and the electrical analogue, an LRC-series circuit (b).

librium of forces acting on the mass to get a differential equation whose solutions will describe its motion. The expression (a), below, equates the sum of the inertial force $\lambda \dfrac{d^2x}{dt^2}$, the frictional force $\rho \dfrac{dx}{dt}$, and the spring restraint $\dfrac{x}{\gamma}$, to the periodic driving force $\mathfrak{F} \cos \omega t$.

$$\lambda \frac{d^2x}{dt^2} + \rho \frac{dx}{dt} + \frac{x}{\gamma} = \mathfrak{F} \cos \omega t \tag{a}$$

Here ω is the angular frequency in radians per second (being 2π times the frequency of oscillation ν, in cycles per second) and t is the time.

Since it is our purpose to use complex numbers, we transform this equation into a complex equation before solving it. We imagine our mass identically constrained along an imaginary but independent direction (y), and driven by an imaginary periodic force in phase quadrature with our former force. The equation expressing force equilibrium for this imaginary case gives a second similar differential equation. We signify the imaginary character of each of its terms, in the expression (b) below, by incorporation of a factor $j = \sqrt{-1}$.

$$\lambda \frac{d^2(jy)}{dt^2} + \rho \frac{d(jy)}{dt} + \frac{(jy)}{\gamma} = j\mathfrak{F} \sin \omega t \tag{b}$$

The sum of the two equations above is

$$\lambda \frac{d^2\tilde{w}}{dt^2} + \rho \frac{d\tilde{w}}{dt} + \frac{\tilde{w}}{\gamma} = \tilde{F} \tag{c}$$

where $\tilde{w}$ is the complex number $\tilde{w} = x + jy$, and where $\tilde{F}$, also complex, is $\tilde{F} = \mathfrak{F} (\cos \omega t + j \sin \omega t)$. $\mathfrak{F}$, a scalar quantity, we call the amplitude of the oscillating force.†

We are now involved in a mathematical "circumlocution." Our real differential equation (a) is buried in the complex equation (c), involving the complex numbers $\tilde{w}$ and $\tilde{F}$. We may recover our real quantities whenever we wish by applying the operation symbolized by R, discussed below. It will develop, as we proceed, that many practical advantages inhere in using this "circumlocution."

---

† Here and subsequently we shall use a super tilde to indicate that a symbol represents a complex number. Examples are $\tilde{w}$, $\tilde{F}$, $\tilde{v}$, $\tilde{\mathfrak{z}}$, $\tilde{V}$, $\tilde{\imath}$, and $\tilde{z}$. An italic symbol without the tilde, such as F or V, represents the real part of the complex number; for example,

$F = R\tilde{F} = R\mathfrak{F}e^{j\omega t} = \mathfrak{F} \cos \omega t$

$V = R\tilde{V} = R\mathcal{V}e^{j\omega t} = \mathcal{V} \cos \omega t$

Finally, script symbols, such as $\mathfrak{F}$ and $\mathcal{V}$, represent scalar or static quantities, such as the amplitude of oscillating or of complex quantities.

We differentiate both sides of equation (c), getting the equation below, which we solve for the velocity, $\tilde{v} = \dfrac{d\tilde{w}}{dt}$.

$$\lambda \frac{d^2\tilde{v}}{dt^2} + \rho \frac{d\tilde{v}}{dt} + \frac{\tilde{v}}{\gamma} = j\omega \mathfrak{F} e^{j\omega t} = j\omega \tilde{F} \tag{d}$$

The steady state solution of (d) which interests us here is

$$\tilde{v} = \tilde{v}_0 e^{j\omega t} \tag{e}$$

Here $\tilde{v}_0$ is a complex constant. If this solution, $\tilde{v}$, and its derivatives are substituted in the differential equation (d), we get

$$\tilde{v}_0 = \frac{\mathfrak{F}}{\rho + j\left(\omega\lambda - \dfrac{1}{\omega\gamma}\right)} \tag{f}$$

Or, multiplying both sides by $e^{j\omega t}$, the ratio of $\tilde{F}$ to $\tilde{v}$ is a complex number, $\tilde{\mathfrak{z}}$, which depends on the constants of the motion and the frequency of the driving force.

$$\frac{\tilde{F}}{\tilde{v}} = \tilde{\mathfrak{z}} = \rho + j\left(\omega\lambda - \frac{1}{\omega\gamma}\right) \tag{g}$$

Now this complex number is quite analogous to the ratio of $\tilde{V}$ to $\tilde{\imath}$ in an L-R-C electrical circuit. When $\tilde{V} = \mathcal{V}_0 e^{j\omega t}$ is an alternating electrical potential, written in complex notation, and $\tilde{\imath}$ is the alternating current which it produces, similarly written, then their ratio, $\tilde{z}$, is a complex number which depends only on the circuit constants and on ω:

$$\frac{\tilde{V}}{\tilde{\imath}} = \tilde{z} = R + j\left(\omega L - \frac{1}{\omega C}\right) \tag{h}$$

$\tilde{z}$ is called the electrical impedance, and analogously, in our oscillator problem, $\tilde{\mathfrak{z}}$ is a mechanical impedance. Corresponding terms and expressions appear in Table 1-1 to illustrate this analogy. R is the electrical resistance and ρ the corresponding mechanical resistance. λ is the mass, which corresponds to the electrical self-inductance L. And finally, the electrical capacity, C, and γ, play similar roles in these analogous problems.

At the resonance frequency ω_0 where $\omega_0 L = \dfrac{1}{\omega_0 C}$, or $\omega_0\lambda = \dfrac{1}{\omega_0\gamma}$, the impedances, $\tilde{z}$ and $\tilde{\mathfrak{z}}$, become real. The frequency of the driving potential, or of the driving force which produces resonance, is $\omega_0 = \dfrac{1}{\sqrt{LC}}$, or $\omega_0 = \dfrac{1}{\sqrt{\lambda\gamma}}$.

The maximum velocity, and its phase or timing, may be written by the product of a scalar coefficient and a complex exponential. Thus the complex

TABLE 1-1

L-R-C	S-H-M
L — inductance	λ — mass
R — resistance	ρ — resistance
C — capacitance	γ — spring compliance
$\tilde{q}$ — charge	$\tilde{w}$ — displacement
$\tilde{\imath}$ — current	$\tilde{v}$ — velocity
$\tilde{z}$ — electrical impedance	$\tilde{\mathfrak{z}}$ — mechanical impedance
$\tilde{z} = R + j\left(\omega L - \dfrac{1}{\omega C}\right)$	$\tilde{\mathfrak{z}} = \rho + j\left(\omega\lambda - \dfrac{1}{\omega\gamma}\right)$
$\left.\begin{array}{l} V\,- \\ v\,- \end{array}\right\}$ electrical potential	$\left.\begin{array}{l} F\,- \\ \mathfrak{F}\,- \end{array}\right\}$ mechanical force

velocity $\tilde{v}$ is written

$$\tilde{v} = v_0 e^{j(\omega t - \varphi)} \tag{i}$$

The value of φ is easily obtained if we write $\tilde{\mathfrak{z}}$ as a complex exponential. We divide and multiply $\tilde{\mathfrak{z}}$ by $\mathfrak{z} = \sqrt{\rho^2 + \left(\omega\lambda - \dfrac{1}{\omega\gamma}\right)^2}$, writing $\dfrac{\rho}{\mathfrak{z}} = \cos\varphi$ and

$\dfrac{\left(\omega\lambda - \dfrac{1}{\omega\gamma}\right)}{\mathfrak{z}} = \sin\varphi$. Then

$$\tilde{\mathfrak{z}} = \mathfrak{z}\,(\cos\varphi + j\sin\varphi) = \mathfrak{z}e^{j\varphi}$$

and from $\tilde{v} = \dfrac{\tilde{F}}{\tilde{\mathfrak{z}}}$, writing v_0 for $\dfrac{\mathfrak{F}}{\mathfrak{z}}$, we get the expression (i) for $\tilde{v}$. φ is the phase angle by which the phase of $\tilde{v}$ lags the phase of $\tilde{F}$. The dependence of φ on ω is

$$\varphi = \tan^{-1}\left(\frac{\omega\lambda - \dfrac{1}{\omega\gamma}}{\rho}\right)$$

φ becomes zero and the velocity and driving force are in phase at resonance.

If we want now to know the mechanical displacement $\tilde{w}$, rather than the velocity $\tilde{v}$, we may integrate $\tilde{v}$ just as we would integrate $\tilde{\imath}$ in an electrical problem to get the electrical charge $\tilde{q}$. This is particularly easy in the case of our exponential exposition. Neglecting the constant of integration,

$$\tilde{w} = \int \tilde{v}\, dt = \frac{v_0}{j\omega}\, e^{j(\omega t - \varphi)}$$

As mentioned above, our real displacement x, or the real velocity $\dfrac{dx}{dt}$, lies buried in the complex displacement $\tilde{w}$, or the complex velocity $\tilde{v}$. To uncover x,

FIG. 1-8 Maximum   displacement
(solid) and phase (dashed) of an under-
damped and critically damped harmonic
oscillator.

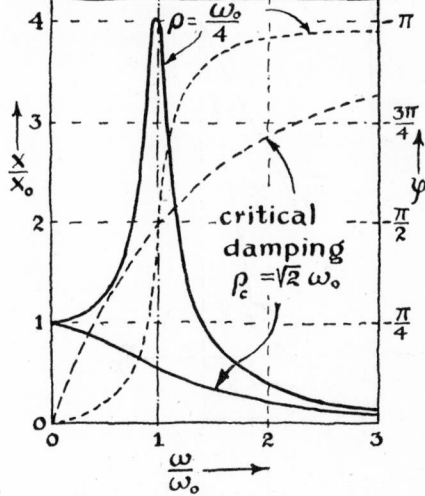

$$\ddot{x} + \rho\dot{x} + \omega_0^2 x \;=\; \omega_0^2 x_0 \cos\omega t$$

or $\dfrac{dx}{dt}$, we apply the operation of taking the real part of the corresponding complex quantity. We symbolize this operation by the druggist's prescription
symbol ℞.

$$x = ℞\tilde{w} = \frac{v_0}{\omega}\sin\,(\omega t - \varphi) = \frac{\mathcal{F}}{\omega\mathfrak{z}}\sin\,(\omega t - \varphi)$$

$$\frac{dx}{dt} = ℞\tilde{v} = v_0\cos\,(\omega t - \varphi) = \frac{\mathcal{F}}{\mathfrak{z}}\cos\,(\omega t - \varphi)$$

The dependences of the real displacement and φ on ω are illustrated in Fig.
1-8.

The integrity of the real motion of a mass, buried in the complex notation,
is uncompromised, while it is thus buried, when we make the regular algebraic
manipulations, and integrate or differentiate.

A special feature of $\tilde{w}$ and $\tilde{v}$, which is of great use in optics, is that they may
be represented as vectors in the complex plane. This representation has important applications for adding oscillations. As an illustration, we represent
two harmonic motions, together with their sum, as vectors in the complex
plane in Fig. 1-9. In the illustration we suppose that a single mass is driven
by two separate forces, $\mathcal{F}_a e^{j(\omega t+\varphi_a)}$ and $\mathcal{F}_b e^{j(\omega t+\varphi_b)}$. The corresponding motions
will be $\tilde{w}_a = \dfrac{\mathcal{F}_a}{j\omega\mathfrak{z}}\,e^{j(\omega t+\varphi_a-\varphi_{\mathfrak{z}})}$ and $\tilde{w}_b = \dfrac{\mathcal{F}_b}{j\omega\mathfrak{z}}\,e^{j(\omega t+\varphi_b-\varphi_{\mathfrak{z}})}$. The real motions in the
steady state will be $x_a = ℞\tilde{w}_a = \dfrac{\mathcal{F}_a}{\omega\mathfrak{z}}\sin\,(\omega t + \varphi_a - \varphi\mathfrak{z})$ and $x_b = ℞\tilde{w}_b =$

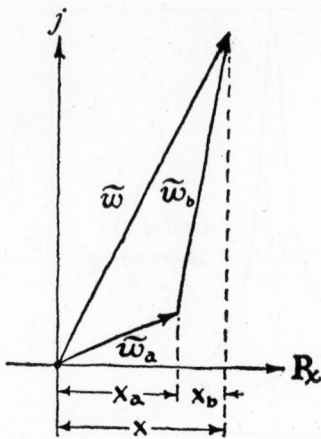

FIG. 1-9 Vector addition of two complex numbers, $\tilde{w}_a$ and $\tilde{w}_b$.

$\dfrac{\mathfrak{F}_b}{\omega\mathfrak{z}}\sin(\omega t + \varphi_b - \varphi\mathfrak{z})$. It is as if we had solved two problems. The sum of the two solutions is of course a solution to our single differential equation and to our problem, so that the total steady state motion due to the two driving forces is $x = x_a + x_b = \mathfrak{R}(\tilde{w}_a + \tilde{w}_b)$.

A cardinal point in the use of complex numbers follows from the fact that the operation $\mathfrak{R}$, and operations of addition, are interchangeable: it is true not only that $x = x_a + x_b = \mathfrak{R}\tilde{w}_a + \mathfrak{R}\tilde{w}_b$; but $x = \mathfrak{R}\tilde{w}$ where $\tilde{w} = \tilde{w}_a + \tilde{w}_b$. Thus, we can add the two motions as complex numbers or complex vectors and later extract their real resultant. Thus Fig. 1-9 shows our $\tilde{w}_a$ and $\tilde{w}_b$ plotted in the complex plane, as well as the closing vector, their sum. The projections of these two vectors on the real axis of abscissa give the real components of motions, x_a and x_b; and projection of $\tilde{w}$ gives the real component of their sum, x. This graphical representation of complex numbers is much employed in optics.

1-5. *Polarized Waves in a Stretched String*

Continuing, we apply complex notation now to wave motions in a stretched string. Such wave motions may be thought of as the coordinated simple harmonic motions of the various elements of the string. We can generalize what we learn of string wave motions to wave motions in a drumhead or on water surfaces, thus introducing the treatment of wave motions in an extended three-dimensional medium.

The description of the wave motions of a stretched string is inherent in the differential equation expressing the equilibrium of forces acting on an elementary length of it. The solution of this equation represents the coordi-

FIG. 1-10 Vertical forces acting on the
element dz of a string, due to tension in it.

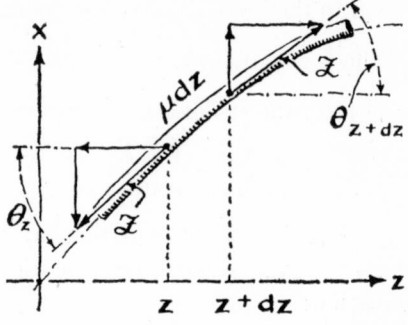

nated simple harmonic motions of its segments. Fig. 1-10 shows a short segment of a vibrating string and the deflected element of it that lies between
z and $z + dz$. Here z is a coordinate that runs along the length of the undeflected string. This coordinate axis is represented by a dashed abscissa line
in the figure. We take the azimuth of the wave motion of the string to be
in the xz-plane, with polarized waves in this azimuthal plane running toward
larger and smaller z's; and we neglect friction for the present, to make our
treatment simple. The deflection in the xz-plane in Fig. 1-10 is represented
by the coordinate x.

    Let the tension along the stretched string be $\mathfrak{F}_0$, and consider that the de-
flections are all small, so that ϑ, the inclination of the string to the z-axis,
is nowhere large, and so that horizontal forces acting on the segment will be
equal and opposite. The horizontal forces are doubly equal, because ϑ is small
so that $\cos \vartheta$ is very nearly unity on the one hand, and because z lies near
$z + dz$, on the other. The vertical force (downward), at z, is

$$\mathfrak{F}_0 \sin \vartheta_z \cong \mathfrak{F}_0 \left(\frac{\partial x}{\partial z} \right)_z$$

At $z + dz$, the vertical force (upward) is

$$\mathfrak{F}_0 \sin \vartheta_{z+dz} \cong \mathfrak{F}_0 \left(\frac{\partial x}{\partial z} \right)_{z+dz}$$

Using the second derivative of x to express this force at $z + dz$:

$$\left(\frac{\partial x}{\partial z} \right)_{z+dz} = \left(\frac{\partial x}{\partial z} \right)_z + \left(\frac{\partial^2 x}{\partial z^2} \right)_z dz$$

we subtract the upward and downward forces to get the net force on the element, arising from the string tension:

$$\mathfrak{F}_0 \frac{\partial^2 x}{\partial z^2} dz$$

    Since we have neither a driving force nor a friction force, we get the differ-
ential equation for the motion of the string by setting this string tension force

equal to the inertial force. The inertial force is, of course, the product of mass of the segment, $\mu\,dz$, and its acceleration, $\dfrac{\partial^2 x}{\partial t^2}$, where μ is the mass per unit length for the string. Thus the differential equation becomes

$$\frac{\partial^2 x}{\partial z^2} = \frac{\mu}{\mathcal{F}_0}\frac{\partial^2 x}{\partial t^2} = \frac{1}{v^2}\frac{\partial^2 x}{\partial t^2}$$

As in § 1-4, where equation (a) was the real part of equation (c), we set $\tilde{w} = x + jy$, and make this equation the real part of the complex differential equation:

$$\frac{\partial^2 \tilde{w}}{\partial z^2} = \frac{1}{v^2}\frac{\partial^2 \tilde{w}}{\partial t^2}$$

This differential equation has two steady state solutions at frequency ω, which are significant here, and they can be shown to be solutions by differentiation and substitution back in the differential wave equation. These two solutions describe polarized waves that propagate along the string with velocity $v = \sqrt{\dfrac{\mathcal{F}_0}{\mu}}$. Together, these two solutions describe the composite total polarized wave motion of frequency ω:

$$\tilde{w} = \tilde{w}_a + \tilde{w}_b = A e^{j\omega\left(t + \frac{z}{v}\right)} + B e^{j\omega\left(t - \frac{z}{v}\right)}$$

$$R\tilde{w} = x_a + x_b = A\,\cos\omega\left(t + \frac{z}{v}\right) + B\,\cos\omega\left(t - \frac{z}{v}\right)$$

Here $\omega = 2\pi\nu$ is angular frequency, and ν is the number of complete vibrations per second at a particular point on the string. The two complex solutions $\tilde{w}_a$ and $\tilde{w}_b$, or their real parts, x_a and x_b, represent two waves, one propagating in one direction along the string, the other propagating in an opposite direction. A and B are the amplitudes of these two waves.

We shall first discuss only one of these solutions, taking $B = 0$ for the other.

$$x = x_a = A\,\cos\omega\left(t + \frac{z}{v}\right)$$

When the phase angle, $\omega\left(t + \dfrac{z}{v}\right)$, is repeated, or when the phase angle is increased or decreased by 2π, then or there the deflection, as well as the neighboring configuration of the string, is likewise repeated.

We interpret $\tilde{w}_a$ as representing a wave traveling to the left because $x_a = A\,\cos\omega\left(t + \dfrac{z}{v}\right)$ defines a configuration at the place z_1, and at the time t_1, which is the same as the configuration at a larger value of z at an earlier time $\left\{\left(t_1 + \frac{z_1}{v}\right) = \left(t_1 - \Delta t + \frac{z_1 + \Delta z}{v}\right)\right\}$. Thus $\tilde{w}_a$ represents a wave con-figuration traveling to the left.

The maximum deflection at z_1 comes when $\omega\left(t + \frac{z_1}{v}\right) = 0$, or when it is equal to an even integral number of π's. Then $x_a = A$ and A is the amplitude of the wave. The minimum deflection at z_1 comes when $\omega\left(t + \frac{z_1}{v}\right)$ is an odd integral number of π's.

The deflection and adjacent configuration is repeated whenever $\omega\left(t + \frac{z}{v}\right)$ changes by 2π or when $\nu\left(t + \frac{z}{v}\right)$ changes by ± 1. First, considering the time fixed, the configuration is repeated wherever $\nu\frac{\Delta z}{v} = \pm 1$. This length incre-ment $\pm\Delta z$ is called the wavelength, and it is symbolized by λ. We previously used λ for mass; but we now abandon this for its customary use, symbolizing wavelength. On substitution of λ for $\pm\Delta z$ we have $\lambda\nu = v$.

Next, considering a fixed point along the string, when $\nu\left(t + \frac{z}{v}\right)$ now changes by ± 1, we have $\nu\Delta t = \pm 1$. If we write τ for this time interval, called the period of the wave motion, we have $\tau = \frac{1}{\nu}$ or $\omega = \frac{2\pi}{\tau}$.

All these considerations apply similarly for $A = 0$ and B finite. Then $x_b = B\cos\omega\left(t - \frac{z}{v}\right)$. This equation describes a wave traveling to the right, since an increase of t yields the same configuration at an increased z.

The kinetic energy of the element of our string, dz, due to such wave motions, is half of the product of the mass, $\mu\,dz$, and the square of the velocity, $\left(\frac{dx}{dt}\right)^2$. Invoking $x = A\cos\omega\left(t + \frac{z}{v}\right)$, we get for this kinetic energy

$$\frac{1}{2}(\mu\,dz)\left(\frac{dx}{dt}\right)^2 = \frac{\mu\,dz}{2}A^2\omega^2\sin^2\omega\left(t + \frac{z}{v}\right)$$

The time average of this kinetic energy is

$$\frac{1}{\tau}\int_{t_1}^{t_1+\tau}\frac{\mu\,dz}{2}A^2\omega^2\sin^2\omega\left(t + \frac{z}{v}\right)dt = \frac{1}{4}\mu A^2\omega^2\,dz$$

In contrast, the energy of configuration of the same element of our string is equal to an integral from $x = 0$ out to x. It is the integral of the product of the restoring force, $-\mathfrak{F}_0\frac{d^2x}{dz^2}\,dz$, and dx. On averaging this integral over one period of oscillation we get the same time average that we got before for the kinetic energy. And the sum of these average kinetic and potential energies gives a constant total energy for the element,

$$\tfrac{1}{2}\mu \, dz \, A^2\omega^2$$

That is, the average linear energy density in the string due to its wave motion is equally divided between the energy of configuration of the string, or potential energy, and the energy of motion of the string, or kinetic energy. These energy densities are each proportional to ω^2, and also *proportional to the square of the amplitude of motion.*

1-6. *Velocities of Mechanical Waves*

The phases for our two solutions were $\omega\left(t \pm \dfrac{z}{v}\right)$—the $+$ sign being associated with waves traveling toward $z = -\infty$, and the $-$ sign with waves traveling toward $z = +\infty$. The velocity of propagation in either case was v, and this velocity was determined by the square root of the ratio of a tension (or force term) and a linear density (or mass term):

$$v = \sqrt{\frac{\mathcal{F}_0}{\mu}}$$

The velocity of transverse waves in an extended isotropic solid medium is also the square root of the ratio of a force term and a mass term. But in this case the force term is the shear modulus of elasticity of the solid, n, and the mass term is the density of the solid, ρ. The velocity of transverse waves is

$$v_t = \sqrt{\frac{n}{\rho}}$$

These transverse waves are polarizable just like waves in our stretched string.

And further, in an extended isotropic solid medium the velocity of longitudinal waves is the square root of the ratio of a force term and a mass term:

$$v_l = \sqrt{\frac{k + \tfrac{4}{3}n}{\rho}}$$

Here the force term is more complex; it includes both n and the bulk modulus of compressibility of the solid, k. Longitudinal waves are, of course, not polarizable.

For gases and liquids the shear modulus n is zero; and, therefore, transverse mechanical waves do not occur. For gases the bulk modulus k is $\dfrac{c_p}{c_v} P$ where P is the pressure and $\dfrac{c_p}{c_v}$ is the ratio of specific heats. For a gas the density is

$\rho = \dfrac{M}{V}$ where M is the molecular weight and V is the gas molecular constant.

We invoke the ideal gas law, $PV = \Re T$ where T is the absolute temperature, and get

$$v_l = \sqrt{\frac{\dfrac{c_p}{c_v}\,\Re T}{M}}$$

for the velocity of the longitudinal waves of sound.

Ripples and surface waves on water, such as originally inspired Huygens, afford a final category of mechanical waves to interest us. When the length of surface waves on water is large, the dominant restoring force is gravity; but when λ is small, it is surface tension. In all cases the velocity is given by

$$v = \sqrt{\frac{\lambda g}{2\pi} + \frac{2\pi\sigma}{\lambda\rho}}$$

Here g is the acceleration of gravity, and σ is the surface tension of the water or other liquid of density ρ.

Chapter **II**

Superposition of
Wave Motions

In the preceding chapter we saw how the density of kinetic or potential energy along a stretched string, arising from wave motion, was proportional to the square of the amplitude of that wave motion. And, also, when we had the superposition of two transverse waves of the same frequency, expressed as complex functions, we saw how the aggregate wave motion was the sum of the complex functions representing the superimposed components. In this chapter we shall be concerned with the methods of determining such sums by vector addition in the complex plane. We shall see in the next and later chapters that these mathematical methods may be applied to light since it also is a transverse wave motion, being oscillations of electric and magnetic fields. The problems of mathematical expression and summation of such electric and magnetic field oscillations are the same as for the transverse mechanical waves in our stretched string.

The superposition illumination that we see when two beams of the same frequency are superimposed is proportional to the square of the aggregate amplitude. Thus, two equal light wave amplitudes of opposite sign could cancel each other, and two equal waves of the same sign would add to give an aggregate amplitude twice as great as either component alone. In the first case we have the adding of light to light to get darkness, as Francesco Maria Grimaldi expressed it; and in the second case, with the doubling of amplitude, or quadrupling of the illumination, we have an apparent violation of the principle of conservation of energy. The explanation of these matters is now also our concern.

Two particular solutions for the differential equation of the stretched string,

22

which we discussed previously, were characterized by the phases $\omega\left(t + \dfrac{z}{v}\right)$

and $\omega\left(t - \dfrac{z}{v}\right)$. There is, however, no restriction on the phases we may take for our solutions, nor on the amplitudes. And, for that matter, azimuths of polarization along the string also are unrestricted.

Although waves in the stretched string must run to the right or the left, waves in a drumhead, for which the equations are similar, can propagate in various directions in its plane. Propagation directions for waves in an extended medium are unrestricted. In a drumhead, all wave motions are restricted to lie perpendicular to the membrane, and are thus mutually parallel, irrespective of the direction of propagation. In contrast with these restrictions, transverse waves in an extended medium may have different azimuths of vibration as well as different directions of propagation.

If, on the one hand, component transverse waves are in phase, we may get their sum from algebraic addition of the geometric vectors involved. On the other hand, when parallel displacements of geometrically parallel component transverse wave motions are not in phase, we can get the resultant wave motion by a vector summation in the complex plane. This is the vector addition of the graphical representations of the complex numbers that describe the geometrically parallel component wave motions. In the general case, however, when component waves at a point are neither in phase nor parallel in displacement, owing to differing directions of propagation or azimuths of polarization, we must use both types of vector summation to get the resultant wave motion. The summation in geometric space of parallel components takes account of different propagation directions and polarization azimuths; the vector summation in the complex plane takes account of different phases. This general case is treated in § 2-5.

2-1. *Sum of Two Cosine Functions of the Same Frequency*

Let us consider the addition of two cosine functions representing two parallel wave motions with amplitudes A and B and with phases α and β. These two cosine functions may be taken as the real components of two complex functions, and their sum may be written

$$\bar{w}_c = \bar{w}_a + \bar{w}_b$$

$$\bar{w}_c = A e^{j(\omega t + \alpha)} + B e^{j(\omega t + \beta)} = C e^{j(\omega t + \gamma)}$$

$$x_c = x_a + x_b = A \cos(\omega t + \alpha) + B \cos(\omega t + \beta) = C \cos(\omega t + \gamma)$$

The expansion of the double arguments of the cosine functions above, for x_a and x_b as well as for x_c, gives the interrelationships

$$\gamma = \tan^{-1} \frac{A \sin \alpha + B \sin \beta}{A \cos \alpha + B \cos \beta}$$

$$C = \sqrt{(A \cos \alpha + B \cos \beta)^2 + (A \sin \alpha + B \sin \beta)^2}$$

$$= \sqrt{A^2 + B^2 + 2AB \cos (\alpha - \beta)}$$

Thus the sum of two cosine functions, x_a and x_b, of the same frequency, is also a cosine function of that frequency, x_c. And it is evident that the sum of two, added to a third, and so on, *ad infinitum*, will give a final sum that is still a cosine function of the same frequency. Thus we have a kind of integrity of frequencies in these summations.

On the physical side, as we proceed into the following chapters, we shall see demonstrations of this integrity of frequency, that the sum of two or many component light wave motions of the same frequency also has that same frequency. And it is also a fact of experience that the *observed* composite amplitude that several superimposed known amplitudes of light of the same frequency produce, when made to pass simultaneously through a given point, conforms to the *calculated* sum of the amplitudes that each component alone would produce. Of course we must take proper account, in the calculated sum, of the component amplitudes, phases, azimuths of polarization, and directions of propagation. The statement of experience, that we have this strict conformance for light between observed and calculated superpositions, is called the *principle of superposition*.

Whereas the expressions above, for the superposition of only two waves, easily yielded the lumped phase and a lumped amplitude (α and β yielding γ, A and B yielding C), when many waves are superimposed the finding of the sum becomes a more complicated procedure. And in addition to this complexity for parallel wave motions, we encounter other complexities arising from the different directions of propagation of components or their different azimuths of polarization. The azimuth of polarization in the case of polarized wave motion in our string is the plane that contains the undeflected string as well as its polarized motion. Previously this plane was the xz-plane.

2-2. *Standing Waves*

In some cases the complex functions representing component waves can be algebraically combined, in other cases not. As a special example of the first, let us consider the following combination of two complex functions:

$$\tilde{w} = \tilde{w}_1 + \tilde{w}_2 = A_1 e^{j\omega\left(t + \frac{z}{v}\right)} + A_2 e^{j\omega\left(t - \frac{z}{v}\right)}$$

This sum represents a combination of two wave motions, both polarized in the xz-plane. These wave motions are propagating in opposite directions in our string, and the sum takes on a special significance when $A_1 = -A_2$. This particular combination describes the case where the string is clamped at $z = 0$, so that there $\bar{w} = 0$, and where $\bar{w}_1$ represents an infinitely long train of waves traveling to the left, with phase $\omega\left(t + \dfrac{z}{v}\right)$. In this case we may consider $\bar{w}_2$ as representing these same waves after reflection at $z = 0$, with both a reversal of propagation direction and a phase change π there. At $z = 0$, the phase of the reflected waves $\bar{w}_2$ becomes $\omega\left(t - \dfrac{z}{v}\right) + \pi$. The change in sign, before the term $\dfrac{z}{v}$, comes from the reversal of direction. And at $z = 0$, our clamping condition requires $A_1 = -A_2$ so that $\bar{w} = 0$. This reversal of the sign of the amplitude of $\bar{w}_2$ is equivalent to $+A_2$ with the phase change π, since $e^{j\pi} = -1$. Adding both incident and reflected waves gives the combination which applies to our string clamped at $z = 0$.

$$\bar{w} = A_1\left[e^{j\omega\left(t+\frac{z}{v}\right)} - e^{j\omega\left(t-\frac{z}{v}\right)}\right] = A_1 e^{j\omega t}\left(e^{j\omega\frac{z}{v}} - e^{-j\omega\frac{z}{v}}\right)$$

Remembering the exponential definitions of the circular functions, the parenthesis above is seen to be $2j \sin \omega \dfrac{z}{v}$. Taking the real component of $\bar{w}$, we get

$$x = \mathrm{R}\bar{w} = -2A_1 \sin \omega t \sin \omega \frac{z}{v}$$

The standing wave pattern which $\bar{w}$ represents has nodes where $\omega \dfrac{z}{v}$ takes on integral values times π; and $\bar{w}$ has antinodes where $\omega \dfrac{z}{v}$ lies midway between these nodes. At the nodes x is always zero; between, at the antinodes, the maximum string deflection x takes on the values $\pm 2A$.

The inherent interest in this expression, which describes standing waves in our string, resides in the fact that standing waves are also encountered in optics.

2-3. *Graphical Summation of Wave Motions*

Above we considered the algebraic sum of two complex functions representing two wave motions. Now we consider the graphical sum of two complex functions that are plotted as vectors in the complex plane; in particular, we consider the graphical sum of the two vectors shown in Fig. 2-1, representing $\bar{w}_c = \bar{w}_a + \bar{w}_b$. Here we take A and B of $\bar{w}_a$ and $\bar{w}_b$ equal; and we represent the $\bar{w}$'s graphically at the instant when $t = 0$, or when $e^{j\omega t} = 1.0$. Obviously the vector sum $\bar{w}_c$ is the bisector of the angle between the equal vec-

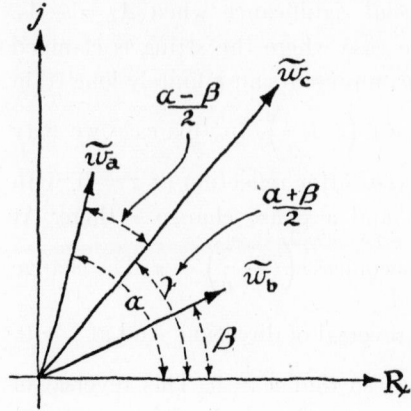

FIG. 2-1 Vector addition of two complex numbers of equal amplitude.

tors $\widetilde{w}_a$ and $\widetilde{w}_b$; and obviously, also, half the length of the vector for $\widetilde{w}_c$ is the projection of either component onto this bisector. Thus $\gamma = \dfrac{\alpha + \beta}{2}$ and $C = 2A \cos \dfrac{\alpha - \beta}{2}$, so that

$$\widetilde{w}_c = 2A \cos \frac{\alpha - \beta}{2} e^{j\left(\omega t + \frac{\alpha+\beta}{2}\right)}$$

Fig. 2-2 shows the vector representations where A and B of $\widetilde{w}_a$ and $\widetilde{w}_b$ are not equal. The vectors make the angles $(\omega t + \alpha)$ and $(\omega t + \beta)$ with the R-axis. Thus, as time flows, both vectors rotate counterclockwise. But at the instant when $t = 0$, or when $e^{j\omega t} = 1.0$, the vector sum is

$$\widetilde{w}_c = Ae^{j\alpha} + Be^{j\beta} = Ce^{j\gamma}$$

The real projections of these rotating vectors will oscillate as time flows; and their projections on the R-axis correspond to our results of § 2-1. We may

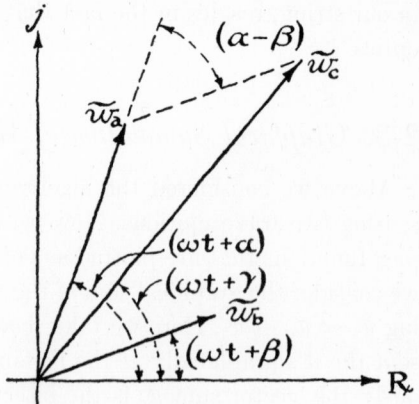

FIG. 2-2 Vector addition of two complex numbers of unequal amplitude.

regard the display in Fig. 2-2 as a snapshot of these rotating vectors. As time flows past the instant this figure represents, the triangle of vectors will rotate, all together, counterclockwise in the complex plane, all with the same angular velocity ω. And although these three vectors all rotate as time flows, they rotate together so that their mutual internal relationships are unchanging.

Going beyond this simple case, we add more than two parallel wave motions of the angular frequency ω as follows: we first specify a convenient time and represent each component for that instant with a complex number $\bar{w}_i$; then we plot these $\bar{w}_i$'s as vectors end to end in the complex plane; and finally we draw the closing side to get their sum $\bar{w}$. And, finally, the operation $R\bar{w}$ gives the real aggregate wave motion. This is the circumlocution we referred to in the previous chapter. Components thus added will give us sums that will be of value for the explanations of optics.

2-4. *Superposition of Polarized Wave Motions*

The above summations in the complex plane relate to wave motions with parallel displacements, such as waves that are polarized in the same xz-plane as they were in our stretched string; or such as wave motions in a drumhead, where all displacements are necessarily perpendicular to the membrane, and hence mutually parallel.

We shall consider the superposition of two wave motions in our stretched string that are mutually perpendicular. Let one motion, say $\bar{w}_x$, be polarized in the xz-plane; and the other, $\bar{w}_y$, be polarized in the yz-plane; with both wave motions traveling together along the z-axis of our string. For reasons which will become evident later we endow these two wave motions with a relative phase difference of $\frac{\pi}{2}$. And further, we endow the waves, $\bar{w}_x$ and $\bar{w}_y$, with equal amplitudes of motion, A. We represent these waves by the complex numbers

$$\bar{w}_x = A e^{j\omega\left(t - \frac{z}{v}\right)}$$

$$\bar{w}_y = A e^{j\omega\left(t - \frac{z}{v}\right) - j\frac{\pi}{2}} = A e^{j\omega\left(t - \frac{z}{v}\right)} e^{-j\frac{\pi}{2}}$$

The corresponding real motions are, of course,

$$R\bar{w}_x = x = A \cos \omega \left(t - \frac{z}{v}\right)$$

$$R\bar{w}_y = y = A \sin \omega \left(t - \frac{z}{v}\right)$$

We must not confuse the real y here with the imaginary part of $\bar{w}_x$, for which we only evanescently used the same symbol y. Here we use y for the real part of $\bar{w}_y$, just as we use x for the real part of $\bar{w}_x$. If we represent the aggregate displacement of any point in our string by the cylindrical coordinates r and θ, and z, our expressions for x and y yield

$$r = \sqrt{x^2 + y^2} = A$$

$$\theta = \tan^{-1}\frac{y}{x} = \omega\left(t - \frac{z}{v}\right)$$

Thus, on adding these mutually perpendicular wave motions, it is apparent that they give, together, at any point z_1 a constant string displacement. At the point z_1 the string segment executes a circular motion at the angular rate ω. This constant displacement, $r = A$, varies with z in such a manner as to give the string, at any instant, an helical configuration. And this helical configuration rotates with time like a rotating screw thread.

These representations of this circular mechanical motion will be of value to us later because the mechanical motions here are easily visualized, and because the complex expressions that represent them have the same form as the complex expressions we shall encounter in the representations of circularly polarized light.

2-5. *General Procedures of Superposition*

Light, as we shall see in the next chapter, can be treated as a polarizable transverse wave motion. Actually light waves are progressing transverse electric and associated magnetic fields. These fields are vector fields, just as the wave displacements in our string are. Therefore, the complex expressions for describing light waves of some prescribed frequency in an extended medium must include the specification of the azimuths of polarization as well as directions of propagation, amplitudes, and phases of wave motion. We can, for example, express the electric displacements at some point in free space, P, for three wave motions, all of the same frequency, as follows:

$$\bar{\tilde{E}}' = (\mathcal{E}_x'\mathbf{i} + \mathcal{E}_y'\mathbf{j} + \mathcal{E}_z'\mathbf{k})e^{j(\omega t + \varphi')}$$

$$\bar{\tilde{E}}'' = (\mathcal{E}_x''\mathbf{i} + \mathcal{E}_y''\mathbf{j} + \mathcal{E}_z''\mathbf{k})e^{j(\omega t + \varphi'')}$$

$$\bar{\tilde{E}}''' = (\mathcal{E}_x'''\mathbf{i} + \mathcal{E}_y'''\mathbf{j} + \mathcal{E}_z'''\mathbf{k})e^{j(\omega t + \varphi''')}$$

In these equations, $\tilde{E}$, with a tilde over it, means the oscillating electric field of a light wave expressed as a complex number. $\bar{E}$, with a bar also over it, means that the direction of the electric field is also expressed. The electric fields parallel to the axes of a Cartesian coordinate system, which we use here, have their respective amplitudes expressed as $\mathcal{E}_x$, $\mathcal{E}_y$ and $\mathcal{E}_z$. Here $\mathbf{i}$, $\mathbf{j}$ and $\mathbf{k}$ are the usual unit vectors of the coordinate system. Single, double and triple primes are used to distinguish the three component wave motions. Although all the wave motions above have the same frequency, they have different characterizing phases, φ', φ'', and φ'''. Unprimed symbols, below, characterize the aggregate wave motion. This example illustrates how we may add any number of superimposed component waves. The procedure is as follows:

First, vector additions of the Cartesian components are made in the complex plane. Each set of the Cartesian components, at a time when $e^{j\omega t} = 1.0$, gives a Cartesian component for the aggregate motion:

$$\tilde{E}_x = \mathcal{E}_x'e^{j\varphi'} + \mathcal{E}_x''e^{j\varphi''} + \mathcal{E}_x'''e^{j\varphi'''}$$
$$\tilde{E}_y = \mathcal{E}_y'e^{j\varphi'} + \mathcal{E}_y''e^{j\varphi''} + \mathcal{E}_y'''e^{j\varphi'''}$$
$$\tilde{E}_z = \mathcal{E}_z'e^{j\varphi'} + \mathcal{E}_z''e^{j\varphi''} + \mathcal{E}_z'''e^{j\varphi'''}$$

Second, after these first vector additions have been made in the complex plane, the resultants are combined vectorially in geometric space to give the final direction and phase of the resultant electric field:

$$\tilde{E} = \tilde{E}_x\mathbf{i} + \tilde{E}_y\mathbf{j} + \tilde{E}_z\mathbf{k}$$

Several exercises for the student to work out are given as problems.

Fortunately, in many of the superpositions we encounter in optics, it is possible to proceed as if all the vectors were parallel, or at least lay in one plane.

2-6. *Single-slit Diffraction*

In Fig. 2-3 a light wave-train of parallel waves is imagined to be incident from the left on a rectangular slit of width a. As each wave front arrives between the slit jaws we imagine that it acts as parent to produce seven daughter Huygens wavelets, one emitted from each of the seven strips into which the slit

---

FIG. 2-3 Single-slit diffraction as treated with seven equal Huygens wavelets originating between the slit jaws and superimposed at P.

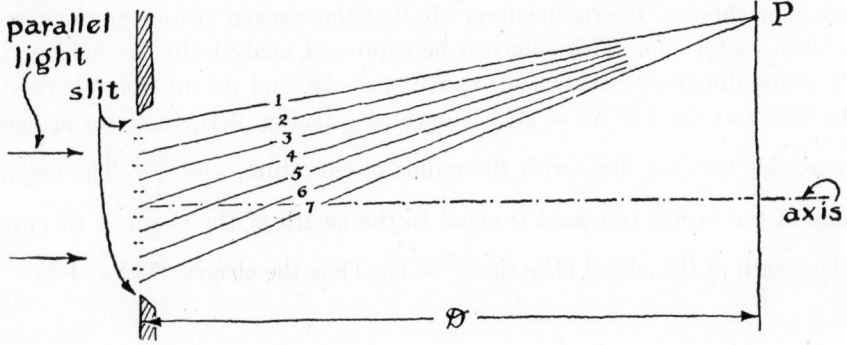

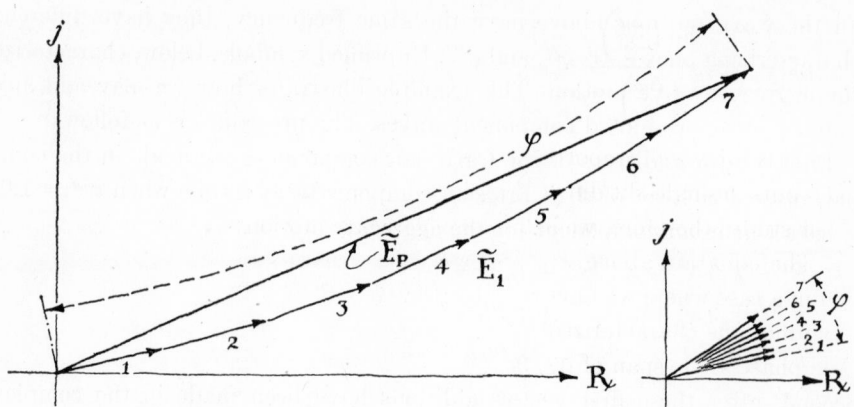

FIG. 2-4 Vector addition of the seven wavelets of Fig. 2-3.

is shown divided. The produced equally strong seven wavelet trains are all in phase in the plane of the slit, and they are polarized in parallel planes. However, because of unequal distances to the point P, as shown, the wavelets will have progressively different phases as they arrive at P. In Fig. 2-3, if $\mathfrak{D} \gg a$ we may assume that the Huygens wave motions at P lie geometrically parallel. Fig. 2-4, at the right, shows seven equal vectors $\mathcal{E}_1$ plotted in the complex plane. These vectors represent the Huygens wavelets at P. We take $\mathcal{E}_1$ as the amplitude of the field at P due to each one of the equally strong Huygens wavelets. The phases of the seven vectors are, however, not the same at P; but they are uniformly distributed throughout the phase interval $\Delta\varphi = \omega \dfrac{\Delta r}{c}$, where Δr is the geometrical path difference. In the same figure, at the left, these seven representative vectors are shown plotted in the complex plane end to end.

If we had divided the slit into a larger number of strips, $\mathfrak{N} \gg 7$, the end-to-end vectors would have closely followed the arc of a circle, of arc length $\mathfrak{N}\mathcal{E}_1$. The chord of the rudimentary arc that these seven vectors approximate is their vector sum. This sum can be expressed analytically, as follows: If the phase difference between successive $\mathcal{E}_1$'s is $\delta\varphi$, and the difference between the first and the last $\Delta\varphi = \mathfrak{N}\delta\varphi$ (shown as φ in Fig. 2-4), then, for $\mathfrak{N}$ very large, this arc is a circle with the radius of curvature, $\rho = \dfrac{\mathfrak{N}\mathcal{E}_1}{\Delta\varphi}$. The amplitude of the vector resultant is equal to the length of the chord of this arc. The length of this chord is $2\rho \sin \dfrac{\Delta\varphi}{2} = \mathcal{E}_P$. Thus the electric field at P is

$$\mathcal{E}_P = \mathfrak{N}\mathcal{E}_1 \frac{\sin\left(\frac{\Delta\varphi}{2}\right)_P}{\left(\frac{\Delta\varphi}{2}\right)_P} \sim \frac{\sin\left(\frac{\Delta\varphi}{2}\right)_P}{\left(\frac{\Delta\varphi}{2}\right)_P}$$

since $\mathcal{E}_1$ is inversely proportional to $\mathfrak{N}$.

To pursue single-slit diffraction further now would disrupt the organization of our study; therefore, we leave the discussion of diffraction for a later chapter. The equation above was developed here as an illustration of superposition in a case where we have many geometrically parallel, and equally strong, wave motions characterized by a regular distribution of their phases over a total phase-angle span of $\Delta\varphi$.

2-7. *Superposition of Waves with Random Phases*

The next case of superposition we treat is different from the one above; for now we take the total phase-angle span as $\Delta\varphi = 2\pi$. And further, we take the phases of many equally strong component amplitudes as distributed throughout this interval according to a different rule: the phase of any particular component wave has an equal probability of falling anywhere in the interval $\Delta\varphi = 0$ to $\Delta\varphi = 2\pi$. If we have a large number of equally strong and parallel components, thus defined, we might expect that the real components of as many of them would be negative as positive, thus giving a negligible resultant. The following analysis, however, shows this not to be the case. $\mathfrak{N}$ such equally strong component wave motions, all parallel, are added as follows: We write the complex number $\mathcal{E}_k = \mathcal{E}_1 e^{j(\omega t + \varphi_k)}$ for each of our components, letting the phase of each wave be described by φ_k, which has equal probability of lying anywhere within the interval 2π. The resultant for such a superposition may be accomplished either graphically, or analytically. Here we add all the representative complex numbers analytically, writing $\tilde{E}$ for the sum.

$$\tilde{E} = \sum_{k=1}^{\mathfrak{N}} \mathcal{E}_1 e^{j(\omega t + \varphi_k)}$$

We evaluate this sum at the instant when $e^{j\omega t} = 1.0$, remembering that the sum vector remains unchanged in length in the complex plane as time flows. This sum $\tilde{E}$, then, is

$$\tilde{E} = \sum_{k=1}^{\mathfrak{N}} \mathcal{E}_1 e^{j\varphi_k}$$

If we multiply this complex number $\tilde{E}$ by its complex conjugate we get the square of the length of its vector in the complex plane, *i.e.*, $\mathcal{E}^2$. This squaring is equivalent to applying the theorem of Pythagoras. The sums of the squares

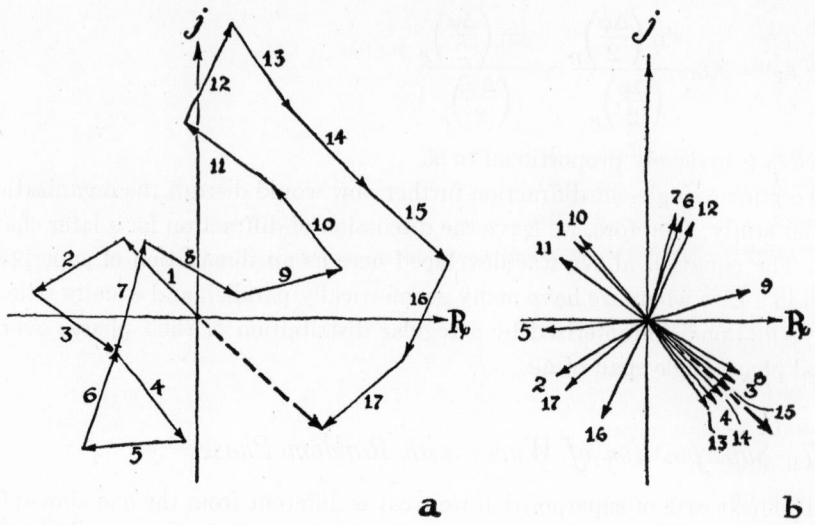

FIG. 2-5 Vector addition of seventeen vectors of equal amplitude and random phase (determined by telephone numbers).

of the real and imaginary components of the resultant vector $\tilde{E}$ are given below.

$$\left[\sum_{k=1}^{\mathfrak{N}} \mathcal{E}_1 \cos \varphi_k\right]^2 = \mathcal{E}_1{}^2 \left[\sum_{k=1}^{\mathfrak{N}} \cos^2 \varphi_k + \sum_{k=1}^{\mathfrak{N}}\sum_{l=1}^{\mathfrak{N}} \cos \varphi_k \cos \varphi_l\right]$$

$$\left[\sum_{k=1}^{\mathfrak{N}} \mathcal{E}_1 \sin \varphi_k\right]^2 = \mathcal{E}_1{}^2 \left[\sum_{k=1}^{\mathfrak{N}} \sin^2 \varphi_k + \sum_{k=1}^{\mathfrak{N}}\sum_{l=1}^{\mathfrak{N}} \sin \varphi_k \sin \varphi_l\right]$$

It is to be noted that there are two kinds of sums on the right: the first single sum, for $k = l$, and the second double sum, for $k \neq l$. The second kind can be taken to vanish, when $\mathfrak{N}$ is large, because the sines and cosines will be as often positive as negative. But the first kind adds terms that are always positive. The square of the length of the hypotenuse, $\mathcal{E}^2$, is thus

$$\mathcal{E}^2 = \mathcal{E}_1{}^2 \left[\sum_{k=1}^{\mathfrak{N}} (\sin^2 \varphi_k + \cos^2 \varphi_k)\right] = \mathfrak{N}\mathcal{E}_1{}^2$$

Fig. 2-5 illustrates these sums at a particular instant for $\mathfrak{N} = 17$ components of equal amplitude.

Here $\mathcal{E}_1{}^2$ is proportional to the illumination that one component alone would produce at P taking the electric fields to be due to light. Our result states that all components together, in random phase interrelation, produce an illumination represented by $\mathfrak{N}\mathcal{E}_1{}^2$, which is equal to the sum of the illuminations which the components alone would produce. The result is the same for the superposition of many unequal components of random phase interrelation, although we shall not prove this here. If $\mathcal{E}_{av}$ is the average component amplitude, for such a case, the superposition sum is such that

$$\mathcal{E}^2 = \mathfrak{N}\mathcal{E}_{av}^2$$

This result applies to light because no means is available to synchronize the emissions of the light from the component atoms of a source. Each component atom emits its fixed frequency with an independent phase. Thus $\mathcal{E}_{av}^2$ might represent, at a point P, the average intensity of emission from sodium atoms in a Bunsen flame. Then if there were $\mathfrak{N}$ atoms in the flame simultaneously emitting toward P, the aggregate intensity at P would be $\mathfrak{N}\mathcal{E}_{av}^2$.

Lord Rayleigh's treatment (*Theory of Sound*, Vol. I, pp. 35–37) of the isoperiodic case of equal component amplitudes is somewhat different from that given above, and it is quoted below:[†]

"We have seen that the resultant of two isoperiodic vibrations of equal amplitude is wholly dependent upon their phase relation, and it is of interest to inquire what we are to expect from the composition of a large number ($\mathfrak{N}$) of equal vibrations of amplitude unity, of the same period, and of phases accidentally determined. The intensity of the resultant, represented by the square of the amplitude, will of course depend upon the precise manner in which the phases are distributed, and may vary from $\mathfrak{N}^2$ to zero. But is there a definite intensity which becomes more and more probable when $\mathfrak{N}$ is increased without limit?

"The nature of the question here raised is well illustrated by the special case in which the possible phases are restricted to two *opposite* phases. We may then conveniently discard the idea of phase, and regard the amplitudes as at random *positive* or *negative*. If all the signs be the same, the intensity is $\mathfrak{N}^2$; if, on the other hand, there be as many positive as negative, the result is zero. But although the intensity may range from 0 to $\mathfrak{N}^2$, the smaller values are more probable than the greater.

"The simplest part of the problem relates to what is called in the theory of probabilities the 'expectation' of intensity, that is, the mean intensity to be expected after a great number of trials, in each of which the phases are taken at random. The chance that all the vibrations are positive is $(\frac{1}{2})^{\mathfrak{N}}$, and thus the expectation of intensity corresponding to this contingency is $(\frac{1}{2})^{\mathfrak{N}}\mathfrak{N}^2$. In like manner the expectation corresponding to the number of positive vibrations being $(\mathfrak{N} - 1)$ is

$$\left(\frac{1}{2}\right)^{\mathfrak{N}} \mathfrak{N}(\mathfrak{N} - 2)^2$$

† From Lord Rayleigh's *The Theory of Sound*. Reprinted with the permission of Dover Publications, Inc., New York.

and so on. The whole expectation of intensity is thus

$$\frac{1}{2^{\mathfrak{N}}}\left\{1\cdot\mathfrak{N}^2 + \mathfrak{N}(\mathfrak{N}-2)^2 + \frac{\mathfrak{N}(\mathfrak{N}-1)(\mathfrak{N}-4)^2}{2\cdot1}\right.$$

$$\left. + \frac{\mathfrak{N}(\mathfrak{N}-1)(\mathfrak{N}-2)(\mathfrak{N}-6)^2}{3\cdot2\cdot1} + \cdots\right\}$$

"Now the sum of the $(\mathfrak{N}+1)$ terms of this series is simply $\mathfrak{N}$, as may be proved by comparison of coefficients of x in the equivalent forms

$$(e^x + e^{-x})^{\mathfrak{N}} = 2^{\mathfrak{N}}\left(1 + \frac{1}{2}x^2 + \cdots\right)^{\mathfrak{N}}$$

$$= e^{\mathfrak{N}x} + \mathfrak{N}e^{(\mathfrak{N}-2)x} + \frac{\mathfrak{N}(\mathfrak{N}-1)}{2\cdot1}e^{(\mathfrak{N}-4)x} + \cdots$$

The expectation of intensity is therefore $\mathfrak{N}$, and this whether $\mathfrak{N}$ be great or small."

2-8. *Superposition of Waves of Different Frequencies*

We have restricted ourselves, heretofore, to the superposition of component waves of the same frequency, such as steady-state solutions to a common second order differential wave equation for a vacuum space. The sum of two expressions that describe component waves of different frequencies, which two waves each satisfy a common differential equation, is also its solution. From this fact it follows that the principle of superposition applies equally to component waves of different frequencies.

2-9. *Transients*

As a matter of casual interest we now consider the transient solutions to our differential equations, such as the one for the driven mass of § 1-4.

Suppose we have a repetitive applied force, or even a repetitive "square wave" applied impulse. We know, from Fourier's theorem, that it is possible to represent such a complicated force, by means of a Fourier series, as a sum of component simple harmonic driving forces. Now for each of these components of the driving force we shall have a component motion of the same frequency as a solution to the differential equation. And the sum of these separate solutions represents the aggregate motion which the complicated driving force produces. Just as Fourier's series representation serves to solve for the aggregate motion that a complicated repetitive periodic driving force produces, Fourier's integral representation serves to solve for the aggregate transient motion that an aperiodic driving force produces. We may understand how this method of determining transient motions goes if we imagine a Fourier series representation that has gone to the limit where the fundamental repetition rate approaches zero. All the successive arguments, then, of the representing trigonometric series, which arguments are multiples of this fundamental repetition rate $(2\pi f_0)$, are successively different by the fundamental amount $2\pi f_0$; and as this amount $2\pi f_0$ approaches zero, the Fourier series changes to the Fourier integral. We may expect a separate motional solution for each one of the infinite number of infinitesimal elements of that driving force integral. And the sum of all these infinitesimal elementary motions, which is also an integral, gives the appropriate overall resultant transient motion. Since the procedures of manipulating Fourier series and integrals are described elsewhere, this is as far as we shall go here with these considerations.

2-10.   *An Apparent Failure of the Principle of Superposition*

In our last chapter we gave a quotation: ". . . the eye can readily single out the smallest [star], whose light is sufficient to affect it, and contemplate it, untroubled by the light of the more powerful luminaries shining in other parts of the heavens." Perhaps the strongest evidence we have to support the principle of superposition is the experience of seeing a star whose rays have come to us during a total eclipse, passing close by the eclipsed sun, without being modified in any way by the strong field of streaming sunlight which it has penetrated.

We shall finish this chapter with a quotation from Rayleigh's *Theory of Sound* which is concerned with the principle of superposition. This example, although unconcerned with optics, does illustrate an interesting complexity encountered in the field of sound where the principle of superposition apparently fails. It represents a complexity from which we are happily free in optics since the principle of superposition does not fail for light. Rayleigh says:†

"The simplest periodic functions with which mathematicians are acquainted are the circular functions expressed by a sine or cosine; indeed there are no others at all approaching them in simplicity. They may be of any period, and admitting of no other variation (except magnitude) seem well adapted to produce simple tones. Moreover it has been proved by Fourier, that the most general single-valued periodic function can be resolved into a series of circular functions, having periods which are submultiples of that of the given function. Again, it is a consequence of the general theory of vibration that the particular type, now suggested as corresponding to a simple tone, is the only one capable of preserving its integrity among the vicissitudes which it

---

† From Lord Rayleigh's *The Theory of Sound*. Reprinted with the permission of Dover Publications, Inc., New York.

may have to undergo. Any other kind is liable to a sort of physical analysis, one part being differently affected from another. If the analysis within the ear proceeded on a different principle from that according to the laws of dead matter outside the ear, the consequence would be that a sound originally simple might become compound on its way to the observer. There is no reason to suppose that anything of this sort actually happens. When it is added that according to all the ideas we can form on the subject, the analysis within the ear must take place by means of a physical machinery, subject to the same laws as prevail outside, it will be seen that a strong case has been made out for regarding tones as due to vibrations expressed by circular functions. We are not however left entirely to the guidance of general considerations like these. In the chapter on the vibration of strings, we shall see that in many cases theory informs us beforehand of the nature of the vibration executed by a string, and in particular whether any specified simple vibration is a component or not. Here we have a decisive test. It is found by experiment that, whenever according to theory any simple vibration is present, the corresponding tone can be heard, but, whenever the simple vibration is absent, then the tone cannot be heard. We are therefore justified in asserting that simple tones and vibrations of a circular type are indissolubly connected. This law was discovered by Ohm.

"According to the literal statement of Ohm's law, the ear is capable of hearing as separate tones all the simple vibrations into which the sequence of pressures may be analysed by Fourier's theorem, provided that the pitch of these components lies between certain limits.

"The principle of superposition assumed in ordinary acoustical discussions, depends for its validity upon the assumption that the vibrations concerned are infinitely small, or at any rate similar in their character to infinitely small vibrations, and it is only upon this supposition that Ohm's law finds immediate application. One apparent exception to the law has long been known. This is the combination-tone discovered by Sorge and Tartini in the last century. If two notes, at the interval for example of a Major Third, be sounded together strongly, there is heard a grave sound in addition to the two others. In the case specified, where the primary sounds, or generators, as they may conveniently be called, are represented by the numbers 4 and 5, the combination-tone is represented by 1.

"In the numerous cases where differential tones are audible which are not reinforced by resonators, it is necessary in order to carry out Helmholtz's theory to suppose that they have their origin in the vibrating parts of the outer ear, such as the drum-skin and its attachments. Helmholtz considers that the structure of these parts is so unsymmetrical that there is nothing forced in such a supposition. But it is evident that this explanation is admissible only when the generating sounds are loud, i.e., powerful as they reach the ear . . . My own experience tends rather to support the view of Helmholtz that loud generators are necessary. On several occasions stopped organ-pipes d''', e''', were blown with a steady wind, and were so tuned that the difference-tone gave slow beats with an electrically maintained fork, of pitch 128, mounted in association with a resonator of the same pitch. When the ear was brought up close to the mouths of the pipes, the difference-tone was so loud as to require all the force of the fork in order to get the most distinct beats. These beats could be made so slow as to allow the momentary disappearance of the grave sound, when the intensities were rightly adjusted, to be observed with some prevision. In this state of things the two tones of pitch 128, one the difference-tone and the other derived from the fork, were of equal strength as they reached the observer; but as the ear was withdrawn so as to enfeeble both sounds by distance, it seemed that the combination-tone fell off more quickly than the ordinary tone from the fork.

"An observation, of great interest in itself, and with a possible bearing upon our present subject, has been made by König and Mayer. Experimenting both with forks and bird-calls they have found that audible difference-tones may arise from generators whose pitch is so high that they are separately inaudible. Perhaps an interpretation might be given in more than one way, but the passage of an inaudible beat into an audible difference-tone seems to be more easily explicable upon the basis of Helmholtz's theory.

"Upon the whole this theory seems to afford the best explanation of the facts thus far considered, but it presupposes a more ready departure from superposition of vibrations within the ear than would have been expected."

Chapter **III**

Electromagnetic Waves

WE HAVE derived the differential equation for a stretched string and discussed its solutions, and the superposition of its solutions. We have compared these and other mechanical waves with light, taking light as a combination of transverse electric and magnetic oscillating fields propagating through space together. Now we shall outline the empirical formulas for electric and magnetic fields from which we derive two fundamental differential wave equations for light, one for electric and one for magnetic oscillations. We shall examine some of the simplest solutions of these differential equations that will reveal the nature of light waves. Our main objective in this chapter is to establish an easy familiarity with the representation of light as electromagnetic waves that can be described with complex numbers and associated rotating vectors in the complex plane. Students who do not yet use the vector calculus notation freely, or who do not care to prepare themselves to follow the swift derivations of the differential wave equations that are possible with it, can either resort to more pedestrian derivations,† or they may take the results we develop here for granted.

Only a small part of the whole of electromagnetic radiation is capable of exciting the retina to produce the sensation of vision. And this small part lies between the part of the total electromagnetic spectrum where radiations are mainly particle-like (such as X-rays or γ-rays) on the one hand, and the part where radiations are mainly wave-like (such as radio waves or radar) on the other. Fig. 3-1 illustrates the narrow interval that light occupies in the total electromagnetic radiation spectrum. Although light, in its complete character, partakes of both these characteristics and simulates the dominant properties of both ends of the spectrum, here we shall be concerned mainly

---

† See N. H. Frank, *Introduction to Electricity and Optics*, 2nd ed. (1950, McGraw-Hill Book Co., New York). See also F. K. Richtmyer, *Introduction to Modern Physics*, 2nd ed. (1934, McGraw-Hill Book Co., New York), Chap. IV.

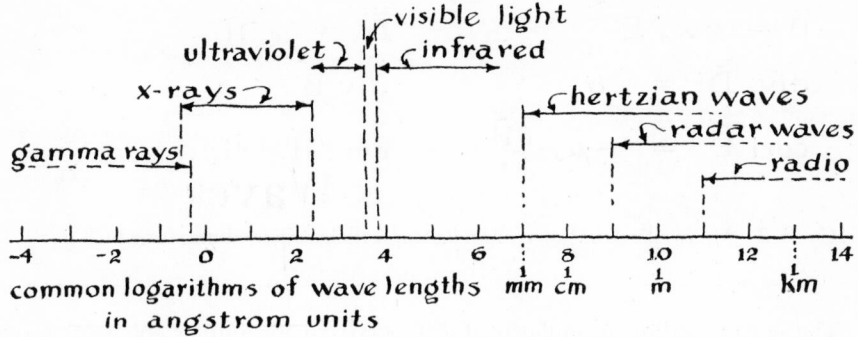

FIG. 3-1 The gamut of the electromagnetic spectrum from gamma rays to radio.

with its wave nature. It is this wave nature of light that affords explanations of the main topics of physical optics: namely, reflection and refraction; velocity and dispersion; polarization and crystals; and, finally, interference and diffraction.

The equations and mathematical formulations we shall derive for light waves provide us the mathematical basis of determining superposition sums for Huygens' wavelets as well as a quantitative basis for discussing the interaction of light and matter. The superposition of Huygens' wavelets explains interference and diffraction. Our formulations will serve to explain the interaction of oscillating electric fields with elastically bound, charged, mass particles, which we shall invoke to represent the atoms or molecules of matter. And our formulations with these representations will serve to explain absorption and emission, velocity and dispersion, reflection and refraction, and, finally, polarization and crystals.

The complex number representation of the transverse electric fields of a plane wave train of light propagating in vacuum in the positive z-direction, polarized in the xz-azimuth, is already somewhat familiar to us:

$$\tilde{E}_x = \mathcal{E}_x e^{j\omega\left(t - \frac{z}{c}\right)}$$

If the same wave were propagating in an absorbing medium of index of refraction N, with velocity $v = \dfrac{c}{N}$, the formulation representing the amplitude would become

$$\tilde{E}_x = \mathcal{E}_x e^{-\alpha z} e^{j\omega\left(t - \frac{Nz}{c}\right)}$$

Here α is the amplitude extinction coefficient, and $\mathcal{E}_x$ is the amplitude of motion at $z = 0$.

$$\bar{\bar{D}} = \kappa \kappa_\circ \, \bar{\bar{E}} \qquad\qquad \bar{\bar{B}} = \mu_\circ \, \bar{\bar{H}}$$

$$\text{div } \bar{\bar{D}} = q_v \qquad\qquad \text{div } \bar{\bar{B}} = 0$$

$$\text{curl } \bar{\bar{E}} = -\mu_\circ \frac{\partial \bar{\bar{H}}}{\partial t} \qquad \text{curl } \bar{\bar{H}} = i_v + \kappa \kappa_\circ \frac{\partial \bar{\bar{E}}}{\partial t}$$

FIG. 3-2   Maxwell's equations for an isotropic dielectric medium, for which $\mu = 1.0$.

The corresponding formulation for a polarized expanding spherical wave in vacuum is

$$\bar{\bar{E}}_t = \frac{\mathcal{E}_1}{r} e^{j\omega\left(t - \frac{r}{c}\right)} \mathbf{t}$$

where **t** is a unit vector tangent to the spherical wave front, and lying in whatever azimuth the light is polarized. In all these formulations we assume that the "clock is set," so to speak, so that it makes the phase, $\omega\left(t - \dfrac{z}{c}\right)$ or $\omega\left(t - \dfrac{r}{c}\right)$, equal to zero *when* $t = 0$, and *where* z or r is zero. The factor $\dfrac{1}{r}$, in the case of the spherical wave, is equivalent to the inverse square law "geometric attenuation factor," $\dfrac{1}{r^2}$. According to this well known law the illumination by a spherical wave expanding about a small source, at a point P, is inversely proportional to the square of the distance from P to the source.

Such expressions as those above are solutions to the electromagnetic differential wave equations that we shall derive from the empirical laws describing the phenomena of electricity and magnetism. Maxwell has condensed these descriptive empirical laws into six differential equations, the so-called Maxwell equations. Fig. 3-2 presents these equations. Four of these equations represent our knowledge of electric and magnetic fields. The fifth represents electromagnetic induction. The sixth relates magnetic fields with electric currents and contains Maxwell's famous invention of the displacement current. We shall derive the wave equations from these six as soon as we have reviewed briefly the history of the electromagnetic nature of light.

3-1. *History of Electromagnetic Waves*

The concept of Hooke and Huygens of light as a wave motion, first put on a sound quantitative basis by the interference and diffraction experiments of

Young and Fresnel, and put in dominance over the corpuscular concept by the velocity measurements of Foucault, was even more firmly established when Maxwell derived the correct velocity of light by his theory from results of measurements that had been done entirely in the laboratory with electricity and magnetism.

In the interval after the experiments of Young and Fresnel, and before Maxwell's theory was published, the burning question was "waves in what?" A hypothetical medium, the luminiferous ether, had been invented by Huygens to fill the philosophic vacuum and serve as the answer to this question. The concept of this medium was, however, encumbered by a dilemma: the medium must be substantial enough, on the one hand, to serve as transient repository of the solar energy, as it streamed from the sun to the planets, and rigid to support transverse waves; and on the other hand, the medium must be mechanically quite tenuous so that the planets could pass through it without suffering any substantial damping.

The beginning suggestion which led to our current theory of light came from Professor Michael Faraday.† In the words of Maxwell,‡ in his famous paper of 1856: "The conception of the propagation of transverse magnetic disturbances to the exclusion of normal ones is distinctly set forth by Professor Faraday in his 'Thoughts on Ray Vibrations.' The electromagnetic theory of light, as proposed by him, is the same in substance as that which I have begun to develop in this paper, except that in 1846 there were no data to calculate the velocity of propagation."

In the paper to which Maxwell referred above, Faraday had said: "The view which I am so bold as to put forth considers . . . radiation as a high species of vibrations in the lines of force which are known to connect particles and also masses of matter together. It endeavors to dismiss the ether but not the vibrations." From the orbit of our moon around the earth, and from Newton's theory of it, as well as from the solar tides, we know that gravitational forces act over long distances. Faraday pictured each mass point of the sun as connected by a line of gravitational force with each mass point of the earth, connected in a manner not unlike two masses connected by a stretched string; and be similarly conceived of long lines of magnetic force. It was along such magnetic lines that Faraday suggested that the waves of sunlight could propagate. This was his conception.

The student will find it quite worth-while, as a lesson both in prejudice and contrasting open-mindedness, to compare an almost ridiculously critical article§ that appears in the same volume of the *Philosophical Magazine* as

---

† Michael Faraday, Phil. Mag., *28*, 345 (1846).
‡ James Clerk Maxwell, Phil. Mag., *29*, 152 (1865).
§ G. B. Airy, Phil. Mag., *28*, 532 (1846).

Faraday's article with the later article by Maxwell which presented a constructive extension of Faraday's conception.

Several attributes make Maxwell's theory masterful. First, by the invention of a displacement current Maxwell made the differential equations for electric and magnetic fields beautifully symmetrical. Second, his theory predicted the velocity of light from electrical and magnetic measurements which had been made entirely in the laboratory. And finally, his theory stimulated Heinrich Hertz to produce waves which were of the same character as light. He produced these waves by means of electric sparks in a resonant electric circuit.

The vector product of the vectors representing Maxwell's electric and magnetic fields gives Poynting's vector, correctly describing the forward flux of radiant energy. This last attribute of Maxwell's theory, coupled with our certain experimental knowledge that a vacuum can indeed support both changing electric and magnetic fields, makes a convincing case for his electromagnetic theory—even though we do not yet have an easy answer to the question "waves in what?"

3-2. *Maxwell's Equations*

The first of Maxwell's six equations defines relations between the electrical fields $\bar{D}$ and $\bar{E}$. The second relates the magnetic quantities $\bar{B}$ and $\bar{H}$ similarly. The third and fourth equations are generalizations of the Gauss theorem relating to electrical and magnetic fields, respectively. The fifth equation is a generalization of Faraday's law of induction. The sixth and last equation relates magnetic fields with the currents that produce them. This last equation is a generalization of the law of Biot and Savart, and it contains Maxwell's invention of the displacement current. This sixth equation connects variable magnetic fields with the variable electric currents that originate them, while the fifth connects the induction of variable currents with oscillating magnetic fields. These last two equations reciprocally interconnect the electric and magnetic waves which together comprise light; since, according to Maxwell, oscillating electric fields have the magnetic attributes of alternating currents.

In the first equation $\bar{E}$ is the electric field which determines force on a charge; while $\bar{D}$ is the electric displacement. $\bar{D}$, like $\bar{E}$, is a vector field. $\bar{D}$ is proportional to the placed charge density, and it is related to $\bar{E}$, in a homogeneous dielectric medium, by the equation

$$\bar{D} = \kappa \kappa_0 \bar{E}$$

Here κ_0 is the electric permittivity of free space, being determined by measurement to be 8.85×10^{-12} in the $MKS\mu_0$ units. κ is the dielectric constant. Although the dielectric constant in a crystal is a complex tensor quantity when we are concerned with varying fields, it is only a complex scalar quantity for varying fields in an isotropic medium; and it is just a plain scalar for static fields. Finally, κ, for vacuum, is unity. Thus, for vacuum, $\bar{D} = \kappa_0 \bar{E}$.

The magnetic induction, $\bar{B}$, determines the force on a moving charge (or current). $\bar{H}$ is the magnetic field. These two vector fields are similarly related: $\bar{B} = \mu\mu_0\bar{H}$. Here μ_0 is the magnetic permittivity of free space, being by international agreement $4\pi \times 10^{-7}$ in $MKS\mu_0$ units. μ is the magnetic permeability which in our optical study here is unity, both for vacuum and for glass. Thus $\bar{B} = \mu_0\bar{H}$.

The theorem of Gauss says that the enclosed charge is equal to the flux of the field $\bar{D}$ integrated over an enclosing surface. $\iint \bar{D} \cdot d\bar{S} = \iiint q_v\, dv$. Here q_v is charge density. In the limit of an enclosing volume of infinitesimally small dimensions, this equation assumes the differential form; div $\bar{D} = q_v$. And in charge-free vacuum, div $\bar{D} = 0$. Since there are no magnetic charges, the Gauss theorem for magnetic induction gives $\iint \bar{B} \cdot d\bar{S} = 0$. And in the limiting case this expression assumes the differential form, div $\bar{B} = 0$.

These four equations are valid for variable fields, which we may express with complex numbers as follows.

$$\bar{D} = \kappa_0\bar{E} \tag{a}$$

$$\bar{B} = \mu_0\bar{H} \tag{b}$$

$$\text{div } \bar{D} = \text{div } \bar{E} = 0 \tag{c}$$

$$\text{div } \bar{B} = \text{div } \bar{H} = 0 \tag{d}$$

Faraday's law of electromagnetic induction, generalized, yields Maxwell's fifth equation. This law states that the electromotive force induced in a closed circuit is the negative rate of change of flux linking that circuit. If the flux through the area enclosed by a circuit is $\Phi = \iint \bar{B} \cdot d\bar{S}$, Faraday's law states,

$$\int \bar{E} \cdot d\bar{l} = -\frac{\partial\Phi}{\partial t} = -\frac{\partial}{\partial t}\iint \bar{B} \cdot d\bar{S}$$

Now if the area enclosed by the circuit has very small dimensions, $\bar{B}$ may be taken as a constant and removed from under the integral. Dividing by the area gives the curl and we get

$$\frac{\int \bar{E} \cdot d\bar{l}}{\iint d\bar{S}} = \text{curl } \bar{E} \qquad \text{and} \qquad \frac{-\dfrac{\partial}{\partial t}\iint \bar{B} \cdot d\bar{S}}{\iint d\bar{S}} = -\frac{\partial \bar{B}}{\partial t}$$

And taking $\mu = 1.0$ for free space,

$$\text{curl } \bar{E} = -\mu_0\frac{\partial \bar{H}}{\partial t}$$

Biot and Savart's law, generalized and combined with Maxwell's invention of the displacement current, yields the sixth equation of Fig. 3-2. This law of Biot and Savart states that the magnetomotive force in a circuit is equal to the enclosed current; writing $\bar{\imath}_v$ for current density, $\int \bar{H} \cdot d\bar{l} = \iint \bar{\imath}_v \cdot d\bar{S}$. Again, we may remove $\bar{\imath}_v$ from the integrals, taking it as a constant in the limit of a vanishingly small closed circuit. On invoking the definition of the curl as a line integral per unit area, we get, similarly,

$$\operatorname{curl} \bar{H} = \bar{\imath}_v$$

Of course, in a charge-free space we must use Maxwell's displacement current for $\bar{\imath}_v$:

$$\bar{\imath}_v = \frac{\partial \bar{D}}{\partial t} = \kappa_0 \frac{\partial \bar{E}}{\partial t}$$

In the section below we explain how this curl $\bar{H}$ equation created the necessity that mothered Maxwell's invention of the displacement current; but first let us write the two last relationships with complex numbers and complete the list of Maxwell's six beautifully symmetric equations:

$$\operatorname{curl} \bar{\bar{E}} = -\mu_0 \frac{\partial \bar{\bar{H}}}{\partial t} \tag{e}$$

$$\operatorname{curl} \bar{\bar{H}} = \kappa_0 \frac{\partial \bar{\bar{E}}}{\partial t} \tag{f}$$

3-3. *Maxwell's Displacement Current*

The necessity that mothered Maxwell's invention of the displacement current, writing $\dfrac{\partial \bar{D}}{\partial t}$ for $\bar{\imath}_v$, arose from the inconsistency illustrated by Fig. 3-3. This figure shows an alternating current generator with a connected condenser. It is well known that an alternating current can flow in such a circuit with no real current flowing across the space between the condenser plates. If we were to integrate the real current density, $\bar{\imath}_v$, over the indicated surface around the upper condenser plate, we should find that our integral does not vanish—

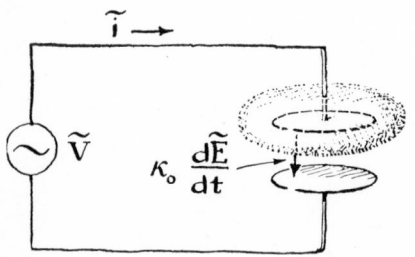

FIG. 3-3 Illustration of Maxwell's principle of no open circuits with alternating currents.

meaning that div $\bar{i}_v$, within, does not everywhere vanish. And yet, in such a circuit, an alternating magnetic field is generated, and $\int \bar{H} \cdot d\bar{l}$ has a real value. These findings contrast sharply with what we would expect from the differential equation, curl $\bar{H} = \bar{i}_v$: if we take the divergence of both sides of this differential equation, since div curl $\bar{H}$ is identically zero for any vector field, we find, mathematically, that div $\bar{i}_v$ must be zero. Thus we have the following embarrassing inconsistency: experience with alternating currents gives div $\bar{i}_v \neq 0$, while mathematical theory gives div $\bar{i}_v \equiv 0$. The inconsistency is embarrassing because the arguments leading to these contrasting results are both valid. Obviously a serious need for interpretation is created. Maxwell removed this embarrassing inconsistency by construing $\dfrac{\partial \bar{D}}{\partial t}$, or $\kappa_0 \dfrac{\partial \bar{E}}{\partial t}$, as a second current to be added to the flow of real charge. This is called the displacement current. In the case of our condenser, the displacement current term, proportional to the changing field between the condenser plates, yields a total current exactly equal to the real current in the connecting wires—a circumstance expressed by the statement that open circuits cannot occur when we are dealing with alternating currents. Thus, according to Maxwell's sixth equation, a changing electric field in a vacuum has the attributes of an electric current, although no real electric charges are involved.

3-4. *Derivation of the Differential Wave Equation*

The equation for electromagnetic waves in free space may be derived from Maxwell's six equations in the following mathematical steps. First we take the curl of both sides of Eq. 3-2e:

$$\text{curl curl } \bar{E} = -\mu_0 \text{ curl } \frac{\partial \bar{H}}{\partial t} = -\mu_0 \frac{\partial}{\partial t} \text{ curl } \bar{H}$$

The left side, by a vector identity, is

$$\text{curl curl } \bar{E} = \text{grad div } \bar{E} - \nabla^2 \bar{E}$$

and since div $\bar{E}$ is zero in free space, from Eq. 3-2c, the left side reduces to $-\nabla^2\bar{E}$. The right side, substituting Eq. 3-2f for curl $\bar{H}$, becomes $-\kappa_0\mu_0 \dfrac{\partial^2 \bar{E}}{\partial t^2}$ after differentiation. Thus we get a wave equation for electric field.

$$\nabla^2\bar{E} = \nabla^2\bar{E}_x\mathbf{i} + \nabla^2\bar{E}_y\mathbf{j} + \nabla^2\bar{E}_z\mathbf{k} = \mu_0\kappa_0 \frac{\partial^2 \bar{E}}{\partial t^2}$$

Similarly, taking the curl of both sides of Eq. 3-2f and substituting Eq. 3-2e, and remembering that div $\bar{H} = 0$, following Eq. 3-2d, we get a differential wave equation for magnetic field.

$$\nabla^2\bar{H} = \nabla^2\bar{H}_x\mathbf{i} + \nabla^2\bar{H}_y\mathbf{j} + \nabla^2\bar{H}_z\mathbf{k} = \mu_0\kappa_0 \frac{\partial^2 \bar{H}}{\partial t^2}$$

Both of these equations are of the same character as the differential equation that we got in § 1-5 for $\bar{w}$. The solutions, too, have the same character, although here they apply to an extended medium and are consequently more complicated.

These two differential equations for $\bar{E}$ and $\bar{H}$ do not yet give us electromagnetic waves; rather, they give us independent electric fields $\bar{E}$ and magnetic fields $\bar{H}$. However, each of these fields has the character of a wave motion, and the two wave motions are indissolubly associated with one another by the reciprocal interconnections that Maxwell's equations impose, Eqs. 3-2e and 3-2f.

3-5. *Nature of Electromagnetic Waves*

We shall illustrate the interrelationship of our $\bar{E}$ and $\bar{H}$ waves by means of two compatible solutions to our two wave equations; namely, two compatible plane waves of infinite extent. We make them compatible by giving them equal angular frequencies, ω, and by taking their directions of wave propagation parallel. These simple solutions may be written

$$\bar{E} = (\mathcal{E}_x \mathbf{i} + \mathcal{E}_y \mathbf{j} + \mathcal{E}_z \mathbf{k}) e^{j\omega\left(t - \frac{z}{c}\right) + j\varphi}$$

$$\bar{H} = (\mathfrak{IC}_x \mathbf{i} + \mathfrak{IC}_y \mathbf{j} + \mathfrak{IC}_z \mathbf{k}) e^{j\omega\left(t - \frac{z}{c}\right) + j\psi}$$

Here we have written these equations without prejudice as to phase interrelationships between the fields $\bar{E}$ and $\bar{H}$ and without prejudice as to field directions relative to the common direction of propagation, or mutual polarization azimuth. The differential equations show that both fields have the same velocity, $\mu_0\kappa_0 = \dfrac{1}{c^2}$.

It is required of two such infinite plane wave motions, since the waves are of uniform strength across their wave fronts, that $\dfrac{\partial \bar{E}}{\partial x}$ and $\dfrac{\partial \bar{E}}{\partial y}$ as well as $\dfrac{\partial \bar{H}}{\partial x}$ and $\dfrac{\partial \bar{H}}{\partial y}$ must be zero. And, furthermore, the requirements of Eqs. 3-2c and 3-2d apply. Therefore, since the divergences of the fields must be zero, $\operatorname{div} \bar{E} = \dfrac{\partial \tilde{E}_x}{\partial x} + \dfrac{\partial \tilde{E}_y}{\partial y} + \dfrac{\partial \tilde{E}_z}{\partial z} = 0$; and similarly $\operatorname{div} \bar{H} = 0$. In combination, all these conditions make $\dfrac{\partial \tilde{E}_z}{\partial z}$ and $\dfrac{\partial \tilde{H}_z}{\partial z}$ both zero, meaning that $\tilde{E}_z$ and $\tilde{H}_z$ are constants. Thus $\tilde{E}_z$ and $\tilde{H}_z$ are either static fields or zero; and because, in radiation, we are not concerned with static fields, the necessary vanishment of $\dfrac{\partial \tilde{E}_z}{\partial z}$ and $\dfrac{\partial \tilde{H}_z}{\partial z}$ means that the oscillating fields of our plane waves can have no

z-components—therefore $\tilde{E}_z$ and $\tilde{H}_z$ are zero. In short, *electromagnetic waves in free space are transverse waves*, as Faraday surmised.

We accordingly set $\mathcal{E}_z$ and $\mathcal{H}_z$ in the equations above equal to zero. And if we set our clock so that $\varphi = 0$, these expressions for our compatible but as yet unrelated wave fronts reduce to

$$\bar{\bar{E}} = (\mathcal{E}_x \mathbf{i} + \mathcal{E}_y \mathbf{j}) e^{j\omega\left(t - \frac{z}{c}\right)}$$

$$\bar{\bar{H}} = (\mathcal{H}_x \mathbf{i} + \mathcal{H}_y \mathbf{j}) e^{j\omega\left(t - \frac{z}{c}\right) + j\psi}$$

To establish the interrelation of $\bar{\bar{E}}$ and $\bar{\bar{H}}$ we take the curls and time derivatives of $\bar{\bar{E}}$ and $\bar{\bar{H}}$ and substitute them in Eqs. 3-2e and 3-2f. Afterward, on equating $\mathbf{i}$ and $\mathbf{j}$ components, we get

$$\mathcal{H}_x e^{j\psi} = -c\kappa_0 \mathcal{E}_y$$

$$\mathcal{H}_y e^{j\psi} = c\kappa_0 \mathcal{E}_x$$

For one thing, these final two relations show that $\psi = 0$; *i.e.*, the *oscillations of electric and magnetic field are in phase*. The second geometrical result of these final two relations is that $\bar{E}$ and $\bar{H}$ are required to be *mutually perpendicular*.

The velocity given by the combination of κ_0 and μ_0 as they appear in our two differential equations is:

$$c = \frac{1}{\sqrt{\kappa_0 \mu_0}} = (2.997930 \pm 0.000005) \times 10^8 \frac{\text{meters}}{\text{sec}}$$

This velocity of light as determined purely from laboratory electrical and magnetic measurements agrees closely with that obtained by travel-time measurements in vacuum. The agreement between this velocity of light calculated from κ_0 and μ_0 and the velocity of light from the astronomical observations of Roemer and Bradley, as it was known in Maxwell's time, provided the original touch of authenticity for Maxwell's theory.

Hertz's discovery of electromagnetic waves in 1886–1888, when he found that the oscillations in a resonant circuit could excite sympathetic oscillations in a distant tuned circuit, provided an additional authenticity for Maxwell's theory.

Finally, a convincing result of Maxwell's theory is manifest by the time average of the vector product of $\bar{\bar{E}}$ and $\bar{\bar{H}}$. These vectors are mutually perpendicular to each other, and perpendicular to the direction of propagation as well, so that Poynting's vector lies in the direction of propagation. This vector has the dimensions of flux of energy per unit area. From the values of $\bar{\bar{E}}$ and $\bar{\bar{H}}$ for waves in vacuum, the average value of Poynting's vector is

$$(\bar{\bar{E}} \times \bar{\bar{H}})_{\text{av}} = \tfrac{1}{2} c\kappa_0 (\mathcal{E}_x^2 + \mathcal{E}_y^2)\mathbf{k}$$

$$\bar{\bar{\mathcal{P}}}_{\text{av}} = \tfrac{1}{2} c\kappa_0 \mathcal{E}^2 \mathbf{k}$$

In the $MKS\mu_0$ system of units the time average of this vector, when $\mathcal{E}$ is expressed in volts per meter, is

$$\bar{\mathfrak{P}}_{av} = 2.65 \times 10^{-3}\mathcal{E}^2\mathbf{k} \text{ watts per square meter}$$

3-6. *Light Waves in Glass*

The index of refraction of glass, or other isotropic transparent medium, is, as we determined from Huygens' principle, $N = \dfrac{c}{v}$, where v is the velocity of light in the medium, and $\mathbf{c}$ is its velocity in vacuum. It will appear in the next chapter that N for a transparent medium, at very long wavelengths, is equal to the square root of its dielectric constant κ. When the Maxwell equations are solved for such a medium, where κ and μ are not unity, just as we solved them for free space, the differential equations turn out to contain $\kappa\kappa_0\mu\mu_0$ where before we had only $\kappa_0\mu_0$. Also, such manipulations, with Maxwell's equations for plane waves, for a medium of index N, yield $\sqrt{\mu\mu_0}\mathcal{H}_g = \sqrt{\kappa\kappa_0}\mathcal{E}_g$. Since $\mu = 1.0$, $v = \dfrac{c}{\sqrt{\kappa}}$, or $N = \sqrt{\kappa}$. Thus the relationship of corresponding magnetic and electric field amplitudes in such a medium becomes

$$\mathcal{H}_g = N\sqrt{\frac{\kappa_0}{\mu_0}}\,\mathcal{E}_g$$

And Poynting's vector in this medium of index N becomes

$$\bar{\mathfrak{P}}_{av} = (\bar{E}_g \times \bar{H}_g)_{av} = \tfrac{1}{2}c\kappa_0 N\mathcal{E}^2\mathbf{k}$$

3-7. *Boundary Conditions*

At a boundary between dielectrics the definitions of $\bar{D}$ and B require that there be no discontinuity of the normal components of these vector fields there. Also, the principle of conservation of energy requires that there be no discontinuity of the tangential components of $\bar{E}$ and $\bar{H}$ there. Application of these boundary conditions to electromagnetic waves gives us a mathematical procedure for deducing reflection and transmission coefficients when light is incident on dielectric interfaces. Such application also gives directions of propagation of the reflected and refracted light. These boundary conditions serve for calculation of the diffraction of light, but the procedures of using them for such results are often very difficult to carry out mathematically. The mathematical *modus operandi* for treating the case where a wave is incident on a dielectric surface, both the incident wave and the dielectric boundary surface being represented mathematically, is to conjure up such mathematical expressions for reflected and refracted or diffracted waves as will give in combination with the mathematical expression for the incident wave, normal continuity of $\bar{D}$ and $\bar{B}$, and tangential continuity of $\bar{E}$ and $\bar{H}$ over all the dielectric boundary surface. Although this task is generally formidable, in some cases it is simple; for example, when we have an infinite plane wave incident on the infinitely large plane boundary between glass and vacuum. This simple case illustrates the physics involved in the intractable ones. In this case we take our incident plane wave of infinite extent to be incident on a vacuum-glass boundary from the vacuum side; and we use a 3-dimensional Cartesian coordinate system for the mathematical formalism that we shall need. We consider the glass surface to be represented by the equation $z = 0$, with vacuum at $z < 0$. Further, we take the incident light wave fronts to be oriented such that their normals are parallel to the plane $y = 0$. At $z > 0$, the index of refraction of the glass, $N = \sqrt{\kappa}$ makes $v = \dfrac{c}{N}$. The boundary conditions, at the glass-vacuum interface, at $z = 0$, must be satisfied at all times, as well as at all places over the boundary surface. These conditions require certain compatibilities between any reflected and refracted waves that we conjure up to combine with the incident waves.

For the associated electric and magnetic vectors of the incident light we take

$$\bar{E}_0 = \mathcal{E}_0 e^{j\omega_0\left(t - \frac{x\sin i + z\cos i}{c}\right)}$$

$$\bar{H}_0 = \sqrt{\frac{\kappa_0}{\mu_0}}\,\mathcal{E}_0 e^{j\omega_0\left(t - \frac{x\sin i + z\cos i}{c}\right)}$$

The wave fronts of this incident light are the surfaces of constant phase represented by the general equation

$$\varphi = \text{constant} = \omega t_1 - \omega\,\frac{x\sin i + z\cos i}{c}$$

For the *refracted* light we conjure up the following plane waves, each of which is a solution for Maxwell's equations.

$$\bar{E}_g = \mathcal{E}_g e^{j\omega_g\left(t - \frac{\alpha x + \beta y + \gamma z}{v}\right)}$$

$$\bar{H}_g = N\sqrt{\frac{\kappa_0}{\mu_0}}\,\mathcal{E}_g e^{j\omega_g\left(t - \frac{\alpha x + \beta y + \gamma z}{v}\right)}$$

Here v is $\dfrac{c}{N}$, and α, β and γ are the as yet unspecified cosines of the direction of propagation within the glass. Also, ω_g is an unspecified angular frequency of the fields in the glass.

For the *reflected* light we conjure up the equations,

$$\bar{\bar{E}}_{r'} = \mathcal{E}' e^{j\omega_{r'}\left(t + \frac{\alpha'x + \beta'y + \gamma'z}{c}\right)}$$

$$\bar{\bar{H}}_{r'} = -\sqrt{\frac{\kappa_0}{\mu_0}}\, \mathcal{E}' e^{j\omega_{r'}\left(t + \frac{\alpha'x + \beta'y + \gamma'z}{c}\right)}$$

Our first requirement of compatibility among these six vectors, that they satisfy the boundary conditions at all times, yields $\omega_0 = \omega_g = \omega_{r'}$.

Our next requirement of compatibility among them is that β and β' both be zero. This requirement stems from the fact that the wave amplitudes at any instant of time are of uniform strength over the surface, with respect to y; and hence, for satisfaction of boundary conditions over the whole surface, the reflected and refracted amplitudes must also be uniform with respect to y; i.e., both reflected and refracted waves must have wave front normals in the $y = 0$ plane.

Furthermore, if the refracted and reflected wave trains are to be compatible with the incident wave train over all of the boundary surface, they must have the same functional dependency on x, as the incident wave; i.e.,

$$-\alpha\frac{x}{v} = -\sin i\,\frac{x}{c}$$

$$+\alpha'\frac{x}{c} = -\sin i\,\frac{x}{c}$$

Writing $\alpha = \sin r$ and $\alpha' = \sin r'$, where r is the angle of refraction and r' is the angle of reflection, we find that these conditions yield Snell's law of refraction and the law of reflection:

$$\frac{\sin i}{\sin r} = \frac{c}{v} = N \qquad \text{and} \qquad i = -r'$$

3-8. *Fresnel Reflection Coefficients*

We use the symbol $\mathfrak{r}$ for the ratio of the amplitude of the electric field in the reflected light, to its amplitude in the incident light; and we use the symbol $\mathfrak{R}$ for the ratio of reflected to incident illumination. The values of these amplitude- and intensity-reflection coefficients, $\mathfrak{r}$ and $\mathfrak{R}$, depend on whether the incident transverse oscillations are polarized with the electric vector in the plane of incidence, $\mathfrak{r}_\pi$ and $\mathfrak{R}_\pi$; or perpendicular thereto, $\mathfrak{r}_\sigma$ and $\mathfrak{R}_\sigma$. The subscripts π and σ mean, respectively, parallel and perpendicular orientation of the electric oscillation with the plane of incidence. If r is the angle of refraction, and i that of incidence, the dependence of the $\mathfrak{r}$'s on i and r can be derived from our boundary conditions. Such a derivation yields the so-called Fresnel equations:

$$\mathfrak{r}_\sigma = -\frac{\sin (i - r)}{\sin (i + r)} \qquad \mathfrak{r}_\pi = \frac{\tan (i - r)}{\tan (i + r)}$$

These expressions are to be derived by the student (as set forth in the problem section). Here, as a guide to a procedure for these derivations, we solve for $\mathfrak{r}_\pi$ and $\mathfrak{r}_\sigma$ in the limiting case where $i = r = 0$, in which case we would expect $\mathfrak{r}_\pi = \mathfrak{r}_\sigma$.

First we write expressions for the incident, refracted and reflected vector fields, keeping the amplitudes, $\mathcal{E}_0$ and $\mathcal{E}'$, unspecified, so that they may be determined from boundary conditions. For normal incidence the $\tilde{E}$'s, and the associated $\tilde{H}$'s, of the incident, the refracted and the reflected fields are related, respectively, as follows:

$$\tilde{E}_0 = \mathcal{E}_0 e^{j\omega\left(t-\frac{z}{c}\right)}\mathbf{k} \qquad \tilde{H}_0 = \sqrt{\frac{K_0}{\mu_0}}\,\mathcal{E}_0 e^{j\omega\left(t-\frac{z}{c}\right)}\mathbf{k}$$

$$\tilde{E}_g = \mathcal{E}_g e^{j\omega\left(t-\frac{z}{v}\right)}\mathbf{k} \qquad \tilde{H}_g = \sqrt{\frac{K_0}{\mu_0}}\,N\mathcal{E}_g e^{j\omega\left(t-\frac{z}{v}\right)}\mathbf{k}$$

$$\tilde{E}' = \mathcal{E}' e^{j\omega\left(t+\frac{z}{c}\right)}\mathbf{k} \qquad \tilde{H}' = -\sqrt{\frac{K_0}{\mu_0}}\,\mathcal{E}' e^{j\omega\left(t+\frac{z}{c}\right)}\mathbf{k}$$

The first two equations describe an incident electromagnetic wave traveling at velocity **c**. The second two equations concern the refracted ray traveling in the same direction, but at velocity v. Here, as set forth in § 3-6, the associated magnetic amplitude is N times larger than it would be in vacuum for the same $\mathcal{E}_g$. The last two equations represent the reflected ray as traveling in the opposite direction, at velocity **c**. We note that $\tilde{H}'$ is given a negative sign, as is necessary in order to reverse Poynting's vector.

We get $\mathfrak{r}$ by equating tangential E's and H's where $z = 0$, and when $e^{j\omega t} = 1.0$. There and then,

$$\mathcal{E}_0 + \mathcal{E}' = \mathcal{E}_g \qquad \text{for the electric fields}$$

$$\mathcal{E}_0 - \mathcal{E}' = N\mathcal{E}_g \qquad \text{for the magnetic fields}$$

Eliminating $\mathcal{E}_g$ gives $\mathfrak{r}$:

$$\mathfrak{r} = \frac{\mathcal{E}'}{\mathcal{E}_0} = \left(\frac{1-N}{1+N}\right)$$

Now $\mathfrak{r}$ is -0.2 for glass of index 1.5. This negative sign of $\mathfrak{r}$ means a change of phase of π for the electric vector on reflection.[†] This is reminiscent of § 2-2 where we saw how the mechanical wave $\bar{w}_1$ was reflected with a change of phase of π to become $\bar{w}_2$.

Poynting's vector gives the illuminations for the incident and reflected light proportional to $\mathcal{E}_0{}^2$ and $(\mathcal{E}')^2$, so that the intensity-reflection coefficient becomes

---

† That it is the electric vector that changes phase, not the magnetic vector, was established by O. Wiener's experiment [Wiedem. Ann., Bd. *40*, 203 (1890)].

$$\mathfrak{R} = \frac{(\mathcal{E}')^2}{(\mathcal{E}_0)^2} = \left(\frac{1 - N}{1 + N}\right)^2$$

3-9. *Conventions of Sign*

If we take limiting values of $\mathfrak{r}_\pi$ and $\mathfrak{r}_\sigma$, when we let i and r go to the limit $i = r = 0$, we get

$$\lim_{i \to 0} \left\{ \mathfrak{r}_\pi = \frac{\tan (i - r)}{\tan (i + r)} \right\} = \left(\frac{N - 1}{N + 1}\right)$$

$$\lim_{i \to 0} \left\{ \mathfrak{r}_\sigma = -\frac{\sin (i - r)}{\sin (i + r)} \right\} = -\left(\frac{N - 1}{N + 1}\right)$$

However, the plane of incidence loses its meaning at $i = 0$. Thus, this disagreement in signs presents a contradiction. But this contradiction is not a real one—it is only an apparent contradiction, as the argument below will show. At grazing incidence, or when $(i + r) > \frac{\pi}{2}$, then $\tan (i + r) < 0$ and both r's are negative, and there is no contradiction. Then

$$\mathfrak{r}_\pi = \mathfrak{r}_\sigma = -1$$

At grazing incidence both r's mean the same thing; namely, a change of phase of π on reflection. It will appear, when $(i + r) < \frac{\pi}{2}$, that the contradiction resides in our convention of signs. The explanation of our difficulty lies in the fact that $\mathfrak{r}_\pi > 0$ means a phase reversal on reflection at normal incidence just as $\mathfrak{r}_\sigma < 0$ does. Their meanings come from our convention of signs: Vectors lying in the plane of incidence are reckoned negative if directed toward the boundary surface, and positive if directed away. Thus, $\mathfrak{r}_\pi > 0$ near normal incidence means that the incident vector lies almost exactly antiparallel to the reflected vector. On the other hand, vectors perpendicular to the plane of incidence are reckoned positive or negative depending on whether the field is directed to the right side or left side of the plane of incidence, as we look toward the light source.

The phase relationships at the surface both just before and just after reflection are illustrated for the parallel or perpendicular azimuths of the electric vector in the incident light, in Figs. 3-4a and 3-4b. These figures represent the vector field directions and Poynting vector fluxes at S, where reflection occurs, both before and after reflection, and for $\mathfrak{r}_\pi > 0$ and $\mathfrak{r}_\sigma < 0$. The Poynting vectors before and after reflection are seen to be in the proper directions—applying the right-hand rule for vector multiplication of $\bar{\bar{E}}$ and $\bar{\bar{H}}$.

The arbitrary negative sign we gave $\bar{\bar{H}}'$ in order to give Poynting's vector

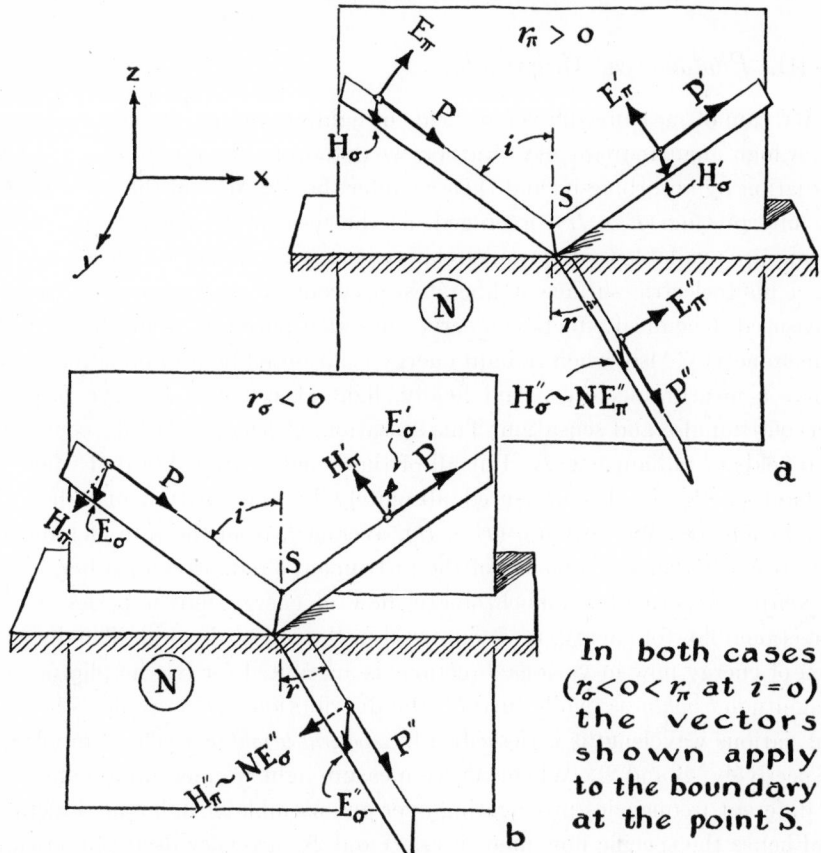

FIG. 3-4 Change of phase on reflection for electromagnetic waves polarized with
the electric vector in the plane of incidence and perpendicular thereto.

a reversal at normal reflection is in agreement with the convention of signs
which we have just described. This convention is an easy one to remember,
for it makes $\bar{\bar{E}}$ the vector that exhibits a phase reversal on reflection. However,
the opposite convention (with $\bar{\bar{H}}'$ not negative, but $\bar{\bar{E}}'$) would have been
equally workable. We must, however, in order to avoid much confusion,
choose a convention and stick to it.

Our main objective of this chapter, so far, has been to get an easy familiarity
with the expressions which describe associated electric and magnetic fields,
and to get a starting fluency in the manipulation of these expressions. This
fluency should now suffice for an introduction to the consideration of the
photometric quantities.

3-10. *Photometric Magnitudes*

We cannot measure the vector fields of light, $\overline{E}$ and $\overline{H}$, directly because of their high angular frequency. Rather, we determine the flux of energy of the radiation by absorbing it; and then we infer the $\overline{E}$ and $\overline{H}$ of the wave motion by interpreting $(\overline{E} \times \overline{H})$, averaged, as the measured energy flow. Various means are available to determine relative energy flow. When light is absorbed on a photoelectric surface it liberates electrons in amounts which may be measured. Chemical effects can serve the same purpose (as in photographic sensitometry). Also, when radiant energy is absorbed by a radiometer, it produces a measurable force. And finally, light absorbed in the eye produces nervous impulse and sensation. This sensation, at least, serves for comparing two fields of radiant energy. But all of these means are calibrated when the radiant energy is also measured absolutely by the instrument called the pyroheliometer. We estimate $(\overline{E} \times \overline{H})$, averaged, from the rate of temperature rise and thermal capacity of the instrument's radiation absorber.

Neither a perfectly monochromatic flow of energy, nor a perfectly unidirectional flow of energy, is encountered in the physical world. The distribution of energy flow in various directions is accounted for by the photometric magnitude *brightness;* while further, the distribution of energy flow throughout various wavelengths is described by *spectral brightness.* The symbols are, respectively, $\mathcal{B}$ and $\mathcal{B}_\lambda$. Whenever we measure light we measure an ensemble of different frequencies propagating over an ensemble of different directions, and hence the specific flow quantities, $\mathcal{B}$ and $\mathcal{B}_\lambda$, are very useful practically. When $\mathcal{B}_\lambda \, d\lambda$ is integrated over wavelength we get total brightness; when $\mathcal{B}$ is multiplied by area we get intensity I; and when $\mathcal{B} \, d\Omega$ is integrated over solid angle we get illumination, $\mathcal{E}$. $\mathcal{E}$ has the same dimensions, watt meter^{-2}, the same as Poynting's vector averaged.

The units of brightness and illumination, as well as intensity and flux, are given in Table 3-1 in the $MKS\mu_0$ system. The several photometric quantities for spectral brightness, spectral illumination, and spectral intensity are designated by adding the subscript λ to the several symbols and by making a corresponding alteration of dimensions. Thus the multiplication of the spectral magnitudes by a length, $d\lambda$, gives the dimensions tabulated.

The table below also gives the names of the corresponding visual photometric magnitudes. The fundamental unit of the physical magnitudes is flux and its basic measure, related back to the $MKS\mu_0$ units, comes from the rate of temperature rise of a pyroheliometer element multiplied by its heat capacity in joules per degree. The fundamental unit of the visual magnitudes is the *candela,* for intensity. The candela is $\frac{1}{60}$ of the normal intensity emitted from one square centimeter of a window that opens into a cavity maintained at the temperature of melting platinum. Within such a cavity, radiation and incandescent matter are in thermal equilibrium. This new candle, the candela, is about 2% less than the former sperm candle standard, but it is a much more definite and reproducible quantity.

TABLE 3-1

Unit	Physical Magnitudes	Visual Magnitudes
$\mathfrak{F}$	Radiant flux [watt]	Luminous flux [lumen]
$\mathfrak{J}$	Intensity; or flux per unit solid angle [watt steradian⁻¹]	[lumen-steradian⁻¹] or [candela]
$\mathfrak{E}$	Illumination; or flux per unit area [watt meter⁻²]	[lumen meter⁻²]
$\mathfrak{B}$	Brightness; or flux per unit area, per unit solid angle	
	Radiance	Luminance
	[watt meter⁻² steradian⁻¹]	[candela meter⁻²]

Because of differences in responsiveness of the normal eye to various wavelengths, as illustrated by the curves of Fig. 3-5, different amounts of flux of different colors are required to produce the same subjective response. The relationship of subjective response to irradiation by different colors, or the spectral sensitivity of the eye, as shown by the visibility curves of Fig. 3-5, is different for the rods and the cones of the retina. At the point where $\mathfrak{V}_\lambda$

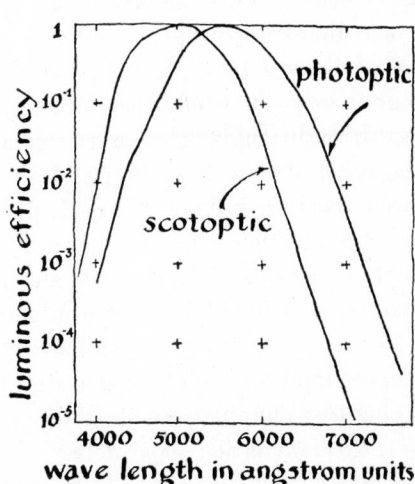

FIG. 3-5 Luminous efficiency of light in producing sensation in the eye at low (scotoptic) and at ordinary (photoptic) levels of illumination.

is 1.0, the eye requires 1467 microwatts of 5550 Å radiation to produce the same subjective response as one lumen of white light. Thus the total subjective response to any colored light, $\mathfrak{F}_V$, is the integral

$$\mathfrak{F}_V = \frac{10^6}{1467} \int_0^\infty \mathcal{V}_\lambda \mathfrak{F}_\lambda \, d\lambda \text{ lumens}$$

where $\mathfrak{F}_\lambda$ is the spectral flux measured in watts.

A comparison of white light with light of $\lambda = 5550$ Å, or the intercomparison of colored lights, which Fig. 3-5 implies, involves heterochromatic visual photometry. The visual comparison of white lights, or comparison of lights of the same color in contiguous fields, requires only that the brighter one be dimmed by a known amount until the two appear equal, or that the same field shall flash between first one of the lights and then the other, with a known attenuation of the stronger light until the sensation of flicker vanishes. But in heterochromatic photometry we must judge a subjective equality of different things. In this connection a quotation from J. W. T. Walsh's *Photometry* is cogent:†

"An observer faced with the problem of making a photometric measurement by the comparison of two surfaces differing markedly in colour is tempted at once to condemn the operation as senseless, and the result obtained as almost without meaning, and he is to a certain extent justified by the physical principle that things which differ in kind cannot be compared in degree except by some quality which is common to both. Thus, in the case of two lights of different colours, while there is no theoretical difficulty in comparing their relative energies expressed in watts, this quantity being common to all forms of radiation, there is very considerable difficulty in comparing their relative effects on the retina, since these effects are different in kind as well as in degree. This argument, however, if pushed to its logical conclusion, would almost deny the possibility of photometry at all. The position has been well expressed by C. Fabry, as follows: 'Confining oneself to the region of pure theory, one would therefore be tempted simply to condemn the problem as, by its very nature, contrary to reason. But it is not only from the theoretical point of view that the problem of heterochromatic photometry must be faced; its interest is pre-eminently practical and even commercial; the problem *demands* solution, even if it be partly by means of a convention.'

"It is self-evidently essential that the results obtained by any method of photometry should be in accordance with the ordinary laws of physical quantities, i.e. that two luminances found to be each equal to a third should also be equal to each other, and that the luminance which results from superposing two illuminations should be equal to the sum of the luminances due to each illumination separately."

† From J. W. T. Walsh, *Photometry* (1953, Constable & Co., London).

We do not go into these esoteric matters here, except to say that the situation is fortunately manageable. The student is referred to Walsh's book for the solutions of questions raised in the quotations above. We pass on to a consideration of the interrelationship of the physical photometric units, which interrelationships, happily, apply equally to both the physical and visual magnitudes.

3-11. *Interrelation of Photometric Units*

Fig. 3-6 shows an elementary black emitting surface of area dS_1. The surface of a black radiation source emits radiations in all directions in amount proportional to its projected area; *i.e.*, proportional to $\mathfrak{B}_1 \, dS_1 \cos \theta_1$. A surface thus characterized (and a window in a cavity is such a surface) is called a Lambert surface. The flux sent from an element of it, dS_1, through a second surface element, dS_2, at the distance r, is proportional to the projected area of the second element, $dS_2 \cos \theta_2$; or more properly to the solid angle it subtends at dS_1; *i.e.*, $d\Omega = \dfrac{dS_2 \cos \theta_2}{r^2}$. Thus the total flux emitted by dS_1 that also passes through a hemisphere above it, is

$$\mathfrak{F} = \int_0^{\frac{\pi}{2}} (\mathfrak{B}_1 \, dS_1 \cos \theta_1) \left(\frac{dS_2 \cos \theta_2}{r^2} \right)$$

If we take a zone on the hemisphere as dS_2, and replace r by the hemisphere radius a, then

$$dS_2 = 2\pi a^2 \sin \theta_1 \, d\theta_1$$

Entering this area element in the integral, we get the total flux:

$$\mathfrak{F} = 2\pi \mathfrak{B}_1 \, dS_1 \int_0^{\frac{\pi}{2}} d(\sin^2 \theta) = \pi \mathfrak{B}_1 \, dS_1 \sin^2 \theta \Big|_0^{\frac{\pi}{2}} = \pi \, \mathfrak{B}_1 \, dS_1$$

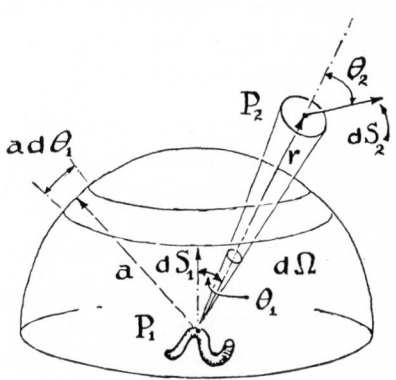

FIG. 3-6 Interrelationships between brightness, flux, and geometric factors.

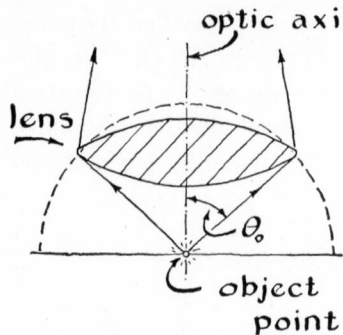

optic axis FIG. 3-7 Flux from a point source collected by a lens.

lens

object point

The illumination $\mathfrak{E}_1$ at dS_1 then is $\dfrac{\mathfrak{F}}{dS_1}$, making

$$\mathfrak{E}_1 = \pi\mathfrak{B}_1$$

Our integral for $\mathfrak{F}$ gives the flux emitted within this angle θ_0, such as might be intercepted by the lens of Fig. 3-7, as

$$\mathfrak{F} = \pi\mathfrak{B}_1 \, dS_1 \sin^2 \theta_0$$

For small angles $\Delta\Omega = 2\pi(1 - \cos\theta_0) \cong \pi\theta_0^2$; so that, for small angles $\mathfrak{F} = \mathfrak{B}_1 \, dS_1 \, \Delta\Omega = \pi\mathfrak{B}_1 \, \theta_0^2 \, dS_1$.

The solid angle subtended by less than the whole hemisphere, say within a cone of angle radius θ_0, as shown, is

$$\Omega = \int_{\theta=0}^{\theta_0} d\Omega = \int_0^{\theta_0} \frac{2\pi a^2 \sin\theta \, d\theta}{a^2} = \left. -2\pi \cos\theta \right|_0^{\theta_0}$$

Substituting $\theta_0 = \dfrac{\pi}{2}$ we get, for the whole hemisphere, $\Omega = 2\pi$. We might expect the brightness of the window in a cavity emitter to be the total emission into a half space, *i.e.*, $\mathfrak{E}_1$, divided by the area of the window and the solid angle into which the energy is emitted, 2π. But our relationship above gives $\mathfrak{B}_1 = \dfrac{\mathfrak{E}_1}{\pi}$. This factor of 2, difference between expectation and calculation, arises from the projection factor $\cos\theta$, in the integral for $\mathfrak{F}$, which does not appear in the integral for Ω.

3-12. *Energy Density of Radiation*

Let us consider an aperture of area S in the walls of a cavity from which radiation characterized by $\mathfrak{B}_1 = \dfrac{\mathfrak{E}_\lambda}{\pi}$ is emerging through each surface element of the aperture equally in each direction. That is to say, we have a homogeneous angular distribution of energy flow with respect to direction within the cavity. From the radiation propagating through S in directions θ relative to the surface normal of S, or in directions adjacent to θ lying within the solid angle $d\Omega$, we may calculate the energy in a layer of the cavity of thickness τ, adjacent to S. Fig. 3-8 illustrates this calculation. Half the radiant energy in this layer τ, propagating at the angle θ with respect to S, will pass through S in the time $dt = \dfrac{\tau \sec \theta}{c}$. This radiation, propagating in a slant direction, occupies a parallelogram of cross section $dS \cos \theta$, and of slant height $\tau \sec \theta$. The energy in the parallelogram propagating in this direction is the product of dt and the illumination $\mathfrak{B}_\lambda \, d\lambda \, d\Omega$. Or, in joules, this energy passing through dS is

$$\frac{\mathfrak{E}_\lambda}{\pi} \, d\lambda \, 2\pi \sin \theta \, d\theta \, \frac{\tau \sec \theta}{c} \times dS \cos \theta = \frac{2\mathfrak{E}_\lambda \, d\lambda}{c} \tau \, dS \sin \theta \, d\theta$$

On multiplying this energy by 2 to account for energy which would leave the layer through its other bounding surface, and then integrating over θ, we get the total energy that would lie between the bounding surfaces if the window were closed. On dividing this total energy by the volume between the bounding surfaces, $\tau \, dS$, we get a relation between the illumination in

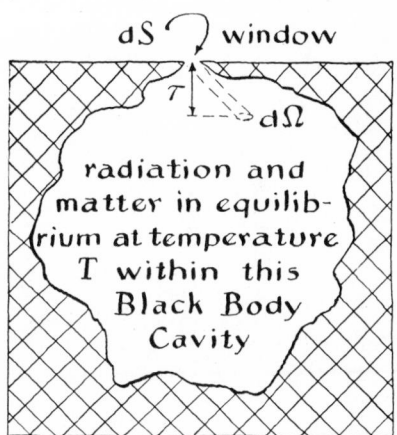

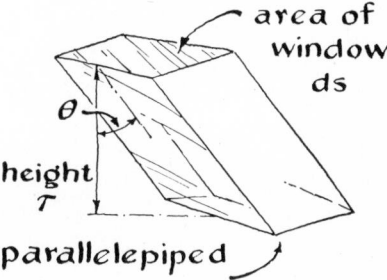

FIG. 3-8 Flux emitted through a window from a blackbody cavity containing energy density of radiation ϵ.

the cavity, $\mathfrak{E}_\lambda$, and its energy density, ϵ_λ.

$$\frac{4}{c}\,\mathfrak{E}_\lambda\,d\lambda = \epsilon_\lambda\,d\lambda \qquad \text{or} \qquad \mathfrak{E}_\lambda = \frac{c}{4}\,\epsilon_\lambda$$

This relationship is both important and useful since the derivations of the spectral distribution of energy in a cavity, for radiation in equilibrium with matter, give ϵ_λ.

Before leaving this subject of interrelationships, it is proper that we should point out here an extra engineering unit for brightness, the foot-lambert. When brightness is expressed as candles per square foot, the illumination, $\mathfrak{E}$, is $\dfrac{\mathfrak{B}}{\pi}$. However, to avoid the factor π, engineers have invented the extra unit, the foot-lambert, for which $\mathfrak{E} = \mathfrak{B}'$.

Chapter IV

Interaction of Light and Matter

WE HAVE just worked through a review of the electromagnetic theory as it pertains to light as a wave motion in free space, with velocity $c = \dfrac{1}{\sqrt{\kappa_0 \mu_0}}$. The waves were described as electric and associated magnetic fields with amplitudes for vacuum related as follows: $\mathcal{H}_\sigma = \sqrt{\dfrac{\kappa_0}{\mu_0}}\, \mathcal{E}_\pi$. Likewise the equivalent theoretical treatment for light as a wave motion in glass, or some such transparent isotropic dielectric, gave the velocity $v = \dfrac{c}{\sqrt{\kappa}}$, with $N = \dfrac{c}{v}$ and $\mathcal{H}_\sigma = N\sqrt{\dfrac{\kappa_0}{\mu_0}}\, \mathcal{E}_\pi$. Finally, we saw how the requirements of field continuity at a glass-vacuum interface, the so-called boundary conditions, yielded the laws of refraction and reflection, with Fresnel's equations for r_π and r_σ. These results afford an introduction to our present topic.

We shall now work through a treatment of the interaction of light and matter in which we determine the action of an oscillating electric field that engulfs the component molecules of matter. There are myriads of these component molecules and each may be considered to be a charged simple harmonic oscillator. When these component oscillators are driven by the engulfing electric field of the light they become excited by that field and emit Huygens-like spherical wavelets, as any oscillating dipole would. Such excitations and stimulated emissions will explain many of the phenomena of optics.

This representation of matter as oriented or random arrays, or dense or tenuous clouds of identical oscillators, may be applied to solids, liquids, and gases. As for the identical oscillators themselves, we shall first consider them as isotropic—with spherically symmetrical friction and restoring forces acting on the charges of each separate oscillator. Later, in Chapter VII, we elaborate this picture and invoke a species of oscillator characterized by anisotropic restoring forces acting on its charges. Although generally a mixture of oscillator species is required to represent optical properties realistically, our simple, single-species representation, with lumped characteristics, will suffice for present purposes, which are mainly pedagogical, and qualitative. Our treatment of liquids and vapors differs especially as regards the population density of the oscillators that we invoke to represent them. By population density we mean the number of oscillators per unit volume or, more meaningfully, the number of oscillators in a cube which is a wavelength on a side. Liquids are pictured as dense clouds, vapors as tenuous clouds; and in both we take the oscillators to be randomly oriented. Glasses are treated as congealed liquids, while crystals have their oscillators located in regular fixed positions with repeating orientations.

When we thus represent matter, and subject it to an engulfing field of light, the constituent oscillators are stimulated by means of excitation energy extracted from the engulfing light. When, then, we shall have taken account of all such stimulated emissions, as well as any residue of the light (not used up in excitation), then we shall have a basis for our optical explanations of reflection and refraction, velocity and dispersion, polarization and crystals, and finally interference and diffraction. Owing to the primitiveness of our representations, using only one species of oscillator, our considerations of these optical topics are incompletely quantitative, but the explanations are the qualitatively correct classical ones. And this classical picture serves admirably to introduce more advanced and sophisticated treatments.

To proceed with our explanations of the optical properties of matter, we first calculate the motion of the excited charge of one component typical oscillator submerged in an engulfing field of light; then, with this motion determined, we can calculate the variable dipole moment that the incident electric field induces. At first we apply this result to describe the optical properties of a tenuous cloud of oscillators, randomly oriented and spaced as in a gas. The individual typical oscillator which we invoke is considered to be comprised of a pair of equal charges, a heavy one and a light one. The lighter is characterized by a mass m, charge q, friction coefficient $m\rho$, and restoring force coefficient $m\omega_0^2$. The heavy charge is considered immobile at light frequencies, due to its large mass, M. To represent a gas by a tenuous cloud of randomly spaced oscillators, we take the population density of oscillators, $\mathfrak{N}_v$, as a relatively small number. When $\mathfrak{N}_v$ is not low, but $m\rho$ is zero, we have a second representation that can be simply treated mathematically.

It affords representation of a transparent dense medium such as a transparent glass or liquid. Finally, when $\omega_0 = 0$, we have a useful model for representation, at low radiation frequencies, of the optical properties of metals.

4-1. *Harmonic Oscillator*

We first solve for the motion of the lighter charge. We characterize our oscillator by opposite charges q, and masses m and M. The lighter charge is considered to be constrained in its motion by a friction coefficient $m\rho$; and it is considered to be bound to a rigid nucleus, of opposite charge, by a binding force $m\omega_0^2$. Fig. 4-1 shows a purely mechanical representation of this typical oscillator. All of the quantities mentioned above are expressed in the $MKS\mu_0$ practical system of units: q is in coulombs; m in kilograms; t in seconds; and the engulfing field E is expressed in volts per meter so that qE, the forcing term, is in newtons. We may think of q and m as approximating the charge and mass of an electron, and ω_0 as its resonant frequency. This frequency may correspond to some wavelength of light between 4000 Å and 7200 Å, say $\lambda = 5892$ Å as for the sodium atom. This particular frequency may be considered as representative of the lumped resonance response of the sodium atom. Actually, sodium has two resonances, one at wavelength 5890 Å and the other at 5896 Å, the first being twice as strong as the second. The motion induced into the heavier oppositely charged sodium

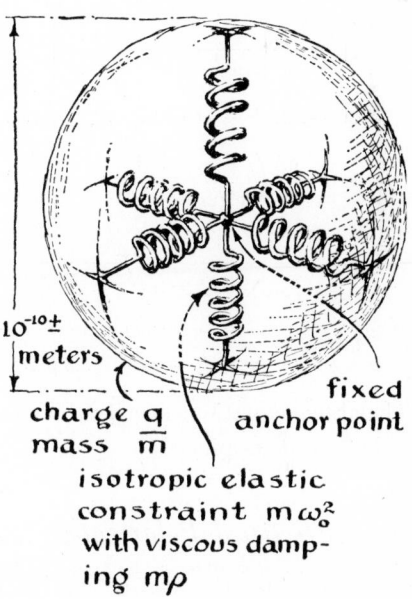

FIG. 4-1 Model of charged harmonic oscillator. A spherical shell of charge q, with associated mass m, is here bound isotropically to a fixed anchor point of opposite charge. Both charges, in the absence of an electric field, here coincide effectively.

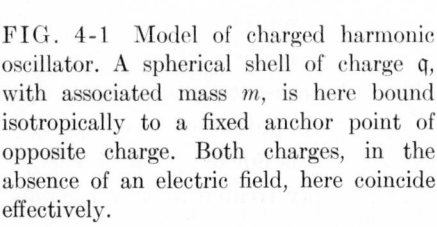

$10^{-10}\pm$ meters

charge q
mass m

fixed anchor point

isotropic elastic constraint $m\omega_0^2$ with viscous damping $m\rho$

nucleus, to which we consider the electron bound, is negligibly small, because it is thousands of times heavier than the lighter electron.

We shall now calculate the motion of the lighter charge when our typical oscillator is subjected to an engulfing field $\bar{E}$. We consider this electric field of the light beam as propagating in the z-direction, and polarized in the xz plane. Thus, with our clock properly set, the field at the oscillator will be $\bar{E} = \mathcal{E}_0 e^{j\omega t}$. This field will produce a real force on the lighter charge of $R_\mathrm{l}q\bar{E} = q\mathcal{E}_0 \cos \omega t$. This real force will lie parallel to the x-axis. The differential equation for the resulting real motion of the charge therefore is

$$m\ddot{x} + m\rho\dot{x} + m\omega_0^2 x = q\mathcal{E} \cos \omega t$$

This equation is similar to Eq. a of § 1-4. The corresponding equation, with the real motion buried in the complex variable, $\bar{w} = x + jy$, becomes

$$m\ddot{\bar{w}} + m\rho\dot{\bar{w}} + m\omega_0^2\bar{w} = q\bar{E}$$

For this equation, as before, we assume an exponential steady-state solution involving a complex constant $\tilde{A}$:

$$\bar{w} = \tilde{A}e^{j\omega t}$$

$$\dot{\bar{w}} = j\omega\bar{w}$$

$$\ddot{\bar{w}} = -\omega^2\bar{w}$$

Substitution of this motion and its derivatives in the differential equation yields the value of $\tilde{A}$, which we may write as $\tilde{A} = \mathcal{Q}e^{j\varphi}$:

$$\tilde{A} = \frac{q\mathcal{E}}{m}\left[\frac{1}{(\omega_0^2 - \omega^2) + j\omega\rho}\right]$$

$$\tilde{A} = \mathcal{Q}e^{j\varphi} = \mathcal{Q}\cos\varphi + j\mathcal{Q}\sin\varphi$$

$$\mathcal{Q} = \frac{q\mathcal{E}}{m}\left(\frac{1}{\sqrt{(\omega_0^2 - \omega^2)^2 + \omega^2\rho^2}}\right)$$

$$\left.\begin{array}{l} \cos\varphi = \dfrac{\omega_0^2 - \omega^2}{\sqrt{(\omega_0^2 - \omega^2)^2 + \omega^2\rho^2}} \\[4ex] \sin\varphi = \dfrac{-\omega\rho}{\sqrt{(\omega_0^2 - \omega^2)^2 + \omega^2\rho^2}} \end{array}\right\} \varphi = \tan^{-1} - \left(\dfrac{\omega\rho}{\omega_0^2 - \omega^2}\right)$$

The above equations specify the motions of the charged mass, both its amplitude $\mathcal{Q}$ and its phase φ.

4-2. *Electrical Polarization and* N

We now use this charge motion to calculate the index of refraction of a dielectric represented by a cloud of these typical oscillators. We get the index from the previously mentioned result from the theory of electromagnetic fields, when $\mu = 1.0$:

$$N = \frac{c}{v} = \frac{\sqrt{\kappa\kappa_0\mu\mu_0}}{\sqrt{\kappa_0\mu_0}} = \sqrt{\kappa}$$

Our oscillator motions are used first to get the dielectric constant of the cloud of the typical oscillators representing the dielectric. The dielectric constant κ is related to the polarization which an applied electric field E produces in a dielectric. From the theory of electric fields, E and the polarization of a cloud of oscillators, $\mathfrak{p}_v$, are related as follows:

$$\kappa\kappa_0 E = \kappa_0 E + \mathfrak{p}_v$$

Here $\mathfrak{p}_v$ is the product of $\mathfrak{p}$, the dipole moment of our typical oscillator, multiplied by the population density of oscillators, $\mathfrak{N}_v$. And of course, for our individual oscillator, $\mathfrak{p}$ is the dipole charge multiplied by the separation from its opposite heavy charge; namely, $\mathfrak{p} = \mathfrak{q}x$. In our complex notation, with the variable x buried in $\tilde{w}$: $\tilde{\mathfrak{p}} = \mathfrak{q}\tilde{w}$. This complex variable $\tilde{\mathfrak{p}}$ gives $\tilde{\mathfrak{p}}_v = \mathfrak{N}_v\mathfrak{q}\tilde{w}$. Substitution of the expressions we have derived for $\tilde{w}$ gives the complex polarization as

$$\tilde{\mathfrak{p}}_v = \mathfrak{N}_v\mathfrak{q}\tilde{w} = \frac{\mathfrak{N}_v\mathfrak{q}^2}{m[(\omega_0^2 - \omega^2) + j\omega\rho]}\tilde{E}$$

From this expression we get a complex dynamic dielectric constant,

$$\tilde{\kappa}\kappa_0\tilde{E} = \kappa_0\tilde{E} + \tilde{\mathfrak{p}}_v \qquad \text{or} \qquad \tilde{\kappa} = 1 + \frac{\tilde{\mathfrak{p}}_v}{\kappa_0\tilde{E}}$$

Finally, taking the square root of this $\tilde{\kappa}$, we get the complex index of refraction for our cloud of oscillators. In this complex index of refraction we write n for the real part, and jk for the negative of the imaginary part:

$$\sqrt{\tilde{\kappa}} = \tilde{N} = n - jk$$

We call $\mathrm{R}\tilde{N} = n$ the *propagation constant;* and we call k in the imaginary part of $\tilde{N}$ the *absorption constant.* The complex $\tilde{N}$ gives a complex velocity $\dfrac{c}{\tilde{N}}$, and on putting this in the expression for $\tilde{E}_g$, we find that n appears in the phase term $\varphi = \left(t - \dfrac{nz}{c}\right)$, and thus determines wave propagation; while jk combines with the exponent factor j to give a negative exponential decay term, $e^{-\omega\frac{k}{c}z}$. These meanings are illustrated if we substitute the complex $\tilde{N}$, for an absorbing medium, where we formerly had a real N for a transparent medium. Formerly, in an expression for $\tilde{E}_g$, we had

$$\tilde{E}_g = \mathcal{E}_g e^{j\omega\left(t - \frac{z}{v}\right)} = \mathcal{E}_g e^{j\omega\left(t - \frac{Nz}{c}\right)}$$

Now, writing $\tilde{N} = n - jk$ where we formerly wrote N, we get

$$\tilde{E}_g = \mathcal{E}_g e^{-\omega \frac{k}{c} z} e^{j\omega\left(t - \frac{nz}{c}\right)}$$

The negative real coefficient of z, $\alpha = \omega \dfrac{k}{c}$, is called the *extinction coefficient*.

To recapitulate the genesis of these relationships: The quantities q, m, ρ, and ω_0 determine the motion of a typical oscillator that is produced by a field $\tilde{E}$. This motion, $\tilde{w}$, and the oscillator charge determine $\tilde{p}$. $\tilde{p}$ combined with the population density of oscillators, $\mathfrak{N}_v$, determines the polarization $\tilde{p}_v$. This polarization gives $\tilde{\kappa}$. And $\tilde{N} = n - jk = \sqrt{\tilde{\kappa}}$. Or, all together,

$$n - jk = \sqrt{1 + \frac{\tilde{p}_v}{\kappa_0 \tilde{E}}} = \sqrt{1 + \frac{\mathfrak{N}_v q^2}{m\kappa_0} \left[\frac{1}{(\omega_0{}^2 - \omega^2) + j\rho\omega}\right]}$$

4-3. *Optical Properties of Sodium Vapor*

When a vapor of sodium is sufficiently tenuous, then $\dfrac{\tilde{p}_v}{\kappa_0 \tilde{E}} = \epsilon$ will be much less than unity, and the square root may be taken simply because, when $\epsilon \ll 1.0$, $\sqrt{1 + \epsilon} = 1 + \dfrac{\epsilon}{2}$, even if ϵ is complex.

$$n - jk = 1 + \frac{\mathfrak{N}_v q^2}{2\kappa_0 m}\left[\frac{1}{(\omega_0{}^2 - \omega^2) + j\omega\rho}\right]$$

Rationalizing the complex fraction, in parenthesis, and equating reals and imaginaries, we get

$$(n - 1) = \frac{\mathfrak{N}_v q^2}{2\kappa_0 m}\left[\frac{\omega_0{}^2 - \omega^2}{(\omega_0{}^2 - \omega^2)^2 + \omega^2\rho^2}\right]$$

$$k = \frac{\mathfrak{N}_v q^2}{2\kappa_0 m}\left[\frac{\omega\rho}{(\omega_0{}^2 - \omega^2)^2 + \omega^2\rho^2}\right]$$

Fig. 4-2 indicates how these values of $(n - 1)$ and k vary with ω in the vicinity of ω_0. The absorption constant exhibits a peak at ω_0, with $(n - 1)$ running through zero there.

In the solar spectrum Fraunhofer's absorption line, D, is due to such a resonance absorption peak for the sodium atomic oscillator at $\lambda = 5892$ Å (as seen in a low-power spectroscope). This dark line in the sun's continuous spectrum is due to the peaked absorption by sodium vapor in the cool outer layers of the sun's atmosphere.

The variation of $(n - 1)$ near ω_0 gives negative values, meaning phase velocities v which are greater than c. This circumstance is particularly inter-

FIG. 4-2 Variation of propagation constant (n) and absorption constant (k) with frequency, near the resonant frequency.

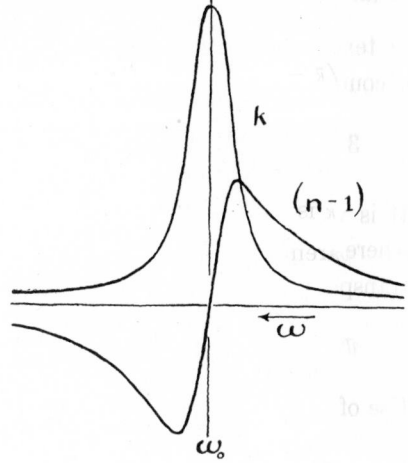

esting; and an experiment with sodium vapor which demonstrates $v > c$ is explained in the next chapter.

4-4. *Optical Properties of Glass*

Above, we have represented matter by means of only one species of typical oscillators. Glasses and liquids are, of course, actually comprised of several species of oscillators, and we should have solved for the polarization due to each of these superimposed species, and then added these polarizations together to get the total polarization. It is polarizations which are to be superimposed, and not indices of refraction, or some other quantity. To illustrate: let us represent a glass as a superimposed mixture of v species of oscillators, individuals of each species being characterized by the magnitudes q_i, m_i, ρ_i and ω_i. If the population densities for each of these species is $\mathfrak{N}_i$, then the resultant total polarization, when the same applied electric field simultaneously engulfs all, will be:

$$\frac{\tilde{\mathfrak{p}}_v}{\kappa_0 \tilde{E}} = \sum_{i=1}^{v} \frac{\mathfrak{N}_i q_i^2}{m_i \kappa_0} \left[\frac{1}{(\omega_i^2 - \omega^2) + j\omega\rho_i} \right]$$

If our oscillators form a dense cloud so that the $\mathfrak{N}_i$'s are large, the oscillators will no longer behave independently in the engulfing field; but they will polarize each other. This interaction is due to the fields that the oscillators themselves emit to engulf each other. In this case, the relationship between $\tilde{\kappa}$ and $\tilde{\mathfrak{p}}_v$, which we used before, is no longer adequate. That relationship,

$$\tilde{\kappa} - 1 = \frac{\tilde{\mathfrak{p}}_v}{\kappa_0 \tilde{E}}$$

must be modified to take account of these interactions between oscillators, by means of the correction factor, $\dfrac{3}{\bar{\kappa} + 2}$. The student will find this factor derived in texts on electricity and magnetism. The relationship corrected to take account of oscillator interactions becomes

$$3\left(\frac{\bar{\kappa} - 1}{\bar{\kappa} + 2}\right) = \frac{\tilde{p}_v}{\kappa_0 \tilde{E}}$$

It is very difficult to solve this relation for $\sqrt{\bar{\kappa}}$ except in the special case where κ is not complex; *i.e.*, when all ρ_i's are negligible as when we have a transparent glass. Neglecting the ρ_i's, the expression for $\tilde{w}_i$ becomes

$$\tilde{w}_i = \frac{q\tilde{E}}{m_i(\omega_i{}^2 - \omega^2)}$$

Use of these $\tilde{w}_i$'s makes $\tilde{p}_v$ real. Then $\bar{\kappa}$ becomes real, and $\tilde{N}$ is real also.

4-5. *Formula of Clausius and Mossotti*

With the above real expression for $\tilde{w}_i$ we now derive the equation of Clausius and Mossotti which gives the index of a transparent homogeneous mixture in terms of the index of each pure component and its concentration in the mixture.

Here we consider a transparent medium as a mixture of ν species of oscillators with characterizing population densities, $\mathfrak{N}_i$, and with the individual oscillators of each species characterized by q_i, m_i and ω_i. The $\mathfrak{N}_i$'s can be expressed in terms of the partial densities of the components δ_i and the total oscillator mass M_i. The sum of polarizations, giving κ and N, is

$$\sum_{i=1}^{\nu} \frac{\delta_i}{M_i}\left[\frac{q_i{}^2}{3\kappa_0 m_i} \frac{1}{(\omega_i{}^2 - \omega^2)}\right] = \frac{\tilde{p}_v}{3\kappa_0 E} = \frac{\kappa - 1}{\kappa + 2} = \frac{N^2 - 1}{N^2 + 2}$$

In practice we solve for N by representing the term in brackets above by R_i. The R_i's are determinable from observations on each pure species i (having density d_i, and index N_i). The term R_i is called the *refractivity*. Substitution of R_i for the brackets, in the sum above, gives the equation of Clausius and Mossotti:

$$\sum_{i=1}^{\nu} \frac{\delta_i}{d_i} \frac{N_i{}^2 - 1}{N_i{}^2 + 2} = \frac{N^2 - 1}{N^2 + 2}$$

4-6. *Cauchy's Formula*

The index of a transparent optical glass varies with the frequency of the light much as the index of sodium vapor varies with frequency; and this variation may be represented throughout the visible spectrum by the equation of § 4-5. When we lump all oscillator characteristics and represent them by a cloud of representative oscillators characterized by $\mathfrak{N}_v$, q, m and ω_0:

$$\frac{N^2 - 1}{N^2 + 2} = \frac{\mathfrak{N}_v q^2}{3\kappa_0 m} \left(\frac{1}{\omega_0^2 - \omega^2}\right) \tag{a}$$

It is possible to lump these characteristics because the significant ω_0's, for all the ν species comprising optical glass, fall in the ultraviolet.

In 1836 Cauchy put forth a simple and very useful formula which expressed the variation of the index of optical glasses with wavelength. Cauchy's formula may be derived from the formula above if we invoke several approximations, all of which depend on the fact that $\omega_0 \gg \omega$. Cauchy's formula has enjoyed much use because of its simplicity—it uses only two empirical constants A and B:

$$N = A + \frac{B}{\lambda^2}$$

First we solve Eq. 4-6a for $N^2 - 1$, writing $\dfrac{\omega}{\omega_1} = \dfrac{\lambda_1}{\lambda}$ and $\omega_1^2 G = \dfrac{\mathfrak{N}_v q^2}{\kappa_0 m}$, with

$\omega_1^2 \cong \omega_0^2 - \dfrac{\mathfrak{N}_v q^2}{3\kappa_0 m}$, giving

$$N^2 - 1 \cong \frac{G}{1 - \dfrac{\lambda_1^2}{\lambda^2}}$$

Because $\lambda \gg \lambda_1$, and since ω_1 is not greatly less than ω_0, the application of the binomial theorem gives

$$N^2 \cong (1 + G) + G\left(\frac{\lambda_1}{\lambda}\right)^2$$

On applying the binomial theorem again, we get Cauchy's formula:

$$N \cong (\sqrt{1 + G}) + \left(\frac{G\lambda_1^2}{2\sqrt{1 + G}}\right)\frac{1}{\lambda^2}$$

4-7. *Optical Properties of Metals at Low Frequencies*

In adapting our model for matter to explain the optical properties of sodium vapor (§ 4-3) by taking a low density for our cloud of oscillators we were able to remove a radical and express $\tilde{N} = n - jk$ explicitly. And in adapting our model for matter to explain transparent solids, to get the theory that culminated above in Cauchy's equation, it was necessary for us to ignore absorption by the oscillators, setting $\rho = 0$. We now adapt our model for matter to explain the optical properties of metals, letting the binding forces become vanishingly weak. Thus, in the limit, our model comes to represent a cloud of free charges. It happens that the computed optical properties of such a cloud do approximately describe the optical properties of metals at low frequencies; and at such low frequencies, the electrical conductivity of metals and these optical properties are simply related.

When we set $\omega_0 = 0$, our differential equation for the motion of a charge becomes

$$m\ddot{w} + m\rho\dot{w} = q\dot{E} = q\mathcal{E}e^{j\omega t}$$

The steady state solution of this differential equation is

$$\dot{w} = -\frac{q}{m}\frac{\mathcal{E}e^{j\omega t}}{\omega^2 - j\rho\omega}$$

This expression for the free charge motion gives $\tilde{\kappa}$, and from $\tilde{\kappa} = \tilde{N}^2$ we get

$$(n - jk)^2 = \tilde{\kappa} = 1 + \frac{\tilde{p}_v}{\kappa_0\tilde{E}} = 1 - \frac{\mathfrak{N}_f q^2}{m\kappa_0}\left(\frac{1}{\omega^2 - j\rho\omega}\right)$$

If the population density of the free or unbound charges is $\mathfrak{N}_f$, the current density which a field $\tilde{E}$ induces will be $\tilde{\imath}_v = q\dot{w}\mathfrak{N}_f = j\omega q\tilde{w}\mathfrak{N}_f$. Thus the direct and alternating electrical conductivities for our cloud are given by the ratio $\dfrac{\tilde{\imath}_v}{\tilde{E}} = \tilde{\sigma}$. For alternating current

$$\tilde{\sigma} = \frac{\tilde{\imath}_v}{\tilde{E}} = +\frac{q^2\mathfrak{N}_f}{m\rho}\left(\frac{\rho}{\rho + j\omega}\right)$$

The direct current conductivity at $\omega = 0$ is

$$\sigma_0 = \frac{q^2\mathfrak{N}_f}{m\rho}$$

Using this value σ_0, the alternating current becomes

$$\tilde{\sigma} = \frac{\sigma_0}{\sqrt{1 + \left(\dfrac{\omega}{\rho}\right)^2}} e^{-j\tan^{-1}\left(\frac{\omega}{\rho}\right)}$$

σ_0 is the handbook value of conductivity. It is useful here because it contains $\mathfrak{N}_f$, q, m, and ρ—quantities that are needed to calculate the optical properties of metals at low frequencies. Writing this handbook value in the expression above, $\tilde{\kappa} = (n - jk)^2$, rationalizing $\dfrac{1}{\omega^2 - j\omega\rho}$, and then equating reals and imaginaries, we get the following expressions:

$$n^2 - k^2 = 1 - \frac{\sigma_0}{\kappa_0\rho}\left[\frac{1}{1 + \left(\dfrac{\omega}{\rho}\right)^2}\right] \cong 1 - \frac{\sigma_0}{\kappa\rho}$$

$$nk = \frac{\sigma_0}{2\kappa_0\rho}\left(\frac{\rho}{\omega}\right)\left[\frac{1}{1 + \left(\dfrac{\omega}{\rho}\right)^2}\right] \cong \frac{1}{2}\frac{\sigma_0}{\kappa_0\omega}$$

At low frequencies where $\omega \ll \rho$, these expressions take on the approximate simpler forms indicated above at the right. And these simpler forms, at very low frequencies, require that $n^2 - k^2$ must remain constant, while nk grows very large. This means that n and k become approximately equal at low frequencies: In the limit, setting $n = k$ makes $k = \sqrt{\dfrac{\sigma_0}{2\kappa_0\omega}}$. At $\omega = 1.9 \times 10^{13}$ (for $\lambda = 100\mu$), taking $\kappa_0 = 8.85 \times 10^{-12}$, and using the handbook value of $\sigma_0 = 5.8 \times 10^7$ mho per meter for copper, we get $k = 850$. Such values of k may be used to calculate $\mathfrak{R}$ as follows.

We calculate $\tilde{r}$ for normal incidence from Fresnel's equation of § 3-8, simply putting $\tilde{N}$ for N:

$$\tilde{r} = \frac{1 - \tilde{N}}{1 + \tilde{N}} = \frac{(1 - n) + jk}{(1 + n) - jk}$$

And the intensity reflection coefficient, $\mathfrak{R}$, is obtained from this amplitude reflection coefficient by multiplying $\tilde{r}$ by its complex conjugate $\tilde{r}^*$. Since $k \gg 1.0$, our assumption $k = n$ gives

$$\mathfrak{R} = \tilde{r}\tilde{r}^* = 1 - \frac{4k}{2k^2 + 2k + 1} \cong 1 - \frac{2}{k}$$

This theory for metals was originally developed by Hagen and Rubens, and its predictions were compared by them with laboratory measurements of $\mathfrak{R}$ at very low frequencies. This theory is valid for metals at $\lambda \geq 100\mu$. Unfortunately it is of only qualitative value for the near infrared and of no value at all for the visible or ultraviolet spectrum.

4-8. *Optical Properties of Metals for Light*

The theory of the interaction between visible light and metals is not simple. Many observations on the optical properties of metals for visible light have been made using thin films deposited on glass plates by means of the process of thermal evaporation. The obtained values of n and k for thermally evaporated silver and aluminum films are given in Table 4-1. The values for

TABLE 4-1   *Optical Constants for $\lambda = 5896$ Å*

Metal	n	k
Silver	0.1	4
Aluminum	1.0	6
Germanium	4.0	0.1

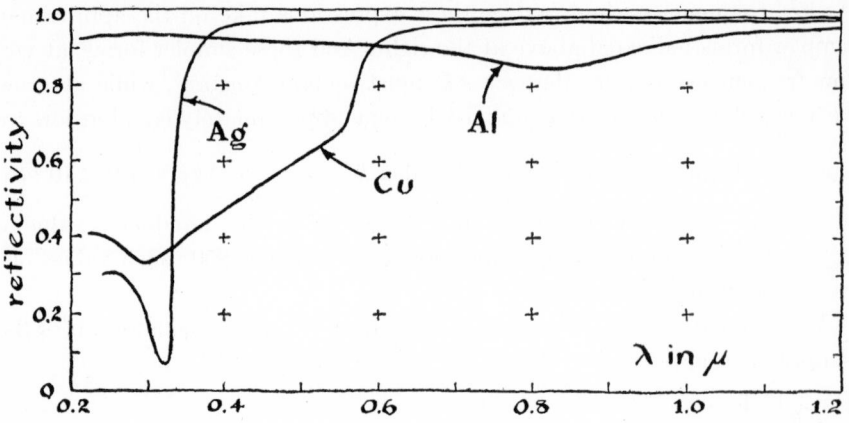

FIG. 4-3   Reflectivity of evaporated metal films, $\Re$.

germanium are also given, since this material contrasts interestingly with silver.

Although the theory of thin absorbing films is far too complex for consideration here, the following formulas for massive metal surfaces giving $\Re_\pi$ and $\Re_\sigma$, and their dependence on angle of incidence i, are interesting:

$$\left.\begin{aligned}\Re_\pi &= \frac{(n - \sec i)^2 + k^2}{(n + \sec i)^2 + k^2} \\[2mm] \Re_\sigma &= \frac{(n - \cos i)^2 + k^2}{(n + \cos i)^2 + k^2}\end{aligned}\right\} \Re_{i=0} = \frac{(n - 1)^2 + k^2}{(n + 1)^2 + k^2}$$

Fig. 4-3 shows the reflectivities of massive silver, aluminum, copper, and gold for wavelengths throughout the visible, and for the adjacent ultraviolet and infrared. Although sufficiently thin films of all metals are found to be partially transparent, silver is so selectively transparent at or near 3160 Å, where it exhibits a minimum of reflectivity, that it is often used as an ultraviolet filter. For example, R. W. Wood used thin silver films as a filter to take his ultraviolet photographs of the moon. Although to the eye the crater Aristarchus appears to be one of the brightest areas on the moon, Wood's ultraviolet pictures showed it to be relatively dark. Because sulfur appears light to the eye, but reflects ultraviolet light poorly, Professor Wood explained the darkness of Aristarchus, in his ultraviolet pictures, as due to lunar deposits of sulfur.

The reflection of light by metals may be qualitatively ascribed to induced alternating currents in the metal surface. The motions of individual free charges in the metal surface, when they are excited by a common engulfing

incident light, are all synchronous. Because of this synchronism, the amplitudes of the wavelets which the free charges emit, rather than their intensities,
will add. These synchronous wavelet amplitudes add up to produce strong
waves. As contrasted with the behavior of Huygens' wavelets, as described
in § 1-1, the re-emitted wavelets here cancel any residue of the original exciting wave in the forward direction. And furthermore, here, because of different phase shifts on re-emission, the back wave is strong and the forward wave
is absent, also contrasting with the results of § 1-1.

The Huygens-like wavelets emitted here by free electrons are the same in
character as emission from an oscillating dipole, as treated below.

4-9. *Dipole Emission*

As we saw above, a harmonically variable dipole moment may be written as
a complex number, $\tilde{p} = q\tilde{w} = p_0 e^{j\omega(t+\varphi)}$, where $p_0 = qx_0$ and $x = x_0 \cos(\omega t + \varphi)$.
The real motion of oscillating charges here is equivalent to a real current
$i = q\dfrac{dx}{dt}$ or to the complex current $\tilde{i} = q\dfrac{d\tilde{w}}{dt}$. This harmonic current may be
expected to produce an associated harmonic magnetic field just as a displace-
ment current $\kappa_0 \dfrac{d\tilde{E}_\pi}{dt}$ produces a magnetic field $\tilde{H}_\sigma$, as defined by Maxwell's
equations. These harmonic magnetic fields and the variable electric fields of
the dipole, together, propagate away from the dipole with velocity **c** as
spherical electromagnetic waves. These fields, at a distance r from the dipole,
and at the time t, are a result of what happened at the dipole at an earlier
time, $\left(t - \dfrac{r}{c}\right)$, where $\dfrac{r}{c}$ is the travel time required for the spherical wave to
get itself expanded to a radius r.

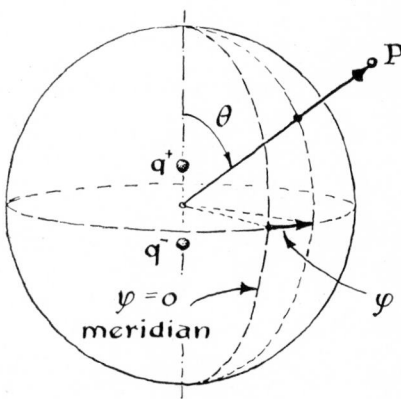

FIG. 4-4   Coordinates used to express
the emission of an oscillating dipole.

When Maxwell's equations are applied to this problem of spherical waves they yield the associated fields $\tilde{E}$ and $\tilde{H}$. These fields are expressed below by equations in polar coordinates. The coordinates are oriented with respect to the direction of the dipole oscillation, $\theta = 0$, as shown in Fig. 4-4. These field equations, just as those we developed earlier for plane waves, are solutions to the differential wave equations for $\tilde{E}$ and $\tilde{H}$; and they have the interrelation required by Maxwell's equations. These fields are:

$$\tilde{H}_r = \tilde{H}_\theta = \tilde{E}_\varphi = 0$$

$$\tilde{E}_r = \frac{2\mathfrak{p}_0\omega^3}{4\pi\kappa_0 \mathbf{c}^3}\left\{\frac{j\mathbf{c}^2}{\omega^2 r^2} + \frac{\mathbf{c}^3}{\omega^3 r^3}\right\}\cos\theta e^{j\omega\left(t-\frac{r}{\mathbf{c}}\right)}$$

$$\tilde{E}_\theta = -\frac{\mathfrak{p}_0\omega^3}{4\pi\kappa_0 \mathbf{c}^3}\left\{\frac{\mathbf{c}}{\omega r} - \frac{j\mathbf{c}^2}{\omega^2 r^2} - \frac{\mathbf{c}^3}{\omega^3 r^3}\right\}\sin\theta e^{j\omega\left(t-\frac{r}{\mathbf{c}}\right)}$$

$$\tilde{H}_\varphi = j\frac{\mathfrak{p}_0\omega^3}{4\pi \mathbf{c}^2}\left\{\frac{j\mathbf{c}}{\omega r} + \frac{\mathbf{c}^2}{\omega^2 r^2}\right\}\sin\theta e^{j\omega\left(t-\frac{r}{\mathbf{c}}\right)}$$

When r is large, so that we may ignore terms involving higher orders of $\left(\dfrac{\mathbf{c}}{\omega r}\right)$, the only remaining components of field are

$$\left.\begin{array}{l}\tilde{E}_\theta = -\dfrac{1}{r\kappa_0 \mathbf{c}}\left\{\dfrac{\mathfrak{p}_0\omega^2}{4\pi \mathbf{c}}\right\}\sin\theta e^{j\omega\left(t-\frac{r}{\mathbf{c}}\right)} \\[3mm] \tilde{H}_\varphi = -\dfrac{1}{r}\left\{\dfrac{\mathfrak{p}_0\omega^2}{4\pi \mathbf{c}}\right\}\sin\theta e^{j\omega\left(t-\frac{r}{\mathbf{c}}\right)}\end{array}\right\} \quad \tilde{H}_\varphi = \sqrt{\frac{\kappa_0}{\mu_0}}\,\tilde{E}_\theta$$

These expressions represent a polarized spherical wave with the plane of polarization of the electric field parallel to the polar axis.

When the above expressions for field vectors are cross multiplied to yield Poynting's vector, we find that illumination is proportional to $\dfrac{1}{r^2}$, as required by the inverse square law, or "geometric attenuation factor" referred to in the beginning of Chapter III. The student can show that the time average of $(\tilde{E} \times \tilde{H})$ for the emission direction θ has the magnitude

$$\mathfrak{E}_{\text{av}} = \frac{1}{2r^2}\frac{\mathfrak{p}_0^2\omega^4}{(4\pi)^2\kappa_0 \mathbf{c}^3}\sin^2\theta \quad [\text{watts per square meter}]$$

Then he can get the total flux of radiation emitted by the dipole if he integrates this illumination over the whole spherical wave front. This integration gives

$$\mathfrak{F} = \iint (\tilde{E} \times \tilde{H})\cdot d\tilde{S} = \frac{\mathfrak{p}_0^2\omega^4}{12\pi\kappa_0 \mathbf{c}^3} \quad [\text{watts}]$$

4-10. *Radiation Pressure*†

Light produces many effects in its interaction with matter. It ejects photo-electrons; it induces the property of electrical conduction in photoconductors; it produces fluorescence and phosphorescence in crystals; it heats any surface which absorbs it; in photography, as one example, it produces photochemical response; and in the retina of the eye it produces nervous response. These topics need not be treated here, for the student may learn of them in atomic physics, photochemistry, and physiology. But it is cogent for us briefly to consider radiation pressure and to review the laws describing the character of radiations when they are in thermal equilibrium with matter. Such equilibrium radiation is called *blackbody radiation*.

When radiation falls on a metallic surface and is reflected, the pressure it exerts may be explained qualitatively as follows: Consider a field of polarized light incident normally on an ideal metal surface (with assumed unit reflectivity). This field induces alternating currents, or oscillating charges, in the metal surface. The oscillating electrons in the metal surface are driven by the incident field $\overline{E}$ in directions that would lie parallel to the surface if they were not subjected also to the associated incident magnetic field, $\overline{H}$, which lies perpendicular to both the induced current and $\overline{E}$. This magnetic field deflects the oscillating electrons in the direction of propagation of the incident light. Of course all electron motions interact eventually with the crystal matrix structure which contains them. The lateral collisions of the electrons with the lattice structure of the metal, however, cancel each other so that there is no net lateral force. But there are unbalanced thrusts in the direction of propagation of the incident light. These unbalanced forces give a net force which is the origin of the radiation pressure.

There is a relationship between $(\bar{\bar{E}} \times \bar{\bar{H}})$ for the incident light and the pressure produced when we have total reflection or absorption. We do not deduce this relationship here; but it may be found elsewhere—as in Harnwell's *Principles of Electricity and Electromagnetism*, McGraw-Hill, 1949, p. 578. It is shown there that the pressure produced is ϵ' when a plane wave is absorbed or $2\epsilon'$ when reflected. Here ϵ' is the energy density of the normally incident monodirectional radiation. Poynting's vector, averaged, has the dimensions of watt meter^{-2}. If we divide $(\bar{\bar{E}} \times \bar{\bar{H}})$ by c we get a quantity that has the dimensions watt sec meter^{-3}—the same as the dimensions for energy density and also the same as the dimensions for pressure. Consider now radiation of energy density ϵ propagating homogeneously at all angles. We

† Richard A. Beth [Phys. Rev., *50*, 115 (1936)] succeeded in measuring the torque produced by the angular momentum of circularly polarized light in interaction with a crystal.

may compute the overall pressure it exerts from an average of the normal pressures produced by its directional components. The normal pressure of each directional component will be proportional to the cosine of its angle of incidence on the absorbing surface. The illumination must be multiplied by an area-foreshortening factor, $\cos \theta$. Finally, the illumination will be given by the brightness multiplied by a solid angle, $d\Omega$, proportional to $2\pi \sin \theta \, d\theta$. Thus the average of the normal pressure components of all incident irradiations is

$$\frac{\int_0^{\frac{\pi}{2}} \cos^2 \theta \, 2\pi \sin \theta \, d\theta}{\int_0^{\frac{\pi}{2}} 2\pi \sin \theta \, d\theta} = \frac{1}{3}$$

And the total pressure exerted on a black surface by such homogeneous radiation, of overall density ϵ, is $\frac{\epsilon}{3}$.

4-11. *Blackbody Radiation Laws*

It is important that the student become familiar with Planck's distribution law for blackbody radiation and with the associated laws presented below. The derivations of these laws are not given here. These derivations will be encountered later in courses on atomic physics.

Planck's distribution law for the density of the radiation in a cavity, in thermodynamic equilibrium with the cavity walls at some absolute temperature T, is

$$\epsilon_\lambda = \frac{8\pi hc}{\lambda^5} \left(\frac{e^{-\frac{hc}{\lambda kT}}}{1 - e^{-\frac{hc}{\lambda kT}}} \right)$$

h is Planck's constant; k is Boltzmann's constant; and T is the absolute temperature of the cavity.

When $\frac{hc}{\lambda kT} \gg 5$, the exponential in the parenthesis above effectively dominates, giving Wien's simpler distribution law as an asymptotic form of Planck's equation.

$$\epsilon_\lambda = \frac{8\pi hc}{\lambda^5} e^{-\frac{hc}{\lambda kT}}$$

At the other extreme, when $\frac{hc}{\lambda kT} \ll 1.0$, expansion of the exponential in

Planck's equation in power series gives the so-called Rayleigh-Jeans simpler distribution law as the other asymptotic form of Planck's equation:

$$\epsilon_\lambda = \frac{8\pi \mathbf{k} T}{\lambda^4}$$

From $\pi\mathfrak{B} = \mathfrak{E} = \frac{c}{4}\epsilon$ we easily convert these energy densities to determine illuminations and brightnesses. Fig. 4-5 shows Planck's equation and its two simpler asymptotic forms plotted as a function of λT.

On differentiation of Planck's equation, and setting $\frac{\partial \epsilon_\lambda}{\partial \lambda} = 0$, we get the condition for determining the wavelength and temperature at which ϵ_λ, or $\mathfrak{E}_\lambda$, has its maximum value. Equating this derivative to zero, and writing $x = \frac{\mathbf{hc}}{\lambda \mathbf{k} T}$, gives

$$5(e^x - 1) = xe^x$$

This transcendental equation has as its root x at approximately 5—actually

---

FIG. 4-5 Planck's radiant energy distribution law for radiation in equilibrium with matter, and the Wien and Rayleigh-Jeans asymptotic forms of that law. Here the ordinate y is proportional to the illumination, $\mathfrak{E}_\lambda$, that is emitted through a window in a cavity.

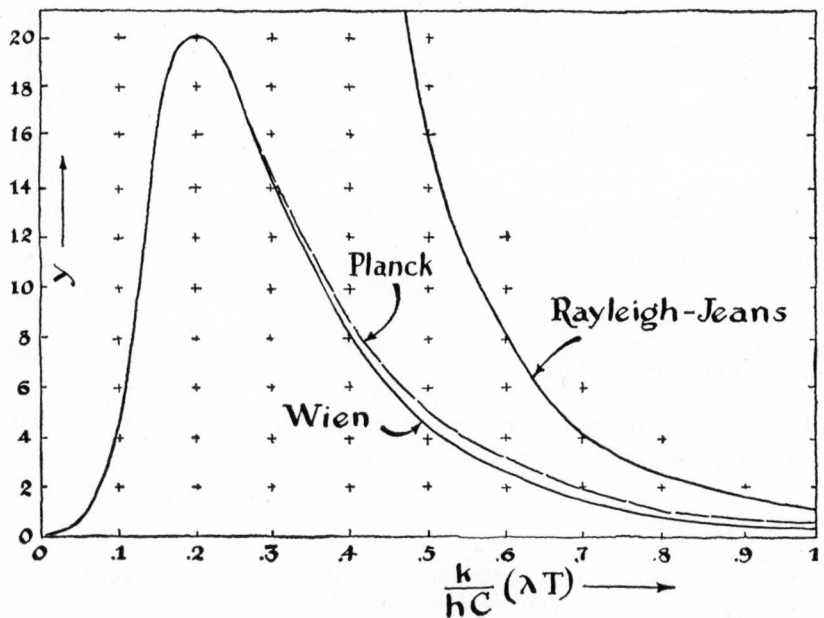

$x = 4.965$. Thus λ_m, the wavelength of the maximum illumination, $\mathfrak{E}_\lambda$, occurs when

$$\lambda_m T = \frac{hc}{4.965\,\mathbf{k}} = 2898 \quad \text{[micron-degree]}$$

It will be useful for the student to be aware of the range of validity of the simpler distribution laws, as illustrated both by Fig. 4-5, and by the following algebraic comparisons between them and Planck's law.†

Wien's equation holds very well for low temperatures and short wavelengths; but it begins to be unsatisfactorily representative of Planck's equation near the blackbody maximum at $\frac{hc}{\lambda_m \mathbf{k} T} = 4.965$. There

$$\mathfrak{E}_{\text{Wien}} = \left\{ \frac{e^{4.965} - 1}{e^{4.965}} \right\} \mathfrak{E}_{\text{Planck}} = \frac{142.3}{143.3} \mathfrak{E}_{\text{Planck}}$$

In contrast with this, the values given by the Rayleigh-Jeans equation at the blackbody maximum are not even of the same order of magnitude as those given by Planck's equation:

$$\mathfrak{E}_{\text{Rayleigh-Jeans}} = \left\{ \frac{e^{4.965} - 1}{4.965} \right\} \mathfrak{E}_{\text{Planck}} = 28.7\,\mathfrak{E}_{\text{Planck}}$$

It is evident that for the Rayleigh-Jeans equation to be a good approximation of Planck's equation we must go to wavelengths very much longer than this maximum. Expanding exponentials when $\lambda \gg 5\lambda_m$, we get

$$\frac{\mathfrak{E}_{\text{Rayleigh-Jeans}}}{\mathfrak{E}_{\text{Planck}}} = \frac{e^{5\frac{\lambda_m}{\lambda}} - 1}{5\frac{\lambda_m}{\lambda}} \cong 1 + \frac{5}{2}\frac{\lambda_m}{\lambda} + \frac{25}{3!}\left(\frac{\lambda_m}{\lambda}\right)^2 + \cdots$$

For example, applying this series to the sun, at $T = 6000°$K, when the blackbody emission maximum lies at $\lambda_m = 0.48\mu$, the Rayleigh-Jeans equation is in error 10%—and the ratio of illuminations, Rayleigh-Jeans to Planck, is 1.10, even at 12μ where $5\frac{\lambda_m}{\lambda} = \frac{1}{5}$.

4-12. *Total Radiation*

The equation of Stefan-Boltzmann relates the total thermal radiation density with the temperature. We may derive it as follows:

$$\epsilon = \int_0^\infty \epsilon_\lambda\, d\lambda$$

---

† Alfred H. Canada, in *General Electric Review*, Dec. 1948, describes a radiation slide rule that the student can acquire. It is very useful in applications of the radiation laws.

Letting $x = \dfrac{hc}{\lambda kT}$, as before,

$$\epsilon = 8\pi \frac{(kT)^4}{(hc)^3} \int_0^\infty \frac{x^3 e^{-x}}{1 - e^{-x}}\, dx$$

The integrand here can be expanded and integrated term by term to give $\sum_1^\infty \dfrac{1}{n^4} = \dfrac{\pi^4}{90}$, as written. Since $\mathfrak{E} = \dfrac{c}{4}\,\epsilon$, we get

$$\mathfrak{E} = \frac{2\pi^5 k^4}{15 c^2 h^3} = 5.67 \times 10^{-5}\, T^4\, \frac{\text{watt}}{\text{meter}^2}$$

Chapter V

Velocities and Scattering of Light

In the last chapter we wrote **c** for the velocity of light in vacuum and $v = \dfrac{c}{N}$ for its velocity in a transparent medium. For an absorbing medium we wrote the velocity complex: $\tilde{v} = \dfrac{c}{n - jk}$. This complex velocity produced two exponents in the wave equation: n, the real part of $\tilde{v}$, produced an imaginary exponent, $j\omega \dfrac{nz}{c}$, that described phase; while k, the imaginary part of $\tilde{v}$, produced the real negative exponent, $\omega \dfrac{kz}{c} = \alpha$, that described attenuation.

We determine v from measurement of N with a refractometer, $\dfrac{c}{N} = v$. But the travel time for the energy in a pulse of light in a dispersive medium, in which N varies with wavelength, is not $\dfrac{z}{v}$, but $\dfrac{z}{u}$, where u is the so-called *group velocity*. The value of u is less than v by the amount $\lambda \dfrac{dv}{d\lambda} = -\dfrac{\lambda}{N^2} \dfrac{dN}{d\lambda}$. Our explanation of group or pulse velocity involves a fictitious concept, the concept of monochromatic light.

When a light source is a white hot incandescent solid, its emission, of course, is not monochromatic but contains all the visible spectrum, and more. By means of a monochromator we may narrow the spectral band width of this emission, and purify it from all component colors but one dominant wavelength band; or we may isolate only one of many nearly monochromatic spectrum lines emitted, such as are emitted by an electrically excited atomic vapor; but none of these radiations will contain only one pure frequency.

Rather, each will be characterized by a continuous band of frequencies—although the characterizing band may be a very narrow band indeed. Thus, available light is never strictly monochromatic. But before we go into the consequences of this universal lack of monochromaticity, let us briefly review an early attempt to measure the velocity of light.

Galileo was among the first to try to measure the travel-time required for light to traverse a known path length. He tried to do this as it had been done successfully for sound, by means of a relay signaling method. He provided himself with a lantern and shutter, and with an assistant similarly outfitted and suitably trained. He and his assistant took up widely separated stations where they could see each other, and they consecutively exposed and eclipsed their lanterns by hand. This procedure was of course too crude, for light has a very great velocity. Although the principle that is inherent in Galileo's method was usable, it was only after great refinements of chopping, returning of the same light by mirrors, and after great finesse of timing had been introduced, that the principle yielded accurate group velocities. Light travels so quickly, in fact, that it would pass through a hole in the earth from north pole to south pole in just less than 24 milliseconds, if such a hole existed. Thus the velocity is so great that there was no possibility of Galileo's method being practical. Even across this vast distance from pole to pole the human eye and the hand would be too sluggish for a precise measurement.

Ole Roemer was the first to know the magnitude of the velocity of light (in 1675). He assumed a finite velocity of light to explain the variations in the observed periodicity of successive eclipses of Jupiter's moon. Taking τ as the period of the repetition of this event, the variations in τ, observed over a year, were explained as due to variations of line of sight components of the distance that the earth moved in the interval τ. Early observers did not agree on the value of τ, for one reason or another. As a result, Roemer's explanations of the variations of eclipses were not at first generally accepted, even by Roemer's own collaborators; and his explanations had to wait half a century for acceptance, until James Bradley had discovered the aberration of light and thus got a confirming value for its velocity.

Bradley discovered the aberration of light while striving to measure stellar parallax. The stellar parallax of a relatively near star is the apparent shift of its position, compared to the distant background stars, that is caused by the earth's motion. Bradley's measurements of star positions revealed a consistent annual shift of angular position of *all* the stars in the heavens. And furthermore, the stars were shifted in a direction that was parallel with the vector of the earth's velocity, in its motion around the sun. Since the earth's velocity, and not its displacement, was found by Bradley to be parallel with this stellar shift, the observed shifts could not be ascribed to parallax; rather,

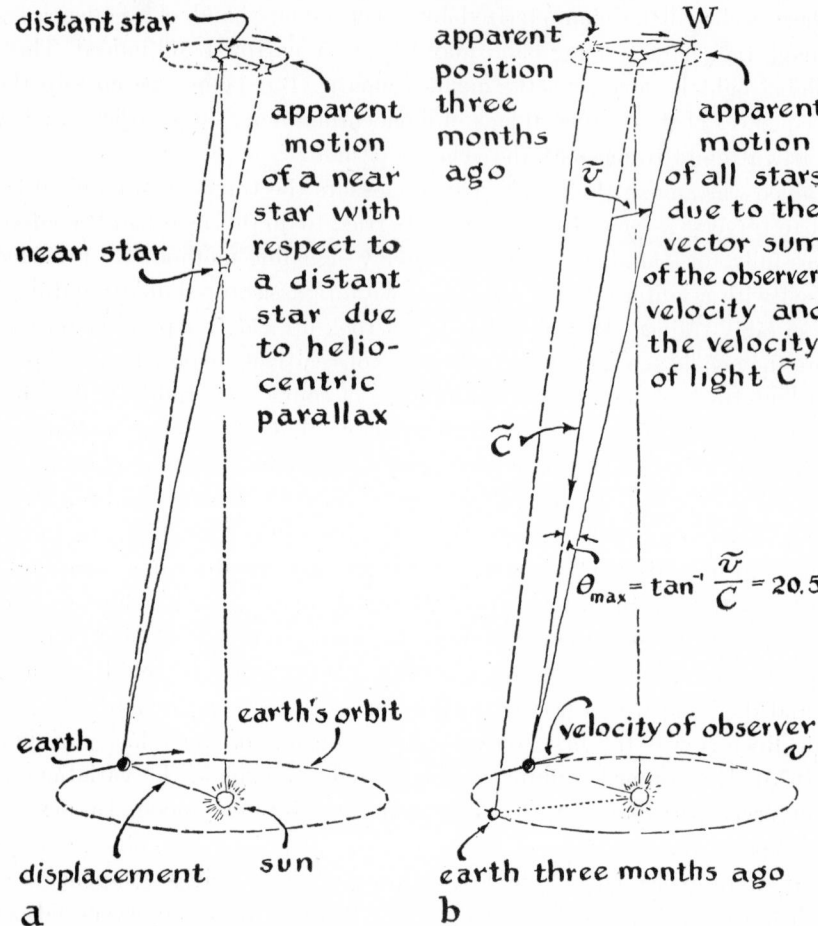

FIG. 5-1 Diagrammatic representation of stellar parallax (a), caused by the earth's variation of position, and of aberration (b), caused by the earth's velocity.

they were produced by a vector combination of the earth's velocity and the velocity of light, as illustrated in Fig. 5-1. From the known maximum velocity of the earth in its orbit, and from the measured maximum shifts (20.5 seconds of arc), Bradley inferred a velocity for light that was in agreement with the one Roemer had gotten a half century earlier. Nowadays the velocity of light is known so very precisely that the situation is reversed; the velocity is now used to yield the astronomical parameters that were formerly used to obtain it.

The methods of measuring the velocity of light are now so refined, in fact, that travel-time measurements will yield the separation of an observing and an associated station that are only a few miles apart on the earth's surface. In these modern applications of the velocity of light for surveying, the error in distance determination is only about one inch.

5-1. *Spectral Lines in Absorption and Emission*

Fig. 4-2 shows how our theory for an absorbing vapor gave k greatest at the frequency where charge displacements were greatest; *i.e.*, at the resonant frequency of the charged harmonic oscillators that we invoked to represent the vapor atoms. In § 4-9 our theory gives the emission of a dipole as proportional to the maximum charge displacement, and hence greatest at the resonant frequency, where the oscillator "wants to oscillate."

When sodium atoms are heated in a Bunsen flame, their kinetic impacts with other atoms would be expected to excite all oscillation frequencies, with the greatest deflections at the resonant frequency. Thus the emission of thermally excited sodium atoms would be expected to be greatest at frequency ω_0. This fact, that a heated gas emits the most light at the frequencies where its absorption constant k is greatest, can be predicted by thermodynamic reasoning. Excitations other than kinetic bombardment, such as electrical discharges, also produce strong emission at or near such resonant frequencies ω_0, and produce spectral lines. And because the resonance is sharp, the light of a spectral line is usually quite monochromatic—but, as stated above, it is not completely so.

The two main wavelengths at which sodium vapor emits strongly when heated in a Bunsen burner lie at 5890 Å and 5896 Å. These two wavelengths are lumped together here and represented by $\lambda = 5892$ Å. The corresponding frequency is $\nu = 5.1 \times 10^{14}$ cycles per second, or 510 million megacycles. From $2\pi c = \lambda \omega_0$, we get $\omega_0 = 32 \times 10^{14}$ radians per second for the angular frequency.

When a Bunsen flame containing sodium vapor is viewed with a spectroscope of low dispersion, we see this lumped resonance as a single emission line at 5892 Å. But when a very bright continuous white light source, such as the old lime light, or a carbon arc crater, is viewed with a spectroscope through such a Bunsen flame, doped with sodium atoms, the 5892 Å spectral component of the brighter white background light is absorbed by the sodium vapor of the cooler Bunsen burner, and the displayed spectrum of the white light shows absorption at 5892 Å, *i.e.* a dark absorption line.

Bunsen and Kirchhoff were the first to make these observations, and to correctly infer that their dark absorption line, at 5892 Å, originated from the same process as that which produced Fraunhofer's D line in the solar spectrum. A paper of Kirchhoff (1859), translated for Magie's *Source Book in Physics*, is quoted below to illustrate this discovery:†

"While engaged in a research carried out by Bunsen and myself in common on the spectra of colored flames, by which it became possible to recognize the qualitative composition of complicated mixtures from the appearance of their spectra in the flame of the blow pipe, I made some observations which give an unexpected explanation of the origin of the Fraunhofer Lines and allow us to draw conclusions from them about the composition of the sun's atmosphere and perhaps also of that of the brighter fixed stars.

"Fraunhofer noticed that in the spectrum of a candle flame two bright lines occur which coincide with the two dark lines D of the solar spectrum. We obtain the same bright lines in greater intensity from a flame in which common salt is introduced. I arranged a solar spectrum and allowed the sun's rays, before they fell on the slit, to pass through a flame heavily charged with salt. When the sunlight was sufficiently weakened there appeared, in place of the two dark D lines, two bright lines; if its intensity, however, exceeded a certain limit the two dark D lines showed much more plainly than when the flame charged with salt was not present.

"The spectrum of the Drummond light generally contains both the bright sodium lines, if the glowing part of the lime cylinder has not been long exposed to the heat; if the cylinder remains unbroken these lines become weaker and finally disappear. If they have disappeared or are very weak, and if an alcohol flame in which salt is introduced is placed between the lime cylinder and the slit, then in place of the bright lines two dark lines appear remarkably sharp and fine, which in every respect correspond with the D lines of the solar spectrum. Thus the D lines of the solar spectrum have been artificially produced in a spectrum in which they do not naturally occur.

"If we introduce lithium chloride into the flame of a Bunsen burner, its spectrum shows a very bright, sharply defined line which lies between the Fraunhofer lines B and C. If we allow the sun's rays of moderate intensity to pass through the flame and fall on the slit, we shall see in the place indicated the lines bright on a darker ground; when the sunlight is stronger there appears at that place a dark line which has exactly the same character as the Fraunhofer lines. If we remove the flame the line disappears completely, so far as I can see.

"I conclude from these observations that a colored flame in whose spectrum bright sharp lines occur so weakens rays of the color of these lines, if they pass through it, that dark lines appear in place of the bright ones, whenever a source of light of sufficient intensity, in whose spectrum these lines are other-

---

† From W. F. Magie, *A Source Book in Physics* (Harvard University Press, Cambridge).

wise absent, is brought behind the flame. I conclude further that the dark lines of the solar spectrum, which are not produced by the earth's atmosphere, occur because of the presence of those elements in the glowing atmosphere of the sun which would produce in the spectrum of a flame bright lines in the same position. We may assume that the bright lines corresponding with the D lines in the spectrum of a flame always arise from the presence of sodium; the dark D lines in the solar spectrum permit us to conclude that sodium is present in the sun's atmosphere."

Some remarks from Piazzi Smyth's *Teneriffe, an Astronomer's Experiment*, referring to the earlier year, 1857, are of some historical interest in this connection:

"We turned therefore with some hope to optical questions, wherein the height on which we were placed [Teneriffe, a mountain in the Canaries] was everything, and the nature of the ground nothing. Foremost among these, came the subject of black lines in the spectrum; primarily discovered by Wollaston, and, secondly, but quite independently, by Fraunhofer, and so much better taken account of practically by him, as to be now generally known by his name. First viewed as so many defects or originally missing portions of the sun's light, suspicions were afterwards raised as to their being caused by the absorption of our own atmosphere. But in that case, even allowing for a moment such cosmical egotism, as the light of our own luminary is perfect in itself, and the earth's atmosphere the sole cause of evil,—how comes it that the light of stars, some of them many times larger than our sun, on passing through the same atmosphere, presents in each of their spectra a different series of black lines?"

Of course the answer came from Kirchhoff: the relatively cool outer absorbing mantle of gas around the sun or a star produces the observed dark lines. And this gaseous layer often has a different chemical composition in different stars. Spectroscopic analysis thus had its beginnings in this far-flung application, yielding from the start qualitative chemical analyses of objects which were millions of miles away, or more, as well as analyses of proximate flames. Later, by such analysis, helium was discovered on the sun before it was found on the earth.

5-2. *Natural Line Width*

Classically, an excited sodium atom will expend its energy by radiating it. This expenditure of radiation occurs at a rate given by the equation of § 4-9 for the flux, $\mathfrak{F}$, emitted by a radiating dipole. This dissipation of energy by radiation will itself dampen the oscillations, even if the mechanical friction term ρ of our oscillator were zero. Such a loss of oscillator deflection is called

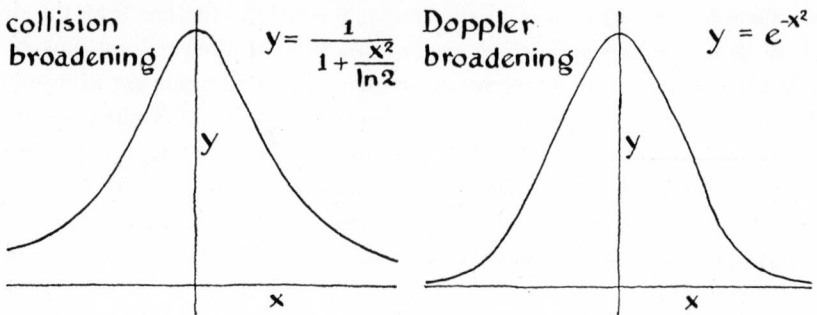

FIG. 5-2 Spectrum illumination contours (y) as functions of wavelength displacement from the line center (x), such as that produced by collision broadening and Doppler broadening.

radiation damping. It is of such a character that the oscillator deflections decay according to an exponential law. The total flux emitted by a radiation-damped oscillator has the following time dependence: $\mathfrak{F} = \mathfrak{F}_0 e^{-\frac{t}{r}}$. Now the Fourier integral representation of a decaying electric field $\bar{E}$, such as we have here, contains not only ω_0, the natural frequency of the oscillator, but frequencies on either side of ω_0 as well. An appendix defines this degradation of monochromaticity. The resultant band of wavelengths, calculated to represent a decaying emission, is shown to have the distribution of Fig. 5-2a, where the half width of the band is

$$\Delta\lambda_N = \frac{2\pi}{3} \frac{q^2}{mc^2}$$

Here q and m have the meanings given in § 4-1. This $\Delta\lambda_N$ is called the natural line width.

5-3. *Doppler Line Width*

When an emitting source of light, like a sodium atom, is advancing, or receding, along a line of sight, then the observed frequencies emitted will all appear increased or decreased. This frequency change is of the same character as the changes in the pitch of sound from a moving whistle, altered by its motion of advance or recession. This effect is, of course, the Doppler effect. For a moving sodium atom emitting the wavelength λ_0, the observed wavelength will be $\lambda_0 \left(1 \pm \dfrac{v}{c}\right)$, where v is the atomic velocity of advance $(-)$ or recession $(+)$. Doppler shifts of frequency afford another reason why the frequencies observed in radiations from sodium atoms in a Bunsen flame do not all lie at the resonant frequency but include frequencies on either side of it. Because of different velocities of sodium atoms, to and fro, the observed light assumes the form of a continuous narrow band of radiations, as shown in Fig. 5-2b, rather than a single spectral line.

Because frequencies of the same atom are the same everywhere, even on stars and planets, the Doppler effect has been much used by astrophysicists to determine the line of sight velocities of astronomical bodies. This use is illustrated in accompanying figures. Fig. 5-3 illustrates spectra of the east and west limbs of the sun, as well as a spectrum of its center. Here the spectra

FIG. 5-3 Doppler effect of the east and west limbs of the sun's disk on the position of spectrum lines, as compared with spectrum lines from the midpoint of the sun's disk.

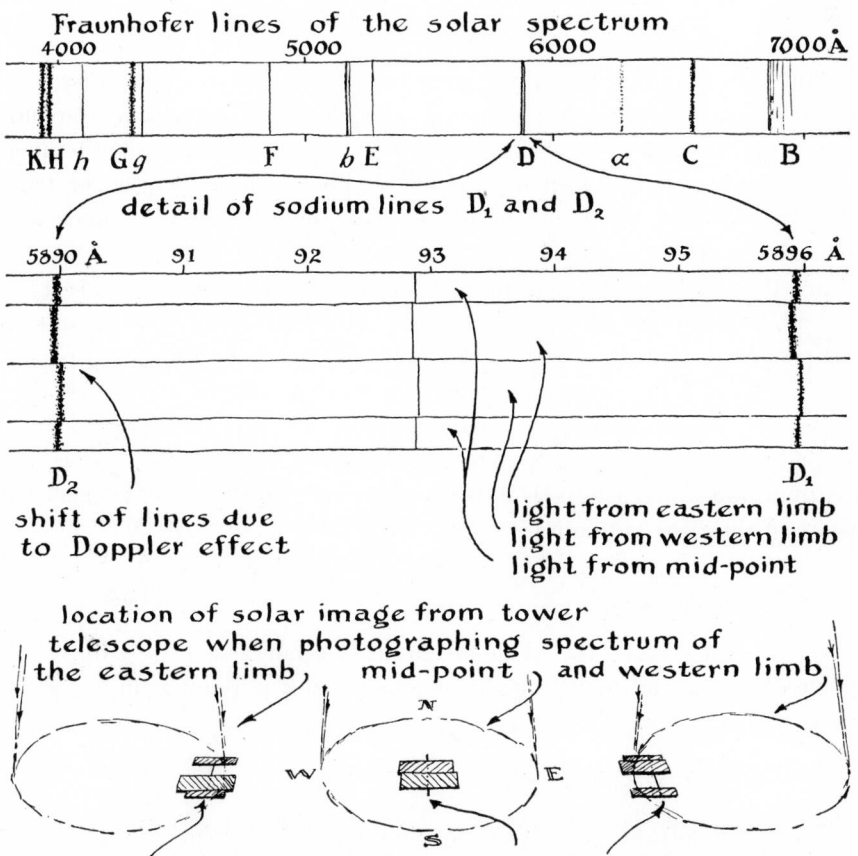

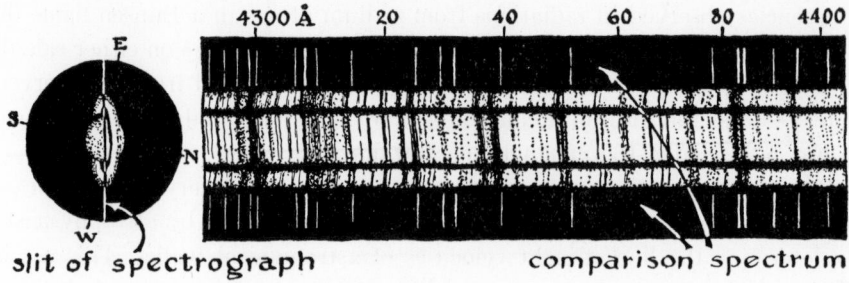

slit of spectrograph   comparison spectrum

FIG. 5-4 Doppler effect in the reflection spectrum of Saturn's ball and rings. [From a photograph by E. C. Slipher in the *Encyclopaedia Britannica*, Vol. 17, p. 998 (1944).]

---

are put in juxtaposition for comparison, to show apparent line shifts in the limb spectra that arise from the recession and advance of these limbs due to the solar rotation (once every 25 days). Fig. 5-4 shows how the spectrum of Saturn and his rings, focused on a spectrograph slit, displays Doppler shifts arising from the rings' orbital rotation. This figure shows comparison spectra to reveal the velocity of the planet as a whole. Relative motions of double stars may be similarly determined. Finally, Fig. 5-5 shows the enormous Doppler shifts of the Fraunhofer H and K lines as found in the spectra of distant nebulae. It is on such observations as this that the hypothesis of an expanding universe is based.

In a Bunsen flame some sodium atoms will be approaching the spectrograph and others receding, and most of the atomic velocities will have at least some component along the line of sight. All these various velocities and components arise because of the various kinetic energies of the sodium atoms in the hot flame. In such a thermally excited source we see the superposition of the myriad of these different Doppler shifted emissions of sodium atoms. The distribution of wavelengths is illustrated also in Fig. 5-2b. The Doppler half width of this distribution is

$$\Delta\lambda_D = \lambda_0 \sqrt{\frac{2kT \ln 2}{Mc^2}}$$

Here M is the mass of the whole oscillator, or atom, as contrasted to m, the reduced mass of the oscillating charge alone, and k is Boltzmann's constant.

5-4. *Pressure Broadening*

Another important broadening effect impressed on the emission of spectrum lines arises from the relatively violent intermolecular collisions which interrupt atomic emissions, thus limiting the free time, τ, during which an oscillator can execute its characteristic frequency, ω_0. For example, an oscillator whose emission is violently interrupted at $t = 0$, and again at $t = \tau$, so that it emits a radiation pulse of duration τ, will contain other frequencies than ω_0. The Fourier integral representation of such a pulse, of duration τ, taking the emission as zero at $t < 0$ and at $t > \tau$, requires side bands as well as the dominant frequency ω_0. It is because of these side bands that the monochro-

FIG. 5-5 Doppler effect observed in the spectra of galaxies by M. L. Humason of the Mount Wilson Observatory.

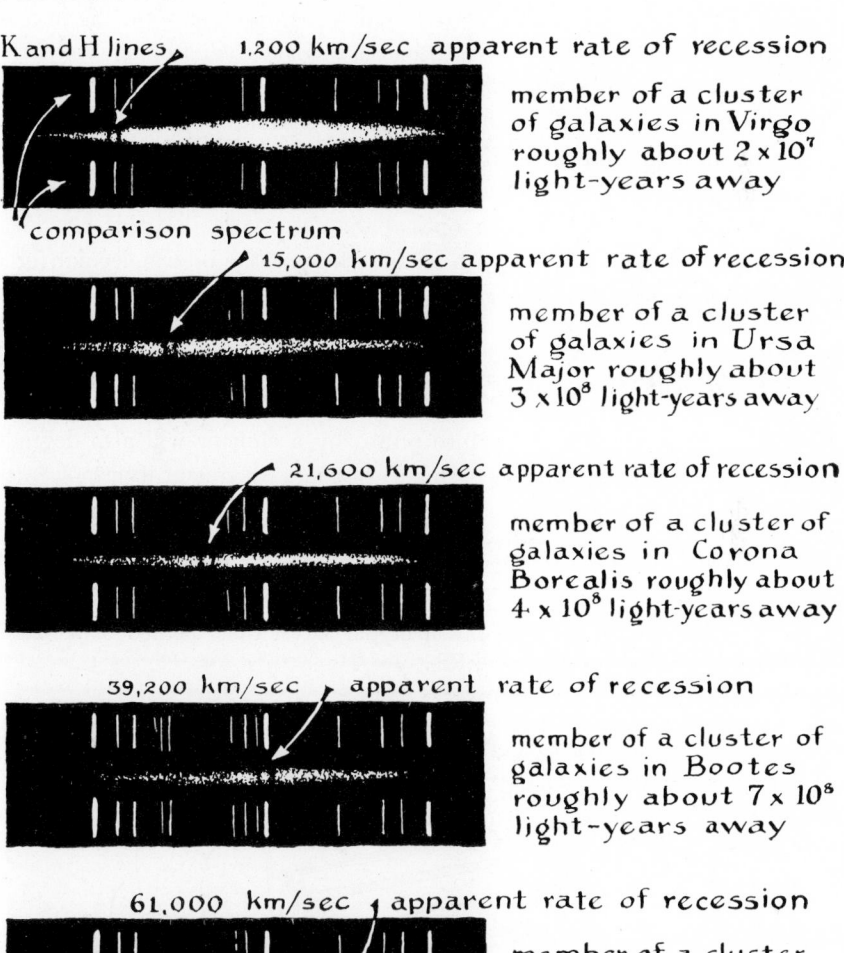

K and H lines,   1,200 km/sec apparent rate of recession

member of a cluster of galaxies in Virgo roughly about 2×10^7 light-years away

comparison spectrum

15,000 km/sec apparent rate of recession

member of a cluster of galaxies in Ursa Major roughly about 3×10^8 light-years away

21,600 km/sec apparent rate of recession

member of a cluster of galaxies in Corona Borealis roughly about 4×10^8 light-years away

39,200 km/sec apparent rate of recession

member of a cluster of galaxies in Bootes roughly about 7×10^8 light-years away

61,000 km/sec apparent rate of recession

member of a cluster of galaxies in Hydra roughly about 10^9 light-years away maticity of oscillator emission is degraded. An appendix defines this degradation of monochromaticity.

The representation of the frequencies introduced by all the different τ's is shown in Fig. 5-2. The over-all line width is determined by the average emission time, τ_{av}.

$$\Delta\lambda_P = \frac{\lambda^2}{c\tau_{av}} \qquad \text{or} \qquad \Delta\nu_P = \frac{1}{c\tau_{av}}$$

Here the average, τ_{av}, may be calculated from the kinetic theory of gases. It depends on the partial pressures and temperature of the gas components and, in addition, on the effective optical collision diameters of the molecules involved. These optical diameters are found to be different from the impact diameters that are involved in the kinetic theory prediction of viscosity; and this difference arises because the range of intermolecular forces involved in an impact collision that is violent enough for significant momentum transfer may be different from the range of forces in an optical collision which is violent enough for a significant wave train interruption. The measurement of pressure broadening, from which optical diameters are derived, provides, therefore, an experimental approach for determining the intermolecular force fields between the molecules in gases.

5-5. *Modulation Broadening*

Chopping of long wave trains into pulses by a shutter will also degrade monochromaticity just as pulse length curtailments by violent intermolecular collisions do. If we were able to have an infinitely long wave train of frequency ω_0, the chopping of this wave train by means of the shutter, as shown

---

FIG. 5-6 Diagrammatic representation of the travel time method of measurement of group velocity through liquid CS_2.

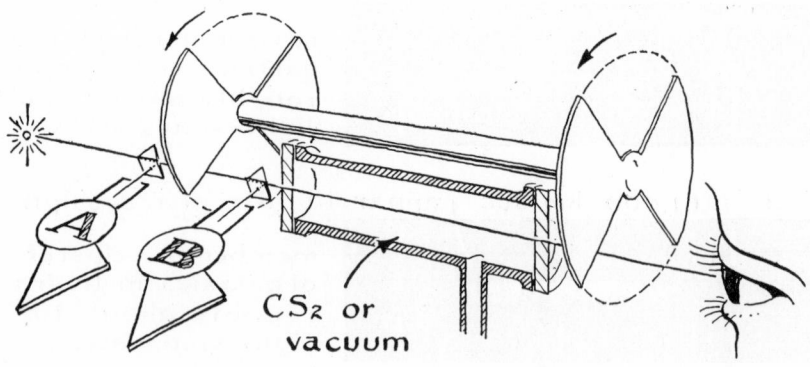

CS_2 or vacuum

FIG. 5-7 Diagrammatic representation of spectral distribution of illumination in a spectrum line before rapid chopping (*A*) and after (*B*).

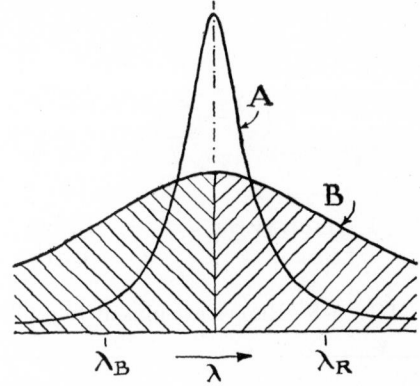

in Fig. 5-6, or even as Galileo tried, would degrade its monochromaticity so that frequencies other than ω_0 would be introduced. An actual source, at best, can only emit the finite wave train that is characterized by a natural line width $\Delta\lambda_N$. This width can be revealed experimentally, as by means of the spectrograph *A* shown in Fig. 5-6. If a spectrum line should appear broadened, as spectrum *A* shows in Fig. 5-7, another spectrograph, after the chopper, would reveal a further broadening of the line, as *B* shows, arising from chopper modulation.

In practice, the distribution *A* will be due to a combination of the natural line width with the Doppler and pressure-broadening effects that were described above. For example, if the source were a cadmium lamp giving the spectrum line 6438 Å (as was used by Michelson to measure the meter), its expected Doppler width would be $\Delta\lambda_D = .004$ Å. The spectrum line Hg 2536 Å pressure-broadened by argon at one atmosphere has a line breadth due to this cause alone of $\Delta\lambda_P = 0.01$ Å. The natural width of spectral lines, due to radiation damping, is much smaller than that due to either of these broadening effects. For example, a spectrum line would exhibit, from this cause alone, $\Delta\lambda_N = 0.0005$ Å. Thus, from such line broadening effects alone, to say nothing of the additional breadth introduced by the chopping, it is abundantly proper for our expression of the pulse-travel-time velocity u to contain the dispersion, $\dfrac{dN}{d\lambda}$.

In communication engineering it is well known that sinusoidal modulation of a carrier of frequency ω_0 at a modulation frequency ω_1 introduces sidebands, $\omega_0 \pm \omega_1$. Our chopping, by means of the rotating sector shown in Fig. 5-6, is in effect a square wave modulation; it would be expected, accordingly, to introduce side-band frequencies here as well. E. Rupp succeeded in demonstrating such modulation broadening. He used the resonance line from a cool thallium vapor discharge tube as his light source. The relatively

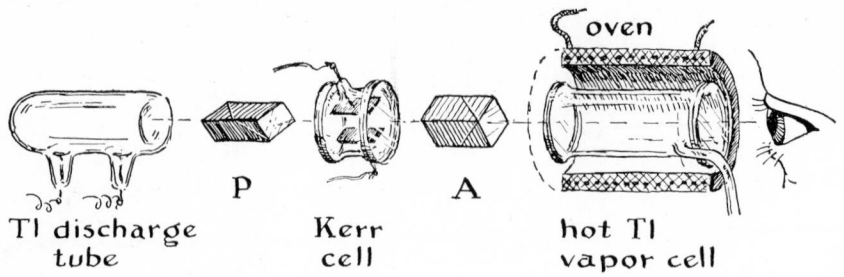

FIG. 5-8 Rupp's experiment to establish the validity of Fig. 5-7 qualitatively.

narrow thallium line, emitted by this vapor source, was totally absorbed by the thallium vapor of a warm cell if the emission of the source was not chopped before entering the absorber cell. But when Rupp chopped that emission at the rate of 10^9 times per second he could see the source through the cell—the side-bands he introduced extended far enough on either side of the central resonant frequency ω_0, and the monochromaticity was thus sufficiently degraded so that light leaked past the resonance absorption peak of the hot thallium vapor cell and made the discharge visible. Rupp chopped the green light at this high frequency with a Kerr cell between crossed Nicol prisms. His combination is shown in Fig. 5-8. When a very high frequency of electrical potential is put on the electrodes of such a Kerr cell combination, the liquid between its electrodes (usually nitrobenzene) is made double-refracting. As a result, the plane of incident polarization is periodically rotated so that the Nicol analyzer periodically fails to extinguish the light emergent from the cell.

5-6. *Observed Group Velocity*

Michelson measured the pulse-travel-time velocity through liquids by means of apparatus such as is represented in Fig. 5-6. This apparatus represents the functioning of all travel-time measuring devices diagrammatically—both the rotating mirror type of device that was used by Foucault and later by Michelson, and the toothed wheel device that was used in the experiments of Fizeau. The tube containing liquid represents the travel-time optical path.

The quadrant wheel of the chopper-shutter system chops up the long wave trains from the source into pulses, or groups of waves. During the time it takes for these groups to travel down the tube, the chopper-shutter system will have rotated through a small angle, and as the rate of rotation of the chopper-shutter is increased, this angle grows. If, finally, a sufficiently high rotation rate is reached, such that this angle is 90°, the light passed by an open quadrant of the chopper will be eclipsed at the other end of the tube by the opaque quadrant of the coupled shutter. When this eclipsing is complete the group velocity u may be calculated from the tube length and the measured time for the chopper-shutter system to rotate 90°. In practice the chopper-shutter system has a great many openings so that its necessary rotation during the travel time is very much less than 90°.

In Michelson's experiment the measured time and tube length gave $\frac{c}{u} = 1.77$ for CS_2; while in separate refractometer measurements he got $N = \frac{c}{v} = 1.64$. It remained for Lord Rayleigh to explain that one would not expect $u = v$, but rather $u = v - \lambda\frac{dv}{d\lambda}$, as Michelson found. As was previously pointed out, u is called the *group velocity*.

5-7. *Explanation of Group Velocity*

A close observation of the expanding ripples that are sent out when a stone is cast into a pond shows the essential phenomenon to be explained. § 1-6 gives the phase velocity of the water waves that the stone produces. The equation there shows that this velocity depends on λ so that the term $\frac{dv}{d\lambda}$ is not zero. We should, therefore, expect the phase velocity v to be different from the group velocity u. Careful observation of a particular wave crest on the water, produced by the stone, will reveal that it travels faster than the group of waves as a whole, as expected. Such a particular crest will advance from the inside through the center of the group to the outer leading edge and disappear. In contrast with such water waves, where we can easily determine v by observing a particular wave crest, in the case of light we cannot "follow" a particular crest, or mark it; we can only "follow" the group of waves as a whole. Thus only u, the group velocity, is observable from travel-time observations.

We cannot determine v from travel-time experiments with "monochromatic light," and avoid the term $\frac{dv}{d\lambda}$, because pulses are inherently not monochromatic, for the reasons that were set forth above in § 5-5.

We use two approximations in our development, below, of Lord Rayleigh's equation, $u = v - \lambda\frac{dv}{d\lambda}$. Firstly, we represent successive pulses of light, passed by the chopper of Fig. 5-6, as successive beats of two hypothetical monochromatic waves of wavelength λ_R and λ_B, as Fig. 5-9a shows. Secondly, we

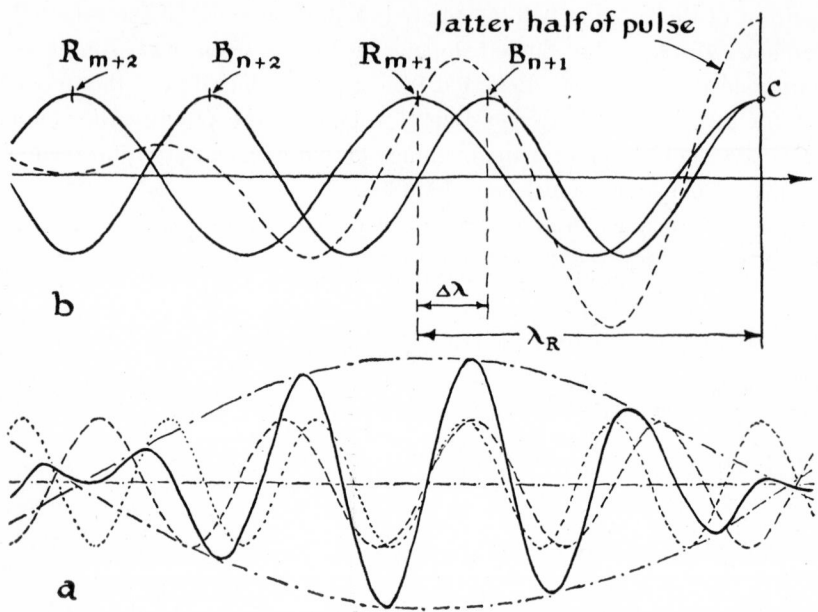

FIG. 5-9 Representation of a pulse as the "beat note" of two monochromatic waves (a) and illustration of how their difference in velocity yields the pulse, or group velocity (b).

represent the distribution B of Fig. 5-7 by these same hypothetical monochromatic waves; λ_R being representative of the cross-hatched right half of B; λ_B being representative of the left half of B. The two wavelengths λ_R and λ_B thus lie at the centroids of the red and blue halves of the distribution of frequencies which represents the chopped light. These approximations give Rayleigh's equation correctly.

Fig. 5-9b shows the two representative component waves, R and B, and their superposition. The superposition sum is represented by the dashed line. This sum has a maximum value at C, corresponding to the group or pulse center, where λ_R and λ_B are, so to speak, in phase. At the instant represented in this figure, the nth crest of λ_B coincides with the mth crest of λ_R. Because λ_B has a lesser phase velocity than λ_R in CS_2, this pulse center, with velocity u, will fall behind the progression of the component waves, with velocities v_B and v_R. Therefore, after the lapse of a certain time, the $(n-1)$th crest of λ_B and the $(m-1)$th crest of λ_R will coincide, and the new point of phase coincidence, marking the pulse center, will have fallen behind the mth crest of λ_R by the distance λ_R. The time required for this to occur, due to $v_R > v_B$, is

$$\Delta t = \frac{\lambda_R - \lambda_B}{v_R - v_B} = \frac{\Delta\lambda}{\Delta v}$$

In the limit we may write $\dfrac{1}{\Delta t} = \dfrac{dv}{d\lambda}$. During this time Δt, while the redder waves will have advanced the distance $v_R\Delta t$, the advance of the center of the pulse is only $v_R\Delta t - \lambda_R$. Division of this pulse advance by Δt gives the group velocity of the pulse:

$$u = \frac{v_R\Delta t - \lambda_R}{\Delta t} = v_R - \frac{\lambda_R}{\Delta t} = v - \lambda\frac{dv}{d\lambda}$$

In a dispersive medium v_R and v_B can be determined from refractometer measurements. Such measurements show that $v_R = \dfrac{c}{N_R}$ is greater than $v_B = \dfrac{c}{N_B}$ by the amount given by the dispersion, $\dfrac{dN}{d\lambda}$, and $(\lambda_R - \lambda_B)$. This derivative can be obtained by differentiating a Cauchy-formula representation of N's as a function of λ (see § 4-6); or $\dfrac{dN}{d\lambda}$ can be obtained approximately from the ratio $\dfrac{N_2 - N_1}{\lambda_2 - \lambda_1}$.

From $v = \dfrac{c}{N}$, we may express $\dfrac{dv}{d\lambda}$ in terms of the dispersion:

$$\frac{dv}{d\lambda} = \frac{d}{d\lambda}\left(\frac{c}{N}\right) = -\frac{c}{N^2}\frac{dN}{d\lambda}$$

So

$$u - v = -\frac{c}{N}\frac{\dfrac{dN}{N}}{\dfrac{d\lambda}{\lambda}}$$

Michelson made his measurements with a white light source. For his CS_2 we may use $N_F = 1.652$ and $N_D = 1.628$ at $\lambda_F = 4861$ Å and $\lambda_D = 5892$ Å, taking 1.64 for the average N, corresponding to an average wavelength $\lambda = 5380$ Å. The substitution of these numbers yields a predicted difference,

$$u - v = -\frac{c}{1.64}\frac{0.024/1.64}{1030/5380} = -\frac{c}{1.64}(0.076)$$

while Michelson observed the difference

$$u - v = c\left(\frac{1}{1.77} - \frac{1}{1.64}\right) = -\frac{c}{1.64}(0.073)$$

Thus Rayleigh's theory is confirmed.

We know that the velocity of light in interstellar space is the same for all

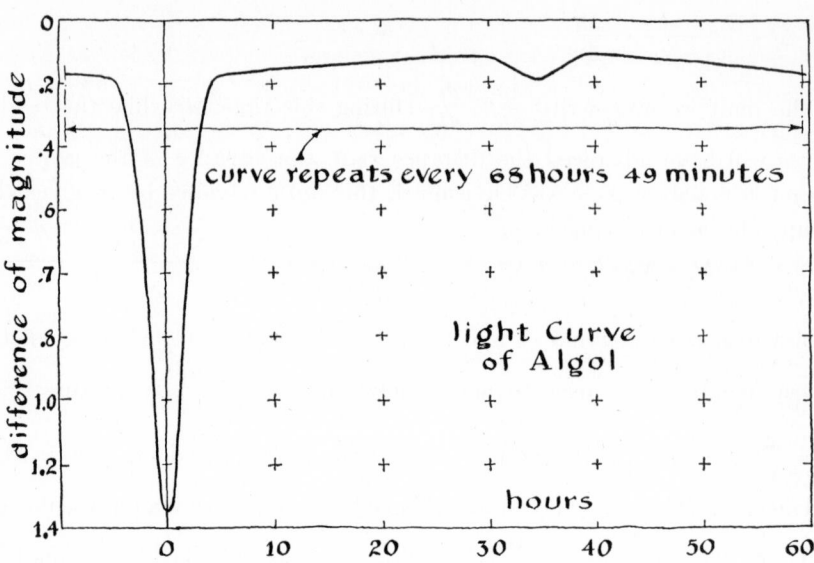

FIG. 5-10   Light curve of the eclipsing binary star, Algol.

wavelengths from the manner in which the light of the star β-Persei varies with time. This star is a binary, and its light variation shown in Fig. 5-10 is due to eclipsing of the brighter star by a darker companion star. This eclipsing is not unlike the chopping of Fig. 5-6 with the darker star of the binary acting as the chopper. If the velocity for red light in interstellar space were as much different as one part in a million from the velocity for the blue then the eclipse as seen in red light should occur at a measurably different time than the eclipse as seen in blue light. This finesse of discrimination, with such a slow chopping rate, arises because of the great distance of the star from us—this binary lies 120 light years away. The most careful measurements on β-Persei show that the light intensity decreases for all colors simultaneously.† We thus infer that $\frac{dv}{d\lambda}$ is zero in interstellar space.

5-8. *Phase Velocity in Sodium Vapor*

Let us now examine the origins of the phase velocity itself and of its dispersion, the medium in which light propagates being considered a cloud of oscillators for which $\tilde{N} = n - jk$. For example, the medium might be sodium vapor. The explanation of how v can differ from or exceed c involves consideration of the individual oscillators; and how they are driven by an incident

† See John S. Hall, J. Franklin Institute, *228*, 411 (1939).

light wave; and especially how the phase of their stimulated motions leads or lags the phase of the forcing fields of incident lights; and finally how the forward emissions of such stimulated oscillators combine with and modify the phase of the residue of the incident light that propagates at velocity **c**. We shall see that v can be inferior to **c**, or exceed it, depending on whether stimulated re-emissions lead or lag the exciting light. This explanation of the origins of N, as the manifestation of a combination of the residue of incident light with the forward re-emitted light, has much in common with explanations of scattering; both are manifestations of re-emitted light: N is explained by forward re-emitted light; scattering is explained by laterally re-emitted light. Although the light laterally re-emitted by oscillators turns out to be vanishingly small if the cloud is dense enough so that a myriad of oscillators occupy the volume of a cube that is only a wavelength on an edge, the forward re-emitted light is never negligible, but makes itself manifest as N.

The maximum and minimum of the $(n - 1)$ curve of Fig. 4-2 demark a very interesting band of oscillator frequencies. Beyond this region on either side n decreases as ω decreases. This variation of n with ω is characteristic of normal dispersion, making u less than v. But within these stationary values we have the opposite dependence of n on ω—with n increasing toward longer wavelengths. This opposite dependence is called anomalous dispersion. It was first discovered for the dye fuchsin by C. Christiansen (of Christiansen filter fame).

Furthermore, where $(n - 1) < 0$, or $n < 1$, we have $v > $ **c**.

The fact that v can exceed **c** may be demonstrated beyond doubt with the apparatus of Fig. 5-11. Here a long horizontal glass cylinder containing only metallic sodium and a residual trace of hydrogen gas has its ends closed by parallel windows. When this tube is heated on its under side, as shown, the sodium metal vaporizes and eventually condenses on the upper cooler cylinder walls. And as this metallic vapor diffuses upward through the residual hydrogen gas, its density decreases. Thus light that penetrates the tube longitudinally passes through a denser cloud of metal vapor atoms at the bottom, and a more tenuous cloud at the top. This density gradient gives the equivalent of a prism of sodium vapor. For the frequency lying red of ω_0, where our formulation predicts that v is less than **c** in sodium vapor, the cell will deflect the light downward toward the thick edge of this equivalent prism. This deflection toward the thick edge is the usual one for an ordinary dielectric prism; it results from the greater phase retardation of the part of the wave front that penetrates the bottom of the tube, or prism. But for the frequency in the region lying just blue of ω_0, where $n < 1.0$ corresponding to $v > $ **c**, it is observed that the light is deflected toward the thin edge of the equivalent prism. Fig. 5-11 shows the optical arrangement that displays these opposite

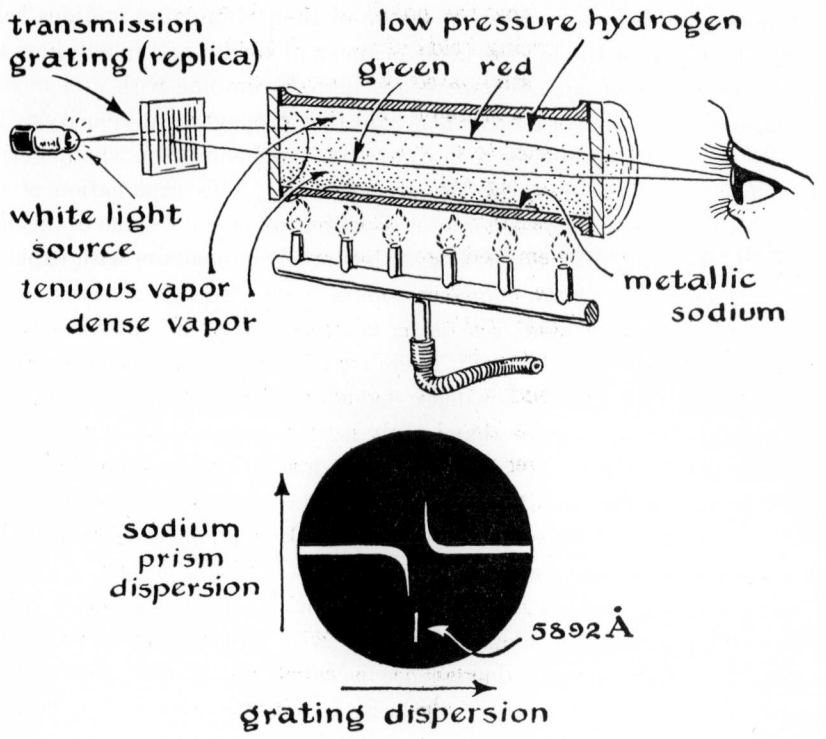

FIG. 5-11 Demonstration of anomalous dispersion in sodium vapor.

deflections. Here the point source of white light is observed through both a transmission grating and the tube, with the grating dispersion arranged horizontal. The insert shows how the grating spectrum appears, and demonstrates the dependence of n on ω in the vicinity of the sodium resonance at $\lambda = 5892$ Å; the maximum of k in Fig. 4-2 is illustrated by the missing segment of the spectrum.

Phase velocities in a dielectric medium that are greater or less than **c** may be understood further from the interaction between light and the oscillators of which we consider dielectrics composed. The incident light loses amplitude by inducing oscillator motions; and these induced oscillations lead or lag the forcing field qE of the incident light, depending on whether ω is just greater or just less than ω_0. These induced oscillations re-emit light in the forward direction of the same frequency. This forward re-radiation, in superposition with the residue of the depleted incident light, gives a resultant wave front that is characterized by a leading or a lagging phase. Thus we account for a phase velocity that may be greater or less than **c**.

Consider the thin slab of vapor represented in Fig. 5-12, bounded by two parallel planes, one at $z = 0$ and the other at $z = \Delta z$. First, considering amplitudes, let the light incident on this slab be described by $\hat{E} = \mathcal{E}_0 e^{j\omega\left(t-\frac{z}{c}\right)}$ giving $\tilde{E} = \mathcal{E}_0 e^{j\omega t}$ at $z = 0$. We suppose $\mathfrak{f}\mathcal{E}_0$ to be the amplitude depletion, on the average, produced by each of the $\mathfrak{N}$ sodium atoms that lie in the path of a light ray. This fraction $\mathfrak{f}$ excites the atomic oscillations which, when all atoms in Δz are considered, explains $v \neq c$. Thus if $\mathfrak{N}\mathfrak{f}$ is small, the residual amplitude of the incident wave at $z = \Delta z$ is $\tilde{E}' = (1 - \mathfrak{N}\mathfrak{f})\mathcal{E}_0 e^{j\omega\left(t-\frac{\Delta z}{c}\right)}$. Now we suppose, further, that a similar fraction $\mathfrak{f}''$ of the induced amplitude from each oscillator is re-emitted in the forward direction. Considering individual phases, the phase of the primary wave when it excites a sodium atom, say a at $z = \delta z_a$, is $\omega\left(t - \dfrac{\delta z_a}{c}\right)$; and the phase of the re-emitted radiation by this atom, after suffering a typical phase lag or lead φ, will be $\omega\left(t - \dfrac{\delta z_a}{c}\right) + \varphi$. The phase of this re-emitted amplitude at $z = \Delta z$ is therefore

$$\omega\left(t - \frac{\delta z_a}{c}\right) + \varphi - \omega\left(\frac{\Delta z - \delta z_a}{c}\right) = \omega\left(t - \frac{\Delta z}{c}\right) + \varphi$$

And for a second sodium atom, say atom b at δz_b, the phase of re-emitted radiation at $z = \Delta z$ will be the same as that due to a:

$$\omega\left(t - \frac{\delta z_b}{c}\right) + \varphi - \omega\left(\frac{\Delta z - \delta z_b}{c}\right) = \omega\left(t - \frac{\Delta z}{c}\right) + \varphi$$

FIG. 5-12   Demonstration of the origins of phase velocity in a dielectric comprised of identical simple harmonic oscillators, a, b, etc.

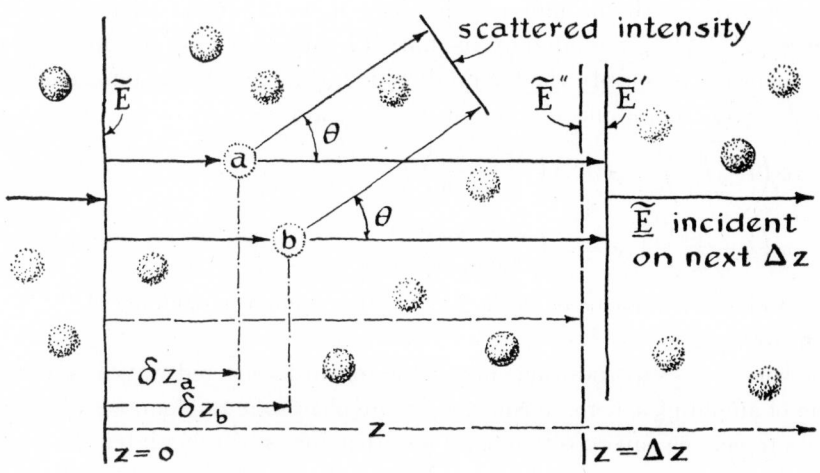

Thus since φ is the same for all the individual sodium atoms, the secondary re-emitted wavelets are in phase at $z = \Delta z$, regardless of the atomic positions.

These secondary wavelets for all oscillators in Δz superimpose at $z = \Delta z$ to form a secondary wave of amplitude which is proportional to

$$\tilde{E}'' \frown \mathfrak{N}\mathfrak{f}\mathfrak{f}''\mathcal{E}_0 e^{j\omega\left(t-\frac{\Delta z}{c}\right)} e^{j\varphi}$$

while the residue of the exciting wave is there proportional to

$$\tilde{E}' \frown (1 - \mathfrak{N}\mathfrak{f})\mathcal{E}_0 e^{j\omega\left(t-\frac{\Delta z}{c}\right)}$$

These two amplitudes $\tilde{E}'$ and $\tilde{E}''$ combine to give the field which is incident on a subsequent slab (say with boundaries at $z = \Delta z$ and $z = 2\Delta z$). These components $\tilde{E}'$ and $\tilde{E}''$ are, of course, polarized in the same plane. And further, in a transparent dielectric it turns out that $\mathfrak{f}''$ is such that the amplitude of $(\tilde{E}' + \tilde{E}'')$ is the same as $\tilde{E}$. But the phase of $(\tilde{E}' + \tilde{E}'')$, on entering the second slab, is not the same as $\tilde{E}$ would have been in vacuum, but it is different by some phase angle φ'. Similarly, the phase of the wave incident on the third slab will be further different by the phase angle $2\varphi'$, lead or lag; and so on. This expected continuing advancement, or continuing retardation of phase, as the atomic vapor is penetrated by the light, is equivalent to a phase velocity that is greater than **c**, if φ' is positive, and less than **c**, if φ' is negative.

A cogent point is this: within our cloud of oscillators we see that no wave passes through the interatomic space with any velocity different from the *signal velocity* of light **c**.

5-9. *Scattering*

In contrast with the mutually in-phase forward re-emitted light vectors above, $\mathcal{E}'' = \mathfrak{f}\mathfrak{f}''\mathcal{E}_0$, the laterally re-emitted light vectors are not in phase for the following reason: In the lateral direction, θ, at $z = \Delta z$ the phases of scattered waves from atoms a and b of Fig. 5-11 are

$$\omega\left(t - \frac{\delta z_a}{c}\right) + \varphi - \omega\left(\frac{\Delta z - \delta z_a}{c \cos \theta}\right)$$

$$\omega\left(t - \frac{\delta z_b}{c}\right) + \varphi - \omega\left(\frac{\Delta z - \delta z_b}{c \cos \theta}\right)$$

These phases cannot be equal because of the random positioning of the oscillator sources.

Whereas the superposition sum of the forward scattered wavelets was the sum of amplitudes, here, in contrast, where phases are random, we add intensities to get the superposition sum; and therefore scattering intensity is proportional to the numbers of scattering oscillators. As we learned in § 2-7, the resultant on adding intensities is much less than the resultant on adding amplitudes. Thus, whereas the forward re-emissions make themselves felt strongly, in determining phase velocity, the lateral re-emissions appear much less strongly, as scattered light.

We have not proven it here, but when the density of oscillators is high (millions in a cube which has the dimensions of one wavelength on an edge), then the summation of lateral emissions goes over into a continuous integral, and the value of that integral for laterally scattered light becomes zero (meaning $f'' = 1.0$).

Professor R. W. Wood has conducted experiments that demonstrate this difference between lateral scattering for a tenuous cloud of oscillators, such as represents ethyl ether vapor, and the evanescence of this scattering when the density of oscillators becomes sufficiently great, such as the denser cloud of oscillators that would represent liquid ethyl ether. Although here the ether molecule, not a sodium atom, is the oscillator, what goes on is quite the same. Professor Wood's experiments showed that the scattering by a given volume of liquid ether was only 50 times greater than scattering by the same volume of vapor. Scattering would have been 1000 times greater if it had been proportional to the density of oscillators. The missing factor of 20, for we do expect

FIG. 5-13   R. W. Wood's laboratory demonstration of scattering of sunlight by clean air.

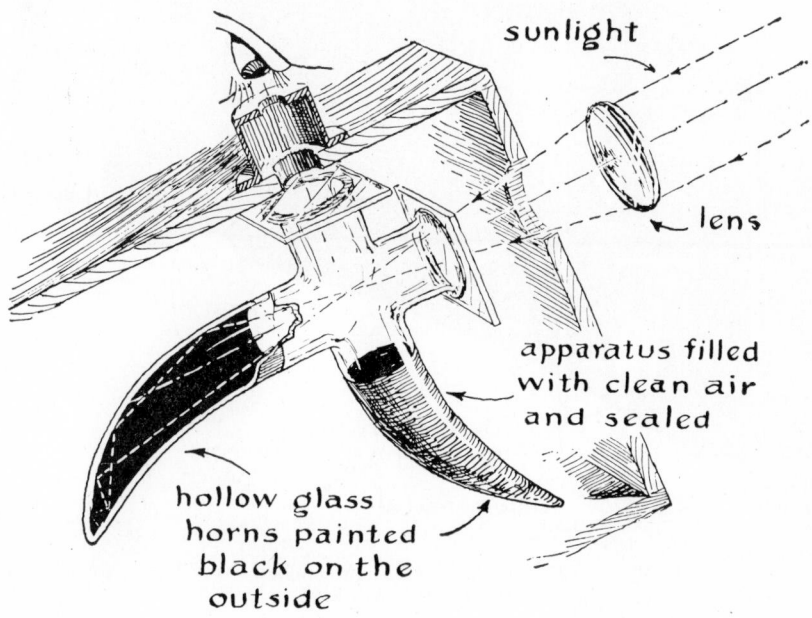

sunlight

lens

apparatus filled
with clean air
and sealed

hollow glass
horns painted
black on the
outside

the intensity of scattered light to be proportional to the number of scatterers, is explained as evanescence of scattering associated with the approach toward a transparent homogeneous medium in the liquid, with myriads of oscillators in a cube with the dimensions of one wavelength on an edge.

When the illumination frequency ω, for a dense cloud of oscillators, lies far from ω_0, the equation of § 4-1 predicts that the motion, $\bar{w}$, will be weak; and hence scattering by them will be weak. In such a case scattering can be seen only on a grand scale, as in the sky when our atmosphere is illuminated by the sun, or, on a smaller scale, when the air is illuminated and viewed by some such special apparatus as is illustrated in Fig. 5-13.

5-10. *Scattering by Sodium Vapor*†

When, on the other hand, the frequency of the illumination, ω, lies near the frequency of resonance, then $\bar{w}$ will be large and scattering is easily ob-

---

FIG. 5-14 R. W. Wood's laboratory demonstration of scattering of sodium resonance radiation by sodium vapor.

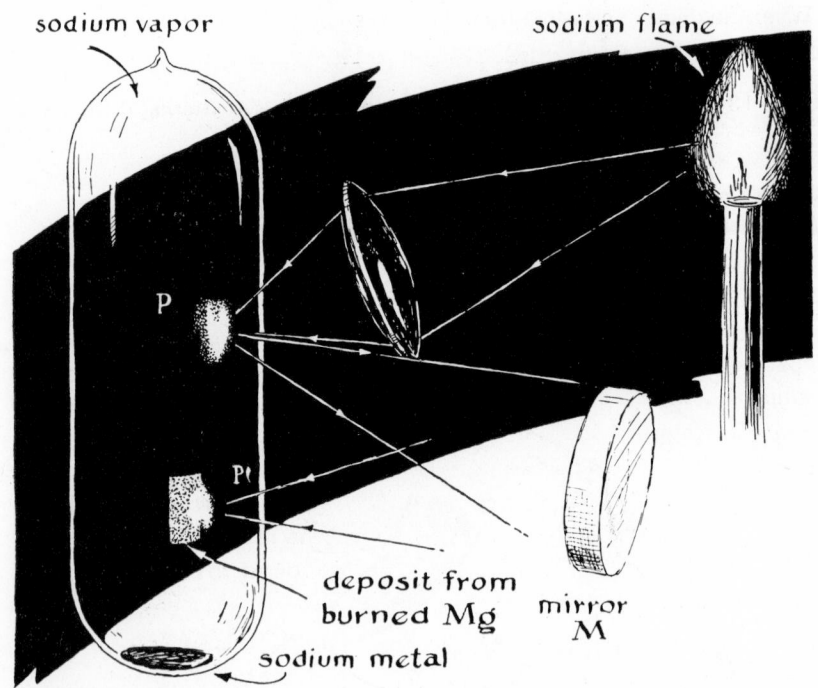

---

† See Robert W. Wood, *Physical Optics* (1934, Macmillan, New York), for additional information on scattering, particularly by sodium vapor.

servable. Such is the case for sodium vapor near $\lambda = 5892\text{Å}$, and for mercury vapor near $\lambda = 2536$ Å. Such sodium scattering is so strong, for example, that the scattering of sunlight just after sundown by the little sodium that exists in our upper air, at 70 kilometers height, is detectable. Fig. 5-14 shows Professor Wood's set-up to demonstrate such strong scattering by sodium vapor near ω_0. The large evacuated glass capsule shown contains metallic sodium. The light from a sodium vapor lamp is focused within the capsule. As the capsule is gradually heated the pressure of sodium vapor within it increases, so that the number of sodium atoms per unit volume, $\mathfrak{N}_v$, also increases. At a moderate temperature scattering by these sodium atoms, as vapor, becomes manifest in the capsule; it appears to be filled with a fine dust. This scattering of sodium light contrasts with what we would observe with white light illumination: when white light is focused within the capsule, scattering is not evident. As $\mathfrak{N}_v$ increases further, at higher temperatures, the penetration of the focused light of the sodium lamp decreases due to absorption by the sodium vapor. And finally, when the temperature is even higher, the incident radiations are so very strongly absorbed that the scattering by sodium atoms within the capsule all occurs very near the inner surface of the enclosing walls.

When this light, scattered by dense vapor at P near the capsule surface, is collected and refocused by a mirror M on the walls at P', part of the light being made to fall on a deposit of magnesia formed there by a burning magnesium ribbon, and the rest being focused on the adjacent clear glass surface, with sodium vapor behind, it is observed that the scattering of this refocused light from the vapor-backed glass is as complete as it is from the MgO-whitened surface. This stronger reflection of the light occurs because frequencies already once scattered by cool vapor lie nearer to ω_0 than the original light from the hot sodium lamp.

When, as with our ether above, the density of sodium vapor here gets high enough, we should expect the phases of the scattered light to begin to lose randomness. And, indeed, in the case of the similar scattering of the 2536 Å mercury resonance radiation, Professor Wood was able to demonstrate the realization of this expectation by the onset of specular reflection, occurring at the surface of dense mercury vapor. He obtained a sufficient density of mercury vapor in a strong fused quartz capsule, with walls transparent to the mercury 2536 Å light, to realize a specular reflection at the inner wall surface for the resonance radiations of 25%.

5-11. *Sky Light*

When the resonant frequency is high relative to the frequency of incident light, $\omega \ll \omega_0$, the expression of § 4-9 simplifies; and if the scattering medium is transparent, the expression simplifies further. When $\omega \ll \omega_0$, and $\rho = 0$, then the maximum dipole moment of each oscillator becomes

$$\mathfrak{p}_0 = \frac{q^2}{m\omega_0^2} \, \mathcal{E}_0$$

The average flux scattered per molecule is $\mathfrak{F} = \dfrac{\mathfrak{p}_0^2 \omega^4}{12\pi\kappa_0 c^3}$ watts. On division of this flux by the average incident illumination, $\mathfrak{E} = \frac{1}{2}c\kappa_0\mathcal{E}_0^2$, we get the flux scattered per molecule per unit illumination, or a kind of scattering coefficient:

$$\frac{q^4\lambda_0^4}{6\pi m^2 \kappa_0^2 c^4} \left(\frac{1}{\lambda^4}\right)$$

The important thing to note here is that the scattering is predicted to be proportional inversely with the fourth power of the wavelength. This dependence explains why the blue light in the incident sunlight is more strongly scattered in the sky than red light.

Lord Rayleigh was able to obtain this inverse fourth power law by means of an interesting application of reasoning, called *dimensional analysis*. He reasoned that the scattered amplitude at a distance r from an air molecule should depend on the following parameters raised to appropriate powers: $\dfrac{1}{r}$ due to the inverse distance; $\mathcal{E}_0$, the incident illumination scattered; λ; N; the velocity of light c; and the volume of the scattering air molecule. Furthermore, the scattered amplitude, $\mathcal{E}_S$, should depend on these parameters in just the way necessary to yield the dimensions of an amplitude of light. Writing $\frac{4}{3}\pi\rho^3$ for the molecular volume, we have

$$\mathcal{E}_S \smile \frac{\mathcal{E}_0}{r} \lambda^\alpha N^\beta c^\gamma \left(\frac{4}{3}\pi\rho^3\right)$$

Dimensional equality requires that γ be zero since $\mathcal{E}_S$ and $\mathcal{E}_0$ have the same dimensions, and since time does not appear elsewhere. Since N is dimensionless, the exponent β gives no help. Balancing the exponents of the length dimensions gives $\alpha = -2$—volume giving $+3$, combined with r raised to the power -1. This balancing yields the inverse fourth power dependence of the ratio of scattered to incident illumination on wavelength. Further exposition on this kind of reasoning is to be found in P. W. Bridgman's *Dimensional Analysis*, Yale University Press, New Haven (1920).

Chapter VI

Polarized Light and Dielectric Boundaries

IN THIS chapter we shall elaborate our considerations of the amplitude-reflection coefficients and intensity-reflection coefficients for a dielectric boundary (including internal incidence and total reflection, as well as external incidence). And in addition we shall set up the corresponding amplitude-transmission and intensity-transmission coefficients.

We describe light incident on a dielectric boundary in terms of the components that are polarized in perpendicular azimuths: one component polarized in the plane of incidence, the π-component (π for parallel), and one component polarized in the plane perpendicular to the plane of incidence, the σ-component (σ for *senkrecht*). This procedure of description is possible since any incident beam of light can be specified in terms of two components of plane polarized light that are polarized in perpendicular azimuths, provided the interrelations of the amplitudes and phases of these components are specified.

In our descriptions we shall invoke polarizers and analyzers. A polarizer such as we invoke may be a suitable plate of tourmaline, or a sheet of Polaroid. For the present we shall ignore full explanations of how polarizers function. Either a plate of tourmaline or a sheet of Polaroid passes only the components of electric fields of incident light that lie in the direction of its azimuth of easy passage; the components perpendicular to its azimuth of easy passage are absorbed. The analyzer, of course, is simply a polarizer used to determine the azimuth of polarization of a polarized beam of light.

Plane polarized light will be our first consideration. Partially polarized light can be found in nature in the blue, haze-free northern sky. Skylight is the sunlight after having been scattered from the sun's beam. Also, natural polarized light is found when the light from a cloud is reflected from a pond or other still water surface at an angle of incidence of 53°. But although the naked eye can just discern when light is polarized by means of Haidinger's brush (described later), this observation has been much too subtle for discovery. Thus, in history, the knowledge of polarized light awaited the discovery of something to make the phenomena associated with polarization observable.

Without further introductory remarks, let us consider the state of polarization of the radiation which a simple source of light emits; namely, waves emitted into one beam by the sodium atoms in a distant Bunsen burner. Each sodium atom will emit polarized light, but the azimuths of different atomic oscillators will be randomly distributed, at every possible azimuth angle. Of all the emissions, only those whose azimuths fall in the direction of easy passage of a polarizer, or only components of emissions parallel to this direction of easy passage, are transmitted by that polarizer. Thus, by use of a polarizer we can get *plane polarized* light from such a light source.

Consider, for a moment, the wave trains of this light before it strikes a polarizer. We have learned that the emission time of a sodium oscillator is only $\tau = 10^{-10}$ second. Accordingly, the sum of all the atomic polarized components from our Bunsen flame will exhibit a substantial persistence of phase over an interval of time that is shorter than this τ; but over a span of time longer than τ, there will be no predictable interrelation of phases. During such a persistence time, and before our light is passed through the polarizer, although it is said to be unpolarized or natural light, we may actually represent the beam by two mutually perpendicular polarized components, say π- and σ-components. At any point the relation of the two phases of these components, $\bar{E}_\pi$ and $\bar{E}_\sigma$, will be always changing, in a purely random fashion. Their phase difference will vary between $\varphi_\sigma - \varphi_\pi = 0$ and 2π. Whenever $\Delta\varphi$, this phase difference between $\bar{E}_\pi$ and $\bar{E}_\sigma$, becomes temporarily zero, or π, the light incident on our polarizer will actually be evanescent plane polarized light. When $\Delta\varphi$ is $\pm \frac{\pi}{2}$ it will be right or left circularly polarized light; and for intermediate values of $\Delta\varphi$ it will be elliptically polarized light. Each of these states of polarization is evanescent—it will persist no longer than $\tau = 10^{-10}$ second. Thus natural light is not unpolarized light but rather a continually changing succession of these four states of polarized light. Natural light is polarized *plane* when the two perpendicular components are in phase, or in antiphase; *circular* during the time that $\Delta\varphi$ persists at $\pm \frac{\pi}{2}$; but generally *elliptical*, with continually varying $\Delta\varphi$. The phase persistence of this light is determined by its monochromaticity. In view of all this, then, we must understand that a set of oscillators, of such a character that each oscillator emitted a continuing or infinitely long wave train of polarized light, could not *en ensemble* produce unpolarized light—the light would be stable plane-, circular-, or elliptical-polarized.

A partial determination of the state of polarization of a beam of light can be made with a Polaroid sheet, used as an analyzer, with the face of the sheet or plate perpendicular to the tested beam. To make this determination we rotate the analyzer in azimuth until the passed illumination exhibits a measured maximum, $\mathfrak{E}_{max}$; and again, we rotate the azimuth until the passed illumination exhibits a measured minimum, $\mathfrak{E}_{min}$. These two quantities are combined as follows to give the degree of polarization, β.

$$\beta = \frac{\mathfrak{E}_{max} - \mathfrak{E}_{min}}{\mathfrak{E}_{max} + \mathfrak{E}_{min}}$$

If such a test of a beam of light yields $\beta = 1.0$, it is, of course, plane polarized light, and this test result is unequivocally interpretable. When the test yields $\beta = 0$, the result is not unequivocal, since both circular and unpolarized light beams yield $\beta = 0$. Then, in order to define the state of polarization, we need to know the phase relationship between orthogonal components of the light. When there is a stable phase difference between orthogonal components, the light is either plane polarized, circularly polarized, or elliptically polarized, as may be determined by tests to be described in § 7-8.

6-1. *Polarization in Nature*

We have pictured the reflected waves at a surface of optical discontinuity as produced by the wavelets that are emitted by the stimulated oscillators of which the reflecting surface is composed. Oscillator charge motions just within the reflecting surface when stimulated give off dipole-like emissions as characterized in § 4-9. Two features of these emissions are of especial interest now. One is that the emissions of the stimulated dipoles are polarized; the other is the absence of emission in the propagation direction which lies parallel to the poles of that dipole motion.

A bound charge in a transparent dielectric has three independent degrees of freedom, if the binding is isotropic. And all these degrees of freedom may be excited in a Bunsen flame by intermolecular collisions. But when the oscillators are not stimulated thus, but stimulated by an incident beam of unpolarized engulfing light, then only two degrees of freedom are excited. And only one degree of freedom is excited when the incident beam of engulfing light is plane polarized. When we have such geometrically restricted excited motions it is not surprising, then, that reflected or scattered light is found to

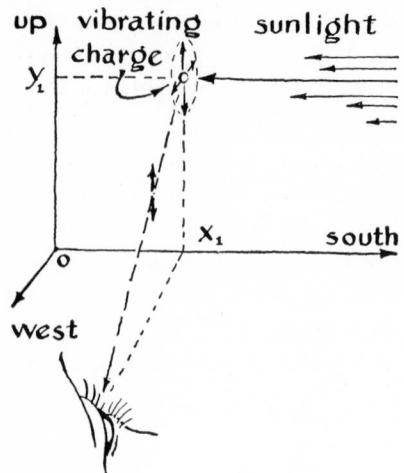

FIG. 6-1 Polarization of singly scattered sky light.

be polarized. In such cases the reflected or scattered light is comprised of the superimposed emissions of identically stimulated oscillators.

In particular, Fig. 6-1 shows how a beam of the sun's light excites the molecules in the atmosphere to scatter polarized sky light, in the perpendicular direction. This polarized emission is due to oriented stimulated oscillations that all lie in a plane perpendicular to the exciting sun's rays, as shown in the figure. Single-scattered sky light which the eye sees in a plane perpendicular to the sun's rays will have its electric vector lying in that plane, and thus be polarized sky light. Actually, even on the clearest days, one fifth of the sky light is not due to single-scattered, but to multiple-scattered light. And of course a much higher proportion than one fifth is multiply scattered on hazy days. Fig. 6-2 illustrates the manner in which multiple scattering

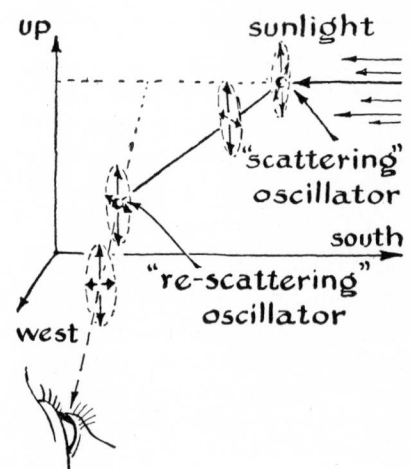

FIG. 6-2 Depolarization of sky light by multiple scattering.

depolarizes the single-scattered sky light. This figure shows one particular multiple scattering. Of course there are many other possible ones. For our particular multiple-scattered ray, it is scattered at 45° to the sun's beam by a molecule south of the perpendicular plane in the south-west direction. Then when it reaches the perpendicular plane we show it re-scattered to the observing eye. Its plane of vibration, as shown, will not be the same as that of the rays singly scattered from the same point in this perpendicular plane— it will contain components perpendicular to those of the single-scattered light. The re-scattered rays thus dilute and depolarize the single-scattered polarized light.

Naturalists claim that the worker bees can perceive this polarized sky light, and that they use it to communicate to other worker bees the directions in which good sources of nectar are to be found. This use of polarization by bees is described in the *Scientific American* for July 1955.

As for reflected polarized light, as from a pond's smooth surface, this light is emitted from stimulated water oscillators just as the 2536 Å radiation specularly reflected from high density mercury vapor, in Professor Wood's experiment, § 5-10, was emitted by mercury vapor atoms. From this experiment of Professor Wood we learned to look on specular reflected light as a limiting case of scattered light, where the density of stimulated oscillators was so great that re-emissions became coherent in phase. And taking this picture as a model, we may ascribe all the reflected beam from the pond's water surface to energy taken from the incident beam and re-emitted by atomic oscillators. On the basis of this picture, therefore, when the re-emitted light lies perpendicular to the direction of the exciting beam within the water, we would expect the re-emitted light to be polarized, just as the scattered sky light is polarized when re-emitted perpendicular to the sunbeams. This reflected light from the pond's surface is polarized because the motions of the stimulated oscillators are oriented. Fig. 6-3 shows the polarized reflected beam when it propagates perpendicular to the direction of the refracted beam within the water—*i.e.*,

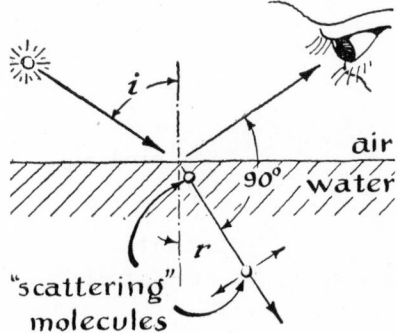

FIG. 6-3 Reflection at Brewster's angle as coherent "scattering." The oscillation of the "scattering" molecule parallel to the plane of reflection does not emit in the direction of the eye, at Brewster's angle.

where $(i + r) = \frac{\pi}{2}$, the condition for the reflected beam to be perpendicular to the refracted beam. This perpendicular reflected light is thus constituted only of electric vibrations perpendicular to the plane of reflection and refraction, because oscillators cannot radiate poleward. The angle of incidence which makes $(i + r) = \frac{\pi}{2}$ is called Brewster's angle, i_B. And since the corresponding refraction angle, r_B, is the complement of i_B, so that $\cos r_B = \sin i_B$, we have, from Snell's law, $\tan i_B = N$.

This polarization by reflection at Brewster's angle was discovered by E. L. Malus in 1808. He did not find polarized light, however, in the reflected sunbeams from a pond's surface, but in reflected light from the surface of a window in the Luxembourg Palace, as illustrated in Fig. 6-4. Malus did not discover this polarization until a means of analyzing a beam of light to determine its polarization was available to him. This means of analysis was, for Malus, a calcite rhomb.

6-2. *Polarizers and Analyzers*

Malus discovered the plane polarized state of the window-reflected light by means of a natural rhomb of the remarkable crystal, calcite, as shown in Fig. 6-4. Fig. 6-5 shows how an incident unpolarized ray is doubled after penetrating such a rhomb. The usual second ray through a calcite rhomb, at certain orientations, is absent after an incident polarized ray of light has penetrated it. It was owing to this absence of a second ray that Malus in-

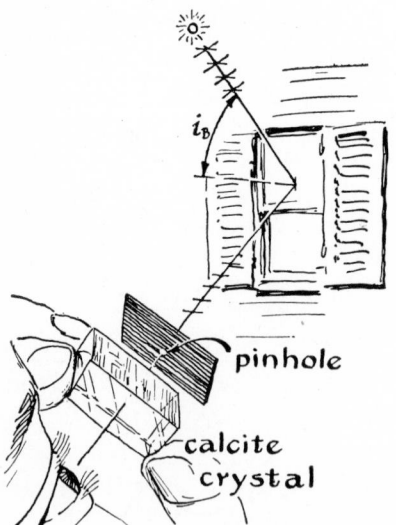

FIG. 6-4 Demonstration of polarization of a reflected beam by means of a calcite rhomb.

FIG. 6-5　Double refraction by calcite.

ferred that the window-reflected light was polarized. Without the calcite rhomb, or some equivalent means to recognize polarization, Malus could not have discerned that his light was polarized. Calcite excited great interest among scientists after it was introduced. This interest is evidenced by the following quotation† of Erasmus Bartholinus, who first described calcite, after the crystal was introduced into Europe from Iceland in the 17th century.

"Greatly prized by all men is the diamond, and many are the joys which similar treasures bring, such as precious stones and pearls, though they serve only for decoration and adornment of the finger and the neck; but he, who, on the other hand, prefers the knowledge of unusual phenomena to these delights, he will, I hope, have no less joy in a new sort of body, namely, a transparent crystal, recently brought to us from Iceland, which perhaps is one of the greatest wonders that nature has produced. I have occupied myself for a long time with this remarkable body and carried out a number of investigations with it, which I gladly publish, since I believe that they can serve lovers of nature and other interested persons for instruction, or at least for pleasure."

It is not entirely correct to say that the naked eye itself cannot discern when light is polarized, but such a discernment constitutes a very subtle observation indeed. The eye can determine both that a field of light is polarized, and that it is polarized in this or that azimuth as well. These determinations are possible from observation of the presence and orientation of Haidinger's brush. But Haidinger's brush, at best, is only just visible in nature in the polarized blue light of the northern sky. And if, long ago, someone had seen it, his descriptions would be taken as "subtlety added to subtlety," and those descriptions would be perhaps only meaningful to one "with the Phaedon instead of Bowditch in his head." This Haidinger's brush is defined by Minnaert as follows:‡

"Many a laboratory physicist is astonished and inclined to disbelieve us when we tell him we are able to see with our naked eye, unaided by any instrument, that the light from the sky is polarized! It does however require a certain amount of practice. . . .

---

† From W. F. Magie, *A Source Book in Physics* (Harvard University Press, Cambridge).
‡ From M. Minnaert, *The Nature of Light and Colour in the Open Air.* Reprinted with the permission of Dover Publications, Inc., New York.

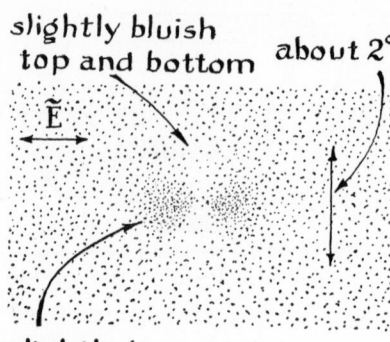

slightly bluish
top and bottom   about 2°

$\tilde{E}$

slightly brownish-
yellow at sides

FIG. 6-6 Haidinger's brush as it appears when a uniform white background is observed through a Nicol prism or a Polaroid sheet.

"If you have a Nicol at your disposal, then look through it at a white cloud —or at an evenly illuminated [white] surface and try to distinguish the figure by the fact that it revolves when the Nicol is rotated.

"After one has observed the uniformly blue sky for a minute or two, a kind of marble effect will begin to appear. This is shortly followed by that remarkable figure known as Haidinger's brush, a figure resembling more or less the one shown [in Fig. 6-6]. It is a yellowish brush with a small blue cloud on either side.

"Haidinger's brush is caused by the dichroism of the yellow spot on our retina. That all observers do not, apparently, see this remarkable figure in the same way no doubt depends on the difference in shape and structure of this yellow spot. . . ."

No wonder then that discoveries of polarized light itself, and full understanding of phenomena relating to it, awaited the introduction of the calcite rhomb as it was used by Malus in 1808; the Nicol prism, introduced later, in 1829; or the prisms by Wollaston and Rochon; or finally, in this century, the plastic polarizer sheets introduced by E. H. Land.

In our descriptions involving polarized light, as already mentioned, we invoke modern Polaroid sheets as representative of polarizers and analyzers. And for simplicity, we shall frequently endow these sheets in our discussions with an ideal property that is hardly realized in practice, of passing light without any attenuation when its electric vector is parallel to the polarizer's azimuth of easy passage; but of absorbing completely all components of electric field perpendicular thereto. This property of a Polaroid sheet to polarize light, like the same in a tourmaline plate, depends on crystal anisotropy of absorption. In contrast the earlier prisms of Nicol, of Wollaston, and of Rochon depended, for their polarizing function, on crystal anisotropy of refraction. But our present concern is not with explanations of the function of polarizers, but primarily with interactions of polarized light and matter at boundaries between different isotropic clouds of charged oscillators.

6-3. *Law of Malus*

Let us consider two Polaroid sheets, ideal as described above, one sheet being used as a polarizer and the other as an analyzer. The polarizer will transmit half of the illumination of an incident beam of unpolarized light; and of this, the transmission by the analyzer depends on how nearly parallel together the plane of polarization and the azimuth of easy passage of the analyzer lie; or how nearly perpendicular. Intermediately, if the azimuth of easy passage of the analyzer lies at angle θ, with respect to that of the polarizer, and if the polarizer transmits light described by the electric field E_1, then a component of E_1, $E_{1\pi}$, will lie parallel to the analyzer; and a component $E_{1\sigma}$ will lie perpendicular. Our ideal analyzer will pass the one component, $E_{1\pi}$, and absorb the other, $E_{1\sigma}$. These components of E_1 are: $E_{1\pi} = E_1 \cos \theta$ and $E_{1\sigma} = E_1 \sin \theta$. The illumination, after passing the first Polaroid polarizer, is thus $\mathfrak{E}_1 = \frac{1}{2}c\kappa_0(\mathcal{E}_{1\pi}{}^2 + \mathcal{E}_{1\sigma}{}^2) = \frac{1}{2}c\kappa_0\mathcal{E}_1{}^2$; while the illumination after passing both polarizer and analyzer will be $\mathfrak{E}_2 = \frac{1}{2}c\kappa_0\mathcal{E}_{1\pi}{}^2$. From $\mathcal{E}_{1\pi} = \mathcal{E}_1 \cos \theta$ the transmission of the polarized light by the analyzer is therefore

$$\frac{\mathfrak{E}_2}{\mathfrak{E}_1} = \cos^2 \theta = \left(\frac{1 + \cos 2\theta}{2}\right)$$

Fig. 6-7 shows these components and their relationships. The double angle dependence of transmission on θ emphasizes the fact that we find two maxima of transmission for each full rotation of the analyzer, and two minima.

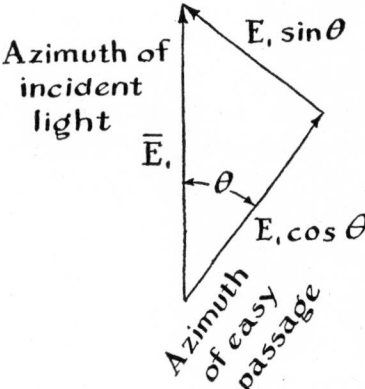

FIG. 6-7 Vector diagram involved in the derivation of Malus' law.

6-4. *Reflection of Polarized Light*

The amplitude reflection coefficients of Fresnel hold as well when the light is internally incident on a dielectric boundary as when it is externally incident. We let primes indicate internal incidence within the glass, and external refraction in the vacuum; i' and r'. In applying Snell's law, the external angle of refraction r' corresponds to our former i, and i' is equal to our former r. But this interchange causes only a reversal of the signs in the corresponding amplitude-reflection coefficients (using primed r's to represent internal incidence), and no change whatever in the intensity-reflection coefficients.

$$\mathfrak{r}_\pi' = \frac{\tan (i' - r')}{\tan (i' + r')} = \frac{\tan (r - i)}{\tan (r + i)} = -\mathfrak{r}_\pi$$

$$\mathfrak{r}_\sigma' = -\frac{\sin (i' - r')}{\sin (i' + r')} = -\frac{\sin (r - i)}{\sin (r + i)} = -\mathfrak{r}_\sigma$$

The ratio of reflection coefficients, in either case, is

$$\frac{\mathfrak{r}_\sigma}{\mathfrak{r}_\pi} = \frac{\mathfrak{r}_\sigma'}{\mathfrak{r}_\pi'} = -\frac{\cos (i - r)}{\cos (i + r)}$$

This reversal of the sign of the r's, on changing from external to internal incidence, can also be established by an interesting proof due to G. G. Stokes. This proof is illustrated by Fig. 6-8. Stokes' proof uses the principle of microscopic reversibility which requires that a reversal of time, involving a reversal of the directions of transmitted and reflected rays, will reproduce the originally incident ray with reversed direction. Fig. 6-8 represents reflection and transmission at a boundary between two transparent dielectrics with indices N_1 and N_2, where N_1 may be either greater or less than N_2. The amplitude-transmission and amplitude-reflection coefficients in this one proof may be applied to the case of either π- or σ-components. The t's and r's are given subscripts 1,2 or 2,1 to discriminate between the case where light incident from medium 1 is reflected by the boundary, or penetrates it into medium 2, and the opposite case, where light from medium 2 is reflected by the boundary,

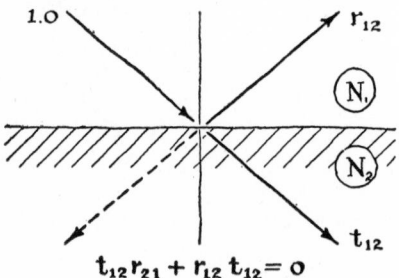

FIG. 6-8 Rays involved in Stokes' proof that $\mathfrak{r}' = -\mathfrak{r}$.

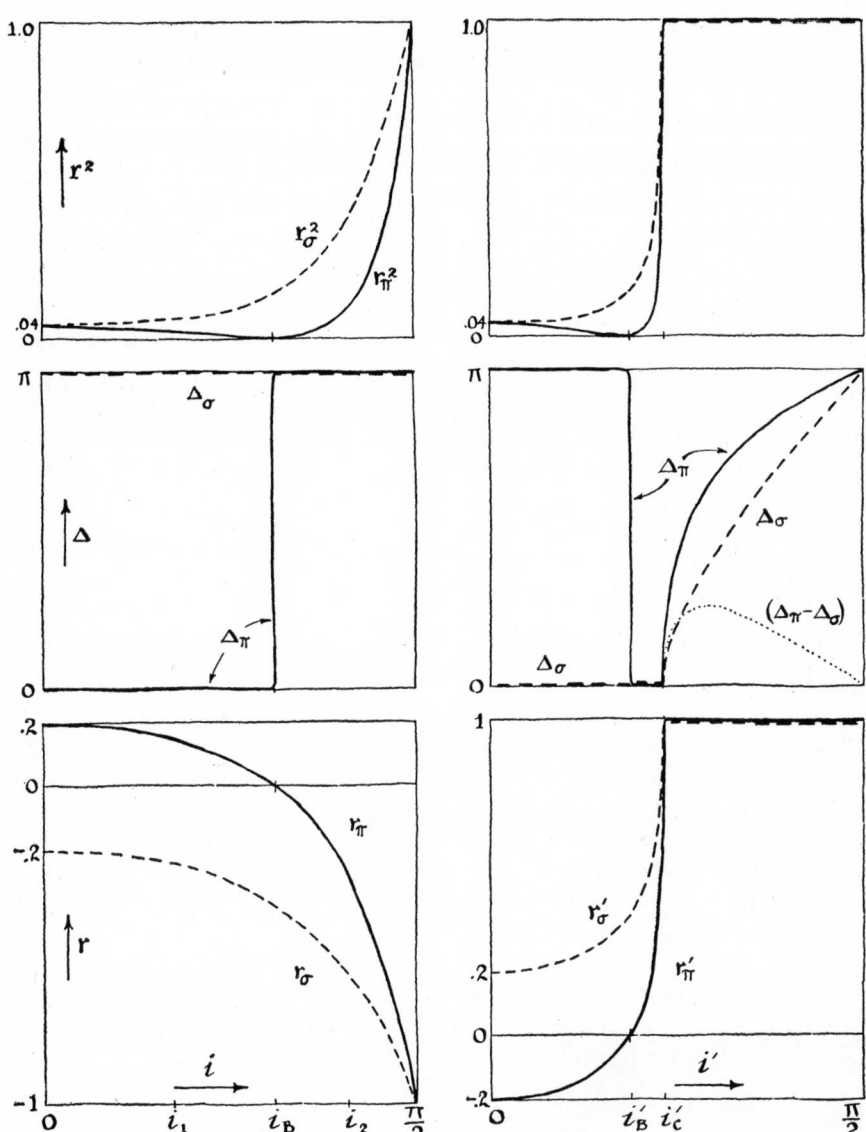

FIG. 6-9 $\Re$, r, and phase shifts for external reflection at a dielectric boundary.

FIG. 6-10 $\Re$, r, and phase shifts for internal reflection at a dielectric boundary.

or penetrates it into medium 1. In Stokes' proof we take the incident beam as having unit amplitude. This beam is divided by the boundary into two beams with the amplitudes: $r_{1,2}$ reflected, and $t_{1,2}$ transmitted. On reversal, each of these is divided into two beams: $r_{1,2}$ is divided into $r_{1,2}t_{1,2}$ and $r_{1,2}{}^2$; and $t_{1,2}$ is divided into $t_{1,2}r_{2,1}$ and $t_{1,2}t_{2,1}$. On reversal, $r_{1,2}t_{1,2}$ and $t_{1,2}r_{2,1}$ must add to zero or we would get the extra beam shown dotted in the figure. Using this condition, $r_{1,2}t_{1,2} + t_{1,2}r_{2,1} = 0$, and canceling out $t_{1,2}$, yields $r_{1,2} = -r_{2,1}$.

Figs. 6-9 and 6-10 illustrates actual values of r_π and r_σ at the surface of ordinary glass for both external and internal reflection as a function of the angles of external or internal incidence, i and i'. For external incidence r_π becomes zero at Brewster's angle when $i = i_B$, and $i_B + r_B = \dfrac{\pi}{2}$ so that $\tan (i + r)_B = \infty$. Then $r_\sigma = -\sin (i_B - r_B)$. At this angle of incidence, the reflected light is plane polarized. It was at just this angle of incidence that Malus discovered polarization in 1808.

For reflection at internal incidence: the variations of the r'''s recapitulate the variations of the r's, with reversed signs. As i goes from 0 to $\dfrac{\pi}{2}$ for the r's, the equivalent values of i' go from 0 to i_C'. Here i_C', the critical angle, is $\sin^{-1}\dfrac{1}{N}$.

Fig. 6-11 shows a simple device that may be used to produce polarized light by means of a glass plate. It is called a Nörremberg doubler. A Nicol prism or Polaroid sheet is usually used as analyzer with it.

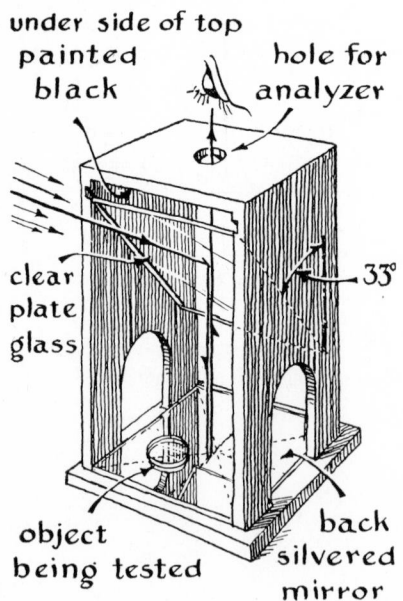

under side of top
painted
black

hole for
analyzer

clear
plate
glass

35°

object
being tested

back
silvered
mirror

FIG. 6-11   Nörremberg doubler.

6-5. *Total Reflection*

The curves of Fig. 6-10 are particularly interesting when i' is equal to the critical angle, i_C', and beyond. This critical angle is $i_C' = 41°50$ when $N = 1.5$. When we calculate r_π' and r_σ', substituting $\sin i' = \dfrac{1}{N}$ at this angle, and $\sin r' = 1.0$, we get $r_\pi' = r_\sigma' = 1.0$, meaning total internal reflection.

We can demonstrate that we also get total internal reflection for angles i' beyond i_C'. However, this demonstration involves a step which will be surprising. In the range where i' lies beyond i_C', Snell's law yields values of $\sin r'$ which are absurd because they exceed unity, and values of $\cos r'$ which are imaginary:

$$(\sin r' = N \sin i') > 1.0$$

$$\cos r' = \pm \sqrt{1 - \sin^2 r'} = \pm j\sqrt{N^2 \sin^2 i' - 1}$$

The step that is surprising is this: if we substitute this absurd sine and imaginary cosine in the Fresnel equations for r_π' and r_σ', we get a complex amplitude-reflection coefficient; and this $\tilde{r}$ not only predicts total reflection correctly, but yields the proper phase jumps that occur at total reflection. These phase jumps are designated Δ_π and Δ_σ, for internal incidence. They are shown in Fig. 6-10 for $i' > i_C'$. The convenient forms of Fresnel's equations in which to substitute this absurd sine and imaginary cosine are

$$r_\pi' = \frac{\tan (i' - r')}{\tan (i' + r')} = \frac{\sin i' \cos i' - \sin r' \cos r'}{\sin i' \cos i' + \sin r' \cos r'}$$

and

$$r_\sigma' = -\frac{\sin (i' - r')}{\sin (i' + r')} = -\frac{\sin i' \cos r' - \sin r' \cos i'}{\sin i' \cos r' + \sin r' \cos i'}$$

First let us make our substitutions in the expression for r_π', getting

$$\tilde{r}_\pi' = \frac{\sin i' \cos i' - jN \sin i'\sqrt{N^2 \sin^2 i' - 1}}{\sin i' \cos i' + jN \sin i'\sqrt{N^2 \sin^2 i' - 1}}$$

On writing this complex $\tilde{r}$ as an exponential we get

$$\tilde{r}_\pi' = e^{-j\left\{2 \tan^{-1}\left(\frac{N\sqrt{N^2 \sin^2 i' - 1}}{\cos i'}\right)\right\}} = e^{-j\Delta_\pi}$$

This exponential is interpreted to mean total reflection because it has unit magnitude; and we interpret the exponent, Δ_π, as the angular phase jump at reflection. This phase jump is the retardation of the π-component.

Similarly, for r_σ', we get

$$\tilde{r}_\sigma' = e^{-j\left\{2 \tan^{-1}\left(\frac{\sqrt{N^2 \sin^2 i' - 1}}{N \cos i'}\right)\right\}} = e^{-j\Delta_\sigma}$$

If we substitute

$$\tan\frac{\Delta_\pi}{2} = \frac{N\sqrt{N^2 \sin^2 i' - 1}}{\cos i'} \qquad \text{and} \qquad \tan\frac{\Delta_\sigma}{2} = \frac{\sqrt{N^2 \sin^2 i' - 1}}{N \cos i'}$$

in the trigonometric equation for the tangent of the difference of two angles,

$$\tan\left(\frac{\Delta_\pi - \Delta_\sigma}{2}\right) = \frac{\tan\dfrac{\Delta_\pi}{2} - \tan\dfrac{\Delta_\sigma}{2}}{1 + \tan\dfrac{\Delta_\pi}{2} \tan\dfrac{\Delta_\sigma}{2}}$$

we get an easily observable quantity,

$$(\Delta_\pi - \Delta_\sigma) = 2\tan^{-1}\left(\frac{\cos i'\sqrt{N^2 \sin^2 i' - 1}}{N \sin^2 i'}\right)$$

$(\Delta_\pi - \Delta_\sigma)$ may be observed when we send a plane polarized beam onto a glass-vacuum interface internally, with the polarization azimuth of incidence at 45° to the plane of incidence. The predictions of theory are confirmed by experiment. At this 45° azimuth of incidence, the π- and σ-components of the electric vector of our incident beam are equal and in-phase. The phase difference of these components, $(\Delta_\pi - \Delta_\sigma)$, after reflection, is easily observable because all other phase retardations suffered by the two components are equal. This phase difference is especially observable after two successive reflections inside the particular glass rhomb shown in Fig. 6-12, giving emergent circularly polarized light. Then an analyzer Polaroid transmits the emergent light equally at all orientations because it is circularly polarized. Fig. 6-10 gives a maximum value $(\Delta_\pi - \Delta_\sigma)$ of $\frac{\pi}{4}$ if $N = 1.496$ and $i' = 52°$. This i' is 10° beyond $i_C' = 42°$. Because the value $\frac{\pi}{4}$ is a stationary value at $i' = 52°$,

---

FIG. 6-12 One Fresnel rhomb makes circularly polarized of plane polarized light, and two rhombs rotate the plane of polarization through 90°.

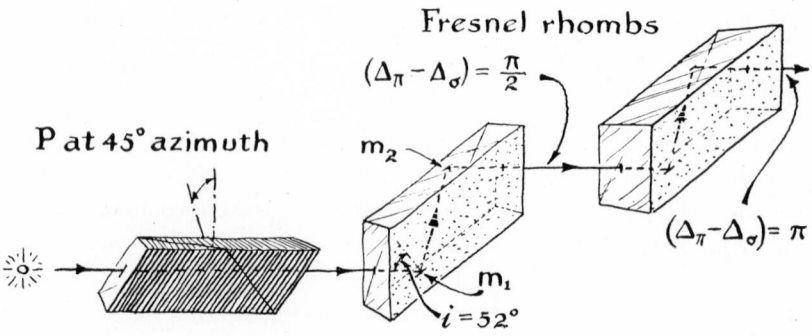

the rhomb's action varies but slowly with i'. And two successive internal rhomb reflections, each at $i' = 52°$, give a total phase retardation of $\frac{\pi}{2}$, changing incident plane polarized light to emergent circularly polarized light.

6-6. *Circularly and Elliptically Polarized Light*

Our equations of § 2-4 taught us that two orthogonal transverse waves of equal amplitude, propagating in the same direction with $\pm \frac{\pi}{2}$ phase difference between them, gave a circular wave motion. And our treatment above shows that Fresnel's rhomb introduces just this difference between the phase of the π-component and that of the σ-component of the incident plane polarized beam. If we orient our rhomb in a Cartesian coordinate system so the emergent beam runs along the z-axis, and so the yz-plane corresponds to the plane of incidence, with σ-components in the xz-plane, we may then represent the components of the emergent beam, after two internal reflections, as

$$\tilde{E}_y = \mathcal{E}_y e^{j\left(\omega t - \omega\frac{z}{c} - \Delta_\pi\right)}$$

$$\tilde{E}_x = \mathcal{E}_x e^{j\left(\omega t - \omega\frac{z}{c} - \Delta_\sigma\right)}$$

Or at some point z_1 along the propagation direction after emergence, writing $\varphi = \left(\omega t - \omega\frac{z_1}{c} - \Delta_\sigma\right)$, the real components of field are

$$E_y = \mathrm{R}\tilde{E}_y = \mathcal{E}_y \cos\left[\varphi - (\Delta_\pi - \Delta_\sigma)\right]$$
$$E_x = \mathrm{R}\tilde{E}_x = \mathcal{E}_x \cos\left[\varphi\right]$$

These two equations represent an ellipse in parametric form, abscissae and ordinates of the points on the ellipse being E_x and E_y. The ellipse generally has its major axis inclined to the x-axis, or to the y-axis; but for the case where the incident beam is plane polarized at $\theta = 45°$, making $\mathcal{E}_y = \mathcal{E}_x = \mathcal{E}$, and when $2(\Delta_\pi - \Delta_\sigma) = \frac{\pi}{2}$, the ellipse degenerates to a circle. Its equally strong emergent components are $E_y = \mathcal{E} \sin\varphi$ and $E_x = \mathcal{E} \cos\varphi$. On superposition of these emergent mutually perpendicular components we get the sum

$$E_r = \sqrt{E_x^2 + E_y^2} = \mathcal{E}.$$ The azimuth of polarization of $\overline{E}_r$ is $\theta = \tan^{-1}\dfrac{E_y}{E_x} = \varphi$.

At a point and time where $\varphi = 0$, the electric field of the emergent light lies in the σ-plane, and $\theta = 0$. But *as time elapses*, θ increases, and the azimuth of the constant strength vector, $\overline{E}_r$, rotates counterclockwise (when viewed looking toward the light source) at the angular rate ω. And *at a fixed time* as z increases, θ decreases. Thus the end of the electric vector describes in space a left-handed helix. It is for this reason that we refer to the emergent light, when $(\Delta_\pi - \Delta_\sigma) = \dfrac{\pi}{2}$, as *left circular polarized light*. Rotating the incident plane polarized light so that θ becomes $-45°$, rather than $+45°$, makes $E_y = -\mathcal{E} \sin \varphi$ while leaving E_x unchanged. The negative sign for E_y makes $\theta = -\varphi$; and such a θ yields *right circular polarized light*.

When two rhombs are in series, as indicated in Fig. 6-12, the total phase difference in the finally emergent beams will be π. When the entrance azimuth for the first rhomb $\theta = 45°$, the emergent light components from the second rhomb will be

$$E_y = \mathcal{E} \cos (\varphi - \pi) = -\mathcal{E} \cos \varphi$$

$$E_x = \mathcal{E} \cos (\varphi)$$

Thus, for two rhombs in series, $\theta = \tan^{-1} \dfrac{E_y}{E_x}$ or $-45°$ for all values of z, and the emergent light is plane polarized at an azimuth perpendicular to the plane of polarization of the incident light.

It will be of use to us later not only to know that two equally strong orthogonal components in phase quadrature give circular polarized light, but to show, also, that plane polarized light can be portrayed as a sum of two equally strong circularly polarized components. We show this now, writing E_x as

$$E_x = \tfrac{1}{2}(\mathcal{E}_x \cos \omega t - \mathcal{E}_y \sin \omega t) + \tfrac{1}{2}(\mathcal{E}_x \cos \omega t + \mathcal{E}_y \sin \omega t)$$

Here the first parenthesis is right circular polarized light if $\mathcal{E}_x = \mathcal{E}_y$; the second parenthesis is left circular polarized light. Thus we see that plane polarized light is a sort of binary compound of two circularly polarized components, just as circular is a binary compound of two plane polarized components.

6-7. *Transmission of Polarized Light*

We now consider the transmission of polarized components of light through a dielectric interface when the light is incident on it at Brewster's angle. For water, with $N = \tfrac{4}{3}$, Brewster's angle is $i_B = 53°$. At this angle $r_\pi = 0$, but at other angles of incidence the surface reflection for the π-component is not zero. In contrast, the σ-component surface reflection is never zero. The reflection of light from the sky or clouds, at a water surface, produces a veiling glare so that the visibility of objects submerged under the water surface is impaired. But at Brewster's angle, looking through Polaroid glasses that are oriented so as to pass only the π-component, we do not see the veiling glare, and the visibility of objects under the water surface is much improved. It is for this reason that Polaroid glasses are popular with sport fishermen. Polaroid glasses are effective not only at this angle but at angles near i_B, since the intensity-reflection coefficient of the π-component, $r_\pi{}^2$, not only is zero at i_B, but has a stationary value there as well. And, for like reasons, Polaroid glasses are popular with motorists when they are driving their automobiles against the sun. Such polarizing glasses filter out the glare due to the σ-component of sunlight or sky light specularly reflected from oil-glazed surfaces on the highway ahead.

We now calculate amplitude- and intensity-transmission coefficients at dielectric boundaries for both π- and σ-components, and their ratios. We define the amplitude-transmission coefficient as $t = \dfrac{\mathcal{E}_t}{\mathcal{E}_0}$, just as we defined $r = \dfrac{\mathcal{E}_r}{\mathcal{E}_0}$, in § 3-8. For the intensity transmission we take $\mathfrak{T} = \dfrac{\mathfrak{E}_t}{\mathfrak{E}_0}$, where $\mathfrak{E}_0$ is incident illumination, and $\mathfrak{E}_t$ is transmitted illumination. We first calculate $\mathfrak{T}$ below, using the principle of conservation of energy. According to this principle, the flux reflected from a transparent dielectric interface, together with that refracted, must be equal to the incident flux. Considering a plane rectangular interface of width w and length l, with light externally incident at angle i, the cross sections of the incident and reflected beams will be $wl \cos i$, while the cross section of the refracted beam will be $wl \cos r$. On equating fluxes,

$$\mathfrak{E}_0 wl \cos i = \mathfrak{E}_r wl \cos i + \mathfrak{E}_t wl \cos r$$

Now the incident, reflected, and transmitted $\mathfrak{E}$'s are, respectively, $\mathfrak{E}_0 = \frac{1}{2} c \kappa_0 \mathcal{E}_0{}^2$; $\mathfrak{E}_r = \frac{1}{2} c \kappa_0 \mathcal{E}_r{}^2$; $\mathfrak{E}_t = \frac{1}{2} N c \kappa_0 \mathcal{E}_t{}^2$. These values come from $\mathfrak{E} = (\bar{E} \times \bar{H})$ averaged; remembering $\mathfrak{K}_0 = c \kappa_0 \mathcal{E}_0$; $\mathfrak{K}_r = c \kappa_0 \mathcal{E}_r$; $\mathfrak{K}_t = N c \kappa_0 \mathcal{E}_t$. Writing in these values for the $\mathfrak{E}$'s in our flux equation, and solving for t, with $\mathcal{E}_t = t \mathcal{E}_0$ and $\mathcal{E}_r = r \mathcal{E}_0$ for either π- or σ-components, we get

$$t = \frac{\mathcal{E}_t}{\mathcal{E}_0} = \frac{1}{\sqrt{N}} \sqrt{1 - r^2} \sqrt{\frac{\cos i}{\cos r}}$$

Although, from the principle of the conservation of energy, we might expect $t_\pi = 1.0$ for incidence at Brewster's angle, where $r_\pi = 0$, we get from this principle, when there is no reflection, $t_\pi = \dfrac{1}{\sqrt{N}} \sqrt{\dfrac{\cos i}{\cos r}}$. We may confirm this unexpected value of t_π by setting $r_\pi = 0$ so that $\mathfrak{E}_r = 0$. Using $\mathcal{E}_t = \dfrac{\mathcal{E}_0}{\sqrt{N}} \sqrt{\dfrac{\cos i}{\cos r}}$ and $\mathfrak{K}_t = N c \kappa_0 \mathcal{E}_t$, canceling out $\frac{1}{2} c \kappa_0 wl$, and equating the fluxes, we get

$$\mathcal{E}_0{}^2 \cos i = \mathcal{E}_t{}^2 \cos r \text{ or } N \frac{\mathcal{E}_0{}^2}{N} \left(\frac{\cos i}{\cos r} \right) \cos r$$

which, as required by the principle of conservation of energy, balances.

For internal incidence a parallel procedure gives t', the internal-to-external coefficient, as

$$t' = \sqrt{N} \sqrt{1 - r'^2} \sqrt{\frac{\cos i'}{\cos r'}}$$

(as the student can easily prove for himself). Since $i' = r$, and $i = r'$, with $r' = -r$, the total amplitude transmission factor for the two surfaces of a plane parallel plate, tt', simplifies to

$$tt' = (1 - r^2)$$

Substituting Fresnel's expressions for r_π and r_σ in this expression for t, and writing $\dfrac{\sin i}{\sin r}$ for N, we get

$$\left.\begin{aligned} t_\pi &= \frac{2 \cos i \sin r}{\sin (i + r) \cos (i - r)} \\[2mm] t_\sigma &= \frac{2 \cos i \sin r}{\sin (i + r)} \end{aligned}\right\} \quad \text{with} \quad \frac{t_\sigma}{t_\pi} = \cos (i - r)$$

6-8. *Transmission of a Pile of Plates*

Fig. 6-13 shows a polarizer made from a stack of glass plates. Several separated plane-parallel dielectric plates are all tipped at Brewster's angle, so that the rays strike the surfaces externally and internally, both at Brewster's angles. When $r_\pi = 0$ we get $tt' = 1.0$, so that no light is lost by reflection from the π-component of an incident beam. Since, however, light from the incident σ-component is lost, the light emergent from such a pile will be partially polarized. The degree of this polarization is low from one plate; but a pile of m plates, if m is large, will act as an effective polarizer.

Because $i_B' = r_B$ and $r_B' = i_B$, the ratios of components in emergent light from one plate will be

$$\frac{t_\sigma t_\sigma'}{t_\pi t_\pi'} = \cos^2 (i_B - r_B)$$

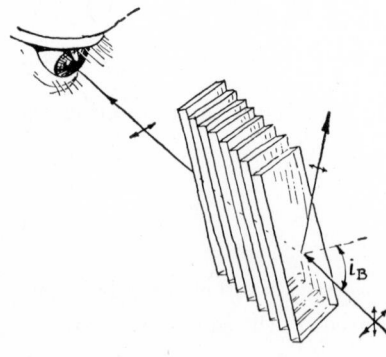

FIG. 6-13 Polarization produced by a stack of glass plates.

This will be $\cos^{2m}(i_B - r_B)$ for a pile of m successive plates. The ratio of emergent illuminations for incident natural light, squaring, becomes

$$\frac{\mathfrak{E}_\sigma}{\mathfrak{E}_\pi} = \cos^{4m}(i_B - r_B) = \left(\frac{2N}{N^2+1}\right)^{4m}$$

The last step above, changing the cosine to a function of N only, is arrived at by dividing $\cos(i_B - r_B) = \cos i_B \cos r_B + \sin i_B \sin r_B$ by $\sin^2 i_B +$ $\cos^2 i_B = 1.0$ and by using $\cos\begin{Bmatrix} i_B \\ r_B \end{Bmatrix} = \sin\begin{Bmatrix} r_B \\ i_B \end{Bmatrix}$ and $N = \dfrac{\sin i_B}{\sin r_B}$. Table 6-1 gives values of this ratio $\dfrac{\mathfrak{E}_\sigma}{\mathfrak{E}_\pi}$ for $m = 1$ and $m = 4$ and for plates with $N = 1.5$ and $N = 2.0$.

TABLE 6-1

Index	$m = 1$	$m = 4$
$N = 1.5$	0.72	0.27
$N = 2.0$	0.41	0.03

The results of this table, however, are not applicable to a polarizer of the type shown in Fig. 6-13 because the σ-component light, which is reflected out of the incident beam by one plate, is not at once lost, as our calculation presumes, but some of it is re-reflected back into the emergent beam. The results of this table would apply if the plates were very thick, and if they were also widely separated. Polarizer performances predicted by Table 6-1 are, therefore, optimistic. Such polarizers as Fig. 6-13 shows are inferior to the Nicol prism, or a sheet of Polaroid, for polarizing visible radiations,

FIG. 6-14 Pfund's method of polarization by reflection from selenium surfaces.

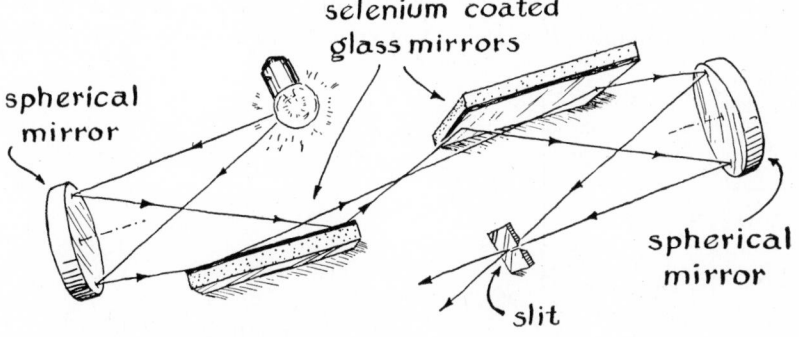

although they have been used in the past. But, in the infrared, parallel plate polarizers are currently used where materials with suitable anisotropy of refraction, or with suitable anisotropy of absorption, for making polarizers, are not available. The plate polarizers are made of AgCl, which material has a high index (greater than 2.0) and also a good transmission in the infrared spectrum out to 20μ, and beyond. Formerly, following Professor A. H. Pfund, reflections at Brewster's angle at the surface of selenium, as shown in Fig. 6-14, were used to produce polarized infrared radiations, or to analyze them, but this arrangement is clumsy to use.

6-9. *Frustrated Total Reflection*

It is instructive to apply our theory and calculate the electric fields, $\tilde{E}$, and displacements, $\tilde{D}$, which occur on the vacuum side of a glass-vacuum interface when light is incident internally on the interface at an angle $i' > i_c'$, so that it is totally reflected there. When we do this, our result may be confirmed by frustrating that total reflection. To frustrate total reflection we put another glass prism hypotenuse surface close by, without touching, as Fig. 6-15 shows, and "tap" off flux from the otherwise totally reflected beam in the first prism. When the second prism is absent, it is to be expected, from considerations of conservation of energy, that no flux can penetrate the vacuum-glass boundary and be thus lost from the reflected beam. Such, in fact, is the case. But at the same time, in order to satisfy the boundary conditions for continuity of fields, there must be electric fields (of the light frequency) in the vacuum beyond the boundary. These flux-less fields are predicted by our theory. To predict them we refer to Fig. 6-16, which shows a plane wave incident internally on a glass-vacuum interface, supposedly at $i' > i_c'$. To apply theory we arrange a Cartesian coordinate system as shown, so that the interface lies at the plane $z = 0$, and so that the internally incident wave normal lies in the $y = 0$ plane. Further, we represent the internally incident electric field at the point $x = z = 0$ by means of the complex number $\tilde{E} = \mathcal{E}_0 e^{j\omega t}$. Now consider the distance from this origin

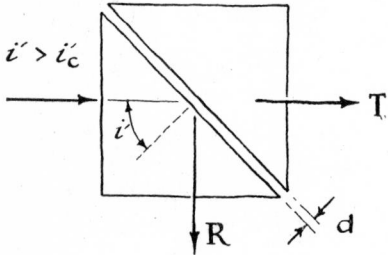

FIG. 6-15   Frustrated total reflection.

FIG. 6-16 Rays and wave fronts involved in total internal reflection.

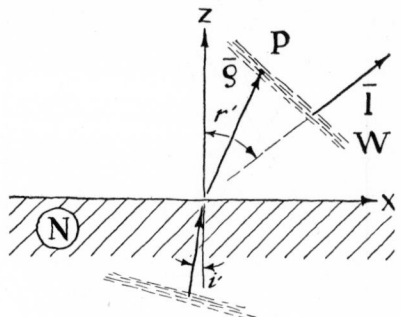

to some point P on W, a supposedly transmitted wave front in the vacuum. This distance may be expressed by a vector $\bar{p} = x\mathbf{i} + z\mathbf{k}$. Consider also the unit vector $\mathbf{1} = \sin r'\mathbf{i} + \cos r'\mathbf{k}$, which lies normal to the supposed wave front W. The normal distance s from the point $x = z = 0$ to W is $s = \bar{p}\cdot\mathbf{1} = x \sin r' + z \cos r'$. The travel time associated with this s should be $\dfrac{s}{c}$. We use this $\dfrac{s}{c}$ to describe the phase of the field at P, as

$$\tilde{E}_P = \mathcal{E}_{t'}e^{j\omega\left(t-\frac{s}{c}\right)}$$

where $\mathcal{E}_{t'}$ is the transmitted electric vector amplitude. To go further with this calculation we lay aside caution and substitute our absurd sine and imaginary cosine in the expression for s, getting the field at P as

$$\tilde{E}_P = \mathcal{E}_{t'}\left[e^{\pm\frac{\omega z}{c}\sqrt{N^2 \sin^2 i'-1}}\right]e^{j\omega\left(t-\frac{Nx \sin i'}{c}\right)}$$

We interpret the term in brackets, with the negative exponent, to mean that the electric field on the vacuum side falls off exponentially with normal distance from the interface. And we interpret the parenthesis as the phase of an electric field on the vacuum side, propagating tangentially in the x-direction, with phase velocity $\dfrac{c}{N \sin i'}$. The validity of this deduced equation is confirmed by experiment and may be demonstrated by the tapping prism mentioned above. The ratio $\dfrac{\mathfrak{R}}{\mathfrak{T}}$ for the gap between the two totally reflecting prisms, as shown in Fig. 6-15, although not deducted here, may be calculated along the same lines as above, for $i' = 45°$. Such a calculation gives

$$\frac{\mathfrak{R}}{\mathfrak{T}} = a\,\frac{(N^2-1)^2}{N^2(N^2-2)}\,\sinh^2\left[\pi\,\frac{\sqrt{2}\,d}{\lambda}\,\sqrt{N^2-2}\right]$$

where $a = 1.0$ for the σ-component, and $a = \frac{1}{4}$ for the π-component. The transmission $\mathfrak{T}$, and the frustrated reflection $\mathfrak{R}$, both vary with d as shown

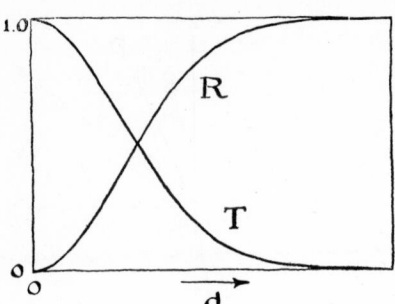

FIG. 6-17 Dependence of $\mathfrak{R}$ and $\mathfrak{T}$ on d in Fig. 6-15.

in Fig. 6-17. Since the flux carried away from the gap by the two beams must be equal to the incident flux, $\mathfrak{R} + \mathfrak{T} = 1.0$. Accordingly we may compute either $\mathfrak{R}$ or $\mathfrak{T}$ from the expression above for $\dfrac{\mathfrak{R}}{\mathfrak{T}}$. Fig. 6-17 shows how $\mathfrak{R}$ and $\mathfrak{T}$ vary with d, $\mathfrak{R}$ becoming zero when $d = 0$, and unity when $d \gg \lambda$.

Chapter VII

Double Refraction—
Calcite and Quartz

IN AN earlier discussion, our charged oscillator model was characterized by isotropic binding. But now, to portray the optical properties of the two crystals, calcite and quartz, we need more elaborately bound model oscillators—we must ascribe to our model oscillating charges anisotropic binding such as is represented by the mechanical model of Fig. 7-1. This figure is to be contrasted with our former mechanical model with isotropic binding, shown in Fig. 4-1. Whereas, before, the three principal elastic restraints on the charge were equal, $\gamma_x = \gamma_y = \gamma_z$, now we have two of them equal, $\gamma_y = \gamma_z = \gamma_\perp$, with the third restraint different, $\gamma_x = \gamma_{||}$. In our two crystals, this third restraint defines a direction that is called the *optical axis* and the $\perp$ and $||$ subscripts on γ mean perpendicular or parallel to this axis, respectively. An array of such oriented anisotropic charged oscillators will explain the optical properties

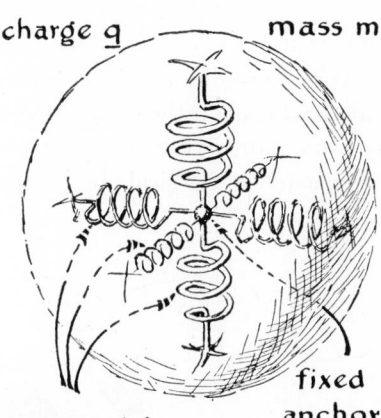

charge q   mass m

asymmetric   fixed anchor
elastic constraints   point

FIG. 7-1 Charged oscillator model, with asymmetrically bound charge, as used to explain optically anisotropic crystals.

of calcite and many of those of quartz. And, with further elaborations, this model can also explain the optical activity of quartz, which is discussed later, in § 7-11.

7-1. *Calcite Double Refraction*

A ray of unpolarized light normally incident on the cleaved crystal face of a calcite rhomb, such as Fig. 7-2 shows, is doubly refracted; it becomes two rays within the crystal. These two rays are constituted of mutually perpendicular polarized components of the incident light. One of these two rays propagates undeviated, normal to the entering surface, and emerges without any lateral shifting; while the other is refracted within the crystal (at the angle $r_{ext} = 6°14'$) and emerges from the back parallel cleaved face of the rhomb propagating again in the same direction as the incident beam. But this other ray emerges shifted laterally, because of the askew internal angle of propagation. Because of the lateral shift of one of the rays, an object viewed through the calcite rhomb is seen double, as Fig. 7-2 shows. A dot on a piece of paper, covered by a rhomb of calcite, is seen double; and if we rotate the rhomb on the paper, one of the double dots remains stationary as the other appears to move around it with the crystal. The fixed dot behaves as it would if the paper were covered by a thick plate of glass. The rays by which it is seen are called the *ordinary rays*. The extraordinary movement of the second dot is due to its being seen by *extraordinary rays*. The extraordinary rays are polarized in the plane which contains the calcite optical axis, while the ordinary rays are polarized perpendicular to this plane.

Huygens did this simple experiment as well as others with two rhombs in cascade. He found "surprising phenomena touching the rays which pass two separated pieces; the cause of which is not explained." Although in 1690 he did not fully understand calcite (because he did not think of light as transverse wavelets), he did adapt his wavelets construction principle to double refraction, and explain the extraordinary refraction exactly as we do today. It was over a hundred years later that Huygens' explanations were fully matured by the masters of classical optics: Malus, Young, Fresnel, Arago, Brewster, Biot, and others.

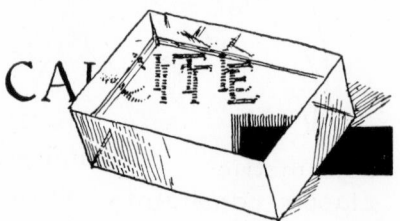

FIG. 7-2 Iceland spar.

7-2. *Spheroidal Huygens Wavelets*

We may speculate that Huygens, realizing that the ordinary spherical Huygens wavelets were inadequate for extraordinary refraction, invoked the next logical elaboration of spherical wavelets; namely, spheroidal Huygens wavelets. We express Huygens' two types of expanding wavelets for calcite, the spheres and the spheroids, as follows:

$$\frac{x^2 + y^2 + z^2}{v^2} = t^2 \qquad \text{for spherical wavelets}$$

$$\frac{x^2 + y^2}{v_{\parallel}^2} + \frac{z^2}{v_{\perp}^2} = t^2 \qquad \text{for the spheroidal wavelets}$$

Here our coordinate system is oriented so that the z-coordinate coincides with the optical axis, or crystal direction of optical symmetry; and for calcite $1.6584\,v_{\perp} = 1.4864\,v_{\parallel} = $ **c**. These equations, when $t^2 = 1.0$, define surfaces which are called the *velocity surfaces*.

Fig. 7-3 shows an xz-section of such velocity surfaces. The circle is the section of Huygens' sphere, and the ellipse is the section of the required circumscribing Huygens spheroid. The sphere represents the phase velocity of ordinary rays in calcite—the rays that have the polarization of their electric vector perpendicular to the plane that contains the optical axis. The spheroid represents the orthogonally polarized *e*-ray velocity in calcite. This spheroid gives the phase velocity at different directions with respect to the axis of optical symmetry.

Fig. 7-4 shows the direction of optical symmetry in a cleaved calcite rhomb; the arrow which lies symmetrical to the three cleavage faces at the blunt corner of the rhomb shows the direction in which the optical axis is found to be oriented in this crystal.

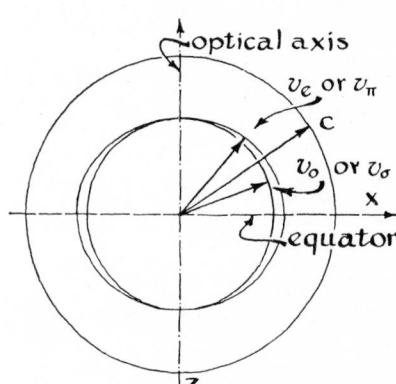

FIG. 7-3 Sections of the velocity surfaces of calcite.

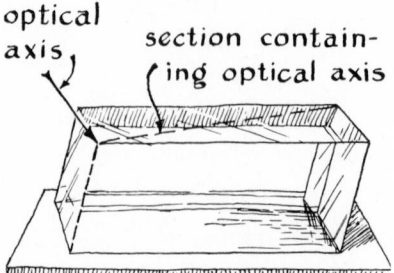

optical axis

section contain-
ing optical axis

FIG. 7-4 Orientation of the optical axis in calcite.

The plane represented by the dotted section of Fig. 7-4, which lies perpendicular to one cleavage face and also contains the optical axis, is shown in Fig. 7-5 with illustrated wave fronts. The wave train, W, is incident on the cleavage face normally. This wave W is shown together with two refracted wave fronts, W_{ord} and W_{ext}, which an earlier incident wave of the same incident wave train has previously excited. W_{ord} and W_{ext} are envelopes for the spheres, and for the spheroids, at a time t after simultaneous excitation of the wavelets at the crystal surface. These envelopes both lie parallel to W; but the extraordinary wave front is propagating in the direction of a vector drawn from the origin of one of the extraordinary wavelets to its point of tangency with W_{ext}. This direction makes an angle with the normal to the crystal face of $r_{ext} = 6°14'$, as is calculated below.

The polarization of ordinary and extraordinary rays is indicated in our figures by dots on the rays, or by short lines transecting them. If the electric vector lies in the plane of the figure, short lines are used; dots, if perpendicular.

7-3. *Calculation of* r_{ext} *for a Calcite Rhomb*

Fig. 7-6 shows the refraction of one extraordinary ray of the incident beam W. It strikes the cleaved surface at O, and is refracted at the angle r_{ext} in the direction OP. At P the wave front and the Huygens wavelet from O are tangent. We may calculate the angle of refraction, r_{ext}, from our equation of the elliptic section of the extraordinary velocity surface. Our coordinate system, with its z-axis coincident with the calcite optical axis, expresses the elliptic

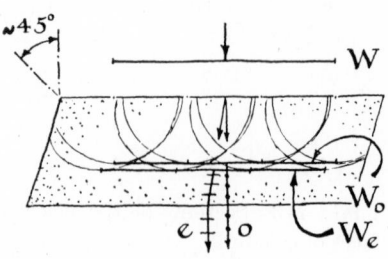

FIG. 7-5 Huygens wavelets in calcite.

FIG. 7-6 Use of the velocity surface in calcite to determine the angle of refraction of the extraordinary ray.

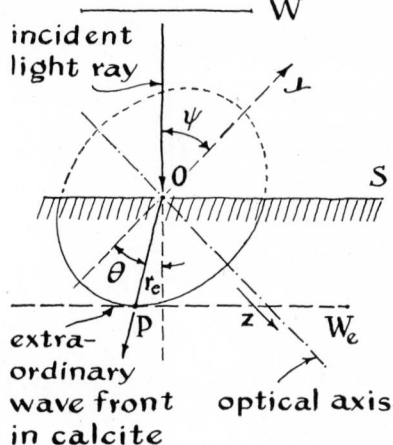

section of the ellipsoid wave surface as

$$\frac{z^2}{v_\perp{}^2} + \frac{y^2}{v_{||}{}^2} = t^2$$

Here $\left(\dfrac{v_{||}}{v_\perp}\right)^2 = \left(\dfrac{1.658}{1.486}\right)^2 = 1.245$. Proceeding to solve the geometry of Fig. 7-6, we first differentiate our equation and get

$$\left(\frac{dy}{dz}\right)_P = -\frac{z_P}{y_P}\left(\frac{v_{||}}{v_\perp}\right)^2 = -1.245\,\frac{z_P}{y_P} = \tan \psi$$

Again, referring to the figure, we see that $\psi = \theta + r_{\text{ext}}$. And $\tan \theta = -\dfrac{z_P}{y_P}$. A student exercise gives the angle of the optical axis relative to the cleavage faces as $\psi = 45°$ so that the combination $\tan \psi$ and $\tan \theta$ gives

$$\tan r_{\text{ext}} = \frac{\tan \psi - \tan \theta}{1 + \tan \psi \tan \theta} = \tan 6°14'$$

7-4. *Refraction by Two Calcite Rhombs in Series*

The "surprising phenomena touching the rays which pass two separated pieces; the cause of which is not explained," which had concerned Huygens, are illustrated by Figs. 7-7 to 10. In these figures the component of the incident unpolarized ray which is polarized in the plane which contains the optical axis of the first rhomb is called e_1, while the component perpendicular is called o_1. These rays are shown traced through the two crystals for the relative rhomb orientations that Huygens used, namely:

The optical axes are parallel in Fig. 7-7.

The optical axes are in the same plane but not parallel in Fig. 7-8.

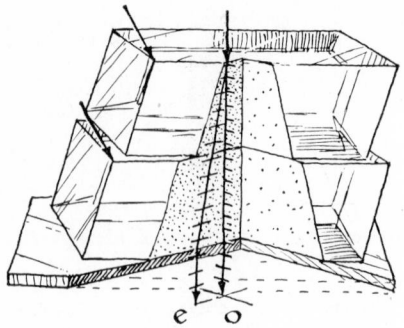

FIG. 7-7 Calcite crystals similarly oriented.

e o

The optical axes are in perpendicular vertical planes in Fig. 7-9.

The optical axes are in 45°-vertical planes in Fig. 7-10.

Huygens did not use the concept of polarized light. This concept is the key to understanding of these experiments. Without this key what would one expect? We may presume that Huygens first expected that the difference between the e-rays and o-rays resided in the light itself. It could, indeed, have been an intrinsic difference, like the difference between the colors which are inherent in white light and not in the prism which displays them, or like the longitudinal and transverse waves from an earthquake epicenter. But his experiments showed that the ordinary ray did not retain its ordinariness in the second crystal; nor did the e-ray also remain extraordinary. We now know that it is the azimuthal planes of transverse vibration of e-rays, or o-rays, which persist here in the second crystal; and if the successive crystals are arranged as in Fig. 7-9, this very persistence causes the o_1-ray in the first rhomb to become an e_2-ray in the second rhomb, while the e_1-ray in the first becomes o_2 in the second, as observed.

Huygens observed that in some instances, *viz.* Fig. 7-7, the displacement of the e-ray was doubled; in other instances, Fig. 7-8, its displacement was canceled. Fig. 7-9 shows his case where e_1 becomes o_2 in the second crystal, and o_1 becomes e_2. Fig. 7-10 shows the 45° relative crystal positions where the

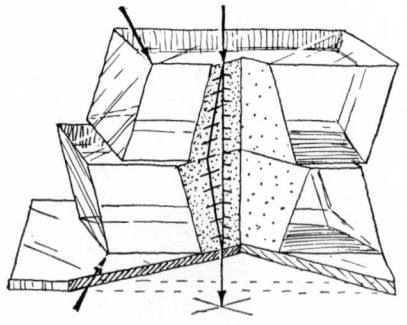

FIG. 7-8 Calcite crystals oppositely oriented.

FIG. 7-9   Calcite crystals with 90° relative rotation,

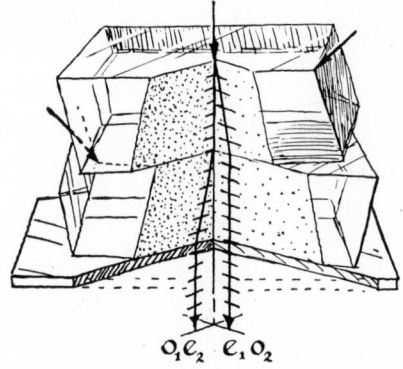

o_1e_2 e_1o_2

e_1 and o_1 rays both have components parallel and perpendicular to the plane of the optical axis in the second crystal—here e_1 becomes e_1e_2 and e_1o_2; and, also, o_1 becomes o_1e_2 and o_1o_2, thus giving four equally strong, finally emergent rays. Once the role of transverse polarization is comprehended, the coexistence of two sets of Huygens' wavelets is seen as natural, and Huygens' "surprising phenomena" are easily understood.

Malus inferred that the reflected window light illustrated in Fig. 6-4 was polarized, because at certain rhomb azimuths he observed only one ray emergent from his rhomb, and because he *always* got two emergent rays when an unpolarized ray was incident normally on a rhomb.

7-5. *Explanation of the e-Wave Surface*

The askew propagation vector, $\overline{OP}$, of Fig. 7-6, represents propagation which does not lie perpendicular to the refracted wave front, W_e. This seems strange. But it can be explained in terms of our model of calcite, if we consider that crystal to be an oriented assemblage of anisotropic oscillators: First we

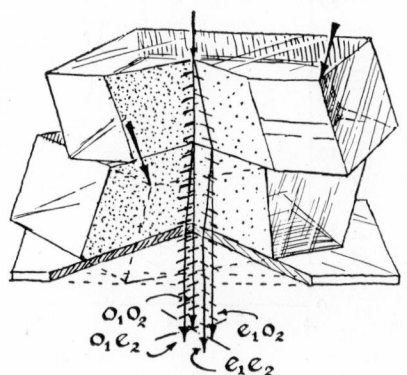

o_1o_2
o_1e_2 e_1o_2
e_1e_2

FIG. 7-10   Calcite crystals with 45° relative rotation.

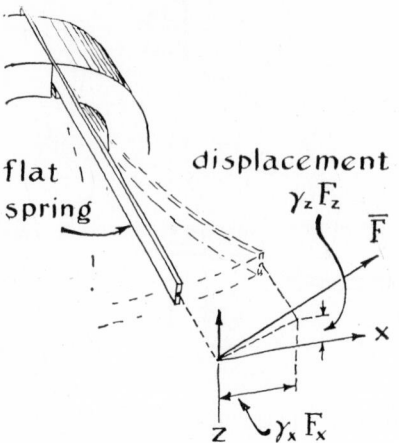

FIG. 7-11 Mechanical analogue of asymmetrically bound charge.

note that the microscopic oscillatory motion of the charge of one of the oscillators may lie in a direction askew to the forcing function, $q\bar{E}$; or to $\bar{E}$, the electric wave front which produces it. The derivative of this askew charge motion when multiplied by the charge, $q\dot{\bar{w}}$, as we have seen, acts as an askew alternating electric current and this askew current produces a magnetic field which lies in a direction also askew to the electric wave front $\bar{E}$. Thus it is clear that the vector product $\bar{\mathfrak{P}} = (\bar{E} \times \bar{H})$ does not lie necessarily perpendicular to the electric wave front, as we know it for vacuum and isotropic media, but as Figs. 7-5 and 7-6 show. Fig. 7-11 shows a flat leaf-spring that illustrates such anisotropic compliance in two dimensions. Here we know from

FIG. 7-12 Reflection of calcite in the infrared region.

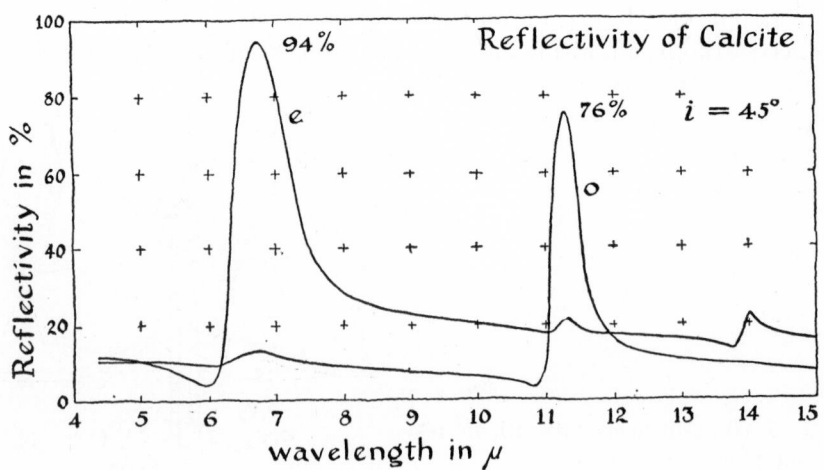

common experience that the deflection of such a leaf spring, where $\gamma_x \neq \gamma_z$, often lies in a direction not parallel to the deflecting force F; but rather, it lies in an askew direction which favors the weakest restraint, as shown.

From the duplicity of γ's for the oscillators in calcite we might expect to have double reflection in the infrared. Fig. 7-12 depicts just such a manifestation. This figure, for near normal incidence, shows the reflectivity curves of a plate cut from calcite so as to contain the optical axis parallel to its face. One curve represents the reflectivity for infrared radiation polarized with the incident electric vector in the azimuth that contains the optical axis, labeled e; the other is for the orthogonal polarization, labeled o. These curves do exhibit resonances at different frequencies such as would be expected for our simple oscillator model with $\gamma_x \neq \gamma_z$. We would expect $\omega_x = \sqrt{\dfrac{1}{\gamma_x m}}$ and $\omega_z = \sqrt{\dfrac{1}{\gamma_z m}}$ if the oscillating masses were equal for these two polarizations, the crystal exhibiting one reflection peak for resonance with radiation polarized parallel, ω_z, and one for radiation polarized perpendicular to the optical axis, ω_x.

Here we do not derive the velocity for extraordinary rays from oscillator bindings and densities, as we did for the ordinary velocity in § 4-2. It is enough for us to get the concept of the e-velocity surface as explained qualitatively by the different magnitudes and directions of $\bar{w}$ that are produced by a given $\bar{E}$ applied in different directions within the crystal.

7-6. *Various Refractions of the* e-*Ray by Calcite*

We have already described and calculated the direction of the e-ray refraction in the case of normal incidence on a cleaved crystal face. The principle for determining other refractions, or refraction in general, is the same. In general we use the same procedure of Huygens' construction, invoking the velocity spheroids for the extraordinary rays. It is possible to use Snell's law for determining extraordinary ray refractions for calcite only when the section of the e-ray velocity surface cut by the plane of incidence is a circle. For example, Snell's law obviously does not apply in our case of normal incidence

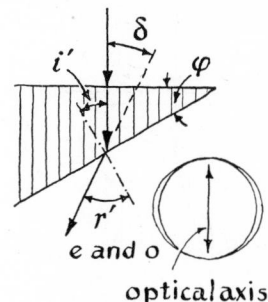

FIG. 7-13   Refraction by a calcite prism.   optical axis

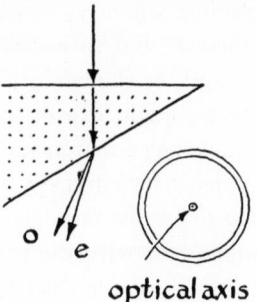

FIG. 7-14 Refraction by a calcite prism.

optical axis

on a calcite rhomb, since the angles, $i = 0$ and $r_{ext} \neq 0$, would give $N = 0$. Refraction angles are easy to determine by the method of Huygens' construction when the optical axis lies in the plane of incidence, or perpendicular to it, and parallel to the refracting surface, or perpendicular. Figs. 7-13, 7-14, and 7-15 show three special cases of internal incidence; Figs. 7-16, 7-17, and 7-18 show three special cases of external incidence; and Fig. 7-19 illustrates a general case. The general case will require a little effort on the part of the student to visualize it, and more to calculate it.

In Fig. 7-13, double refraction is not manifest because, when the rays are propagating parallel to the optical axis within the prism, we have $v_{\parallel} = v_{\perp}$. The e- and o-rays are therefore equally retarded within the crystal, and thus equally deviated through the angle δ at the emergent face. The angles i' and r', for a half prism of angle φ, are $i' = \varphi$ and $r' = \varphi + \delta$. Applying our Huygens relation of § 1-1, we get

$$\frac{c}{v_{\perp}} = \frac{c}{v_{\parallel}} = \frac{\sin (\varphi + \delta)}{\sin \varphi}$$

Double refraction is manifest in Figs. 7-14 and 7-15. In both cases $v_{\perp} < v_{\parallel}$; so the o-ray is refracted more than the e-ray. Our Huygens relation of § 1-1 gives

$$\frac{c}{v_{\perp}} = \frac{\sin (\varphi + \delta_{ord})}{\sin \varphi} \quad \text{and} \quad \frac{c}{v_{\parallel}} = \frac{\sin (\varphi + \delta_{ext})}{\sin \varphi}$$

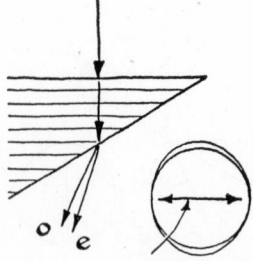

optical axis FIG. 7-15 Refraction by a calcite prism.

FIG. 7-16   Refraction by a calcite plate.

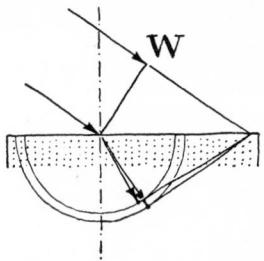

The refractions are independent of the azimuthal orientation of the optical axis as long as that axis lies parallel to the entering face of the prism, and as long as this face is entered normally by the light.

In Fig. 7-16 the extraordinary ray lies in a plane of incidence perpendicular to the optical axis and the sections of the wave surfaces are circles, as shown. In this case Snell's law applies and we may describe the crystal double refraction with two indices, $N_{ext} = \dfrac{c}{v_{||}} = 1.486$ and $N_{ord} = \dfrac{c}{v_{\perp}} = 1.658$. This figure illustrates also how we may apply Huygens' construction.

Figs. 7-17 and 7-18 show how we can determine other refraction angles for external incidence when the optical axis lies parallel or perpendicular (rather than at $\psi = 45°$) to the refracting surface, and in the plane of incidence. In Figs. 7-16, 7-17, and 7-18 external wave fronts (W) are shown when their bottom edge first touches the refracting surface; and the Huygens wavelets in the crystal are shown at the instant when the upper edge of the incident wave front arrives at the refracting face. From Huygens' construction a line drawn from a point of late wavelet origin to a point of tangency with the earliest wavelet in the crystal gives the internal wave front and a ray from origin to point of tangency gives the propagation direction. The reality of the ellipsoid velocity surface is experimentally demonstrated by such simple confirming refraction observations as are indicated by the figures.

Fig. 7-19 illustrates the general case as contrasted to these special ones. In this figure the optical axis lies neither parallel nor perpendicular to the refracting surface, nor in the plane of incidence. Nevertheless, the graphical

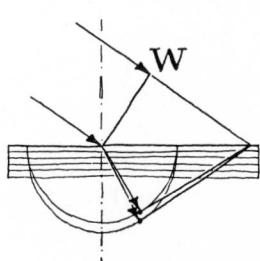

FIG. 7-17   Refraction by a calcite plate.

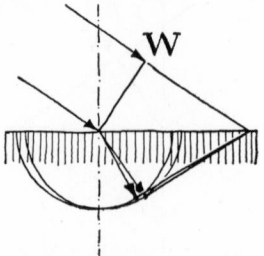

FIG. 7-18 Refraction by a calcite plate.

procedure for finding the extraordinary refracted wave fronts, as well as propagation directions, is the same in principle as above; although much more difficult of execution. The *e*-ray in general is refracted out of the plane of incidence, as indicated. Of course, Snell's law always applies for determining refractions of the *o*-ray.

In this chapter we do not treat biaxial crystals, which are more complex than calcite. The properties of biaxial crystals are, in fact, complex enough, and different enough, from crystal to crystal, to provide the basis in mineralogy for the precise diagnostic power of the petrographic microscope.

FIG. 7-19 Generalized refraction by a calcite surface.

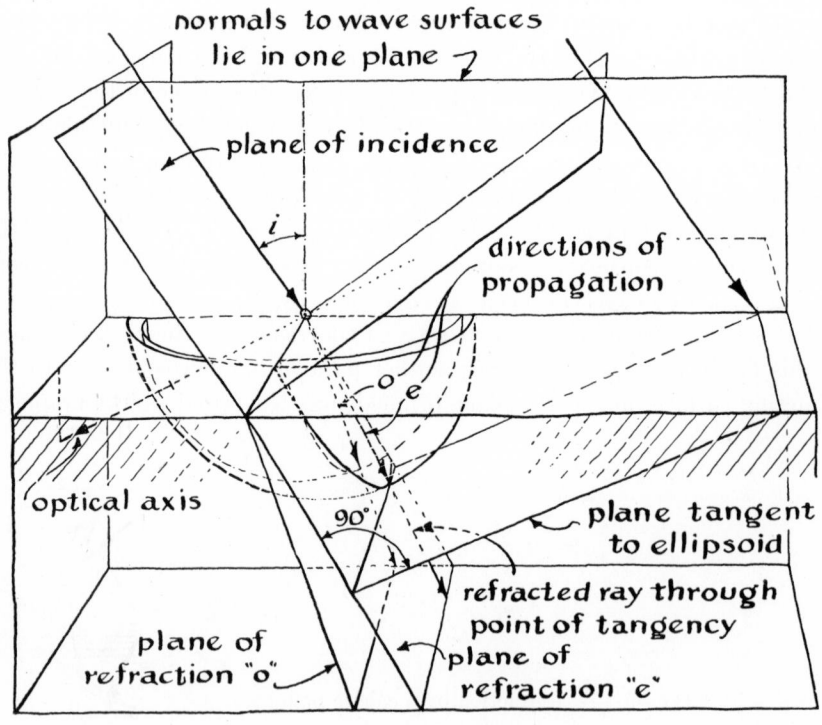

Geologists and mineralogists are able to determine the complex optical properties of mineral samples by means of the petrographic microscope. They then use these properties for mineral identification. Their analysis procedure is possible because their petrographic microscopes are fitted with components made of calcite, quartz, and mica. Although we shall not explain all the complexities of the use of the petrographic microscope, because mineral identifications require a full knowledge of crystal optics, we do explain the microscope components from which the precise diagnostic powers derive: polarizers, analyzers, quarter wave plates, and the Babinet and Soleil compensators.

7-7. *Uses of Calcite*

After Malus initiated the first useful application of a cleaved rhomb of calcite, for determining that light was polarized, and in which azimuth, this crystal enjoyed much use in analysis of light. It was much used throughout the early fertile days in optics before W. Nicol, in 1829, introduced his polarizing prism (or more properly his polarizing parallelepipedon). After 1829, Nicol's prism and other devices fabricated from calcite (and quartz) competed successfully with usage of the simple cleaved rhomb. The Rochon and Wollaston composite prisms were among these other devices. As contrasted to Malus' calcite rhomb, from which these rays emerge parallel, these last two composite prisms separate the o-ray from the o-ray in angle.

Fig. 7-20 shows the construction of Nicol's prism which eliminates the o-ray entirely. To make a Nicol, two prismatic wafers are sliced off the ends of a suitably long and narrow cleaved calcite rhomb; and the crystal is cut diagonally, as shown, into two halves. These two main parts then have their sliced and cut surfaces ground and polished, and the two cut surfaces are reunited, face to face, and cemented with a balsam cement in which the velocity of light is intermediate between $v_\perp$ and $v_\parallel$. When natural light is incident along the axis of a composite prism thus constructed, the o-ray is refracted down at the entering face and strikes the balsam surface at an angle i', which is greater than the critical angle, i_c', there. Thus this ray is totally reflected out of the optical path, and absorbed. On the other hand, the e-ray, which is substantially undeviated, is transmitted through the balsam layer with but little loss of light; and it finally emerges through the exit face, again substantially without angular deviation.

Fig. 7-21 shows Rochon and Wollaston prisms. Although these can be made of either calcite or quartz, here we describe them as made of calcite. This material is a more appropriate one for us now, pedagogically, although, in practice, quartz is easier to fabricate, and it is more permanent in use. The orientations of optical axes are indicated in Fig. 7-21 for these two composite

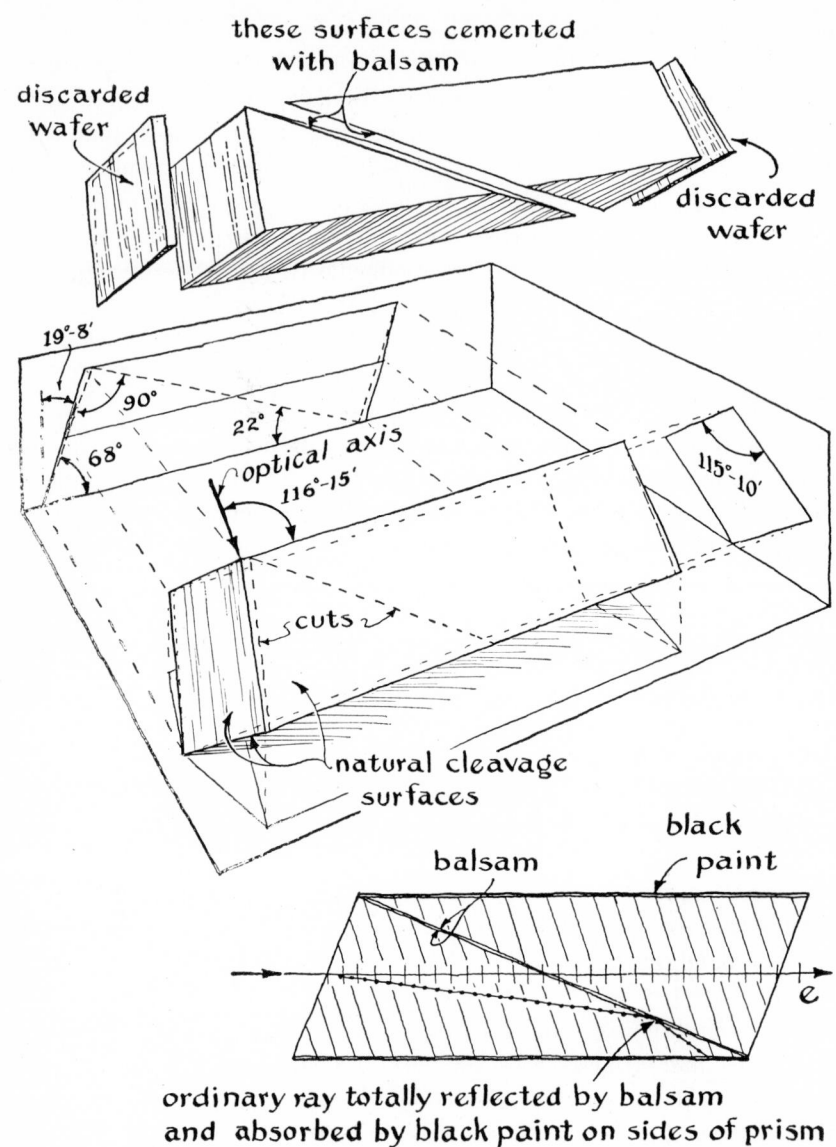

FIG. 7-20 Nicol prism.

prisms, made of calcite. In the Rochon, the interface between prism halves affords no optical discontinuity at all for the normally incident ray polarized perpendicular to the optical axis in the second component. This o-ray is therefore undeviated at the inclined interface. However, the other component, the e-ray, suffers a discontinuity of index of $(N_{ext} - N_{ord})$ at the interface,

FIG. 7-21   Polarization beam splitters.

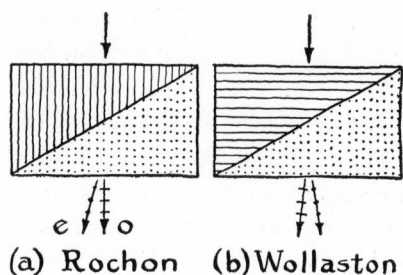

(a) **Rochon**   (b)**Wollaston**

and it is therefore deviated. In the Wollaston, a discontinuity $(N_{ext} - N_{ord})$ occurs for both polarizations, since the *e*-ray in the first prism is the *o*-ray in the second, and *vice versa*. Thus the angle deviation is doubled.

Fig. 7-22 illustrates a *phase retardation plate*. We consider this plate to be a very thin parallel plate of calcite, of thickness *d*. It is cut and polished from a crystal so that its surfaces lie parallel to the optical axis. Consider a beam of light striking this plate normally with its electric vector, $\overline{E}_1$, polarized in a plane which makes an azimuth angle θ with the optical axis, as shown in the figure. Since $v_{ord} < v_{ext}$, or $N_{ord} > N_{ext}$, such a plate will retard the perpendicular component of $\overline{E}_1$ more than the parallel component. A glass plate of the same thickness would introduce a phase retardation of $\dfrac{\omega(N_g - 1)}{c} d$ in the emergent light, compared to the phase it would have had with the plate absent. But here we are interested in differences of retardation for *e*-rays and *o*-rays. Perhaps such a plate of calcite as is described above should be called a *relative phase retardation plate,* or something like that. Such a name would distinguish it from a simple glass plate, but the name is too clumsy. Actually, for such phase retardations as are wanted in a petrographic microscope, the calcite would be much too thin for construction, considering its brittleness. The biaxial crystal, mica, affords a more appropriate material than calcite,

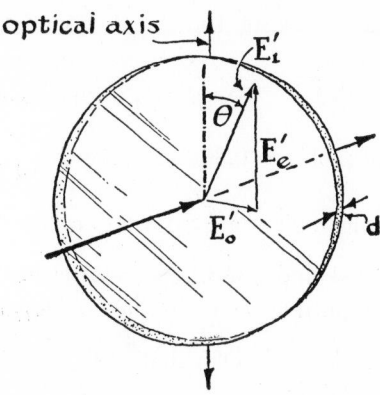

FIG. 7-22   Calcite relative-retardation plate.

because by splitting it is easy to make it thin enough; and because it is flexible, and thus not so easily broken. Nevertheless, for pedagogic reasons, we shall consider our relative phase retardation plate here as made of calcite.

Let $\bar{E}_1 = \mathcal{E}_1 e^{j\omega t}$ be the electric vector of the incident light polarized at angle θ to the optical axis; and let $\bar{E}_1' = \mathcal{E}_1' e^{j\omega t}$ be the field just *inside* the crystal boundary, also at $z = 0$. The component fields inside, at any point or time within the crystal, can be written in terms of the components of $\bar{E}_1'$: parallel and perpendicular to the optical axis, $\mathcal{E}_{ext}'$ and $\mathcal{E}_{ord}'$.

$$\bar{E}_{ext}' = \mathcal{E}_1' \cos\theta \, e^{j\omega\left(t - \frac{z}{v_{ext}}\right)} \qquad \text{and} \qquad \bar{E}_{ord}' = \mathcal{E}_1' \sin\theta \, e^{j\omega\left(t - \frac{z}{v_{ord}}\right)}$$

Neglecting reflection losses at both entering and emergent faces of the plate, the component fields just *outside* the emergent face of the crystal at $z = d$ will be

$$\bar{E}_{ext} = \mathcal{E}_1 \cos\theta \, e^{j\omega\left(t - \frac{d}{v_{ext}}\right)} \qquad \text{and} \qquad \bar{E}_{ord} = \mathcal{E}_1 \sin\theta \, e^{j\omega\left(t - \frac{d}{v_{ord}}\right)}$$

Thus the incident and emergent illuminations, neglecting reflection losses, are equal:

$$\tfrac{1}{2}c\kappa_0\mathcal{E}_1^2 = \tfrac{1}{2}c\kappa_0(\mathcal{E}_1^2 \cos^2\theta + \mathcal{E}_1^2 \sin^2\theta)$$

However, the components of the emergent light are no longer synchronous. Our explanations are simplified if we reset the clock so that $\omega t = \left(\omega t' + \dfrac{\omega d}{v_{ext}}\right)$, if we write $\dfrac{2\pi c}{\lambda}$ for ω and express $\dfrac{1}{v_{ext}}$ as $\dfrac{N_{ext}}{c}$, and if we write $\dfrac{1}{v_{ord}}$ as $\dfrac{N_{ord}}{c}$ (as we may, since the propagation is perpendicular to the optical axis). Then the expressions for emergent fields are

$$E_1 \cos\theta \, e^{j\omega t'} \qquad \text{and} \qquad E \sin\theta \, e^{j\left\{\omega t' + \frac{2\pi d}{\lambda}(N_{ext} - N_{ord})\right\}}$$

Our calcite plate will be called a *quarter wave plate* when the relative phase lag of these orthogonally polarized emergent beams is $\dfrac{\pi}{2}$, as when $-d(N_{ext} - N_{ord}) = \dfrac{\lambda}{4}$. Since $N_{ext} - N_{ord} = -0.172$ for calcite, a plate thickness of $d = 0.86\mu$ will give this $\dfrac{\lambda}{4}$ retardation for $\lambda = 5892$ Å. Such a plate of calcite gives a $\dfrac{\pi}{2}$ retardation of o-rays, relative to e-rays, and as we learned in § 2-4 and § 6-6, this retardation between orthogonal components of equal amplitude can produce emergent circularly polarized light. When the thickness is such that $d(N_{ext} - N_{ord}) = \dfrac{\lambda}{2}$, yielding $\Delta\varphi = \pi$, we then call our plate a *half wave plate*. And a calcite plate of twice this thickness would be a *full wave plate*, also called a *tint plate*. Although it is pedagogically proper for us to make these plates of calcite of 0.86, 1.72, and 3.44 microns thickness respectively, they are, in reality, much too thin to be of interest to the experimental physicist. Quartz, a positive uniaxial crystal (since $N_{ext} > N_{ord}$), requires a much thicker plate ($d = 16\mu$) for $\frac{\pi}{2}$ retardation, since ($N_{ext} - N_{ord}$) $= .0091$. Actually, most quarter-wave plates are made of mica. Cleaved thin sheets of this material have effective indices for orthogonal components of light, incident normally on the sheets, of $(N_2 - N_1) = 1.5997 - 1.5941 = .0056$. Thus $d = 26.3\mu$ yields a quarter-wave plate. This thickness is easily obtained by splitting; in fact, with care one can split out sheets of mica thin enough to give $\frac{\lambda}{16}$ relative phase retardation.

The full-wave plate is called a tint plate because, although it may be a full-wave plate for green light, it will be less than a full-wave plate for red, and more than a full-wave plate for blue. If such a plate is put between crossed Nicols so that white polarized light from the first Nicol strikes it at $\theta = 45°$, then the green component of the white light will be extinguished by the analyzer Nicol; but the red and blue components will penetrate it, giving transmitted light of a magenta tint. The eye is particularly sensitive to variations of this hue, such as are produced when a further positive or negative increment of retardation comes into play. More retardation changes the magenta toward blue; less retardation changes it toward red. One use of a tint plate is shown in Fig. 7-36, where it exaggerates the effects of slight retardations introduced in a beam by stressed glass. Retardations produced by the stressed glass, which alone would not be apparent to the observing eye, are easily measured when added to the tint plate retardations. This use of a tint plate is described in § 7-12.

Fig. 7-23 shows a *Babinet compensator*. A compensator can be made of two calcite prisms, with their optical axes mutually perpendicular, as shown in Fig. 7-23. The composite introduces relative phase retardation because of increasing thickness of one component prism, and decreasing thickness of the other. If the relative phase retardation introduced by the upper prism

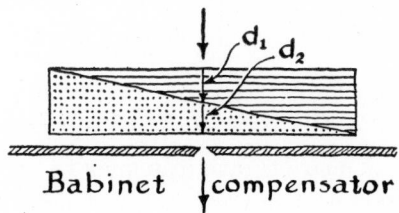

FIG. 7-23   Babinet compensator.

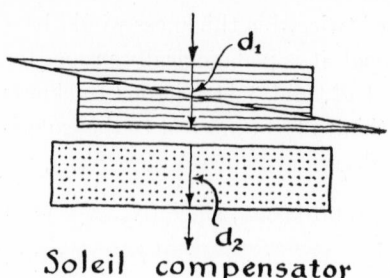

FIG. 7-24 Soleil compensator.

Soleil compensator

in this figure is $\dfrac{2\pi d_1}{\lambda}\,(N_{ext} - N_{ord})$, that produced by the second prism will be

$-\dfrac{2\pi d_2}{\lambda}\,(N_{ext} - N_{ord})$. The negative sign arises because the rays that are e-rays

or o-rays in the first prism become, respectively, o-rays and e-rays in the second
prism. Thus the total phase difference introduced is

$$\Delta\varphi = \frac{2\pi}{\lambda}\,(d_1 - d_2)(N_{ext} - N_{ord})$$

$\Delta\varphi$ is zero at the center of the compensator, where $d_1 = d_2$, but $\Delta\varphi$ increases
to one side of the center and decreases to the other side. Actually, Babinet
compensators are usually made of quartz, but the principle is the same.

Soleil invented a modification of the Babinet compensator that has the advantage of producing the same phase retardation over all the area of component prism overlap, as contrasted to the changing retardation from side to side
of the Babinet. In the Soleil, made of calcite, two prisms are used, as in the
Babinet, except that these component prisms have the same optical axis
orientations as that of the upper component of the Babinet. Soleil added a
third component under these prisms, which is a plane parallel plate of calcite
as Fig. 7-24 shows. The optical axis orientation of the third component is the
same as that of the lower component of the Babinet. To describe the net
retardation of these three Soleil components, we set d_1 as the effective total
thickness of the two Soleil prisms. This thickness, although variable under
control, is constant over the area of overlap. Further, we set d_2 as the thickness of the third plate. Thus the net retardation is given by the equation above
for $\Delta\varphi$; and the d_1 in this equation can be varied by shifting the upper prism
relative to the lower prism. Although the Soleil compensator is here described
as made of calcite, it too is usually made of quartz.

7-8. *Analysis of Polarized Light*

A quarter-wave plate may be used in conjunction with a Nicol or Polaroid
sheet to analyze a beam of light more completely than is possible with a

calcite rhomb, or Nicol alone. In order to understand such analysis it is necessary to keep in mind that a beam of light can be described in terms of the amplitudes and phase relationships of two components having vibrations in mutually orthogonal azimuthal planes. Furthermore, we must keep in mind that the description of a beam can sometimes be expressed in more than one way. For example, the statement that a beam is circularly polarized light is equivalent to saying that its orthogonal plane polarized components are equally strong, and in phase quadrature. And, conversely, the statement that a beam is plane polarized light is equivalent to saying that it is comprised of two equally strong right and left circularly polarized components.

When the phase relationship of orthogonal plane polarized components is unstable, as in the case of *natural light*, so that their relative phase difference, $\Delta\varphi$, wanders over the whole range, $\Delta\varphi = 0$ to $\Delta\varphi = 2\pi$, in a random way, then the instantaneous state of polarization can assume all possibilities—each possible state lasting for an average time that is determined by the degree of monochromaticity of the light (see the beginning of Chapter VI). When, however, relative phase relationships of orthogonal components are stable, then the state of polarization of the light is stable, and may be characterized as *plane, circularly,* or *elliptically polarized*. For example, if two components are not equally strong, or if they are equally strong but their relative phase relationship is not 0, $\pm\frac{\pi}{2}$ or $\pm\pi$, then we shall have elliptically polarized light.

And elliptically polarized light may (equally well) be expressed as a coherent combination of plane polarized light with circular. Finally, any beam of light can be represented as a combination of plane polarized light $\mathscr{P}$, circularly polarized light $\mathscr{C}$, elliptically polarized light $\mathscr{E}$, with more or less unpolarized light $\mathscr{N}$; that is, $(\mathscr{N} + \mathscr{P})$, $(\mathscr{N} + \mathscr{C})$, or $(\mathscr{N} + \mathscr{E})$. Using these symbols $\mathscr{P}$, $\mathscr{C}$, $\mathscr{E}$, and $\mathscr{N}$, as defined above, we shall review some of the tests that can be made on an unknown beam to determine the state of its polarization: first we test the light with a Nicol or Polaroid analyzer alone; then this test is followed by tests of the light with a Nicol and $\frac{\lambda}{4}$-plate. The function of the $\frac{\lambda}{4}$-plate in such tests as we shall describe is not to change plane polarized light to circular, but, conversely, to change circularly or elliptically polarized light to plane. The $\frac{\lambda}{4}$-plate transforms circular to plane at any azimuthal orientation, but it transforms elliptical to plane only at certain azimuthal orientations. With all this in mind, the following tests can be made significant.

Consider first that a tested beam passes no light at one azimuth with the Nicol alone, and all at another, thus yielding $\beta = 1.0$. In such a case we know certainly then that this analyzed beam can be described as pure $\mathscr{P}$.

But, on the other hand, the test with the Nicol alone may give no change of transmission as the Nicol azimuth is varied, thus yielding $\beta = 0$. In such a case we can make no unequivocal inference: the analyzed beam may be pure $\mathfrak{N}$; pure $\mathfrak{C}$; or a mixture, $(\mathfrak{N} + \mathfrak{C})$. To discriminate further, we put a $\frac{\lambda}{4}$-plate ahead of the Nicol. If now, where we formerly got $\beta = 0$, rotation of the Nicol still gives $\beta = 0$ for all orientations of the $\frac{\lambda}{4}$-plate, we know certainly that the analyzed beam is pure $\mathfrak{N}$. And in contrast with this test result, if all orientations of the $\frac{\lambda}{4}$-plate give $\beta = 1.0$, where before we had $\beta = 0$, we know certainly that the analyzed beam is pure $\mathfrak{C}$.

Intermediately, if the test with the Nicol alone gives some variation of transmission, but yields $0 < \beta < 1.0$, then we test with the Nicol and a preceding $\frac{\lambda}{4}$-plate. If certain orientations of this $\frac{\lambda}{4}$ plate yield $\beta = 1.0$ in the transmitted light, as tested with the following Nicol, then we know certainly that the analyzed light is pure $\mathcal{E}$.

All other test results are ambiguous.

7-9. *Uses of the Half-wave Plate*

It is often useful to cross an analyzer with the azimuth of a beam of plane polarized light in order to extinguish it. In this manner we may determine its azimuth of polarization. If we introduce something between polarizer and analyzer which will rotate the plane of polarization, we may determine the angle through which the plane of polarization is rotated by a rotation of the analyzer until we re-establish extinction. For example, an introduced sugar solution will produce such a rotation, and the introduced rotation will depend on the sugar concentration and on the optical path length in it. Such a measurement of rotation by a known solution path length may be employed to determine the concentration of the sugar solution. This application of polarized light is called *saccharimetry*.

The determination of analyzer azimuth angles which give extinctions of plane polarized light is not inherently precise. This lack of precision arises because the amount of light transmitted by the analyzer has a stationary value at extinction, and this makes the transmission insensitive to small increments of analyzer rotation. The precision of azimuth determination may be greatly enhanced, however, if a half-wave plate is properly attached to the analyzer Nicol—that is, preceding the Nicol and covering *only half* of its field. The

FIG. 7-25 Transmissions of analyzer for polarized light—through uncovered part (solid line) and through part covered by $\frac{\lambda}{2}$-plate.

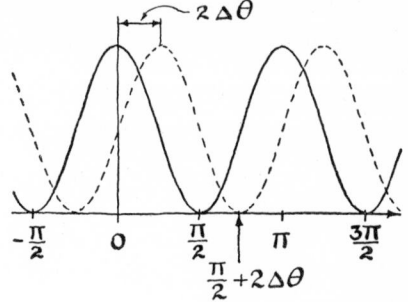

proper attachment requires that the optical axis of the $\frac{\lambda}{2}$-plate be oriented at the small angle, $-\Delta\theta$, relative to the azimuth of extinction of the analyzer Nicol. A $\frac{\lambda}{2}$-plate, thus attached, has the property of transmitting incident plane polarized light as emergent plane polarized light, but with its plane of vibration rotated to the symmetrical azimuthal position on the opposite side of the optical axis of the $\frac{\lambda}{2}$-plate. Figs. 7-25 and 7-26 illustrate this property of the $\frac{\lambda}{2}$-plate. In Fig. 7-25 the solid curve represents the idealized transmission through an uncovered area of the Nicol analyzer as a function of analyzer angle, while the dotted line represents the transmission through the area of the analyzer covered by the attached $\frac{\lambda}{2}$-plate, as a function of the same angle. Both curves are plots of Malus' double-angle law, where θ is the angle between the plane of the polarized light and the azimuth of easy passage of the analyzer Nicol. The first stationary value of zero transmission, referred to above, occurs at $\theta = \frac{\pi}{2}$ for the uncovered analyzer, and at $\frac{\pi}{2} + 2\Delta\theta$ for the covered area of its field.

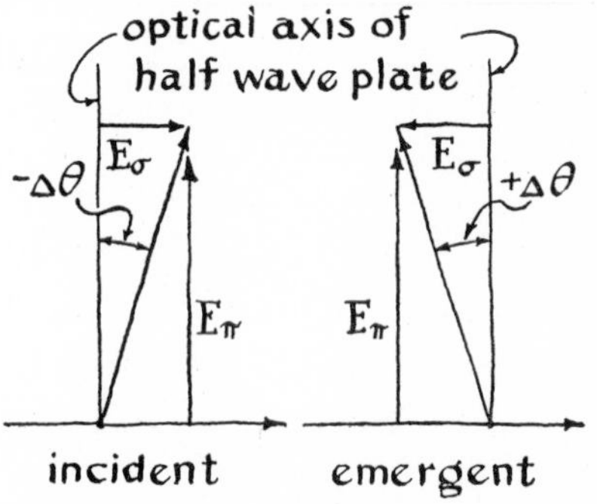

FIG. 7-26   Vector diagram for $\frac{\lambda}{2}$-plate.

The explanation of the above $2\Delta\theta$ rotation is facilitated by referring to Fig. 7-26. When the polarized light lies at an azimuth such that it is extinguished by the uncovered half of the analyzer, and lies at the azimuth angle $-\Delta\theta$ in respect to the optical axis of the $\frac{\lambda}{2}$-plate, the part of that polarized light which falls on the $\frac{\lambda}{2}$-plate of amplitude E will have synchronous components $E_{\parallel}$ and $E_{\perp}$. On emergence from the $\frac{\lambda}{2}$-plate, the component $E_{\perp}$ will be retarded in phase by the angle π, relative to the phase of the component $E_{\parallel}$. Now since $e^{-j\pi} = -1$, this retardation is equivalent to synchronism, but with a reversal of the sign of $E_{\perp}$, and this reversal, as shown in the figure, is equivalent to a rotation of azimuth of E from $-\Delta\theta$ incident, to $+\Delta\theta$ emergent. Thus the curve representing the law of Malus in Fig. 7-25, for the covered half, has its angle of extinction not at $\frac{\pi}{2}$, but at $\left(\frac{\pi}{2} + 2\Delta\theta\right)$.

We take advantage of this $2\Delta\theta$ rotation by setting the $\frac{\lambda}{2}$-plate analyzer combination at the angle $\left(\frac{\pi}{2} + \Delta\theta\right)$. The criterion of setting here is to match the light transmitted by the half covered Nicol with that transmitted through the uncovered half. This matching affords an angular setting that is more precise than an extinction setting on either of the adjacent minima; a slight incremental analyzer rotation, $\delta\theta$, away from $\frac{\pi}{2} + \Delta\theta$, brightens one half of the field and darkens the other.

Fig. 7-27 shows two other means of achieving this same end. By one, at the left, a Nicol is cut longitudinally and a small angle wedge is removed; then parts are reassembled. Fig. 7-27 at the right shows the other, the Lippich combination of a large and small Nicol prism.

FIG. 7-27   Cornu-Jellett and Lippich analyzers.

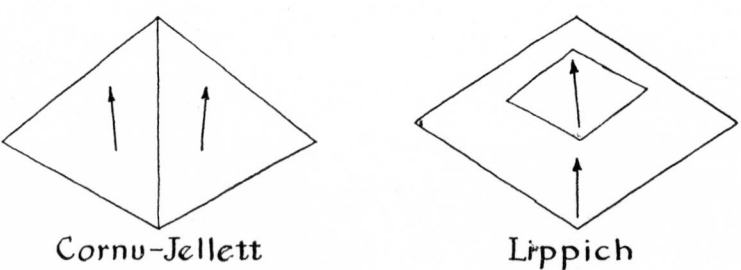

Cornu–Jellett   Lippich

7-10. *Calcite and Quartz Contrasted*

Fig. 7-28 compares the velocity surfaces of both calcite and quartz, both uni-axial crystals. But whereas $(N_{ext} - N_{ord})$ is -0.172 for calcite, $(N_{ext} - N_{ord})$ is $+.0091$ for quartz. Calcite is called a negative uniaxial crystal; quartz is therefore positive. For calcite, the extraordinary velocity surface is an oblate spheroid that circumscribes the spherical, ordinary velocity surface. For quartz, the ordinary velocity surface circumscribes the extraordinary velocity surface, which is a prolate spheroid. A further and more subtle difference is that the two velocity surfaces for quartz do not touch where each intersects the optical axis; while for calcite, the *e*-ray becomes perpendicular to the axis, and takes on the character of the *o*-ray where its velocity surface intersects the optical axis. But for quartz the *e*-ray and *o*-ray remain different in charac-ter and different in velocity. At the equator the *o*-ray for quartz is polarized perpendicular to the optical axis, as with calcite, but as the *o*-ray inclines more and more toward the pole, or optical axis, and is finally parallel to it, this ray's characterizing property is finally that of circular polarized light. On the other hand, at the equator the *e*-ray, polarized parallel to the optical axis, becomes also a circular polarized ray finally at the pole, but circular in an opposite sense. If the *o*-ray, propagating in the poleward direction, becomes a right circularly polarized ray, the *e*-ray will be a left circularly polarized ray. The possibility of right or left for the *o*-ray, with the opposite for the *e*-ray, suggests that there might be, in nature, two kinds of quartz crystals, one with the *o*-ray right circularly polarized, and one with the *o*-ray left circularly polarized; and, indeed, there are two such kinds of quartz. If the *o*-ray is right circularly polarized in a quartz crystal it is called right-handed quartz, or *R*-quartz; while the other kind is left-handed, or *L*-quartz. Fig. 7-29 shows the outward

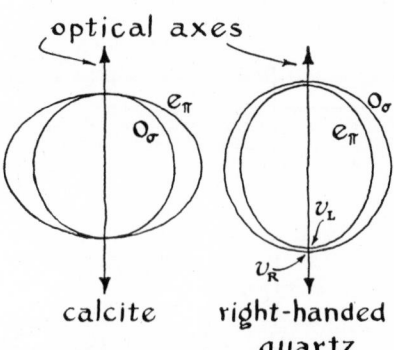

FIG. 7-28 Velocity surface sections of cal-cite and quartz contrasted.

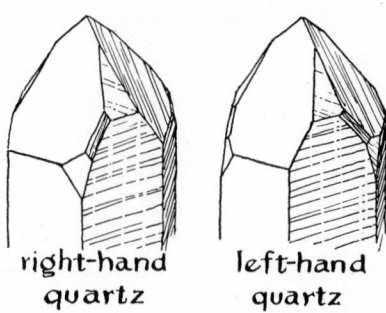

FIG. 7-29 Right- and left-hand quartz crystals.

right-hand quartz　　**left-hand quartz**

forms of these two kinds of naturally occurring quartz crystals; one is the mirror image of the other; and they are called isomers.

Ascribing indices for the propagation direction along the optical axis $N = \dfrac{c}{v}$, the difference of wave velocities for R and L-quartz, at $\lambda = 5892$ Å, where $N_{\text{ord}} = 1.544$, gives

$$\frac{c}{v_L} - \frac{c}{v_R} = (N_L - N_R) = \begin{cases} +71 \times 10^{-6} & \text{for } R\text{-quartz} \\ -71 \times 10^{-6} & \text{for } L\text{-quartz} \end{cases}$$

Fig. 7-30 shows the composite prism that was used by A. J. Fresnel to demonstrate this property of quartz. In this composite the light propagates substantially parallel to the optical axes of the quartz from which the prism components are made. Fresnel's composite consisted of a 152° prism of R-quartz with its optical axis oriented as shown. This symmetrical prism was buried optically between two half prisms of L-quartz, of 76° half-prism angle. Either natural or plane polarized light entering, say, on the left face of the

FIG. 7-30 Fresnel's demonstration that plane polarized light can be resolved into right- and left-circularly polarized components.

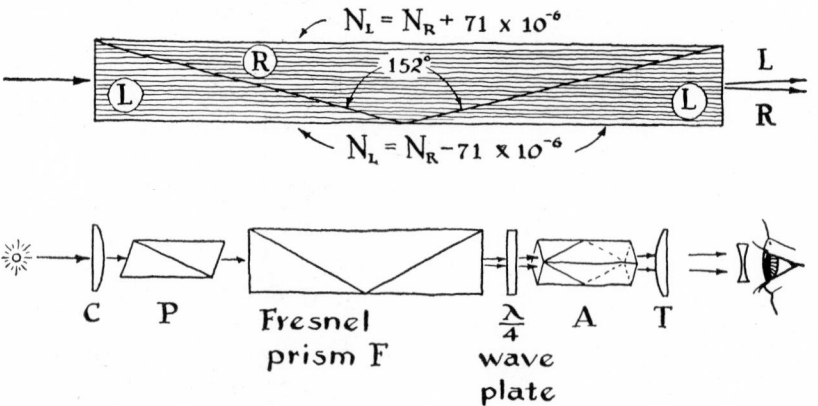

$N_L = N_R + 71 \times 10^{-6}$

$N_L = N_R - 71 \times 10^{-6}$

152°

C　P　Fresnel prism F　$\dfrac{\lambda}{4}$ wave plate　A　T

combination, as shown, finds the velocity of its left circularly polarized component less in the central prism than in the two end half-prisms and, therefore, it is refracted upward. Contrarily, the right circularly polarized component is refracted downward. A student exercise poses the problem of calculating the total expected angular separation of right and left circularly polarized components produced by these refractions. The optical train below in Fig. 7-30 shows how this angular separation is measured. A beam of incident plane polarized light, thus broken up into two beams, is observed with the telescope *T*. The character of the two separated emergent beams can be identified as being right and left circularly polarized by means of the $\frac{\lambda}{4}$-plate and Nicol (at orientations $\pm 45°$). If the entrance beam is only right-polarized, or only left-polarized, there will be only one emergent beam, and this will be deviated by a very small angle. But when both states are present the differential refraction separates them.

Fig. 7-31 shows how two half quartz prisms, one right and one left, were combined by A. Cornu, for use in a crystal quartz prism spectrograph. Cornu's combination avoids double refraction and a consequent doubling of spectral lines. This figure also shows how two half lenses of quartz were combined by Cornu to avoid double imaging.

We have so far considered only the rays propagating along the axis of the velocity surface. As rays in the quartz become more and more inclined to this pole, the *e*-rays and *o*-rays change their character. They change from circular to elliptical polarizations; and this change is followed, not far from the poles, by a further degeneration of the elliptically polarized light to plane polarized light. For most purposes, therefore, using quartz near the equator, we may ignore the elliptical nature of the *e*-ray and *o*-ray and simply consider quartz as if it were a weakly positive calcite crystal. Thus quartz can be used, just like calcite, for making Wollaston and Rochon prisms and for compensators.

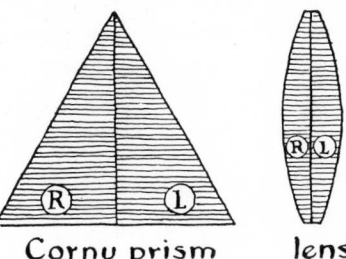

FIG. 7-31   Cornu prism and Cornu lens.   **Cornu prism**   **lens**

7-11. *Optical Activity*

Plane polarized light, after it penetrates a plane parallel quartz plate cut with the optical axis perpendicular to the plate faces, is observed to have its azimuth rotated. The amount of this rotation is 21.7° of angle per millimeter of travel along the optical axis. No matter what the thickness of such a plate may be, the light that penetrates the quartz parallel to its optical axis emerges as plane polarized light. The property of a dielectric medium to thus rotate the plane of polarization is called *optical activity*. Some liquids show this property, like the sugar solution we mentioned in § 7-9. Also, some vapors, like turpentine vapor, show optical activity.

The optical activity of quartz, along its axis, is a consequence of the difference of phase velocities, v_L and v_R. Let us consider such a quartz plate mathematically: For a plate of R-quartz cut perpendicular to the optical axis, as described above, we write expressions for entering light, which is taken to be plane polarized in the xz-azimuth. We write these expressions in terms of a total amplitude E' just within the entering crystal face. We consider the entering face to be located in the $z = 0$ plane of a Cartesian coordinate system, and we further consider the light to be propagating in the $+z$-direction inside the quartz. The plane polarized light vector at $z = 0$ may be written as the sum of two circularly polarized components of equal amplitude. Following § 6-6:

$$E_x' = \tfrac{1}{2}\mathcal{E}' \cos \omega\!\left(t - \frac{z}{v_R} \right) + \tfrac{1}{2}\mathcal{E}' \cos \omega\!\left(t - \frac{z}{v_L} \right)$$

$$E_y' = -\tfrac{1}{2}\mathcal{E}' \sin \omega\!\left(t - \frac{z}{v_R} \right) + \tfrac{1}{2}\mathcal{E}' \sin \omega\!\left(t - \frac{z}{v_L} \right)$$

Thus at $z = 0$, $E_x' = \mathcal{E}' \cos \omega t$ and $E_y' = 0$. At any point beyond this, in the quartz, the instantaneous azimuth of the electric field is $\theta = \tan^{-1} \dfrac{E_y'}{E_x'}$. Now if we reset the clock for any point z_1 beyond, so that $\omega t = \omega\!\left(t' + \dfrac{z}{v_R} \right)$, and if we write $2\vartheta = \omega z\!\left(\dfrac{1}{v_R} - \dfrac{1}{v_L} \right) = -2\pi \dfrac{z}{\lambda} (N_L - N_R)$, then, as in the case of Problem 2-4, the application of trigonometry gives θ at this point z_1, within the quartz, as independent of time—that is, the light is plane polarized there. And further, the azimuth of polarization is $\theta = \vartheta$. For R-quartz $\vartheta = -\dfrac{\pi z}{\lambda} (71 \times 10^{-6})$. Thus θ decreases as z increases so that the rotation produced is clockwise as seen looking toward the light source. A substance that produces a rotation of the plane of polarization in this sense is called a

FIG. 7-32 Rotation of plane of polarization of microwaves by twisted wave guide.

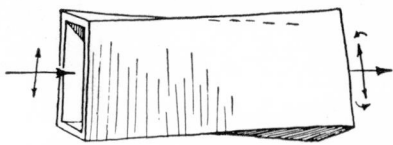

dextrorotatory substance, as contrasted to one that produces a counterclockwise rotation, which is called a levorotatory substance. The ratio $\frac{\theta}{z}$ is called
the *specific rotatory power;* and for quartz it is $\pm217°$ per cm.

This rotation by quartz of the plane of polarization was discovered by
D. F. Arago in 1811. In 1815 Biot discovered the same type of rotation by
both liquid and vaporous turpentine. The specific rotatory power, $\frac{\theta}{z}$, for all
optically active substances shows a dispersion that may be described by a
Cauchy-type equation:

$$\frac{\theta}{z} = A + \frac{B}{\lambda^2}$$

The ratio $\frac{\theta}{z}$ also varies with temperature.

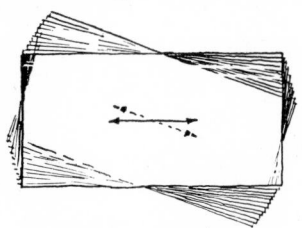

FIG. 7-33 Rotation of light by twisted stack of
mica plates.

Optical activity can be explained as a progressive rotation of the azimuth
of easy constraint of a complex oscillator in the quartz. X-ray studies show
that the atoms in the quartz crystal are arranged in rudimentary right- or
left-handed helices. In analogy, a bag of short right-handed coiled springs,
each wrapped separately in paper, will rotate transmitted polarized microwaves one way while left-handed coils rotate them oppositely, just as in
saccharimetry.

Fig. 7-32 shows how a twisted wave guide rotates the plane of polarization
of the microwaves it transmits.

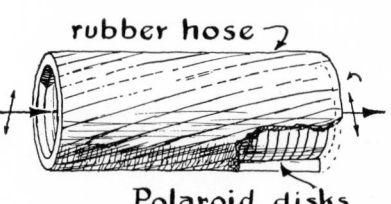

FIG. 7-34 Rotation of plane of polarized light by succession of Polaroid sheets
with 90° between first and last sheets.

A stack of oriented mica sheets and a stack of oriented Polaroid sheets when twisted as shown in Fig. 7-33 or 7-34 will also each rotate the azimuth of incident plane polarized light.

7-12. *Induced Double Refraction*

Our considerations of crystals show how double refraction is manifest. Now we shall mention several methods of inducing double refraction in normally isotropic dielectric media.

In 1814 Sir David Brewster discovered that glass under mechanical compression became double refracting. The explanation of his experience is that compression partially orients otherwise randomly oriented anisotropic oscillators. The birefringence thus induced is that of a negative crystal with the optical axis parallel to the direction of compression, and with $(N_{\text{ext}} - N_{\text{ord}})$ proportional to the stress. This discovery of Brewster currently enjoys wide application in the art of photoelasticity. By its application, stress distributions are predicted for proposed designs of mechanical load bearing structures in which the concentration of stress is to be avoided or minimized. The prediction is based on experiments with a model of the mechanical structure whose design is to be optimized. This model is made out of a sheet of calibrated transparent plastic. It might be, for example, the model of a gear tooth. The model is stressed in the same manner as the structure it represents will be stressed in service, and the birefringence produced by the two dimensional distribution of these stresses is measured. The distribution of measured birefringence then gives the stress distribution in the model. The transparent model material may be calibrated as shown by Fig. 7-35 by a combination of prisms. Even-numbered prisms of the substance in question are made longer than the odd-numbered prisms, and they are arranged so that pressure can be put only on them in a press. The function of the odd, unstressed prisms, of the same material, is to cancel out the average deviation produced by the stressed prisms. The difference in *e*-ray and *o*-ray deviations, because of the pressure, gives the birefringence calibration. Fig. 7-36 shows a simple set-up used to test immersed irregular glass shapes for internal stress, such as the metal-to-glass seal shown. A tint plate is used. With it a relative retardation

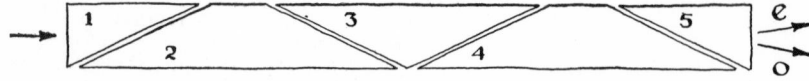

FIG. 7-35   Procedure for determining the coefficient of forced double refraction in an isotropic medium.

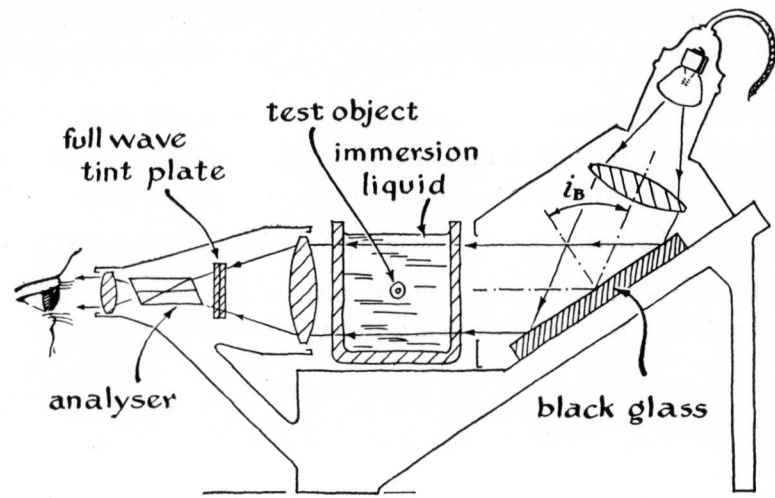

FIG. 7-36 Photoelasticity gadget for studying stress in glass by observation of induced double refraction.

of as little as $\frac{\lambda}{50}$ (10 millimicrons) is readily noticed. The tested glass shapes are immersed in a liquid of the same index to cancel out all but the differential refraction due to birefringence.

There are other ways of inducing birefringence in an otherwise homogeneous medium—for example, by magnetic and electric fields. In 1845 Michael Faraday successfully induced optical activity in glass by a strong magnetic field oriented parallel with the propagation direction of the light. In 1907 Cotton and Mouton induced double refraction in liquid nitrobenzene by using a strong transverse magnetic field.

In 1875 John Kerr discovered that a strong static electric field induced double refraction in glass; and it has been later determined that the induced double refraction is much stronger in liquid nitrobenzene than in glass. Furthermore, with the colloid called bentonite, the effect is exhibited a million times stronger than in nitrobenzene.† The relative Kerr phase retardation in such media as glass, nitrobenzene, and bentonite colloids is proportional to the square of the static electric field that is applied to the medium, ε; and it is also proportional to the length of optical path in that medium, l. Thus the expression for relative phase retardation is

$$\Delta\varphi = 2\pi B l \varepsilon^2$$

† Hans Mueller, J. Opt. Soc. Am., *31*, 286 (1941).

Here B, called the Kerr coefficient, is characteristic of the medium. When a Kerr cell is placed between a polarizer and analyzer, with azimuths of easy passage parallel, and a properly strong static electric field, such as will produce a phase difference of $\Delta\varphi = \pi$, is applied at 45° inclination, then the plane of polarization of the light transmitted by the Kerr cell will be rotated 90°, or perpendicular to the analyzer. Such a rotation gives extinction, just as a $\frac{\lambda}{2}$-plate would do. Such a combination of Kerr cell and two Nicols was used by Rupp to get the very quick chopping that we have described in § 5-7. He applied a high frequency rather than a static field.

Chapter VIII

Interference of Two Sources Laterally Separated

WE HAVE considered several optical topics that prepare us to take up superpositions of light amplitudes and explain the classical *two-beam interference* experiments of Young, Fresnel, and others.

Young's experiments with two-beam interference established the wave nature of light and first gave us the wavelengths of the different colors. Soon after, the experiments of Fresnel and Arago on the interference of polarized light established the transverse character of the waves.

Interference, as we are about to consider it in this chapter, and diffraction, as we shall consider it in the next chapter, lie in the very heartland of optics. First we consider interference patterns as they are produced by superposition of light from two synchronous and persistent sources of waves which are separated laterally. Many applications grow out of the manifest phenomena of this two-beam interference. Later we shall consider two-beam interference obtained when two synchronous and persistent sources are separated in depth.

In Chapter II we treated superposition of wave amplitudes generally, including two waves polarized in mutually perpendicular planes to give circular and elliptical polarization. Our theory there required wave trains of only a few wavelengths' persistence; and that theory was supported by experiments using quarter- and half-wave plates and the Fresnel rhomb.

Here, in contrast, we consider two superimposed beams that must be synchronous and polarized in the same azimuthal plane; *i.e.*, beams derived from the same atomic oscillators. Furthermore, since these superpositions will add amplitudes emitted by atomic sources at one time, with those that were or will be emitted by the same particular atomic sources, but at a previous or future time, the phase of the composite of such atomic emissions must be stable during several tens, hundreds, thousands or even millions of charge oscillations in order to get observable superpositions. Two overlaid beams of light always produce a superposition pattern, but the phases of the overlaid beams must be stable long enough to be sensed by the eye if the characteristic patterns of superposition are to be observable. When two beams of the same frequency have this requisite phase stability, so that superposition patterns are stable long enough to be observable as interference patterns, we say the two beams are *coherent*.

8-1. *Coherence of Light*

Our explanations of the laws of reflection and refraction of natural light of § 1-1, using Huygens' construction, only required synchronism over a very limited area of the wave front. We call such lateral *spatial* synchronism, over the wave front, *lateral coherence*. Lateral coherence, or synchronism, is inherent in the definition of a wave front; so that the merit of formally defining lateral coherence resides mainly in its clarification of the meaning of *longitudinal coherence*, which depends on the *temporal* stability of phase in succeeding wave fronts of a wave train.

Even single pulses of limited lateral coherence suffice for Huygens' explanation of the laws of reflection and refraction because all the daughter waves that are required for those explanations come from a common parent wave front, and the waves that superimpose came from immediately adjacent surface elements of that parent wave front. In contrast with this, the explanations of interference, which we now take up, require the addition of daughter Huygens wavelets which derive from earlier and later wave fronts of a common wave train, thus requiring a *temporal* persistence of phase in the parent wave train. This temporal persistence is required if superposition of amplitudes rather than intensities is to be observed. When such persistence of phase obtains, for two overlaid light amplitudes, we say that they are coherent, and we call this specifically *longitudinal coherence;* and just as lateral coherence is inherent in the definition of a wave front, longitudinal coherence is inherent in the definition of a wave train. We shall sometimes call a succession of Huygens wavelets, with temporal persistence of phase, *Huygens wave trainlets*.

As we learned in Chapter V, the length of wave trains, or degree of longitudinal coherence, is a measure of monochromaticity. And as we shall see later, increased monochromaticity means increased observability of interference bands. The superposition pattern of two beams of white light is less observable by eyes that are color-blind than by normal eyes. This is equivalent to saying that the color sensitivity of the normal eye acts as a virtual monochromator to make superposition patterns apparent where a color insensitive eye would not see them.

We can illustrate the coherence in the light emitted by an ensemble of $\mathfrak{N}$ atomic oscillators by considering an analogous situation. Consider an ensemble of $\mathfrak{N}$ equally powerful radio stations, all emitting the same frequency of radiation. Imagine that each station radiates for an epoch of approximately one-second and then rests an instant before radiating again. And also, consider that each station follows the rule that the phase of its emission during each radiation epoch is to be unrelated with the phase of its emission during the previous epochs; that is, the persistent phase during any radiation epoch is to be determined by pure chance. Finally, if the instants when the different radio stations begin and end their epochs of emission are unrelated, the component radiations will be relatively constant in intensity, but they will be mutually incoherent. Following the teachings of § 2-7, if the superposition sum of such component emitted waves is determined at a distant observing station, say one on the moon, the aggregate amplitude will be only $\sqrt{\mathfrak{N}}$ times greater there than the amplitudes that would be determined for each station separately. The phase observed at our distant station on the moon will be reasonably constant over a time interval that is short compared to the radiation interval, but phases will be unrelated after time lapses which are longer than that interval. Thus temporal stability of phase is described by the time duration of the radiation epochs of the component sources; *i.e.*, by one second. And the degree of longitudinal coherence will be described by the distance light travels in this epoch; *i.e.*, $\dfrac{1}{c} = 3 \times 10^8$ meters.

This analogy supplies the picture for the D-line emissions of a sodium doped Bunsen flame. In such a flame, each thermally excited sodium atom begins its emission at a time that is unrelated to the activities of other atoms, and the duration of its atomic radiation epoch is 10^{-10} seconds. At any instant some sodium atoms will be firing-up while others are burning-out; and the overall phase of the aggregate radiations, if it were observable at a distant observing station, would be always varying as time flows—although that phase would be reasonably constant over time intervals short compared to 10^{-10} seconds. Here the degree of longitudinal coherence is 3 cm.

With these conceptions of independent radiation sources of the same frequency, and the degree of longitudinal coherence among them, we are prepared to proceed with explanations of interference and diffraction, where Huygens wave trainlets of coherent radiation are overlaid.

8-2. *Young's Experiment*

Thomas Young was the first physicist both to arrange apparatus in the manner which yields two laterally separated white light sources of sufficient synchronism and persistence of phase to produce an observable superposition pattern, and, further, to correctly interpret the observed interference pattern and determine from it the wavelengths of the colors of which white light is comprised.

Fig. 8-1 illustrates Young's arrangement of apparatus (above) and the significant geometric parameters (below) that he used for interpreting his observations. As the figure shows, a white light source L illuminates a single slit in an opaque baffle S_1. Huygens wave trainlets, which are considered to originate between the jaws of this slit, spread out by diffraction on both sides of the incident direction of propagation. Wave trainlet wave fronts arriving synchronously at the two parallel slits in a second baffle S_2 excite the generation of two sets of secondary Huygens wave trainlets between the jaws of these slits, one set of trainlets from each slit. The wave fronts of these secondary wave trainlets also spread out by diffraction and fall, overlaid, on the observing screen at the distance $\mathfrak{D}$ beyond S_2. In such a Young experiment the lateral

---

FIG. 8-1 Diagram of Young's double-slit interference experiment (above) and of the geometry it involves (below).

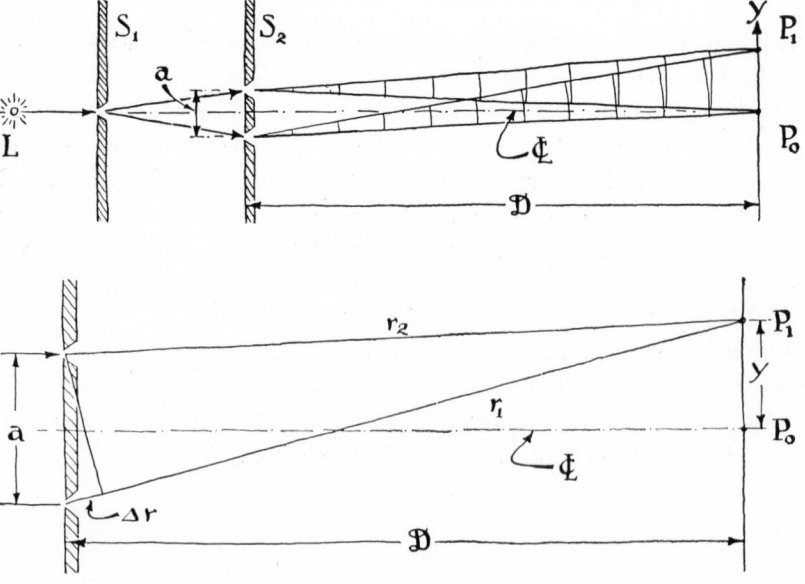

separation of the two slits is of the order of a fraction of a millimeter while $\mathcal{D}$ is of the order of a meter. On the observing screen the superimposed amplitudes of wave trainlets from the two slits S_2 add or subtract according to their relative phases. The dimensions in Fig. 8-1 are exaggerated. Although the experiment works with a narrow source at S_1, say a modern tungsten straight filament lamp, a diffraction source such as the slit that Young used is best. For now it may be assumed that the two slits, S_2, lying perpendicular to the plane of Fig. 8-1, are short. Interference bands appear on the observing screen. The produced bands lie perpendicular to the plane of the figure along P_0P_1.

The wave trainlets emitted from S_1, when they arrive at either one of the slits at S_2, may be represented by the expression

$$\tilde{E} = \frac{\mathcal{E}_1}{r_{12}} e^{j\left\{\omega\left(t - \frac{r_{12}}{c}\right) + \phi(t)\right\}}$$

Wave trainlet lengths depend on the persistence of phase; *i.e.*, on the epoch over which $\phi(t)$ in the expression above is substantially constant. Here r_{12} is the distance from the single slit S_1 to either one of the slits at S_2; $\mathcal{E}_1$ is the amplitude at unit distance from S_1; and the phase at S_1 is $\omega t + \phi(t)$. Young's method of getting synchronism at the slits S_2, so that the same expression $\tilde{E}$ applies to either slit, is called *wave front division*. The resultant amplitude beyond S_2 is the sum for two segments of successive wave trainlet fronts from S_1; and before that, the wave trainlets derive from the same atomic oscillators at L.

In order to appreciate Young's experiment, which gave the first wavelengths of light, we must understand the qualitative nature of the observed pattern, together with its quantitative interpretation. The quantitative analysis below applies to Young's experiment and, to some extent, to two-beam interference generally.

8-3. *Analysis of Two-beam Interference*

The amplitudes of the wave trainlets diffracted at S_2, at unit distance beyond S_2, are approximately equal. We write $\mathcal{E}_0$ for these magnitudes. The resultant of superposition on the observing screen at the point P at distances r_1 and r_2, respectively, from the double slits, as shown in Fig. 8-1b, is the sum of the expressions

$$\tilde{E}_1 = \frac{\mathcal{E}_0}{r_1} e^{j\omega\left(t - \frac{r_1}{c}\right)} e^{j\phi(t')} \qquad \tilde{E}_2 = \frac{\mathcal{E}_0}{r_2} e^{j\omega\left(t - \frac{r_2}{c}\right)} e^{j\phi(t'')}$$

Amplitudes at P are equal to the degree of approximation that $\dfrac{\mathcal{E}_0}{r_1} = \dfrac{\mathcal{E}_0}{r_2} = \mathcal{E}_2$.

They are equal to a high degree of approximation because the distance difference $(r_1 - r_2) = \Delta r$ is small compared to the values of r_1 or r_2.

The terms $\phi(t')$ and $\phi(t'')$ must vary together if the interference pattern around P is to be stable, and thus observable. Here t' refers to the time when the parent waves yielding $\tilde{E}_1$ originated from the atomic source, at L; while t'' refers to the time when the parent waves yielding $\tilde{E}_2$ originated from the same sources. In order for the sum, $\tilde{E}_1 + \tilde{E}_2$, to have a stable value the terms $\phi(t')$ and $\phi(t'')$ must be substantially equal, although they may vary together slowly. Thus the wavelets added must originate from the same atoms during a same epoch of their atomic emission times. For a sodium emission, $(t' - t'')$ must therefore be small compared to 10^{-10} seconds; or, $(r_1 - r_2) = \Delta r$ must be small compared to 3 cm. In so far as $\phi(t') = \phi(t'')$, and we continually keep setting our clock as phases change so that $e^{j\phi(t)}$ is continually unity, then our two amplitudes may be more simply represented by

$$\mathcal{E}_2 e^{j\omega\left(t - \frac{r_1}{c}\right)} \qquad \mathcal{E}_2 e^{j\omega\left(t - \frac{r_2}{c}\right)}$$

Writing $\Delta\varphi = \dfrac{\omega}{c}(r_1 - r_2)$, our formula of § 2-3 gives the superposition sum of these two wave motions, $\tilde{E}_P$, as follows:

$$\tilde{E}_P = \left[2\mathcal{E}_2 \cos \frac{\Delta\varphi}{2}\right] e^{j\left\{\omega t - \left(\frac{\varphi_1 + \varphi_2}{2}\right)\right\}}$$

$$= \left[2\mathcal{E}_2 \cos \frac{\omega}{2c}(r_1 - r_2)\right] e^{j\omega\left(t - \frac{r_1 + r_2}{2c}\right)}$$

The geometry of Fig. 8-1b gives $\Delta r = \dfrac{ay}{\mathfrak{D}}$. On writing $\dfrac{\omega}{c} = \dfrac{2\pi}{\lambda}$, we get the amplitude at P as

$$\mathcal{E}_P = 2\mathcal{E}_2 \cos \pi \left(\frac{ay}{\lambda \mathfrak{D}}\right) \tag{a}$$

This superposition sum gives the observed illumination,

$$\mathfrak{E}_P = \tfrac{1}{2}c\kappa_0 \mathcal{E}_P^2 = 2c\kappa_0 \mathcal{E}_2^2 \cos^2 \pi \left(\frac{ay}{\lambda\mathfrak{D}}\right) = c\kappa_0\mathcal{E}_2^2 \left[1 + \cos 2\pi \left(\frac{ay}{\lambda\mathfrak{D}}\right)\right] \tag{b}$$

These equations for $\mathcal{E}_P$ and $\mathfrak{E}_P$ are plotted as functions of $\Delta\varphi = 2\pi\left(\dfrac{ay}{\lambda\mathfrak{D}}\right)$ in Fig. 8-2.

When one of the added amplitudes has an additional phase advance or retardation of $\pm\pi$, making the total phase difference $\Delta\varphi = 2\pi\left(\dfrac{ay}{\lambda\mathfrak{D}}\right) \pm \pi$, the electric field and illumination expressions become

$$\mathcal{E}_P = 2\mathcal{E}_2 \cos \left[\pi\left(\frac{ay}{\lambda\mathfrak{D}}\right) \pm \frac{\pi}{2}\right] = \pm 2\mathcal{E}_2 \sin \pi \left(\frac{ay}{\lambda\mathfrak{D}}\right) \tag{c}$$

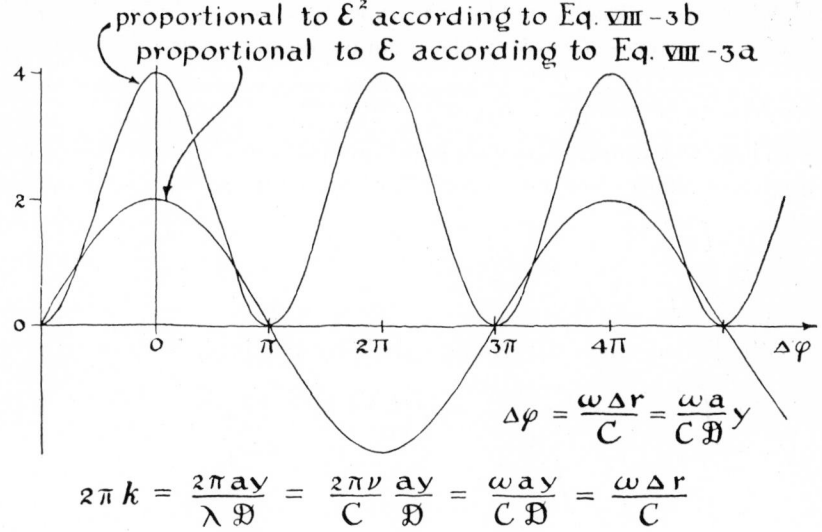

FIG. 8-2   Amplitude and intensity in Young's experiment.

$$\mathfrak{E}_P = 2c\kappa_0\mathcal{E}_2^2 \sin^2 \pi \left(\frac{ay}{\lambda\mathfrak{D}}\right) = c\kappa_0\mathcal{E}_2^2 \left[1 - \cos 2\pi \left(\frac{ay}{\lambda\mathfrak{D}}\right)\right] \tag{d}$$

Eqs. 8-3a and b apply to Young's interference pattern; Eqs. 8-3c and d apply to Lloyd's interference pattern, to be described further in § 8-4, where we have this phase advance or retardation.

To understand Young's and Lloyd's interference patterns we need to define a quantity that we call the order of interference, k. The *order of interference* here is $\frac{\Delta r}{\lambda} = \frac{ay}{\lambda\mathfrak{D}} = k$. The extent of the interference pattern that is embraced by Δy_c, such that $\left(\frac{a}{\lambda\mathfrak{D}}\right)\Delta y_c = 1.0$, is called a cycle of interference. The phase difference, $\Delta\varphi$, corresponding to a cycle of interference is 2π, and the corresponding path difference is $\Delta r = \lambda$.

In constructive two-beam interference $\mathcal{E}_P = 2\mathcal{E}_2$ and $\mathfrak{E}_P = 2c\kappa_0\mathcal{E}_2^2$. This constructive interference occurs for Eqs. 8-3a and b when k assumes an even integer value; $0, 1, 2 \cdot \cdot \cdot$ etc. In destructive two-beam interference $\mathcal{E}_P = \mathfrak{E}_P = 0$. This destructive interference occurs for these equations when k assumes a half integer value; $\frac{1}{2}, 1\frac{1}{2}, 2\frac{1}{2} \cdot \cdot \cdot$ etc. For Eqs. 8-3c and d we have destructive interference when k assumes even integer values, and constructive interference when k assumes half integer values.

In either case, when $\mathcal{E}_P = \mathfrak{E}_P = 0$ and we have no illumination we have, as Grimaldi expressed it, light added to light to produce darkness. On the other hand, where $\mathfrak{E}_P = 2c\kappa_0\mathcal{E}_2{}^2$, we have the appearance of a violation of the principle of conservation of energy: The illumination for one slit alone is $\frac{1}{2}c\kappa_0\mathcal{E}_2{}^2$ and we might expect twice this when the second slit is opened, rather than the fourfold increase that our equations predict. These paradoxical results disappear, however, if we compare averaged illuminations, averaged over a complete cycle of interference. For one slit alone, we get the averaged illumination,

$$\mathfrak{E}_{av} = \frac{\int_y^{y+\Delta y_e} \frac{1}{2}c\kappa_0\mathcal{E}_2{}^2 \, dy}{\int_y^{y+\Delta y_e} dy} = \frac{1}{2}c\kappa_0\mathcal{E}_2{}^2$$

For both slits together the average, similarly obtained, is twice as great.

$$\mathfrak{E}_{av} = \frac{\int_y^{y+\Delta y_e} c\kappa_0\mathcal{E}_2{}^2 \left[1 + \cos 2\pi\left(\frac{ay}{\lambda\mathfrak{D}}\right)\right] dy}{\int_y^{y+\Delta y_e} dy} = c\kappa_0\mathcal{E}_2{}^2$$

Thus, when we compare averaged illuminations, the principle of conservation of energy does not appear to be violated. The deficiencies of illumination in interference minima, where $\mathfrak{E}_P = 0$, are compensated by excesses in the interference maxima, where $\mathfrak{E}_P = 4\mathfrak{E}_{av}$.

This result is a strange one, because the flow of energy appears to be determined here by the sum of $\mathcal{E}$'s at P; but causality at P seems too late. It would seem, in a proper intuitive theory, that $\mathfrak{E}_P$ should be determined from calculations applied nearer the doublt slit, rather than at P.

There were objections to Young's experiment when his results were publicly announced in 1802; and it remained for other two-beam experiments to overcome those objections and clear up interpretations. Some objections were valid, but others were simply polemical. For example, one politically prominent contemporary, reviewing Young's work, wrote: "We wish to raise our feeble voice against innovations that can have no other effect than to check the progress of science, and renew all those wild phantoms of the imagination which Bacon and Newton put to flight from her temple. This paper contains nothing which deserves the name of either experiment or discovery."

Young took such criticism seriously, and it was primarily because of professional antagonism that he withdrew from optics. Afterward, he applied himself to deciphering the Rosetta stone, and thus made important contributions to Egyptology. All together, Young established himself as one of the most versatile and effective creative intellects of all time. William C. Gibson's excellent thumbnail biography in the *Scientific Monthly* of July 1955† recapitulates Young's career. He wrote, in part, as follows:

---

† "Significant Scientific Discoveries by Medical Students," p. 22.

"Thomas Young is remembered chiefly for his theory of light and of color vision and for Young's modulus of elasticity, which is used daily in engineering. Yet, the investigation that he made on the lens of the eye remains a monument to a medical student, aged 20. For this work he was elected a fellow of the Royal Society of London at age 21.

"Young had come up to St. Bartholomew's Hospital Medical School from his Quaker home in Somerset, full of ancient languages, the calculus, botany, meteorology, and lens grinding. He had scarcely begun his anatomy course when he was invited to dissect a very recently removed ox eye. The ciliary muscles responded to stimulation, thereby compressing the lens and changing its shape. At once Young saw that the alleged lengthening and shortening of the eyeball to accommodate for distance was a myth. Before long Young set out for Edinburgh to join Joseph Black. Within a year he went to Göttingen, where he graduated as doctor of medicine in 1795. His thesis, on the human voice, was remarkable for a 47-character alphabet that he had designed. Young's 'universal alphabet' was to permit, by its combinations, the rendering of every sound expressible by the human voice.

"From this graduation thesis of a 22-year-old medical student came a development, truly Newtonian in its import. Young's work on the combinations of sounds, and especially on the propagation of sound, led him into experiments in the analogous field of light. His wave theory of light and his theory of accommodation in vision were only two of his contributions made as a result of his investigations during his medical-student days. In addition, he became a pioneer in the study of the tides, in the construction of life-expectancy tables, in the measurement of 'ultimate particles,' or atoms, and finally in the deciphering of 'the Rosetta Stone.' "

8-4. *Other Two-beam Interference Experiments*

Young's rational critics argued properly that his experiment depended on a poorly understood phenomenon—diffraction. And since diffraction was poorly understood, they argued that an interference experiment where diffraction was involved could not be fully understood. These valid objections to Young's experiment were first overcome by experiments of Fresnel some ten years later. Fresnel, like Young, superimposed beams from two coherent sources that were laterally separated. But in his experiments Fresnel achieved lateral doubling of the source by reflection and by refraction, rather than by diffraction. Fresnel not only devised these unobjectionable experiments, but he also introduced more precise methods of measuring the interference pattern; namely, by the use of a micrometer eyepiece, observing the interference patterns directly in space, rather than indirectly, by means of light scattered from an observing screen, as had been done previously. Fresnel's experimental arrangements using reflections by a double mirror or refractions by a double prism achieved wave front division and lateral source doubling as shown in Fig. 8-3. Felix Billet later achieved lateral source doubling by means of the split lens shown (below) in the same figure, and thus produced Young-type interference bands. In all the experiments of Fig. 8-3, if we interpret a and $\mathfrak{D}$ as they are indicated in the figure, Eqs. 8-3a and b apply.

Fig. 8-4 shows the experimental arrangement of Lloyd in which a source is laterally doubled by use of only one mirror. By Lloyd's arrangement the source S is doubled at S' because of light reflected at grazing incidence from the surface of black glass. A distinguishing characteristic of Lloyd's experiment is that a phase change π is introduced into this reflected beam, from S', and not in the direct beam (because the grazing incidence reflection coefficients are $\mathfrak{r}_{\pi} = \mathfrak{r}_{\sigma} = -1.0$). It is this characteristic of Lloyd's experiment that requires Eqs. 8-3c and d, rather than a and b, to describe the superposition pattern of amplitudes, and the illumination in the interference bands. In applying Eqs. 8-3c and d, we use $\mathfrak{D}$ as the distance from the source S to the observing screen, and for a we use twice the normal distance of S from the plane reflector surface.

FIG. 8-3 Double-slit experiments by means of Fresnel's double mirror (above), Fresnel's double prism (middle), and Billet's split lens (below).

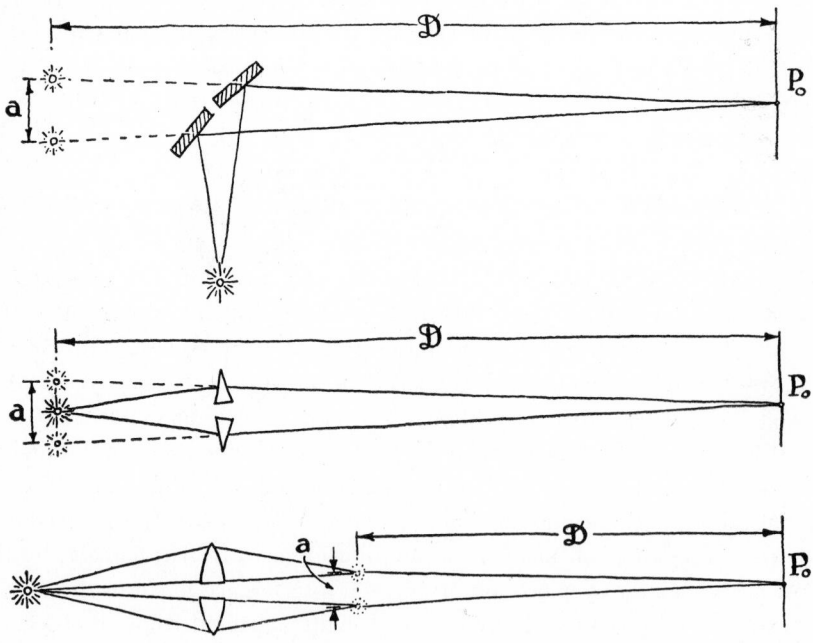

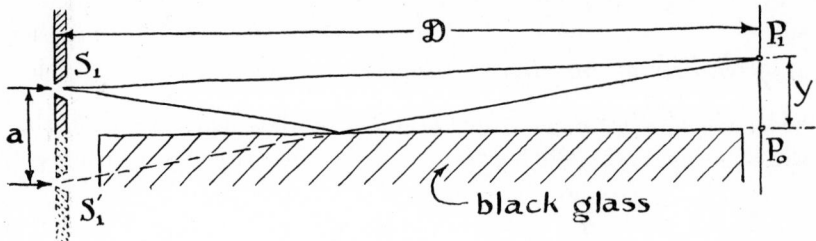

FIG. 8-4   Lloyd's experiment: source doubling by reflection from a black glass surface.

The Young and Lloyd interference bands are complementary, as is easily shown by adding Eq. 8-3b for the illumination of one set of bands to Eq. 8-3d for the other, giving an aggregate illumination that is constant, or independent of y. Thus, if we added the light for a Young experiment on a common observing screen to that for an equivalent Lloyd experiment, with equivalent a's and $\mathfrak{D}$'s, we would get constant illumination, or white light at all y's. The interference colors for a white light source in one experiment, given in Table 8-1 below, are the complementary colors of those for the other experiment.

TABLE 8-1   *Colors for Superposition of Two Equal Amplitudes of White Light (from Bruhat)*

Path Difference Δr	Young's Interference Colors	Lloyd's Interference Colors
$.000\mu$	white	black
.158	brownish white	blue grey
.259	bright red	greenish white
.332	blue	vivid yellow
.565	clear green	purple
.664	orange	sky blue
.747	red	green
.866	violet	yellow green
1.101	green	magenta
1.376	violet	green

8-5. *Visibility of Interference Bands*

We have mentioned the impaired visibility of white light interference bands for the color-blind eye—and how the color-sensitive eye acts as a virtual monochromator to increase the observability of interference. Now we shall consider, in more detail, how the visibility of the interference bands depends on monochromaticity. In particular, we shall intercompare interference bands of Young's and Lloyd's arrangements when the slit S_1 is illuminated with both red and blue light; when S_1 is illuminated by the D-lines of a sodium-doped Bunsen burner; when S_1 is illuminated by one isolated D-line; and finally when S_1 is illuminated by the highly monochromatic mercury light at $\lambda = 5461$ Å, using monoisotopic mercury, Hg^{198}. This mercury is obtained by transmutation of gold in a cyclotron, and its emission may be facetiously dubbed the green mercury radiation from gold.

Fig. 8-5 shows the superposition $\mathfrak{E}_P$'s for the simultaneous red and blue light illuminations, calculated by the appropriate equations: the plot of $\mathfrak{E}_P$ for $\lambda = 6400$ Å is labeled R, with B for the blue light, $\lambda = 4800$ Å. These illuminations are plotted as ordinates with y as abscissa. The subscripts on the labeled maxima and minima of $\mathfrak{E}$ give the corresponding values of order of interference, k. Since here the ratio of the λ's is 4:3, $k = \left(\dfrac{ay}{\lambda \mathfrak{D}}\right)$

FIG. 8-5 Red and blue interference bands in Young's experiment (above) and in Lloyd's experiment (below).

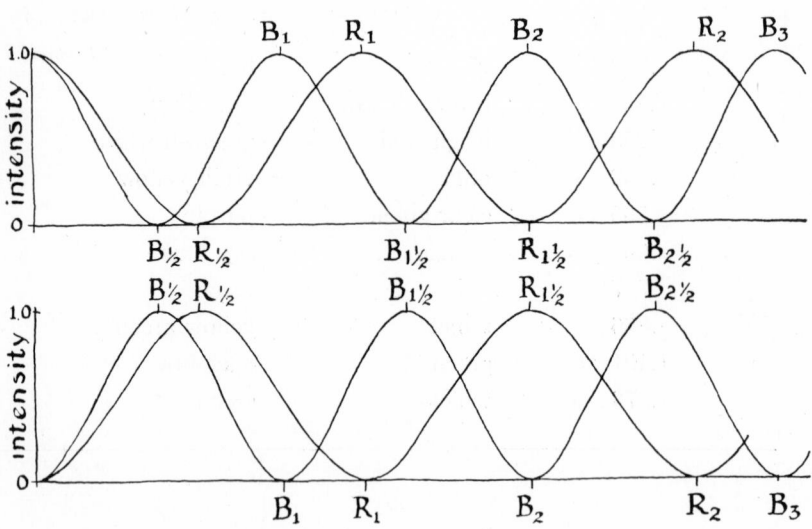

will be 2 for B at the same point P, on the screen, where k is $1\frac{1}{2}$ for R. Comparing Young's and Lloyd's experiments, with equivalent a's and $\mathfrak{D}$'s: in Young's experiment we only see blue light at the point P where y_P gives $k_B = 2$; whereas, in Lloyd's experiment, at the equivalent point, we see red light only. In either experiment, if the two colors separately appeared equally strong to a color-blind eye, then, together, on the observing screen at P, that eye would notice little or no variation of illumination; for the color-blind eye the blue maximum would fill in for the red minimum, and *vice versa*. In contrast, the normal color-sensitive eye not only sees the colored interference bands, but sees them beautifully.

Consider now our next category of monochromaticity for the illumination of S_1—the situation where the two wavelengths are $\lambda_B = 5890$ Å and $\lambda_R = 5896$ Å, representing illumination by means of light from a sodium-doped Bunsen burner. These two sodium wavelengths provide a situation similar to our previous one, but on a different scale. Just as the maximum of illumination for $\lambda_B = 4800$ at y_P giving $k = 2$ filled in the minimum for $\lambda_R = 6400$ at $k = 1\frac{1}{2}$, we now expect the maximum for $\lambda_B = 5890$ at y_P giving $k = 491$ to fill in the minimum for $\lambda_R = 5896$ at $k = 490\frac{1}{2}$. But here the maximum for λ_B does not fill in the minimum for λ_R completely since the $\lambda_R = 5896$ Å spectrum line is only half as strong as the one at 5890 Å. Actually the color sensitivity of the normal eye is not acute enough to discriminate the two sodium wavelengths as it could the red and blue (above). Thus the interference bands at and around $k = 491$ lead to the type of impairment of the visibility of the sodium interference bands that the color-blind eye experiences with R and B at $k = 2$.

When y_P is such that $k = 982$ for the bluer sodium line at $\lambda_B = 5890$ Å, and $k = 981$ for the redder one at $\lambda_R = 5896$ Å, then both lines have coincident maxima and minima, and the visibility of the interference bands is unimpaired again, as it was around $k = 0$. And so it goes, with another visibility minimum at $k = 1473$, and with good visibility again around $k = 1964$, and so forth. There is, nevertheless, a slight impairment at each of these recurrent maxima of visibility, until the visibility becomes zero and remains so, at and beyond $k = 50,000$.

Since the two sodium lines are incoherent, intensities rather than amplitudes add to describe the superposition pattern on the screen. At $k = 0$, 982, 1964, etc. the maxima and minima of $\mathfrak{E}$ for the 5890 Å and 5896 Å lines are, respectively, proportional to 2 and 1. Thus for these k's the intensities of maxima in concert are proportional to 3, while the adjacent minima are zero. The visibility of interference bands is defined quantitatively by a certain combination of the illuminations observed at a maximum and at an adjacent minimum, $\mathfrak{E}_{max}$ and $\mathfrak{E}_{min}$, of the interference pattern. The visibility $\mathfrak{v}$,

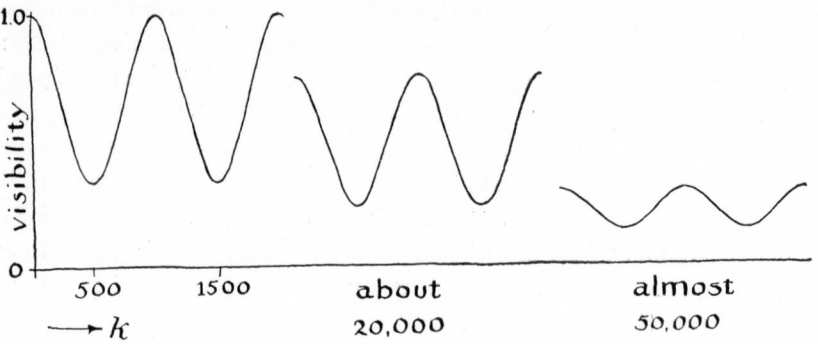

FIG. 8-6   Visibility of interference bands for the sodium D-lines.

expressing this combination, is

$$\mathcal{U} = \frac{\mathfrak{C}_{\max} - \mathfrak{C}_{\min}}{\mathfrak{C}_{\max} + \mathfrak{C}_{\min}}$$

This expression gives the visibilities of the interference bands, for the sodium lines together, at $k = 0, 982, 9164$, etc., as

$$\mathcal{U} = \frac{(2+1) - 0}{(2+1) + 0} = 1$$

At $k = 491, 1473$, etc. the aggregate values of $\mathfrak{C}_{\max}$ are $(2+0)$, and those of $\mathfrak{C}_{\min}$ are $(1+0)$. Thus the calculated visibilities are

$$\mathcal{U} = \frac{2-1}{2+1} = \frac{1}{3}$$

Fig. 8-6 illustrates Michelson's measures of $\mathcal{U}$ for the sodium light. Around $k = 50,000$ and beyond $\mathcal{U}$ decays to zero. This decay is a manifestation of the decay that interference bands by either one of the D-lines alone would manifest, due to lack of monochromaticity. This final disappearance of interference bands may be explained in two ways, as follows:

First, the path difference $\Delta r \cong 3$ cm, corresponding to $k = 50,000$, is so great that no sodium atom that contributes to the wave trainlet from the one of the double slits that is the closer to P is still "burning" when the wave trainlet from the other slit arrives simultaneously at P; *i.e.*, Δr for the superimposed two trainlets is greater than the longitudinal coherence length, 3 cm.

Second, the monochromaticity of either one of the D-lines under high spectral resolving power is comprised of a distribution of frequencies such as B represents in Fig. 5-7. Imagine the halves of such a broadened line as represented by two monochromatic component lines lying at the wavelengths, λ_R and λ_B, each at the centroid of one of the halves, as shown. A separation

$(\lambda_R - \lambda_B) = \frac{1}{17}$ Å would cause the first visibility minimum to occur at the k value of 50,000:

$$(5896 - \tfrac{1}{34})k = (5896 + \tfrac{1}{34})(k - \tfrac{1}{2})$$

And $\frac{1}{17}$ Å is not an unreasonable value for the half width of the sodium D-lines. (*Cf.* § 5-5.) The difference between this first loss of the interference visibility for the representative monochromatic components, λ_R and λ_B, and our previously considered minimum of $\mho$ at $k = 491$, for both lines, is this: with the continuous distributions of wavelengths that our two centroids represent here, once the interference bands are lost, they remain lost; while beyond $k = 491$ the visibility recovers again. It is the same, after 50,000 cycles of interference, as it is for Young's white light bands after 5 cycles, as explained below.

The fringes of the mercury radiation from gold are visible until $\Delta r = 30$ cm, which indicates that line 5461 Å is about ten times narrower than either one of the sodium lines.

8-6. *White Light Interference Bands*

Consider the use of a white light source to illuminate S_1 in a Young or Lloyd experiment, with equivalent a's and $\mathfrak{D}$'s. At a point on the observing screen where the red was formerly missing and the blue was in full strength, or *vice versa*, intermediate colors will now appear in the interference pattern at intermediate strengths. Table 8-1 gives the resulting colors for white light interference in both the Young and Lloyd patterns, and for various values of the geometrical path difference between the interfering beams, Δr, expressed in microns. The colors of one arrangement are complementary to those of the other arrangement for the reasons already set forth in § 8-4.

When the path difference Δr grows greater than 1.4μ, the colors become diluted and disappear and the illumination on the observing screen goes white. The reason for this disappearance of color is apparent from Fig. 8-5. Already at $k = 1\frac{1}{2}$ and 2, for red and blue, the maximum of one color tends to fill up the minimum of the other for the color-blind eye. But at higher k's the maxima for one wavelength compensate the minima for nearer adjacent wavelengths, so that, finally, even the color discrimination of the normal eye is inadequate to discern the interference. After Δr is so large that the sensitivity of the eye cannot discern any pattern of color and the interference is lost, it can be recaptured if the superimposed beams are observed through a spectroscope.

To summarize, we can see about five orders of interference on either side of $k = 0$ in two-beam interference, when the illumination is with white light; we can see 500 orders before the minimum of $\mho$ when both D-lines comprise the illumination; we can see 50,000 orders with one or the other of the D-lines, and finally we can see 500,000 orders of two-beam interference with the monochromatic green mercury radiations from gold. In terms of the span of longitudinal coherence this is 3μ for white light, 3 cm for one D-line, and 30 cm for the 5461 Å line.

If, in Lloyd's experiment, we use a short grating spectrum of a white light slit at S_1 with the blue end of the spectrum near the glass, and the red end away; and if the distances of the colors are such that $\dfrac{\lambda}{a}$ is the same for red and blue, and for all colors between; then the y_P positions giving a certain order of interference are the same for all colors, and maxima or minima of all colors of the same order coincide; thus, many orders of white interference bands become visible, even with white light illumination. Such colorless fringes are called achromatized fringes.

8-7. *Spectral Analysis with Two-beam Interference*

In the early days of spectroscopy H_α, the spectral line of hydrogen, at $\lambda = 6563$ Å, was regarded as a single line until A. A. Michelson studied the visibility of the two-beam interference bands using this red line from a hydrogen discharge for illumination. He measured the decay curve of visibility of the bands, and his result is reproduced in Fig. 8-7. This curve was obtained with

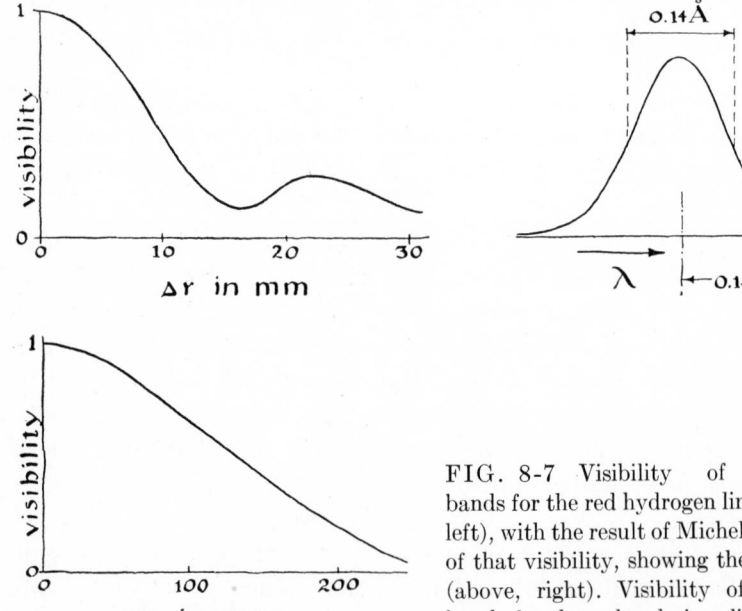

FIG. 8-7 Visibility of interference bands for the red hydrogen line, H_α (above, left), with the result of Michelson's analysis of that visibility, showing the satellite line (above, right). Visibility of interference bands for the red cadmium line (below).

the two-beam interference of Michelson's interferometer, but the principle is the same as if it had been obtained with Young's apparatus. Just as the character of the curve of $\mho$ in Fig. 8-6 reveals the nature of the D-lines, their mutual separation, relative strengths, and individual widths; so also, the peculiar decay of $\mho$ in Fig. 8-7 indicated that H_α was not a simple line. Lord Rayleigh was able to supply the quantitative theory by which the $\mho$ curve in this figure was analyzed. The analysis showed the H_α line to have a satellite. Separation, relative strength, and individual widths of the main line and its satellite are also shown in Fig. 8-7. Later on, when more powerful instruments became available to spectroscopists, the presence of this satellite was seen directly and confirmed; and also, when more sophisticated theory superseded Niels Bohr's first explanation of H_α, the satellite was theoretically predicted.

As another example of the visibility of fringes of a single line, Fig. 8-7 shows the result of Michelson for the cadmium red line, which he used to measure the meter—but more about that measurement in Chapter XI.

8-8. *Stellar Interferometers*

Two-beam interference was first used by Michelson to determine stellar diameters (in 1920) much as it had been used previously to determine the separation of components of binary stars. The determinations in both cases use a modification of Young's experimental arrangement. In order to understand these two applications of Young's double-slit interferometer we first consider a doubling of the light source L of a Young arrangement as shown in Fig. 8-8a. Let two point sources of illumination, S and S', be mutually incoherent and of equal strength, so that they subtend an angle α at the double slit. These two point sources may be thought of as representing components of a double star. Their angular separation α will correspond to a relative phase difference of $\dfrac{2\pi a\alpha}{\lambda}$ for the particular arrangement involved; if the phase difference in the two-beam interference for one source at some point P on the observing plane is $\Delta\varphi = 2\pi\left(\dfrac{ay}{\lambda\mathfrak{D}}\right)$, that for the other source, at the same point, is $\Delta\varphi \pm \dfrac{2\pi a\alpha}{\lambda}$. Thus, if the interference pattern for S is described by $\mathfrak{E} = 2c\kappa_0\mathcal{E}_2{}^2 \cos^2\left(\dfrac{\Delta\varphi}{2}\right)$, the illuminations for S' will be given by

$$\mathfrak{E}' = 2c\kappa_0\mathcal{E}_2{}^2 \cos^2\left(\frac{\Delta\varphi}{2} + \frac{\pi a\alpha}{\lambda}\right)$$

And if $\dfrac{\pi a\alpha}{\lambda} = \dfrac{\pi}{2}$, as when $\alpha = \pm\dfrac{\lambda}{2a}$, then $\mathfrak{E}' = 2c\kappa_0\mathcal{E}_2{}^2 \sin^2\left(\dfrac{\Delta\varphi}{2}\right)$. And since

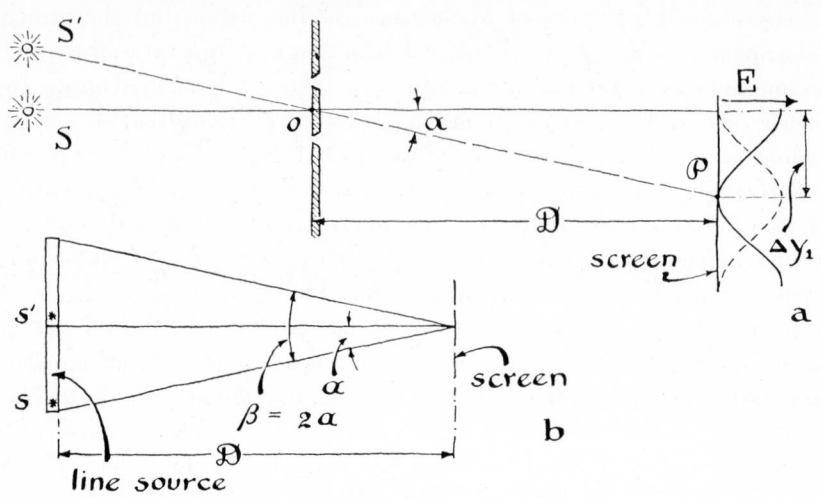

FIG. 8-8   Illustration of $\upsilon = 0$ for two stars (a) and for a line source (b).

our two sources are incoherent, the aggregate illumination of both patterns on an observing screen or in an eyepiece is the sum of separate pattern illuminations rather than the sum of separate amplitudes. And this sum, for $\alpha = \pm\dfrac{\lambda}{2a}$, is a constant: $\mathfrak{E}_2 = \mathfrak{E} + \mathfrak{E}' = 2c\kappa_0\mathcal{E}_2{}^2$. Thus when the separation α is increased until the Young-type interference bands are first lost and $\upsilon = 0$, the angular separation of two equal point sources is $\alpha = \pm\dfrac{\lambda}{2a}$.

Next let us consider these two sources replaced by a line source of length 2α, as shown in Fig. 8-8b. This line may be imagined as equivalent to a string of many equal stars, all of equal strength, and all arranged in linear juxtaposition. Now consider that this line subtends an angle at the double slits of $\beta = 2\alpha = \dfrac{\lambda}{a}$. If we pair off component stars in one half of the line with corresponding or mating stars in the other half of the linear array at the angular distance $\alpha = \dfrac{\lambda}{2a}$ away, then the interference pattern for each pair of stars is a constant illumination; thus the interference patterns for the whole line of stars combine to give constant illumination, and no interference bands are seen.

Finally, consider the line source of Fig. 8-8b to be replaced by a uniformly bright disk of light. The interference bands produced by a short strip element of this disk, just as for a point, may be described by the circular function of Eq. 8-3b. And if we represent one half of the bright disk by a series of short

line strips of increasing length, the interference bands that each element produces may be thus described. And the totality of this series of trigonometric
functions, of various strengths and phases, but of the same period, is itself a
trigonometric function, or a trigonometric function plus a constant, like that
representing a single star. Of course the same may be said for the representation of the illumination produced by the other half of the disk. And when
the two representative trigonometric functions have the relative phase difference of π, due to the angular separation of the disk halves they represent, then,
again, the illumination in the interference pattern will be constant, and no
interference bands for the disk as a whole will be manifest. A calculation we
can understand intuitively, but which we do not make here, shows that the
visibility of interference bands in the Young pattern for such a disk-shaped

source is zero when the diameter of the disk subtends the angle $\gamma = 1.22\dfrac{\lambda}{a}$

at the double slit.

The first and last of these formulas, for α or γ, are the ones used in astrophysics for determining the separation of double stars α, or for determining
the diameter of a large single star γ. Astrophysical measurements by interferometry of stellar separations and star diameters are made as follows:

A telescope aperture is covered by a baffle with two cat's-eye apertures cut
through it as shown in Fig. 8-9. These apertures are separated by a distance,
a, and they become the Young-type double slits of the interferometer. The
interference bands produced by the two beams these cat's-eye apertures transmit are observed at the focus of the telescope by means of an eyepiece. If

---

FIG. 8-9   Anderson's stellar interferometer.

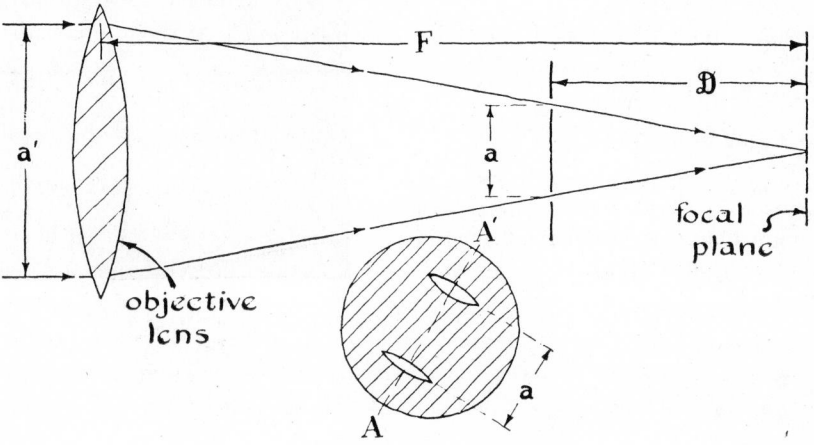

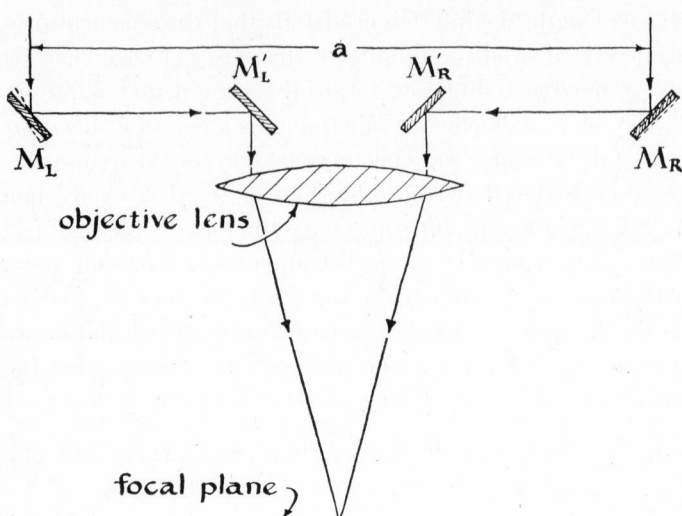

objective lens

focal plane

diffraction pattern
due to M_L and M_R

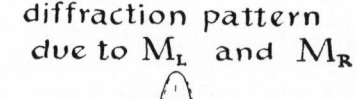

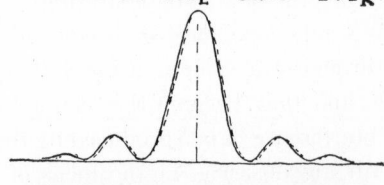

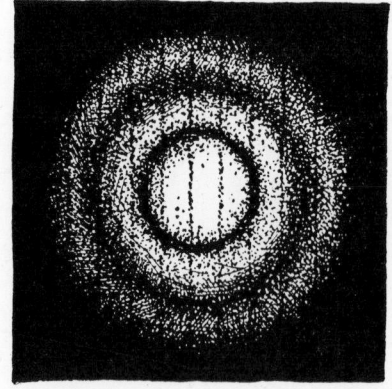

FIG. 8-10 Michelson's stellar interferometer (above), with the overlapping stellar diffraction images, and the Young interference bands (below).

these apertures are over the telescope objective, its focal length corresponds to our Young's factor $\mathfrak{D}$.

The arrangement of Fig. 8-10 illustrates Michelson's stellar interferometer attachment for the 100″ telescope at Mt. Wilson Observatory.

For doub e-star measurements the Fig. 8-9 arrangement uses a less than the objective diameter. In contrast, to have a larger than 100″, and large enough for stellar diameter measurements, the Fig. 8-10 arrangement attaches an outrigger to the telescope.

The Young apertures of Fig. 8-9 are located in the diaphragm at the distance $\mathfrak{D}$ ahead of the telescope focus, and they are separated a distance a, so that their virtual separation at the objective is $a' = a\dfrac{F}{\mathfrak{D}}$. In practice this diaphragm is mounted so that it may be rotated in azimuth, and so that its position may be varied. Adjustments of azimuth and $\mathfrak{D}$ are made until the visibility of the interference bands is a minimum, or zero if the stellar components have equal magnitude and color. The orientation gives the azimuth AA' along which the stellar components are separated in the heavens, and $\mathfrak{D}$ gives their angular separation, $\alpha = \dfrac{\lambda}{2a'} = \dfrac{\lambda\mathfrak{D}}{2aF}$. Some such value as $\lambda = 5500$ Å is taken to represent visible light. The visibility seldom goes to zero because the stellar components seldom have equal magnitudes and colors.

Fig. 8-10 represents the 100″-telescope reflector diagrammatically as a lens. The outrigger beam with four auxiliary mirrors extends beyond the lens (or 100″-mirror). On a night in December, 1920, Michelson and Pease successfully reduced the visibility of the Young-type bands for the single star Betelgeuse to zero by setting $a = 120$ inches. This separation of the outrigged mirrors, $a = 120''$, yields an angular stellar diameter of .047 second of arc. One of the more recently measured stars, Antares, whose parallax and hence distance are well known, has an actual diameter larger than that of the orbit of our planet Mars; namely, 141 million miles.

Fortunately for the operation of the stellar interferometer, the interference bands are less disturbed by the optical inhomogeneities of our atmosphere than astronomers had first anticipated. Fortunately, also, the interference bands are immune to small tippings of the mirrors M_L and M_R (as shown for M_L), or to small displacements (such as is shown for M_R). It is only required that the two diffraction patterns of the stars shall overlap, as shown in the insert.

8-9. *Interference of Polarized Light*

A long and complicated set of experiments was conducted in the early years of the last century by Arago and Fresnel on the interference of polarized light. It was only after these experiments that plane, circular, and elliptical polarized light was fully understood. These experiments by Arago and Fresnel were done with a Young-type interference apparatus. For polarizers they used stacks of parallel mica plates tipped at Brewster's angle. They covered the two slits of their experimental arrangement, similar to that of Fig. 8-1, by separate and independent polarizers, and were thus able to examine the Young interference bands under different conditions. Their results are usually summarized by three statements:

1. When the azimuths of the two coherent polarized light beams emerging from the two slits were arranged so that these azimuths were parallel, then the interference bands produced were observed to be similar in all respects to those produced by unpolarized light.

2. When the azimuths were arranged perpendicular, no interference bands were observed.

3. They were never able to produce observable interference bands with two beams, derived from perpendicular components of natural light, even when these beams were subsequently rotated to the same azimuthal plane.

These results are all consistent with the hypothesis that light is a transverse wave motion. The following quotation from Ernst Mach's *The Principles of Physical Optics* is interesting in this connection:†

"As we have seen, great difficulties had to be overcome before the nature of polarization could be thoroughly explained. This was, no doubt, the reason why workers of such renown worked on this problem, and the cause of the tardy nature of its solution. It would thus be worth our while to go into the nature of these difficulties. Here it is best to adopt the naive attitude of the beginner, to which even the great investigator must succumb with respect to an entirely new problem. It was while attending the grammar school (gymnasium) that I first heard of the transverse vibrations of which light must consist. This view made a very strange, phantastic, and unsympathetic impression on me, without my knowing the actual cause. When I tried, with the aid of my memory, to obtain a clearer view of the matter, I had to confess that I felt instinctively the impossibility of transverse vibrations in so readily movable (displaceable) a medium as air, and thus more so for the ether, which

---

† From Ernst Mach, *The Principles of Physical Optics*. Reprinted with the permission of Dover Publications, Inc., New York.

I considered must be still more rarefied and easily displaceable. Malus' mirror, which when turned about the polarized ray as axis at one time reflects and at another not, soon brought me a steadfast conviction of the existence of transverse waves capable of propagating light. The question of mechanical possibility was quite forgotten, being altogether insignificant compared with the resulting explanation of actual facts. The sequel to, and usage of, this conception alone finally seemed to justify it. Others, as students, may probably have had similar experiences. This, it seems to me, has also occurred in the history of optics. It is one question, what actual property of light it is that makes itself evident in polarization, and quite a different one, whether this property may be (mechanically) explained or further reduced. The fact that these two questions have not always been clearly distinguished has often had a retarding action on the progress of optics. This is obvious even with Huygens. Newton clearly set forth the facts of polarization, as far as they were known in his time, and distinctly separated from them the explanation. Malus did the same, but in Fresnel's time the necessity for such a separation was no longer so clearly recognized by investigators. The Fresnel-Arago interference experiments indicate, as an actual fact, *that the periodic properties of light behave like geometrically additive elements of a two-dimensional space* (a plane perpendicular to the ray), and this is true whether it can be explained or not. There was no need for Fresnel to keep back this important discovery since he thought that he had found a mechanical explanation, this being of a very hypothetical nature, whereas the matter to be explained is in no wise hypothetical. Newton behaved in a similar manner when he did not announce the law of inverse squares in the case of gravity because he was unable to explain it."

Chapter IX

Fresnel Diffraction

FROM Huygens' construction we can predict a pattern of light and shadow such as is cast on an observing screen by an obstruction placed in the light of a distant point source, as described in § 1-1. We get the prediction that the shadow will be a true geometrical projection of obstruction contours: that is, darkness within the silhouette border, and an abrupt change to uniform illumination outside the border. But whenever an experimental test of this prediction is made, and we observe the border closely, we find some illumination on the side predicted to be the dark, and we find the border fringed with bands of greater and less illumination rather than an abrupt transition to uniform illumination. Huygens' construction fails to predict these observed details because it considers that the wavelet trains can be ignored everywhere that they do not touch a common wave envelope—Huygens' construction is tantamount to ignoring all of each Huygens wavelet except its exactly forward segment.

Fresnel elaborated Huygens' procedure by taking account of all of each wavelet surface. He considered the aggregate amplitude at any point on the observing screen as the superposition sum of all the wave trainlets, one from each surface element of the unobstructed wave front, taking due account of relative phases and relative amplitudes. This procedure, a "corrected Huygens construction," gave predictions in much more satisfying conformity with observations. But Fresnel's theory predicted, as did Huygens', an unrealistic back wave. And furthermore, his theory gave resultant phases at all points on the observation screen for the superposition sums, which were in error by

the phase angle $\frac{\pi}{2}$.

Kirchhoff's theory provided refinements over Huygens' construction, and over Fresnel's theory as well. Gustav Kirchhoff developed a theory which

180

introduced two new factors. One of these, j, removed the $\frac{\pi}{2}$ error of Fresnel; while the other $\left\{\frac{1+\cos\theta}{2}\right\}$, called the obliquity factor, freed the superposition sum from the prediction of a false back wave. But even this Kirchhoff refined theory is only a scalar theory—that is to say, it is only valid for representing scalar waves, such as sound, where a scalar quantity describes the amplitude; it fails to represent light completely, because the electric field of light is a transverse vector quantity. As an example, Kirchhoff's theory when applied to diffraction by a narrow slit in a metallic sheet does not discriminate between the electric light fields polarized parallel to the slit and those polarized perpendicular, and certainly these different orientations of the electric field, with respect to conducting slit jaws, in diffraction, may be expected to interact differently—the jaws will tend to be short circuits for fields parallel to their metallic edges, and condensers for perpendicular fields. Nevertheless Kirchhoff's theory is very useful except when the slit widths are very narrow (of the same magnitude in width as λ). And predictions by Kirchhoff's theory are useful for many other applications. In this chapter we make several such applications of the theory of Kirchhoff, although it falls beyond our present presumptions to derive the obliquity factor and the factor j that distinguish it.

In contrast with Young's interference, where our theory summed only two Huygens wave trainlets, in diffraction, which we now take up, we represent the wave trainlet from each unobstructed element of a wave surface as an infinitesimal spherical wavelet source; and we integrate all these many infinitesimals as they successively arrive at any predetermined point at which we wish to find a resultant field. These integrations are in general difficult to carry out. However, for two classes of diffraction the appropriate infinitesimals are easily integrable. These classes are the so-called *Fresnel diffraction* and *Fraunhofer diffraction*. The first is mainly concerned with fringes of the illumination around the borders of silhouettes; the second is concerned with fringes of illumination around images at the focus of a lens or mirror.

9-1. *Kirchhoff's Differential and the Case of No Diffraction*

Referring to Fig. 9-1, Fresnel's expression for the infinitesimal field at a point P, due to the Huygens wavelet from an unobstructed surface element at S, is

$$d\tilde{E}_P = \frac{\mathcal{E}_0}{r}\, e^{j\omega\left(t-\frac{r}{c}\right)}\, dS$$

Fresnel integrated all the myriad of such differentials, each representing a

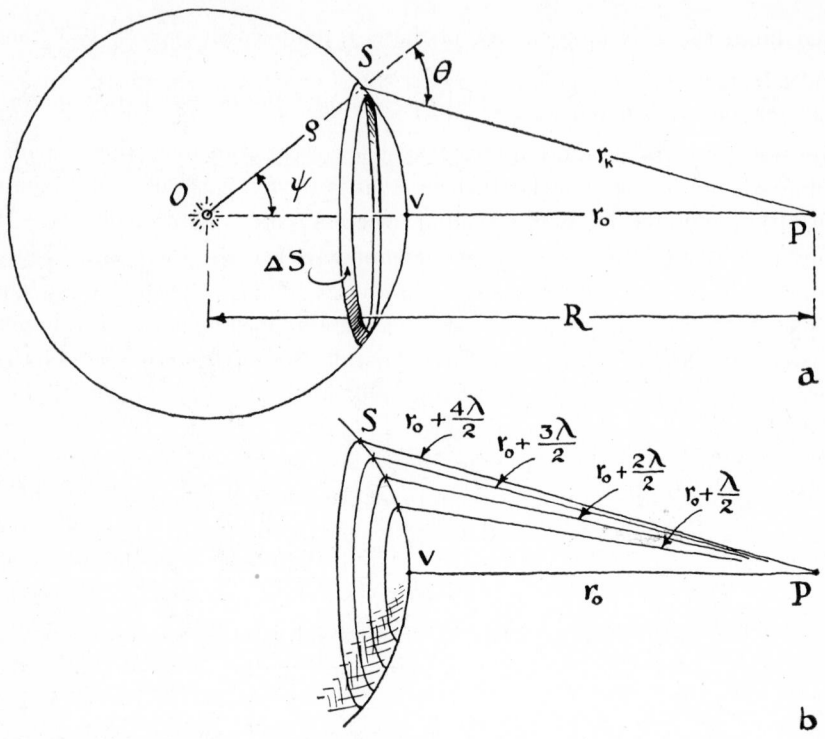

FIG. 9-1 Geometry involved in calculation of the case of no diffraction.

wave trainlet generated by a parent wave train; the factor $\mathcal{E}_0 \dfrac{dS}{r}$ giving relative amplitudes, and the factor $e^{j\omega\left(t-\frac{r}{c}\right)}$ giving relative phases. Fresnel's differential, as modified by Kirchhoff by the factors mentioned above, becomes

$$d\tilde{E}_P = \frac{j\mathcal{E}_0}{\lambda r} e^{j\omega\left(t-\frac{r}{c}\right)} \left\{\frac{1 + \cos\theta}{2}\right\} dS \tag{a}$$

Here a factor λ, not already mentioned, comes from the derivation. It is needed to give both correct dimensions and correct physical meaning.

As a first example, we shall apply this differential in a simple case where its integral should predict no diffraction. In Fig. 9-1a, O is a point source of light which at some earlier time has emitted the parent wave train of which one wave is presently at S. To get the future field at P we integrate $d\tilde{E}_P$ over this parent spherical wave surface S. It lies at the distance ρ from the source. The integral, made with respect to P at a distance R from the source, determines the field at P due to the wave train. From prior considerations, and particularly from the equations of § 4-9, we can anticipate the value of the integral: We should expect the inverse square law and the usual phase retardations with distance to apply; so that, if $\mathcal{E}_1$ is the amplitude at unit distance from O, and ωt is the phase of the field there, the fields at V on the surface S, and at P beyond, should be, respectively,

$$\tilde{E}_V = \frac{\mathcal{E}_1}{\rho} e^{j\omega\left(t - \frac{\rho}{c}\right)}$$

$$\tilde{E}_{P'} = \frac{\mathcal{E}_1}{R} e^{j\omega\left(t - \frac{R}{c}\right)}$$

We write this field at P with a prime to indicate that it is the anticipated field; and our task will be to integrate $d\tilde{E}_P$ over all the surface S to determine if the Kirchhoff differential does indeed give this anticipated resultant field.

At first we ignore the obliquity factor, and the fact that the Huygens wavelet propagation directions are not parallel as they pass through P. We integrate $d\tilde{E}_P$ over the surface S one zone at a time as Fig. 9-1b suggests. These zones at b are chosen so that θ, the diffraction angle at zone boundaries between the parent wave normal and the direction of propagation of the wave trainlet toward P, is constant. Furthermore, the zones at b are chosen so that the distance from dS to P varies over any one zone from $\left[r_0 + (k - 1)\frac{\lambda}{2}\right]$ to $\left[r_0 + k\frac{\lambda}{2}\right]$, where k is an integer. The zones on the surface S are thus all symmetrical about V, as shown. The first zone has an inside radius of zero— it is a disk. Its center at V lies at the distance r_0 from P and its outer border lies at the distance $\left[r_0 + \frac{\lambda}{2}\right]$. This outer border is the inside border of the second zone, and so forth. Thus the inner and outer borders of any zone are loci, respectively, of constant distances from P, $\left[r_0 + (k - 1)\frac{\lambda}{2}\right]$ and $\left[r_0 + k\frac{\lambda}{2}\right]$. Such zones are called Fresnel's half-period zones because it takes light the time of a half period of oscillation to travel the distance $\frac{\lambda}{2}$.

We calculate the area of the kth zone, referring to Fig. 9-1a. This area, ΔS_k, is given by the following mathematics:

$$\Delta S_k = (2\pi\rho \sin \psi)(\rho\Delta\psi) = -2\pi\rho^2\Delta(\cos \psi)$$

Applying the cosine law to $\triangle$ SOP, we get

$$r_k^2 = \rho^2 + R^2 - 2\rho R \cos \psi$$

Since ρ and R are each constant, differentiation gives

$$\Delta(\cos \psi) = -\frac{r_k}{\rho R}\, \Delta r_k$$

Substituting this differential in our expression for ΔS_k, and setting $\Delta r_k = \dfrac{\lambda}{2}$, we find that the areas of the zones are constant, in so far as r_k is constant:

$$\Delta S_k = \frac{\pi \rho r_k \lambda}{R}$$

Next, we divide up each of our k zones into m subzones of a similar character, with successive subzone path differences of only $\dfrac{\lambda}{2m}$. The successive subzone areas $\delta S = \dfrac{\Delta S}{m}$ are all equal for the reasons set forth above relative to ΔS_k. In so far as r_k is constant over all the surface of one of our subzones, contributions over the facets of a subzone need no integration. Substituting in Kirchhoff's Eq. 9-1a, for $\delta \tilde{E}_P$, and ignoring the obliquity factor, we get

$$\delta \tilde{E}_P = \frac{j \mathcal{E}_S}{\lambda r_{km}} e^{j\omega\left(t-\frac{\rho}{c}-\frac{r_{km}}{c}\right)}\, dS$$

On writing $\mathcal{E}_S = \dfrac{\mathcal{E}_0}{\rho}$ and $\delta S = \dfrac{\Delta S}{m}$, the contribution of the last subzone becomes

$$\delta \tilde{E}_P = j\,\frac{\pi}{m}\frac{\mathcal{E}_0}{R} e^{j\omega\left(t-\frac{\rho}{c}-\frac{r_{km}}{c}\right)}$$

$$= -j\,\frac{\pi}{m}\frac{\mathcal{E}_0}{R} e^{j\omega\left(t-\frac{R}{c}\right)} e^{-jk\pi} = -\left[\frac{\mathcal{E}_0}{R} e^{-j\omega\left(t-\frac{R}{c}\right)}\right] j\,\frac{\pi}{m}\, e^{-jk\pi}$$

Although contributions of the other subzones, adjacent to the one expressed above, will all have the same amplitude, they will have phases differing successively by $\dfrac{\pi}{m}$. We may neglect the expression in brackets above, for the moment, and add the last vector $j\dfrac{\pi}{m}\, e^{-jk\pi}$ to those of the preceding inner subzones. We do this addition for the m subzones vectorially. Although all the m subzone vectors in the complex plane will have the same length, successive vectors will have their inclination increasing by the angle increments $\dfrac{\pi}{m}$. Therefore, when they are placed successively end-to-end in the complex plane, as taught in Fig. 2-4, they will approximate the semicircle of Fig. 9-2a. The vector sum of these subzone components for a half-period zone is a vector parallel to the j-axis of length equal to the diameter of this semicircle, or equal to the fraction $\dfrac{2}{\pi}$ times its arc length. Multiplying our neglected expression in

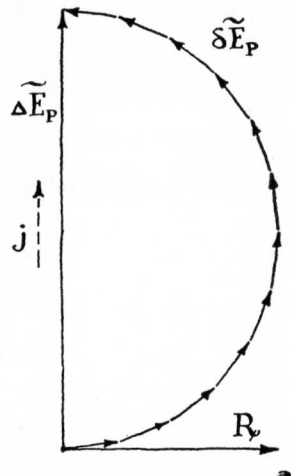
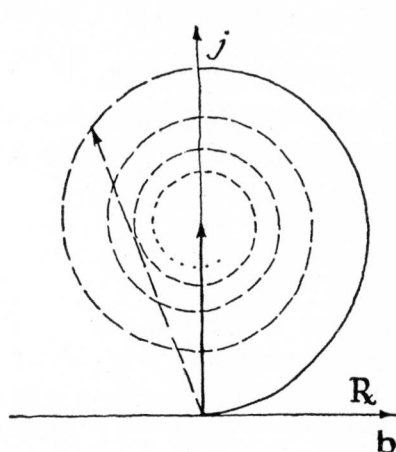

FIG. 9-2   Vector diagram for the first Fresnel half-period zone comprised of $m = 12$ subzones (a) and for subsequent outer zones (b).

brackets by the fraction $\dfrac{2m}{\pi}$ makes the superposition sum for all the subzones of a half-period zone equal to

$$\Delta\tilde{E}_k = \Sigma\,\delta\tilde{E}_P = -2\frac{\mathcal{E}_0}{R}e^{j\omega\left(t-\frac{R}{c}\right)}e^{jk\pi}$$

$$\Delta\tilde{E}_k = (-e^{jk\pi})2\tilde{E}_P'$$

This sum is $+2\tilde{E}_P'$ or $-2\tilde{E}_P'$ accordingly as k for the half-period zone is odd or even. The sum of an odd number of half-period zones, $\tilde{E}_0 = \Sigma\Delta\,\tilde{E}_k$, is larger than the expected by the factor 2, and zero if we take an even number of terms. However, we must remember that we ignored the obliquity factor, also that the electric fields due to zone contributions, as k increases, will not lie perpendicular to the $\mathcal{C}$ at P, but progressively inclined. These factors, together, make the diameter of successive vector semicircles less and less, as indicated in Fig. 9-2b. Finally, as this exaggerated spiral of the subzone vectors shows, the sum for all the $\Delta\tilde{E}_k$'s is just half of $\Delta\tilde{E}_1$. And this is just our expected value $\tilde{E}_P'$.

The sum of all Fresnel's half-period zones can be written as an alternating series, since the ΔE_k's alternate in sign.

$$\tilde{E}_P = \Delta\tilde{E}_1 + \Delta\tilde{E}_2 + \Delta\tilde{E}_3 \cdots$$

$$|\tilde{E}_P| = \tfrac{1}{2}|\Delta\tilde{E}_1| + \tfrac{1}{2}(|\Delta\tilde{E}_1| - |\Delta\tilde{E}_2|) - \tfrac{1}{2}(|\Delta\tilde{E}_2| - |\Delta\tilde{E}_3|)$$

$$+ \tfrac{1}{2}(|\Delta\tilde{E}_3| - |\Delta\tilde{E}_4|) - \cdots \quad \text{(b)}$$

If all the half-period zones contributed equally, the dominating terms of this sum would be the first and the last. But, as Lord Rayleigh put it, "We may usually suppose that a large number of the outer rings are incomplete, so the last term at the upper limit [of the above series] may be taken to vanish." Also, a lack of longitudinal coherence would bring about this incompleteness. And such a lack of longitudinal coherence is expected when $(r - r_0)$ gets large. In either case, we are left with half the amplitude of the first half-period zone, as expected.

Our treatment here, of this case of no diffraction, has demonstrated the significance of Kirchhoff's j without which we would have had Fresnel's $\frac{\pi}{2}$ phase error.

We can modify this treatment of no diffraction and get explanations for several interesting cases of diffraction, such as arise with circular obstructions, or apertures. We shall treat such cases with P lying on the symmetry axis before we take up diffraction by a square or rectangular aperture at points off the axis of symmetry.

9-2. *Circular Obstructions and Apertures*

Fresnel's essay on his diffraction theory contained essentially the above analysis, but without the factor j. It was submitted to a Prize Essay Committee of the French Academy in 1818. S. D. Poisson, a member of the Committee, was an adherent of the corpuscular theory of light, and thus, like Young's critic, he was antagonistic toward a wave theory of light. From Fresnel's theory Poisson deduced that a spot of light should appear at the center of the shadow of a circular obstruction, and that its illumination should be as great as if there were no obstruction. Poisson offered this deduction as evidence that Fresnel's theory was absurd and unrealistic, not knowing that the predicted spot had been discovered and announced by Maraldi over a half century earlier. When Poisson communicated his deduction to Arago, Arago promptly tried the implied experiment with a 2 mm circular obstruction, and immediately rediscovered Maraldi's spot, which had become lost to science. Thus Poisson's hostility to Fresnel's wave theory only served to establish it. Maraldi's spot, rediscovered by Arago, is known today as Poisson's bright spot. When one sees it, it appears as if a mischievous research assistant had bored a small hole through the center of the obstructing ball. Fig. 9-3a illustrates how Poisson's spot may be observed; and Fig. 9-3b shows its imaging properties.

The quantitative prediction of the bright spot comes at once from the considerations that led to the alternating series of Eq. 9-1b; a circular obstruction eliminates a segment of a spherical wave front out to a ring on the sphere S,

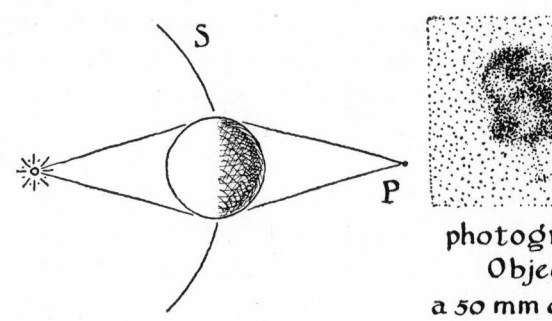
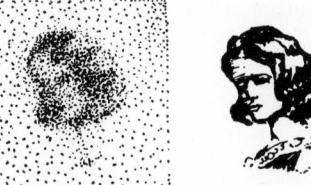

photograph   object
Object photographed with
a 50 mm disc. object distance =
a   image distance = 35 m   **b**

FIG. 9-3 Diagrammatic arrangement for a circular obstacle, giving Poisson's bright spot (a), and a photograph taken with this arrangement (b). (From photographs by E. von Angerer.) Here the ball diameter is greatly exaggerated.

which lies at a distance r_0' from P. It is this distance now, rather than the distance r_0 from V to P, that we define as the inner border of the first Fresnel half-period zone. The outer border of the first zone is at $r_0' + \dfrac{\lambda}{2}$, and so forth.

Thus we get the same alternating series as before. Only now the phase of the disturbance at P is further retarded by the amount

$$\tilde{E}_P = \tilde{E}_{P}'e^{j\omega\left(\frac{r_0 - r_0'}{c}\right)}$$

The illumination in the Poisson spot is thus due to half the amplitude of the first unobstructed Fresnel half-period zone, and this illumination is the same as we would get without any obstruction.

When a circular aperture, rather than obstruction, is put at S, then the amplitude at P will be determined by its radius, or where the aperture rim, symmetrical about V, cuts off an integral of the type of § 9-1b. The field will vary from zero to $2\tilde{E}_{P}'$—being zero if an even number of Fresnel half-period zones are exactly unobstructed, and $2\tilde{E}_{P}'$ if an odd number are exactly unobstructed. If the aperture, in addition to passing an integral number of zones, also passes a fractional zone in addition, the vector representing the final term of the series, to be added for the fractional zone, will be the one represented by the vector chord of the fractional part of the appropriate representative semicircle of Fig. 9-2b (the dashed vector).

9-3. *Zone Plates*

A zone plate consists of a central circular obstruction together with a set of combined ring obstructions. These obstructions may be arranged so that, together, they just cover the central $k = 1$ zone, about V of Fig. 9-1, and the encircling odd zones. These zonal obstructions remove terms from our alternating series of § 9-1 so that it reduces to

$$\tilde{E}_P = \tilde{E}_{even} = +\Delta\tilde{E}_2 + \Delta\tilde{E}_4 + \Delta\tilde{E}_6 - \cdots$$

$$= -2\tilde{E}_P' - 2\tilde{E}_P' - 2\tilde{E}_P' - \cdots$$

Such a set of opaque zones is illustrated in Figs. 9-4 and 9-5. When such a figure as Fig. 9-5 is photographed on a fine-grain emulsion and reversed, at such demagnification that the outer ring is reduced to only a centimeter or so in diameter, then we have an ordinary *zone plate*.

The reduced but unreversed photograph also acts as a zone plate. Fig. 9-4 illustrates the central areas of both types of plates. Before photographic reversal the central zone appears clear, and clear zones are photographed opaque. The series of § 9-1, representing the unreversed photograph illustrated at the center in Fig. 9-4, is

$$\tilde{E}_P = \tilde{E}_{odd} = \Delta\tilde{E}_1 + \Delta\tilde{E}_3 + \Delta\tilde{E}_5 + \cdots$$

$$= 2\tilde{E}_P' + 2\tilde{E}_P' + 2\tilde{E}_P' + \cdots$$

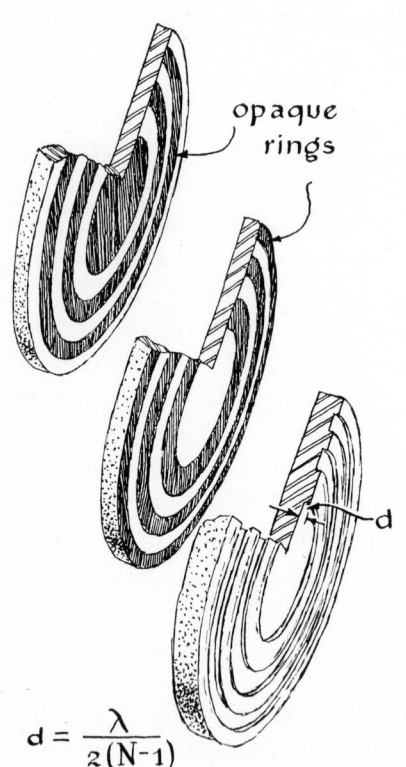

opaque rings

$$d = \frac{\lambda}{2(N-1)}$$

FIG. 9-4 Two types of obstruction zone plates and the phase reversal zone plate (below).

FIG. 9-5   Zone plate. Each black ring has the same area as the central spot.

A third type of zone plate, suggested by Lord Rayleigh, is also shown in Fig. 9-4 (bottom of the figure). It is called a *phase reversal zone plate.*† Here either the odd or even half-period zones are covered with transparent rings of thickness d, as shown in the figure with even zones covered. The ring thickness, d, is such that $(N - 1)d = \dfrac{\lambda}{2}$. Now a path difference of $\dfrac{\lambda}{2}$ represents a π phase change, *i.e.* a relative phase reversal. If contributions of these rings are added over the even zones, as above, all together the even zones yield $-\tilde{E}_{\text{even}}$, not $+\tilde{E}_{\text{even}}$; and for all zones together we get

$$\tilde{E}_P = +2k\tilde{E}_P'$$

This sum makes the illumination for a phase reversal zone plate $4k^2$ times

---

† R. W. Wood, Phil. Mag., *45*, 511 (1898).

greater at P than illumination with no zone plate. A problem is given which shows that these zone plates have an equivalent focal length, and some of the properties of lenses, just as the Poisson ball of Fig. 9-3b also has.

9-4. *Diffraction Beyond a Rectangular Aperture*

Fig. 9-6 illustrates the coordinate systems we shall use to find the diffraction pattern produced by rectangular apertures. Here two planes, $z = 0$ and $z = r_0$, locate two parallel flat screens, each provided with a coordinate system as shown. The screen at $z = r_0$ is an observing screen; or it may be a photographic plate. The flat screen at $z = 0$ is an opaque screen except for the central rectangular aperture. The sides of the aperture are at $\pm x_0$ and $\pm y_0$. P is an observation point on the observing screen at $z = r_0$, with coordinates (x_P, y_P, r_0). This figure represents a case where we have plane waves of a parent wave train propagating in the z-direction toward $z = \infty$, with the electric field in the $z = 0$ plane described by $\mathcal{E}_0 e^{j\omega t}$. To get a predicted field at P we consider each surface element of the aperture dS, at $(x, y, 0)$, to be the source of a Huygens wave trainlet. And we integrate Kirchhoff's differential over the open rectangular area of the diffraction screen.

It will turn out that there are two limiting cases where the double integration over the aperture area is separable into two integrals, one over x, and the other over y. The surface element in setting up the Kirchhoff differential, $d\tilde{E}_P$, will be $dS = dx\, dy$, with r as the travel distance from this origin of a Huygens

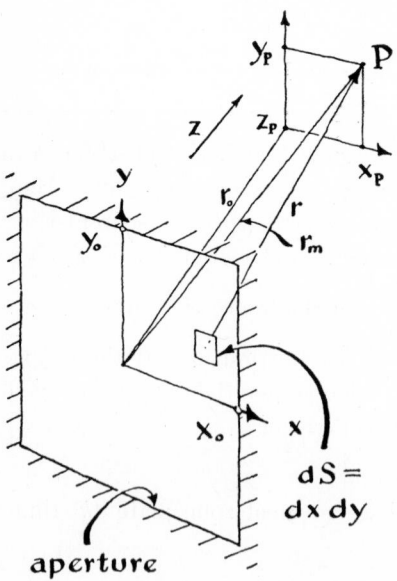

FIG. 9-6 Geometry involved in calculation of diffraction by a rectangular aperture, with parallel incident light.

wave trainlet to P. Here

$$r = \sqrt{(x_P - x)^2 + (y_P - y)^2 + r_0{}^2} \tag{a}$$

The integral that predicts the field at P, $\tilde{E}_P$, if we ignore the obliquity factor, is

$$\tilde{E}_P = \int d\tilde{E}_P = \frac{j\mathcal{E}_0}{\lambda} \iint e^{j\omega\left(t - \frac{r}{c}\right)} \frac{dx\,dy}{r} \tag{b}$$

In the two limiting cases mentioned above, where this integral is separable, it assumes the form

$$\int d\tilde{E}_P = \left\{\frac{j\mathcal{E}_0}{\lambda r_c} e^{j\omega\left(t - \frac{r_c}{c}\right)}\right\} \int e^{j\frac{2\pi}{\lambda}f(x)} \int e^{j\frac{2\pi}{\lambda}f(y)} \, dx\,dy \tag{c}$$

where r_c is taken constant. This form is assumed in either one of the two following cases: First, in the limiting case of a diffraction situation where $r_0{}^2 \gg (x_P - x)^2 + (y_P - y)^2$, Eq. 9-4a reduces to

$$r_F \cong r_0 + \frac{(x_P - x)^2}{2r_0} + \frac{(y_P - y)^2}{2r_0} \tag{d}$$

Whereas Huygens' construction ignored all the wave front except the almost exactly forward part, which touched the envelope, here our Fresnel theory takes $\dfrac{x_P - x}{2r_0}$ and $\dfrac{y_P - y}{2r_0}$ to be small as compared to unity, but not so small as Huygens' construction implies; thus the Fresnel integral takes account of a more extensive segment of the wavelets than Huygens' construction does.

Second, in the limiting case where $x_P \gg x$ and $y_P \gg y$, Eq. 9-4a reduces to

$$r = r_\infty \cong r_m - x \sin \alpha - y \sin \beta \tag{e}$$

Substituting Eqs. (d) and (e) for r in (b) gives (c), where r_c represents either r_0 or r_m. Eqs. (d) and (e) come from Eq. (a) by applying the binomial theorem for taking a square root: $\sqrt{1 + \epsilon} \cong 1 + \dfrac{\epsilon}{2}$ when $\epsilon \ll 1.0$. Eq. (d) is obtained by first dividing Eq. (a) by r_0, then taking the square root, and finally multiplying by r_0. Eq. (e) is similarly obtained by setting $r_m{}^2 = r_0{}^2 + x_P{}^2 + y_P{}^2$, then neglecting x^2 and y^2 in the expression for $r_0{}^2$, writing $\sin \alpha = \dfrac{x_P}{r_m}$ and $\sin \beta = \dfrac{y_P}{r_m}$, dividing by r_m, then taking the $\epsilon \ll 1.0$ square root, and finally multiplying by r_m, as before.

The two types of diffraction we describe by these limiting expressions, of which one is the subject of this chapter, are *Fresnel diffraction*, when $r = r_F$, and *Fraunhofer diffraction*, when $r = r_\infty$. Fig. 9-7 illustrates the illumination distribution that is to be explained. This figure shows the illumination along

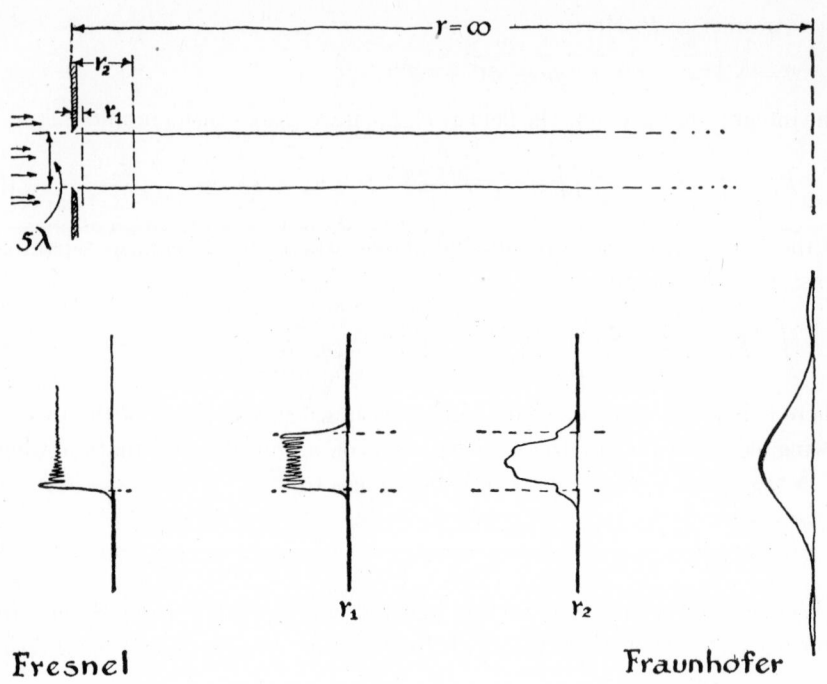

FIG. 9-7 Transition from Fresnel to Fraunhofer diffraction patterns with parallel incident light. (Redrawn from Slater's *Introduction to Theoretical Physics*.)

the x_P-axis that is produced by a long rectangular slit at various observing screen distances, r_0's; and finally at the focus of a lens where r_0 is effectively at infinity. The distribution at small r_0's illustrates Fresnel diffraction; at large r_0's, and especially at the focus of the lens, we have an illustration of Fraunhofer diffraction. In the case of far Fraunhofer diffraction the image by the lens on the observing screen will be the image of a point source surrounded by diffraction fringes (or the image of an extended source, fringed by Fraunhofer diffraction fringes). Fig. 9-8 shows such a focused diffraction pattern for a point source of light, as produced by a square aperture. In Fresnel diffraction, the pattern on the observing screen appears as the silhouette of the aperture, fringed by our other type of diffraction fringes.

In the case of $r = r_F$, each integral of Eq. (c) further may be broken up into two integrals if we write the exponential with its imaginary argument as an ordinary complex number $\mathcal{C} + j\mathcal{S}$. These integrals are called Fresnel's integrals; and Cornu devised a beautiful method of describing these integrals by means of the geometrical properties of a spiral in a complex plane.

FIG. 9-8   Fraunhofer diffraction pattern of a square
aperture at the focus of a lens beyond the aperture.

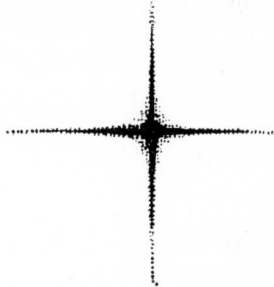

9-5. *Cornu's Spiral*

Fig. 9-9 shows Cornu's spiral in the complex plane. Here Fresnel's integrals
are represented by $\mathcal{C}$ plotted along the $\mathcal{R}$-axis against $\mathcal{S}$, plotted along the
j-axis. The definitions of $\mathcal{C}(u)$, $\mathcal{S}(u)$, and the parameter u itself, as well as
optical meanings, are developed below.

FIG. 9-9   Cornu's spiral as conventionally plotted.

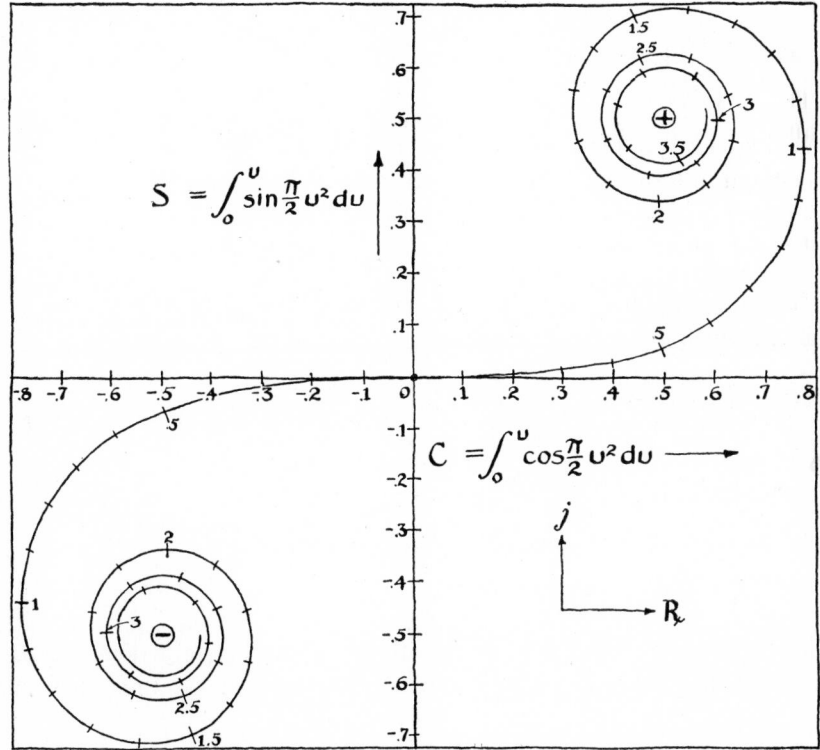

$$S = \int_0^u \sin \frac{\pi}{2} u^2 \, du$$

$$C = \int_0^u \cos \frac{\pi}{2} u^2 \, du \longrightarrow$$

Our integration, when $r = r_F$ in Eq. 9-4c, may be written

$$\int d\tilde{E}_P = \tilde{X}\tilde{Y}\tilde{Z}$$

where

$$\tilde{X} = \sqrt{\frac{2}{\lambda r_0}} \int_{-x_0}^{+x_0} e^{-j\pi\frac{(x-x_P)^2}{\lambda r_0}}\, dx$$

$$\tilde{Y} = \sqrt{\frac{2}{\lambda r_0}} \int_{-y_0}^{y_0} e^{-j\pi\frac{(y-y_P)^2}{\lambda r_0}}\, dy$$

$$\tilde{Z} = \frac{j\mathcal{E}_0}{2} e^{j\omega\left(t-\frac{r_0}{c}\right)}$$

We first treat the integration $\tilde{Y}$; and whatever we say of it is analogously true of the integration $\tilde{X}$. The parameter u is defined as $\dfrac{y - y_P}{\sqrt{\dfrac{\lambda r_0}{2}}}$ or

$u^2 = \dfrac{2(y - y_P)^2}{\lambda r_0}$. In $\tilde{Y}$, the exponent becomes $-j\dfrac{\pi u^2}{2}$. The limits of integra-

tion are $u_1 = -\dfrac{y_0 + y_P}{\sqrt{\dfrac{\lambda r_0}{2}}}$ when $y = -y_0$; and $u_2 = \dfrac{y_0 - y_P}{\sqrt{\dfrac{\lambda r_0}{2}}}$ when $y = +y_0$.

And $du = \dfrac{dy}{\sqrt{\dfrac{\lambda r_0}{2}}}$.

Writing out $\tilde{Y}$ in terms of the limits u_1 and u_2, and du, and expressing the exponential in terms of its trigonometric equivalent, we get

$$\tilde{Y} = \int_{u_1}^{u_2} e^{-j\frac{\pi u^2}{2}}\, du = \left[\int_{u_1}^{u_2} \cos\frac{\pi u^2}{2}\, du - j\int_{u_1}^{u_2} \sin\frac{\pi u^2}{2}\, du\right] \tag{a}$$

These integrals may now be expressed in terms of the Fresnel integrals $\mathcal{C}$ and $\mathcal{S}$, which are defined as follows:

$$\mathcal{C}_1 = \int_0^{u_1} \cos\frac{\pi u^2}{2}\, du \quad \text{and} \quad \mathcal{S}_1 = \int_0^{u_1} \sin\frac{\pi u^2}{2}\, du$$

Our $\tilde{Y}$, expressed in terms of $\mathcal{C}$ and $\mathcal{S}$, becomes

$$\tilde{Y} = (\mathcal{C}_2 - \mathcal{C}_1)_y - j(\mathcal{S}_2 - \mathcal{S}_1)_y = \Delta\mathcal{C}_y - j\Delta\mathcal{S}_y$$

Or, in the equivalent exponential notation, for this complex number:

$$\tilde{Y} = \sqrt{(\Delta\mathcal{C}_y)^2 + (\Delta\mathcal{S}_y)^2}\, e^{-j\tan^{-1}\frac{\Delta\mathcal{S}}{\Delta\mathcal{C}}}$$

Referring to the spiral of Fig. 9-9, it will appear from a demonstration below that the parameter u is the integral of the line element along the spiral, from the origin out to the point defined by $\mathcal{C}$ and $\mathcal{S}$: $u = \int_0^u du$. Thus the parameters u_1 and u_2 represent two points on the spiral separated by the arc length $(u_2 - u_1)$, so that $\tilde{Y}$ is the complex number representing the vector chord that extends from u_1, where it has its feather, to u_2, where it has its flint. And similarly for $\tilde{X}$:

$$\tilde{X} = (\mathcal{C}_2 - \mathcal{C}_1)_x - j(\mathcal{S}_2 - \mathcal{S}_1)_x = \Delta\mathcal{C}_x - j\Delta\mathcal{S}_x$$

$$\tilde{X} = \sqrt{(\Delta\mathcal{C}_x)^2 + (\Delta\mathcal{S}_x)^2}\, e^{-j\,\tan^{-1}\frac{\Delta\mathcal{S}}{\Delta\mathcal{C}}}$$

where u_x is defined as u_y was.

Our remark about $u = \int_0^u du$ follows from the definition of the line element:

$$\sqrt{(d\mathcal{C})^2 + (d\mathcal{S})^2} = \sqrt{\cos^2\frac{\pi u^2}{2} + \sin^2\frac{\pi u^2}{2}}\, du = du$$

A geometric feature of the spiral is apparent when we write out its slope:

$$\frac{d\mathcal{S}}{d\mathcal{C}} = -\frac{\sin\dfrac{\pi u^2}{2}}{\cos\dfrac{\pi u^2}{2}} = -\tan\frac{\pi u^2}{2}$$

Here, when $\frac{\pi u^2}{2}$ increases by 2π, the slope of the spiral changes from φ to $\varphi + 2\pi$. The increment of u^2 that represents one turn of the spiral is therefore given by $(u_{\varphi+2\pi}{}^2 - u_\varphi{}^2) = 4$. On factoring we get $(u_{\varphi+2\pi} - u_\varphi) = \dfrac{4}{(u_{\varphi+2\pi} + u_\varphi)}$. And, when u is large, $\Delta u = \dfrac{2}{u_{\mathrm{av}}}$. This result is associated with the fact that the curvature of the spiral is proportional to u.

The optical meaning of the parameters u_1 and u_2, as they relate to our Fresnel diffraction problem, is developed below. These u's, and the arc length, $(u_2 - u_1)$, define a chord $\tilde{Y}$ of length $\sqrt{(\Delta\mathcal{C})^2 + (\Delta\mathcal{S})^2}$ and of inclination $-\tan^{-1}\left(\dfrac{\Delta\mathcal{S}}{\Delta\mathcal{C}}\right)$. Our rectangular aperture gives a fixed arc length,

$$(u_2 - u_1) = 2\sqrt{\frac{2}{\lambda r_0}}\, y_0 \qquad \text{or} \qquad 2\sqrt{\frac{2}{\lambda r_0}}\, x_0$$

The location of this arc on the spiral, however, depends on the value of y_P (or x_P). Thus: if $y_P = -y_0$ then $u_1 = 0$ and the arc length lies all in the positive spiral arm; if $y_P = +y_0$ then $u_2 = 0$ and the length lies all in the negative arm; if $y_P = 0$ the length lies equally in positive and negative arms, with $u_1 = -u_2$. If $y_P < -y_0$ all the length lies on the positive arm, and $y_P > +y_0$ puts it all on the negative arm. When $y_P = +\infty$ the arc is wound tightly about the negative eye of the spiral, when $y_P = -\infty$ it is wound tightly about the positive eye, so that the chord of this arc, $\tilde{Y}$, is zero at $y_P = \pm\infty$. In between, where the arc length is not coiled tightly, but is more or less open, chord length may be determined graphically from the ends on the spiral. Also, the chord may be determined from numerical values of $\mathfrak{C}$ and $\mathfrak{S}$ such as are to be found in the tables of Jahnke and Emde.

Once we have made an evaluation of $\tilde{Y}$ and $\tilde{X}$, for a particular optical problem, and a point P, either graphically from the spiral, or from numerical tables, and when $\tilde{Z}$ is determined from λ_0, $\mathcal{E}_0$, and r_0, the product of the three complex numbers $\tilde{X}$, $\tilde{Y}$, and $\tilde{Z}$ gives the illumination on the observing screen.

Following an application of these procedures, we shall discuss the case of Fresnel diffraction produced by a straight edge, by a narrow slit, and by a narrow strip. But first, in order to become familiar with such uses of Cornu's spiral, we shall apply our integrals, again, to a case of no diffraction.

9-6. *No Diffraction, by Cornu's Spiral*

If the incident parallel field at $z = 0$ is $\mathcal{E}_0 e^{j\omega t}$, we expect an infinitely large aperture to give a field at $z = r_0$ on the observing screen opposite the center of our aperture, of $\mathcal{E}_0 e^{j\omega\left(t - \frac{r_0}{c}\right)}$; and we may use Cornu's spiral also to determine this field. We take the aperture so large that we may write $\pm x_0 = \pm y_0 = \pm\infty$. Cornu's spiral should give $\tilde{E}_P = \mathcal{E}_0 e^{j\omega\left(t - \frac{r_0}{c}\right)}$ at $x_P = y_P = 0$ on the screen. Thus, a wide aperture should produce no diffraction pattern opposite its center. The associated fixed arc lengths involved here, $(u_2 - u_1)_x$ and $(u_2 - u_1)_y$, are both infinitely long and symmetrically laid off on the spiral. At $x_P = y_P = 0$ these arcs reach from the negative eye to the positive eye of the spiral, and their ends are tightly wound up about these eyes. Thus, for both $\tilde{X}$ and $\tilde{Y}$, we get $\sqrt{2}$ for the chord lengths and 45° for the chord inclinations: $\tilde{X} = \tilde{Y} = \sqrt{2} e^{-j\frac{\pi}{4}}$. Writing $e^{j\frac{\pi}{2}}$ for j in our expression for $\tilde{Z}$, the aggregate $\tilde{E}_P$ given by Cornu's spiral is

$$\tilde{E}_P = \tilde{X}\tilde{Y}\tilde{Z} = (\sqrt{2} e^{-j\frac{\pi}{4}})(\sqrt{2} e^{-j\frac{\pi}{4}})\left(\frac{j\mathcal{E}_0}{2} e^{j\omega\left(t - \frac{r_0}{c}\right)}\right) = \mathcal{E}_0 e^{j\omega\left(t - \frac{r_0}{c}\right)}$$

Thus our theory gives the expected field at P. It is not an unsullied result, however. In this particular problem our integrations involved values of r that may not be expressed by the approximations of Eq. 9-4d. The expression r_F for r is valid only when the quantities $\left|\dfrac{x - x_P}{2r_0}\right|$ and $\left|\dfrac{y - y_P}{2r_0}\right|$ are small compared to unity, and here we have allowed these quantities to become infinite. The violation of these conditions for representation of r by r_F is discussed by R. W. Ditchburn in his *Light*:[†] "The above discussion suffers from the defect

† From R. W. Ditchburn, *Light* (1954, Interscience Publishers, New York).

that we have included a portion of wave front corresponding to large angles of diffraction, whereas the equations from which the spiral was constructed are exactly valid only for small angles of diffraction. The effect of zones remote from the center is very small. The factors we have omitted would, if included, merely deform the spiral a little in the region where it is near the asymptotic point [eye]."

Even when the conditions which validate r_F are satisfied, this theory of diffraction is not exact, for it is a scalar theory. And although the theory does greatly expand the forward area of the Huygens wavelets which are accounted for, it must be admitted that, in the end, it begs the question much as Huygens' construction did. And, also, our neglect of the obliquity factor must not be overlooked. But this neglect is not serious if we only apply the theory to the forward areas of the wavelets. In spite of these impairments, the Cornu spiral treatment has been useful in many applications.† And besides, pedagogically, it affords the student an example of a typical theory in physics which has an impressive neatness, inspiring awe; which makes necessary compromises, requiring prudence; and which is blemished by a lack of complete validity, requiring understanding.

Now, with an awareness of these impairments, let us apply Cornu's spiral to a treatment of diffraction by the straight edge and by the slit and strip.

9-7. *The Straight Edge*

As in our treatment of § 9-6, we again take a very large diffraction aperture and examine $\tilde{E}_P$ at $y_P = y_0$, leaving $x_P = 0$. In the environs of $y_P = y_0$, on the observing screen, we will have a diffraction pattern equivalent to that of the straight edge.

With $x_P = 0$ our integration over x is the same as above, yielding $\tilde{X} = \sqrt{2}e^{-j\frac{\pi}{4}}$. And the integration over y at $y_P = y_0$, for which $u_2 = 0$, yields a chord with the same slope as above, but of only half the former length; namely, $\tilde{Y} = \frac{\sqrt{2}}{2} e^{-j\frac{\pi}{4}}$. Here the infinite fixed length arc is mostly wound tightly about the negative eye of the Cornu spiral, and its positive end just reaches out to $u_2 = 0$. Thus $\tilde{E}_P$ has the same phase but only half the magnitude it has at $y_P = 0$; and $\mathfrak{E}_P$ gives only one quarter of the unobstructed illumination.

Outside the border, where, as y_P increases beyond y_0, u_2 decreases, the chord length of the fixed length arc will decrease monotonously until finally, in full

---

† C. L. Andrews, Phys. Rev., *71*, 777 (1947).

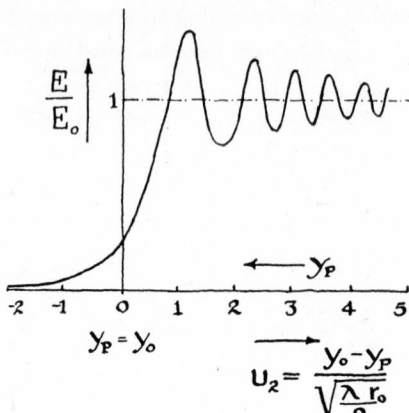

FIG. 9-10 Illumination near edge of silhouette of a straight edge—Fresnel diffraction.

shadow, where $y_P - y_0 \gg 1.0$, all the infinite fixed length arc is wound tightly around the negative eye of the spiral, and $\tilde{E}_P = 0$. As u_2 decreases, the phase retardation also increases, monotonously.

Inside the border, where, as y_P decreases, u_2 increases, the chord length first increases from its length $\dfrac{\sqrt{2}}{2}$ to a maximum, when $u_2 = 1.2$ (see Fig. 9-9); then it decreases to a minimum at $u_2 = 1.9$; and so on, oscillating with smaller and smaller fluctuations until, finally, the flint of the vector lies at the positive eye, with the feather still at the negative eye. Finally the chord takes on the value $\tilde{Y} = \sqrt{2}e^{-j\frac{\pi}{4}}$, the same as its value at $y_P = 0$. The corresponding oscillations of the illumination are shown in Fig. 9-10. These oscillations of illumination inside the border of the silhouette represent the typical Fresnel display of diffraction fringes.

Of course, eventually, as y_P goes to even larger negative values, the feather will come out of its eye and reach $u_1 = 0$ at $y_P = -y_0$.

9-8. *Fresnel Diffraction by Slit and Strip*

The diffraction patterns for the slit and strip are predicted by an adaptation of this treatment, when P is almost opposite the edges of the large aperture, and by the treatment of § 9-6 for opposite its center. Since $(u_2 - u_1)$ will not be infinite for a narrow slit, neither end of the fixed length arc will lie at an eye of the spiral when P lies opposite the slit in the vicinity of $x_P = y_P = 0$. In such a case we determine $\tilde{E}_P$ and $\mathfrak{E}_P$ by first finding $\tilde{Y}_P$ graphically, with the spiral. We get $\tilde{Y}_P$ from the location of u_1 and u_2, or u_{av}, as determined by the coordinate y_P. From the fixed arc $(u_2 - u_1)$, thus located on the spiral, to get $\tilde{Y}_P$ we determine the chord length and its inclination. Or, as mentioned above, we may use tables of $\mathcal{C}$ and $\mathcal{S}$ to determine $\tilde{Y}_P$.

Let ω in the expression $\tilde{E} = \mathcal{E}_0 \, e^{j\omega t}$, describing the incident light at $z = 0$, correspond to $\lambda = 5 \times 10^{-5}$ cm. If the slit width is $2y_0 = 1$ mm, and if $r_0 = 1$ meter, then $(u_2 - u_1) = 2$. Fig. 9-11 shows this arc on the spiral, and its chord $\tilde{Y}_{\text{slit}}$, for y_P giving $u_{\text{av}} = 0$.

If the slit is long, then $\tilde{X} = \sqrt{2}e^{-j\frac{\pi}{4}}$, and the illumination pattern, as y_P varies, is given by the electric field, $\tilde{E}_P = (\sqrt{2}e^{-j\frac{\pi}{4}}\tilde{Z})\,\tilde{Y}_{\text{slit}}$. For the purposes of the following discussion we take the expression in parenthesis above as unity, so that $\tilde{E}_P = \tilde{Y}_{\text{slit}}$. Thus $\mathcal{E}_P$ is proportional to $\tilde{Y}\tilde{Y}^*$. When P lies on either side of $y_P = 0$, where u_{av} is large, the illumination will be a minimum whenever $\dfrac{2}{u_{\text{av}}}$, the length of a turn, divides into $(u_2 - u_1) = 2$ an integral number of times; then the ends of the arc will lie very near each other.

The diffraction pattern of a strip may be determined from $\tilde{Y}_{\text{slit}}$ for the slit by applying the principle of superposition. We think of the slit jaws as two large rectangular apertures filled with fitting opaque rectangles. Let $\tilde{Y}_{\text{strip}}$ represent the field for a strip, complementary to our slit above, of width $2y_0 = 1$ mm, so that it will just fit in the slit. When the strip is fitted in the slit, we have complete opacity. When the strip only is removed, we have $\tilde{Y}_{\text{slit}}$. When the slit jaws only are removed, we have $\tilde{Y}_{\text{strip}}$. And when all obstructions are removed, we have $\tilde{Y}_{\text{open}}$. In the light of the principle of superposition,

$$\tilde{Y}_{\text{slit}} + \tilde{Y}_{\text{strip}} = \tilde{Y}_{\text{open}} = \sqrt{2}e^{-}$$

Fig. 9-11 shows the complementarity feature of these complex numbers graph-

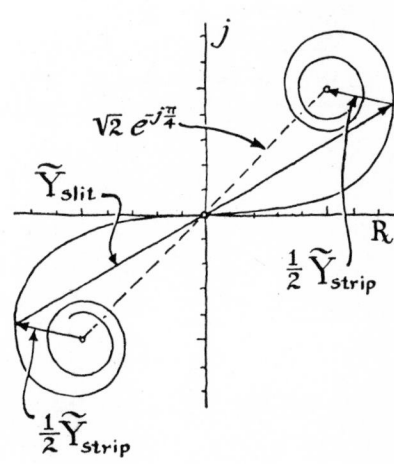

FIG. 9-11   Use of Cornu's spiral to predict Fresnel diffraction by slit and strip. In Fig. 9-9, and here, we have drawn the spiral as it usually appears in optics books. As a consequence, all vectors, as well as the spiral itself, appear as reflected in the $\mathcal{R}$-axis. This is because the imaginary term in Eq. 9-5a carries a negative sign.

ically for y_P, where $u_{av} = 0$ and remembering that complex vectors can represent these complex numbers: $\tilde{Y}_{slit}$, $\tilde{Y}_{strip}$, and $\tilde{Y}_{open}$.

9-9. *Babinet's Principle*

The above summing of the diffraction pattern amplitudes for two complementary sets of obstacles, which yields no resultant for the case where there are no obstacles, provides the basis for Babinet's Principle. Complementary sets of obstacles are those for which one set has openings where the other set is opaque, and *vice versa*. Imagine an unobstructed wave front focused on an observing screen, and consider a point P where the field $\tilde{E}_P = 0$ as shown in Fig. 9-12. If complementary screens are alternately interposed and give diffraction fields represented by $\tilde{Y}_1$ and $\tilde{Y}_2$ at P, then with neither screen at points where $\tilde{E}_P = 0$ we must get, from the principle of superposition,

$$\tilde{Y}_1 + \tilde{Y}_2 = 0 \qquad \text{or} \qquad \tilde{Y}_1 = -\tilde{Y}_2 \qquad \text{and} \qquad \mathfrak{E}_1 = \mathfrak{E}_2$$

FIG. 9-12   Application of Babinet's theorem.

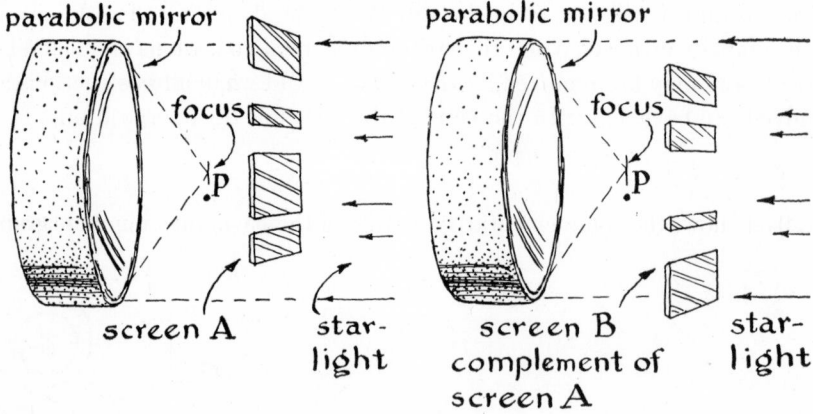

Chapter X

Fraunhofer Diffraction

THE INTEGRATION of Eq. 9-4c with r_∞ rather than r_F for r gives expressions describing Fraunhofer diffraction. In this chapter we shall make such integrations for a rectangular aperture. Our result will describe the Fraunhofer diffraction produced by the single slit. Then we shall elaborate our integrations to describe diffraction produced by multiple slits to give the elementary theory of diffraction gratings. The Fraunhofer diffraction produced by other openings is predicted similarly, although integrations may be more difficult.

10-1. *The Rectangular Aperture*

As before, we divide the integral of Kirchhoff's differential into three parts:

$$\tilde{E}_P = \tilde{X}\tilde{Y}\tilde{Z}$$

In this Fraunhofer case, where $r_0 = \infty$, the terms have the following meaning:

$$\tilde{X} = \frac{1}{2x_0} \int_{-x_0}^{x_0} e^{j\frac{2\pi x \sin \alpha}{\lambda}} \, dx$$

$$\tilde{Y} = \frac{1}{2y_0} \int_{-y_0}^{y_0} e^{j\frac{2\pi y \sin \beta}{\lambda}} \, dy$$

$$\tilde{Z} = (4x_0 y_0) \frac{j\mathcal{E}_0}{\lambda r_m} e^{j\omega\left(t - \frac{r_m}{c}\right)}$$

We require r_0 to be very large by our present approximation. The illumination on the observing screen then will be correspondingly small. This is because r_m^2 gets large in the denominator of $\mathfrak{E}$, from $\tilde{Z}\tilde{Z}^*$. In practice, however, if we put the screen at the distance $r_0 = z_P = f$, so that it lies at the focal plane of a thin lens, it is equivalent, as far as our approximation is concerned, to putting the observing screen at infinity. With such a lens $\mathfrak{E}$ does not get too small to see.

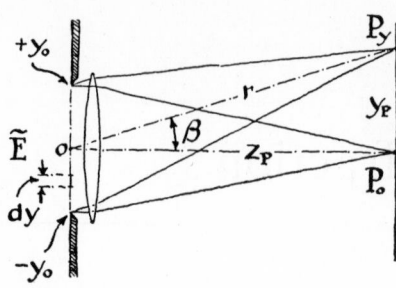

FIG. 10-1   Fraunhofer diffraction.

Fig. 10-1 illustrates this modification of Fig. 9-6. We shall now analyze this modification. Although in Fig. 10-1 the source is focused on the observing screen by a lens, the expression for r_m (if we neglect the thickness of the lens) remains $\sqrt{x_P{}^2 + y_P{}^2 + r_0{}^2}$. Setting $\xi = \dfrac{2\pi x_0 \sin \alpha}{\lambda}$, our integral over x becomes

$$\tilde{X} = \frac{1}{2x_0} \int_{-x_0}^{x_0} e^{j\frac{2\pi x \sin \alpha}{\lambda}}\, dx = \frac{e^{j\xi} - e^{-j\xi}}{2j\xi} = \frac{\sin \xi}{\xi}$$

And similarly, setting $\eta = \dfrac{2\pi y_0 \sin \beta}{\lambda}$, our other integral becomes

$$\tilde{Y} = \frac{1}{2y_0} \int_{-y_0}^{y_0} e^{j\frac{2\pi y \sin \beta}{\lambda}}\, dy = \frac{e^{j\eta} - e^{-j\eta}}{2j\eta} = \frac{\sin \eta}{\eta}$$

Thus both $\tilde{X}$ and $\tilde{Y}$ are pure real numbers. The phase of $\tilde{E}_P$, except for sign, is represented in the term $\tilde{Z}$. Here $\sin \alpha = \dfrac{x_P}{r_m}$ and $\sin \beta = \dfrac{y_P}{r_m}$. In § 2-6 we had already obtained the equivalent of the above formulas for $\tilde{X}$ and $\tilde{Y}$. In Fig. 2-4 we had $\mathfrak{N}$ component Huygens wavelets represented and plotted as vectors in the complex plane, with their phases distributed uniformly over the total phase interval 2ϕ. Here the corresponding total phase interval is 2ξ or 2η. Our former finite $\mathfrak{N}$ has here gone to the limit required by integration; but otherwise the summation is parallel to that of § 2-6. Each Huygens wave trainlet that we sum here comes from an elementary area $(dx\, dy)$ at (x, y), and its phase, due to its coordinate x, relative to that of the wave trainlet from $(x = 0, y)$, is $\dfrac{2\pi x \sin \alpha}{\lambda}$. With the dx's all equal, our integration is here equivalent to the limit of an infinite series of $\mathfrak{N}$ complex numbers with uniformly distributed imaginary phase exponents, and with their equal infinitesimal amplitudes proportional to $\dfrac{2x_0}{\mathfrak{N}}$. Thus $\tilde{X}$ gives the average contribution of all surface elements in the phase interval from $-\xi$ to ξ. The effect of wavelets with leading phases on the final phase cancels that of those with retarded phases, so that the phase of the average is the same as that of the wavelet from the center of the aperture at $x = 0$; hence $\tilde{X}$ is real.

When x or y is zero, then $\tilde{X}$ and $\tilde{Y}$ becomes 1.0:

$$\text{Lim}_{\epsilon \to 0} \left(\frac{\sin \epsilon}{\epsilon} \right) = \text{Lim}_{\epsilon \to 0} \left(1 - \frac{\epsilon^3}{\lfloor 3} + \cdots \right) = 1.0$$

When ξ or η is π, 2π, 3π, $\cdots$ etc., then $\tilde{X}$ or $\tilde{Y}$ is zero. When ξ or η is $\dfrac{\pi}{2}$, $\dfrac{3\pi}{2}$, $\dfrac{5\pi}{2}$, $\cdots$ etc., then $\tilde{X}$ and $\tilde{Y}$ have very nearly extreme values. These extreme values lie exactly where the transcendental equation, $\tan (\xi \text{ or } \eta) = (\xi \text{ or } \eta)$, has its roots; and for our purposes we assume that these roots, as mentioned above, lie where ξ or η takes on values equal to odd integral multiples of $\dfrac{\pi}{2}$.

When $\mathcal{E}_0 e^{j\omega t}$ is taken as the field at $z = 0$, the field at a point P on the observing screen, which subtends the angles α and β at the center of the rectangular aperture, is

$$\tilde{E}_P = \tilde{X}\tilde{Y}\tilde{Z} = 4(x_0 y_0)\frac{j\mathcal{E}_0}{\lambda r_m}\frac{\sin \xi}{\xi}\frac{\sin \eta}{\eta} e^{j\omega\left(t - \frac{r_m}{c}\right)}$$

The illumination at the center of the diffraction pattern, $\xi = \eta = 0$, corresponding to the electric field given above, is

$$\mathcal{E}_C = \frac{1}{2} c\kappa_0 \frac{(4x_0 y_0)^2 \mathcal{E}_0^2}{\lambda^2 r_m{}^2}\frac{\text{watt}}{\text{meter}^2}$$

Thus, at other points P defined by ξ and η, we get the illuminations

$$\mathcal{E}_P = \mathcal{E}_C \frac{\sin^2 \xi}{\xi^2}\frac{\sin^2 \eta}{\eta^2}\frac{\text{watt}}{\text{meter}^2}$$

10-2. *Square and Circular Apertures Contrasted*

Figs. 10-2 and 10-8 illustrate light distributions produced on a screen by a square and by a circular aperture. In Fig. 10-3, $\dfrac{\mathcal{E}_P}{\mathcal{E}_C}$ is plotted for each pattern. In this figure the circular aperture diameter is taken to be equal to the width of the square aperture, $2x_0$. In either case, without diffraction, the image would be a point.

In the case of the circular aperture the central disk of the pattern, which contains 84% of the light, is surrounded by a succession of diffraction rings. The central disk is called Airy's disk. Only the first diffraction ring is conspicuous around it. The fields and illuminations for the circular aperture are not calculated here, but merely quoted.

In the case of the plot of $\dfrac{\mathcal{E}_P}{\mathcal{E}_C}$ for a square aperture, or a slit, along $y_P = 0$

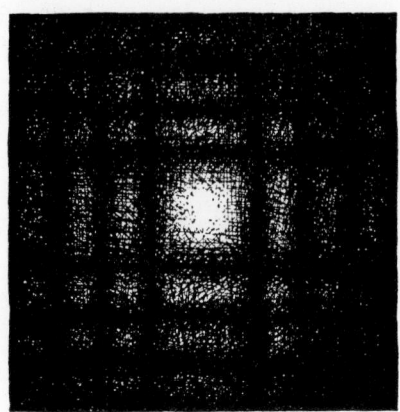

FIG. 10-2 Fraunhofer diffraction by a square aperture.

on the observing screen, the path difference for corresponding points in the two halves, separated by the distance x_0, is $x_0 \sin \alpha$, the corresponding number of cycles of interference being $\dfrac{x_0 \sin \alpha}{\lambda}$, and the corresponding phase retardation being $\xi = 2\pi \dfrac{x_0 \sin \alpha}{\lambda}$. And when ξ takes on the values $\dfrac{3\pi}{2}, \dfrac{5\pi}{2} \cdots$ etc., $\mathfrak{E}_P$ has the maximal values of $\mathfrak{E}_{\max} = \mathfrak{E}_C \left(\dfrac{2}{n_{\mathrm{odd}}\pi} \right)^2$. Here n_{odd} is an odd integer. If ξ and η simultaneously take on such values, for the square aperture, then, along the diagonal where $\alpha = \beta$, the illumination has the maximal values

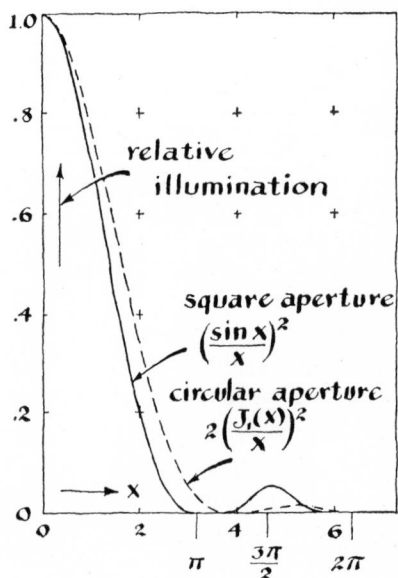

FIG. 10-3 Fraunhofer diffraction by a circular and a square aperture, or slit, contrasted.

of $\mathfrak{E}_{max} = \mathfrak{E}_C \left(\dfrac{2}{n_{odd}\pi} \right)^4$. Along this diagonal the distances, ρ, of maxima are farther from the pattern center, by the factor $\sqrt{2}$, than the distances of the corresponding maxima along $y_P = 0$, or $x_P = 0$. Relative illuminations, and relative distances from the pattern center, are given for successive maxima and minima in Table 10-1. For the square aperture it is apparent that the $\mathfrak{E}_{max}$ values fall off most rapidly on the observing screen along the diagonal. And they fall off least rapidly for the square-on square. For the circular aperture, the $\mathfrak{E}_{max}$ values fall off with distance on the observing screen at an intermediate rate. Also, it is apparent that we have the widest central maxima for the most rapid fall-off of illumination—the maximum for the circular aperture, as measured by the distance to the first minimum, is 22% greater than for the square-on square, but narrower than for the diagonal square.

In some instances where we may wish to see two close equally strong stars resolved, so that the duplicity of sources is apparent in their composite image, we must have the central maxima narrow even if outer maxima are not the weakest possible. But, on the other hand, there are instances, as in the case of the star Sirius and its dark companion, where one of the sources is very much stronger than the other. There the narrowness of the central maximum may be less important than suppression of diffraction images due to the stronger star.

TABLE 10-1   *Successive Diffraction Maxima*

Square-on Square		Square Diagonally		Circular Aperture	
$n_{odd} = \dfrac{4x_0}{\lambda z_P} x_P$	$\mathfrak{E}_P/\mathfrak{E}_C$	ρ/x_P	$\mathfrak{E}_P/\mathfrak{E}_C$	ρ/x_P	$\mathfrak{E}_P/\mathfrak{E}_C$
0	1.00	0	1.00	0	1.00
1	0	1.41	0	1.22	0
1.5	0.045	2.12	.002	1.64	.0174
2	0	2.82	0	2.22	0
2.5	0.016	3.53	.00026	2.69	.0041
3	0	4.24	0	3.24	0
3.5	0.008	4.95	.000069	3.72	.0016
4	0	5.64	0	4.24	0
4.5	0.005	6.36	.000025	4.72	.0008
5	0	7.06	0	5.24	0
5.5	0.003	7.77	.000011	5.72	.0004

FIG. 10-4 Satellite of the helium line, by Prof. P. Jacquinot.

In the case of Sirius and its dwarf companion their separation, 12 seconds of arc, is not excessively small; but their difference in strength is excessive, being 10,000 to one. Here for a telescope with circular aperture, the diffraction rings of the main star, overlying the companion star, can mask the view of it. At a separation of the dwarf of 12 seconds, or 6.6×10^{-6} radians of arc, the value of $\frac{\rho}{x_P}$ in Table 10-1, for $x_0 = 4$ inches, is 5.5. This angular distance on the observing screen gives the overlying diffraction maximum of Sirius as $33/10,000$ $\mathfrak{E}_C$ for the square-on square aperture; $4/10,000$ $\mathfrak{E}_C$ for the circular aperture, at the same angular distance; but only between $0.70/10,000$ $\mathfrak{E}_C$ and $0.25/10,000$ $\mathfrak{E}_C$ for the diagonal square aperture. Thus, where a square-on aperture of 8″ on a side would present the dwarf submerged under diffraction maxima that were 33-fold stronger, and where a circular aperture of 8″ diameter would present it submerged under rings that were 4-fold stronger, along the diagonal of an 8″ square aperture the central image of the companion would dominate the diffraction pattern.

Another example where the fainter of two objects is confused by diffracted light with the stronger, with relative strengths also of 10,000:1, is found in spectroscopy in the case of the yellow spectrum line of helium and its satellite. Professor P. Jacquinot of Paris has separated this helium line and its satellite in his spectrograph by ingeniously using the collimated light emerging from a pinhole rather than from a slit, so that, actually, he got spectrum "spots" rather than spectrum "lines." This spectrum line and its satellite are not ordinarily separated in photographs of the spectrum because of diffraction by the prism, which constitutes not only a rectangular aperture stop, but one that is arranged square-on. In order to avoid veiling diffraction, Professor Jacquinot stopped his prism aperture with a square aperture that was rotated to the diagonal, or at 45° azimuth position. He was then able to resolve the satellite helium "spot." Fig. 10-4 shows how his photograph looked.

Dr. Jacquinot and his associates have studied other means of achieving the suppression of high order diffraction (at the expense of broader central diffraction). All such procedures are called *apodizing* (meaning *to make without feet*). They have exploited the use of circular aperture stops with symmetrically variable transmission screens—screens that are clear in the center with opacity increasing toward their rims. Such screens remove from play the diffraction that is produced by sharp borders, so to speak; for they have no sharp border. In this country Dr. W. M. Sinton used an aluminized circular disk of gradually increasing opacity to thus apodize the Johns Hopkins 9″ telescope and make the otherwise submerged dwarf companion of Sirius observable.

10-3. *Interference and Diffraction Combined*

Fig. 10-5 shows $\mathfrak{N}$ identical parallel slits each of width $2x_0$. These slits are separated by equal intervals a. This figure may be thought of as representing Fig. 10-1 with the single aperture here replaced by a regular array of slits. Wave trainlets from the successive surface elements of each one of the separate slits superimpose, as described previously. Thus, all together, our array produces $\mathfrak{N}$ superimposed single slit diffraction amplitudes on the observing screen. And these $\mathfrak{N}$ superimposed diffraction amplitudes produce interference effects. We can easily check the analysis of these interference effects, that we are about to develop, because our theory must reduce to our earlier analysis of Young's double-slit interference pattern when we set $\mathfrak{N} = 2$. When $\mathfrak{N}$ is small, but greater than 2, multiple slits produce interference maxima of illumination such as those illustrated in Fig. 10-6. When $\mathfrak{N}$ is large, say from 10^4 to 10^5, the theory we are about to develop applies to diffraction gratings.

The mathematical result describing these combined diffraction-interference effects is obtained by integrating Kirchhoff's differential over the open area of all the separate slits. We have already gotten the result of integration of Huygens' wave trainlets over one slit. This can guide us in obtaining the

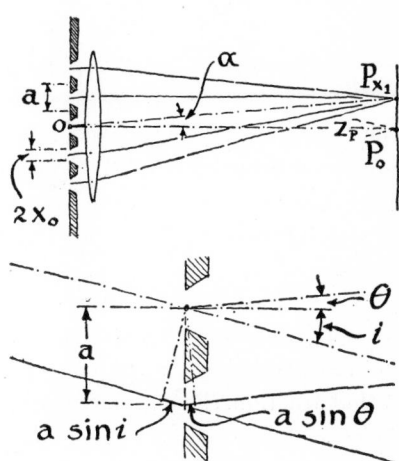

FIG. 10-5 Multiple-slit interference and diffraction.

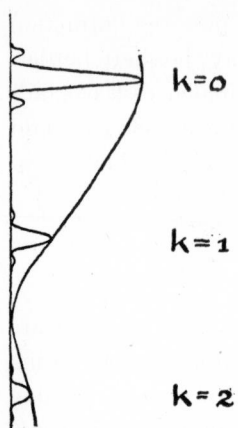

FIG. 10-6 Multiple-slit interference enveloped by single-slit diffraction.

k=o

k=1

k=2

superposition sum of $\mathfrak{N}$ diffraction patterns, thus compounding interference with diffraction. If $\tilde{X}_i$ is the integral for any one particular slit, and $\tilde{X}$ is the superposition sum for all, then

$$\tilde{X} = \sum_{i=1}^{\mathfrak{N}} \tilde{X}_i$$

or

$$\tilde{X} = \frac{1}{2x_0} \int_{-x_0}^{x_0} e^{j\frac{2\pi x \sin \alpha}{\lambda}} \, dx + \frac{1}{2x_0} \int_{a-x_0}^{a+x_0} e^{j\frac{2\pi x \sin \alpha}{\lambda}} \, dx$$

$$+ \cdots + \frac{1}{2x_0} \int_{(\mathfrak{N}-1)a-x_0}^{(\mathfrak{N}-1)a+x_0} e^{j\frac{2\pi x \sin \alpha}{\lambda}} \, dx$$

We shall use our earlier result for the first integral, here labeled $\tilde{X}_S$. It appears at once that the second integral contains $\tilde{X}_S$ if we change the integration variable to $x' = x - a$; thus

$$\frac{1}{2x_0} \int_{a-x_0}^{a+x_0} e^{j\frac{2\pi x \sin \alpha}{\lambda}} \, dx = e^{j\frac{2\pi a \sin \alpha}{\lambda}} \frac{1}{2x_0} \int_{-x_0}^{x_0} e^{j\frac{2\pi x' \sin \alpha}{\lambda}} \, dx' = \tilde{X}_S e^{j\frac{2\pi a \sin \alpha}{\lambda}}$$

The treatment of this second term indicates how the third and subsequent terms may also be treated. On writing $k\lambda$ for $a \sin \alpha$, where k becomes the order of interference between corresponding parts of adjacent slits, we get, for all slits, the series

$$\tilde{X} = \tilde{X}_S[1 + e^{j(2\pi k)} + e^{2j(2\pi k)} + \cdots + e^{(\mathfrak{N}-1)j(2\pi k)}]$$

The series, in the brackets above, is equivalent to the polynomial fraction in the brackets below, as can be shown by applying simple polynomial division to that fraction. Thus, remembering the integration over y, and the term $\tilde{Z}$,

$$\tilde{E}_P = \tilde{X}_S\{\tilde{Y}\tilde{Z}\} \left[\frac{1 - e^{\mathfrak{N}j(2\pi k)}}{1 - e^{j(2\pi k)}} \right]$$

Representing the complex number of the brackets above by $\tilde{B}$, the illumination on the observing screen at P for the interference-diffraction pattern becomes

$$\mathfrak{E}_P = \tfrac{1}{2}c\kappa_0\tilde{X}_S\tilde{X}_S*\{\tilde{Y}\tilde{Z}\}\{\tilde{Y}\tilde{Z}\}*\tilde{B}\tilde{B}*$$

We evaluate $\tilde{B}\tilde{B}*$ as follows: Multiplying our complex fraction by its conjugate:

$$\tilde{B}\tilde{B}* = \left\{\frac{1 - e^{2j\mathfrak{N}\pi k}}{1 - e^{2j\pi k}}\right\}\left\{\frac{1 - e^{-2j\mathfrak{N}\pi k}}{1 - e^{-2j\pi k}}\right\} = \frac{2 - (e^{2j\mathfrak{N}\pi k} + e^{-2j\mathfrak{N}\pi k})}{2 - (e^{2j\pi k} + e^{-2j\pi k})}$$

On writing the exponential terms in this expression as cosines, and then the cosines as sines, squared, we get

$$\tilde{B}\tilde{B}* = \frac{1 - \cos 2\mathfrak{N}\pi k}{1 - \cos 2\pi k} = \frac{\sin^2 \mathfrak{N}\pi k}{\sin^2 \pi k}$$

We now check ourselves to see if we get $\mathfrak{E}_P = 2c\kappa_0\mathcal{E}_2^2 \cos^2 \pi\left(\dfrac{ay}{\lambda\mathfrak{D}}\right)$ (Eq. 8-3b) for $\mathfrak{N} = 2$. In $\tilde{B}\tilde{B}*$ we set $\mathfrak{N} = 2$ and write $\dfrac{y}{\mathfrak{D}}$ for $\dfrac{y_P}{r_0}$ to express $\sin \alpha$, as in our earlier terminology for Young's double slit. Thus $k = \dfrac{a \sin \alpha}{\lambda} = \dfrac{ay}{\lambda\mathfrak{D}}$. Trigonometry gives $\dfrac{\sin^2 2x}{\sin^2 x} = 4 \cos^2 x$, so that $\tilde{B}\tilde{B}*$ now gives $4 \cos^2 \pi\left(\dfrac{ay}{\lambda\mathfrak{D}}\right)$, as expected.

When the order of interference is zero, then $\tilde{B}\tilde{B}*$ is $\mathfrak{N}^2$. Also, when k is positive or negative, and an integer, the value of $\tilde{B}\tilde{B}*$ is again $\mathfrak{N}^2$, as demonstrated below. Taking k as zero, or as any integer, and going to the indicated limit,

$$\operatorname*{Lim}_{\delta \to 0} \frac{\sin^2 \mathfrak{N}(\pi k + \delta)}{\sin^2 (\pi k + \delta)} = \operatorname*{Lim}_{\delta \to 0} \frac{\sin^2 \mathfrak{N}\delta}{\sin^2 \delta} = \mathfrak{N}^2$$

These k values represent the constructive interference of all the $\mathfrak{N}$ single-slit diffraction amplitudes taken together, giving an aggregate illumination that is $\mathfrak{N}^2$ greater than any one of the single slits alone would produce. When $\mathfrak{N}$ is large, an illumination maximum takes on the character shown in Fig. 10-7.

When $\mathfrak{N}$ is very large, this interference shape is geometrically similar to the shape for a single slit diffraction pattern, such as the one described in Fig. 10-3 for a slit. To show this geometrical similarity, let $k = k_1 \pm \dfrac{n}{2\mathfrak{N}}$. Then

$$\tilde{B}\tilde{B}* = \frac{\sin^2 (\mathfrak{N}\pi k_1 \pm \tfrac{1}{2}\pi n)}{\sin^2 \left(\pi k_1 \pm \dfrac{1}{2} \pi \dfrac{n}{\mathfrak{N}}\right)} = \frac{\pm\sin^2 \pi \dfrac{n}{2}}{\sin^2 \dfrac{1}{2} \dfrac{\pi n}{\mathfrak{N}}} \cong \mathfrak{N}^2\left\{\frac{\sin \dfrac{\pi n}{2}}{\dfrac{\pi n}{2}}\right\}^2$$

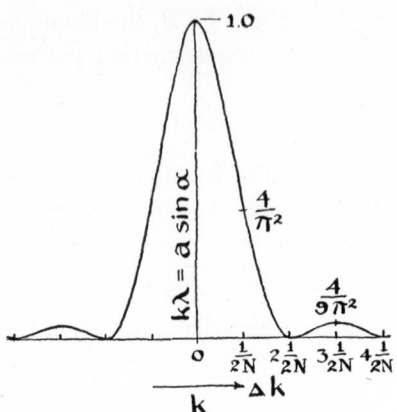

FIG. 10-7 Multiple-slit interference pattern.

When k_1 and n are small integers, and $\mathfrak{N}$ is a large one, we may simplify the expression for $\tilde{B}\tilde{B}^*$ by writing $k = k_i \pm \dfrac{n}{2\mathfrak{N}}$. Then on solving for the extreme values and zeros, we get

$$
\tilde{B}\tilde{B}^* = \begin{cases} \sin^2 \mathfrak{N} \left(k_1 \pi \pm \dfrac{n\pi}{\mathfrak{N}2} \right) \\[2ex] \sin^2 \left(k_1 \pi \pm \dfrac{n\pi}{\mathfrak{N}2} \right) \end{cases} = \begin{cases} 0 & \text{if } n \text{ is even} \\[2ex] \left(\dfrac{\pm 1}{\sin \dfrac{\pi n}{2\mathfrak{N}}} \right)^2 \simeq \dfrac{4\mathfrak{N}^2}{n^2\pi^2} & \text{if } n \text{ is odd} \end{cases}
$$

These values of $\tilde{B}\tilde{B}^*$ give an interference contour similar to the single slit diffraction pattern, as shown in Fig. 10-7.

10-4. *Diffraction-limited Resolving Power*

The diffraction patterns of Figs. 10-2 and 10-8 are used below to explain angular resolving power and the spectral resolving power of a prism. The geometrically similar interference pattern of Fig. 10-7 is involved in the spectral resolving power of a diffraction grating.

Fig. 10-8 shows photographs of two artificial double stars of equal magnitude variously separated. These photographs were taken by J. A. Anderson with a telescope of circular aperture. They illustrate the angular separation that is necessary in order to make the duplicity of a double star source apparent. If the component separations are too small, the composite image appears like that of the single star. When the separation is great, the Airy disks and their encircling diffraction rings for the two images are completely separated. Intermediately, the stars are said to be just resolved at that separation where the composite image is judged first to manifest the duplicity of

FIG. 10-8 Circular aperture diffraction of stars—to illustrate resolving power. The top pattern is essentially the same as that of a single star.

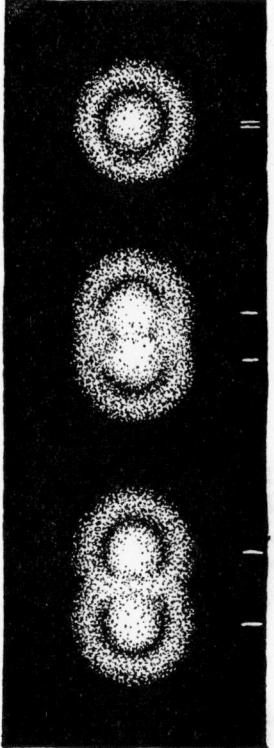

the source. This judgment occurs near a separation that puts the center of one disk over the first dark ring of the other. And the necessary separation is much the same for the resolution of two parallel equally strong fine line sources— they first appear double near where their angular separation is such that the center of one diffraction pattern overlies the first minimum of the other. Partly for simplicity, but mainly because it represents experience, Lord Rayleigh chose these simply expressed necessary angular separations as the criterion for resolvability. For a circular aperture this necessary angular point source separation is $\alpha_R = 1.22 \dfrac{\lambda}{w}$, where w is the aperture diameter, while the separation is $\alpha_R = \dfrac{\lambda}{w}$ for a rectangular aperture square-on, where w is the rectangular width.

Fig. 10-9 shows some calculated composite illuminations for superimposed square aperture patterns of two equal parallel line sources. Here, of course, it is the illumination of the components which we add to get the composite illumination, since such separated component line sources are incoherent.

In the case of light from two equal lines diffracted by a slit or rectangular

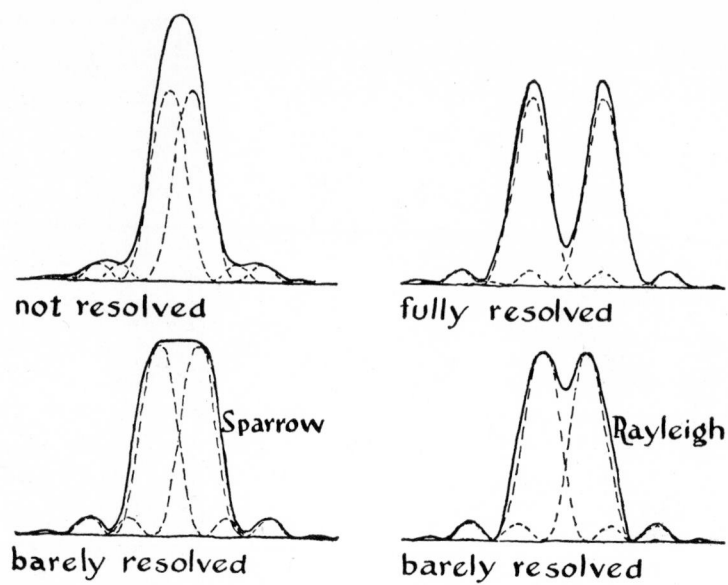

FIG. 10-9   Composites of equal diffraction patterns at various slit separations.

aperture, when they are separated according to this Rayleigh criterion, the composite illumination gives a pattern exhibiting a pair of maxima, of illumination strength $\mathfrak{E}_c$, with a central minimum between, of illumination strength

$$2\mathfrak{E}_c \frac{\sin^2 \dfrac{\pi}{2}}{\left(\dfrac{\pi}{2}\right)^2} \cong 0.8\mathfrak{E}_c.$$

A student exercise is concerned with Sparrow's criterion of resolution for slits or such rectangular apertures. This criterion for equal lines calls for the narrower separation, that which just removes the central minimum of illumination, as shown in Fig. 10-9.

10-5. *Spectroscopic Resolving Power*

Before we take up the treatment of spectral resolving power, as contrasted to angular resolving power, let us define the typical spectroscopic equipments to which our spectral resolving powers will apply. Fig. 10-10 shows a spectrograph, a spectroscope, and a spectrometer, all diagrammatically.

In the spectrograph, shown at a, the light to be analyzed passes through an entrance slit e. C is either a lens or mirror, represented here diagrammatically as a lens. C, called the collimator, makes the light from each point of the entrance slit (or the entrance spot of § 10-2) parallel. P is a prism or diffraction grating. This element causes the different wavelengths to be deviated in angle by different amounts. And finally T is a lens or mirror to focus the parallel beams of different wavelengths, propagating in different directions after penetrating P, onto the surface of a photographic plate. Here T and the photographic plate holder constitute a camera.

We observe the spectrum visually in a spectroscope such as is shown at b. We have a telescope now instead of the camera lens or mirror and plate holder above.

In a spectrometer, the eyepiece of the telescope is replaced by an exit slit (in its focal plane) as is shown at c. Behind this exit slit x a radiation thermopile, bolometer, photocell, or other photometric device measures the different wavelength radiations that the exit slit passes when the telescope is pointed in different directions.

In a monochromator, white light is passed through an entrance slit e, and light of substantially one wavelength emerges from the exit slit x. A double monochromator is arranged in the manner shown in Fig. 10-15b. It is comprised of two monochromators in series, with the exit slit of the first x_1 being

FIG. 10-10 Diagrammatic representation of spectrograph (a), spectroscope (b), and monochromator or spectrometer (c).

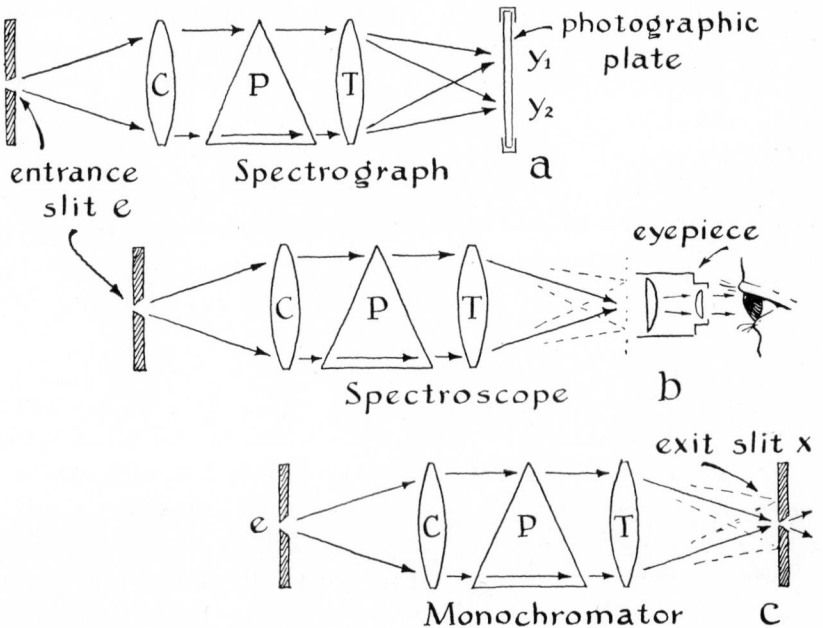

also the entrance slit of the second e_2. The three slits are usually designated thus: e for entrance slit, m for middle slit, and x for final exit slit.

In all of these instruments the spectral resolving power is defined by $\dfrac{\lambda}{\delta\lambda}$, where λ is an average wavelength and $\delta\lambda$ is the difference in wavelength between two adjacent equal spectrum lines that can just be discriminated. This discrimination is determined by the interplay of the angular size of the spectrum-line images with their angular separation. The image sizes which determine resolving power may arise from aberration or from a wide entrance slit setting, because the entrance slit is imaged through P by C and T. But here we take it that the image sizes are determined by $\delta\lambda_R$, due predominantly to diffraction. In any case, in order for two adjacent spectrum lines to be resolved, their angular separation, $\delta\alpha$, must exceed their equal angular widths, whatever those widths may be due to.

We calculate this needed separation $\delta\alpha_R$, when the equal angular widths are determined by diffraction alone. We determine the wavelength difference, $\delta\lambda_R$, corresponding to the necessary angular separation, from Rayleigh's criterion by means of the relationship $\delta\lambda_R = \delta\alpha_R \Big/ \left(\dfrac{d\alpha}{d\lambda}\right)$. Here $\left(\dfrac{d\alpha}{d\lambda}\right)$ is the angular dispersion of P—prism or grating.

Fig. 10-11 illustrates the calculation of the resolving power of a prism when it is worked at minimum deviation. Here we have a prism of apex angle 2ϵ. Its reflecting face has a length f, and its total base length is b. A light beam of wavelength λ, and width w as shown, is refracted symmetrically through the prism, and emerges after an angular deviation of 2δ. The relationships between δ, ϵ, and N, the index of the prism, if we take i and i', and r and r', for the external and internal angles of incidence and refraction, are

$$i' = r = \epsilon \qquad i = r' = \epsilon + \delta$$

and, from Snell's law,

$$N \sin \epsilon = \sin (\epsilon + \delta)$$

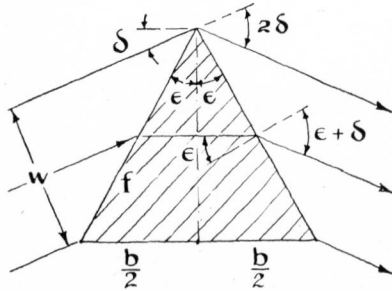

FIG. 10-11 Prism worked at minimum deviation.

We get $\left(\dfrac{d\alpha}{d\lambda}\right)$ from $\left(\dfrac{d\alpha}{dN}\right) \times \left(\dfrac{dN}{d\lambda}\right)$, as follows: First, differentiating Snell's law,

$$\frac{d\alpha}{dN} = \frac{d(2\delta)}{dN} = \frac{2 \sin \epsilon}{\cos (\epsilon + \delta)}$$

From the geometry of Fig. 10-11, $\sin \epsilon = \dfrac{b}{2f}$ and $\cos (\epsilon + \delta) = \dfrac{w}{f}$. Introducing $\dfrac{dN}{d\lambda}$ to describe the prism material, we get

$$\frac{d\alpha}{d\lambda} = \frac{d\alpha}{dN} \cdot \frac{dN}{d\lambda} = \frac{b}{w}\frac{dN}{d\lambda}$$

We use the Rayleigh separation, $\delta\alpha_R = \dfrac{\lambda}{w}$, to get $\delta\lambda_R$. This gives a resolving power:

$$\frac{\lambda}{\delta\lambda_R} = \frac{\lambda \left(\dfrac{d\alpha}{d\lambda}\right)}{\delta\alpha_R} = b \left(\frac{dN}{d\lambda}\right)$$

It is notable that this resolving power is independent of the prism's refracting angle, 2ϵ.

If P in Fig. 10-10 is a grating, normally illuminated, we may use our analysis for multiple slits of § 10-3 to get $\delta\lambda_R$. An equation of § 10-3 gave the light as concentrated in directions defined by $k\lambda = a \sin \alpha$, where k was an integer. In these directions the amplitude of the electric field was predicted to be $\Re X_s\{\tilde{Y}\tilde{Z}\}$; and the distribution of monochromatic light adjacent to these exact directions of concentration, as shown in Fig. 10-7, falls off exactly as in a diffraction pattern. To get the angular dispersion of a set of slits, we simply differentiate the grating equation. For normal incidence, differentiating $k\lambda = a \sin \alpha$ gives $\dfrac{d\alpha}{d\lambda} = \dfrac{k}{a \cos \alpha}$. The first minimum of the pattern of Fig. 10-7 corresponds to $k = k_1 \pm \dfrac{1}{\Re}$. From $\lambda\delta k = a \cos \alpha \, \delta\alpha$, setting $\delta k = \dfrac{1}{\Re}$, we get

$$\delta\alpha_R = \frac{\lambda}{\Re a \cos \alpha}$$

Translating this angle to $\delta\lambda_R$, we get the resolving power of the so-called transmission amplitude grating at normal incidence,

$$\frac{\lambda}{\delta\lambda_R} = \frac{\lambda \left(\dfrac{d\alpha}{d\lambda}\right)}{\delta\alpha_R} = \frac{\dfrac{\lambda k}{a \cos \alpha}}{\dfrac{\lambda}{\Re a \cos \alpha}} = \Re k$$

10-6. *Diffraction Gratings*

A set of multiple slits constitutes an *amplitude transmission grating*, but such a set is not very useful as a grating because it is impossible to make the elements—alternate openings and obscurations—fine enough. In diffraction gratings it is desirable to have up to 30,000 diffracting elements per inch.

Our treatment of § 10-3 may be applied directly to an *amplitude grating*, where the amplitude is modulated across the wave front. It also applies to the more common grating, the *phase grating*, where phase is modulated across the wave front, and where the amplitude is only slightly modulated, if at all. There are two general subtypes of each of these two types: *transmission gratings* and *reflection gratings*. Reflection gratings are almost always phase gratings; they are made by embossing equal parallel straight grooves on a smooth metal surface, by means of the sharp diamond point of a ruling engine. A replica of such an embossed surface on one side of a transparent plate can be used as a transmission phase grating.

Transmission phase gratings may be explained by consideration of the Huygens wave trainlets from our familiar set of $\mathfrak{N}$ multiple slits. Let the slit separations be a, as before, with the kth order of interference occurring at $\alpha = \sin^{-1}\dfrac{k\lambda}{a}$. We consider each one of the $\mathfrak{N}$ openings now divided into m equal sub-strips, which are in juxtaposition as is illustrated in Fig. 10-12. Furthermore, at first, let us consider that the width of each slit is equal to the separation a. Then the m complex numbers, representing the m Huygens wavelets from the m sub-strips, of which our separate slits of width a are each comprised, will add as vectors in the complex plane to form a complete circle (just as they do in the case of two adjacent half-period Fresnel zones in Fig. 9-2b). The vector resultant of a circle of m vectors is, of course, zero. Thus

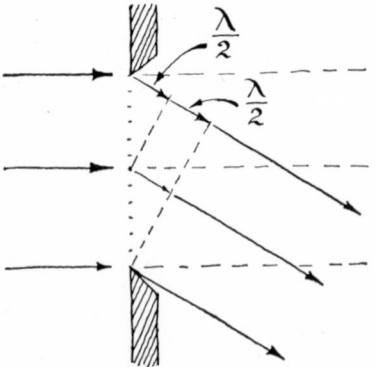

FIG. 10-12 Single-slit diffraction.

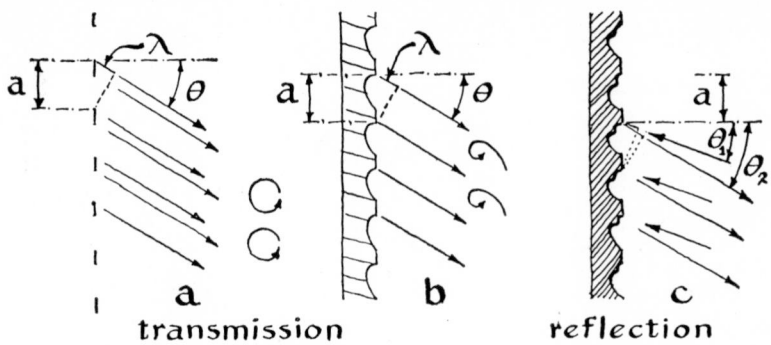

FIG. 10-13   Various diffraction gratings.

this particular set of multiple slits, each of width a, can only give an interference pattern of zero amplitude in the direction α. This amplitude is zero
because in fact we do not have a grating at all, but only a clear aperture of
width $\mathfrak{N}a$. This particular grating, which is really here only an open window,
becomes an ideal *amplitude grating* if the near half, say, of the m vectors for
each set of sub-strips is eliminated by obstructions. Then the vector resultant
of the remainder becomes the diameter of our aforementioned circle of m vectors. But even better, our imaginary grating becomes an ideal *phase grating*,
for the direction α, when we add a suitable tiny transparent prism to each slit
of width a. The prisms are suitable when they introduce a progressive phase
retardation $(N - 1)d$ which varies from zero, for the farthest of the m sub-
strip wavelets of each slit, to $(N - 1)d = k\lambda$, for those that are nearest. This
progressive phase retardation transforms our circle of m vectors, one vector
for each sub-strip, into a straight line of m vectors. Thus the length of the
vector resultant, which was first zero for the full circle and then the circle
diameter, becomes the circle circumference—the m vectors are now uncurled
and arranged in a straight line.

Fig. 10-13a contrasts the multiple-slit equivalent of the amplitude grating
with the phase grating, Fig. 10-14a. Of course any periodic disturbance of
phase, such as is shown in Fig. 10-13b, would in general prevent the array
of m vectors from closing, and giving zero resultant. Thus, almost any
periodic disturbance will produce a spectrum in the direction α. Fig. 10-13c
shows a reflection grating.

In Fig. 10-14 the transmission phase grating gives a strong spectrum in the
direction δ_g. The reflection grating shown in the figure at b has a mirror
facet of width a which diffracts the light incident on the whole face most
intensely in the direction α. This grating, also, is a phase grating. The gratings at b and c are called blazed gratings.

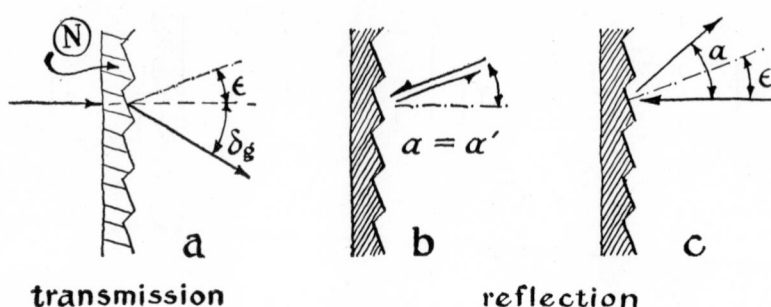

transmission   **reflection**

FIG. 10-14   Blazed gratings.

The prospective usefulness of such gratings was recognized by Lord Rayleigh in his 1888 *Encyclopaedia Britannica* article on "The Wave Theory of Light." In this article, Rayleigh suggested the possibility of the blazed gratings as follows:

"If it were possible to introduce at every part of the aperture [*a*] an arbitrary [phase] retardation, all the light might be concentrated in any desired spectrum [i.e., all the diffracted monochromatic radiation by our multiple apertures could be sent in the direction α]. By supposing the retardation to vary uniformly and continuously we fall upon the case of an ordinary prism; but there is then no diffraction spectrum in the usual sense. To obtain such it would be necessary that the retardation should gradually alter by a wavelength in passing over any element [*a*] of the grating, and then fall back to its previous value, thus springing suddenly over a wavelength. It is not likely that such a result will ever be fully attained in practice; but the case is worth stating, in order to show that there is no theoretical limit to the concentration of light of assigned wavelength in one spectrum."

In 1882 H. A. Rowland achieved the production of precision diffraction gratings, and, by 1910, R. W. Wood had advanced the art of making them, and realized what Rayleigh had allowed theoretically: reflection gratings, such as those of Fig. 14b and c, that put a substantial part (although not all) of the incident radiation (in this instance infrared radiation) in one particular order. The same result was later achieved for visible radiations, largely by the ingenuity, initiative, and skill of Dr. John A. Anderson. More recently, since the introduction of the use of aluminum films on glass grating blanks, the production of blazed gratings for visible and ultraviolet light has become routine.

Fig. 10-13a represents an amplitude grating, and b represents a phase grating, both in transmission. And Fig. 10-14 at a represents a blazed transmission grating. Fig. 10-13c represents a phase grating in reflection, with

Figs. 10-14b and c representing blazed reflection gratings, c being normally illuminated and b illuminated so that the incident and diffracted angles are equal; $\alpha' = \alpha$.

Each of the blazed gratings in Fig. 10-14 introduces the progressive retardation which Rayleigh's suggestion called for; that is, just that retardation necessary to uncurl the circle of complex vectors in the complex plane into a straight line.

The achievement of making precision diffraction gratings, first accomplished by Professor Rowland at Johns Hopkins University, exemplifies the highest manifestation of precision mechanism. The mechanical-minded student will find Dr. John A. Anderson's article in the *Dictionary of Applied Physics*, Professor H. A. Rowland's article on the screw in the *Encyclopaedia Britannica*, the articles by Dean George R. Harrison† in the *Journal of the Optical Society of America*, and the popular articles by A. G. Ingalls‡ in the *Scientific American*, all interesting.

10-7. *Images of Coherent Sources*

In our treatment of angular and spectral resolving power we assumed that the two sources that were to be resolved (white stars or spectrum lines) were incoherent sources. If these sources are coherent, then their superimposed images have a different appearance. The treatment of the appearance of images when sources are coherent is of importance for understanding the microscope. The problems of the imagery of coherent sources by a system of lenses will not be treated generally here; rather, we shall only discuss one particular example, the image of the diffraction pattern of a point source as it is produced by a relay lens. This example is particularly germane to the understanding of the resolving power as manifest by a double monochromator.

Fig. 10-15a shows a single monochromator (with two prisms), and Fig. 10-15b shows two such single monochromators (each with one prism) combined to form a double monochromator. The diffraction patterns of the entrance slit e, for monochromatic illumination, are shown diagrammatically. C_1 is the collimator lens of the single monochromator, which makes the light from each part of the entrance aperture, e_1, parallel. The components P_1 and P_2 are prisms, and T_1 is the telescope lens. This figure, being diagrammatic, does not show the deviation of the rays. e is considered to be illuminated by light of wavelength λ. The image by T_1 if it acts as a circular aperture of diameter w shows a diffraction pattern of width $1.22\,\dfrac{\lambda}{w}$. This diffraction pattern is illus-

---

† George R. Harrison, J. Opt. Soc. Am., *39*, 413 and 522 (1949).
‡ A. G. Ingalls, Scientific American, June 1952.

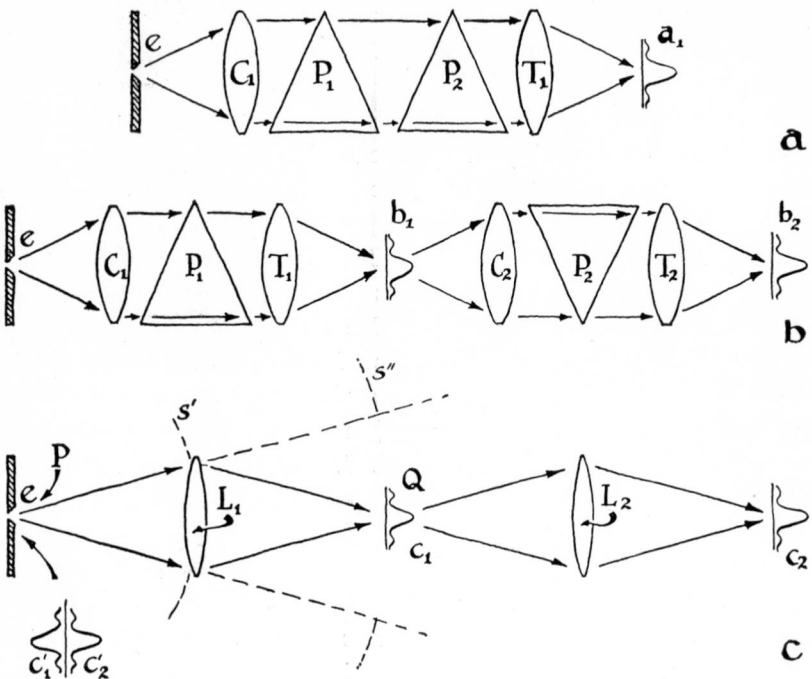

FIG. 10-15 Invariance of diffraction pattern to relaying by a lens.

---

trated diagrammatically at a_1. In the double monochromator C_2, P_2, and T_2 are identical, respectively, with C_1, P_1, and T_1. The resolving power of the single monochromator is determined by the diffraction pattern a_1 in combination with the dispersion of two prisms. The resolving power of the double monochromator is determined in the same manner. It is well known that the resolving power of a double monochromator, compared with one of its single monochromator components, is doubled due to the doubled prism dispersion; it follows therefore that the pattern b_2 must be the same as b_1, or a_1. It is our purpose now to show that this equality, $b_2 = b_1$ or a_1, is indeed expected. To demonstrate this expectation we invoke the simplified diagram of Fig. 10-15c. In this figure a lens L_1 replaces the ensemble of C_1, P_1, and T_1, giving the same pattern, c_1, as b_1. We expect the patterns c_1 and c_2 to be equal if b_1 and b_2 are equal. The demonstration† that this is a valid expectation hinges on applying the principle of microscopic reversibility, together with the principle of Babinet, of § 9-9.

In Fig. 10-15c the segment of the train of waves S' from e which L_1 inter-

---

† This demonstration was developed by Drs. W. M. Sinton and W. C. Davis when they were students.

cepts, is focused at Q. When the rays in this segment arrive at their focus, they form the pattern c_1. Now consider that these rays are reversed at c_1 and that the rays in the unfocused wave front S'' are also reversed at the same instant. The hole in this reversed wave front, S'', is of course fringed by Fresnel fringes, while at c_1 we have rays coming, in reverse, from an image of a point source that is surrounded by Fraunhofer fringes. After simultaneous reversal of all these rays, the principle of microscopic reversibility requires that the reversed rays, when they arrive again at L_1, will reform a complete spherical wave. And later, at e, the returned rays, all together, must converge onto their original source without any surrounding diffracted light; the whole spherical wave must give a field that is everywhere zero except at the point source e. But, applying Babinet's principle, the diffraction of the wave S'' alone, without the returned segment through the lens, would give the pattern of a circular obstruction $(-c_1')$; and, further, this pattern is the negative of the pattern of the returned segment through the circular aperture, c_2'. Therefore, since $c_1' - c_2' = 0$, the refocused returned waves alone c_2' give the same pattern as the circular aperture originally, c_1, except for sign. We therefore expect the pattern c_2, of c_1 imaged by the further lens L_2, to be the same as c_2'. Thus, $c_2 = c_1$. And analogously, in Fig. 10-15b, we should expect $b_2 = b_1$.

Fig. 8-10 shows the focal pattern, in the stellar interferometer, of two coherent diffraction images superimposed. This imagery is to be compared with the focal pattern of two incoherent diffraction images, shown at the top in Fig. 10-8.

Chapter XI

Coherent Sources Separated in Depth

ALTHOUGH Young's great interference experiment (1802) first gave us the wavelength of light, it did not afford the first interference bands that were observed. Those first bands were observed by Robert Boyle (1663) as concentric rings around the contact between a long-radius convex lens surface and a flat glass plate. Because Newton (1666 *et seq.*) studied these interference bands further, they are now known as Newton's rings.

The third quarter of the seventeenth century was a fertile time for optics. For example, Grimaldi, in 1665, discovered diffraction and described diffraction experiments. Studies of diffraction were extended by Newton, who called the inward diffraction by an obstacle *infraction*. Robert Hooke (1665) first suggested a wave concept of light, and Huygens (1678) gave some substance to his longitudinal wave-pulse theory. Neither of these wave theories was acceptable, mainly because of the unaccountable straight line propagation of light in free space, and particularly because the experiments with double refraction of light by a calcite crystal were not explained in terms of longitudinal waves. The concept of transverse waves came later, at an equally fertile time in optics—during the first quarter of the nineteenth century. But to return to Newton's rings—they afford an example of the interference of light from coherent sources that are not separated laterally, as is required to explain Young's experiment: they are separated in depth. Before we take up the wave theory explanations of Newton's rings we shall first endeavor to understand the more general case of the interference manifested by a thin dielectric plate such as is shown in Fig. 11-1.

222

11-1. *Thin Dielectric Plate*

Fig. 11-1 shows a light source S and a transparent plane parallel plate with two reflected beams, one reflected at its front surface and the other at its back surface. The index of this plate is N, and its thickness is d. The plate is immersed in an index of $N = 1.0$. It returns, of course, more than two reflected beams by multiple reflections. For example, a thrice internally reflected beam $\tilde{E}_3$ is shown by dots. But when the amplitude reflection coefficients at the plate interfaces are low, the first two beams, $\tilde{E}_1$ and $\tilde{E}_2$, shown solid in the figure, are the dominant beams. We shall therefore, for beginning simplicity, assume that the three times reflected beam and others are negligible. And furthermore, we shall assume that the incident and reflected beams are normal to the plate, and equal—our figure shows the beams at inclined incidence only for the purpose of distinguishing them. The two beams we consider here are not gotten by *wave-front division*, but rather by *amplitude division*. The amplitude of the parent beam $\tilde{E}_0$ is partly reflected at the first surface to give a reflected beam of amplitude $\tilde{E}_1 = r_{1\rho}\tilde{E}_0$; and the remainder of the parent beam is partly reflected at the second surface of the transparent plate to give a reflected beam of amplitude $\tilde{E}_2 = r_{\rho 1}(1 - r^2)\tilde{E}_0 e^{-j\varphi}$. Here $r_{1\rho} = -r_{\rho 1} = \dfrac{1 - N}{1 + N}$; and $\varphi = \dfrac{2Nd}{c}\,\omega$. The third ignored beam, as Problem 3-3 taught, is $\tilde{E}_3 = r^3{}_{\rho 1}(1 - r^2)\tilde{E}_0 e^{-2j\varphi}$. The reason we may ignore $\tilde{E}_3$ is that it is weak. Here, in the case $N = 1.5$, $\mathcal{E}_1 = -0.2\mathcal{E}_0$ and $\mathcal{E}_2 = +0.192\mathcal{E}_0$; while $\mathcal{E}_3 = 0.008\mathcal{E}_0$ is twenty-five-fold less. Later, in § 11-8, we shall take account of all the multiple internally reflected beams; but now we assume that $\mathcal{E}_1 = -\mathcal{E}_2$, making true two-beam interference. The discrepancy this assumption introduces in amplitudes is: We take $2\mathcal{E}_1 = 0.40$ for the maxima where $|\mathcal{E}_1| + |\mathcal{E}_2| = 0.392$. And, we take $|\mathcal{E}_1| - |\mathcal{E}_2| = 0$ rather than .008 for the minima. Thus, our assumed reflected beams, mathematically, are

$$\left.\begin{aligned}\tilde{E}_1 &= r\mathcal{E}_0 e^{j(\omega t + \pi)}\\[4pt]\tilde{E}_2 &= r\mathcal{E}_0 e^{j\left(\omega t - \omega\frac{2Nd}{c}\right)}\end{aligned}\right\} \text{ where } \tilde{E}_0 = \mathcal{E}_0 e^{j\omega t}$$

$$\tilde{E}_3 = \tilde{E}_4 = 0, \text{ etc.}$$

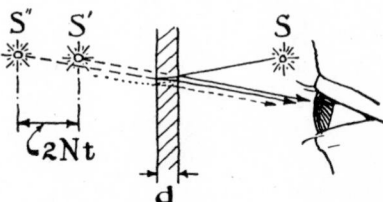

FIG. 11-1 Sources separated in depth because of reflection by the front and back surfaces of a plane parallel transparent plate.

The two reflected beams superimpose as beams from two coherent sources located at S' and S'', separated in depth by the optical distance $2Nd$. These sources are not synchronous at S' and S'' but oscillate out of phase. Their phase difference arises partly from the phase change of π for first surface external reflection, compared to no phase change for the second surface internal reflection. When the monochromaticity of our light is sufficient, reflected beams superimpose as in our formula, Eq. 8-3c, giving

$$\tilde{E} = r\mathcal{E}_0 e^{j\omega t}\left[e^{j\pi} + e^{-j\omega\frac{2Nd}{c}}\right]$$

$$= -2r\mathcal{E}_0 e^{j\omega t}e^{j\left(\frac{\pi}{2}-\omega\frac{Nd}{c}\right)}\sin\frac{2\pi N\acute{a}}{\lambda}$$

with the illumination

$$\mathfrak{E} = 2c\kappa_0 r^2\mathcal{E}_0{}^2\sin^2\left(\frac{2\pi Nd}{\lambda}\right) = c\kappa_0 r^2\mathcal{E}_0{}^2\left[1 - \cos\frac{4\pi Nd}{\lambda}\right]$$

A soap film affords a beautiful example of the above analysis. Such a film can easily be produced thin enough to show white light interference colors. As optical path differences, $\Delta r = 2Nd$, vary (because of variation of d), the white light interference colors of such a film recapitulate the sequence given for Lloyd's experiment, in Table 8-1. Fig. 11-2 shows Boys' spinning top by means of which such a sequence of film thicknesses, and colors, may be

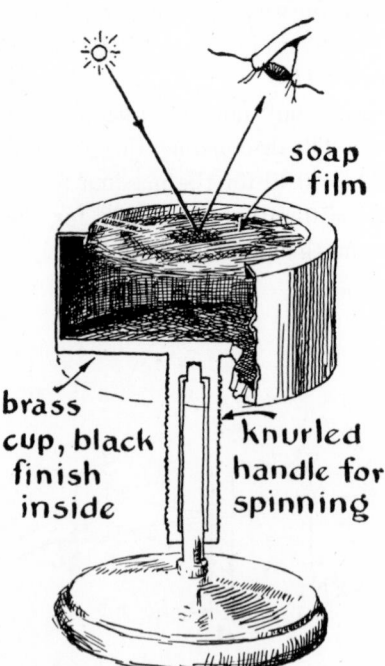

soap
film

brass
cup, black
finish
inside

knurled
handle for
spinning

FIG. 11-2 C. V. Boys' rainbow cup for spinning out thin soap films.

demonstrated. A soap film drawn across the top of Boys' device will spin
out thin as the top rapidly rotates, due to centrifugal force. Finally, the soap
film will become so thin, at the center of the film, that the two reflecting
surfaces of the film are effectively in optical contact, and the reflected illumination, $\mathfrak{E}$, vanishes. This black corresponds to the $\Delta r = 0$ black band of
Lloyd's experiment.

Similarly, the center of Newton's rings, where the lens surface and flat are
close enough to be in optical contact, exhibits a central black spot, called
Newton's black spot.

In contrast to this loss of reflectivity for white light when the film is excessively thin, color is lost when the soap film is thick, so that $k = \dfrac{2Nd}{\lambda} \geq 5$;
then the overall reflected light appears white to the naked eye. Although, in
this case, the two beams from the two virtual sources, separated in depth,
are still in superposition, the interference is lost to observation because the
incident light is insufficiently monochromatic. However, interference in these
superimposed beams can be recaptured by looking at the reflected light with
a pocket spectroscope. The spectrum of the reflected light is then observed to
be crossed by dark interference bands. Such a spectrum is called a *channeled
spectrum*, and the interference bands are called Edser-Butler bands. For example, the light reflected by a cellophane wrapper, which has the thickness
$d = 25\mu$, will manifest about 75 such dark interference bands across the visible
spectrum. Although the color sensitivity of the eye fails to see interference
colors (due to differing interference effects for different wavelengths), the more
powerful color discrimination obtained with the pocket spectroscope recap-
tures the differing interference effects for different wavelengths.

The overall illumination $\mathfrak{E}$ that is reflected by our thin plate, for a wavelength λ, depends on the order of interference between the two superimposed
amplitudes. In the special case of normal incidence, this order of interference
is $k = \dfrac{2Nd}{\lambda}$. However, when the angle of incidence is increased from zero to
the angle i, giving the angle of refraction r, the order of interference is reduced
to $\dfrac{2Nd}{\lambda} \cos r$. Fig. 11-3 illustrates the derivation of this relationship. We derive it from the optical path difference between rays reflected at the first
and second surfaces of a plate as shown in the figure. The optical path difference is readily seen to be

$$\Delta r = N(BC + CD) - BD'$$

and we proceed to show this is $\dfrac{2Nd}{\lambda} \cos r$. The factor N above changes the
geometrical path length within the plate, represented by the parenthesis, to

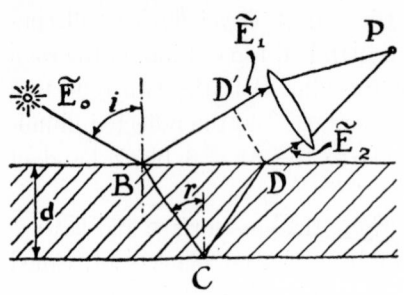

FIG. 11-3 Geometry of path difference on reflection at oblique incidence by front and back surfaces of a plane parallel plate.

an optical path length. We subtract BD' to get the optical path length difference. The elements in this expression for Δr, in terms of i, r and d, are

$$BC = CD = \frac{d}{\cos r}$$

$$BD' = (BD) \sin i = N(BD) \sin r$$

$$BD = \frac{2d}{\cos r} \sin r$$

The substitution of these elements in the expression for Δr gives

$$\Delta r = 2Nd \left\{ \frac{1}{\cos r} - \frac{\sin^2 r}{\cos r} \right\} = 2Nd \cos r$$

When this path difference is divided by λ, we get the order of interference; and further multiplication by 2π gives the relative phase retardation:

$$\Delta\varphi = \frac{4\pi Nd}{\lambda} \cos r = 2\pi k$$

The illumination produced by the superposition of two amplitudes with such a phase difference may be varied by three partial variations. These are: (1) variation with respect to the optical path difference, Nd; (2) variation with respect to the inclination angle, r; (3) variation with respect to λ. These mathematical partial variations of our expression for $\Delta\varphi$ represent three distinctive experimental manifestations of interference bands:

1. Fizeau interference bands, where $\Delta\varphi$ is dominated by a variation of Nd.
2. Haidinger interference bands, where $\Delta\varphi$ is dominated by a variation of r.
3. FECO bands, or the channeled spectrum, where $\Delta\varphi$ is dominated by a variation of λ.

The last designation, "FECO," was introduced by Professor S. Tolansky to mean *fringes of equal chromatic order*. Here we retain Professor Tolansky's designation, although generally we use the word "bands" instead of "fringes" for describing the variations of illumination, or color, produced by two-beam interference—reserving the word "fringes" to describe the variation of illumination around the *edge* of a silhouette, due to Fresnel diffraction, or around the edge of the image of a light source, due to Fraunhofer diffraction.

11-2. *Fizeau Bands*

If we cause monochromatic light to fall on a very thin dielectric plate at normal incidence, as illustrated in Fig. 11-4a, then the observed variations of reflected light over its surface, and especially the bands of maximum and minimum reflected illumination, are dominated by variations of Nd. A locus of constant illumination over the face of the plate is, of course, a locus of constant phase difference, or constant k, and, here, of constant Nd. These Fizeau bands are called "fringes of equal thickness" because the value of k is dominated by d and is insensitive to variation of i or of r. For example, a soap film of 1μ thickness will exhibit a change of k of 29% viewed at normal incidence when d changes to 1.29μ or 0.71μ. But at constant d, a corresponding change of k, due to a change of incidence, requires that r change from zero to 45°, or that i change from zero to 70°. It is because of this stronger dependence on d, and weaker dependence on i, that such superposition bands are called "fringes of equal thickness." The dominance of d is greatest when the illumination on the surfaces is nearly normal, when $\cos r$ is insensitive to changes of the angle r.

The longitudinal coherence necessary in the incident light to yield observable superposition, or to manifest interference bands, is greatest for the largest

---

FIG. 11-4   Method of observing Fizeau interference bands (a) and Newton's interference rings (b)—sometimes called fringes.

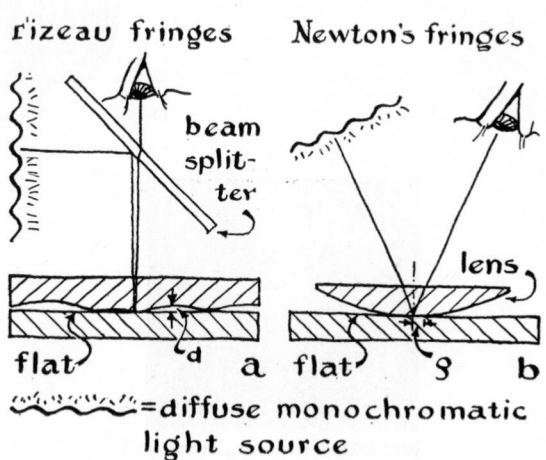

Fizeau fringes   Newton's fringes

beam split-ter

lens

flat   d   a   flat   s   b

=diffuse monochromatic light source

k's. For observers who are not color blind, white light interference may be detected for k's up to about $k = 5$. (See § 8-5.)

When monochromatic light is used to observe the Fizeau bands, they are interpreted in exactly the same manner as the contour lines of a topographical map. There is, however, this difference: whereas a height difference of adjacent lines on the map may be 6 meters, the height difference of adjacent Fizeau interference bands on a thin plate, is only 2.5×10^{-7} meters (for $\lambda = 5000$ Å). Newton's rings, observed as Fig. 11-4b illustrates, afford the classical manifestation of Fizeau bands. Fig. 11-5a shows their normal appearance with monochromatic illumination. The dielectric plate involved here is to be thought of as an air film ($N = 1.0$) immersed in glass of index N—this air film is the thin interspace between the convex lens surface and the flat plate that it contacts. A little algebra will show that the loci of the equal thicknesses, that give integral values of k and hence give dark rings, are circles of radii $\rho = \sqrt{k\lambda R}$, where R is the radius of curvature of the contacting convex glass surface.

When two inclined surfaces produce Fizeau bands, the bands are localized

---

FIG. 11-5 Newton's rings in reflection, with contacting glass surfaces unsilvered (a); Newton's rings in reflection (b) and transmission (c), with contacting glass surfaces silvered; FECO interference bands for the same contacting silvered surfaces (d). (Redrawn from Tolansky.) Monochromatic light—a, b, and c. White light below.

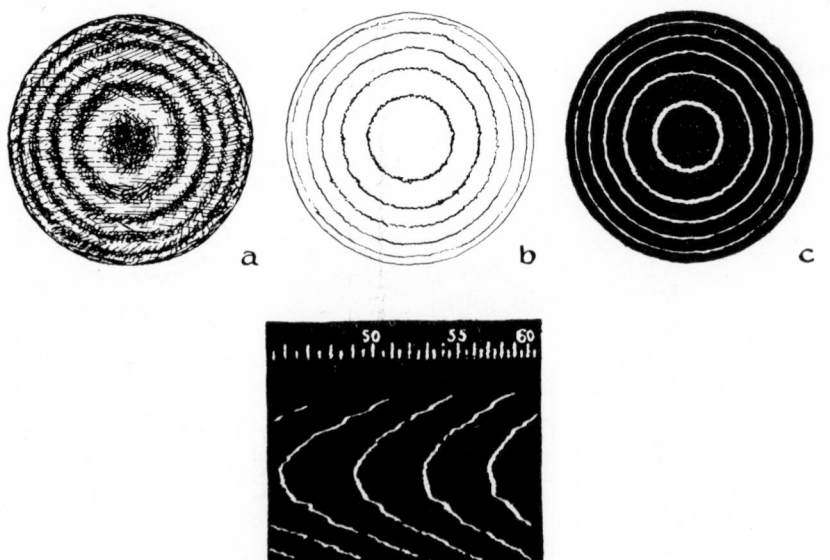

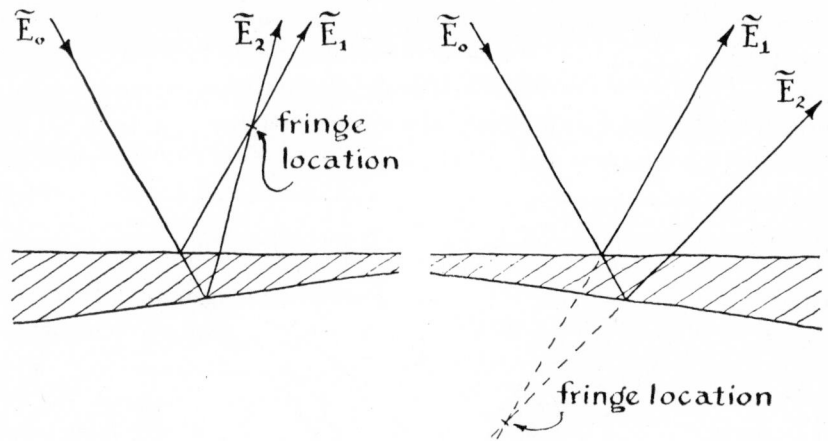

FIG. 11-6 Dependence of "fringe" location on wedge angle and incidence angle.

at the point where one daughter ray, the ray that was produced by amplitude division at the first surface, intersects the other daughter ray from the second surface, as illustrated in Fig. 11-6. This intersection point will lie above or below the reflecting surfaces, depending on the wedge angle, and on the angle of incidence. When the light is incident from the thick side of the wedge, the two reflected rays intersect above the wedge; when it is incident from the thin side, the daughter rays appear to intersect below the wedge. The interference bands have their maximum visibility when the eye focuses on these ray intersection points. Thus Newton's rings are localized above the flat on one side of the central contact, and below it on the other.

Fizeau interference bands have many practical uses. They are, for example, used for proving the quality of manufactured optical surfaces; a surface to be proved flat is laid over a transparent test flat, or it is covered by such a flat, and the separations of surfaces are determined by means of the displayed interference band contours. If these bands are viewed from a sufficiently great distance, and if they appear parallel, straight, and equally spaced, then the tested surface is flat. But if the fringes are curved, or unequally spaced, the tested surface is not flat. The tested shape can be easily deduced from such band contours and spacings. Angle observations may be used to determine if successive interference bands represent an increasing or decreasing d. If r is increased $\cos r$ will decrease and all bands will shift over the surface in such directions as will increase d, so that $\dfrac{2Nd}{\lambda} \cos r$ for the band remains constant.

Fizeau interference bands are used for proving curved lens surfaces in factory production. A convex or concave lens surface is proved by contact with a transparent concave or convex matching master surface, made to specifications. Here displayed bands instantly reveal work quality. (See Prob. 13-2.) Also, Fizeau bands and test flats find many uses in the machine shop in comparing work dimensions with gauge blocks, as illustrated by Fig. 11-7. These bands compare thickness and ball or pitch diameters, and measure taper.

FIG. 11-7 Uses of Fizeau interference bands in mensuration.

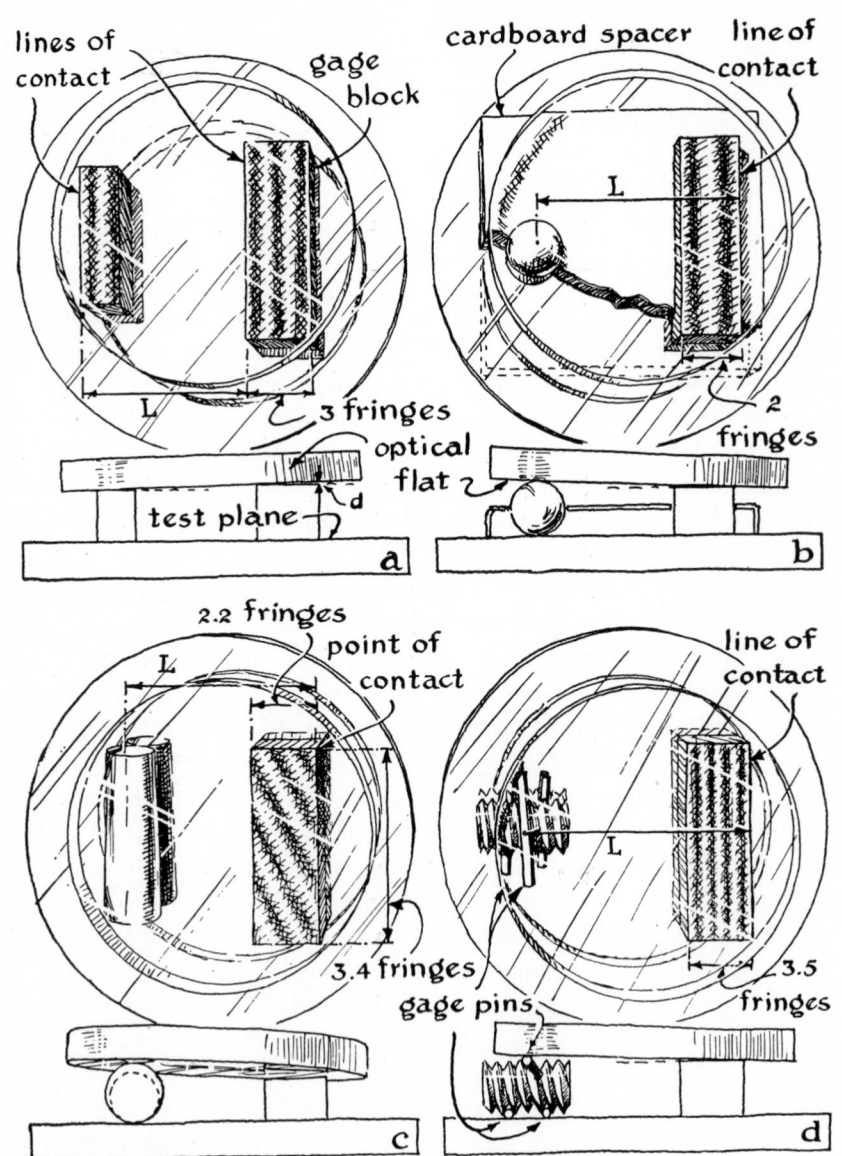

11-3. *Haidinger Bands*

Fig. 11-8 shows the experimental arrangement which manifests the interference bands of equal inclination that are produced by plane parallel reflecting surfaces. They are called "fringes of equal inclination." A monochromatic light, or source with great longitudinal coherence, is usually necessary to produce these bands, because the Nd of the plate makes k so great that the interference is not otherwise observable. Here i, which determines $\cos r$ in

$k = \dfrac{2Nd}{\lambda} \cos r$, is the dominant variable since N, d, and λ are all constants.

In the arrangement shown, the observing eye sees a spot centered around the normal of the parallel surfaces, with encircling rings of light and darkness.

The central spot is a dark one if $\dfrac{2Nd}{\lambda}$ is an integer. Then the surrounding

dark rings occur at angles that give $\dfrac{2Nd}{\lambda} \cos r$ integral values. The Haidinger

bands are not localized, but lie at infinity.

A pair of parallel reflecting surfaces that have been made highly reflecting by applied metal films constitutes a Fabry-Perot interferometer. The Fabry-Perot type bands that are produced when the surfaces are endowed with high reflectivity by deposited films of silver are described later in this chapter (§ 11-10). The characteristic features of the Fabry-Perot interferometer are described in the next chapter (§ 12-7).

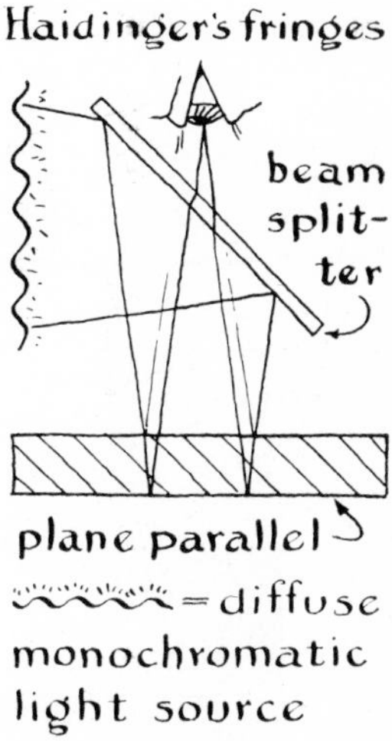

Haidinger's fringes

beam split-ter

plane parallel

$\wedge\wedge\wedge\wedge$ = diffuse monochromatic light source

FIG. 11-8   Method of observing Haidinger's interference bands.

11-4. *FECO Bands and the Channeled Spectrum*

Fig. 11-9 shows parallel light incident and transmitted by the two almost parallel silvered surfaces of a very thin dielectric plate. Here the slit of the spectroscope is focused on the plate, or on the interspace. The superposition sums, although they may not be manifest to the naked eye as interference if k is large, are displayed in the spectrum at the wavelengths λ that make $\dfrac{2Nd}{\lambda} \cos r$ an integer. Fig. 11-10 reproduces two FECO spectrograms from Tolansky's book.†

The transmission bands become very narrow when the surfaces are silvered; and the dark reflection bands also become narrow. Fig. 11-5a shows the two-beam Fizeau rings in reflection for unsilvered surfaces; while b and c show the reflection and transmission rings as they appear after silvering.

Newton's rings in transmission, with unsilvered glass, are invisible. The visibility of these bands in transmitted light is extremely low because of the inequality of the interfering beams. Table 11-1 gives values for the unequal dielectric-air reflected amplitudes $\mathcal{E}$, and the unequal transmitted amplitudes $\mathcal{E}'$. $\mathcal{E}_0$ is taken as unity. These values of $\mathcal{E}$ and $\mathcal{E}'$ are given for a parallel transparent plate of index 1.5; and for one of index 1.3. The first two transmitted amplitudes are $\mathcal{E}_1' = \mathcal{E}_0(1 - r^2)$ and $\mathcal{E}_2' = \mathcal{E}_0 r^2(1 - r^2)$, following the results

---

FIG. 11-9   Set-up for observing FECO bands. (What is wrong in this picture?)

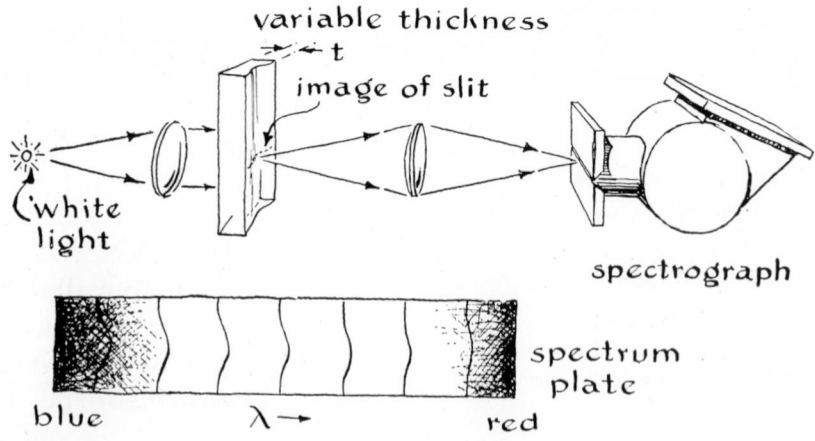

† Samuel Tolansky, *Multiple-beam Interferometry of Surfaces and Films* (Clarendon Press, Oxford).

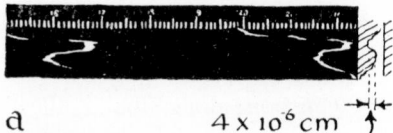

a

4×10^{-6} cm

FIG. 11-10 Actual FECO bands. (Redrawn from Tolansky.)

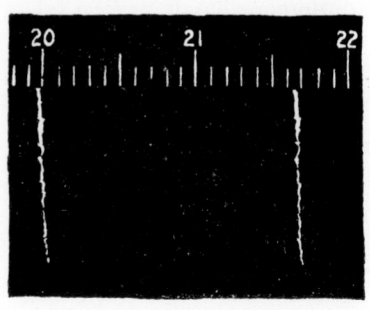

b

of Problem 3-3. Table 11-1 also gives the visibility of the interference bands, as calculated from the formula of § 8-5, for these unequal beams, both in reflection, $\mathcal{V}_R$, and in transmission, $\mathcal{V}_T$.

TABLE 11-1   *Visibility of Interference Bands*

Index	$\mathcal{E}$	$\mathcal{V}_R$	$\mathcal{E}'$	$\mathcal{V}_T$
	$-.200$		0.96	
1.50		.9992		0.08
	$+.192$		0.04	
	$-.142$		0.98	
1.33		.9998		0.04
	$+.138$		0.02	

Fig. 11-5d shows a photograph of the FECO transmission bands obtained with the same Newton lens-plate combination, but with the contacting surfaces partially silvered. The scale and numbers at the top of this spectrum yield wavelengths. Here the incident illumination is parallel white light; and the center of the spectrograph slit is focused on the contact point of the lens. If we have the same phase shifts on reflection at both silvered surfaces, the overall reflection is a maximum, and transmission a minimum when $\dfrac{2Nd}{\lambda} \cos r$ is a half integer. The wavelength along any one of the curved spectrum bands gives the variation of λ along the diameter of the interspace between lens surface and plate, where the slit is focused. The variation of λ along such a band is that necessary to keep the order of interference constant. Thus knowledge of the variation of λ in $\dfrac{2Nd}{\lambda} \cos r = $ constant, along a band, allows inference of the variation of d along the diameter of the Newton ring pattern. The explanations of why silvering narrows interference bands will be taken up later.

Returning to the channel spectrum produced on reflection from an unsilvered transparent parallel plate, the overall light reflected by the parallel surfaces may be used to determine the inter-surface index of refraction, N, and also the inter-surface separation, d. When we know the index of the interference-producing plate, and when we may ignore changes of its N with λ, because its thickness, d, is not large, then we may determine d simply by counting the number of dark bands displayed by the spectrum between two wavelengths, say λ_1 and λ_2. If this counted number is Δk, then

$$\Delta k = \Delta \left(\frac{2Nd}{\lambda} \cos r \right) = 2Nd \cos r \left(\frac{1}{\lambda_2} - \frac{1}{\lambda_1} \right)$$

This expression gives

$$d = \frac{\Delta k \lambda_1 \lambda_2}{2N(\lambda_1 - \lambda_2) \cos r}$$

Taking the limits of the spectrum as $\lambda_1 = 7000$ Å and $\lambda_2 = 4100$ Å, for $N = 1.5$, we get $d = \dfrac{\Delta k}{3}$. It is from this formula that we previously expected 75 dark bands in the reflected light from a cellophane wrapping of 25μ, or .001" thickness. This method of determining d from the number of channels in the spectrum is useful for measuring plastic, glass, or mica films which are too thick, on the one hand, to be estimated from their produced interference colors; and when they are too thin, on the other hand, for easy measurement with micrometer calipers—say films of thickness from 2 to 20μ.

We may determine N from the channel spectrum if we observe how a particular band changes wavelength as i changes. By measurement we find that the band occurs at the wavelengths λ_1 and λ_2 for the angles of incidence i_1 and i_2. The corresponding values of $\cos r_1$ and $\cos r_2$ are given by

$$\cos r = \sqrt{1 - \sin^2 r} = \frac{\sqrt{N^2 - \sin^2 i}}{N}$$

Now, insofar as we can neglect changes of N with λ, $\dfrac{\cos r}{\lambda} = \dfrac{k}{2Nd}$ is a constant, and we may equate $\dfrac{\cos r_1}{\lambda_1} = \dfrac{\cos r_2}{\lambda_2}$, and solve:

$$N = \sqrt{\frac{\lambda_2^2 \sin^2 i_1 - \lambda_1^2 \sin^2 i_2}{\lambda_1^2 - \lambda_2^2}}$$

Again, we can determine Nd from $Nd = \dfrac{k\lambda}{2\cos r}$ if we know k; and we can determine k from the wavelengths of adjacent dark bands, λ_k, and the wavelength of the adjacent band on the blue side, λ_{k+1}. Insofar as $2Nd\cos r$ may be taken constant,

$$k\lambda_k = (k+1)\lambda_{k+1} \qquad \text{giving} \qquad k = \frac{\lambda_{k+1}}{\lambda_k - \lambda_{k+1}}$$

We shall return to various aspects of these Fizeau, FECO, and Haidinger interference phenomena later. But now it is appropriate to treat Michelson's interferometer, which exhibits the Fizeau and FECO interference bands (on a strictly two-beam interference basis).

11-5. *Michelson Interferometer for Monochromatic Light*

Fig. 11-11 shows Michelson's interferometer. Fig. 11-12a shows its use with monochromatic light; and Fig. 11-12b shows its use with a compensator plate for either monochromatic or white light. In either case, the central feature of the interferometer is a plane parallel glass plate with one surface coated with a beam-splitting silver film to divide light incident on it at 45° into two equal beams, one transmitted and one reflected. The remainder of the optical system consists of two fully silvered flat plates, M_1 and M_2, as shown. We

FIG. 11-11   Michelson interferometer.

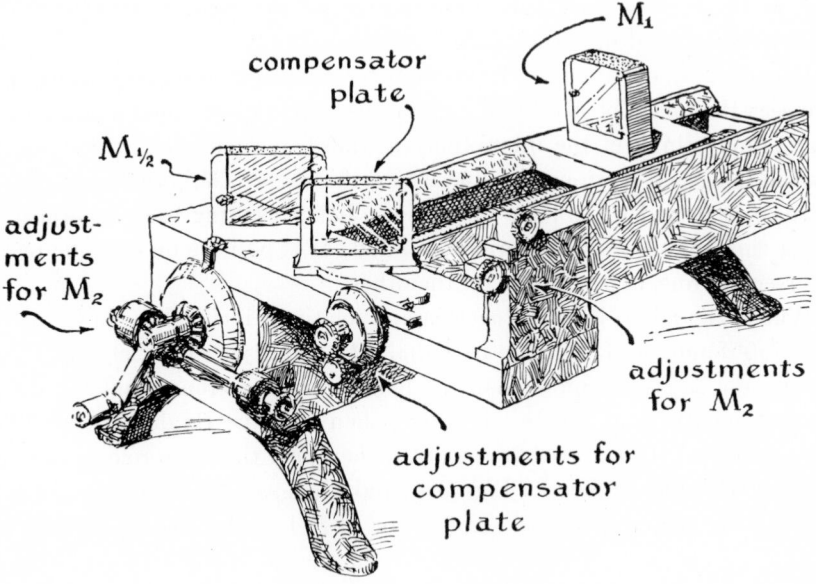

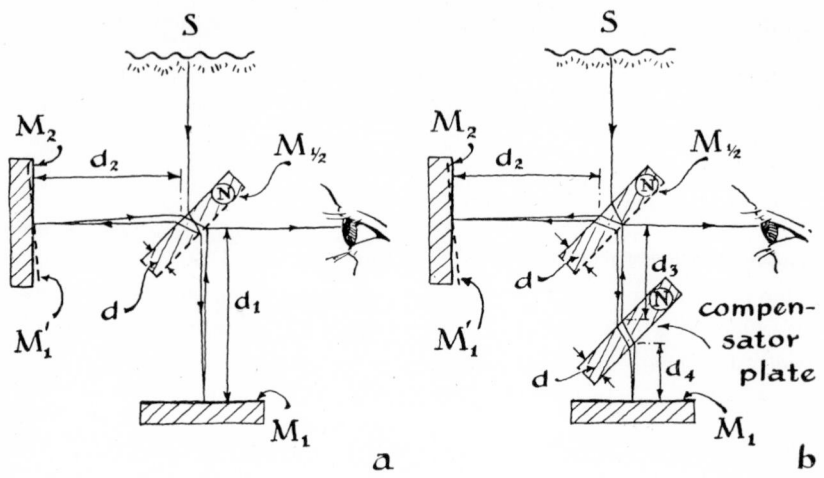

FIG. 11-12 Michelson interferometer without compensator plate, for mono-chromatic light (a), and with compensator, for white light (b).

use an extended light source such as may be realized by back lighting on a diffusion screen. The equally strong beams produced by the beam-divider mirror, $M_{1/2}$, that are returned by M_1 and M_2, are respectively reflected and transmitted to the observing eye by $M_{1/2}$. These equal amplitudes superimpose to give the observed illumination. If the phases are such as to produce con-structive interference, the interferometer returns no light toward the source; and in the case of destructive interference all the light is returned to the source or absorbed in the $M_{1/2}$ film.

In order to understand this interferometer easily, we note that the observing eye sees the image of M_1, by $M_{1/2}$, lying near M_2 or even intersecting it. We may simply think of the interference as due to beams reflected by M_2 and by the image, M_1'. Along the line where this image, M_1', intersects M_2, we have $\Delta r = 0$, and we call the interference band resulting from superposition along this line the $k = 0$ interference band. Δr is positive on one side of this intersection line, and negative on the other.

When M_1' and M_2 lie close together and relatively inclined, we have Fizeau-type interferences. If, on the other hand, M_1' and M_2 are parallel, and if their separation, Δr, is not more than the longitudinal coherence that the diffuse monochromatic light source will allow, then we have Haidinger-type inter-ference. When the surfaces are inclined, they play the same role as the non-parallel reflecting surfaces of Fig. 11-4; and when parallel, they play the same roles as the arrangements of Fig. 11-1 or 11-8. In either case the image and mirror, M_1' and M_2, give virtual sources separated in depth, just like S' and S'' of Fig. 11-1.

In Fig. 11-12a the path difference for rays reflected from M_1' and M_2 when they are parallel is

$$\Delta r = 2nd_1 - (2nd_2 + 2Nd/\cos r)$$

Here n is the index of air, and N that of the beam-splitter support plate. The distances d_1 and d_2 are indicated in the figure, while d is the support plate thickness. r is $\sin^{-1} \dfrac{1}{\sqrt{2N}}$ for $i = 45°$ on the support plate. If there is a differential phase jump $\Delta\varphi$ due to internal reflection at $M_{1/2}$ for one beam and external reflection for the other, then the two-beam superposition at the observing eye produces a resultant amplitude proportional to $\cos \dfrac{1}{2} \left(\dfrac{\omega \Delta r}{c} + \Delta\varphi \right)$.

11-6. *Michelson Interferometer for White Light*

The compensator plate differentiates Fig. 11-12a from 11-12b. Its function is not far to seek. In the expression above, applying to Fig. 11-12a, Δr cannot be made zero except at one wavelength. This is because of the dispersion of the $M_{1/2}$ support plate, making Δr vary with wavelength. For Fig. 11-12b, however, with the compensator plate, which is identical with the support plate except for its metal $M_{1/2}$ coating, the corresponding expression for the optical path difference is

$$\Delta r = \{2n(d_3 + d_4) + 2Nd/\cos r\} - \{2nd_2 + 2Nd/\cos r\}$$

Here, also, the meanings of the various d's are indicated in Fig. 11-12b; and n is the index for air. With the compensator plate the effects of dispersion in the two wavelength dependent terms, $2Nd/\cos r$, cancel. Thus, if $d_2 - (d_3 + d_4) = 0$, we have $\Delta r = 0$ for all wavelengths. When M_1' intersects M_2 at a small angle, and the differential phase jump at reflection is $\Delta\varphi = \pi$, the manifest interference bands will appear exactly as they do in Lloyd's experiment, with the central band, which we shall call the $k = 0$ band, black.

In their tipped position, if M_1' is moved forward or backward, the $k = 0$ white light or monochromatic light band will move laterally across the face of M_2. The unique appearance of the $k = 0$ white light band may be used to ascribe the proper zero order to a monochromatic band when light sources are interchanged. Otherwise one monochromatic band looks too much like another to ascribe the appropriate k's.

11-7. *Michelson's Measurement of the Meter*

Michelson used this interferometer to measure the meter in terms of wavelengths of the monochromatic 6438 Å cadmium spectrum line. The basic operation involved comparing the meter with etalon mirrors which themselves were separated a known number of wavelengths. In his interferometer, Michelson used an etalon of the type shown in Fig. 11-13a and b and he moved it so that M_1' covered first one mirror of the etalon and then the other. He used the $k = 0$ white light band for identifying the $k = 0$ monochromatic light band at the coincidences of M_1' with the etalon mirrors; and he counted the number of monochromatic bands that crossed a fiducial mark on the first mirror as M_1 moved from one coincidence, with the first etalon mirror, to the other coincidence, with the second mirror. After certain intercomparisons of etalons he compared the separation of mirrors of one of the etalons with the meter. Thus Michelson determined the length of the international prototype meter in terms of wavelengths of the 6438 Å cadmium line, and thus made its length available to everyone, for all time.

The particular cadmium line which Michelson chose is indeed remarkable; although cadmium lines have hyperfine structure, which makes them broad and thus degrades the visibility of their interference bands at large path differences, this one line is unique in the spectrum of cadmium in being free of hyperfine structure. Its wavelength width, compared to its wavelength, is only one or two parts in a hundred million. Until recently no other line approached its utility as the primary standard of wavelength or of length. Recently,

---

FIG. 11-13 First and last etalons used by Michelson for measuring the meter.

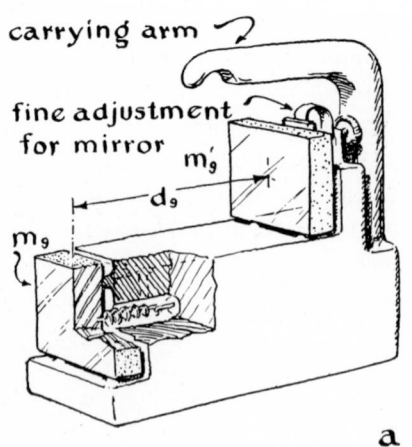

carrying arm

fine adjustment
for mirror m_9'

d_9

m_9

a

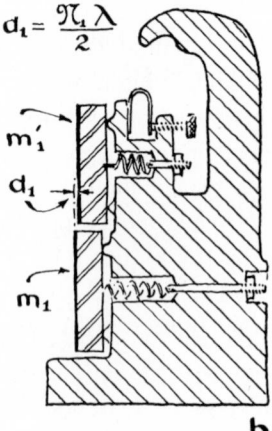

$d_1 = \dfrac{\mathfrak{N}_1 \lambda}{2}$

m_1'

d_1

m_1

b

FIG. 11-14 Appearance of interference bands on Michelson's first etalon mirrors at beginning and end of his interference band count.

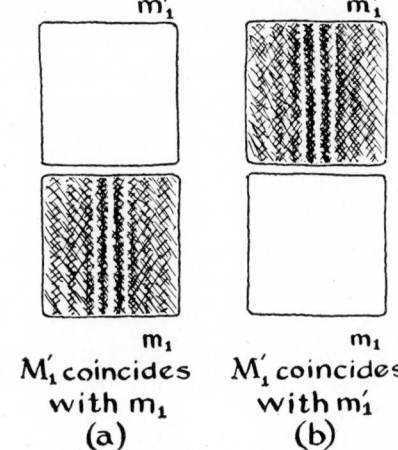

M_1' coincides with m_1

(a)

M_1' coincides with m_1'

(b)

however, the monoisotopic mercury line (Hg[198]) and a monoisotopic krypton line are competitors (with Cd 6438 Å as the current official standard—although a final international selection has not yet been made).

Any great experimental work takes full advantage of all available circumstances that can give it significance; and in the present case the ingenuity of Michelson's procedures certainly made his work great. His procedures almost automatically avoided systematic and accidental errors—making his final count of the wavelengths in the meter correct to within a half wavelength, or one interference band. The main errors which inhered in his count were those in transferring mirror positions to the terminal meter bar engravings. An outline of Michelson's counting procedures follows: First, he counted fringes only across an etalon of 0.4 mm spacing—the etalon of Fig. 11-13b. Next he repetitively doubled this spacing 8 times, in 8 other etalons, until his count, now indirect, corresponded to the 10 cm etalon spacing—the etalon of Fig. 11-13a. Finally he determined the number of times this 10 cm spacing would go into the meter. This determination, taking account of residuals, gave

1 meter = 1,553,163.5 wavelengths of the cadmium light

He used the $k = 0$ white light band to identify the $k = 0$ cadmium 6438 Å interference band in making his optical coincidences. The procedures used were such that an accumulation of error, when it became a substantial part of an interference band, was automatically noticed and compensated.

Michelson counted 1212.35 bands of the 6438 Å light across his first etalon, shown in Fig. 11-13b, as M_1' was shifted from the coincidence with m_1 of that etalon to coincidence with m_1'. For this count the etalon took the place of M_2 in his interferometer. The appearances of defining white light bands at the beginning and end coincidences of this 1212.35 red band count are illustrated in Fig. 11-14a and b. When the $k = 0$ white bands showed on m_1, they were of course invisible on m_1'; but the red bands were visible on both mirrors since the monochromaticity of the 6438 Å line was abundantly sufficient to give good visibility of bands over a much larger spacing than the $\Delta r = 0.4$ mm involved here. Fig. 8-7 shows the visibility of the 6438 Å bands as a function of Δr. The monochromaticity of this line, however, is not sufficient to reach over an optical path difference $\Delta r = 1$ meter. Even if Michelson had had a spectrum line of sufficient monochromaticity to span such a great spacing, and could have attached a microscope to the carriage of M_1, focused its cross hairs on one set of the terminal engraving of the meter bar, and counted bands until he had reached the other terminal engraving on the meter, such a count would have been very tedious. Such a count, of over 3 million bands, counting at the rate of one band a second, 8 hours per day, 6 days per week, would have taken 18 weeks. In contrast with this tedious task, Michelson counted the actual bands over only the 0.4 mm of his etalon #1. He then compared twice this with etalon #2; twice #2 with #3; and so forth until he knew the count for etalon #9, which had a separation of its mirrors, m_9 and m_9', of about 10 cm. Finally, he compared 10 times this etalon with the meter.

In doing the count with the first etalon in the place of M_2, the micrometer carriage carrying M_1 was moved to bring M_1' into coincidence with m_1 for the $k = 0$ red band; then Michelson counted the bands crossing the face of m_1 until the $k = 0$ red band arrived at m_1'. This count was $\mathfrak{N} = 1212$ bands, plus a residual of $n_1 = 0.35$ band. After adjusting the separation of mirrors of etalon #2 to almost exactly twice the separation of those of #1, Michelson determined the exact count for #2 to the fraction of a fringe as follows: He put m_1 and m_2 coplanar in the M_2 arm of the interferometer by means of M_1', using the $k = 0$ white band. He then brought M_1' to coincidence with m_1', and moved etalon #1 until m_1 was where m_1' had been, using the $k = 0$ white band. And finally he determined the number of red bands residual between the second position of m_1' and m_2'. This residual would, of course, have been zero if etalon #2 had been exactly twice #1. Here we do not elaborate on details, such as the need for two fiducial marks, and the need for corrections for the M_1' tip, etc.—these details, of little pedagogic interest, are of course important practically.

Fig. 11-15 shows how the count for etalon #5 is derived from that of #4, and how the count for #6 is derived from #5. Both illustrate how the count of #2 was derived from #1. In Fig. 11-15, M_1' first serves to make m_4 and m_5 coincident. Then it serves to mark the position of m_4' until m_4 can be moved to that position. Finally M_1' serves to measure the residual number of red bands necessary to shift from coincidence with m_4' to m_5', this being the re-

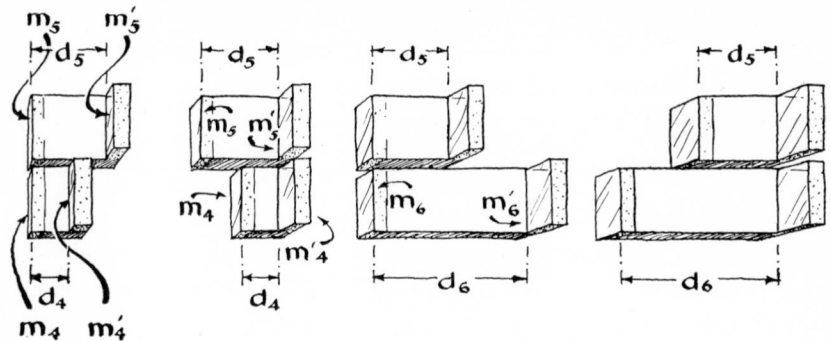

FIG. 11-15   Management of etalons in the doubling of a band count.

sidual $n_{4,5}$. This residual would be zero if etalon #5 were exactly twice #4. Thus the count for etalon #5 is

$$\mathfrak{N}_5 = 2\mathfrak{N}_4 + n_{4,5}$$

In this notation, the previous counts were

$$\mathfrak{N}_4 = 2\mathfrak{N}_3 + n_{3,4}$$
$$\mathfrak{N}_3 = 2(\mathfrak{N}_2 + n_2) + n_{2,3} = 2^2\mathfrak{N}_1 + 2n_2 + n_{2,3}$$
$$\mathfrak{N}_2 = 2(\mathfrak{N}_1 + n_1) + n_{1,2} = \mathfrak{N}_2 + n_2$$

This notation, extended, gives

$$\mathfrak{N}_9 = 2^8\mathfrak{N}_1 + n_9$$

After having thus indirectly counted the interference bands in his etalon #9, which had the nominal mirror separation of 0.4×2^8, or about 100 mm, Michelson attached a microscope to M_1, and by means of the etalon #9 in the other arm of the interferometer he repetitively advanced the microscope a distance corresponding to $\mathfrak{N}_9$ bands, ten times. The starting position was defined by both an interferometric coincidence of m_9 and M_1', as well as by a microscopic coincidence of the cross hairs with an image of one set of the meter bar engravings. After these ten advances of M_1, and its attached microscope, the residual number of red bands, n_{10}, were counted as M_1' moved from interferometric coincidence with m_9' to a position giving microscopic coincidence between the cross hairs and an image of the terminal meter bar engravings. Thus, in wavelengths of 6438 Å, the meter was found to be

$$\tfrac{1}{2}\{10(2^8\mathfrak{N}_1 + n_9) + n_{10}\} = 1{,}553{,}163.5$$

It might be supposed that the many residuals which enter here, n_1 to n_{10}, would have involved a manifold multiplication of the errors that inhere in determining them, but this supposition of large accumulated errors is wrong; the character of all the interferometric coincidences was such that the accumulated errors, when they became a substantial part of a band, or of half a wavelength, were noticed and compensated. Thus Michelson's final count was good to $\frac{\lambda}{2}$, or less. The main errors were those in the microscopic settings on the terminal engravings.

As an analogy to Michelson's procedures, imagine a short red machine screw with a thread near each end painted white, and a series of screws also with white threads at approximately 2^n greater separation, the $n = 8$ separation being nominally 256 times greater than that of the short $n = 1$ screw. If we count the threads and the residual fraction of a thread in the short screw, and proceed to lay it off twice in the 2^1 screw (using red threads extending beyond the white ones on our screw if necessary) and so forth, we can easily visualize how accumulating error will become noticeable when it is comparable in amount to one screw thread.

11-8. *Multiple Sources Separated in Depth*

When the r's for two parallel surfaces of a plate such as the one illustrated in Fig. 11-1 are not small but rather approach unity, as when the surfaces are partially silvered, then the beams we have previously ignored because r was small, $\tilde{E}_3, \tilde{E}_4 \cdots$ *etc.*, the so-called multiply internally reflected beams, as in § 11-1, now take on a dominating importance. The profound changes these multiple beams produce in observed interference bands, as illustrated in Fig. 11-5, are analyzed below.

Fig. 11-16 shows a plate of thickness d and index N_2. It is bounded on the side of incidence of the light by a medium of index N_1 and on the other side by a medium of index N_3. All three dielectrics here are assumed to be transparent. We represent the incident light as plane parallel electric wave fronts, polarized in the π or σ planes, and described at some point on the 1,2 interface by $\tilde{E}_0$. This light is taken to be incident at the angle i. The phase retardation

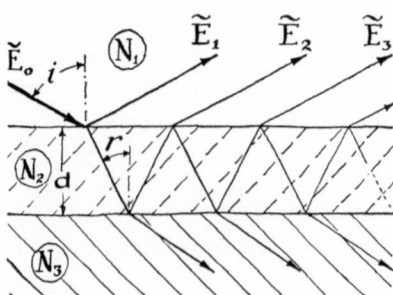

FIG. 11-16 Geometry, indices, and beams involved in deduction of overall reflection and transmission.

of $\tilde{E}_2$, reflected from the second surface of our plate, relative to $\tilde{E}_1$, will be $\Delta\varphi = \dfrac{4\pi N_2 d}{\lambda}\cos r$; for $\tilde{E}_3$ the retardation will be $2\Delta\varphi$; etc. These incident, reflected, and multiply reflected beams, illustrated in Fig. 11-16, may be described by the following complex expressions:

$$\tilde{E}_0 = \mathcal{E}_0 e^{j\omega t}$$

$$\tilde{E}_1 = \mathfrak{r}_{1,2}\tilde{E}_0$$

$$\tilde{E}_2 = \mathfrak{r}_{2,3}(1 - \mathfrak{r}_{1,2}{}^2)\tilde{E}_0 e^{-j\Delta\varphi}$$

$$\tilde{E}_3 = \mathfrak{r}_{2,3}(1 - \mathfrak{r}_{1,2}{}^2)(\mathfrak{r}_{2,3}\mathfrak{r}_{2,1})\tilde{E}_0 e^{-2j\Delta\varphi}$$

$$\tilde{E}_4 = \mathfrak{r}_{2,3}(1 - \mathfrak{r}_{1,2}{}^2)(\mathfrak{r}_{2,3}\mathfrak{r}_{2,1})^2\tilde{E}_0 e^{-3j\Delta\varphi}$$

$$\cdot \qquad \cdot \qquad\qquad\qquad \cdot$$
$$\cdot \qquad \cdot \qquad\qquad\qquad \cdot$$
$$\cdot \qquad \cdot \qquad\qquad\qquad \cdot$$

$$\tilde{E}_m = \mathfrak{r}_{2,3}(1 - \mathfrak{r}_{1,2}{}^2)(\mathfrak{r}_{2,3}\mathfrak{r}_{2,1})^{m-2}\tilde{E}_0 e^{-(m-1)j\Delta\varphi}$$

Here the $\mathfrak{r}$ with numerical subscripts represents either $\mathfrak{r}_\pi$ or $\mathfrak{r}_\sigma$ at the indicated boundary; and from Stokes' theorem $\mathfrak{r}_{1,2} = -\mathfrak{r}_{2,1}$.

To get the overall reflected amplitude, $\tilde{E}$, we must sum all the separate reflected amplitudes above, $\tilde{E}_m$. We may either add separate representative vectors graphically in the complex plane or add the complex numbers themselves algebraically. By the latter method of superposition, we can reduce the sum, in the present case, to a simplified closed algebraic form. This sum is the following series:

$$\tilde{E} = \sum_{m=1}^{\infty}\tilde{E}_m = \tilde{E}_0\{\mathfrak{r}_{1,2} + \mathfrak{r}_{2,3}(1 - \mathfrak{r}_2{}^2)[e^{-j\Delta\varphi} + \mathfrak{r}_{2,3}\mathfrak{r}_{2,1}e^{-2j\Delta\varphi}\cdots]\}$$

Polynomial division shows that the value of the series in brackets above is equivalent to

$$\left[\frac{e^{-j\Delta\varphi}}{1 - \mathfrak{r}_{2,3}\mathfrak{r}_{2,1}e^{-j\Delta\varphi}}\right]$$

Substituting the latter expression in brackets, and solving for the complex overall amplitude reflection coefficient, $\mathfrak{r}$, we get

$$\tilde{\mathfrak{r}} = \frac{\tilde{E}}{\tilde{E}_0} = \mathfrak{r}_{1,2} + \mathfrak{r}_{2,3}(1 - \mathfrak{r}_{1,2}{}^2)\left[\frac{e^{-j\Delta\varphi}}{1 - \mathfrak{r}_{2,3}\mathfrak{r}_{2,1}e^{-j\Delta\varphi}}\right]$$

$$= \frac{\mathfrak{r}_{1,2} + \mathfrak{r}_{2,3}e^{-j\Delta\varphi}}{1 - \mathfrak{r}_{2,3}\mathfrak{r}_{2,1}e^{-j\Delta\varphi}} = \mathfrak{r}e^{j\psi} \tag{a}$$

This expression for $\tilde{\mathfrak{r}}$ is real when $\Delta\varphi$ is an integral number of π's. When $\Delta\varphi$ is an even number of π's, then $e^{-j\Delta\varphi} = +1$, and

$$\mathfrak{r}_{\text{even}} = \frac{\mathfrak{r}_{1,2} + \mathfrak{r}_{2,3}}{1 - \mathfrak{r}_{2,3}\mathfrak{r}_{2,1}} = \frac{\mathfrak{r}_{1,2} + \mathfrak{r}_{2,3}}{1 + \mathfrak{r}_{1,2}\mathfrak{r}_{2,3}} \tag{b}$$

And, similarly, when $\Delta\varphi$ is an odd number of π's, $e^{-j\Delta\varphi} = -1$, and

$$r_{odd} = \frac{r_{1,2} - r_{2,3}}{1 + r_{2,3}r_{2,1}} = \frac{r_{1,2} - r_{2,3}}{1 - r_{1,2}r_{2,3}} \tag{c}$$

To find the predictions of theory, we shall apply these three equations to dielectric films deposited on glass and to partially silvered parallel reflecting surfaces.

11-9. *Dielectric Films on Glass*

In the notation of Fig. 11-16, we let $N_1 = 1.0$ represent air and $N_3 = 1.52$ represent glass, while N_2 is the index of a deposited thin transparent film. N_2 may be lower than N_3, such as we have in a deposit of MgF_2 ($N_2 = 1.38$) produced by thermal evaporation; or it may be higher, as with deposited films such as ZnS or TiO_2 ($N_2 \cong 2.25$).

If $N_2 = 1.38$, we get $r_{1,2} = -0.160$ and $r_{2,3} = -0.048$. And when $N_2d = \frac{\lambda}{4}$, we get the sum of all reflected beams as $r_{odd} = -0.113$. This is to be compared with the algebraic sum of the first two beams, which is -0.112. Thus the neglect of $\tilde{E}_3$, etc., in this instance, leads to a 1% error in calculated overall amplitude. The overall reflectivity of such overcoated glass, in this case, is r_{odd}^2, or $\mathfrak{R} = 1.28\%$. This overall reflectivity is 70% less than the reflectivity of naked glass, $\mathfrak{R} = 4.3\%$. And, furthermore, such an overcoat results in a $(4.3\% - 1.3\%)$ or 3% greater transmission. Such $\frac{\lambda}{4}$ films of FMg_2 or other transparent low index material are called *reflection reducing films*. They are nowadays applied to all good camera lens components to increase their overall transmission and to reduce the inter-component reflection which can produce undesirable ghosts in photographs.

On the other hand, if an overcoat film has a higher index, like $N_2 = 2.25$ for example, then $r_{1,2} = -0.385$ and $r_{2,3} = 0.194$. When $N_2d = \frac{\lambda}{4}$, all beams together for such a film yield $r_{odd} \backsim -0.538$. This is to be compared with the sum of the first two beams, which is -0.579. Here a neglect of $\tilde{E}_3$, etc., leads to an error of 7.5% in reflected amplitude. Such high index overcoat films have an intensity reflection coefficient of 29%.

It is of interest to note that a MgF_2 film of $N_2d = \frac{\lambda}{2}$ gives $r_{even} = \frac{-0.208}{1.0077}$ or $-.2065$, which is exactly equal to $\left(\frac{N_3 - 1}{N_3 + 1}\right)$, the same as naked glass. And for an $N_2d = \frac{\lambda}{2}$ film with $N_2 = 2.25$, we also get $r_{even} = -\frac{0.191}{0.925} = -0.2065$.

Professor Pfund first produced such reflection-enhancing films by thermal evaporation; and the author first produced such reflection-reducing films by the same process.

11-10. *Metallic Films*

Fig. 11-17 shows calculated transmissions for four pairs of ideal non-absorbing reflecting surfaces, where the $\mathfrak{R}$'s for the single surface reflection of each parallel pair are, respectively, $\frac{1}{40}$, $\frac{1}{10}$, $\frac{1}{2}$, and $\frac{9}{10}$. In order to apply Airy's formula to predict such results, we simply set $r_{2,3} = -r_{1,2} = r_{2,1} = \sqrt{\mathfrak{R}_1}$, where $\mathfrak{R}_1$ is the intensity reflection-coefficient of one surface of the pair. Results thus predicted apply approximately to actual interferences by two partially silvered surfaces. From Eq. 11-8a, these predicted results are given by

$$\tilde{r} = \sqrt{\mathfrak{R}_1}\left\{\frac{1 - e^{-j\Delta\varphi}}{1 - \mathfrak{R}_1 e^{-j\Delta\varphi}}\right\}$$

Multiplying $\tilde{r}$ by $\tilde{r}^*$, we get the overall reflectivity, $\mathfrak{R}_2$, for the two surfaces:

$$\mathfrak{R}_2 = \mathfrak{R}_1\left[\frac{2 - (e^{-j\Delta\varphi} + e^{j\Delta\varphi})}{1 - \mathfrak{R}_1(e^{-j\Delta\varphi} + e^{j\Delta\varphi}) + \mathfrak{R}_1^2}\right]$$

Finally, writing the parenthesis in terms of its cosine equivalent, $2\cos\Delta\varphi$:

$$\mathfrak{R}_2 = \mathfrak{R}_1\left[\frac{2(1 - \cos\Delta\varphi)}{1 - 2\mathfrak{R}_1\cos\Delta\varphi + \mathfrak{R}_1^2}\right]$$

Now for non-absorbing films, even when $\mathfrak{R}_1$ is nearly unity, we get

---

FIG. 11-17   Multiple-beam bands—dependence of band sharpness on reflectivity of surfaces.

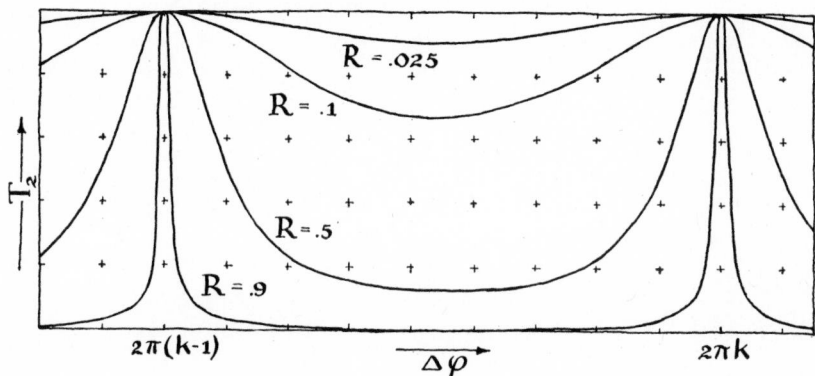

$$\mathfrak{R}_2 = 0 \qquad \text{if } \cos \Delta\varphi = +1$$

$$\mathfrak{R}_2 = \frac{4\mathfrak{R}_1}{(1 + \mathfrak{R}_1)^2} \qquad \text{if } \cos \Delta\varphi = -1$$

And for the total transmission, setting $\mathfrak{T}_2 = (1 - \mathfrak{R}_2)$, we get $\mathfrak{T}_2 = 1.0$ and zero, respectively. Or, generally:

$$\mathfrak{T}_2 = \frac{(1 - \mathfrak{R}_1)^2}{1 - 2\mathfrak{R}_1 \cos \Delta\varphi + \mathfrak{R}_1^2}$$

which gives $\mathfrak{T}_2 = \left(\dfrac{1 - \mathfrak{R}_1}{1 + \mathfrak{R}_1}\right)^2$, when $\cos \Delta\varphi = -1$, and $\mathfrak{T}_2 = 1.0$, when $\cos \Delta\varphi = +1$.

Fig. 11-18 illustrates graphically the interrelations of $\mathfrak{R}_1$, $\Delta\varphi$, and $\mathfrak{T}_2$. In the triangle, $\triangle ABC$, we let the side AB represent the incident amplitude, and we let the base AC represent $\mathfrak{R}_1$. If the heavy line $\mathfrak{R}$ represents $(1 - \mathfrak{R}_1)$, and the dashed line BC is $\mathfrak{D} = \sqrt{1 - 2\mathfrak{R}_1 \cos \Delta\varphi + \mathfrak{R}_1^2}$, then the transmission is $\mathfrak{T}_2 = \left(\dfrac{\mathfrak{R}}{\mathfrak{D}}\right)^2$. $\mathfrak{T}_2$ decreases very rapidly as $\Delta\varphi$ increases, as is apparent from the figure; and it decreases most quickly when $(1 - \mathfrak{R}_1)$ is smallest.

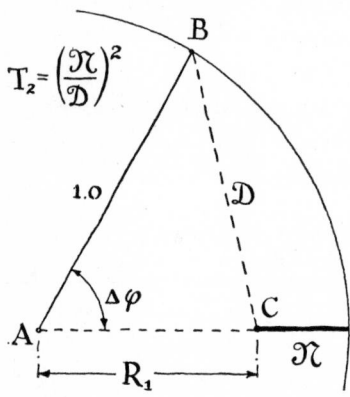

FIG. 11-18 Geometric explanation of beam sharpness with a circle about A.

Chapter **XII**

Applications of
Physical Optics

ALTHOUGH the advent of new principles into the science of physical optics is relatively rare nowadays, new applications of old and established principles are being made regularly, and such applications are exploited with great vigor. For example: reflection-reducing and reflection-enhancing deposits on the surfaces of lenses and beam splitters; periodic film structures for producing color filters; new and improved diffraction gratings, such as Professor G. R. Harrison's echelle; large interferometers, such as the Mach-Zenger interferometer used in aeronautical research; birefringence interferometers, and interferometry in infrared spectroscopy; Professor Frits Zernike's phase microscope; and so forth.

We have already discussed some applications of physical optics, such as Young's interferometer for measuring stellar separations and star diameters and the use of Michelson's interferometer to measure the meter. And in this chapter we shall continue this discussion of applications. We have chosen several applications for their pedagogic value. In our selection, however, many applications have had to be omitted for want of space, but not for want of interest or importance.[†]

In this chapter we shall be concerned with interference bands that extend over a broad wavelength span, such as the $\frac{\lambda}{4}$-films required in reflection-reducing overcoats, where the lowest possible order of destructive interference encompasses the whole visible spectrum, and an entire lens surface. We shall be concerned, also, with interference bands that are less wide, but not

---

† For example, see A. C. S. van Heel, "High Precision Measurements with Simple Equipment," J. Opt. Soc. Am., *40*, 809 (1950).

sharp either, such as are used in color filters to isolate one or two of the seven colors of the spectrum. And finally, we shall treat those applications in multiple-beam interferometry that involve interference bands that are to be made as sharp as possible, in both low and high orders of interference, such as are obtained by partially silvering the opposed surfaces.

12-1. *Reflection-reducing Overcoats*

When the surface on glass of index N_3 is overcoated by a film of $\frac{\lambda}{4}$ optical thickness, and of index $N_2 = \sqrt{N_3}$, the overall reflection for wavelength λ is zero. Such a *reflection-eliminating* film is called an *L*-film. It is also called an *L*-film, as long as $N_2 < N_3$, if its optical thickness is $\frac{\lambda}{4}$. An *H*-film is a $\frac{\lambda}{4}$-film with $N_2 > N_3$. We shall designate these films hereafter simply by *L* and *H*. Thus, for example, a certain three-film combination may be designated *HLH*; etc.

An $Nd = \frac{\lambda}{4}$ film introduces $\frac{\pi}{2}$ phase retardation between reflected amplitudes $\tilde{E}_1$ and $\tilde{E}_2$, for each film traversal by $\tilde{E}_2$; or a total phase shift of π for $\tilde{E}_2$ relative to $\tilde{E}_1$. Such a $\frac{\lambda}{4}$-film contrasts with our $\frac{\lambda}{4}$-plates of § 7-7 (which were of calcite, of quartz, or of mica) as follows: the $\frac{\lambda}{4}$-plates of calcite, etc., give a $\frac{\pi}{2}$ phase retardation of the *o*-ray relative to the *e*-ray for one film traversal; *i.e.* $(N_{\text{ord}} - N_{\text{ext}})d = \frac{\lambda}{4}$.

The index of refraction of water (1.33) is lower than that of ordinary glass (1.52); thus, a film of water on a wetted glass surface may form an *L*-type overcoat when $Nd = \frac{\lambda}{4}$. When a film of water on glass evaporates until its thickness becomes $Nd = \frac{\lambda}{4}$, the reflectivity of the wet glass is reduced, relative to dry glass, by an order of magnitude. Since countless people have looked at this phenomenon, as they have cleaned their eyeglasses, it is surprising that it was so late when its practical implications were first recognized, and the reflection reduction was permanently realized by thermally evaporated *L*-films of CaF_2.

We approach an understanding of *reflection-reducing* overcoats by considering first the properties of *reflection-elimination* films; for although reflection-elimination films are seldom realized in practice, their properties, being

FIG. 12-1   Reflection of glass coated with a reflection-eliminating film.

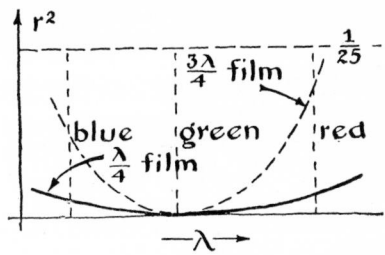

simple, are of interest pedagogically. First, to produce a reflection-elimination film, it is apparent from Eq. 11-8c that we must have $r_{1,2} = r_{2,3}$. On writing $N_1 = 1.0$ in this equality of r's we get the required index for a reflection-elimination film as $N_2 = \sqrt{N_3}$:

$$\frac{1 - N_2}{1 + N_2} = \frac{N_2 - N_3}{N_2 + N_3} \qquad \text{yielding} \qquad N_2 = \sqrt{N_3}$$

For ordinary glass, its index 1.52 calls for an overcoating film of index $N_2 = \sqrt{1.52} = 1.23$. Although no stable solid of such low index is known, thermally evaporated overcoats of MgF_2, for which $N_2 = 1.38$, produce substantial reflection reduction, as was pointed out in the last chapter, and they are extensively used. As is often the case with new inventions, the annoying effects of surface reflection were appreciated, and the materials and operational facilities for practical realization of reflection reduction were well known, or available, before their combination was invented.

In a reflection-reducing overcoat it is desirable to have the desired interference effect extend over as broad a wavelength band as possible. Thus we wish $\Delta\varphi = \dfrac{4\pi N_2 d}{\lambda} \cos r$ to vary as little as possible with wavelength. This variation is least when $N_2 d$ has its least possible value; that is, when it is approximately $\dfrac{\lambda}{4}$. And yet, a $\dfrac{\lambda}{4}$-film of correct thickness for green light, such as to

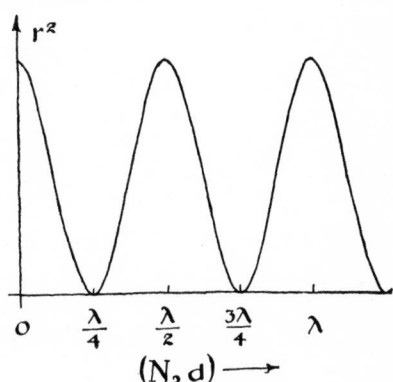

FIG. 12-2   Reflection of glass as a function of film thickness.

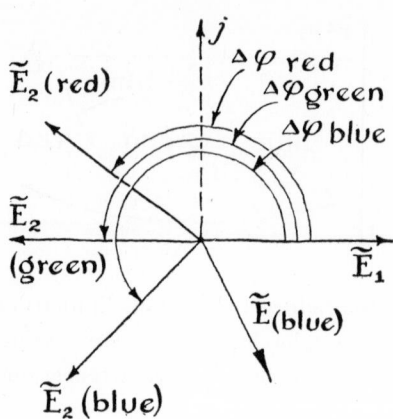

make $\Delta\varphi = \pi$, gives a retardation $\Delta\varphi > \pi$ for blue light, and $\Delta\varphi < \pi$ for red. Fig. 12-1 shows how the overall intensity-reflection coefficient for a reflection-eliminating film varies from its green zero with increasing or decreasing wavelength. Here, taking N_2 constant, the changes of λ control the value of $\Delta\varphi$. Fig. 12-2 shows how the reflectivity of overcoated glass changes as $N_2 d$ changes, with λ fixed. And Fig. 12-3, which is to be compared with Fig. 12-1, shows how the representative vectors add in the complex plane. Here we take $\mathcal{E}_{1,2} = \mathcal{E}_{2,3} = 0.1\mathcal{E}_0$ for red, green and blue light. When $\Delta\varphi = \pi$ for green light, the retardations for blue and red are $\Delta\varphi = \frac{5}{4}\pi$, and $\frac{4}{5}\pi$, if $\lambda_{\text{blue}} = \frac{4}{5}\lambda_{\text{green}}$ and $\lambda_{\text{red}} = \frac{5}{4}\lambda_{\text{green}}$. Thus the overall reflected amplitude, taken as the superposition of the two main beams only, is $\tilde{E} = \tilde{E}_1 + \tilde{E}_2$. This sum is zero only

---

FIG. 12-4 Vector diagram for glass coated with a $3\dfrac{\lambda}{4}$ reflection-eliminating film.

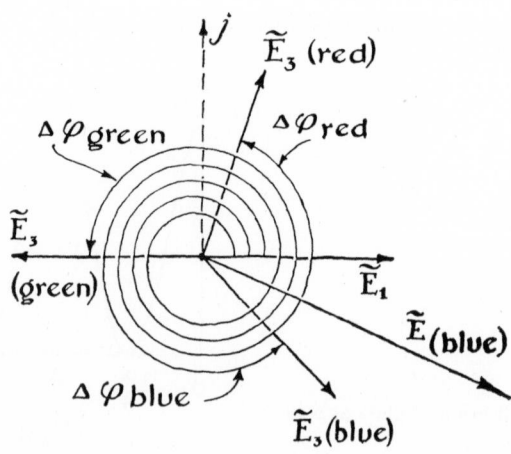

for λ_{green}. Thus white light, after reflection from such films as are represented in Fig. 12-1, will appear magenta in color.

Fig. 12-4 shows the comparable representative vectors in the complex plane for a $3\frac{\lambda}{4}$ film, and Fig. 12-1 also shows (in dots) the plot of reflectivities against wavelength for glass overcoated with this film. It is apparent that the thinner $\frac{\lambda}{4}$-film gives reflection-elimination over a wider spectrum band than the $3\frac{\lambda}{4}$-film, just as we anticipated. In Fig. 12-4, when $\Delta\varphi_{\text{green}} = 3\pi$ for the green, $\Delta\varphi_{\text{blue}} = \frac{5}{4}\Delta\varphi_{\text{green}}$ and $\Delta\varphi_{\text{red}} = \frac{4}{5}\Delta\varphi_{\text{green}}$.

For overcoats thicker than these, the spectrum of the reflected light takes on the character of a channel spectrum. Although the reflected light may appear white for very thick coats ($N_2 = \sqrt{N_3}$) the superposition sums of amplitudes of the light, returned from the front and from the back surfaces of the overcoat film, can be made manifest, and interference fringes can be recaptured, by means of a spectroscope. For such thick films the averaged overall reflection is one half of that of naked glass.

We do not treat here the cases of non-normal incidence. They are mathematically involved but not difficult. There are, of course, two problems to be solved, one for the incident light polarized in the π-plane, and one for light polarized in the σ-plane. In either case, inasmuch as $\Delta\varphi$ is proportional to $\cos r$, an increase of the angle of incidence decreases $\Delta\varphi$, just as an increase of λ or a decrease of d would.

Although overcoats with non-uniform optical properties have had very little application, they afford interesting possibilities. Fig. 12-5 illustrates how the overall reflectivity depends on the thickness of an overcoating film whose index varies linearly from $N_2 = N_3$, at the 2-3 interface, to $N_2 = N_1$, at the 2-1 interface. The author has made such graded films on glass by

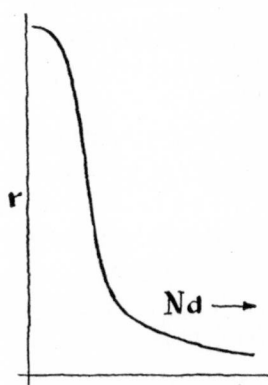

FIG. 12-5 Amplitude reflection coefficient as a function of the thickness of a variable density film.

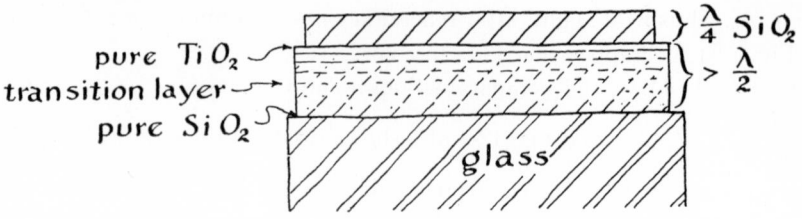

FIG. 12-6 Reflection-eliminating structure made from hard materials.

thermal evaporation. In one such film the index at its outer surface terminated with the value $N_2 = N_3^2$. This structure is illustrated in Fig. 12-6. To accomplish this film he started with the thermal evaporation of pure silica deposited at a rapidly decreasing rate while evaporation of titania was started and continued at a rapidly increasing rate. These rates of evaporation, of silica and titania, were managed so that the composite film had an optical thickness of about $\frac{\lambda}{2}$ when terminated. The film was substantially pure titania outside and pure silica inside. The inside film surface, with its index N_3 approximating that of glass, eliminated reflection at the glass boundary by "immersion"; while the outside surface of this film with its index N_2, which was approximately N_3^2, made the reflectivity the same as that of pure titania. And this altered surface was ideal to receive a pure quartz L-film reflection-elimination overcoat. The advantage of this procedure, of first increasing the reflectivity and then eliminating it, lies in the hardness of the films that can be used. Since titania is not entirely free from absorption some other material, like cerium oxide, might be preferable for the high index material.

12-2. *Multiple Overcoats*

Although H and L films are separately useful, they are especially useful in combination. A carpenter's rule is ideal for visualizing the representative vectors in the complex plane, when several such films, compounded, are overcoated on glass with all interface r's equal. This manner of showing the changes of resultant amplitude due to simultaneous changes of all the $\Delta\varphi$'s must not be disdained, if we are to judge fairly from the fact that it has been much used by those whose creativeness in inventing useful H-L combinations has been greatest. Two applications of the carpenter's rule method (of Dr. Francis Turner) are suggested in Figs. 12-7 and 12-8. Such films, HLH, $HLLH$, etc., producing high reflectivity, or transmission, over a limited wavelength region, without absorption, make useful color filters. And even

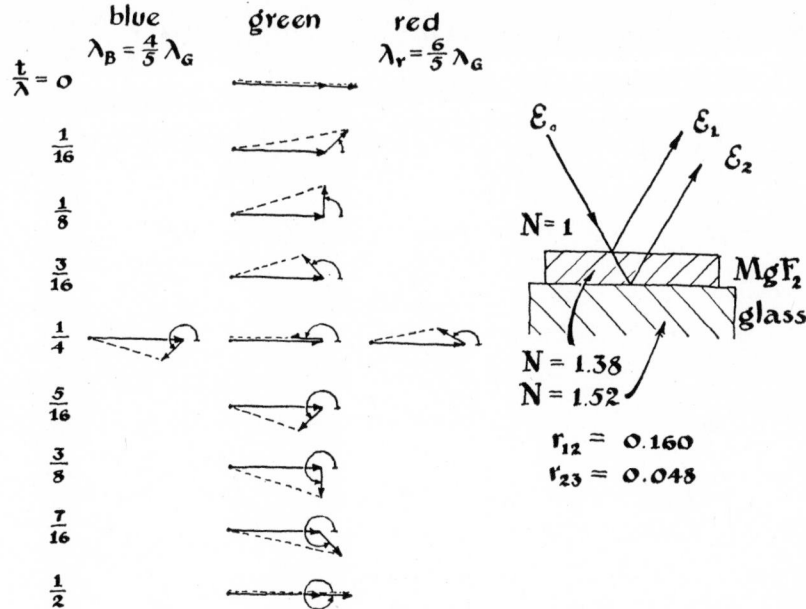

FIG. 12-7   Vector diagrams for a practical reflection-reducing film of MgF₂.

more elaborate periodic structures are used for getting the high-order multiple-beam fringes in Fabry-Perot interferometry.

12-3. *Interference Filters*

An interference filter may be made of two transmitting silver films separated by a dielectric layer of MgF₂. We may deposit this combination by thermal evaporation as follows: First a silver film of such thickness as to have a reflectivity alone of $\Re_1 \cong 0.9$ is deposited on a glass plate. Next follows the deposition of a spacer layer of thickness d, to give interference of the order $k = \dfrac{2Nd}{\lambda}$. Following this, a second $\Re_1 \cong 0.9$ silver film is deposited on the MgF₂ spacer layer. Finally, for protection, a glass cover plate is cemented over all with a plastic cement. The intensity-reflection coefficient of such a combination is given in § 11-10. If the phase shifts at both silver films are the same, overall reflection will be zero when k is a half integer; and the filter will then exhibit a transmission peak. If we set $k = \dfrac{2Nd}{\lambda} = \left(1 + \dfrac{1}{2}\right)$ for $\lambda = 5000$ Å, since $N = 1.38$ for MgF₂, we get $d = 2720$ Å for the required spacer layer thickness. Such a filter will transmit green light. The transmission

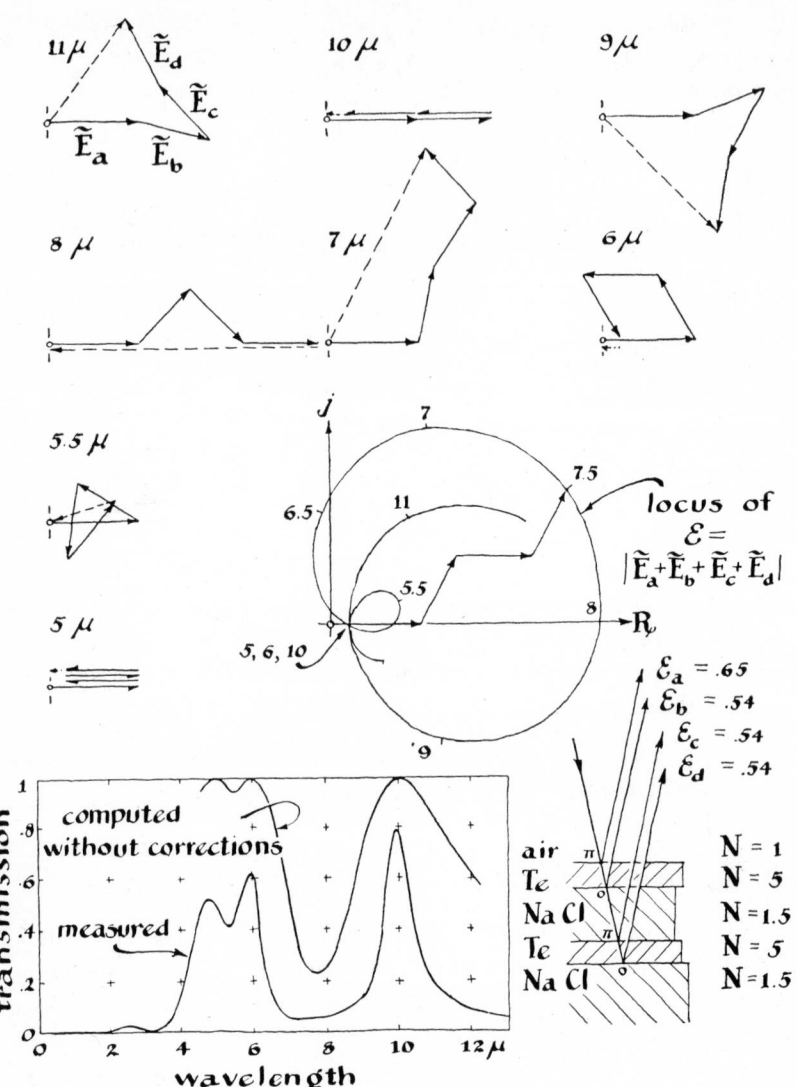

FIG. 12-8 Vector diagram study of a transmission interference filter, neglecting multiple reflections, and actual measured transmission. (Prepared by Dr. Robert Greenler.)

band it exhibits is very narrow. It is so narrow that, in reflection, the filter looks like a simple single silver coating, and not negative green (magenta) as one might expect. However, if we look at the spectrum of the reflected light with a pocket spectroscope, it will show a narrow dark band in the green where the transmitted light is missing.

As another example, consider a similar filter which is to have the thickness that will transmit in the $k = (1 + \frac{1}{2})$ order near $\lambda_1 \cong 7000$ Å, and in the $k = (2 + \frac{1}{2})$ order near $\lambda_2 \cong 4000$ A. Assuming that N for the dielectric spacer layer is 1.38 at both wavelengths, and $\lambda_1 = 7000(1 - \epsilon)$, and $\lambda_2 = 4000(1 + \epsilon)$, we get $\dfrac{\lambda_1}{\lambda_2} = \dfrac{5}{3} = \dfrac{7000(1 - \epsilon)}{4000(1 + \epsilon)}$; or $(1 + 2\epsilon) = \dfrac{21}{20}$, and $\epsilon = .025$. Thus $\lambda_1 = 6825$ Å and $\lambda_2 = 4100$ Å.

And as still another example, consider a similar filter with a wedge-shaped MgF_2 spacer film. Let the wedge be such as to transmit in the $k = 1 + \frac{1}{2}$ order at both the above wavelengths, λ_1 and λ_2. The wedge thickness then will be $d_1 = 3710$ Å for λ_1, and $d_2 = 2230$ Å for λ_2. Such a filter will transmit a narrow band at the ends of the spectrum, and when a white light is viewed through this wedge, the ordered transmission bands for the other spectrum colors will be seen between. If a mercury light is viewed through this wedge filter, the mercury lines will show up in transmission in a display just as they do with a pocket spectroscope.

12-4. *Low-order Multiple-beam Interferometry*

Professor S. Tolansky describes many interesting applications of multiple beam interferometry in his book, *Multiple-beam Interferometry of Surfaces and Films*.† One of these applications is concerned with the determination of the surface structure of terraced mica cleavages. According to Bragg, mica cleaves along surfaces coincident with the planes in the crystal which contain the potassium atoms. And these planes are separated in mica by its c-spacing, which is equal to the molecular dimension, 20 Å, or $\frac{1}{250}$ of the wavelength of green light. In Tolansky's words:

"The interference data can now be compared with these conclusions. The smallest steps recorded from mica interferograms are 20 Å in height, i.e., exactly the c spacing (effectively one 'molecule').

"The following typical set of measurements show five steps in which the precision of measurement is high, since they occur in areas nearly plane. The fractional order displacement is known to 0.001, which corresponds to an error of the very small value of 3 Å. It is a striking fact that *these steps are*

---

† Samuel Tolansky, *Multiple-beam Interferometry of Surfaces and Films* (Clarendon Press, Oxford).

exact integral multiples of 20 Å (within the error). In Ångstrom units they are:

TABLE 12-1 *Mica Cleavages*

$$
\begin{array}{rcl}
& \text{Å} & \\
41 &=& 2 \times 20.5 \\
100 &=& 5 \times 20.0 \\
158 &=& 8 \times 19.8 \\
180 &=& 9 \times 20.0 \\
341 &=& 17 \times 20.0 \\
\end{array}
$$

Furthermore, these steps all being *small* multiples of 20 Å, there is no question about the certainty of the integral ratios. *It is clear that the steps are simple multiples of whole 'molecules'.* There is amongst these five, no evidence of 10 Å being the fundamental unit. This might only be a matter of chance, for clearly if 10 Å happens to be the unit, it is simply a question of whether amongst five random steps all will have an even number of units. Other evidence seems to incline to favour 20 Å as the unit.

"Some mica samples cleave true to a single molecular plane over areas exceeding 20 sq. cm."

Fig. 12-9 illustrates the sharp Fizeau bands that have been obtained by Tolansky with a silvered mica surface and a flat silvered surface. It shows their displacement at mica cleavage terraces. These and other recent applications of low-order multiple-beam interferometry by Dr. Tolansky in Britain, and by Dr. W. F. Koehler in this country, applied to surface contour problems (such as the nature of optical polish) all depend on the very sharp interference bands obtained with two highly reflecting surfaces. This sharpening was illustrated academically by Fig. 11-17. Fig. 11-5a shows Newton's rings as normally seen; and b and c show them after silvering the interfering surfaces, to illustrate how the interference bands are sharpened. The interferences of

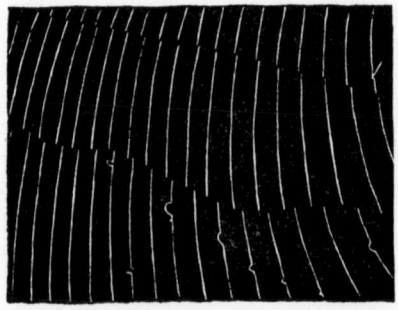

FIG. 12-9 Mica terraces revealed by Fizeau interference bands (produced with silvered surfaces and monochromatic light). (Redrawn from Tolansky.)

a, b, and c in Fig. 11-5 are Fizeau bands; while the FECO interference bands for the Newton lens-plate setup, with silvered surfaces, are illustrated at d. Many other applications of low-order multiple-beam fringes are described in Tolansky's book. This type of interferometry offers a fruitful field of research in applied optics with relatively simple equipment.

12-5. *High-order Multiple-beam Spectroscopy*

Figs. 12-10, 12-11, and 12-12 illustrate Haidinger's transmission fringes applied to spectroscopy; and Fig. 12-13 shows a spectrometric application of interference. In each case the Haidinger transmission bands are sharpened by use of transparent silver films. These Haidinger arrangements display the spectrum by virtue of the variation of their overall transmission, of the parallel surfaces, with angle of inclination of the light. Because the transmission around a ring occurs for a wavelength λ at the angle of inclination r which makes the order of interference an integer, this inclination angle acquires a wavelength meaning. Fig. 12-10 shows the way the structure of a monochromatic line can be photographed, and Fig. 12-11 shows the appearance of the photograph of a spectrum of lines.

Fig. 12-12 illustrates juxtaposed half-photographs taken with the photographic plate in the position where the slit is, above, and with a mercury discharge as source. One of these half-photographs shows the Haidinger bands

---

FIG. 12-10   Use of Fabry-Perot interferometer in spectroscopy.

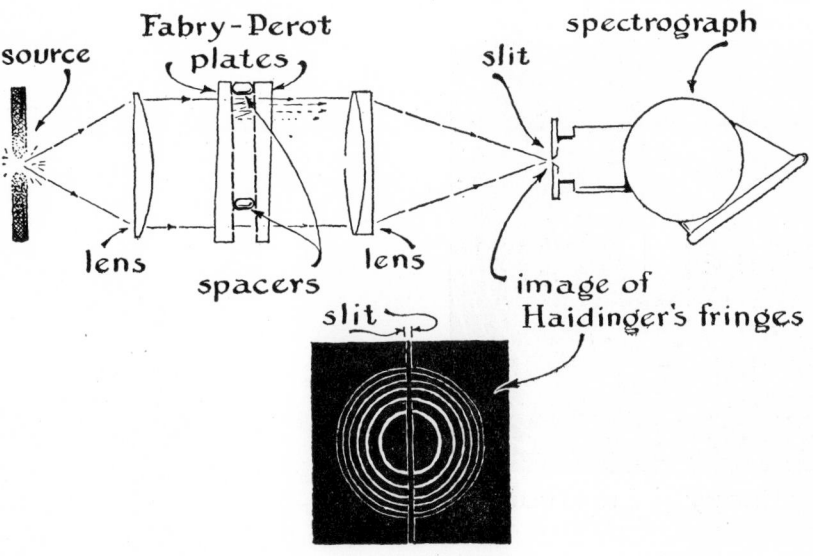

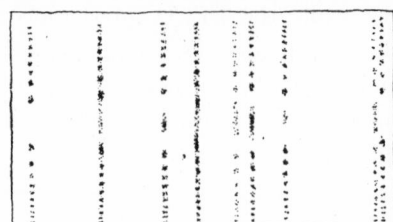

FIG. 12-11 Character of spectrum obtained with the set-up of Fig. 12-10.

of the spectrum line λ = 5461 Å from a natural mercury source; and the other half-photograph shows the bands for the same spectrum line from a mono-isotopic mercury source, using mercury transmuted from gold. The successive rings of the second half-photograph, on the right, correspond to successive orders k. The light from this source is substantially monochromatic. The wider and more complex rings in the first half-photograph, on the left, are representative of the more complex 5461 Å emission of natural mercury. The illumination here is not wide enough in wavelength span so that the complex rings of one order, k, overlap those of adjacent orders, $k \pm 1$. But when such overlapping occurs, the interpretation of the photograph gets confused. The spectral range over which we have no overlapping, and which is thus unconfused, is called the *free spectral range*. This free spectral range is very narrow in high-order spectro-interferometry. Fig. 12-11 shows how the overlapping of fringes is avoided by the use of the spectrograph of Fig. 12-10. The center of the ring pattern is focused on the center of the entrance slit of the spectrograph; and the dispersion of the spectrograph acts as a filter to restrict the spectrum lines in any one photographed interference pattern to less than one free spectral range.

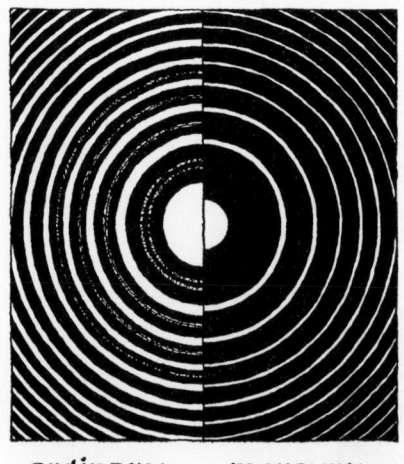

ordinary   mercury
mercury   198

FIG. 12-12 Fabry-Perot ring system for natural mercury and for the 198 isotope of mercury, made from gold.

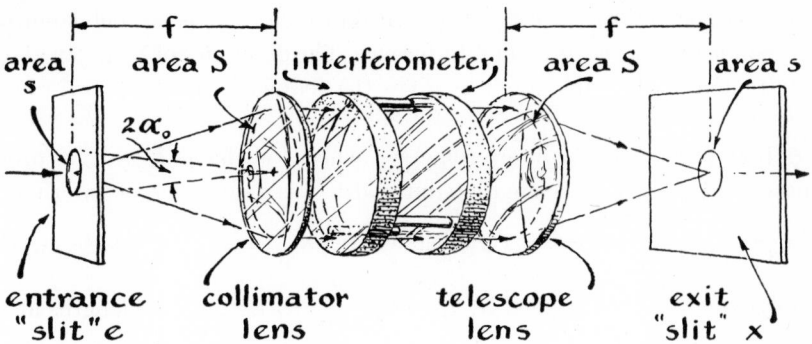

FIG. 12-13 Fabry-Perot spectrometer.

Fig. 12-13 shows an interference spectrometer. This spectrometer uses the central ring or, more properly, the central spot of the Fabry-Perot ring system. The entrance and exit "slits" are here circular apertures. This spectrometer must be used in conjunction with a filter or another spectrometer, so that it is illuminated with only one free spectral range. Otherwise the flux passed by it would have an ambiguous meaning. We shall return to this spectrometer later; but now let us examine the matter of unambiguous display, or free spectral range.

12-6. *Free Spectral Range*

We may introduce the concept of free spectral range by discussing the just overlapping orders that are obtained with a diffraction grating when $k_1\lambda_1 = k_2\lambda_2$. In the field of a diffraction grating we can have, for example, a third order 6000 Å spectrum line; a fourth order 4500 Å line; and a sixth order 3000 Å line all lying at the same diffraction angle α. This superposition of spectrum lines, or near if not exact superposition, was used by Professor Rowland to establish an interrelated set of spectrum lines for use as secondary wavelength standards. But whereas Professor Rowland welcomed and used this confusion of spectra, such overlapping, when working for other ends, is to be avoided by means of filters.

TABLE 12-2

λ	6000		4500		3600		3000
k	3		4		5		6
$\Delta\lambda_F$		1500		900		600	

The free spectral range of a grating within which there is no such confusion corresponds to just one order of interference. The difference of λ's given above, all diffracted at the same angle α, gives the sequential free spectral ranges, $\Delta\lambda_F$.

In high-order spectro-interferometry, where we have $k \gg 1$, a proper filter should pass a wavelength band width just one order of interference wide. This is because $k_0 + 1$ multiplied by $\lambda_-' = \left\{ \dfrac{k_0}{k_0 + \frac{1}{2}} \right\} \lambda_0$ is approximately equal to k_0 multiplied by $\lambda_+' = \left\{ \dfrac{k_0}{k_0 - \frac{1}{2}} \right\} \lambda_0$: the wavelengths at the limits of the free spectral range are λ_-' and λ_+', with λ_0 at the center. The limits above are approximately equal since, for large k,

$$\frac{k_0 + \frac{1}{2}}{k_0 - \frac{1}{2}} \simeq \frac{k_0 + 1}{k_0}$$

The free spectral range required of a Fabry-Perot spectrograph filter in the $k = 5000$ order of interference is 1 Å for green light. Thus a resolving power of at least $\dfrac{\lambda}{\delta\lambda} = 5000$ is needed in the filter or auxiliary spectrograph in order to give an unambiguous display of the spectrum.

12-7. *Fabry-Perot Resolving Power*

Spectral resolving power in the Fabry-Perot ring system may be determined from the lack of angular sharpness of the rings that are produced. When the system is illuminated by two equally strong monochromatic lines (whose wavelengths are different by an amount $\delta\lambda_R$), the angular dispersion must be sufficient to dominate over lack of angular sharpness so that the two lines are separated sufficiently to show the duplicity. In other words, the two rings must have an angular separation, $\delta\alpha_R$, that is determined by the interplay of ring sharpness with wavelength dispersion, $\dfrac{d\alpha}{d\lambda}$.

For unit index between the parallel silver reflecting films, we have $\alpha = r$ and $\Delta\varphi = \dfrac{4\text{v}d}{6} \cos r$ (§ 11-8 and 10). The transmission of the pair of parallel plates is

$$\mathfrak{T}_2 = \frac{(1 - \mathfrak{R}_1)^2}{1 - 2\mathfrak{R}_1 \cos \Delta\varphi + \mathfrak{R}_1^2}$$

The illuminations, due to two close spectrum lines, transmitted near the same angle α, will be additive since spectrum lines are incoherent. Thus the pattern as seen or photographed will be proportional to the sum of two transmission patterns. Fig. 12-14 (repeating part of Fig. 10-9) shows two diffraction patterns and their sum when the two diffraction patterns are separated

FIG. 12-14 Diffraction pattern at the Rayleigh separation for two equally strong lines.

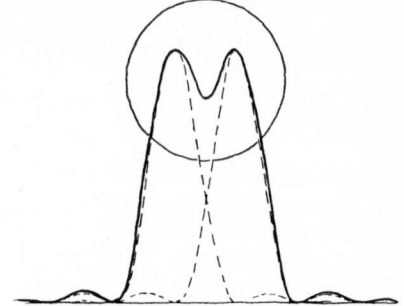

by the amount required to just satisfy the Rayleigh criterion. We cannot, of course, apply Rayleigh's criterion directly to our interferometer here. This is because our Haidinger transmission patterns do not have adjacent minima. But we can apply Rayleigh's criterion indirectly. We can arrange the separation of our interferometer transmission patterns so that a similar minimum occurs between two maxima, shown encircled in Fig. 12-14.

Suppose now that this Fig. 12-14 pattern, produced by two equally strong monochromatic lines, requires the angular separation $\delta\alpha_R$. From $\dfrac{\lambda}{2\pi k}\,\delta\varphi_R = \delta\lambda_R$, by differentiating, $\Delta\varphi = \dfrac{4\pi N d}{\lambda}\cos\alpha$. Let us designate the transmission for one of our two equally strong spectrum lines, where the other has a maximum transmission, by $\mathfrak{T}_{\delta\varphi}$; while we designate the transmission half way between the lines by $\mathfrak{T}_{\frac{\delta\varphi}{2}}$. The central minimum of the composite pattern of Fig. 12-14 will display an illumination proportional to $2\mathfrak{T}_{\frac{\delta\varphi}{2}}$, while the adjacent maxima on either side will display an illumination very nearly proportional to $(1 + \mathfrak{T}_{\delta\varphi})$. Simulation of the Fig. 12-14 pattern requires a central minimum that is equal to 0.8 of the adjacent maxima. Therefore we write

$$2\mathfrak{T}_{\frac{\delta\varphi}{2}} = 0.8(1 + \mathfrak{T}_{\delta\varphi})$$

Here $\delta\varphi_R$ will be a small angle, and we may, accordingly, write

$$\cos(2\pi k \pm \delta\varphi)_R = 1 - \frac{1}{2}(\delta\varphi_R)^2 \quad \text{and} \quad \cos\left(2\pi k \pm \frac{\delta\varphi_R}{2}\right) = 1 - \frac{1}{2}\left(\frac{\delta\varphi_R}{2}\right)^2$$

Making this substitution for both $\cos(\delta\varphi)$ and $\cos\left(\dfrac{\delta\varphi}{2}\right)$, we get

$$\mathfrak{T}_{\delta\varphi} = \frac{1}{1 + \dfrac{\mathfrak{R}_1}{(1 - \mathfrak{R}_1)^2}(\delta\varphi_R)^2} \quad \text{and} \quad \mathfrak{T}_{\frac{\delta\varphi}{2}} = \frac{1}{1 + \dfrac{\mathfrak{R}_1}{(1 - \mathfrak{R}_1)^2}\left(\dfrac{\delta\varphi_R}{2}\right)^2}$$

Letting $x = \dfrac{\mathfrak{R}_1}{(1 - \mathfrak{R}_1)^2}(\delta\varphi_R)^2$ and solving the resulting quadratic equation, taking only the positive root, we get $x = 4.45$. This value of x makes $\delta\varphi_R = \dfrac{(1 - \Re_1)}{\sqrt{\Re_1}} \sqrt{4.45}$. Using $\delta\lambda_R = \dfrac{\lambda}{2\pi k} \delta\varphi_R$, we get finally the resolving power:

$$\frac{\lambda}{\delta\lambda_R} = \frac{2\pi\sqrt{\Re_1}}{\sqrt{4.45}(1 - \Re_1)} k \cong \frac{3\sqrt{\Re_1}}{(1 - \Re_1)} k$$

In this expression we interpret $\dfrac{3\sqrt{\Re_1}}{(1 - \Re_1)}$ as the effective number of interfering beams and write $\Re_{\text{eff}}$ for $\dfrac{3\sqrt{\Re_1}}{(1 - \Re_1)}$. When $\Re_1$ is nearly unity, this number is approximately $\dfrac{3}{(1 - \Re_1)}$. Expressing the resolving power as a *frequency resolving limit*, $\delta\nu_R$, with ν in waves per centimeter, since $k = \dfrac{2d}{\lambda}$ at $\alpha \cong 0$, we get

$$- \delta\nu_R = \frac{\delta\lambda_R}{\lambda^2} = \frac{1}{2\Re_{\text{eff}}d}$$

The free spectral range for the interferometer may be determined by differentiating $k = \dfrac{2d}{\lambda} \cos r$ and setting $\Delta k = \pm 1$ since k is large. At $\alpha = r$ and $\alpha \cong 0$,

$$\Delta\nu_F = \frac{1}{2d} = \Re_{\text{eff}}\delta\nu_R$$

$\Re_{\text{eff}}$ may be interpreted as the number of equal monochromatic spectrum lines that may be equally spaced in wavelength, and just fill in one free spectral range of frequency, with each line still separate enough from its neighbors to be discernible according to Rayleigh's criterion.

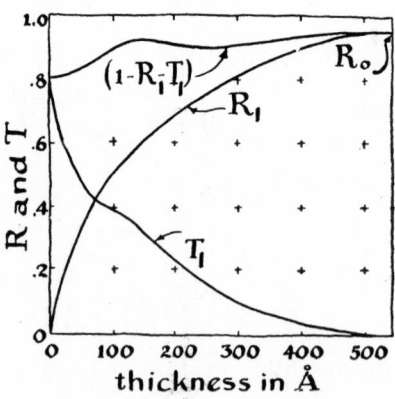

FIG. 12-15 Reflectivity, transmission, and efficiency of silver films.

Fig. 12-15 gives the values of $\mathfrak{R}_1$ and $\mathfrak{T}_1$ for single silver films of increasing thickness. If there were no absorption in the silver films we would have $(1 - \mathfrak{R}_1) = \mathfrak{T}_1$, and $\mathfrak{T}_1$ would not be asymptotic to zero as $\mathfrak{R}_1$ increased; the overall transmission of two parallel films at the peak, for a spectrum line, would be unity. Because of absorption, however, $(1 - \mathfrak{R}_1)$ is asymptotic to $(1 - \mathfrak{R}_0)$, where $\mathfrak{R}_0$ represents the reflectivity of massive silver. And with absorption, $\mathfrak{T}_2$ does not have a maximum of unity for our interferometer; but

we get $\mathfrak{T}_2 = \left(\dfrac{\mathfrak{T}_1}{1 - \mathfrak{R}_1}\right)^2$. We get the ratio of maximum to minimum transmis-

sion of the two ideal interferometer films as $\dfrac{4}{(1 - \mathfrak{R}_1)^2}$. The table below gives $\mathfrak{R}_0$ and $\mathfrak{R}_{\text{eff}}$ for ordinary opaque silver films at $\lambda = 4500$ and 6000 Å. The values for $\lambda = 5461$ Å represent special silver films prepared by Dr. Frank Mooney.

TABLE 12-3   *Silver Reflecting Films*

λ	$\mathfrak{R}_0$	$\mathfrak{R}_{\text{eff}}$
4500 Å	.92	38
6000	.96	75
5461	.985	200

There is a fundamental difference between an amplitude-division type spectrometer, such as the interferometer of Fig. 12-13 for which interference is the basis of operation, and a wave front-division type spectrometer, such as the grating with diffraction as the basis of operation. The difference is apparent when we set about to account for the incident light. In the case of the Fabry-Perot etalon, the energy absent in the narrow transmission bands is largely present in the spectrum of the reflected light. In case of the Haidinger interference bands, the sum of the reflected and transmitted light would be unity if the silver films did not waste light by absorption. In the case of the diffraction gratings, interference effects merely cause a redistribution in angular projection of the diffracted light.

The Fabry-Perot interferometer gives the highest resolving power of any spectroscopic device. Professor K. W. Meissner has used $d = 20$ cm with 31 beams. Here the order is $k = 800,000$. Applying this order in combination with $\mathfrak{R}_{\text{eff}} = 31$ gives a theoretical resolving power of 25 million.

12-8. *The Transmission and Reflection Echelons*

Fig. 12-16 shows the transmission echelon of Michelson and the reflection echelon of Williams. These instruments are quite like blazed transmission and reflection diffraction gratings, except for the grossness of the phase steps. A typical diffraction grating having 14,400 lines per inch, $5\frac{1}{2}$ inches of ruling makes a total of $\mathfrak{N} = 80,000$ interfering beams. Such a grating may be worked in the $k = 5$ order. A Fabry-Perot interferometer, or the echelons of Fig. 12-16, worked in the $k = 10,000$ order, with $\mathfrak{N} = 40$ beams will have the same resolving power—$k\mathfrak{N} = 400,000$.

In either the transmission or reflection echelon of Fig. 12-16 we have steps of rise a, and width d, which are typically 1 mm and 1 cm. Although the transmission echelon is easier to construct, it is very difficult to get sufficiently

FIG. 12-16 Michelson transmission echelon (above) and Williams reflection echelon (below).

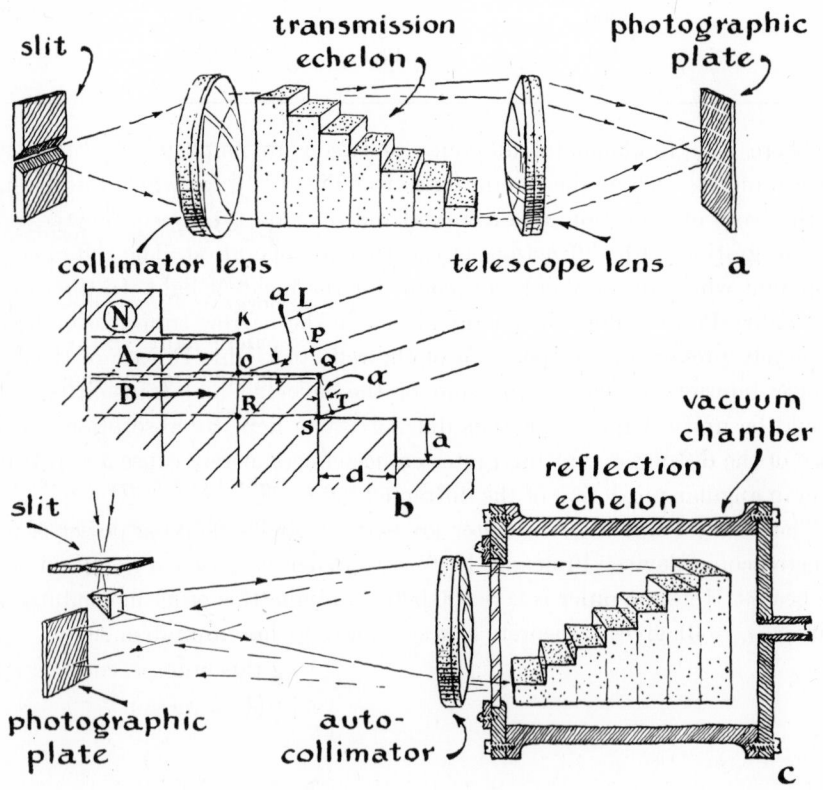

homogeneous glass from which to make it. In the reflection echelon the plates are not penetrated by the light, and this difficulty with glass inhomogeneity is absent. Although Michelson foresaw the advantages of the reflection echelon, its requirement of four times greater precision of construction precluded its realization until 1929, when Williams discovered he could achieve the precision with optically contacted clean fused quartz. Such contacted quartz surfaces can only be parted after several days' soaking in kerosene.

The manner in which echelons are used, and the way they compare with gratings and with the Fabry-Perot interferometer, are set forth hereunder.

In the *transmission echelon* in Fig. 12-16 we have wave front division of internally incident plane waves. We get a retardation, in a direction α, which, in part, is of the same character as that with which we are familiar from our multiple slits; but, in addition, we have a progressive retardation for the beams by virtue of their passage through the successively thicker layers of glass. Consider the adjacent steps A and B in the diagram of Fig. 12-16b, and the emergent beams, which are shown there diffracted upward at the angle α. Along LT these beams weld with each other to form a diffracted wave front, with the total retardation of the diffracted beam B, relative to A, say at the corresponding points, T and P, given by the difference of optical paths $(RT - OP)$. Because the angle α will be small, we may write α for $\sin \alpha$, and

$$RT = Nd + a \sin \alpha \cong Nd + a\alpha$$

$$OP = d \cos \alpha \cong d$$

In order for the diffracted beams to weld into the diffracted wave front LT, the path difference $(RT - OP)$ must be an integral number of wavelengths. Thus

$$(RT - OP) = (N - 1)d + a\alpha = k\lambda$$

In the forward direction the order of interference for $d = 1$ cm will be 10,000, as cited above.

The resolving power of this echelon comes from $\delta\lambda_R$ in $\dfrac{\lambda}{\delta\lambda_R}$; and, in turn, $\delta\lambda_R$ comes from $\left(\delta\alpha_R \div \dfrac{d\alpha}{d\lambda}\right)$, as before. Within this parenthesis the width of the $\mathfrak{N}$ diffracted beams is approximately $\mathfrak{N}a$. Following § 10-5, this makes $\delta\alpha_R = \dfrac{\lambda}{\mathfrak{N}a}$. The dispersion comes from differentiating the equation for $k\lambda$ above, giving $\dfrac{d\alpha}{d\lambda} = \dfrac{k}{a} - \dfrac{d}{a}\dfrac{dN}{d\lambda}$. Thus

$$\frac{\lambda}{\delta\lambda_R} = k\mathfrak{N} - \mathfrak{N}d\frac{dN}{d\lambda}$$

If we express the power of this instrument, for spectroscopy, in waves per centimeter as the *resolving limit*, $\delta\nu_R = \dfrac{\delta\lambda_R}{\lambda^2}$, we get

$$\delta\nu_R = \frac{\lambda}{\mathfrak{N}\left(k - d\dfrac{dN}{d\lambda}\right)}$$

The *free spectral range* here is most meaningfully expressed in wave numbers. The wave number width of one order can be obtained by setting $\Delta k = \pm 1$, for constant α, as follows:

$$\Delta k = -[(N - 1)d + a\alpha]\frac{\Delta\lambda_F}{\lambda^2} = \pm 1$$

giving

$$\Delta\nu_F = -\frac{\Delta\lambda_F}{\lambda^2} = \frac{2}{(N - 1)d + a\alpha} \cong \frac{2}{(N - 1)d}$$

In the case of the *reflection echelon*, with normally incident white **light**, corresponding approximations hold, and we have

$$k\lambda = 2d + a\alpha \qquad \text{with} \qquad \frac{d\alpha}{d\lambda} = \frac{k}{a} \text{ and } d\alpha_R = \frac{\lambda}{\mathfrak{N}a}$$

giving, as before,

$$\frac{\lambda}{\delta\lambda_R} = \lambda\frac{\left(\dfrac{d\alpha}{d\lambda}\right)}{\delta\alpha_R} = \lambda\left[\frac{k}{a}\bigg/\frac{\lambda}{\mathfrak{N}a}\right] = k\mathfrak{N}$$

The free spectral range at $\alpha = 0$, setting $\Delta k = \pm 1 = 2d\dfrac{\Delta\lambda_F}{\lambda^2}$, is given by

$$\Delta\nu_F \cong \frac{1}{2d}$$

Now, although the wavelengths are given by separate equations,

$$k\lambda = \begin{cases} \pm a\alpha + (N - 1)d & \text{for transmission} \\ \pm a\alpha + 2d & \text{for reflection} \end{cases}$$

the illumination in the direction α, in either case, is given by the formula that we have already developed for multiple slits. It involves the phase difference ξ of § 10-1; and illumination is proportional to

$$\frac{\sin^2\xi}{\xi^2} \qquad \text{where} \qquad \xi = \frac{\pi a\alpha}{\lambda}$$

To pursue the question of spectral illuminations further, we invoke the principle that the sum of illuminations in all orders, in transmission, is equal to the incident illumination—thus neglecting reflection losses at glass surfaces. Now $\dfrac{\sin^2\xi}{\xi^2}$, as in Fig. 10-6, envelops the orders, thus relating their illuminations. These illuminations are proportional here to the single-slit diffraction pattern for the facet of width a. When a certain wavelength appears in the direction $\alpha = 0$, we get $\xi = 0$ and $\dfrac{\sin^2 \xi}{\xi^2} = 1.0$ for its order. This order will be $k_0 = (N - 1)\dfrac{d}{\lambda}$. For the two adjacent orders $k = k_0 \pm 1$, we get $\xi = \pm\pi$, making their illuminations zero, as they are for all other orders. In this manner we deduce that all the monochromatic light of this wavelength should appear in the k_0 order at $\alpha = 0$. Now we deduce that a different wavelength, such as would appear at the angle α that makes $\xi = \pm\dfrac{\pi}{2}$, will exhibit an illumination proportional to $\dfrac{\sin^2 \xi}{\xi^2} = \dfrac{4}{\pi^2}$. Other adjacent orders of this wavelength will lie at the angles that make $\xi = \pm\dfrac{3\pi}{2}; \pm\dfrac{5\pi}{2};$ *etc.* And for these angles, illuminations will be proportional to $\dfrac{4}{9\pi^2}; \dfrac{4}{25\pi^2};$ *etc.* The sum of all of the orders of this wavelength will be proportional to the sum

$$2\left(\frac{4}{\pi^2} + \frac{4}{9\pi^2} + \frac{4}{25\pi^2} + \cdots\right)$$

If we divide the light appearing in one strongest order by this sum, we get a spectrometric efficiency of 0.44. Thus no matter what the wavelength may be, our theory predicts that a strongest order may lie at the angles α, corresponding to $\xi = \pm\dfrac{\pi}{2}$; or between one of them at angles α giving $|\xi| < \dfrac{\pi}{2}$. And furthermore, the strength of this order is such as to contain at least 44% of the incident illumination. At $\alpha \leq \dfrac{\lambda}{2a}$ we get, for visible light, $a = 1$ mm, $\alpha \leq \dfrac{1}{4000}$. Thus our approximation, $\sin \alpha = \alpha$, is abundantly valid. Similar arguments apply for the reflection echelon. These arguments are qualitatively valid, also, for blazed diffraction gratings (and the echelle below).

But in the case of diffraction gratings the facets are often not flat; and also, a is usually not large compared to λ, so that Kirchhoff's scalar theory is only an approximation. The fact that Kirchhoff's theory is not valid is irrefutably evinced when we find that π- and σ-components of a grating spectrum of natural light are unequal.

Professor George R. Harrison's echelle is neither a narrow-grooved diffraction grating nor a wide-stepped echelon such as we have just described. Its facet width lies between that of the grating, a few microns, and that of the echelon, thousands of microns—the step of Harrison's echelle is typically some hundreds of microns wide. The free spectral range, $\frac{1}{2d}$, is also intermediate. The advantage of his echelle is that, when it is used with a suitable crossed dispersion, the whole echelle spectrum can be photographed on a single photographic plate.

12-9. *The Lummer-Gehrcke Plate*

The reflectivity of silver gets as low as 4% in the ultraviolet; and it is typically 30% there. This reflectivity is much too low to be efficient on Fabry-Perot plates. Even $\mathfrak{N}_{\text{eff}}$ for aluminum is only about 12; and aluminum certainly gives the most favorable values of $\mathfrak{N}_{\text{eff}}$. Before the advent of aluminum films, the Lummer-Gehrcke interferometer plate was used for the highest ultraviolet resolving power. Fig. 12-17 shows the cross section of such a plate of thickness d and length l. It is a plane parallel slab of quartz provided with the prism shown at the left end, for introducing light. The introduced light is multiply reflected internally at the slab surfaces. At each reflection, which is arranged to be very near the critical angle, a beam emerges from the slab at an almost grazing angle. Thus, by multiple internal reflection, we get amplitude division. Here the wavelets are arranged laterally, as well as retarded in depth. The emerging beams at the angle α lie side by side; and if their relative longitudinal retardation is an integral number of wavelengths, k, they then weld and give an interference wave front.

Because the reflection is near to that giving total internal reflection, the

---

FIG. 12-17   Lummer-Gehrcke interferometer.

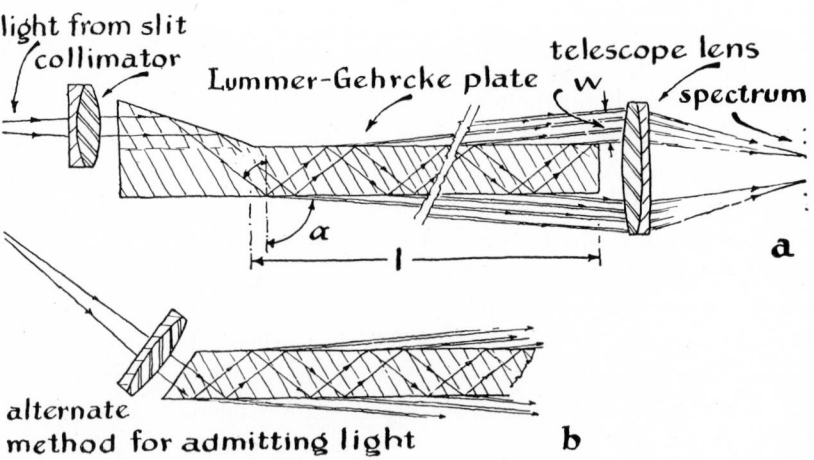

parent beam loses so little amplitude to each of the emergent daughter beams that we can make the assumption, at least for an elementary theory of the Lummer-Gehrcke plate, that the daughter beams are equally strong. From this assumption we may determine the resolution limit, as before, by consideration of the dispersion, together with the diffraction width of spectral lines proper to the width of the welded emergent wave front. This width is $w = l \cos \alpha$, where l is the slab length; and the angular line width proper to it is $\delta \alpha_R = \dfrac{\lambda}{l \cos \alpha}$. The dispersion is obtained by eliminating i's from the square of a Snell's law expression and from the square of an expression for the path difference:

$$\sin r' = N \sin i'$$

$$k\lambda = 2Nd \cos i'$$

On squaring and eliminating i', we get

$$4d^2(N^2 - \sin^2 r') = k^2\lambda^2$$

Differentiating this last equation with respect to r', keeping d and k constant, gives the dispersion since $\alpha = r'$:

$$\frac{dr'}{d\lambda} = \frac{\lambda N \dfrac{dN}{d\lambda} - (N^2 - \sin^2 r')}{\lambda \sin r' \cos r'}$$

It turns out that the term $\lambda N \dfrac{dN}{d\lambda}$ is negligible. Neglecting this term and setting $\sin r' = 1.0$, the dispersion simplifies to

$$\frac{dr'}{d\lambda} = - \frac{(N^2 - 1)}{\lambda \cos r'}$$

On combining this dispersion with the necessary Rayleigh separation, $\delta \alpha_R$, we get

$$- \delta\lambda_R = \frac{\delta r_R'}{\left(\dfrac{dr'}{d\lambda}\right)} = \frac{\lambda}{l \cos r'} \cdot \frac{\lambda \cos r'}{(N^2 - 1)} = \frac{\lambda^2}{l(N^2 - 1)}$$

And the resolving limit in waves per centimeter, from $\delta\nu_R = - \dfrac{\delta\lambda_R}{\lambda^2}$, is

$$\delta\nu_R = \frac{1}{l(N^2 - 1)} \text{ cm}^{-1}$$

We can get the free spectral range by differentiating, as before, setting $\Delta k = \pm 1$. With $\sin \alpha = 1.0$, this operation for the Lummer-Gehrcke plate gives

$$\Delta\nu_F = \frac{1}{2d\sqrt{N^2 - 1}}$$

12-10. *Energy-limited Resolving Power*

The resolving power of a spectroscopic device is not always limited by interference and diffraction effects. A spectrometer in which the attainable resolving power is not limited by diffraction or aberration, but rather by an inadequate flux of energy falling on a too insensitive detector is said to be *energy limited*. This limitation exists, barring aberrations, when the entrance and exit slits cannot be set narrow enough to enjoy the resolving power that diffraction, or freedom from aberration, affords—but when the slits must be kept wide because light sources are not bright enough and/or detectors are not sufficiently sensitive. We shall calculate the energy available at the detector both for a spectrometer, using a prism or grating as the dispersing element P (as shown in Fig. 10-10c) and for the interference spectrometer shown in Fig. 12-13. This calculation will illustrate the significance of energy limitation. For simplicity we shall take the collimator and telescope focal lengths, f, for the prism or grating, and for interferometry, as equal. We shall let s represent the area of the entrance slit, and also the area of the equal exit slit, of the conventional spectrometer; or the area of the round entrance hole, and also the area of the equal exit hole, of the interference spectrometer. In our deductions we shall let S represent the useful projected area of the disperser P, or of the interferometer plates. In both cases, the entrance slit of the conventional spectrometer, or entrance hole of the interference spectrometer, is considered to be flooded with light from a source of spectral brightness $\mathfrak{B}_\lambda$ (expressed in watts per unit area, per steradian solid angle, per unit spectral band pass). The flux emergent through the exit slit, or exit hole, will be $\mathfrak{B}_\lambda$ multiplied by an area; by a solid angle; and by a wavelength band pass; as well as a transmission factor $\mathfrak{T}$:

$$\mathfrak{F} = \mathfrak{B}_\lambda \, s \, \frac{S}{f^2} \, \Delta\lambda\mathfrak{T} \text{ watts}$$

Here $\mathfrak{T}$, the appropriate transmission factor for the system, for monochromatic radiation is given a subscript P, to represent the transmission of the conventional spectrometer; or a subscript I, to represent the transmission for the interference spectrometer. In these cases $\Delta\lambda$ will not be determined by diffraction but by the size of the equal entrance and exit apertures, and by dispersion.

In the case of the conventional spectrometer we introduce a symbol to represent the ratio of the slit length l to the focal length f. This ratio is called the L-number: $L = \dfrac{l}{f}$. If w represents the equal entrance and exit slit widths, then

$$\Delta\lambda = \frac{d\lambda}{d\alpha} \cdot \frac{w}{f} \quad\text{and}\quad s = l\cdot w = (Lf)\left(f\frac{d\alpha}{d\lambda}\Delta\lambda\right)$$

In the case of the interference spectrometer the hole radius subtends the angle α_0, which is related to the spectrometer pass band by the relation $\alpha_0{}^2 = 2\left(\dfrac{\Delta\lambda}{\lambda}\right)$. We get this relation from $\sin r\ \Delta r = \dfrac{k\Delta\lambda}{2d}$ by differentiating $\cos r = \dfrac{k\lambda}{2d}$. We take $r = \alpha$; we call $\cos r = 1.0$; we use $\dfrac{\alpha_0}{2}$ for the average value of $\sin\alpha$; and finally, we take $d\alpha = \alpha_0$. These approximations give $\alpha_0{}^2 = \dfrac{k}{d}\Delta\lambda$. Thus s, for an interference spectrometer that passes the band $\Delta\lambda$, is equal to $\pi\alpha_0{}^2 f^2 = \dfrac{2\pi}{\lambda}f^2\Delta\lambda$.

We are now prepared to intercompare a conventional spectrometer and an interference spectrometer. If, for the first, $\mathfrak{F}_P^*$ is the minimum operable flux in the detector, eliminating s and solving for $\Delta\lambda_P$, the allowed pass band, we get

$$\Delta\lambda_P = \sqrt{\frac{\mathfrak{F}_P{}^*}{\mathfrak{B}_\lambda\left(L\dfrac{d\alpha}{d\lambda}\right)S_P\mathfrak{T}_P}}$$

Because $\dfrac{d\alpha}{d\lambda}$ for a conventional grating is 10-fold, or so, greater than for a conventional prism (excepting crystal quartz prisms in the ultraviolet, which are excellent), we would expect the resolving power of the grating spectrometer, under energy-limited conditions, to be 3-fold better than that of a prism.

A similar calculation for the interference spectrometer, using a minimum flux $\mathfrak{F}_I^*$, and eliminating s, gives

$$\Delta\lambda_I = \sqrt{\frac{\mathfrak{F}_I{}^*}{\mathfrak{B}_\lambda\left(\dfrac{2\pi}{\lambda}\right)S_I\mathfrak{T}_I}}$$

We shall compare energy-limited resolving powers by taking the ratio of the $\Delta\lambda$'s under the following realistic conditions: We strike off $\mathfrak{B}_\lambda$'s, assuming the same source brightness for both spectrometers. For simplicity we strike off the $\mathfrak{T}$'s and take the $\mathfrak{F}^*$'s and S's as equal. We find that $L\dfrac{d\alpha}{d\lambda}$ for the grating compares with $\dfrac{2\pi}{\lambda}$ for the etalon. For a grating normally illuminated $\lambda\left(\dfrac{d\alpha}{d\lambda}\right)_P$ becomes $\tan\alpha_P$, which we may realistically take as unity. Thus the ratio of Δλ's becomes

$$\sqrt{\frac{L_P}{2\pi}}$$

A representative value for L_P is $\dfrac{1}{18\pi}$, making the ratio of the Δλ's, under energy-limited conditions, 1 to 6π.

Professor Stanley Rupert first fully recognized this, and Dr. Robert Greenler first exploited this advantage for interferometry in the infrared. Greenler used evaporated tellurium films for the H layers, and evaporated KBr or NaCl films for the L layers, in his HLH coats to give high $\Re_1$. Tellurium is an appropriate material for the H layers because it has the high index of 5.3, and it is quite transparent in the infrared at 10μ.

Chapter **XIII**

Images of Points by Single Surfaces

IN THE preceding twelve chapters we have treated those topics in optics which fall in its division called physical optics. Our study began with the wave nature of light, and it was primarily concerned with the interaction of waves with matter as manifest in the phenomena of reflection, refraction, polarization, dispersion, and interference and diffraction. We concluded this division by discussions of some practical applications of physical optics, particularly spectroscopic applications.

Now we come to the division of our subject called geometrical optics. Our explanations here will be based primarily on the oldest concept of light, as rays. We combine this concept with the laws of reflection and refraction, and geometrical considerations, to study the confluences or unions of the rays into optical images. But before we get into these applications of geometry to rays of light, we shall briefly consider the geometrical transformations of plane and spherical wave fronts of monochromatic light, that occur when the light passes through plane or spherically contoured boundaries between transparent media of different indices of refraction. We consider these transformations at first under the simplest possible conditions: namely, that the incident wave front and the reflecting or refracting interface are symmetrically oriented with respect to a common optical center line. This line, $\mathbb{C}$, contains the point source, the final image, and the center of curvature of the intermediate surface. And we restrict ourselves to such rays as are nearly parallel to this $\mathbb{C}$, both before and after refraction. In the next chapter we shall enlarge our considerations: Point light sources will still lie on an axis of symmetry; but we shall consider two successive refractions. These two refractions may, for example, be those produced by the two faces of thin or thick lenses.

And we shall extend these considerations to the successive transformations of rays by systems of lenses and mirrors. After this, we shall be prepared to enlarge our considerations still further, and include point light sources that do not lie on the axis of symmetry. Finally, we shall remove or mitigate, successively or simultaneously, other simplifying restrictions.

There is little need to emphasize the practical significance of the study of optical images. In principle, our subject encompasses both the images of the outside world that are formed on the sensing retina of the eye, and corrections of faulty image-forming powers of the eye, achieved by means of spectacles. Geometrical optics concerns itself with magnifiers, microscopes, and telescopes—devices that increase the power of the eye. Also, it concerns itself with cameras and projectors—devices that extend what is available to be seen.

13-1. *General Considerations*

Fig. 13-1 illustrates image formation by reflection and by refraction, and represents the light both by wave fronts and by rays. This figure serves to define new terminology that we shall need, and it illustrates our beginning simplifying restrictions.

Either a point object or its image point is said to be real when the light propagates through it. In this figure a and b represent reflections and refractions, respectively, in cases where the point object and its image are both real. Below this, c and d represent reflections and refractions where the point object is virtual, and its image real. An object point is said to be virtual when the light propagates toward it, but is refracted or reflected in front of it. Finally, e and f represent reflections and refractions where both the image and object points are virtual. An image point is said to be virtual when the light propagates away from it, after reflection or refraction, and never actually propagates through it.

Our beginning simplifying conditions, mentioned above, which we impose on first considerations, are defined in reference to Fig. 13-1 in terms of the ₵ on which object and image points lie, and on which the centers of curvature of the reflecting or refracting surfaces fall. The rims of the reflecting or refracting spherical surfaces, S, have outside radii h_0. Our simplifying conditions are:

1. The light is monochromatic.
2. We ignore diffraction.
3. The object is taken to be a point source on the ₵.
4. The area of the optical surface of radius of curvature r is small; $h_0 \ll r$.
5. The object is at such a distance from the surface on the ₵ at V as to

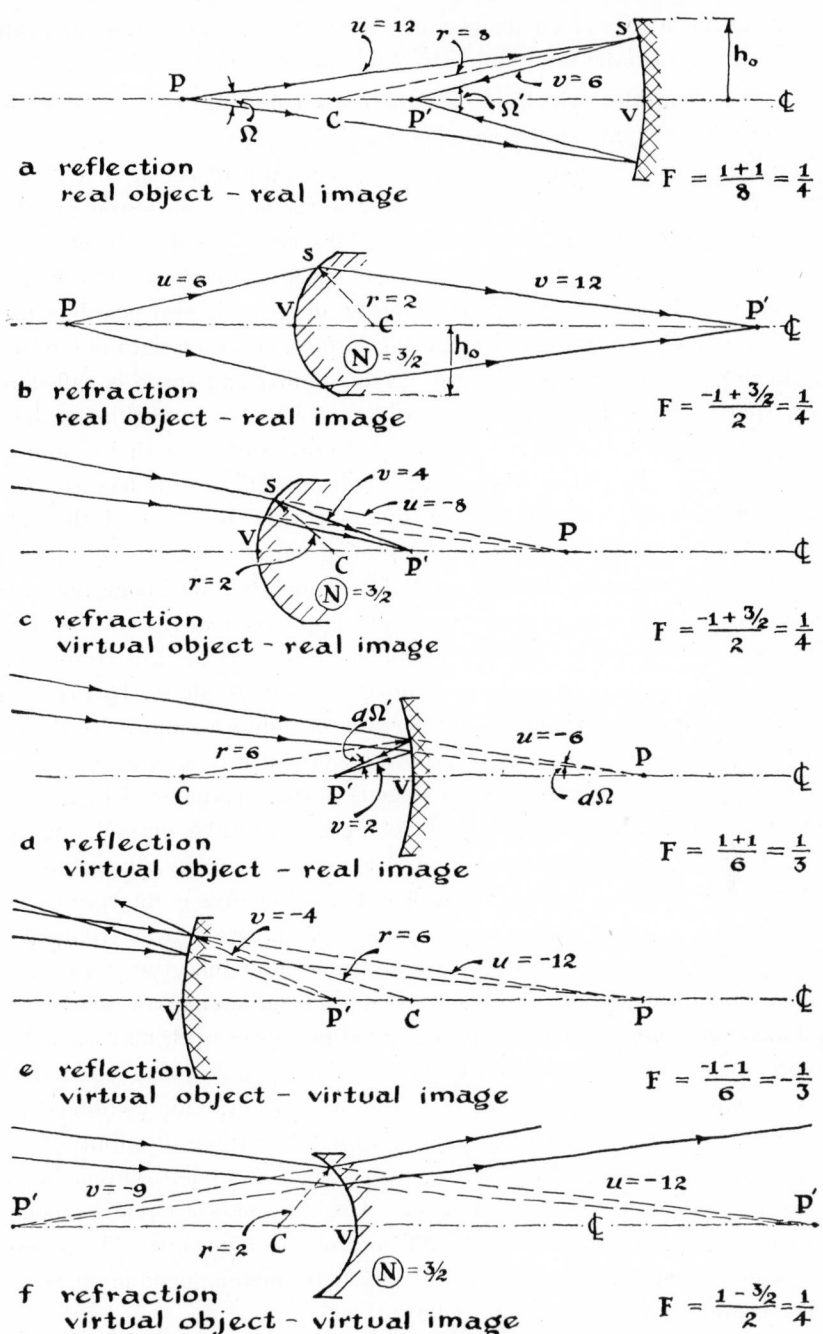

 is the full-page figure. The labels within:

a reflection
real object – real image

$u = 12$ $r = 3$ s h_o $v = 6$ Ω' V

$F = \dfrac{1+1}{8} = \dfrac{1}{4}$

b refraction
real object – real image

$u = 6$ s $r = 2$ $v = 12$ $(N) = \dfrac{3}{2}$ h_o

$F = \dfrac{-1 + \frac{3}{2}}{2} = \dfrac{1}{4}$

c refraction
virtual object – real image

$v = 4$ $u = -8$ s $r = 2$ $(N) = \dfrac{3}{2}$

$F = \dfrac{-1 + \frac{3}{2}}{2} = \dfrac{1}{4}$

d reflection
virtual object – real image

$d\Omega'$ $r = 6$ $u = -6$ P' V $v = 2$ $d\Omega$

$F = \dfrac{1+1}{6} = \dfrac{1}{3}$

e reflection
virtual object – virtual image

$v = -4$ $r = 6$ $u = -12$ V P' C P

$F = \dfrac{-1-1}{6} = -\dfrac{1}{3}$

f refraction
virtual object – virtual image

$v = -9$ $u = -12$ $r = 2$ C V $(N) = \dfrac{3}{2}$

$F = \dfrac{1 - \frac{3}{2}}{2} = \dfrac{1}{4}$

FIG. 13-1   Images by single-surface reflection and refraction.

make the angles of incidence and reflection, or refraction, so small that their sines can be represented by the approximation, $\sin x = x$.

The rays we consider, which lie near the optical $\mathcal{C}$, and are inclined at only a small angle to it, are called *paraxial rays*.

The light is conventionally taken as incident on a reflecting or refracting surface from the left. Thus an object point on the left of the surface is a real object; and one on the right of it is virtual. All space is possible object space—the half-space to the left is called real object space; the half-space to the right is called virtual object space. Concurrently all space is also possible image space—an image on the left of a reflecting surface, or on the right of a refracting surface is a real image, and the corresponding half-space is called real image space. The complementary half-space is, of course, virtual image space.

Fig. 13-1a shows a concave spherical reflecting surface with its center of curvature at C, the radius of curvature being r. The mirror has a circular rim centered about V of radius h_0. The optical $\mathcal{C}$ runs through both the center of this mirror surface, at V, and its center of curvature at C. By virtue of symmetry, if the object point lies on this center line, its image does also. The mirror collects a solid angle segment of spherical waves emitted from, or through, P. This spherical wave segment, of solid angle Ω, is transformed by the mirror into waves convergent on P', of solid angle Ω'. In Fig. 13-1b the refracting surface also has a circular rim, of radius h_0, and it handles the solid angle Ω of the waves from P; the transformed waves, of solid angle Ω', converge here on P'. Matters are analogous in the remainder of Fig. 13-1.

We may preliminarily define a ray of light as a wave normal, but it is appropriate also to think of it as a flow line of radiant energy, like a Poynting vector. In geometrical optics the surface density of rays is interpreted as a measure of image illumination. If we imagine that the wave front from a point source, of solid angle Ω, is handled by a refracting or reflecting surface, and that we divide the surface into equal parts, each subtending an increment of solid angle $\Delta\Omega$, then these equal parts, called pencils of rays, may each have a representative ray ascribed to them. After determining the confluence of these rays, by means of the laws of reflection and refraction combined with geometry, it is only a matter of display to determine the distribution of illumination in their union, if their confluence is not at a point. Often the rays do not come together perfectly, but only within a smallest circle, perpendicular to the $\mathcal{C}$, called the circle of least confusion, or blur circle. The union of rays becomes imperfect, and they fail to pass through a single point as soon as our beginning restrictions are removed.

Paul Drude in his *Theory of Optics* says: "The concept of light rays is . . . introduced for convenience. It is altogether impossible to isolate a single ray and prove its physical existence. For [owing to diffraction], the more one

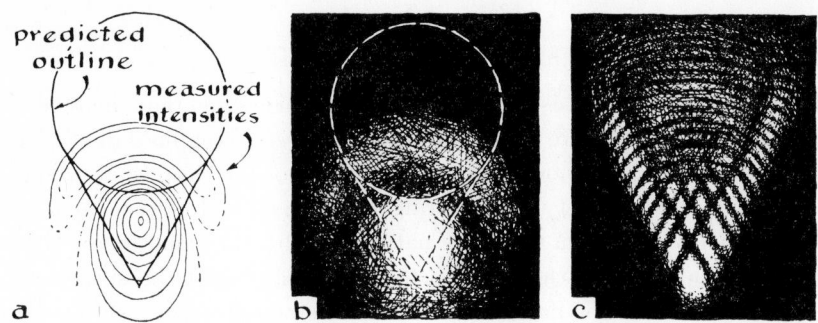

FIG. 13-2 Coma as predicted by ray tracing and as measured (a), and as photographed (b) (redrawn from Martin, *Technical Optics*); coma manifesting diffraction details (c) (redrawn from Kingslake, *Photographic Lenses*).

tries to attain this end by narrowing with restricting apertures, the less does light proceed in straight lines, and the more does the concept of light rays lose its physical significance."

Although the geometry of rays may predict a point image of a pinhole source, we know from diffraction theory that the image size cannot be less than Airy's disk. It is not surprising when the images that are predicted by considerations of rays and geometry to be larger than Airy's disk are found by experimental tests of the distribution of illumination in them to be predicted correctly. It is surprising, however, that the agreement between such predicted distributions of illumination, and test results, should agree as well as they do when predicted images are not much larger than Airy's disk. Fig. 13-2a shows a comparison of predicted and realized illuminations in the photograph of an image that is dominated by the aberration coma. Curves of equal illumination represent the measured pattern. Fig. 13-2b shows a photograph of this image. Here we have a borderline case where geometric results are invalidated by diffraction. And Fig. 13-2c shows the mischief of diffraction in full play. The geometrical prediction of blur circles has been used by lens designers since the time of Sir Isaac Newton and John Flamsteed. This art of geometrical prediction is called *ray tracing*, and it is of great practical value because it costs less to trace a representative set of rays for a proposed lens design than to construct and test prototype models. The full-blown ray tracing treatment of a proposed lens, however, may be a formidable task.

Before we apply ourselves further to geometrical optics, let us consider the required microscopic nature of optical surfaces (§ 13-2) and establish some simple mathematical descriptions by which we can describe macroscopic contours (§ 13-3).

13-2. *Optical Surfaces—Polish*

In our geometrical optics considerations we shall assume that the reflecting and refracting surfaces are smooth and that the refracting material of lenses is both isotropic and homogeneous. If this were not the case, there would be both surface- and body-type scattering of incident and transmitted light. From the point of view of the wave nature of light, the smoothness requirement, to avoid surface scattering, is that irregularities of the surface must be sufficiently small so that the reflected or refracted wave fronts shall be smooth. Tolerable wave front irregularities must be of an order of magnitude less than one wavelength, at least. The required smoothness of refracting or reflecting surfaces occurs naturally on liquids, and it is found in crystal surface cleavages, over small areas. Also, smooth surfaces can be produced on glass by optical polishing, using procedures which are old, and which have been studied extensively (but which are, even now, not conclusively understood).

Fig. 13-3 shows the effect of scattering of radiation by a rough surface and how it detracts from measured apparent specular reflectivity. If the surfaces of the figure had been here microscopically smooth, the samples represented in Fig. 13-3 would each have reflected infrared light almost completely. This figure shows the reflectivity deterioration observed with rough surfaces. The three brass plates represented in the figure were ground

FIG. 13-3   Dependence of specular reflectivity on roughness.

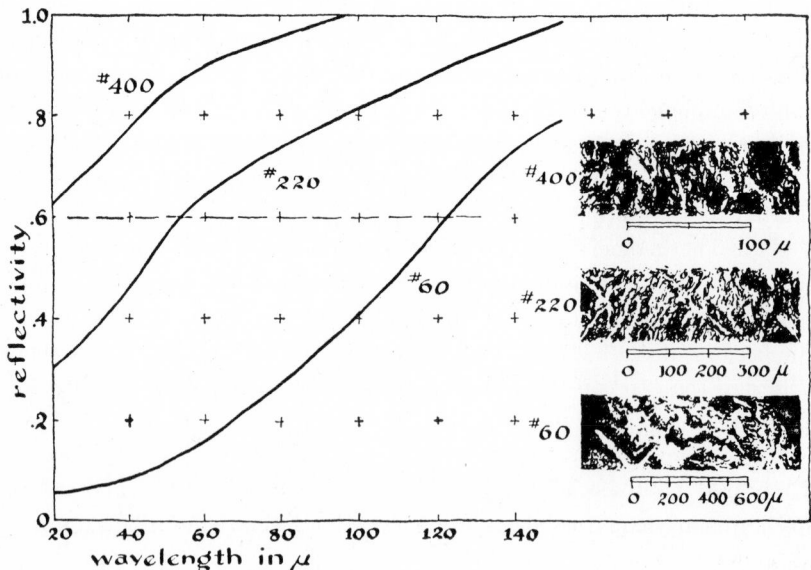

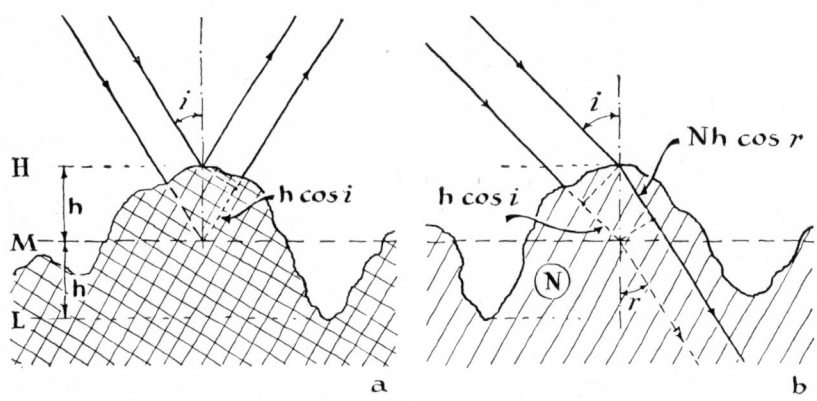

FIG. 13-4 Path differences on reflection (a) and on refraction (b) by rough surfaces.

with #60, #220, and #400 abrasives respectively. Their apparent specular reflectivities were measured over the wavelengths region from 20 to 150μ. Microphotographs of these three rough plates are shown variously enlarged; and on the various scales of these final enlargements, the roughnesses significantly appear similar.

Below, we calculate the relative significance of roughness for producing scattering, both in reflecting and refracting surfaces. Figs. 13-4a and 13-4b represent microscopic surface undulations relative to an average surface, shown as a dashed line. We calculate the optical path difference between a ray reflected or refracted from a peak on this surface, or a depression, relative to another ray that is reflected or refracted by the average surface. Referring to Fig. 13-4a, we consider the two rays reflected from an incident wave front, one at H, the other at M. These two rays travel over different paths, and their final path difference on superposition at a focus is

$$\Delta r = 2h \cos i$$

This equation gives $\Delta r = 2h$ when $i = 0$, $\Delta r = 0$ when $i = \frac{\pi}{2}$. Referring to Fig. 13-4b, the corresponding final path difference for refraction is:

$$\Delta r = Nh \cos r - h \cos i$$

giving $\Delta r = (N - 1)h$ when $i = 0$ and $\Delta r = h\sqrt{N^2 - 1}$ when $i = \frac{\pi}{2}$.

For a fine ground surface, there iş an angle where the red rays from an incandescent filament are first reflected specularly, as i increases toward $\frac{\pi}{2}$, while the surface is still rough for blue rays. This is because increasing i is equivalent to decreasing h, making it small compared to λ. This transition angle is used by opticians to judge the roughness of a surface. If this angle for a worked surface is 80° or less (10°, or more, away from grazing) the surface is deemed to be ready for pitch and rouge polishing.

13-3. *Optical Surfaces—Geometry*

Lenses almost universally have been manufactured with spherical surfaces. This has been mainly because spherical surfaces are the easiest to make, but partly because spherical surfaces are best for reducing certain aberrations. Mirrors, however, often have aspheric contours. These contours are usually conic sections of revolution, symmetrical about an optical ₵. For example: a mirror surface with the shape of a parabola of revolution gives a geomet-

FIG. 13-5   Analytical geometry of conic sections.

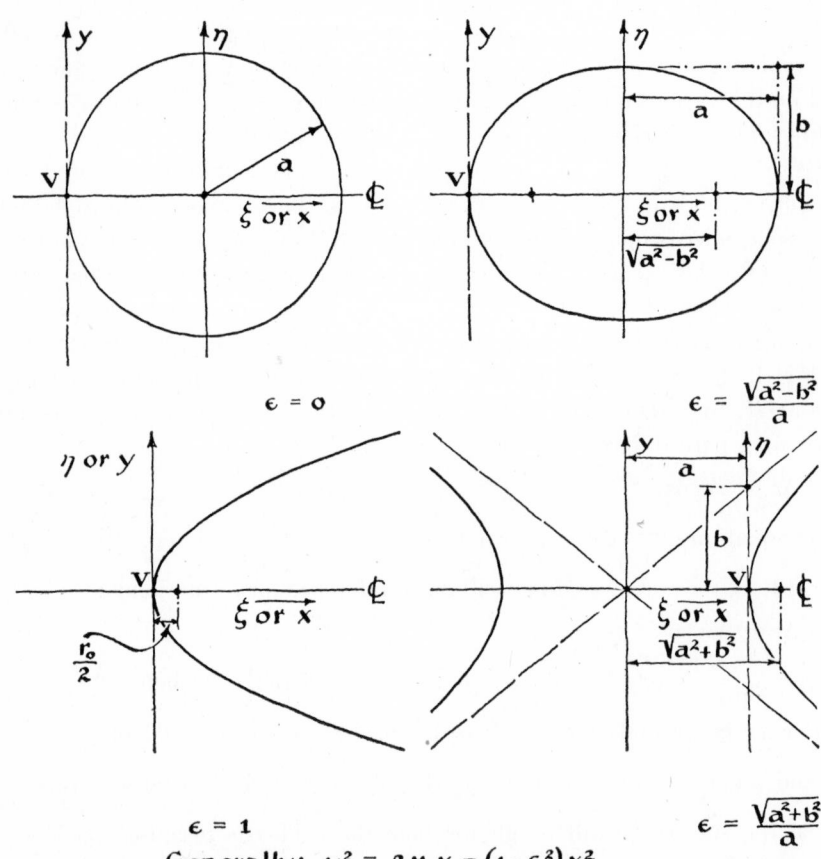

Generally: $y^2 = 2r_0x - (1-\epsilon^2)x^2$

rically perfect confluence of reflected parallel rays at its focus. Rays from a real point object can be united in a geometrically perfect point union by a mirror surface having the shape of an appropriate ellipse of revolution. A properly oriented convex hyperbolic mirror diverges rays from a point object as if they came from a geometrically perfect virtual image. And a spherical mirror is ideal for reflecting rays from a point, at or near its center of curvature, back to a nearby image. It will be shown here that all these conic sections can be mathematically represented by simple and convenient expressions with sufficient accuracy for our purposes. These expressions, developed below, are valid when $h_0 \ll r_0$. Here r_0 is the radius of the sphere which, at V, osculates the conic under consideration. Fig. 13-5 shows all these conic sections and their representation in the V-centered coordinates (x,y). In the usual system of coordinates (ξ,η) the equations, and eccentricities ϵ, are:

Circle: $\qquad \xi^2 + \eta^2 = r_0{}^2 \qquad \epsilon = 0$

Ellipse: $\qquad \dfrac{\xi^2}{a^2} + \dfrac{\eta^2}{b^2} = 1 \qquad \epsilon = \dfrac{\sqrt{a^2 - b^2}}{a} \quad r_0 = \dfrac{b^2}{a} \text{ and } \dfrac{a^2}{b}$

Parabola: $\qquad \eta^2 = 2r_0\xi \qquad \epsilon = 1.0$

Hyperbola: $\qquad \dfrac{\xi^2}{a^2} - \dfrac{\eta^2}{b^2} = 1 \qquad \epsilon = \dfrac{\sqrt{a^2 + b^2}}{a} \quad r_0 = \dfrac{b^2}{a}$

A student exercise shows that the (x,y) coordinates, having their origin at V, may be written in the general form

$$(1 - \epsilon^2)x^2 - 2r_0x + y^2 = 0$$

Solving this equation for x, we get

$$x = \frac{r_0}{(1 - \epsilon^2)}\left\{ 1 \pm \sqrt{1 - (1 - \epsilon^2)\left(\frac{y}{r_0}\right)^2} \right\}$$

Now, when $\dfrac{y}{r_0} \ll 1.0$, the radical above may be expanded by means of Newton's binomial theorem, giving

$$x = \frac{y^2}{2r_0} + \frac{(1 - \epsilon^2)}{8r_0{}^3} y^4 + \cdots$$

In many of the cases we shall treat, even after dropping all but the first two terms of the above series, we may still represent the x of the surface of our conic with sufficient accuracy.

These approximate expressions for the curves of shape of the circle and the parabola, and derivatives for their slopes, are:

Parabola: $\quad x = \dfrac{y^2}{2r_0} \qquad\qquad \dfrac{dx}{dy} = \dfrac{y}{r_0}$

Sphere: $\quad x = \dfrac{y^2}{2r_0} + \dfrac{y^4}{8r_0{}^3} \qquad \dfrac{dx}{dy} = \dfrac{y}{r_0} + \dfrac{1}{2}\left(\dfrac{y}{r_0}\right)^3$

The local radius of the curvature of any curve is given by the calculus as

$$r_1 = \frac{\left[1 + \left(\frac{dx}{dy}\right)^2\right]^{3/2}}{\frac{d^2x}{dy^2}}$$

For the parabola, derivatives of our approximate expression for x, with $\epsilon = 1.0$, give this radius of curvature at y as

$$r_1 = r_0 \left\{ 1 + \left(\frac{y}{r_0}\right)^2 \right\}^{3/2}$$

And for the circle (although we already know its radius everywhere), derivatives of our expression for x with $\epsilon = 0$ gives

$$r_1 \cong r_0 \frac{\left\{ 1 + \left(\frac{y}{r_0}\right)^2 + \left(\frac{y}{r_0}\right)^4 + \cdots \right\}^{3/2}}{1 + \frac{3}{2}\left(\frac{y}{r_0}\right)^2} \cong r_0$$

The curvatures of a parabola, and of its osculating sphere, are approximately constant over a substantial area about the point of contact, V. In the sphere it is this constant curvature which reflects or refracts the incident pencils equally, and it is the tip or inclination of this curvature, so to speak, by an amount proportional to y, which directs the equally focused pencils to a common image.

13-4. *Longitudinal Equations for a Single Surface*

In our expressions we shall frequently, although not always, write h for the distance from the ₵ to a point S at which reflection or refraction occurs. And for the x coordinate of the surface contour, in our V-centered coordinates, we shall often write σ. The sagitta of a sphere or parabolic surface to first approximation is $\sigma = \frac{h^2}{2r_0}$. Fig. 13-6a illustrates this formula. The sagitta $\sigma = BV$ of the circle S, about P, is a side of $\triangle BVS$. This right triangle is similar to $\triangle BSA$ because both triangles have acute angles on the circle which cut out equal arcs from it. Thus the ratios of corresponding sides, taking $BS = h$, give

$$\frac{\sigma}{h} = \frac{h}{2u - \sigma} \qquad \text{or} \qquad \sigma = \frac{h^2}{2u - \sigma} \cong \frac{h^2}{2u}$$

At Fig. 13-6b we see the transformation of a wave front of radius u into one of radius $-u$. This transformation is effected by the plane mirror illustrated. We see the corresponding transformation, effected by a plane refracting surface, illustrated at c. The figure at d shows the transformation of the

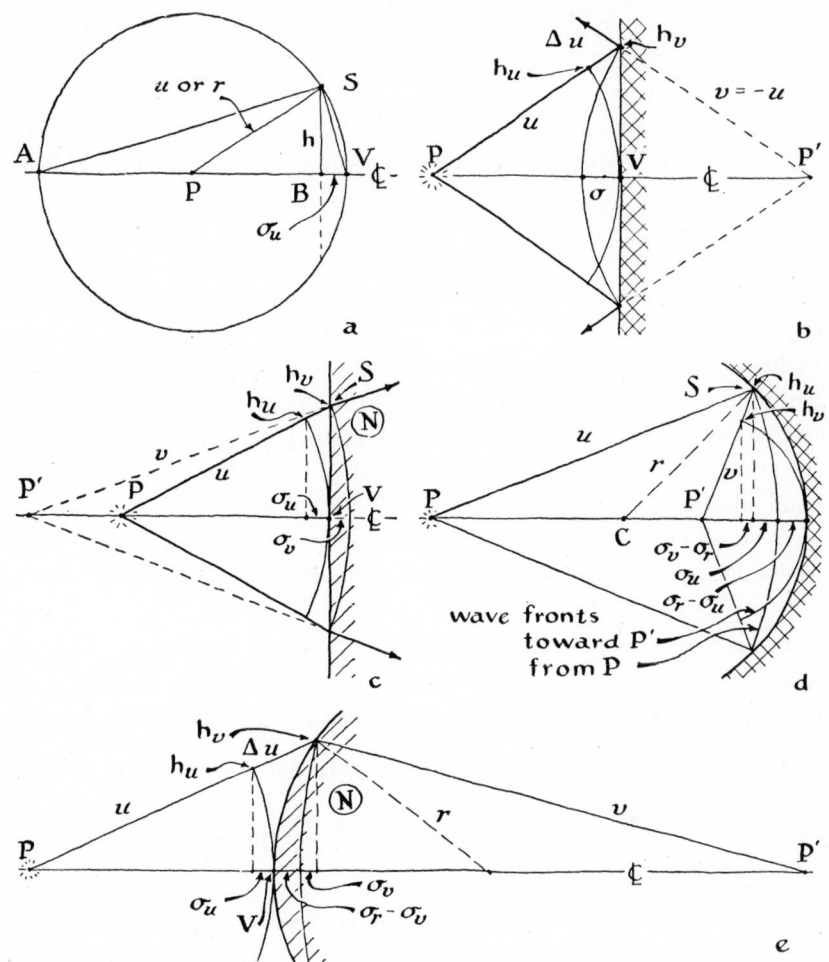

FIG. 13-6   Sagitta method of calculating image positions from wave and surface contours.

wave front of radius u to one of radius v, as produced by a spherical mirror, and, at e, as produced by a curved refracting surface.

In Fig. 13-6b, during the time that the marginal rays are traveling the indicated distance to the plane mirror, Δu, the segment of the wave front which is tangent to the mirror at V will have been returned an equivalent distance, σ. Since we take h as small, σ and Δu have the same length, and the reflected wave has the same radius of curvature. The image of P lies behind the plane mirror at P'.

In Fig. 13-6c, a spherical wave from P is shown when it has progressed to the point where it is just tangent to a refracting surface, at V. It will take a time $\Delta t = \dfrac{\sigma_u}{c}$ for the marginal rays to travel the further Δu distance and strike the surface at S. But this Δu is small since h is small. During this time the central ray, which penetrated the surface at V, will have progressed the distance $\sigma_v = \dfrac{c}{N} \Delta t = \dfrac{\sigma_u}{N}$. From the σ's, since h's are small and $h_u \cong h_v$, we get the descriptive formula $Nu = v$. We shall return to this formula later, and show how it may be used to measure the index of refraction of a glass plate.

In Fig. 13-6d, we have a concave mirror of radius r with its center at C. A marginal ray is shown just progressed to the point where it is striking the mirror at S. It will have been returned toward P' a distance Δv during the time it takes the central ray to reach further to V. For paraxial rays, and h small, this Δv is approximately $(\sigma_r - \sigma_u)$. Thus $\sigma_v = \Delta v + \sigma_r = 2\sigma_r - \sigma_u$; or $\sigma_u + \sigma_v = 2\sigma_r$. Writing in the expressions for σ's, and using $h_u = h_v$, this equation leads to the well-known mirror formula,

$$\frac{1}{u} + \frac{1}{v} = \frac{2}{r}$$

In Fig. 13-6e, we have a curved refracting surface. During the time it takes a marginal ray to travel the indicated distance Δu, in the time interval $\Delta t = \dfrac{\Delta u}{c}$, the central ray at V will have penetrated the refracting medium a distance $(\sigma_r - \sigma_v)$, where $(\sigma_r - \sigma_v) = \dfrac{c}{N} \Delta t$. From the figure it is evident that $\Delta u = \sigma_r + \sigma_u$. Equating Δt's, we get $N(\sigma_r - \sigma_v) = \sigma_r + \sigma_v$, or $\sigma_u + N\sigma_v = (N - 1)\sigma_r$. On entering the appropriate expressions for the σ's, taking $h_u = h_v$ as before, the last equation leads to the important Gauss lens formula for a single refracting surface:

$$\frac{1}{u} + \frac{N}{v} = \frac{(N - 1)}{r}$$

Figs. 13-7 and 13-8 show corresponding cases where, in contrast, we deduce these object and image distance relationships for plane and curved surfaces by application of geometry to rays, rather than by reasoning with wave fronts. Fig. 13-9 illustrates the geometric treatment of the general case of refraction of any ray at a curved surface. When the equation for the general case is specialized for a ray from a point on the $\mathcal{C}$ we get the formula immediately above.

In Fig. 13-7a, paraxial rays from P that are reflected near V come away from the mirror as if from P'. This is because the right triangles, $\triangle PSV$ and $\triangle P'SV$, are congruent. Here $PV = P'V$.

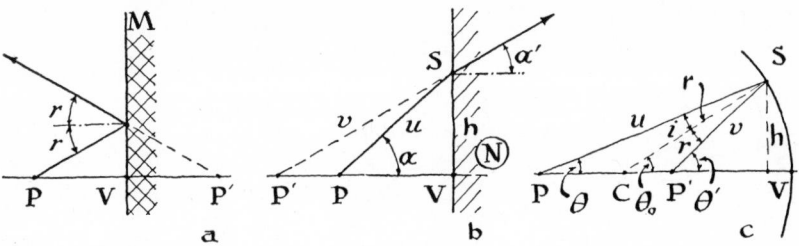

FIG. 13-7   Simple refractions and reflections.

In Fig. 13-7b, we have the case of refraction by a plane dielectric boundary. Rays from P that are refracted at S proceed as if they came from P'. Here we consider the right triangles, $\triangle PSV$ and $\triangle P'SV$, with their common side, h. In these right triangles $u \tan \alpha = v \tan \alpha'$. Thus

$$\frac{v}{u} = \frac{\sin \alpha \cos \alpha'}{\cos \alpha \sin \alpha'} = N \frac{\cos \alpha'}{\cos \alpha} \cong N$$

This descriptive formula is the basis for the method of determining the index of refraction of a plane parallel plate, of thickness v, to which we alluded above. This method, using a microscope, is illustrated in Fig. 13-8. First we focus the microscope on a spot on its stage. Then, after interposing the glass plate of thickness v, as shown in the figure, we raise the microscope lens and tube through the height $\Delta y = (v - u)$ to refocus on the same spot. If the index of the plate is N, using our descriptive formula $v \cong Nu$, we get, on eliminating u,

$$N = \frac{v}{v - \Delta y}$$

Fig. 13-7c illustrates the ray solution for the concave mirror. Here three right triangles in the figure all have the common side h. The hypotenuses of these triangles are u, v, and r. If the angles θ, θ_0, and θ' are small, so that $\tan \theta = \sin \theta = \theta$, then $h = u\theta = v\theta' = r\theta_0$. And since the difference between

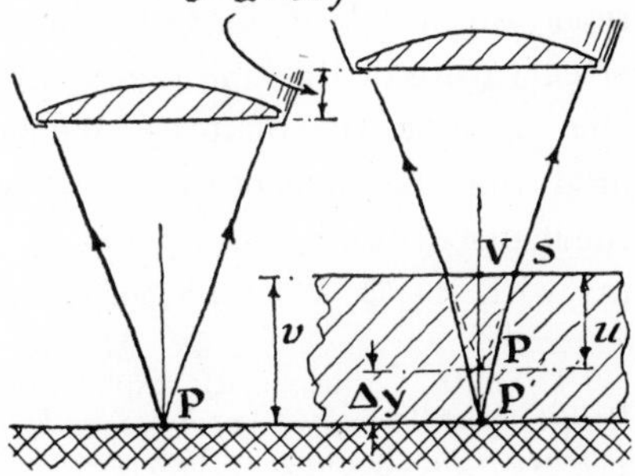

FIG. 13-8 Method of measuring the index, N, of a plane parallel plate with a microscope.

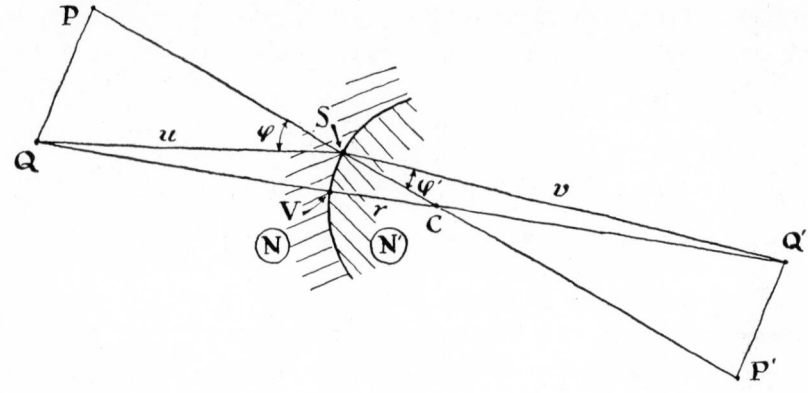

FIG. 13-9 Geometry for deducing the longitudinal equations for a single refracting surface. (Redrawn from T. Smith.)

two angles of any two right triangles, with a common side, is equal to the difference of the two angles opposite that common side, the angles of incidence and reflection are $(\theta_0 - \theta)$ and $(\theta' - \theta_0)$. Equating these angle differences, and writing the angles as $\dfrac{h}{u}$, $\dfrac{h}{v}$, and $\dfrac{h}{r}$,

$$\frac{1}{u} + \frac{1}{v} = \frac{2}{r}$$

Fig. 13-9 shows an object point Q and its image by a refracting surface Q', both object and image lying off the line PP'. This case gives the important general longitudinal equation of Gauss for a single surface; and for the purposes of this chapter, we shall specialize it for the simplest case of paraxial rays.

Before we proceed further, we must settle certain conventions of sign for r, for u and v, and for a quantity called the power of the refracting surface, F. Otherwise we could not proceed without confusion. The conventions we shall arbitrarily adopt are those recommended as Group II Case (1) in T. Smith's report on the "Teaching of Oeometrical Optics."† We always take r the radius of curvature of the considered surface—either a reflecting or refracting surface—as a positive quantity. We ascribe signs to u, the object distance measured from the point object Q to the refracting or reflecting facet S; and to v, the image distance measured from S to the point of cross-over of the refracted ray with the ₵. We take both object and image distances as positive if the

† T. Smith, "Appendix B: Explanatory Notes on the System of Group II, Case (1)," *Report of the Committee Appointed by the Physical Society to Consider and Make Recommendations on the Teaching of Geometrical Optics* (1934, The Physical Society, Cambridge).

light rays actually travel along the lines that u or v measure; otherwise they are taken negative. Distances measured from a real object point, and to a real image point, are thus positive. Distances from Q to P or S, or from S to Q' or P', are negative when the object or image is virtual.

The quantity F, called the power of a surface, takes on the form

$$F = \frac{\pm N \pm N'}{r} \tag{a}$$

Here N is the index of the medium of the incident light rays, and N' is the index of the medium of the emergent or reflected light. We ascribe *positive* or *negative* signs to N's according to the following rules: N for the incident medium is given a positive sign if the incident light is on the concave side of the considered surface at S; and N is given a negative sign if the incident light is on the convex side; and similarly for N', the index of the emergent medium. It is given a positive sign when the emergent or reflected light is on the concave side, and negative for the emergence convex side.

13-5. *The Gauss Equation*

In Fig. 13-9 we consider two rays from Q to its image Q' to get Gauss' important equation. One ray from Q, through C, penetrates the surface normally and is undeviated; the refraction of the ray at S is described by Snell's law relating the angles of incidence and refraction, φ and φ'. Invoking the right triangles $\triangle CQP$ and $\triangle CQ'P'$, which are similar because of equal angles at C, we have

$$\frac{CP}{PQ} = \frac{CP'}{P'Q'}$$

We next write the ratios of sides of these right triangles as

$$\frac{u \cos \varphi + r}{u \sin \varphi} = \frac{v \cos \varphi' - r}{v \sin \varphi'}$$

Applying Snell's law, to remove the ratio of sines by substitution of $\dfrac{N}{N'}$, we get

$$N \left(\cos \varphi + \frac{r}{u} \right) = N' \left(\cos \varphi' - \frac{r}{v} \right)$$

We get the Gauss equation on rearranging terms:

$$\frac{N}{u} + \frac{N'}{v} = \frac{N' \cos \varphi' - N \cos \varphi}{r} \tag{a}$$

This is the first important longitudinal equation of geometrical optics. The right side of this equation is called the power of the surface:

$$F = \frac{\pm N' \cos \varphi' \pm N \cos \varphi}{r} \tag{b}$$

We associate signs, plus and minus, with N and N' according to our rules of § 13-4—*i.e.* the N's on the left side of Eq. 13-5a are always taken positive; but the N's on the right side of this equation have associated signs in agreement with our convention rule. In Fig. 13-9, N on the right side of a is negative since the light is incident from the negative or convex side of the surface; and N' on the right side of a has a positive sign, also in agreement with our convention rule since the refracted light is emergent on the positive concave side. Eqs. 13-4a and 13-5b are quite general, and they apply equally to reflection and refraction. When S lies near V on the line through C, so that φ and φ' are both small, Eq. 13-5a reduces to the form we have already deduced by means of transformed spherical waves:

$$\frac{N}{u} + \frac{N'}{v} = \frac{N' - N}{r}; \qquad \text{or for} \qquad N = 1.0 \qquad \frac{1}{u} + \frac{N'}{v} = \frac{N' - 1}{r}$$

13-6. *Newton's Equation*

Fig. 13-10, intended to get Newton's equation, shows a refracting surface with five rays or lines to or through its center of curvature, C. By means of geometrical considerations applied here we arrive at the second general longitudinal equation for a single refracting surface:

$$(u - f)(v - f') = dd' = \frac{NN'}{F^2} \tag{a}$$

In Fig. 13-10, four of these five lines penetrate the dielectric boundary surface normally. Of these four, one goes through both P and P'. The three lines through S pass, respectively, through P, P' and C. Two lines through S and C are parallel, running through P and Q, while two other lines through S and C are parallel, running through P' and Q'. The ray from P, which

---

FIG. 13-10 Geometry for deducing the principal foci, as well as Newton's equation, for a single refracting surface. (Redrawn from T. Smith.)

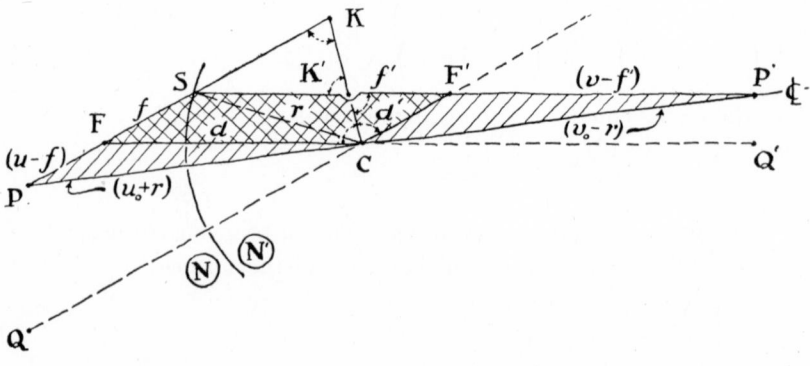

is refracted at S, intersects the one through Q and C at F', whereas the ray through P', which is refracted at S, intersects the one through Q' and C at F. These intersection points F and F' are called the principal foci of the refracting surface.

We make our geometrical deductions from this figure from the parallelogram formed by the intersection of the above two sets of parallel lines. They form the parallelogram $\square CFSF'$, shown cross hatched. Opposite sides of this parallelogram are equal, so that $d = f'$ and $d' = f$. Now if we set $u = \infty$, using Eqs. 13-5a and 13-5b, we get $v = f' = \dfrac{N'}{F}$, for the point F'; and similarly for the other point, setting $v = \infty$, we get $u = f = \dfrac{N}{F}$. Applying the angle relationships of parallel lines, triangles, and supplementary angles, it is easy to show that the triangles, $\triangle PFC$ and $\triangle CF'P'$, are similar. And the ratios of corresponding sides of these similar triangles give the Newton equation

$$\frac{u - f}{d} = \frac{d'}{v - f'} \qquad \text{or} \qquad (u - f)(v - f') = dd' = ff' = \frac{NN'}{F^2}$$

This Newtonian equation, like the Gauss equation, is concerned with longitudinal image positions.

13-7. *Nodal Equation*

There is still a third general longitudinal equation for a single refracting surface. This equation is called the nodal equation:

$$\frac{1}{N\mathbf{u}} + \frac{1}{N'\mathbf{v}} = \frac{F}{NN'} \tag{a}$$

In Fig. 13-10, the deduction of the nodal equation uses the parallel lines through P' and Q', and the parallel lines through P and Q. The transector line CK is drawn so that the triangles, $\triangle KFC$ and $\triangle K'F'C'$, are isosceles and similar, with $KF = d$ and $K'F' = d'$. An examination of the geometry of these figures reveals that the triangle $\triangle KSK'$ is also isosceles and similar, and

$$SK = d - f = SK' = f' - d'$$

The two points K and K', thus located, are called the nodal points; and we designate object and image distances, as measured to or from them, by $\mathbf{u}$ and $\mathbf{v}$. Thus

$$\mathbf{u} = PK = u + SK = u + d - f$$
$$\mathbf{v} = P'K' = v - SK' = v + d' - f'$$

Now if we multiply $u - f = \mathbf{u} - d$ by $v - f' = \mathbf{v} - d'$, one product, according to Eq. 13-6a, is dd', which gives

$$\frac{d}{u} + \frac{d'}{v} = 1$$

On introducing our former values for d and d' $\left(d' = \frac{N}{F} = f \text{ and } d = \frac{N'}{F} = f' \right)$,
we get the Eq. 13-7a above. This nodal equation, like the Gauss and Newton
equations, has the function of predicting the longitudinal location of images.
For paraxial rays the bisector of the parallelogram angle at C, on which the
points K and K' lie, will be nearly perpendicular to the optical ₵; consequently, the nodal points for paraxial rays lie approximately at C.

13-8. *Mirrors*

We shall now apply these new results to the concave and convex reflecting
surfaces of Fig. 13-11. These cases will further illustrate our convention of
signs. In Fig. 13-11a, the positive sign is associated with both N and N', in
the expression of § 13-4 for F, because both incident and emergent rays lie
on the concave or positive side of the considered surface. Thus with these
signs for $N = N' = 1.0$ we get

$$\frac{1}{u} + \frac{1}{v} = \frac{2}{r} = F_a \qquad \text{(a)}$$

In Fig. 13-11b, the negative sign is associated with both N and N', in the
expression for F, because both incident and emergent rays lie on the convex
or negative side of the considered surface. Thus these signs yield

$$\frac{1}{u} + \frac{1}{v} = -\frac{2}{r} = F_b \qquad \text{(b)}$$

And the power of this convex mirror is negative as compared to a concave
mirror of equal radius of curvature.

A negative sign enters Eq. 13-7a for reflection. This negative sign enters

---

FIG. 13-11   The same as Fig. 13-9, but for reflecting surfaces. (Redrawn from
T. Smith.)

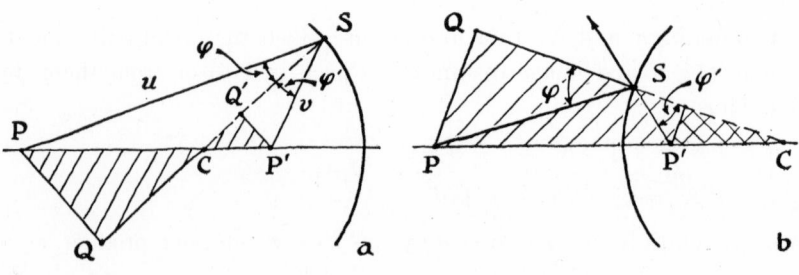

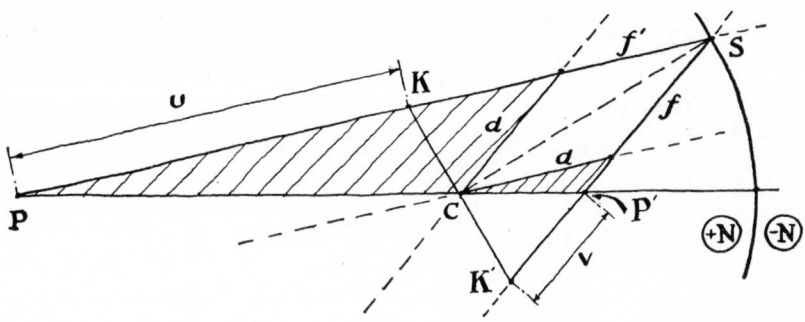

FIG. 13-12   Geometry for deducing the nodal equation for a reflecting surface. (Redrawn from T. Smith.)

---

because our line, CK of Fig. 13-10, that was a bisector of the internal parallelogram angle at C, is now, in Fig. 13-12, the bisector of an external parallelogram angle at C. The nodal distances **u** and **v** are taken positive or negative, according to the same rules that we apply to u and v; namely, positive if the light actually passes over the measured distance to the image, etc. In Fig. 13-12 **v** is negative. Using geometry, and considerations based on two sets of parallel lines similar to those we applied to Fig. 13-10, we get the following expression from Fig. 13-12:

$$\frac{1}{\mathbf{u}} + \frac{1}{\mathbf{v}} = -F \tag{c}$$

And this equation, for reflection, has a sign which is opposite to that of Eq. 13-7a for refraction.

13-9.   *Parabolic Telescope Mirror*

· The above equations are concerned with spherical mirrors. In a telescope used for the purpose of focusing starlight, an aspheric parabolic mirror has a more appropriate shape. Parabolic mirrors also have important applications in projection, such as for searchlight mirrors and automobile headlamps. We shall compare the spherical and parabolic mirror as regards appropriateness for these applications. In particular, we shall compare mirrors of a spherical and a parabolic contour characterized by 6 inches diameter and $r_0 = 100$ inches. These contours are typically produced by amateur telescope makers, successively, on a glass disk during the process of making it into a telescope mirror. A study of such contours, and of their optical performance, will exemplify several optical principles.

Fig. 13-13 shows a typical amateur's telescope using such a parabolic mirror.

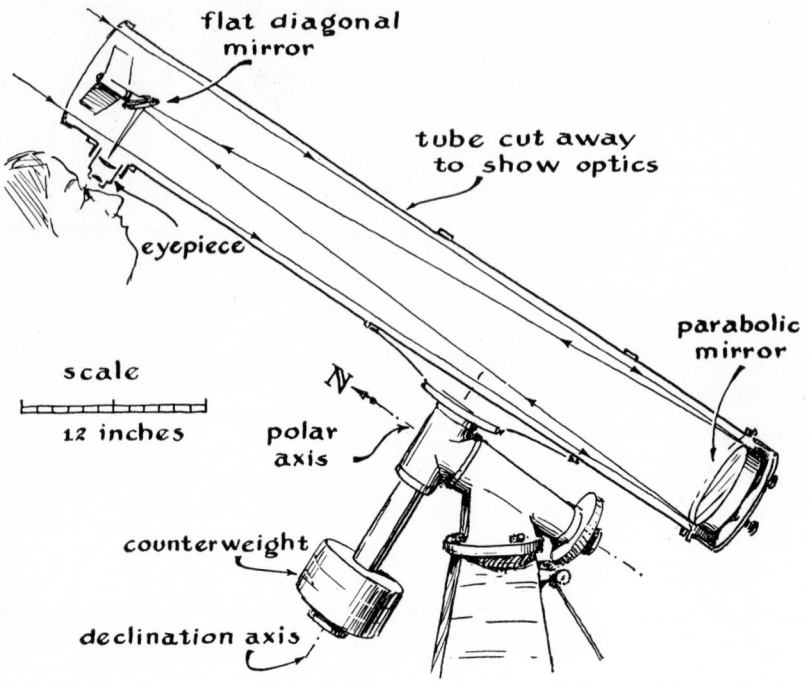

FIG. 13-13 Typical amateur's telescope.

Its optical parts consist of the six-inch diameter primary mirror, a smaller diagonal flat mirror, and an eyepiece for viewing images formed by the primary mirror. These optical parts are held in proper relative alignment by means of a telescope tube. This tube, in turn, is mounted on axes so that it may be trained on any heavenly object that is to be observed. The mounting is arranged so that the tube may be clamped to a driving mechanism which will track an observed star, and keep its image centered in the field of the eyepiece as the star marches across the heavens.

The means used to test these contours—namely, the Foucault knife-edge test and Gaviola's caustic test—are essentially the experimental inverse of mathematical ray tracing: In ray tracing we have a mathematical procedure of predicting where rays from a point go. In these two experimental tests we have means of determining where the rays from a point source actually do go. Such experimental tests can be used either to confirm the predictions of theory, or to test and control the perfection of constructions.

The amateur telescope maker prepares his primary mirror as follows: First he cuts out a 6-inch diameter glass disk of about one-inch thickness. On one face of the disk, by grinding, he produces a concave spherical surface of $r_0 = 100$ inches. This disk is ground first with coarse abrasive grits and then with finer and finer abrasives until a final grinding, with emery flour, produces a surface that will give the transition to specular reflection, at $i \leq 80°$, that we described at the end of §13-2. Then he polishes this surface with a fitting convex pitch tool that is charged with a suspension of rouge in water. And finally he *figures* his mirror to remove zonal errors, and to achieve a true spherical contour. In this *figuring* stage he controls the polishing by optical test results and polishes selectively on those areas of the whole glass surface that are revealed by test to be relatively high, in respect to an imagined perfect sphere. After the surface has been made spherical, it is modified to the parabolic contour by the same process of selective polishing.

It is the amateur's first task to make the spherical contour described by the expression

$$x_C \cong \frac{1}{2r_0} y^2 + \frac{1}{8r_0^3} y^4 + \cdots$$

Once this sphere is achieved, his next task is to remove glass selectively, and modify this sphere into the desired parabolic contour. The parabolic contour finally desired is described by

$$x_P = \frac{1}{2r_0} y^2$$

The equation $\tau = (x_C - x_P) = \dfrac{y^4}{8r_0^3}$ defines the glass that must be selectively removed. At the rim of the sphere $\tau = \dfrac{h_0^4}{8r_0^3} = \dfrac{81}{8 \times 10^6}$, giving 10 microinches as the glass to be removed. This 10 microinches is only half of a wavelength of green light, and yet its removal is very important. We shall determine how important it is by comparing the reflected parallel rays united by the sphere, before this 10 microinches is removed, with the perfect union of them that geometrical optics predicts when they are united by the parabola. Although the predicted geometric union by a parabola is perfect, we know from physical optics that the radius of the produced blur circle can be no less than the radius of Airy's disk. Setting $h_0 = 3$ inches for the rim of the mirror and $\lambda = 20 \times 10^{-6}$ inches for green light, the radius of Airy's disk is predicted by our diffraction theory as

$$1.22 \frac{20 \times 10^{-6}}{6} = 4 \text{ microradians}$$

This Airy disk gives a minimum blur circle of 200 microinches, when $v = 50$ inches, for $r_0 = 100$ inches. And this blur radius describes the best union that can be expected of the parabolic contour. The character of the union of parallel rays reflected by the sphere may be determined from twice the difference between the slope of the sphere and the slope of the parabola, $\left(\dfrac{y}{r_0}\right)^3$. Multiplied by $f = 50$ inches, this doubled slope difference gives a predicted deviation of the marginal rays, reflected by the sphere, of 1350 microinches to one side of the paraxial focus. Actually the rays reflected by the sphere, at a point defined by the coordinate y, cross over the ₵ at a distance $\dfrac{y^2}{4r_0}$ in front of the paraxial focus. A student exercise shows from this that there is a fourfold smaller, and minimum geometric blur circle for this sphere, of radius 337 microinches, lying intermediately between the ₵ crossover point for rim rays, and the focal point for the paraxial rays. Even so, this 337 microinches blur circle radius is larger than the 200 microinches Airy disk radius. It points to the need of parabolizing the sphere by selective polishing, or figuring.

F. Twyman says· that a cloth polishing tool of $1\frac{1}{4}$ inches diameter, when charged with moist rouge, will normally remove about $\frac{1}{10}$ microinch per stroke. This experience of Twyman's illustrates what may be expected in the action of a pitch polisher, such as is used for figuring. Thus the 10 microinches to be removed will require a hundred strokes. Although the work to be done is very delicate, the means to do it is equally delicate; and, as we shall soon see, the test for controlling the work is also delicate.

13-10. *Foucault's Knife-edge Test*

Fig. 13-14a shows a testing arrangement with a pinhole light source near the center of curvature of the mirror to be tested. Here a knife-edge is shown in partial eclipse of the image at the union of rays that are reflected by the tested mirror. The eye is located close enough to the image so that all the uneclipsed rays enter its pupil. Such a view of a mirror or lens surface is called Maxwell's view. Fig. 13-15a illustrates the uniform illumination that is to be expected over the face of the tested mirror when it is a sphere.

The view of a lens or mirror with the eye at or near the union of focused rays, so that the pupil receives them all, is called a Maxwellian view because Maxwell, in some of his color mixing experiments, caused several objectives to be thus viewed simultaneously and superimposed.

Fig. 13-14b illustrates the disposition of the eclipsing knife, the tested mirror, and an auxiliary test flat, as used when the tested mirror has a true parabolic contour. Fig. 13-15b shows the uniform illumination in the Maxwellian view that a true parabolic mirror presents by this arrangement.

Figs. 13-14c and 13-15c relate to the testing arrangement and test result for an untrue spherical mirror with a raised intermediate ring zone, symmetrical about the ₵. In a geometrically perfect image, as in Fig. 13-15a, the

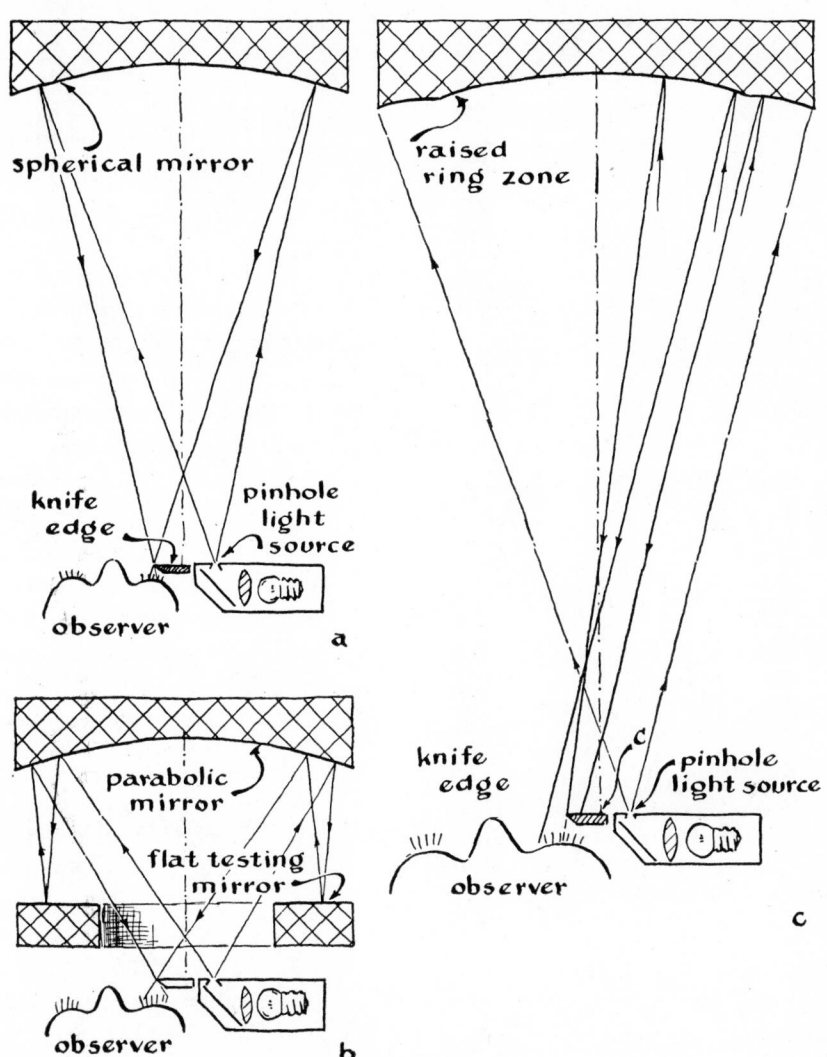

FIG. 13-14 Foucault test of spherical mirror at its center of curvature (a); test of parabolic mirror with auxiliary testing flat (b); test of mirror with symmetrical, raised, intermediate error zone (c).

rays from the different local areas of the mirror are equally eclipsed when the knife-edge is brought into their union. However, for a mirror with an error ring zone, such as that in Fig. 13-14c, the rays are less eclipsed where the mirror facets tip away from the knife-edge, and these facets on the face of the mirror appear relatively brighter when the knife edge is brought into the union; and oppositely tipped facets give a relatively darker appearance.

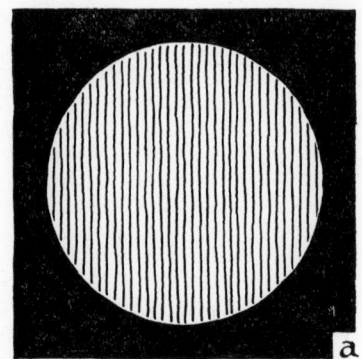

spherical mirror tested
at center of curvature

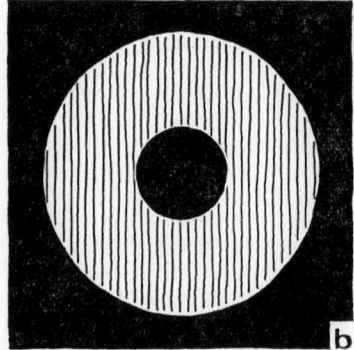

parabolic mirror tested
with a flat testing mirror

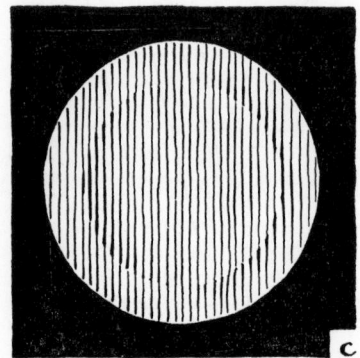

spherical mirror with raised annular ridge as tested
at center of curvature

FIG. 13-15 Appearance of mirrors under the Foucault test of Fig. 13-14.

These appearances are indicated in Fig. 13-15c. While Fig. 13-15a shows the view when the mirror is properly spherical, Fig. 13-16b shows it as viewed with the arrangement of Fig. 13-14a, after it is parabolized. The difference between the parabola and sphere at the best focus appears as a broad intermediate raised zone, qualitatively like that above at c. Fig. 13-16a shows the Maxwellian view appearance for a spherical mirror when tested with the setup of Fig. 13-14b. When the sphere is viewed in this parabola testing setup it appears to have a raised intermediate zone somewhat like that exhibited by the parabola at b. The setup of Fig. 13-14b requires a true testing flat. Without such an auxiliary mirror, the amateur is reduced to figuring his mirror so that it appears qualitatively as shown at Fig. 13-16b and determining if it is quantitatively correct by means of knife-edge test measurements.

A "coincident" pinhole light source and knife-edge is convenient for carrying out Foucault's test, as well as for use with the Gaviola test described in

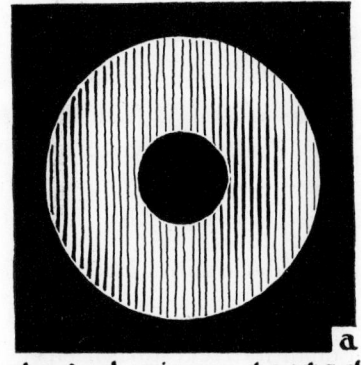 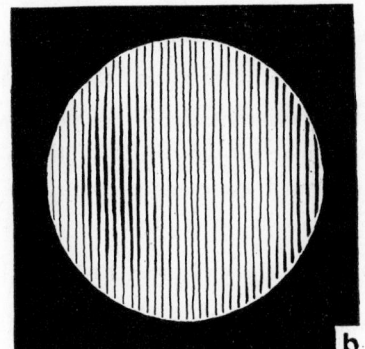

spherical mirror tested   parabolic mirror tested at
with a flat testing mirror   mean center of curvature

FIG. 13-16   Appearance of mirrors under the Foucault test of Fig. 13–14.

the next section. This coincidence is achieved by means of a beam-splitter mirror, such as is shown in Fig. 13-17, the whole device being mounted so that its motion is measurable.

By means of a measurement of longitudinal aberration, the amateur may determine, semi-quantitatively, from the appearance of his mirror as shown in Fig. 13-16b, when he has the correct quantitative departure from a sphere. The longitudinal aberration of a parabola is the difference between the ₡

FIG. 13-17   Coincident source and knife edge, to avoid parallax.

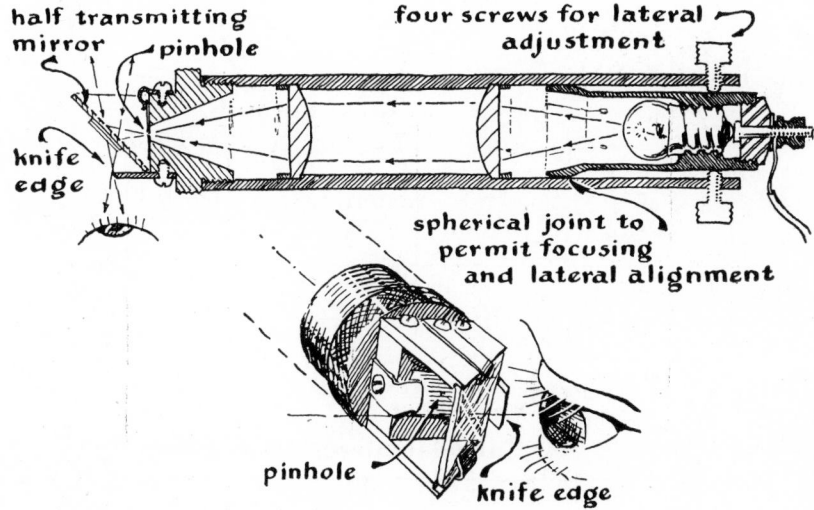

half transmitting mirror   pinhole

four screws for lateral adjustment

knife edge

spherical joint to permit focusing and lateral alignment

pinhole   knife edge

cross-over for paraxial rays at $\frac{r_0}{2}$, and the cross-over point on the ₵ for parallel rays returned by the zone of the parabolic mirror at y. This cross-over
point, experimentally, is the point where the returned rays from the zone y,
on each side of the center of the mirror, are simultaneously cut off by the
"coincident" knife-edge. From geometry these two cross-over points for a
true parabola are found to be separated, longitudinally, by the amount
$x_i = \frac{y^2}{2r_0}$, as shown in Fig. 13-18. Although a parabolic mirror can be surveyed
by determining the cross-over points for a set of contiguous zones laid out
across the face of the mirror, a much better quantitative test has been devised
by Dr. Enrique Gaviola, and it is illustrated by Fig. 13-18, and described
below.

13-11. *Gaviola's Caustic Test*

Dr. Gaviola's test applies to the survey of deviations of a mirror surface
from a true parabolic shape. The surveyed deviations are assumed to be
symmetrical about the ₵, which is a realistic assumption. Gaviola's method
is based on the fact that centers of curvature, in the plane of a diametrical
section through the center of a true parabolic mirror shape, do not lie on the ₵.
Rather, they lie off the center line on a curve called the caustic. The facet
of the parabola, Δy_i, of Fig. 13-18, for example, has its center of curvature

---

FIG. 13-18   Geometry of Gaviola's test with knife edge on the caustic.

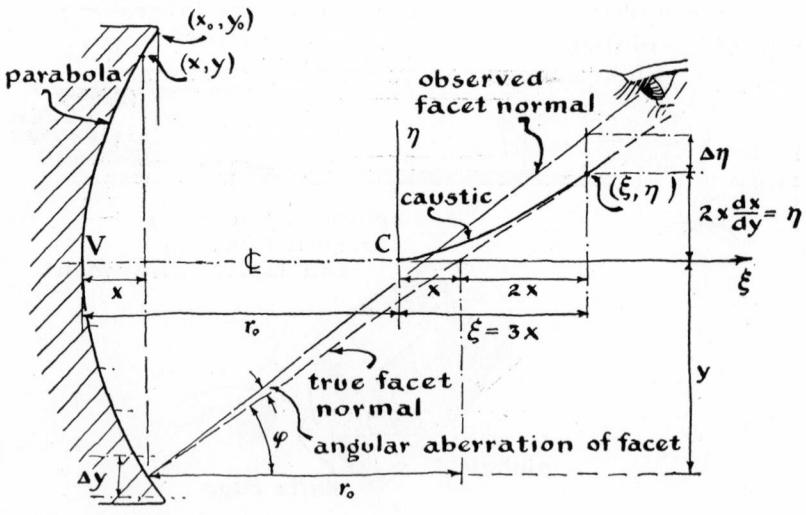

at the point (ξ_i, η_i) in the auxiliary coordinate system shown. The relationships between these caustic coordinates (ξ_i, η_i) and the parabola facets, (x_i, y_i), are given by analytic geometry as

$$\xi_i = \frac{3y_i^2}{2r_0} \qquad \eta_i = 2\frac{y_i^2}{2r_0}\left(\frac{y_i}{r_0}\right) = 2x_i\left(\frac{dx}{dy}\right)_i,$$

A Gaviola survey of the parabolic mirror, using these caustic coordinates, involves the following steps to determine improper inclination of such zones as that shown, Δy_i: We divide a strip across the center of the parabola into a succession of facets of equal width, Δy_i. For a particular one of these facets, centered at y_i as shown, we set the source-and-knife gadget where its center of curvature should lie if the parabola were true, that is at (ξ_i, η_i). If this zonal facet at y_i has the correct inclination then the source-and-knife should lie on its center of curvature. If, however, the mirror is not a true parabola, then the source-knife gadget must be moved, say, to η_i'. Here the light reflected at y_i will be returned on itself, and $(\eta_i' - \eta_i)$ determines the angle of improper inclination, ϵ_i. After the improper inclinations of all facets have been determined, we may calculate the curve of shape of the mirror surface. At each facet the linear increment of deviation is $\epsilon_i \Delta y_i$; and the accumulation of such linear deviations determines the curve of shape. The aggregate deviations are

$$\tau_{y_i} = \sum_{l=1}^{l=\frac{y_i}{\Delta y_i}} \epsilon_l \Delta y_l$$

The procedure of determining the ϵ_i's consists of the following steps. First, by means of the "coincident" pinhole light source and knife-edge of Fig. 13-17, we determine r_0 for the central zone of the parabola. Second, by means of this r_0, and formulas given above, we calculate the coordinates of the caustic, ξ and η. Thirdly, we move the knife-edge device from the center of curvature just determined to the calculated point (ξ_i, η_i), for testing the uncovered facet at y_i. Then we make the "further shift" to η_i', at ξ_i, so that the knife-edge device is brought to lie perpendicular to the facet. The improper inclination, ϵ_i, is proportional to the perpendicular component, *i.e.*, $(\eta_i' - \eta_i) \cos \varphi$. Finally, this component is divided by the parabola's local radius of curvature to give the inclination ϵ_i. The local radius, r_i, is

$$r_i = r_0\left\{1 + \left(\frac{y_i}{r_0}\right)^2\right\}^{3/2}$$

This procedure is repeated for all the facets of the mirror, so that the curve of shape may be determined by the sum given above.

The zones, Δy_i, should be taken narrow compared to any residual errors of figure which are to be measured and removed. Very narrow zonal errors can be seen with much greater sensitivity when they are examined from the caustic than when they are examined with the Fig. 13-17 source-and-knife-edge device located on the optical ₵.

Chapter XIV

Images of Points by Systems of Surfaces

IN THIS chapter we extend our treatment of images to include images formed by two or more reflecting or refracting surfaces in cascade—their centers of curvature being all on the common optical center line, ₵. We consider that successive refractions by successive surfaces give successive images of the rays coming from an object point source on this ₵. In this chapter our concern is with the so-called longitudinal equations, giving the successive and final longitudinal image locations on the ₵. In the next chapter we shall continue these considerations but with the point source moved off the ₵, bringing us to the concepts of magnification and field.

We first consider refractions produced by the two plane surfaces of a parallel plate, and then deviations produced by the two plane surfaces of prisms. Then we pass on to images produced by the curved surfaces of lenses—first the two surfaces of a thin lens; then thick lenses; and finally combinations of lenses.

14-1. *Plane Parallel Plate, and Prisms*

Two plane surfaces successively refracting light rays from an object point produce virtual images. Fig. 14-1 shows a tipping plane parallel plate of thickness d and illustrates how its tipping is used to produce a lateral displacement of light rays (and images). This plate may be used for measuring small angular deflections of a light ray, such as deflections of a mirror galvanometer produce. In this particular usage, the beam after the galvanometer response is returned to its original undeflected position by a measured rotation of the tipping plane parallel plate. Such a return can be accomplished far more

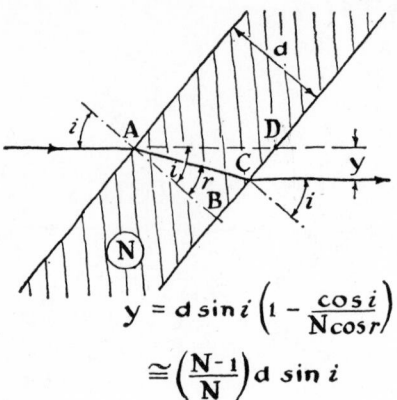

FIG. 14-1 Geometry of tipping-plate ray-displacement device.

$$y = d \sin i \left(1 - \frac{\cos i}{N \cos r}\right)$$

$$\cong \left(\frac{N-1}{N}\right) d \sin i$$

precisely than the galvanometer deflection can be read on the scale directly. In Fig. 14-1, the displacement is seen to be

$$(BD - BC) \cos i = d \cos i \, (\tan i - \tan r) = d \sin i \left(1 - \frac{\cos i}{N \cos r}\right)$$

and if i is a small angle, the expression reduces to

$$\left(\frac{N-1}{N}\right) d \sin i$$

Fig. 14-2 shows several other considerations of the imaging effects of a plane parallel plate for rays diverging from a point P. In a the image of P_1, after one refraction, is P_1'; and the image of this image, after a second refraction, lies at P_2'. P_2' is separated from P_1' by the longitudinal distance $\Delta x = d \left(\frac{N-1}{N}\right)$. The student can derive this equation easily; but we know

---

FIG. 14-2 Geometric optics of plane parallel plate.

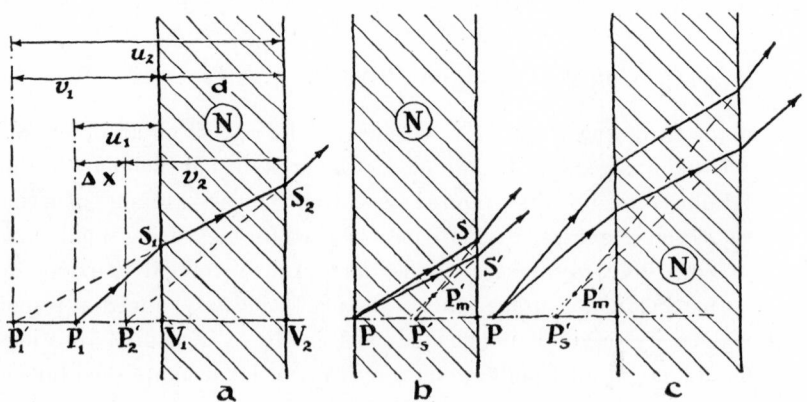

it is correct since the distance from P_2', perpendicular to the ray from P_1, is $\Delta x \sin i$. This is just the lateral displacement we found, in the last paragraph, to be produced by our tipping plate.

In Fig. 14-2b, two rays of an oblique pencil of rays from P, which lie in the plane of the figure, are refracted at S and S', respectively. And the refracted rays appear to diverge from the point P_m'. However, a third ray in the pencil, which lies at the same angle to the ₵ as the one refracted at S, but above or below S, will appear to originate from the same point as the ray through S, at the point P_s' which lies on the ₵. The rays in a plane that are bounded by a pair of rays is called a fan of rays. Thus a fan of rays in the plane of the figure appears to come from P_m', while a fan of rays in a plane perpendicular to the plane of the figure appears to come from P_s'. Here we have a simple example of astigmatism, an aberration we shall encounter in further detail in Chapter XVI.

Fig. 14-2c shows the two astigmatic virtual images of two such fans of rays when the object point P does not lie on the surface of the parallel plate.

In the case of the prism, we shall treat one ray that is deviated successively by its two surfaces. Fig. 14-3 shows the prism penetrated symmetrically at a, and unsymmetrically at b. For convenience we call the refracting angle of the prism $2\varphi_0$. Also, we call the total angular deviation for case (a) $2\delta_0$. In this symmetrical case the angle of refraction at the first face, or the angle of internal incidence at the second, is $r = i' = \varphi_0$. The angle of incidence at the first face, or of refraction at the second face, is $i = r' = (\delta_0 + \varphi_0)$. Applying Snell's law,

$$N = \frac{\sin (\varphi_0 + \delta_0)}{\sin \varphi_0}$$

When φ_0 is small this expression reduces to $\delta_0 = (N - 1)\varphi_0$.

FIG. 14-3   Geometric optics of prism worked on (a) and off (b) minimum deviation.

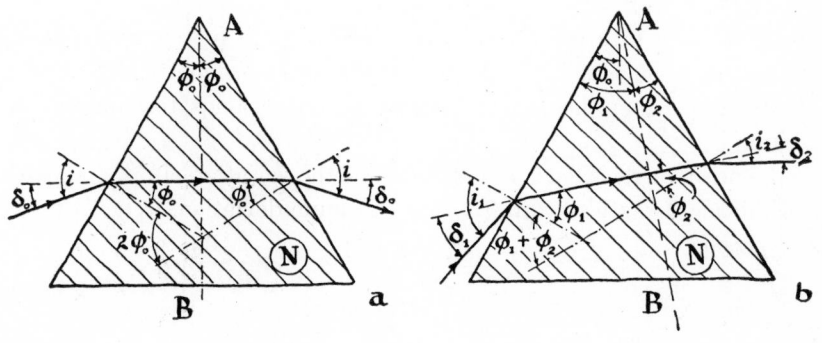

The non-symmetrical penetration of this prism is best considered by taking the total prism angle, $2\varphi_0$, as being comprised of two unequal half prisms. For one reason, we already have the expression which covers each half prism. Each is penetrated symmetrically since the internal ray in Fig. 14-3b is perpendicular to the line AB which divides the total prism angle, now unequally. The line AB divides the refracting angle into angular parts: φ_1, half of a prism of angle $2\varphi_1$; and φ_2, half of a prism of angle $2\varphi_2$. We represent the total deviation in this case, similarly, by two unequal half deviations: δ_1 and δ_2. This procedure makes it possible to treat this non-symmetrical case as the composite of two parts: one with $r_1 = \varphi_1$; and the other with $r_2 = \varphi_2$; and with $i_1 = (\delta_1 + \varphi_1)$ and $i_2 = (\delta_2 + \varphi_2)$. Since our total prism angle is $2\varphi_0$, it follows that the half prism angles of the parts are $\varphi_1 = \varphi_0 + \Delta\varphi$ and $\varphi_2 = \varphi_0 - \Delta\varphi$. To get the total deviation for an only slightly non-symmetrical penetration we use the following first and second derivatives of the component deviations: From $\dfrac{di}{dr} = \dfrac{\tan i}{\tan r}$, obtained from Snell's law,

$$\frac{d(\delta)}{d\varphi} = \frac{d}{dr}(i - r) = \frac{\tan i}{\tan r} - 1$$

$$\frac{d^2(\delta)}{d\varphi^2} = N\frac{d}{dr}\left(\frac{\cos r}{\cos i}\right) = N\frac{\sin r}{\cos i}\left(\frac{\tan^2 i}{\tan^2 r} - 1\right) > 0$$

With these derivatives we may express the component deviations, δ_1 and δ_2, as Taylor series. Recalling that $(\Delta\varphi)$ for prism #1 is $(-\Delta\varphi)$ for prism #2, these deviations become

$$\delta_1 = \delta_0 + \frac{d(\delta)}{d\varphi}(\Delta\varphi) + \frac{1}{2!}\frac{d^2(\delta)}{d\varphi^2}(\Delta\varphi)^2 + \cdots$$

$$\delta_2 = \delta_0 + \frac{d(\delta)}{d\varphi}(-\Delta\varphi) + \frac{1}{2!}\frac{d^2(\delta)}{d\varphi^2}(-\Delta\varphi)^2 + \cdots$$

The total deviation for the non-symmetrical case, adding, becomes

$$\delta_1 + \delta_2 = 2\delta_0 + \frac{d^2(\delta)}{d\varphi^2}(\Delta\varphi)^2$$

This deviation is greater for any penetration which is not symmetrical, since $(\Delta\varphi)^2$ and its coefficient are both positive. Thus the symmetrical penetration, with its simplest mathematical description, corresponds to a minimum of deviation.

If the prism angle is fixed and we examine the variation of $(\delta_1 + \delta_2)$ with λ, using this mathematical expression for minimum deviation, we get the *angular dispersion*

$$\frac{d}{d\lambda}(\delta_1 + \delta_2) = \frac{d}{dN}(\delta_1 + \delta_2) \cdot \frac{dN}{d\lambda}$$

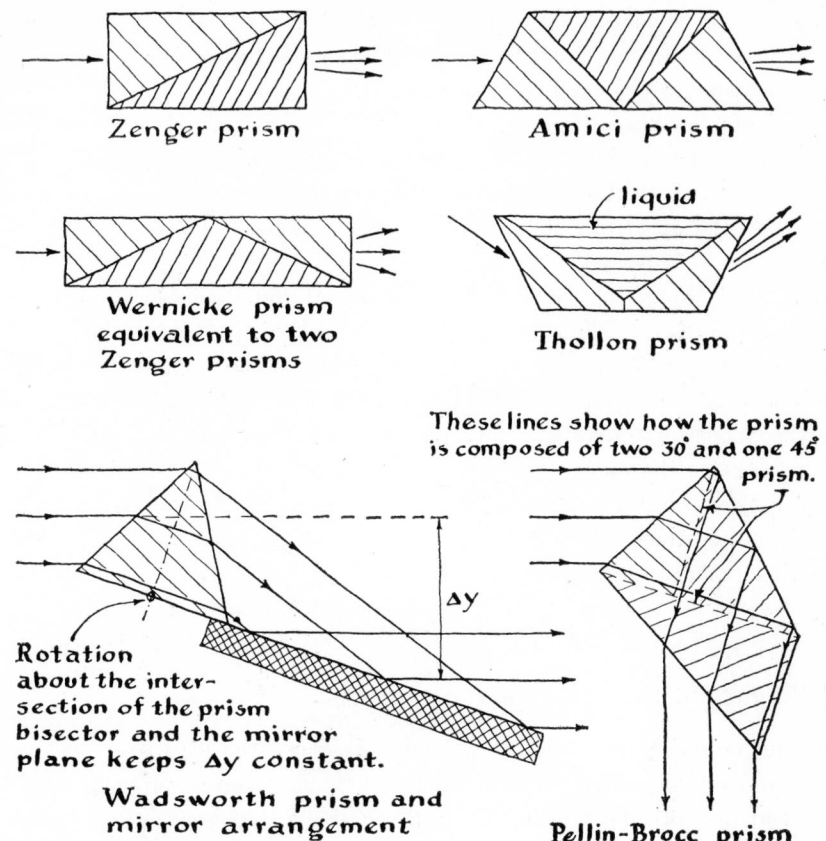

FIG. 14-4   Various composite prisms and various methods of working prisms.

$\dfrac{dN}{d\lambda}$ is an intrinsic property of the prism material, and it can be obtained from Cauchy's equation of § 4-6. Differentiating $N \sin \varphi_0 = \sin (\delta_0 + \varphi_0)$, for the symmetrical and near symmetrical cases, we get

$$\frac{d}{dN} (\delta_1 + \delta_2) = \frac{2 \sin \varphi_0}{\sqrt{1 - N^2 \sin^2 \varphi_0}} \quad \text{or} \quad \frac{1}{\sqrt{1 - \tfrac{1}{4}N^2}} \quad \text{if} \quad \varphi_0 = 30°$$

Sometimes composite prisms are used as shown in Fig. 14-4. For example, the Zenger prism is a composite of two right angle prisms, made of different glasses, with the components arranged hypotenuse to hypotenuse to form a right parallelepiped of glass. Glasses of nearly the same index for the yellow light, but of different dispersion, are chosen for this composite. This composite prism is often used for making a direct vision spectroscope. The Wernicke or Amici prism is a composite of three prisms, as shown. With such a 3-component prism J. Duclaux and G. Ahier [*cf. Rev. d'Optique 17* 417 (1938)] achieved an angular dispersion equal to that of 6 prisms of heavy flint glass, and, at the same time, with higher transmission. The transmission was higher because dispersion grows rapidly as the angle of incidence at a solid-solid interface increases, whereas light loss by reflection at the solid-solid interface grows more slowly. A further advantage in using this type of prism lies in the fact that the spectrum is less polarized by differential σ and π interface reflections.

Fig. 14-4 shows the Pellin-Broca prism and the Wadsworth prism-and-mirror combination. These arrangements are used in constant deviation monochromators. In both, $\varphi_1 = \varphi_2 = \varphi_0$ is invariant at $30°$. The total constant deviation is $90°$ for the Pellin-Broca prism. The total constant deviation is $0°$ for the Wadsworth arrangement. The light that emerges through the exit slit penetrates the entering and emerging prism refracting surfaces equally in both cases. Only the Pellin-Broca prism, or the Wadsworth arrangement, is rotated in order to vary the wavelength of the light passed by the exit slit. In either case, as the wavelengths are varied by rotation, this light is always dispersed as by a $60°$ prism used at minimum deviation.

14-2. *Longitudinal Gaussian Equations*

In our considerations so far, with paraxial rays, the rays from the point source have been considered limited so that they strike the refracting surface only near the optical center line $\mathbb{C}$. In the case of a single refracting or reflecting surface, or thin lens, the radius of the rim h_0 sets this limit on the extent of the wave front which is focused. Half the angle that this rim subtends at the object is called the *angular aperture*. Correspondingly, half the angle that this rim subtends at the image is called the *angle of projection*. When an iris aperture, centered on the $\mathbb{C}$, is in front of the lens it may determine the angular aperture. And correspondingly, the half angle that the image of such a limiting iris rim subtends in image space at the image point is the angle of projection. For the present we shall ignore the relationship which exists between these angles and simply take both the angular aperture and the angle of projection as small.

Fig. 14-5a shows two rays from the object point P_1, located on the $\mathbb{C}$ of a lens. One of the rays runs along the $\mathbb{C}$ itself to V_1, and the other, refracted at S_1 and S_2, is returned to the $\mathbb{C}$ at the cross-over point P_2'. There is an intermediate virtual cross-over point, at P_1' or P_2. The refractions at S_1, and at S_2, are illustrated separately in the figure at b and c. For paraxial rays, if P_1 is a point source of light, P_2' is its final image. P_1', or P_2, is both an image for first surface refraction at S_1, and an object for the second surface refraction at S_2.

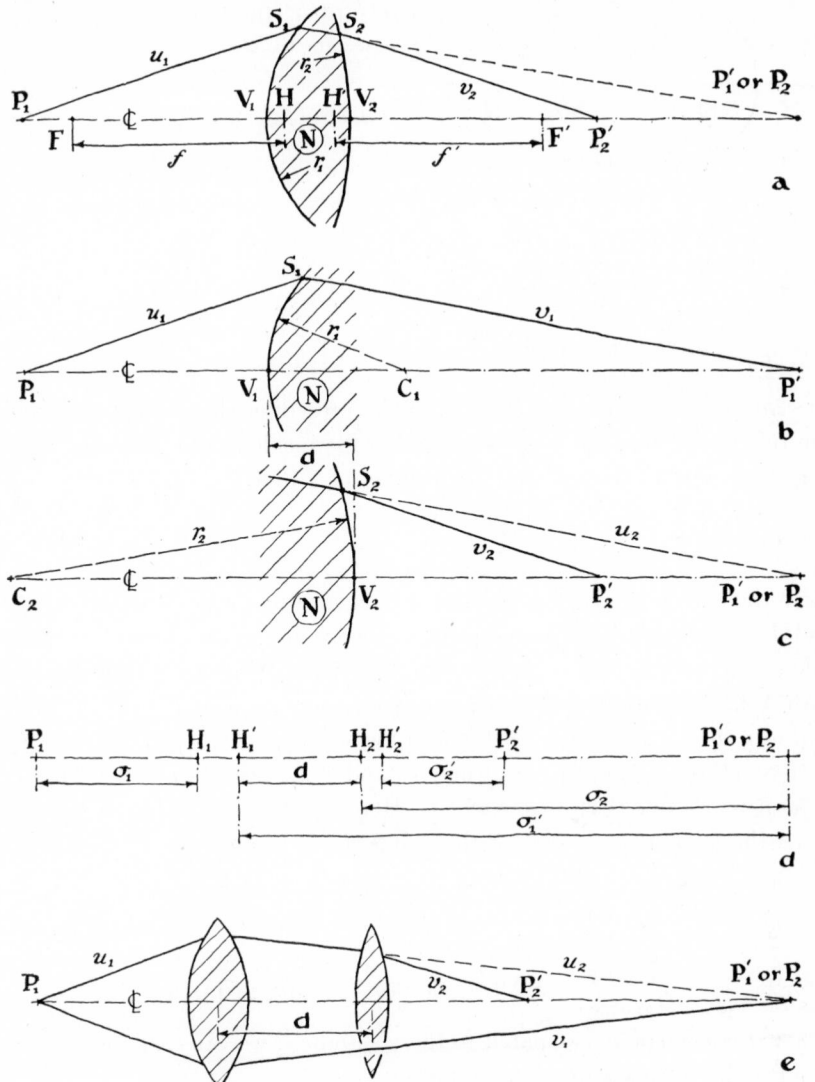

FIG. 14-5   Image location for a thick lens and for combinations of lenses.

The distances u_1 and v_1, measured to and from S_1, as well as u_2 and v_2, measured to and from S_2, are reckoned positive or negative according to our sign conventions of the previous chapters. Thus, in Fig. 14-5, u_2 is a negative quantity, while v_1 and v_2 are positives. These distances for paraxial rays become substantially equal to the distances along the ₵ from V_1 and V_2. Taking $\cos \varphi$ and $\cos \varphi'$ as unity, the powers for the two refracting surfaces of our

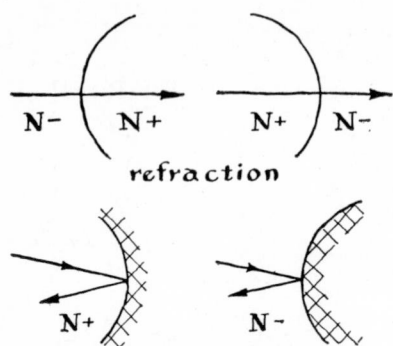

FIG. 14-6 Illustration of convention to assign signs to N's.

lens are obtained from our formula of the preceding chapter for F, invoking the rule given there for ascribing signs to the incident and emergent indices. Fig. 14-6 repeats the rule for the formula of § 13-4.

$$F_1 = \frac{N - 1}{r_1}; \qquad F_2 = \frac{-1 + N}{r_2}$$

Here the emergent rays for the first surface are the incident rays for the second; both emergence and subsequent incidence are on the concave or positive side of their respective refracting surfaces. A positive sign is thus associated with N in both inside cases, with a negative sign associated with the outside index on the convex side of each refracting surface.

If we take the limiting case where S_1 and S_2 lie near V_1 and V_2, since $-u_2$ is positive we may write $u_2 = d - v_1$. Here d is the separation of V_1 and V_2 along the ₵. The equations for the two surfaces become

$$\frac{1}{u_1} + \frac{N}{v_1} = F_1 \qquad \text{giving} \qquad v_1 = \frac{Nu_1}{u_1F_1 - 1}$$

$$\frac{N}{d - v_1} + \frac{1}{v_2} = F_2 \qquad \text{giving} \qquad v_1 = d - \frac{Nv_2}{v_2F_2 - 1}$$

When d is negligible, compared to the u's and v's, we can easily eliminate v_1, and get the so-called thin lens equation:

$$\frac{1}{u_1} + \frac{1}{v_2} = F_1 + F_2 = \mathfrak{F} \tag{a}$$

Here F_1 and F_2 are the powers of the cascaded lens surfaces, and $\mathfrak{F}$ is the power of the lens as a whole.

If d is not negligible, writing $\tau = \dfrac{d}{N}$, we can eliminate v_1, although not quite so quickly. Then we get the so-called thick lens equation. After first eliminating v_1, we get

$$u_1 v_2 + u_1 \left(\frac{F_1 \tau - 1}{F_1 + F_2 - F_1 F_2 \tau} \right) + v_2 \left(\frac{F_2 \tau - 1}{F_1 + F_2 - F_1 F_2 \tau} \right)$$

$$= \frac{\tau}{F_1 + F_2 - F_1 F_2 \tau}$$

It will turn out that the power of this thick lens as a whole is the recurrent denominator in the expression above. This power is represented by $\mathfrak{F}$, and

$$\mathfrak{F} = F_1 + F_2 - F_1 F_2 \tau$$

Rearranging terms,

$$u_1 v_2 + \frac{F_1 \tau}{\mathfrak{F}} u_1 + \frac{F_2 \tau}{\mathfrak{F}} v_2 = \frac{1}{\mathfrak{F}} (u_1 + v_2 + \tau)$$

If we add $\dfrac{F_1 F_2}{\mathfrak{F}^2} \tau^2$ to both sides of the above equation, the left side can then be factored:

$$\left(u_1 + \frac{F_2 \tau}{\mathfrak{F}} \right) \left(v_2 + \frac{F_1 \tau}{\mathfrak{F}} \right)$$

And, since adding $\dfrac{F_1 F_2}{\mathfrak{F}} \tau^2$ to τ yields $\left(\dfrac{F_1 \tau + F_2 \tau}{\mathfrak{F}} \right)$, the right side becomes

$$\frac{1}{\mathfrak{F}} \left\{ \left(u_1 + \frac{F_2 \tau}{\mathfrak{F}} \right) + \left(v_2 + \frac{F_1 \tau}{\mathfrak{F}} \right) \right\}$$

Now, on dividing both the right and the left side by the two factors on the left, while multiplying both by $\mathfrak{F}$, we get

$$\frac{1}{u_1 + \dfrac{F_2 \tau}{\mathfrak{F}}} + \frac{1}{v_2 + \dfrac{F_1 \tau}{\mathfrak{F}}} = \mathfrak{F} \tag{b}$$

From this equation it appears that if we measure object and image distances not from V_1 and V_2, as before, but from points at certain distances inside the V's, these certain distances being independent of u_1 and v_2, then the equation takes on the familiar Gaussian form that our thin lens equation had:

$$\frac{1}{\sigma} + \frac{1}{\sigma'} = \mathfrak{F} \tag{b'}$$

The two points from which object and image distances are now to be measured lie inside the thick lens surfaces by the amounts $\dfrac{F_2 \tau}{\mathfrak{F}}$ and $\dfrac{F_1 \tau}{\mathfrak{F}}$. Surfaces through these two points, perpendicular to the $\mathbb{C}$, are called principal surfaces and sometimes unit surfaces. As T. Smith puts it, ". . . it is evident that the distance of a unit surface from the corresponding refracting surface is approximately a constant fraction of the lens thickness in that neighborhood, so that the unit surfaces are necessarily curved like the refracting sur-

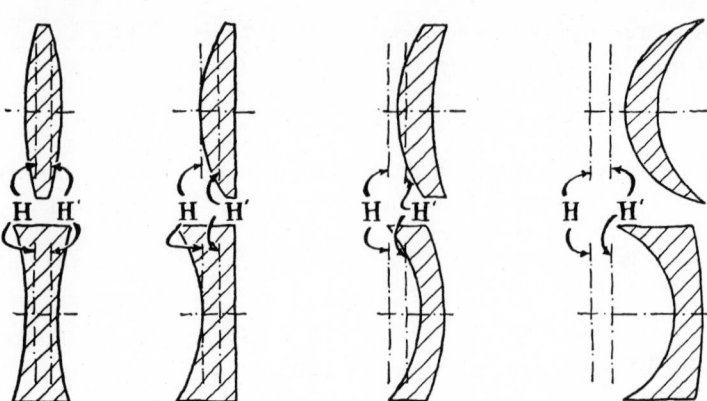

FIG. 14-7 Location of principal planes for various lenses. (Redrawn from Morgan.)

faces themselves. If, however, we are only dealing with rays refracted at a small distance from the axis of a lens, the longitudinal error that we shall make by treating the unit surfaces as planes will be very small. It is only under these conditions that we are justified in using the conception of unit planes." Fig. 14-7 shows some examples of the positions of these unit planes, labeled H and H', for various lenses. We can combine formulas for two thick lenses, mathematically, in the same manner that we have just combined them for two single refracting surfaces. We combine

$$\frac{1}{\sigma_1} + \frac{1}{\sigma_1'} = \mathfrak{F}_1 \qquad \text{with} \qquad \frac{1}{d - \sigma_1'} + \frac{1}{\sigma_2'} = \mathfrak{F}_2$$

where d here is the distance from the last principal plane for the first lens, H_1', to the first principal plane of the second, H_2. These four planes are shown in Fig. 14-5d, and the two corresponding thick lenses in 5e. Now since the equations for these two thick lenses are of the same mathematical form as the equations which represented refractions of the two single surfaces of our thin lens, or of a single thick lens, similar algebraic operations will lead to a similar result; namely,

$$\frac{1}{\sigma_1 + \dfrac{\mathfrak{F}_2 d}{\Phi}} + \frac{1}{\sigma_2' + \dfrac{\mathfrak{F}_1 d}{\Phi}} = \Phi$$

Here Φ is the focal power of the two thick lenses, *en ensemble,* and

$$\Phi = \mathfrak{F}_1 + \mathfrak{F}_2 - \mathfrak{F}_1\mathfrak{F}_2 d \tag{c}$$

And the principal planes of this combination, $\mathfrak{IC}$ and $\mathfrak{IC}'$, lie at the distances $\dfrac{\mathfrak{F}_2 d}{\Phi}$ and $\dfrac{\mathfrak{F}_1 d}{\Phi}$ inside of H_1 and H_2'.

Of course we can extend such considerations to combinations of combinations, all in an entirely analogous mathematical manner.

14-3. *Longitudinal Newtonian Equations*

In addition to the principal or unit planes, there are other fiducial points from which object and image positions on the ₵ can be located. Object and image positions may be located by measuring from the principal foci (of a thin lens, a thick lens, or a combination). These principal foci on the ₵ are labeled F and F'. The point F indicates the object position that gives an image at infinity, while F' is the image formed by parallel light from an object at infinity. We are already familiar with the Newtonian equation from our previous deduction with the equations for the single surface in § 13-6. There we measured the object position by $(u - f)$ and the image position by $(v - f')$, where f and f' were slant distances measured to the refracting surface. For a thick lens, the locations of the principal foci are obtained first by setting $\sigma' = \infty$ and then $\sigma = \infty$. These foci are $f_0 = \dfrac{1}{\mathfrak{F}}$ as measured from the principal planes. With $\dfrac{1}{f}$ written for $\mathfrak{F}$, Eq. 14-2b' can be written

$$\frac{1}{(\sigma - f) + f} + \frac{1}{(\sigma' - f) + f} = \frac{1}{f}$$

Now we multiply the terms of this expression by the least common denominator and represent $(\sigma - f)$ by x, and $(\sigma' - f)$ by x'.

$$(\sigma - f)(\sigma' - f) = f^2 = xx' \tag{a}$$

This is in the Newtonian form of the thick lens equation.

Fig. 14-8 illustrates a procedure by which the focal points and principal points for a thick lens may be experimentally determined. Here two screens with pinhole apertures in them, P and P', are separated by the fixed distance X. Three operations are involved: Firstly, we set the lens at A so that light from the first pinhole is focused back to a sharp image alongside P when returned through the lens by a tipped plane mirror M. This returned light is in focus, or autocollimated, when the first principal point F is at P. Then the distance, $u_F = PV_1$, may be measured. Secondly, we remove M and set the lens at B so that the pinhole at P is sharply focused on P'. The shift from A to B determines a Newtonian distance x. Thirdly, we replace M and move the lens to C, where we autocollimate the pinhole source at P' back on

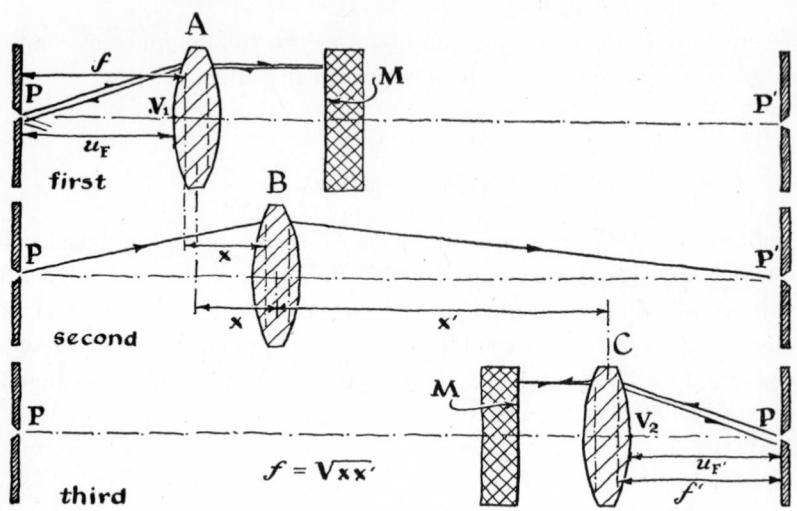

FIG. 14-8 Procedures to determine the principal planes and principal foci of a thick lens.

itself, and measure the distance, $u_{F'} = V_2 P'$. The shift from B to C determines the second Newtonian distance x'. The focal length of the lens is

$$f = \sqrt{xx'}$$

The experimental measurements may be checked by the relation

$$X - (u_F + u_{F'} + x + x') = d$$

where d is the overall thickness of the lens. The principal planes are at the distances $(f - u_F)$ and $(f - u_{F'})$, behind V_1 and V_2. This procedure works for a compound lens as well as a single thick or thin lens.

14-4. *Immersion Lenses*

In our consideration of systems of surfaces, above, we have taken the indices of the initial object space and final image space as both being unity. The index of refraction of object and image space is always different for a single, or an odd number of surface refractions. In optical systems with more than one refracting surface, the initial and final indices are thus often different— the most notable case, of course, being the eye itself. Equations similar to those developed above for the thick lens may be set up for these cases; and they may be manipulated with a similar algebra. For a thick lens of index N_2 separating two media of indices N_1 and N_3, we have $F_1 = \pm \dfrac{N_1 \pm N_2}{r_1}$

and $F_2 = \pm \dfrac{N_2 \pm N_3}{r_2}$. Appropriate signs are, of course, to be prefixed to the N's according to our curvature convention of § 13-4. The equation for such a thick lens, with different indices of object and image space, derives in a familiar manner.

$$\frac{N_1}{u_1} + \frac{N_2}{v_2} = F_1 \qquad \text{giving} \qquad \frac{v_1}{N_2} = \frac{u_1}{u_1 F_1 - N_1}$$

$$\frac{N_2}{d - v_1} + \frac{N_3}{v_2} = F_2 \qquad \text{giving} \qquad \frac{v_1}{N_2} = \frac{d}{N_2} - \frac{v_2}{v_2 F_2 - N_3}$$

If now we write $\dfrac{d}{N_2} = \tau$, and let $\dfrac{u_1}{N_1} = \mathfrak{U}$ and $\dfrac{v_2}{N_2} = \mathfrak{V}$, then our expressions have the same form as before, and we may write their solution by analogy, using $\mathfrak{F} = F_1 + F_2 - F_1 F_2 \tau$. Finally, returning from $\mathfrak{U}$ and $\mathfrak{V}$ to u_1 and v_2, we get

$$\frac{1}{\dfrac{u_1}{N_1} + \dfrac{F_2 \tau}{\mathfrak{F}}} + \frac{1}{\dfrac{v_2}{N_3} + \dfrac{F_1 \tau}{\mathfrak{F}}} = \frac{N_1}{u_1 + \dfrac{N_1 F_2 \tau}{\mathfrak{F}}} + \frac{N_3}{v_2 + \dfrac{N_3 F_1 \tau}{\mathfrak{F}}} = \mathfrak{F} \qquad \text{(a)}$$

Setting first v_2 and then u_1 infinite, we get the focal distances above:

FH as $f = \dfrac{N_1}{\mathfrak{F}}$, and $H'F'$ as $f' = \dfrac{N_3}{\mathfrak{F}}$. Here the f's are measured from principal planes.

The normal eye with its characteristic optical parameters is illustrated in Fig. 14-9. Here we have three refracting surfaces. The index of object space is $N_1 = 1.0$; and of the final image space, $N_3 = 1.336$; while the index of the intermediate immersed crystalline lens is 1.413. Typical radii of curvature at dielectric interfaces are indicated in the figure.

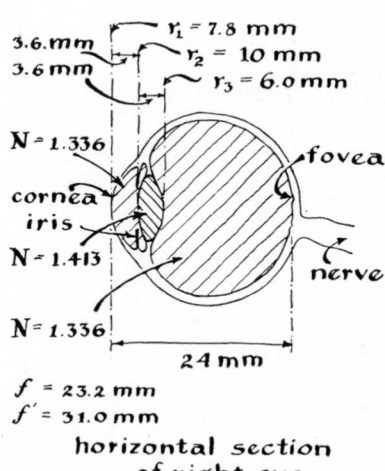

FIG. 14-9 Gullstrand's eye. (Redrawn from Martin.)

14-5. *Freedom from Spherical Aberration*

Although we shall consider the aberrations for off-axis objects in the next chapter, it is appropriate here for us to touch on spherical aberration. A lens begins to show imperfect union of the marginal rays that it collects from an object on the ₵ when the aperture is enlarged so that these marginal rays are no longer refracted at points which lie close to the optical axis. We have seen, in the last chapter, how this imperfect union, called longitudinal spherical aberration, enters at the rim of a spherical mirror used to focus parallel light.

Certain systems are free from spherical aberration when they are worked at certain conjugate points. Fig. 14-10 shows two examples of such systems; one is worked in refraction, and the other is worked in reflection. The first is a sphere worked in refraction at the conjugate points P and P'. The second is comprised of two mirrors—a spherical reflector at S_1, and a cardioid reflector at S_2. In the refraction example, converging monochromatic light incident on the sphere "from" the virtual object point P is refracted without longitudinal aberration to a real image point P'. P lies at a distance Nr to the

right of the center of curvature C, while P' lies at a distance $\dfrac{r}{N}$ to its right.

In the reflection example, rays of white light parallel to the ₵, but displaced variable distances h from it, are all focused after two reflections at a common image point P'.

In Fig. 14-10a, if we consider the triangles, $\triangle PCS$ and $\triangle P'CS$ in the figure, it becomes apparent that all incident rays directed toward P are refracted toward P'. These triangles are similar because they have a common side and

---

FIG. 14-10 Aplanatic points of the sphere (a) and of the Siedentopf cardioid mirror (b).

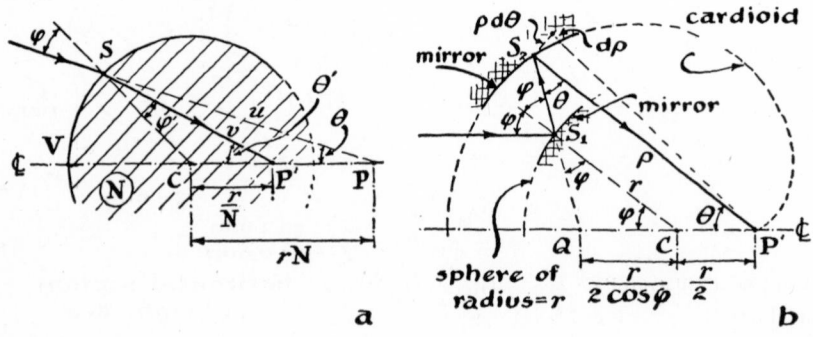

angle at C, and because the ratio of the sides which enclose this angle, for either triangle, is N. Therefore the angle φ' at S of the smaller triangle is equal to the angle at P of the larger triangle. Applying the law of sines,

$$\frac{Nr}{\sin \varphi} = \frac{r}{\sin \theta}; \quad \frac{\dfrac{r}{N}}{\sin \varphi'} = \frac{r}{\sin \theta'} \quad \text{giving} \quad \frac{\sin \varphi}{\sin \varphi'} = N$$

Since these relationships fulfill the requirements of Snell's law, the refractions deviate all rays toward P' that are incidentally directed toward P to give an image without spherical aberration.

In Fig. 14-10b parallel rays from infinity are first reflected by the convex sphere, centered at C, like the one at S_1. Then they are again reflected by the concave cardioid mirror, as at S_2. This mirror has the special shape of a cardioid of revolution—a section of it is expressed in a polar coordinate system by the equation $\rho = r(1 + \cos \theta)$, where r is here the radius of the sphere centered at C. The twice reflected rays are focused at the origin of the polar coordinate system, as we shall show, and at a distance $\dfrac{r}{2}$ behind C. The angles labeled φ in the figure, for the representative ray, are all equal; the triangle $\triangle QS_1C$ is therefore an isosceles triangle; and Q, from which the ray appears to come after reflection at S_1, lies at a distance $\dfrac{r}{2 \cos \varphi}$ to the left of C on the ₵.

Now consider this ray reflected at S_2, a point defined by the radial coordinate ρ and angle θ. By the law of reflection, this ray must be deviated at S_2 through twice the angle which lies between the line perpendicular to $P'S_2$ and the tangent to the cardioid surface; that is, through twice the angle whose tangent is $\left\{ \dfrac{d\rho}{d\theta} \Delta\theta \div \rho\Delta\theta \right\}$. Differentiation of ρ gives this tangent as $\dfrac{\sin \theta}{1 + \cos \theta}$, which is $\tan \dfrac{\theta}{2}$. Thus the total deviation at S_2 is θ; and the triangle $\triangle QS_2P'$ is also isosceles. Now if $\triangle QS_2P'$ and $\triangle QS_1C$ are both isosceles, with a common angle at Q, then θ and φ are equal, and the ray in question goes through P'.

14-6. *Spherical Aberration of a Lens*

Having described two systems that are free of spherical aberration, let us turn to the consideration of the spherical aberration of a simple lens. This aberration depends both on object and image distances, and on the two lens surface curvatures. Although spherical aberration of a simple lens may be minimized by choice of a suitable shape, it cannot be eliminated. The longitudinal spherical aberration of a simple lens has been described analytically in the Paper No. 461 of the Bureau of Standards. It is described in terms of four index of refraction parameters, and two others; an object and image position parameter τ, and a lens shape parameter σ. These last two parameters are:

$$\tau = \frac{v_2 - u_1}{v_2 + u_1}$$

$$\sigma = \frac{r_2 + r_1}{r_2 - r_1}$$

Here, in this σ formula alone we ascribe signs to the r's, and use the convention that r is positive if light is incident on the convex side, and negative if it is incident on the concave side.

The four index of refraction parameters are:

$$a = \frac{N + 2}{8N(N - 1)^2} \quad \text{for } N = \frac{3}{2}: \quad a = 1.166$$

$$b = \frac{N + 1}{2N(N - 1)} \quad\quad b = 1.666$$

$$c = \frac{3N + 2}{8N} \quad\quad c = 0.542$$

$$d = \frac{N^2}{8(N - 1)^2} \quad\quad d = 1.125$$

Spherical aberration will be described in terms of how far the cross-over point, for rays refracted at a distance h from V, misses the paraxial focal point. This difference, Δv_2, is given by the Bureau of Standards expression as follows:

$$\frac{\Delta v_2}{v_2^2} = -\Delta \left(\frac{1}{v_2}\right) = \frac{h^2}{f^3} (a\sigma^2 + b\sigma\tau + c\tau^2 + d)$$

As an example, consider that a thin lens focuses parallel light; *i.e.*, $u_1 = \infty$. Then we have $\tau = -1.0$. Mathematical manipulations soon show that the expression above cannot be made zero by any possible choice of σ, although we can find a σ which will give a minimum of longitudinal aberration. This minimum is gotten from the derivative of $\Delta \left(\dfrac{1}{v_2}\right)$ with respect to σ. On setting this derivative equal to zero we get

$$\sigma_{\min} = -\frac{b}{2a}\tau = -2(N - 1)\left[\frac{N + 1}{N + 2}\right]\tau$$

For parallel light we find that the incident and emergent surfaces of a lens of $N = 1.5$ should both be convex, with the ratio of radii of emergent and incident surfaces -6 to 1. But for a lens of index $N = 2.0$, we find that the incident surface should be convex with $\frac{1}{5}$ the radius of an emergent concave surface. Fig. 14-11 not only illustrates these two cases diagrammatically,

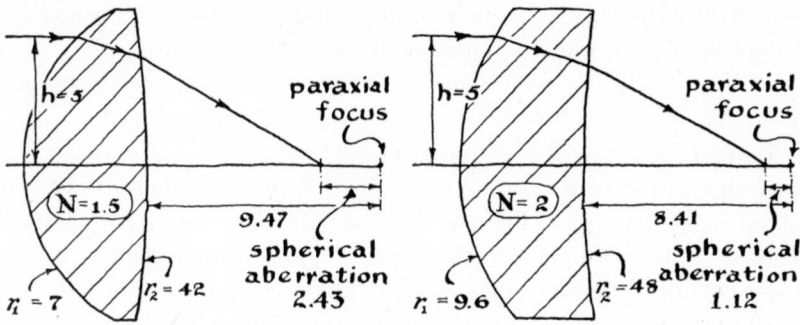

FIG. 14-11   Longitudinal spherical aberration minimized for $N = 1.5$ and $N = 2$.

but shows that our minimum of spherical aberration is considerably smaller for the higher index lens, even when it is a thick lens.

In the preceding chapter we saw that the longitudinal aberration of a spherical mirror had the value $\dfrac{h^2}{4r_0}$, or $0.125 \dfrac{h^2}{f_{\text{mirror}}}$. This value is less than the minimum spherical aberration of a lens. For a lens characterized by $\tau = -1.0$ and $\sigma_{\min} = \dfrac{5}{7}$, of index $N = 1.5$, we get the value $1.07 \dfrac{h^2}{f_{\text{lens}}}$ for the minimum spherical aberration. This eightfold larger value shows that a lens with negative f_{lens} could easily have the same magnitude of longitudinal spherical aberration as a mirror, but with opposite sign.

A. F. E. Mangin (1876) was the first to use a compensating lens for mirror aberrations. He did this by silvering the convex side of a diverging meniscus lens, such as is shown in Fig. 14-12a. The curvatures of this lens,

FIG. 14-12   Mangin mirror and Bouwers' catadioptric system.

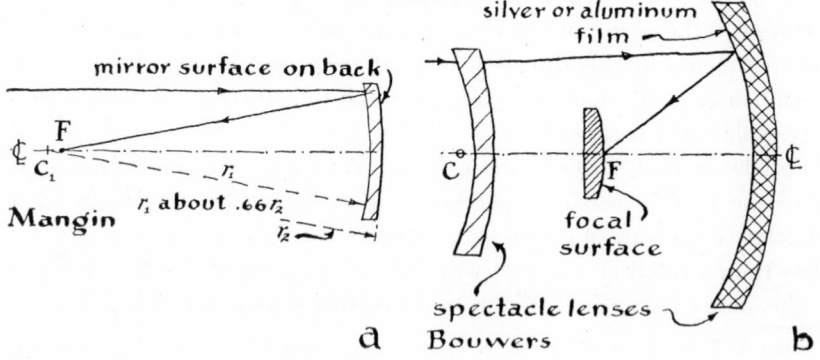

when correctly chosen, give freedom from both spherical aberration and coma. But the field of this combination is small. The center of curvature of the un-silvered face of the meniscus lens lies approximately at the focus of the combination.

A. Bouwers was the first to recognize and use these compensating longi-tudinal aberrations in a separated combination, with the lens at the center of curvature of the mirror. Fig. 14-12b shows his use of a relatively large lens aberration, with relatively small power. The quotation below from his *Achievements in Optics* describes this use. His use retains most of the desir-able achromatic power of a concave spherical mirror while correcting its undesired spherical aberration with a weak negative lens. In the words of Bouwers:†

"It is more or less obvious that a spherical mirror, combined with a suit-able lens or system of lenses, should produce reasonable images. In recent literature there are some interesting examples of such combinations in which the negative curvature of the mirror image is compensated by the positive curvature of the lens system. It is possible to treat systems so far published along the usual lines with the application of the well-known theories of aberrations.

"Insated of following this 'classical' course, we have tried to retain as many as possible of the excellent properties of the spherical mirror, accepting the curved images produced by it. We have reasoned approximately along the following lines. Let us place before the mirror and centred with it a single negative lens of a given small power, say a twentieth of the power of the mirror, its focal distance being thus twenty times the focal distance of the mirror. This lens does not materially alter the power of the mirror nor the curvature of the image, but the system will probably show a considerable amount of spherical aberration, although less so than the mirror would alone.

"We may now replace the negative lens by another one of the same power but another degree of 'bending'; perhaps such that its spherical aberration, increasing with the degree of bending, is comparable with the spherical aberration of the mirror of opposite sign and thus more or less compensating it.

"Experiments in this direction were at once surprisingly successful, and we shall show in the course of this paragraph that these good results may be easily explained *a posteriori*.

"For the sake of curiosity we have reproduced in [Fig. 14-11] a sketch of one of the first successful experimental systems, which, used as a camera with relative aperture of very nearly $f/1.0$, produced [an excellent photograph] in less than a tenth of a second without any special illumination. The result was the more striking as both mirror and corrector were commercial spectacle-

---

† From A. Bouwers, *Achievements in Optics* (1946, Elsevier Publishing Co., New York).

glasses of the meniscus type, bought at an optician's. The concave surface of one of the glasses was covered with aluminium by evaporization in vacuo, and used as mirror, and the other meniscus was used as corrector."†

14-7. *Longitudinal Chromatic Aberration*

It is the property of a mirror to redirect the propagation of all wavelengths equally; hence the images that mirrors form are free from color, or chromatic aberration. The power of a lens is determined by the powers of its surfaces, $F = \pm \dfrac{N_1 \pm N_2}{r}$, and since these F's contain indices of refraction, they consequently vary with wavelength. This variation causes the so-called longitudinal chromatic aberration of a lens when it forms images of white light objects.

In this section we shall consider two simple combinations of lenses, made of properly chosen glasses, and so contrived as to give minimal longitudinal chromatic aberration. The power of the separate lens components may be expressed as

$$(N_1 - 1)S_1 \qquad \text{and} \qquad (N_2 - 1)S_2$$

where S is a shape factor determined by the r's of the component surfaces. The power of a doublet combination of these components, Φ, when the components are separated by a distance, d, will be

$$\Phi = \mathfrak{F}_1 + \mathfrak{F}_2 - \mathfrak{F}_1\mathfrak{F}_2 d$$

In geometrical optics we define the variation of N with λ to first approximation, for a particular glass, by the quantity ν, called the *reciprocal dispersion*. If N_D, N_F, and N_C are indices of refraction for a particular glass at the wavelengths of the Fraunhofer lines D, F, and C, at $\lambda_D = 5892$ Å, $\lambda_F = 4861$ A, and $\lambda_C = 6563$ Å, the reciprocal dispersion is defined as

$$\nu = \frac{N_D - 1}{N_F - N_C}$$

Intermediate partial dispersions define the details of the manner in which N varies through the intermediate wavelengths, when the above ν value, alone, does not define the glass adequately.

If a doublet lens has a minimal longitudinal chromatic aberration, its Φ will satisfy the equation $\dfrac{\partial \Phi}{\partial \lambda} = 0$. We shall examine two possible ways of satisfying this equation.

First, let us consider that we have a doublet made of two different glasses

----

† This experiment was carried out in August 1940 at Eindhoven.

in contact, so that we can set $d = 0$. If $d = 0$, then the partial derivative of Φ is zero when

$$S_1 \frac{\partial N_1}{\partial \lambda} + S_2 \frac{\partial N_2}{\partial \lambda} = 0$$

We now introduce ν into this equation, using $\left(\dfrac{N_F - N_C}{\lambda_F - \lambda_C}\right)_1$ for $\dfrac{\partial N_1}{\partial \lambda}$. Writing

$$S \frac{\partial N}{\partial \lambda} = \left(\frac{S(N_D - 1)}{\lambda_F - \lambda_C}\right)\left(\frac{N_F - N_C}{N_D - 1}\right) = \frac{\mathfrak{F}}{(\lambda_F - \lambda_C)\nu}$$

the condition for $\dfrac{\partial \Phi}{\partial \lambda} = 0$ becomes

$$\frac{\mathfrak{F}_1}{\nu_1} + \frac{\mathfrak{F}_2}{\nu_2} = 0$$

It is possible to satisfy this relationship when one component is made of crown glass and the other is made of flint glass. The achromatic doublet is the realization of this possibility. Below we quote the early history of such doublets as that history appears in Louis Bell's *The Telescope*:†

"Chester Moor Hall, Esq. (1704–1771) a gentleman of Essex, designed and caused to be constructed the first achromatic telescope, with an objective of crown and flint glass. He is stated to have been studying the problem for several years, led to it by the erroneous belief (shared by Gregory long before) that the human eye was an example of an achromatic instrument.

"Be this as it may, Hall had his telescopes made by George Bast of London at least as early as 1733, and according to the best available evidence several instruments were produced, one of them of above 2 inches aperture on a focal length of about 20 inches (F/8) and further, subsequently such instruments were made and sold by Bast and other opticians.

"These facts are clear and yet, with knowledge of them among London workmen as well as among Hall's friends, the invention made no impression, until it was again brought to light, and patented, by the celebrated John Dollond (1706–1761) in the year 1758.

"Physical considerations give a clue to this singular neglect. The only glasses differing materially in dispersion available in Hall's day were the ordinary crown, and such flint as was in use in the glass cutting trade,— what we would now know as a light flint, and far from homogeneous at that.

"Out of such material it was practically very hard (as the Dollonds quickly found) to make a double objective decently free from spherical aberration, especially for one working, as Hall quite assuredly did, by rule of thumb. With the additional handicap of flint full of faults it is altogether likely that these first achromatics, while embodying the correct principles, were not

---

† From Louis Bell, *The Telescope* (McGraw-Hill Book Co., New York).

good enough to make effective headway against the cheaper and simpler spy-glass of the time.

"Dollond, although in 1753 he strongly supported Newton's error in a Royal Society paper against Euler's belief in achromatism, shifted his view a couple of years later and after a considerable period of skilful and well ordered experimenting published his discovery of achromatism early in 1758, for which a patent was granted him April 19, while in the same year the Royal Society honored him with the Copley medal. From that time until his death, late in 1761, he and his son Peter Dollond (1730–1820) were actively producing achromatic glasses.

"The Dollonds were admirable craftsmen and their early product was probably considerably better than were Hall's objectives but they felt the lack of suitable flint and soon after John Dollond's death, about 1765, the son sought relief in the triple objective which, with some modifications, was his standard form for many years.

"Other opticians began to make achromatics, and, Peter Dollond having threatened action for infringement, a petition was brought by 35 opticians of London in 1764 for the annulment of John Dollond's patent, alleging that he was not the original inventor but had knowledge of Chester Moor Hall's prior work. In the list was George Bast, who in fact did make Hall's objectives twenty five years before Dollond, and also one Robert Rew of Coldbath Fields, who claimed in 1755 to have informed Dollond of the construction of Hall's objective.

"This was just the time when Dollond came to the right about face on achromatism, and it may well be that from Rew or elsewhere he may have learned that a duplex achromatic lens had really been produced. But his Royal Society paper shows that his result came from honest investigations, and at worst he is in about the position of Galileo a century and a half before.

"The petition apparently brought no action, perhaps because Peter Dollond next year sued Champneys, one of the signers, and obtained judgment. It was in this case that the judge (Lord Camden) delivered the oft quoted dictum: 'It was not the person who locked up his invention in his scrutoire that ought to profit by a patent for such invention, but he who brought it forth for the benefit of the public.'

"This was sound equity enough, assuming the facts to be as stated, but while Hall did not publish the invention admittedly made by him, it had certainly become known to many. Chester Moor Hall was a substantial and respected lawyer, a bencher of the Inner Temple, and one is inclined to think that his alleged concealment was purely constructive, in his failing to contest Dollond's claim.

"Had he appeared at the trial with his fighting blood up, there is every reason to believe that he could have established a perfectly good case of public use quite aside from his proof of technical priority. However, having clearly lost his own claims through *laches*, he not improbably was quite content to let the tradesmen fight it out among themselves. Hall's telescopes were in fact known to be in existence as late as 1827.

"As the eighteenth century drew toward its ending the reflecting telescope, chiefly in the Gregorian form, held the field in astronomical work, the old refractor of many draw tubes was the spy-glass of popular use, and the newly introduced achromatic was the instrument of 'the exclusive trade.' No glass of suitable quality for well corrected objectives had been produced, and that available was not to be had in discs large enough for serious work. A 3-inch objective was reckoned rather large."

A second way to satisfy $\dfrac{\partial \Phi}{\partial \lambda} = 0$, which we now examine, applies to two separated components made from the same glass. Differentiating Φ, and canceling out equal $\dfrac{\partial N}{\partial \lambda}$'s, as we may if both components have the same dispersion, we get

$$\frac{1}{\mathfrak{F}_1} + \frac{1}{\mathfrak{F}_2} = 2d$$

Huygens was the first to make such two-component achromatic combinations. He invented eyepieces with components made from just one kind of glass— the two components being separated by the amount prescribed above.

When two lens components, in contact, satisfy the relation $\mathfrak{F}_1 \nu_1^{-1} + \mathfrak{F}_2 \nu_2^{-1} = 0$, the aggregate power Φ will be the same at two wavelengths; but it may vary at wavelengths between. Then an image in white light will show secondary longitudinal chromatism, called secondary spectrum. Optical materials with

---

FIG. 14-13 All-calcite telescope devised by Dr. J. A. Anderson. Dr. Anderson used the index for the extraordinary ray for the positive component, and the index for the ordinary ray for the negative component, of this unusual lens, which is also a polarizer. The focus at F_2 is absorbed by an opaque paddle.

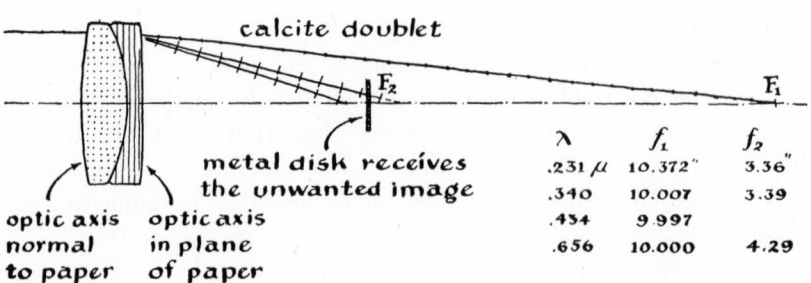

λ	f_1	f_2
.231 μ	10.372"	3.36"
.340	10.007	3.39
.434	9.997	
.656	10.000	4.29

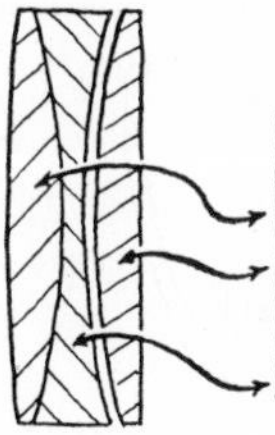

glass	N_D	v	dis-persion	relative partial dispersions					
				C-F	A-C	D-F	E-F	F-G'	F-H
O-543	1.564	50.7	0.01115	0.3354	0.7085	0.3309	0.5830	1.1857	
O-374	1.511	60.8	0.00844	0.3507	0.7026	0.3247	0.5675	1.1564	
mean				0.3420	0.7059	0.3282	0.5163	1.1730	
O-656	1.546	50.1	0.01090	0.3425	0.7052	0.3278	0.5167	1.1145	

FIG. 14-14   Cooke triplet lens. (Redrawn from Martin.)

appropriate partial dispersions can be employed to make compound lenses which are quite free from such secondary spectrum. As an example, crystalline CaF_2 and vitreous silica have appropriate dispersions so that, together, they make doublets with very little secondary spectrum. Again, Fig. 14-13 shows a novel telescope devised by Dr. John A. Anderson which uses the ordinary index of a calcite crystal for one lens component, and the extraordinary index of the same crystal material for the other component. This telescope is designed to work only over a narrow field, and with only one azimuth of polarization of incident light—a little paddle, shown in the figure, removes the other orthogonal component of polarization. The remarkably small variation of the focal length of this telescope, throughout the visible and ultraviolet, is evident from the tabulation of f_λ's in the figure.

Three different glasses, properly chosen for their partial dispersions, can be combined in components to yield a composite lens with the same aggregate power at three wavelengths and with small secondary spectrum between. Fig. 14-14 gives a tabulation of the indices for the D-line, the reciprocal dispersions v, and the relative partial dispersions, all for the three glasses used in the Cooke photo-visual triplet lens. These numbers come from the Jena Catalog and the mean relative dispersions show how the nearly equal positive components of the Cooke lens, shown in Fig. 14-14, act like a lens with a partial dispersion that almost exactly matches the partial dispersion of the negative flint lens.

14-8. *Optical Materials*

The tabulation in Fig. 14-14 suggests why lens designers are always interested in new transparent materials, such as new glasses, crystals, plastics, *etc.*, from which to make their lens components. They want, particularly, to be able to have a wide range of $\dfrac{N}{v}$ values from which to contrive their designs. For example: to make an achromatic doublet, their components must

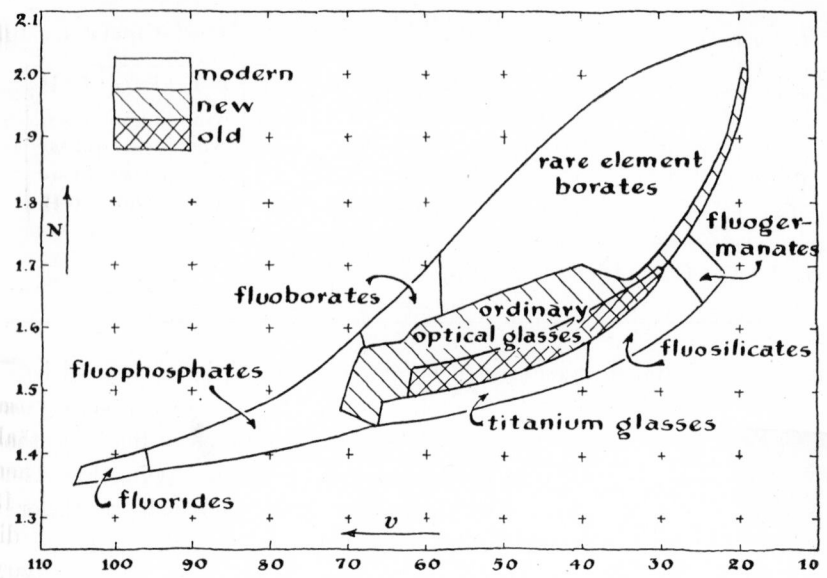

FIG. 14-15   The old, new, and modern glasses. (Redrawn from Kingslake.)

satisfy the equation $\dfrac{\mathcal{F}_1}{\nu_1} + \dfrac{\mathcal{F}_2}{\nu_2} = 0$; and to manifest minimum stigmatism and curvature of field, the glasses must also satisfy Petzval's condition, which is $\dfrac{N_1}{\mathcal{F}_1} + \dfrac{N_2}{\mathcal{F}_2} = 0$. Multiplying these equations gives $\dfrac{N_1}{\nu_1} = \dfrac{N_2}{\nu_2}$ as a condition to be satisfied if two glasses are to be doubly satisfactory. Ernst Abbe, speaking of his interest in glasses, says, "For years we combined with sober optics a species of dream optics, in which combinations made of hypothetical glass, existing only in our imagination, were employed to discuss the progress which might be achieved if the glass makers could only be induced to adapt themselves to the advancing requirements of practical optics." Finally despairing of cooperation from existing manufacturers, Abbe and Schott (in 1880) themselves undertook together to produce the desired glasses. Fig. 14-15 shows the *old glasses* as they existed before this undertaking and the *new glasses* which Abbe and Schott produced. Finally, this figure also shows the latest *modern glasses*. Prior to the work of Abbe and Schott the available *old glasses* were characterized by $\dfrac{N}{\nu} = .024$ for silicate crown glass and $\dfrac{N}{\nu} = .045$ for flint glass.

The *new glasses* of Abbe and Schott led to several famous lenses, like the Dagor cemented triplet and the Cooke air-space triplet; but the available range of N's was still not as great as now, so that curvatures of components in those lenses were necessarily high, to effect achromatism—and these large curvatures meant that reduction of other aberrations was made difficult. Whereas the *new glasses* of Abbe and Schott defined a very important second period in optical history, the present, characterized by the *modern glasses,* offers much wider choice to the designer. These latest glasses have resulted from a determined effort to find *all* the glass-making combinations that are possible with *all* the elements of the periodic table. G. W. Morey has been a dominant personality in the successes of this effort. As one example of the importance of these developments: a lens is now made with three modern glasses that gives performance comparable to that of one of the best four-element lenses made from predecessor glasses.

Other new optical materials that have recently appeared are useful for work in the infrared and ultraviolet as well as in the visible spectral region.

Chapter XV

Magnification, Aperture, and Field

W<small>HEN</small> <small>WE</small> begin to consider images of point objects, formed by a lens or other optical system, which lie off the optical ₵, we find the concept *magnification* useful. And when refracting or reflecting surfaces of an optical system collect more than an infinitesimal solid angle of rays, we find the concept *aperture* useful. Finally, when a lens system has separated reflecting or refracting components, or one such surface and a separated iris diaphragm, or aperture stop, then we find the concept *field* useful. We shall first discuss these concepts, *magnification, aperture,* and *field,* and image defects associated with them, all in context with the properties of Maxwell's perfect optical system. Although the correspondence between such a perfect system and a real system is not valid, even as a limiting case, the ideas of Maxwell do facilitate our thinking.

When the obliquity of monochromatic rays is sufficiently increased, real optical systems fail to focus rays at a point in image space. Rather, the final union will exhibit the defects which are characteristic of point imagery: *spherical aberration, coma,* and *astigmatism,* as well as *longitudinal* or *lateral chromatic aberration,* the defects of white light images. When two or more objects lie in a plane at P perpendicular to the ₵, but off that axis sufficiently far, their images may (and usually do) fail to fall in the conjugate plane which lies at P' perpendicular to the ₵. Rather, the conjugate images will fall on a curved surface, thus exhibiting the image defect *curvature of field.* Also, the array of conjugate image points in the image surface may fail to display themselves in geometrical similitude with parent object points, thus exhibiting the image defect *distortion.*

At first, let us consider the geometrical coordinates of both off-axis object points and their conjugate image points and also the mathematical trans-

326

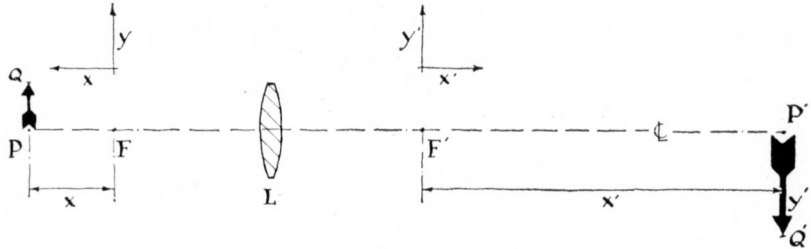

FIG. 15-1 Coordinates of collinear transformations for an ideal optical system.

formations which relate such points in object space to conjugates in image space. Fig. 15-1 shows an azimuthal section containing the $\mathbb{C}$ of a lens, and the coordinate systems we shall use for object and image space, with their origins at F and F'. We describe an object point in the plane of the figure by the coordinates (x,y) of the infinite object space, centered at F. And we describe its image point by (x',y') anywhere in the infinite image space, centered at F', which image space is coexistent with the infinite object space. These coordinates, referred to the principal focal points F and F', measure x positive to the left in object space and x' positive to the right in image space. The coordinates y and y' are reckoned positive upward and negative downward. The transformations of coordinates of an object point (x,y) that yield the coordinates of its image (x',y') are the so-called collinear transformations of Maxwell's ideal optical system:

$$xx' = f^2$$

$$l = \frac{dx'}{dx} = -\frac{x'}{x}$$

$$m = \frac{y'}{y} = -\frac{f}{x} = -\frac{x'}{f}$$

$$l = -m^2$$

The first equation above has the form of Newton's longitudinal equation. The second equation is obtained from the first by differentiation, giving the longitudinal magnification, l. This equation is applicable when dx' and dx are larger than differentials, but still small. The third equation gives the similitude factor, or lateral magnification, m. This equation is derived in the next section, using the geometry illustrated by Fig. 13-10.

It is noteworthy that an ideal Maxwellian system yields $m = l$ only if $m = 1$.

15-1. *Magnification by a Single Surface*

The longitudinal magnification derives from the equations of our previous chapter simply by differentiation:

$$\frac{N}{u} + \frac{N'}{v} = F \qquad \text{gives} \qquad l = \frac{dv}{du} = -\frac{Nv^2}{N'u^2}$$

$$\frac{1}{N\mathbf{u}} + \frac{1}{N'\mathbf{v}} = \frac{F}{NN'} \qquad \text{gives} \qquad l = \frac{d\mathbf{v}}{d\mathbf{u}} = -\frac{N'\mathbf{v}^2}{N\mathbf{u}^2}$$

$$(u - f)(v - f') = \frac{NN'}{F^2} \qquad \text{gives} \qquad l = \frac{dv}{du} = -\left(\frac{v - f'}{u - f}\right)$$

In considering the lateral magnification, $\frac{y'}{y}$, we reckon PQ as positive if Q lies above P and $P'Q'$ as negative if Q' lies below P'. This convention of signs is equivalent to the rule that the magnification of a single surface is negative if the conjugate planes, at P and P', lie on opposite sides of the surface center, and positive if these planes lie on the same side of C.

Considering the rays through S in Fig. 13-9, we see that $QP = y$ is a side of $\triangle PQS$, while $Q'P' = y'$ is a side of $\triangle P'Q'S'$. Thus

$$m = \frac{y'}{y} = -\frac{P'Q'}{PQ} = -\frac{CP'}{CP} = -\frac{v \sin \varphi'}{u \sin \varphi} = -\frac{Nv}{N'u}$$

From the geometry of Fig. 13-10, using the fact that $\triangle F'P'C'$ is similar to $\triangle FPC$, and remembering that $d = f'$ and $f = d'$,

$$m = -\frac{CP'}{CP} = -\frac{v - f'}{f'} = -\frac{f}{u - f}$$

And further, remembering from § 13-6 that $f' = \frac{N'}{F}$ and $f = \frac{N}{F}$, we get

$$m = \left(1 - \frac{vF}{N'}\right) = \left(\frac{1}{1 - \frac{uF}{N}}\right) \tag{a}$$

In Fig. 13-10, CP and CP' are corresponding sides of the triangles $\triangle PKC$ and $\triangle P'K'C$, with $PK = \mathbf{u}$ and $P'K' = \mathbf{v}$. These triangles have equal angles i at K and K', and because the sides PK and $P'K'$ are opposite supplementary angles at C, if we apply the laws of sines, remembering that sines of supplementary angles are equal, we get

$$m = -\frac{CP'}{CP} = -\frac{P'K'}{PK} = -\frac{\mathbf{v}}{\mathbf{u}}$$

As the student may establish for himself, the corresponding lateral magnification for the mirror of Fig. 13-11 $m = -\frac{v}{u} = +\frac{\mathbf{v}}{\mathbf{u}}$. Thus, whereas the expression for m is the same for reflection as for refraction, in terms of u and v, in terms of the nodal distances, $\mathbf{u}$ and $\mathbf{v}$, the expressions for reflection, and for refraction, have opposite signs.

15-2. *Extended Sine Relationship*

The extended sine relationship is an equation of great importance in optics. Here we derive it with the help of Fig. 15-2. In this figure $\wp$ and $\wp'$ are conjugates, as well as P and P'. This figure is much like Fig. 13-10, and with the same markings, except that here we have $\wp$ and $\wp'$ on the $\mathcal{C}$. The perpendicular distances of P and P' from the $\mathcal{C}$ are y and y', respectively. Referring to this figure:

$$PQ = +y = \wp P \sin \theta; \quad F\wp \sin \theta = d \sin \theta'$$

$$P'Q' = -y' = \wp'P' \sin \theta'; \quad F'\wp' \sin \theta' = d' \sin \theta$$

From the Newtonian equation, remembering from § 13-6 that $d' = \dfrac{N}{F}$ and $d = \dfrac{N'}{F'}$,

$$\frac{NN'}{F^2} = (u - f)(v - f') = (F\wp - \wp P)(F'\wp' + \wp'P)$$

$$\frac{NN'}{F^2} = \left(\frac{N'}{F} \frac{\sin \theta'}{\sin \theta} - \frac{y}{\sin \theta} \right)\left(\frac{N}{F} \frac{\sin \theta}{\sin \theta'} - \frac{y'}{\sin \theta'} \right)$$

Multiplying the last equation throughout by $\sin \theta \sin \theta'$, we get the extended sine relation:

$$Ny \sin \theta + N'y' \sin \theta' = yy'F \tag{a}$$

When y and y' are both small, compared to the reciprocal of F, we neglect

---

FIG. 15-2   Geometry involved in deduction of extended sine relationship. (Redrawn from T. Smith.)

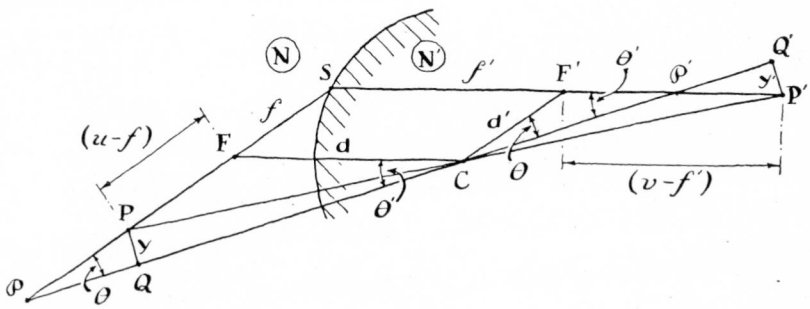

the right side and get the Abbe sine relation:

$$m = \frac{y'}{y} = -\frac{N \sin \theta}{N' \sin \theta'} \tag{b}$$

And when the object is at infinity, then $y \gg y'$, and we may divide Eq. 15-2a by y and neglect the term in which $\frac{y'}{y}$ enters, giving

$$N \sin \theta = y'F \tag{c}$$

15-3. *Magnification by a System of Surfaces*

The equations for magnification of a thick or thin lens, or of a system of surfaces, are obtained by cascading expressions for m that are appropriate to each component surface. First, we consider a thick lens for which the aggregate or cascaded magnification is $m = m_1 m_2$. Suppose that the lens is bounded on the object side by a medium of index N_1, that the lens itself has an index N_2, and that on the image side the lens is bounded by a medium of index N_3. Writing the appropriate expressions for m_1 and m_2, we get

$$m = \left\{ -\frac{N_1 v_1}{N_2 u_1} \right\} \left\{ -\frac{N_2 v_2}{N_3 u_2} \right\} = \frac{v_1 v_2}{u_1 u_2} \frac{N_1}{N_3}$$

Now $v_1 + u_2 = d$ for a thick lens. In the case of a thin lens we may take d as negligible and write $v_1 = -u_2$, getting

$$m = -\frac{N_1 v_2}{N_3 u_1} \qquad \text{or} \qquad m = -\frac{v_2}{u_1} \qquad \text{if} \qquad N_1 = N_3$$

It is often convenient to use the property that the conjugates of points lying in the perpendicular unit plane H are in the unit plane H'; and further, that we have unit magnification in the case of such conjugates. We prove these properties as follows: Firstly, a point lying on the plane H is characterized by $u_1 = -\frac{F_2 \tau}{\mathcal{F}}$. Substitution in Eq. 14-3a,

$$\left(u_1 + \frac{F_2 \tau}{\mathcal{F}} - f \right) \left(v_2 + \frac{F_1 \tau}{\mathcal{F}} - f \right) = f^2$$

yields the longitudinal position of the conjugate point as $v_2 = -\frac{F_1 \tau}{\mathcal{F}}$. Thus the conjugate point lies in the plane at H'. Secondly, substitution of these values for u_1 and v_2 in Eq. 15-1a,

$$m = \left\{ m_1 = \left(\frac{1}{1 - u_1 F_1} \right) \right\} \left\{ m_2 = (1 - v_2 F_2) \right\} = \frac{1 - v_2 F_2}{1 - u_1 F_1}$$

yields $m = 1.0$.

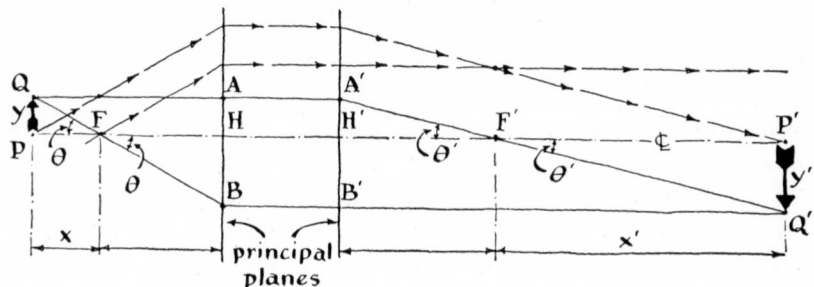

FIG. 15-3   Magnification of a thick lens as inferred from principal planes and rays.

We may use these properties to demonstrate, graphically, the equations for the thick lens, or for a system of lenses. In Fig. 15-3, let P and Q be imaged at P' and Q' by a thick lens or system of surfaces, represented here with reference only to H and H'. Consideration of two rays running from Q to Q' will suffice—one parallel to the ₵ in object space, the other parallel to the ₵ in image space. The ray from Q parallel to the ₵ strikes H at A. Since A and A' are conjugates with $m = 1.0$, this ray "emerges" at A' at a distance y from the ₵. It then passes through F' to Q', making the angles θ' at F'. These angles θ' are angles of the similar right triangles, $\triangle H'F'A'$ and $\triangle P'F'Q'$. Thus, representing the paraxial refraction by the two equivalent H-surfaces, we have for our similar triangles $m = \dfrac{y'}{y} = -\dfrac{x'}{f'}$. Similarly, for the ray from Q which passes through F to B in the plane H and thence, with $m = 1.0$, to B', "emerging" at the distance y' to pass parallel to the ₵ through Q': since $\triangle HFB$ is similar to $\triangle PFQ$ we have $m = \dfrac{y'}{y} = -\dfrac{f}{x}$, as expressed in the collinear equations, since $f = f'$ here.

Fig. 15-4 illustrates the equivalent refracting surface of a compound lens, determined by Eq. 15-2c. Here that equivalent surface is not a plane, but a sphere centered on F'.

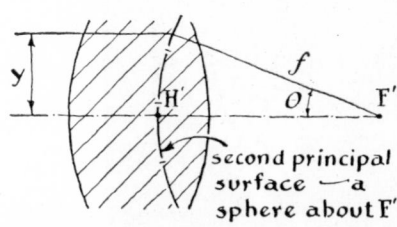

FIG. 15-4 Second principal spherical surface for representation of focusing by a compound lens that is free of coma. (Redrawn from Kingslake.)

15-4. *Aplanatic Points*

Geometrical optics predicts that rays parallel to the ¢ are reflected by a parabolic mirror in a manner such that they unite perfectly on the parabolic axis to give an image free of spherical aberration. When the parabolic mirror's axis is inclined to the incident parallel rays, however, so that the union of reflected parallel rays does not lie on the ¢, then the reflected rays no longer unite with geometrical optics perfection. Rather, their union manifests the image defect called *coma*, illustrated in Fig. 15-5. The bands of Fig. 15-5 are due to diffraction, which geometrical optics ignores.

Whereas spherical aberration is a result of improper deviations produced by different ring zones of a lens or mirror, which lie symmetrical with the ¢, the defect coma, in contrast, is a result of improper deviations, produced by reflection or refraction, at different facets of the same ring zone. A lens or optical system which unites rays perfectly in an image on the ¢ must be free of spherical aberration, while a lens or optical system which also forms a point image off the ¢ must be free from both spherical aberration and coma. A lens or system that is free from both spherical aberration and coma, when worked at certain conjugate planes, is said to be an *aplanatic* system; and the conjugate object and image planes, over which the imagery is free of both aberrations, cut the ¢ at the so-called *aplanatic points*. In § 14-5 we encountered a lens, and also a two-mirror reflecting system, both of which were free from spherical aberration. Both systems are also free from coma, as we shall demonstrate after we have established the necessary condition for perfect point imagery, both on the ¢ and off. It can (and will) be shown that there is, at most, for any one optical system, only one set of conjugate aplanatic points.

We invoke Fermat's law to establish the necessary condition for the existence of conjugate aplanatic points, and to prove the uniqueness of such

FIG. 15-5 Coma pattern, showing diffraction. (Redrawn from Kingslake.)

conjugate aplanatic points for an optical system. Fermat's law relates to a wave front progressing through a lens or optical system, off-axis or on, from a point source to an image. The wave front, at any intermediate position in its travel from source to image, determines the locus of positions along which the various ray trajectories have required equal times to travel from the point source. Hence, at the final image, if this locus becomes a point, the time of travel for each ray in going from source to image must be the same. This consideration affords an explanation of Fermat's law, even before we state it—Fermat's law says that the total optical path length for any ray, from a common source to a common image, is the same as for any other ray. We may express Fermat's law mathematically as a variation:

$$\delta \int_Q^{Q'} N(s)\, ds = 0$$

Here s is the measure of geometrical path length, and $N(s)$, a function of s, describes the dependence of index of refraction, along the path, on s. The first four problems for this chapter illustrate applications of this law, to yield the laws of reflection and refraction.

Fig. 15-6 illustrates an interesting second deduction of Abbe's sine relation, here using Fermat's law. Abbe's relation, already deduced in § 15-2, is the conditional relation which is to be satisfied if we are to have aplanatic points. Fermat's law for point images at P' and Q' requires that the integrals of $N(s)\, ds$ for the two paths between ℂ conjugates in Fig. 15-6, $PAG'P'$ and $PHF\cdot P'$, must be equal; and also that the integrals be equal for the two off-ℂ paths, $QA'G'Q'$ and $QB'F'Q'$. These equalities express the facts that the union of rays on the ℂ at P is free from spherical aberration, and that the union at Q', off the ℂ, is free from both spherical aberration and coma. Thus

$$\left. \begin{array}{l} PAP' = PHP' \\ QA'Q' = QB'Q' \end{array} \right\} \tag{a}$$

FIG. 15-6   Geometry for deduction of Abbe's sine relationship.

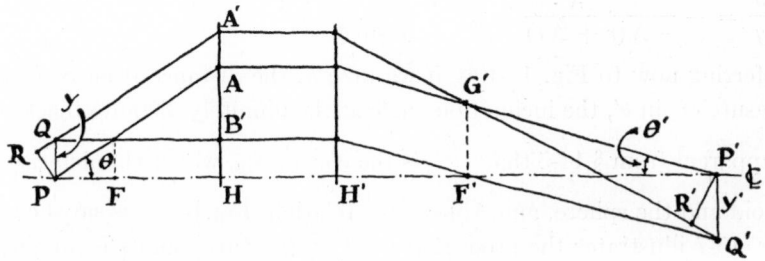

For paraxial rays, when the angles at F' and G' are small,

$$\left.\begin{array}{l} F'Q' = F'P' \\ G'R' = G'P' \end{array}\right\} \tag{b}$$

The points F' and G' lie in the second focal plane. Accordingly, at these points the focused parallel rays, from P and from Q or R, have equal paths,

$$\left.\begin{array}{l} QB'F' = PHF' \\ RA'G' = PAG' \end{array}\right\} \tag{c}$$

Equating the left sides of (b) and (c) to their right sides, we get

$$QB'F' + F'Q' = PHF' + F'P'$$
$$RA'G' + G'R' = PAG' + P'G'$$

or

$$\left.\begin{array}{l} QB'Q' = PHP' \\ RA'R' = PAP' \end{array}\right\} \tag{d}$$

Now, on combining this result (d) with the Fermat equations (a), we get

$$RA'R' = PAP' = PHP' = QB'Q' = QA'Q'$$

and from $RA'R' = QA'Q'$ we get

$$RQ = R'Q'$$

This is the result we seek; on writing the corresponding optical distances which the geometrical distances RQ and $R'Q'$ represent, and invoking N for object space and N' for image space, we get the familiar form of Abbe's relation, Eq. 15-2b:

$$\left.\begin{array}{l} RQ = Ny \sin \theta \\ R'Q' = -N'y' \sin \theta' \end{array}\right\} \quad \text{giving} \quad \frac{y'}{y} = -\frac{N \sin \theta}{N' \sin \theta'}$$

This Abbe sine relation is satisfied by the refracting sphere when it is worked as shown in Fig. 14-10a, where one index is unity. Substitution of N and the distances u and v indicated, in $m = -\dfrac{v}{Nu}$, gives us Abbe's sine relation for paraxial rays,

$$\frac{y'}{y} = -\frac{r + \dfrac{r}{N}}{-N(r + Nr)} = \frac{1}{N^2} = \frac{\sin \theta}{N \sin \theta'}$$

Referring now to Fig. 14-10b, if we take h, the distance from S_1 to ₵, as a measure of $\sin \theta'$, the inclination angle at the infinitely distant object point, it is apparent from § 14-5 that $\dfrac{h}{\sin \theta}$ is the constant r, which characterizes the cardioid and the sphere, and Abbe's sine relation, Eq. 15-2c, is satisfied.

Fig. 15-7 illustrates the proof that a set of aplanatic points is unique, if a

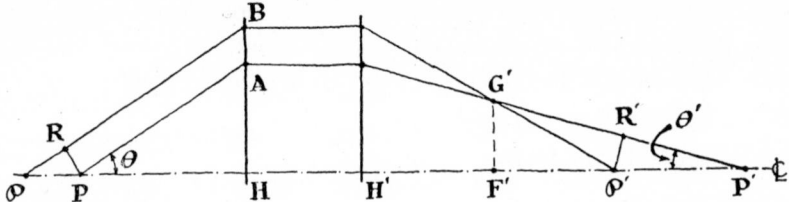

FIG. 15-7 Geometry for proof that a set of aplanatic points of a lens is a unique set.

---

system has such a set. In this figure we presume that the conjugate points, $\wp$ and $\wp'$, which are close to the aplanatic set P and P', respectively, are also free from spherical aberration. We show below that the presumption that $\wp$ and $\wp'$ are free from spherical aberration is untenable, and therefore the conjugate perpendicular planes there can hardly be aplanatic planes.

We start our demonstration, as before, by writing the presumed and necessary optical path equalities. A consequence from Fermat's law for presumed freedom from spherical aberration, at both sets of points, is that

$$\wp H \wp' = \wp B \wp' \qquad \text{and} \qquad PHP' = PAP'$$

And terms in the first equation above are

$$\wp H \wp' = \wp P + PHP' - P'\wp'$$

$$\wp B \wp' = \wp R + RBG' + G'\wp'$$

In Fig. 15-7 $\wp'R'$ is perpendicular to $G'P'$, but since we shall take $\wp'R'$ as being small, we may write $RBG' = PAG'$, and $G'\wp' = G'P' - R'P'$. Combining these equalities gives

$$\wp B \wp' = \wp R + PAP' - P'R'$$

$$\wp P - \wp R = \wp'P' - P'R'$$

Using the angles θ and θ', shown in the figure, $\wp R = \wp P \cos \theta$ and $\wp'R' = \wp'P' \cos \theta'$. These trigonometric expressions give

$$\frac{\wp'P'}{\wp P} = \frac{1 - \cos \theta}{1 - \cos \theta'} = \frac{\sin^2 \dfrac{\theta}{2}}{\sin^2 \dfrac{\theta'}{2}}$$

Now since this ratio must be the same for all rays handled by our system, the ratio of sines of the half angles, $\dfrac{\theta}{2}$ and $\dfrac{\theta'}{2}$, must be constant. But Abbe's relation for the points P and P' requires the ratio of sines of the full angles, θ and θ', to be constant. Thus our proof, based on a presumed freedom from spherical aberration for two sets of conjugate points, is incompatible with

Abbe's condition. Since Abbe's condition "takes precedence over whatever statements it contravenes," we can only conclude that just one set of aplanatic points is possible, and that the possible set is therefore unique.

15-5. *Magnifiers and Microscopes*

Although geometrical optics may predict perfect union of rays when a lens or system of surfaces is worked at aplanatic points, in general we expect a blurring at the union due to physical-optics diffraction. However, when the aperture is enlarged, and diffraction blurring is decreased, or when the object point is at a sufficient distance off-axis, the dominant blurring at the union becomes geometrical-optics aberration. For the normal eye this physical-optics limitation is equally important with the geometrical-optics aberration-limitation when the iris of the eye is about 2.5 mm in diameter (as seen in object space).

In the case of the eye, these two limitations, as well as limitations due to the inherent coarseness of the retinal sensing structure, all together, make its resolving power not far different from that computed from Rayleigh's diffraction criterion alone—namely, $1.22\frac{\lambda}{w} = 0.27$ mils or milliradians, for $\lambda = 0.55\ \mu$. Since there are 206 seconds in a mil, this calculated limit is 56 seconds of arc, or approximately 1 minute. Although those who have exceptionally keen vision may approach this limit under the most favorable observing conditions, we shall take one mil (rather than 0.27 mil) as the normal angular separation of points which can be comfortably resolved.

A suitable lens system may be used to enlarge the linear image separation of close objects that are to be observed, or to enlarge the apparent angular separation of the images of distant objects. A device to achieve the first is called a *magnifier*; a device to achieve the latter is a *telescope*. The magnifier is essentially a lens with a short focal length; and the *microscope* is a magnifier suitably elaborated to give the system an even shorter focal length. The simple lens used as a magnifier forms a magnified image of the observed object at the distance from the eye of closest distinct vision, which we conventionally take as 250 mm. And when its focal length is much less than 250 mm, the lens is worked at $u \cong f$, giving the magnification

$$|m| \cong \frac{250}{f} \tag{a}$$

A simple microscope is comprised, essentially, of two lenses (or magnifiers)—an objective of focal length f_{obj}, and an eyepiece of focal length f_{eye}. These components, when separated by an amount d (usually 160 mm) will have, in aggregate, an equivalent objective focal length, f. We may calculate the aggregate focal power from the powers of the components by $\Phi = \mathcal{F}_{obj} + \mathcal{F}_{eye} - d\mathcal{F}_{obj}\mathcal{F}_{eye}$. Or we may use the equivalent formula for aggregate focal length:

$$f = \frac{f_{obj}f_{eye}}{f_{obj} + f_{eye} - d}$$

Thus, for example, a focal length of the objective of $f_{obj} = 4$ mm, and a focal length of the eyepiece of $f_{eye} = 20$ mm, with $d = 160$ mm, give the reduced effective focal length of the combination as $f = 0.6$ mm. This reduction, f_{obj} to f, is greater than six-fold.

Under favorable conditions of illumination the resolving power of the microscope is determined by the angle of obliquity of the marginal rays that the objective collects, θ_{obj}, and by the immersion index, N. The effective diameter of the objective is $2f_{obj}N \sin \theta_{obj}$; and this gives a Rayleigh resolving power of $1.22 \left(\dfrac{\lambda}{2f_{obj}N \sin \theta_{obj}} \right)$ radians. The quantity $N \sin \theta_{obj}$, represented below by the symbol A, is called the numerical aperture. N enters here because the optical path difference of two rays, one to each opposite edge of the objective, is N-fold greater in an immersion medium of index N than it would be in air, making the diffraction pattern of the objective aperture N-fold narrower.

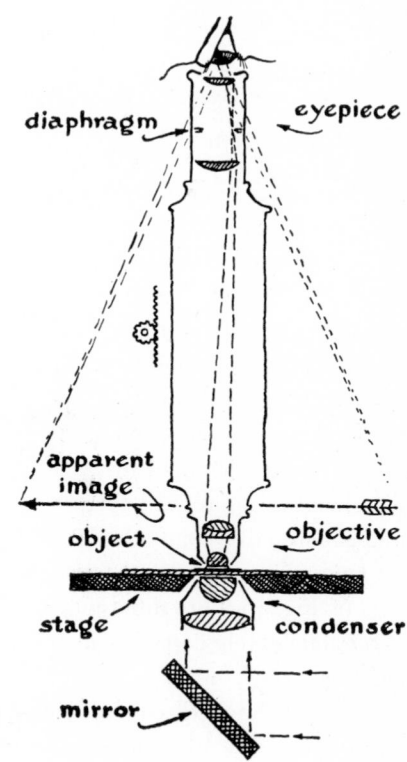

FIG. 15-8   Optics of the microscope.

If two points resolved by the objective of the microscope are to be separated in angle by 1 mil, so they may be resolved comfortably by the eye, then the proper working magnification for the microscope as a whole must be such that $1.22 \dfrac{\lambda}{2f_{obj} A} m = 10^{-3}$ or 1 mil. This magnification is realized when $m = 750 A$. Or, in round numbers, we may take it that the microscope magnification should be $m = 1000 A$.

Objectives and recommended eyepiece magnifications for typical manufactured microscopes, as given by Dr. A. C. S. van Heel, are reproduced in Table 15-1. Δu is the depth of focus (see § 15-8). Fig. 15-8 shows the optical system of a typical compound microscope.

TABLE 15-1

Type of Objective	A	m_{obj}	m_{eye}	Δu
dry	0.2	8	25	$14\,\mu$
dry	0.3	10	30	6
dry	0.4	20	20	1
oil	0.65–0.85	40	16 to 21	0.40
oil	1.3	90	15	0.25

15-6. *Telescopes*

Whereas the microscope serves to cast an image of objects lying close to its objective at the distance of closest distinct vision for the eye, 250 mm, the telescope, in contrast, serves to magnify the angular separation of distant objects. A telescope may, in fact, magnify in angle with actual linear demagnification. And the typical telescope optical system has its components arranged so that its aggregate power as an optical system is zero; $\Phi = 0$. Conversely, a system with $\Phi = 0$ is called a telescopic system. Fig. 15-9 shows two examples of telescopic systems: the astronomical and the Galilean telescope. These telescopes are comprised of a large diameter objective lens of focal length f_{obj}, and an eyepiece lens of focal length f_{eye}. These components are each usually compound lenses. Their separation, $d = f_{obj} + f_{eye}$, makes the aggregate power of the telescope vanish, $\Phi = 0$. The two telescope systems of Fig. 15-9 are different in that the focal length of the eyepiece of the astronomical telescope is positive, while that of the Galilean telescope is negative.

Fig. 15-9 shows three parallel rays, from the left, traced through the two lenses. They are incident on the objective parallel to the ¢; and because

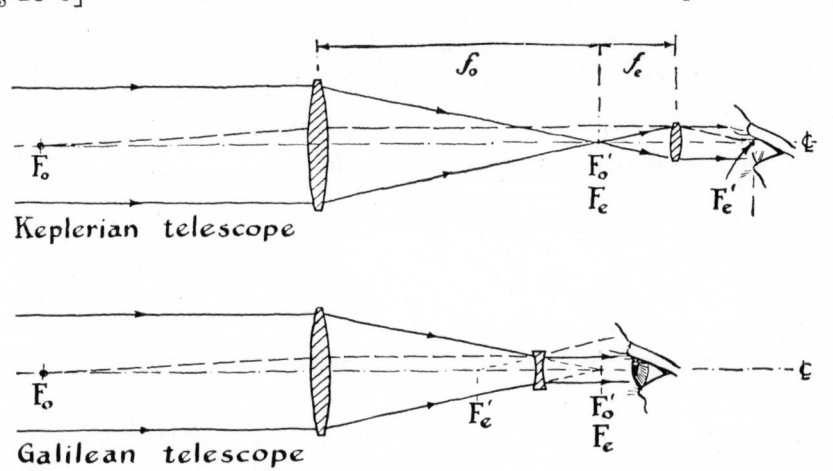

Keplerian telescope

Galilean telescope

FIG. 15-9   Keplerian and Galilean telescopes.

$\Phi = 0$, they emerge parallel to it. The dashed line represents an inclined ray from an off-axis distant object which passes through the first principal focal point of the objective and propagates parallel to the ₵ between the components, to emerge from the second lens in such a direction as to pass through the second principal focal point of the eye lens. From the geometry of these rays, the angular magnification of the image is obviously $-\dfrac{f_{\text{obj}}}{f_{\text{eye}}} < 0$ for the astronomical telescope and $-\dfrac{f_{\text{obj}}}{f_{\text{eye}}} > 0$ for the Galilean telescope. The positive angular magnification, in the second case, means an erect image; the negative angular magnification means an inverted image.

FIG. 15-10   Use of a Galilean telescope to convert a lens to a telephoto lens.

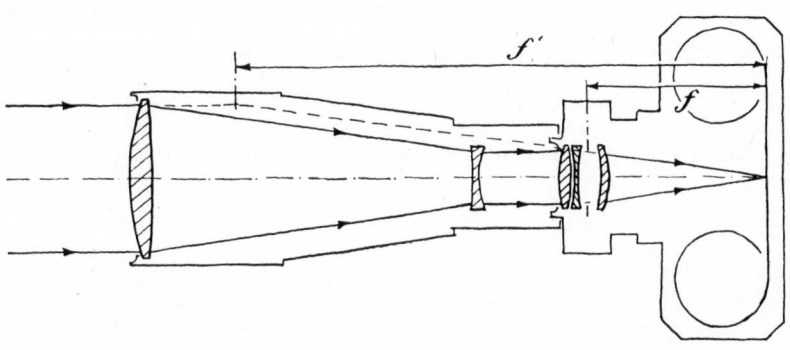

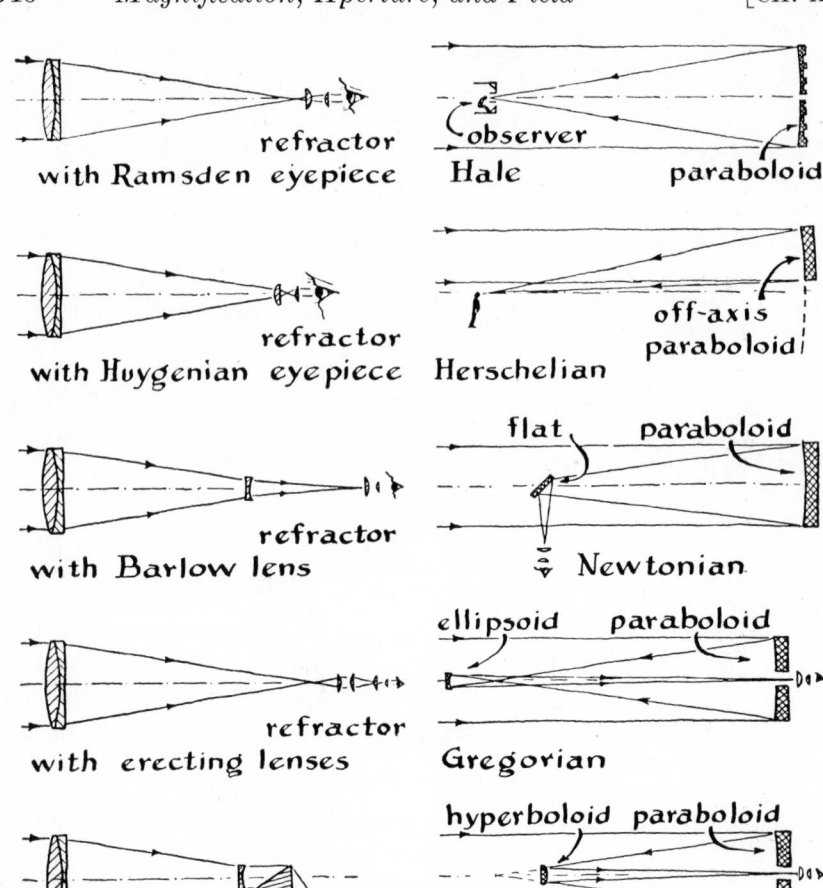

FIG. 15-11 Various types of astronomical telescopes.

The angular resolving power of the objective of a telescope is $1.22 \frac{\lambda}{w_0}$, apply-ing Rayleigh's criterion. For $\lambda = 0.55\,\mu$ this angular resolving power is $\frac{0.67}{w_{obj}}$ mils, or $\frac{13.8}{w_{obj}}$ seconds of arc, when w_{obj} is the objective diameter expressed in centimeters, or about $\frac{5}{w_{obj}}$ seconds when w_{obj} is expressed in inches. The angular resolving power of the 200″ telescope of Mt. Palomar is 0.027 second of arc. Actually, in practice, owing to the ever present optical inhomogeneities of the atmosphere through which the stars are seen, or for other reasons, the resolving power of an astronomical telescope is much less than this.

If two optically resolved stars with the angular separation of $\dfrac{13.8}{w_{obj}}$ seconds are to be comfortably seen by the eye, this angle must be magnified to 1 mil, or to 206 seconds. A magnification if 15 w_{obj} is thus necessary, making the proper eyepiece focal length $f_{eye} = \dfrac{f_{obj}}{15\, w_{obj}}$.

Fig. 15-10 shows a Galilean telescopic system used in front of a camera lens. The Galilean telescope is focused on infinity, making, in effect, a "tele-photo lens."

Fig. 15-11 shows some of the elaborations of simple telescopes, and the main telescope arrangements that use mirror objectives. These designs have various specialized functions.

15-7. *Aperture*

In the case of a compound lens or optical system with several components, or when a simple lens is used in combination with an adjustable iris diaphragm or fixed stop, the solid angle of rays from a source point which the lens can handle will be limited. We may determine which one of the various lens rims and stops sets this limit as follows: each is imaged into object space by all the lens components which precede it in the system; then we determine that one, in object space, which subtends the smallest angle at the object point. The lens rim or stop which subtends the smallest angle is called the *entrance pupil*, and the angle which its radius subtends at the object point is defined

---

FIG. 15-12   Illustration of dependence of entrance aperture on position of object.

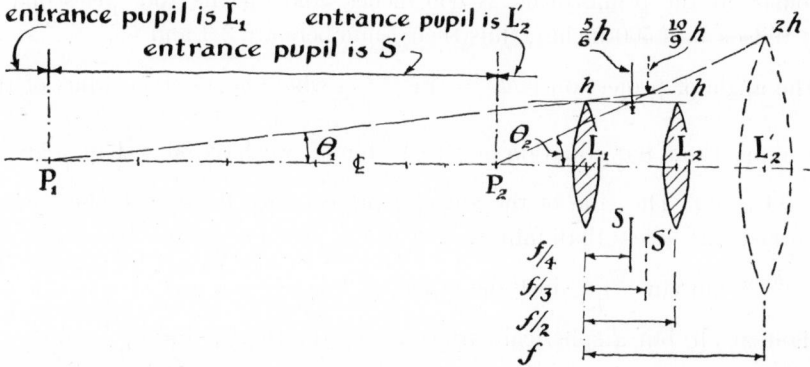

as the *angular aperture*, θ_0. Fig. 15-12 illustrates the location of entrance pupil, and angular aperture, for a particular case. In this case we have two equal thin lenses of radius h separated by the distance, $d = \dfrac{f}{2}$, or by half the focal length of either. Also, we have an iris diaphragm midway between these lenses, and we take its stop radius as $\frac{5}{6}h$. The rim of the first lens, L_1, is, of course, already in object space. The images of other components in object space are shown dotted. The image of the second lens, L_2, lies at L_2'; and that of the stop, S, lies at S', as seen through L_1. Both L_1 and S' subtend the same angle at P_1; and this angle is smaller than the angle subtended there by L_2'. At a point on the ₵ to the left of P_1 the smallest subtended angle is that subtended by L_1; thus it is the entrance pupil. At a point to the right of P_1, and until we reach P_2, S' is the entrance pupil. At P_2 both S' and L_2' subtend equal angles. At a point to the right of P_2, the smallest angle is subtended by L_2', and thus it is the entrance pupil.

The light gathering power of the microscope is determined by its numerical aperture, A. We have previously defined A above as the product of the immersion index and the sine of the angular aperture:

$$A = N \sin \theta_0$$

The exit pupil is the image of the lens rim or stop in image space that subtends the smallest angle at P', each lens rim or stop being imaged into image space by all lens components which follow it in the system. The *exit pupil* of an optical system (called the Ramsden disk in a telescope) is conjugate to the *entrance pupil*. The angle which the radius of the exit pupil subtends at the final image P' is defined as the *angle of projection*, θ_0'.

The angle of projection of a camera lens is measured by a quantity called the *f*-number, which is the ratio of the effective focal length to the diameter of the exit pupil, $2y_{exit}$. The *f*-number is the reciprocal of twice the sine of the angle of projection. For example, f for the 200″ Hale telescope of Mt. Palomar, at the prime focus, is 666 inches; and for the 100″ telescope at Mt. Wilson it is 500 inches; thus their *f*-numbers are 3.3 and 5.

The angle of projection is $\theta_0' = \sin^{-1} \dfrac{y_{exit}}{v}$, where y_{exit} is the radius of the exit pupil. From our equation of § 15-1a for m, we have $m = 1 - \dfrac{v}{f}$, giving $v = f(1 - m)$. The sine of the angle of projection is decreased when a lens is worked with u less than infinite. For $u = \infty$, then m is zero. Here θ_0' is not $\tan^{-1} \dfrac{y_{exit}}{f}$, but $\sin^{-1} \dfrac{y_{exit}}{f}$ since the principal "plane" of a corrected lens is not a plane at all, but a spherical surface about the focal point P, as shown in

Fig. 15-4. When the object gets near the lens the decreased θ_0' becomes approximately $\sin^{-1} \dfrac{y_{\text{exit}}}{f(1 - m)}$.

The reciprocal of the square of the f-number of a lens is a measure of its "speed"; and the iris diaphragm control on a camera is indexed in effective f-numbers for convenience in controlling exposure (and depth of field, as explained in § 15-8). In recent lenses $\mathfrak{T}$-numbers are used instead of f-numbers for closer control of exposure. The $\mathfrak{T}$-number is the f-number corrected for the transmission of the lens, $\mathfrak{T}$. Thus

$$\mathfrak{T}\text{-number} = \frac{f\text{-number}}{\sqrt{\mathfrak{T}}}.$$

Rudolf Kingslake gives $\mathfrak{T} = 0.5$ for certain complex anastigmats, making a $\mathfrak{T}$-number of 15.5 correspond to an f-number of 11.0.

15-8. *Depth of Field*

In cameras the depth of field, which is controlled by the f-number, depends also on the angular resolving power of the eye, which again we take as 1 mil. In most cameras the resolving powers of lenses, and of the photographic materials they use, will usually support a 1 mil limit of angular resolving power when each photograph is viewed normally, from its center of perspective. The depth of field in a photograph is defined as the limits, reckoned longitudinally along the lens ₵, within which we have images with the out-of-focus circle of least confusion, on the photographic surface, subtending an angle of 1 mil or less. The center of perspective is the first principal plane of the lens; the developed and printed photograph is later viewed under normal perspective when the eye is located exactly where the second principal point of the camera lens was when the photograph

FIG. 15-13 Illustration of depth of field.

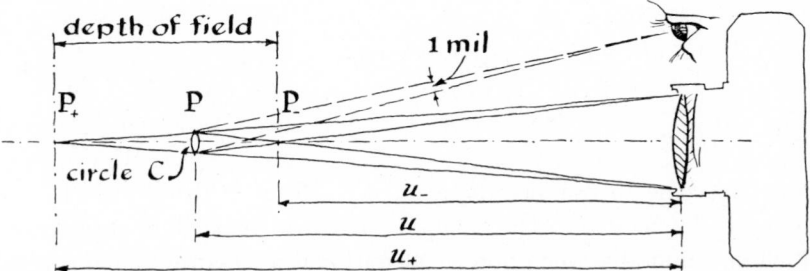

was exposed. Then perspective is the same for the eye, as if the photographed field were viewed with the eye at the position where the first principal point of the lens was during the exposure. If a lens has an entrance pupil diameter w_{ent}, and is focused on a plane at P, then a point beyond, at P_+, will have its rays focused within the image of a circle C at P, as illustrated in Fig. 15-13. This is the circle cut from the plane by the marginal rays from P_+ at P. At a certain near point, P_-, the rays will be focused within the image of the same circle (at P) as illustrated. This point P_- lies where its marginal rays, projected back through P_-, fall on C. The points P_+ and P_- define the limits of sharp field, when the angle the blur circle subtends is just 1 mil. If we take the entrance pupil and the first principal point of the camera lens as coincident, and write u for the distance from lens to the plane which is in sharp focus at P, we see, referring to Fig. 15-13, that

$$\frac{w_{ent}}{u_+}(u_+ - u) = \frac{w_{ent}}{u_-}(u - u_-) = \frac{u}{1000}$$

or

$$u_+ = \frac{u}{1 - \dfrac{u}{1000w_{ent}}} \quad \text{and} \quad u_- = \frac{u}{1 + \dfrac{u}{1000w_{ent}}}$$

These distances, u_+ and u_-, give the difference $(u_+ - u_-)$ that is called the depth of field. If the object photographed is at a distance $u_0 = 1000w_{ent}$, which is called the hyperfocal distance, we have $u_+ = \infty$ and $u_- = \frac{u_0}{2}$. When a camera is focused at this distance all objects it photographs from $\frac{u_0}{2}$ to infinity will be in apparent sharp focus. A 50 mm camera lens stopped to $w_{ent} = 9$ mm (approximately 5.6 f-number) has its hyperfocal distance at 9 meters so the depth of sharp focus extends from 4.5 meters to infinity. When viewed with a magnifying lens or projector of 50 mm focal length, then, no photographed object within the 4.5 meters to infinity range will appear significantly blurred to the eye.

The above two formulas can be solved for u_+ and u_- for any u, if w_{ent} is given; but usually manufacturers of camera lenses provide a scale on the focusing drum which provides this solution.

15-9. *Field*

We now come to consider the angular width of field over which optical systems work effectively. The points in the image plane of an optical system will become unusable when they lie at too great a distance from the ₵. Such images become unusable because of aberrations on the one hand; or, in the

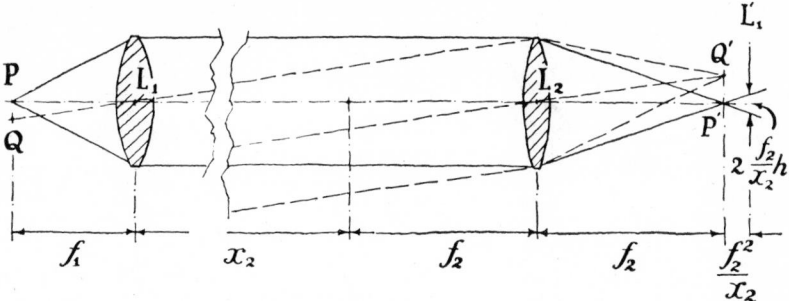

FIG. 15-14   Illustration of lateral field.

---

case of a compound lens, or of a simple lens with a stop, because of deficient light; or both. The light deficiency limitation, which concerns us now, is illustrated in Fig. 15-14. Here we have, as an example, an optical system such as has been used for measuring the transmission of the atmosphere over a long air path. The function of this optical system is to collect radiations emitted by the source PQ and deliver them to a distant spectrometer slit at Q'. This system first makes these rays parallel by a collimating component lens (or it could be a mirror); and then, after penetrating a long modifying path, the modified radiations handled by the system are focused by a component telescope lens (or mirror) at $P'Q'$ on the slit of a spectrometer. It is obvious from the dashed lines in Fig. 15-14 that the field of this optical system does not extend to the point Q', and that the length of spectrometer slit that can be illuminated is narrowly delimited. In this example we take the radii of the rims, $h_C = h_T$, for collimator and telescope. The field is determined by the *exit window*, here $L_C' = L_1'$. The exit window is the lens rim or stop which, imaged into image space, subtends the smallest angle at the center of the exit pupil, here $L_T = L_2$. The exit window radius is $\dfrac{f_T}{x_T} h_C$, where h_C is the radius of the colli-

mator component. The image of C in T lies at $x' = \dfrac{f_T}{x} f_T$ behind the focal plane, or spectrometer slit. The angular field is defined quantitatively as the tangent whose angle, φ, is the ratio of the radius of the field defining rim image to its separation from the exit pupil. Here,

$$\varphi = \tan^{-1}\left[\frac{\dfrac{f_T}{x_T} h_C}{f_T\left(1 + \dfrac{f_T}{x_T}\right)}\right]$$

In this instance, if $x_T = 1000$ meters and $f_T = 1$ meter, then $\varphi \cong \dfrac{h_C}{x_T}$. If h_C is 50 cm, the total length of slit illuminated within the ascribed field is 1 mm;

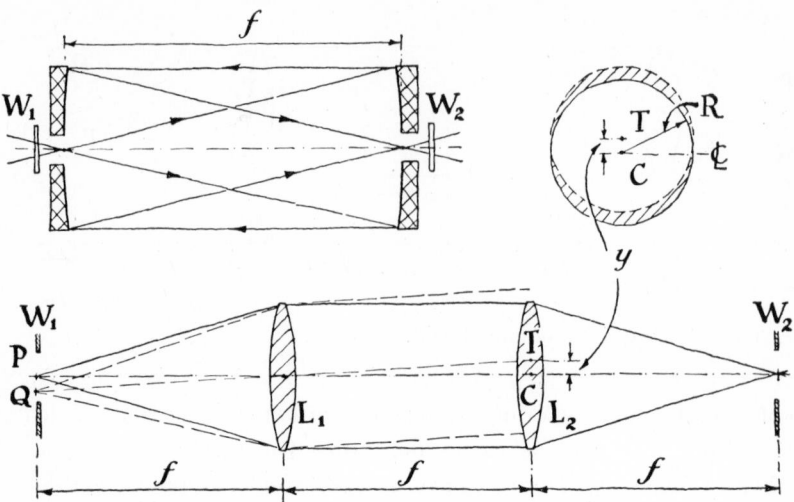

FIG. 15-15 Pfund absorption cell, and the geometry involved in its vignetting characteristics.

with the transmission of the rays collected by the system approximately $\frac{1}{3}$ when Q' is $\frac{1}{2}$ mm from P'.

Fig. 15-15 shows another example of field—a Pfund absorption cell. The function of this optical system is the same as that described above, except here the radiation modifying gases or vapors are enclosed, and the optical path is folded; and furthermore, $f_C = f_T = d$. An equivalent lens schematic of the Pfund cell is also shown in Fig. 15-15. The rays enter the cell through window W_1 and emerge through W_2. It is instructive to calculate the geometrical transmission of this cell. For this calculation we refer to the equivalent lens schematic. Here the rays from Q near P are not delimited by W's. The decrease in geometrical transmission as Q moves away from the ₵ in the focal plane is called *vignetting*—the word comes from *vignette*, which meant originally a miniature portrait which fades out beyond the image of the face. The insert in Fig. 15-15 shows how vignetting is calculated in this instance. The dashed circle shows the geometrical projection of the lens rim L_1 by collimated rays from Q, in respect to the rim of the other lens L_2. The shaded area is lost, and the area of this lune is $2y_Q h_1$, as application of a little geometry will easily prove. The ratio of the area enclosed by the circle L_1, less the $2y_Q h_1$, when divided by πh_1^2, gives the transmission of the cell for rays from Q, i.e.

$$\mathfrak{T} = 1 - 2\left(\frac{y_Q}{\pi h_1}\right)$$

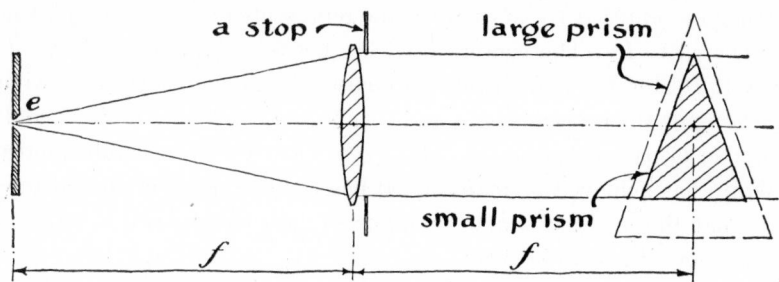

FIG. 15-16 Vignetting in a spectrometer.

If PQ is a uniform line source of light, such as an evenly illuminated narrow slit of length $2y_s$, the average transmission will be

$$\mathfrak{T} = 1 - \frac{y_s}{\pi h_1}$$

Similar formulas apply to a mirror spectrometer where the distance from collimator to prism or grating is f, as shown in Fig. 15-16. Vignetting in these cases is given by the above formula suitably amended; in particular, if the prism or grating acts as a circular stop of equal diameter, $h_C = h_P$, the above formula applies as is; while, if the prism or grating is large and we have equal lens diameters, $h_C = h_T$, the average transmission is $1 - 2\,\dfrac{y_s}{\pi h_T}$.

Fig. 15-17 shows John U. White's absorption cell in which vignetting is minimized. Light enters an entrance window e and is focused at P_3' by M_1,

FIG. 15-17 White's absorption cell.

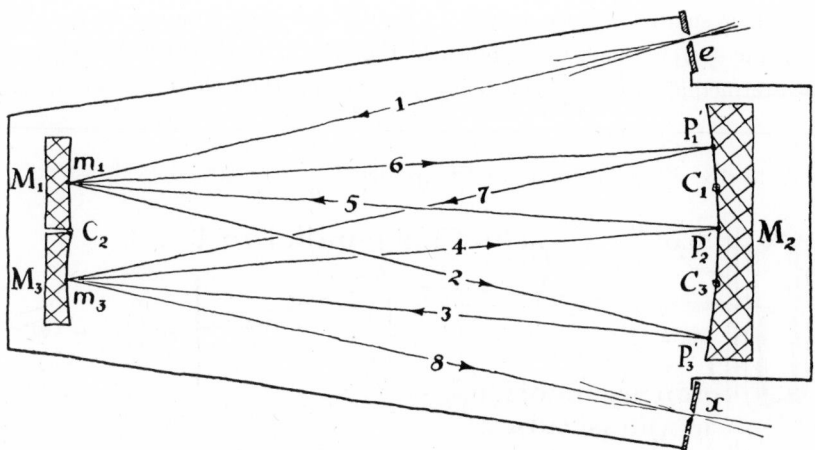

which has its center at C_1. M_2 returns the represented chief ray to m_3, whence it is reflected to P_2'. The center of M_2 lies between M_1 and M_3 at C_2. P_2' is imaged at P_1' by M_1; and finally M_3 focuses the rays onto the exit window, x. By changing the positions of C_1 and C_2, effected by tipping both M_1 and M_2, the number of passes, before the light is sent out through x, can be increased. Dr. White has achieved as many as 90 passes. The point of special interest here is that the mirror M_2 acts to suppress vignetting, since it repetitively focuses M_1 on M_2, and *vice versa*.

There is no gradual vignetting when the dominating stop of an optical system, which determines the field of view, coincides with the object or image plane; for then the field of view is sharply delimited. This is the case for the entrance and exit windows of the Pfund absorption cell, W_1 and W_2.

In a telescope, in order to minimize vignetting, the exit pupil is usually arranged to fall on, and overlie, the iris of the eye. If the magnification is so great that the exit pupil, which is the image of the objective in the eyepiece, is smaller than the iris of the eye, then the telescope image will be unduly dim; but if, on the other hand, the magnification is too small, the iris of the eye will act as the aperture stop rather than the objective. In such a case it would have been more economical to use a smaller objective. But when a magnification is used which gives the exit pupil a diameter of 2 mm, called *normal magnification*, neither limitation is dominant.

Stops are often used to control image illumination as in the ordinary camera. Stops may be used for other ends; for example, to control parallax, as described below. Stops may also be used to control aberration, as we shall see in the next chapter.

15-10. *Telecentric Systems*

When a small aperture stop acts as the entrance window of an almost telescopic system, such as is shown in Fig. 15-18, and when this stop lies at the intermediate focus of the telescopic system, then the entrance aperture

---

FIG. 15-18 Use of a stop to control parallax in a telecentric optical system.

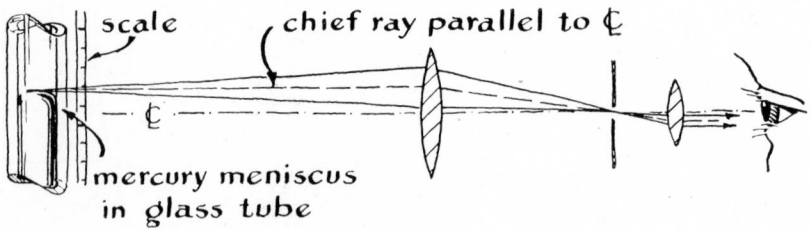

is projected at infinity. The chief ray, defined as the ray which passes through the center of the entrance pupil, accordingly lies parallel to the ₵ of the system, as illustrated. Such a system, called a telecentric system, has useful properties. For example, Fig. 15-18 shows this system used for reading the position of a mercury meniscus located behind a transparent scale. The telecentric optical arrangement avoids errors arising from parallax—all chief rays, for all observable meniscus positions, are horizontal.

15-11. *Field Lens*

Fig. 15-19 shows two optical systems fitted with field lenses, to reduce vignetting. These lenses are located at intermediate real focal planes. These focal positions lie between the components of lens systems, and the field lenses are chosen with just the power to focus the exit window of the preceding part of the system onto the entrance window of the following part. The field lenses cause very little disturbance of the final image of the system because of their location at focal positions.

In the monochromator the field lens focuses a first prism or grating on the following one. Because prisms or gratings are more expensive than lenses or mirrors, they usually act as the aperture stop in spectroscopic equipments.

In the telescope, the field lens, located at the intermediate focal plane, focuses the objective onto the eye ring. The periscope is an example of a specialized telescope, and it affords an example where vignetting would be extreme if field lenses were not used between the relay lenses. Specks of dirt on such field lenses are troublesome since the field lenses are all imaged in the final focal plane. Actually, however, the field lenses can themselves be compound lenses which are only virtually at the positions shown. Such field

FIG. 15-19   Field lenses in a spectrometer and in a periscope.

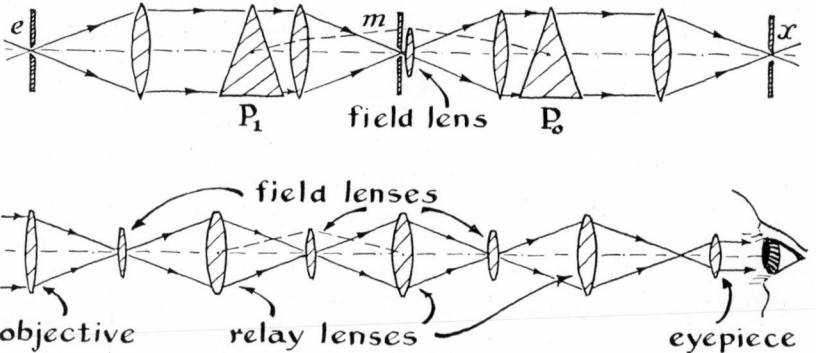

lenses make it possible, for example, to have a large working field through a periscope tube 20 feet long and of only 4 inches inside diameter. The field is of 40° diameter with the field lenses while without them the periscope would have a field of only 1°.

Chapter **XVI**

Image Defects

WHEN THE angular aperture of a lens is widened, spherical aberration and longitudinal chromatism enter. When the field is widened, the monochromatic aberrations—coma, astigmatism, curvature, and distortion—and lateral chromatism enter. The first defects are defects of aperture; the latter are defects of field. We have discussed some of these aberrations and how the image defects they produce may be reduced or eliminated. We found that spherical aberration of a simple lens was controlled by "bending" it. And we found that certain systems were aplanatic, with both spherical aberration and coma absent. We demonstrated the control of longitudinal chromatism by dividing a lens into two components made from suitably different glasses; or by separating two components made from the same glass. Now, in this chapter, we extend these discussions of image defects.

Very few optical systems give images that are free from all defects. Fortunately, certain defects, unimportant in one particular application, can often be tolerated, if the produced images have the specified quality in other respects. Thus, experimentalists can solve optical problems which would be frustrating otherwise. Designers control important defects by the "bending" of components (to minimize aperture defects), or by the position of stops (to minimize field defects), or by the choice of glasses, thicknesses, and spacings for components (to manage chromatism, *etc.*). Recently, designers have been beginning to use aspheric surfaces more extensively.

Mirror components are of course free of chromatism, but the off-axis coma of mirrors may be severe. In a spectrograph, where the camera focuses spectrum lines on a photographic plate, astigmatism is generally unimportant and the photographic plate may be tipped to compensate for longitudinal chromatism. Although a large flat field is of first importance in an ordinary camera, in a monochromator, where the required field is very small, field

351

defects, compared to those of aperture, are of secondary importance. In a photoengraver's lens freedom from distortion is the most important lens specification. Here chromatism is tolerable because photoengraving can be done with more or less monochromatic light. And, within limits, lack of lens speed in a photoengraver's lens may be compensated by long exposure. And so it goes; lenses and optical systems are designed or selected according to the specialized functions they must perform.

16-1. *Third-order Theory*

Our Gauss theory for longitudinal and lateral image positions was based on the approximate analytic series representation of the transcendental circular functions:

$$\sin x = x - \frac{x^3}{3!} + \frac{x^5}{5!} \cdots$$

$$\cos x = 1 - \frac{x^2}{2!} + \frac{x^4}{4!} \cdots$$

The Gauss theory using $\sin x = x$ and $\cos x = 1$ is called first-order theory. The values of x, representing angles of incidence, of refraction, or of obliquity, are constrained by the Gauss theory to have such modest magnitudes that all terms in x of order higher than the first, in the expansions, are negligible. The next analytic approximation to the transcendental circular functions retains the second and third order terms in x in $\sin x$ and $\cos x$. Optical theory which retains these terms is called *third-order theory*. The equations for longitudinal and lateral image positions, taken to the third order of approximation, were first developed a century ago by Ludwig von Seidel.

If we solve for the $\mathbb{C}$ cross-over point of a parallel marginal ray, according to the Gauss theory, where $u = \infty$, and where the rays are refracted by a single convex glass surface of radius r, we get

$$\frac{1}{v} = \frac{N - 1}{Nr}$$

If, however, we use the third-order theory, the cross-over point for a ray incident at a distance h off the $\mathbb{C}$, lies a distance v beyond the glass surface. This distance is given by the above expression amended:

$$\frac{1}{v} = \frac{N - 1}{Nr} + \frac{(N - 1)h^2}{2N^3 r^3}$$

A formula for objects not lying at infinity, using third-order theory, is deduced in Joseph Morgan's *Introduction to Geometrical and Physical Optics*, § 7.1, p. 86. Morgan, in his book, compounds that formula for air-to-glass refraction, at the first surface of a lens, with a similar formula for glass-to-air refraction, at the second surface, to get the expression for the spherical aberration of thin lenses (in terms of an image position parameter τ and a lens shape parameter σ) that we saw in § 14-6.

Von Seidel's third-order theory includes the above amendment to the Gauss first-order theory, together with a series of four more such correction terms:

$$\frac{N}{u} + \frac{N'}{v} = \frac{N' - N}{r} + \sum_{n=1}^{5} S_n$$

These corrections are reminiscent of Van der Waals' virial coefficient corrections to the gas law,

$$PV = RT + \sum_{n=1}^{4} \frac{C_n}{V^n}$$

The five terms of von Seidel may be thought of as expressing the theoretical aberrations of the simple Gauss theory. They have a logical relation to the five categories into which we have classified the monochromatic image defects: The first Seidel expression corrects the theory of Gauss to take account of spherical aberration. The second is concerned with coma, and so on. The character of the Seidel terms as they apply to a compound lens is such that the second amendment does not have meaning unless the first is zero— or more generally, all the amendments preceding any one of the five must be zero for it to have meaning. In an aplanatic system Seidel's first and second amendments are both zero. The third amendment then has meaning relative to astigmatism. In an anastigmat worked at aplanatic points we have a system without spherical aberration, coma or astigmatism; and then the fourth Seidel term relates to curvature of field; while the fifth Seidel amendment, significant if the preceding four are all zero, is concerned with distortion.

16-2. *Image Testing*

We do not presume to deal here with topics of optical engineering. The optical engineer is concerned with the practice of design and construction of optical systems intended for specified functions so that significant defects will be absent or tolerable. Successful engineering combines engineering art and intuition, acquired from education and experience, as well as the science itself. This requirement of art and intuition arises because designs are largely empirical—the radii, thicknesses, dispersions, and separations of component lenses are not explicitly determined from Seidel's theory—and anyway, chromatism is the most confounding image defect.

Once a lens is designed, and suggested compromises have been made, they may be evaluated in prospect, before construction, by ray tracing; and once the lens is constructed, it may be further evaluated by experimental testing. Nowadays ray tracing is done with modern high-speed computing machines so that, for each object point in the object plane, several hundred rays may be calculated, and the points where they penetrate the image plane may be plotted in prospect, to predict the character of their union. These plots are called dot diagrams. The surface density of these points in the image plane is construed as the predicted distribution of image illumination. The dot diagram predictions are often found by subsequent experimental test to have been pessimistic—that is, they predict a poorer quality of the image of a point source than experiment manifests. But in spite of this lack of full concordance (which is understandable with the absence of a tight theoretical basis for ray tracing as set forth in § 13-1), these prospective evaluations, by computation, have enjoyed elaborate use. Ray tracing, to guide design, has been practiced since it was introduced in the time of Newton and Flamsteed.

Although, in this book, we do not take up optical design, we treat, briefly, the different methods of experimental testing. First of all, there is the eyepiece testing method. By means of a suitable magnifier the array of rays

FIG. 16-1 The Hartmann test illustrated by a lens with a raised intermediate zone.

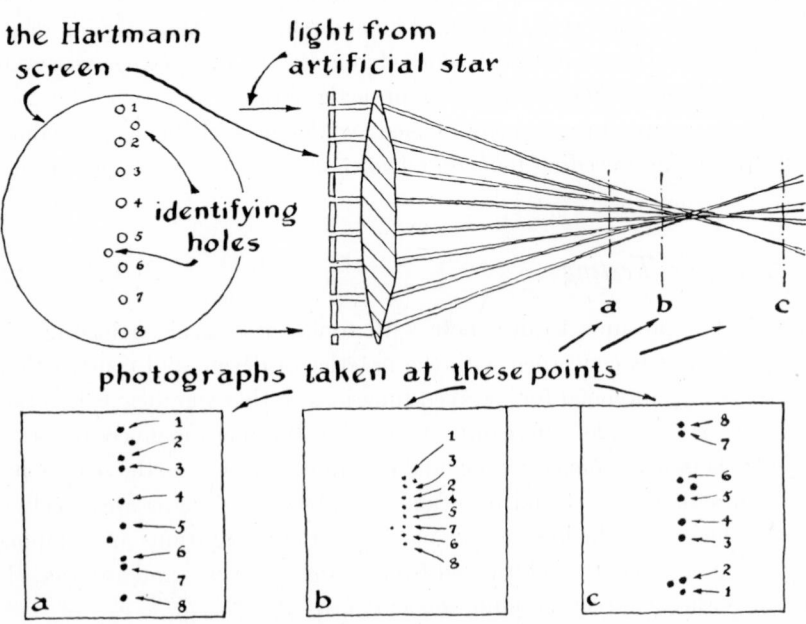

making up an image of a point may be relayed onto the retina of the eye. If the angular aperture of the magnifier is greater than that of the image-forming system, all the rays will be thus relayed; and, by visual examination, the union of rays both inside and outside the best focus can be determined. This eyepiece method becomes a particularly sensitive test for astigmatism. Also, the image may be photographed, and the photographs, rather than the relayed images, may be examined (see Fig. 13-2).

Fig. 16-1 illustrates the Hartmann test. This experimental method for testing an optical system has much in common with theoretical ray tracing. In the Hartmann test a screen obscures all the rays collected by the lens except a set of narrowly defined beams. These sampling beams are uniformly distributed over the aperture, as illustrated. The trajectory of each of these beams may be traced by means of a series of photographs made at regular intervals along the ₵ in front of, and behind, their best union, as shown.

The knife-edge test is also a kind of a ray tracing in reverse. From our previous explanations of the knife-edge test in § 13-10, it is evident that by this test we may determine where rays from each part of the aperture penetrate the focal plane. This determination follows from the position of the knife-edge and the observed location of its shadow, as seen in the Maxwellian view of the lens or mirror.

In addition to these tests, and Gaviola's test of § 13-11, there are other optical tests, which we ignore here; but all have as their purpose to determine the nature of images, and to provide information from which defects due to design or construction defects can be diagnosed or corrected.

16-3. *Spherical Aberration, Coma, and Achromatism*

When points of white light are to be imaged, chromatism is as important as monochromatic spherical aberration in determining if a lens will be satisfactory, or not. In a compound lens worked with white light, both these aberrations enter, and both can be mathematically eliminated (in the cases where Gauss' theory applies). Our expression for the spherical aberration of a simple lens, of § 14-6, may be abbreviated as follows:

$$\frac{h^2}{f^3}\left(a\sigma^2 + b\sigma\tau + c\tau^2 + d\right) = \frac{h^2 S'}{f^3}$$

In case of a compound lens comprised of a pair of close thin lenses, the aggregate spherical aberration will be zero if the following expression vanishes:

$$-\Delta\left(\frac{1}{v}\right) = h^2\left\{\frac{S_1'}{f_1^3} + \frac{S_2'}{f_2^3}\right\}$$

Our condition, from § 14-7, that two thin lenses in contact shall be free from longitudinal chromatism, was: $\dfrac{\mathfrak{F}_1}{\nu_1} + \dfrac{\mathfrak{F}_2}{\nu_2} = 0$. On combining this condition with the spherical aberration condition, $\Delta\left(\dfrac{1}{v}\right) = 0$, we get the condition for a simultaneous freedom from both spherical aberration and longitudinal chromatism, namely:

$$S_1' \nu_1^3 = S_2' \nu_2^3$$

In a lens the off-axis image defect coma may be pictured as a manifestation of differences in lateral magnification for different rays from a single off-axis object point. Lateral chromatism is to longitudinal chromatism somewhat as coma is to spherical aberration. When a lens manifests lateral chromatism, its magnification is different for different colors. In a case where we have freedom from coma, as Fig. 15-4 shows, the second principal surface is a sphere. This is required mathematically by Abbe's sine condition: $\dfrac{h}{\sin\theta'} =$ (constant f). Similarly, freedom from lateral chromatism requires that the second principal surfaces be coincident for all different colors. Fortunately, lateral chromatism is usually a small aberration in a well-corrected, cemented achromatic doublet. For this reason we do not treat it further here.

The coma of a simple lens may be described along the lines of our treatment of spherical aberration, in § 14-6. In that section spherical aberration was described in terms of τ, an image position factor; σ, a lens shape factor; and four constants involving N. We defined these four constants, a to d, in § 14-6. We need two more such constants, e and g, for describing coma:

$$e = \frac{3(N+1)}{4N(N-1)} \qquad g = \frac{3(2N+1)}{4N}$$

Fig. 16-2 illustrates the character of the coma pattern for a simple lens. The third-order theory of Seidel gives the angular length of the coma pattern for a simple thin lens, from its head to the central extremity of its tail. This figure shows where the rays from 12 points on a circular zone penetrate the focal plane. The angular length of the coma pattern, not derived here, is

$$\left(\frac{h^2}{f^2}\tan\varphi\right)(e\sigma + g\tau)$$

Here φ is the angle which the chief ray makes with the ₵ of the lens axis. It is apparent, for a given set of conjugate points defined by τ, that bending the lens to change σ may reduce or even eliminate coma, just as spherical aberration was minimized in § 14-6. Actually, in contrast with spherical aberration, the elimination of coma is physically realizable. Coma vanishes

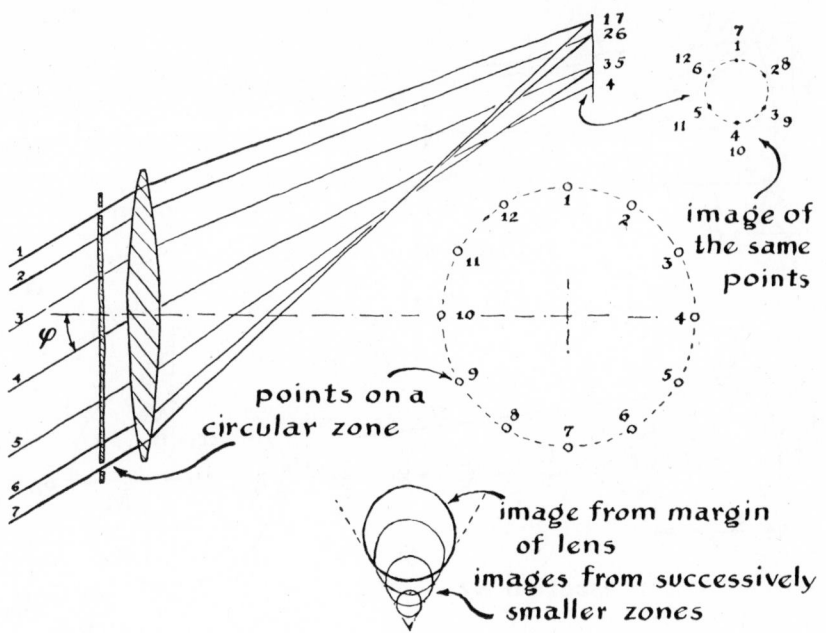

FIG. 16-2   Nature of coma in a simple lens.

when $(e\sigma + g\tau) = S'' = 0$. And, for two thin lenses in contact, since we may cancel out $h^2 \tan \varphi$, the aggregate coma vanishes when

$$S_1''f_2^2 + S_2''f_1^2 = 0$$

16-4. *Coma of a Parabolic Mirror*

We now take up the description of coma as it is manifested by a parabolic mirror worked in parallel light. Fig. 16-3 shows the parabolic mirror used to focus a distant point object (which we shall take in the limit to be at infinity). This figure also contains auxiliary illustrations pertinent to our considerations. The image of the point P, by the parabola, will exhibit coma if the chief ray is inclined at an angle φ to the ₵. We describe this coma here for the limiting case where φ is small. The condition that φ is small allows our mathematics of modest rigor to come into play. With this mathematics of modest rigor we develop a qualitative description of the comatic image that becomes increasingly valid, quantitatively, as the field and the angular aperture decrease. Fig. 16-3, at a, shows the object point at P on the ₵ at a distance u from a mirror facet M. This facet lies on a zone of the mirror of radius h, and its position is further defined on this zone by the hour angle,

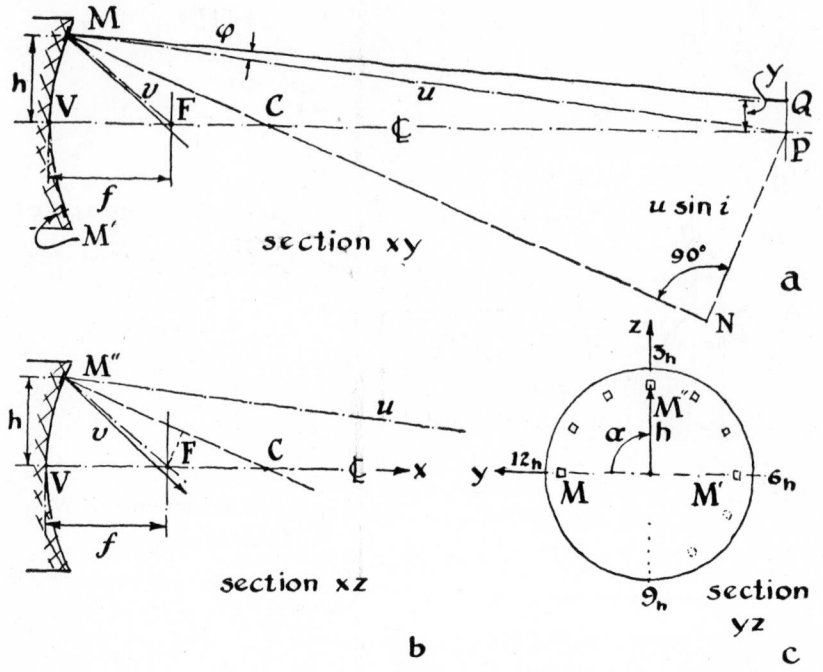

FIG. 16-3 Geometry of rays for calculation of the coma of a parabolic mirror.

or by the angle α at which M lies, when the zone is viewed from P, as shown at c. Thus M, as shown, is at 12^h, and $\alpha = 0$ or 2π. And M', as shown, is at 6^h, and $\alpha = \pi$. M'' at 3^h (or 9^h) is defined by $\alpha = \dfrac{\pi}{2}\left(\text{or } 3\,\dfrac{\pi}{2}\right)$. This definition of the angular facet position on the mirror face is illustrated in Fig. 16-3 at c. At b we see a horizontal section, corresponding to the vertical section at a.

Fig. 16-4 shows three magnified views of the focal plane as seen from P looking toward V. In this figure at a we see the point P', where an on-axis object P is focused. Here coordinates (y', z') in the focal plane serve to define the point where a reflected ray penetrates this focal plane. Q_V' shows where a paraxial ray, reflected by the mirror at V, from an off-axis point Q, penetrates the focal plane. A ray from the same point Q, reflected by a mirror facet located at the radial distance h from the ₡, and at an azimuth angle α, penetrates the focal plane at $Q_{h\alpha}'$. We also define $Q_{h\alpha}'$ by the vectors $\vec{R}$ and $\vec{A}$, in the focal plane, as shown. Fig. 16-4b shows the observed characteristic double comatic circle, of radius ρ, representative of reflections of rays from all the facets of one zone h. The positions of $Q_{h\alpha}'$ for facet hour angles on the

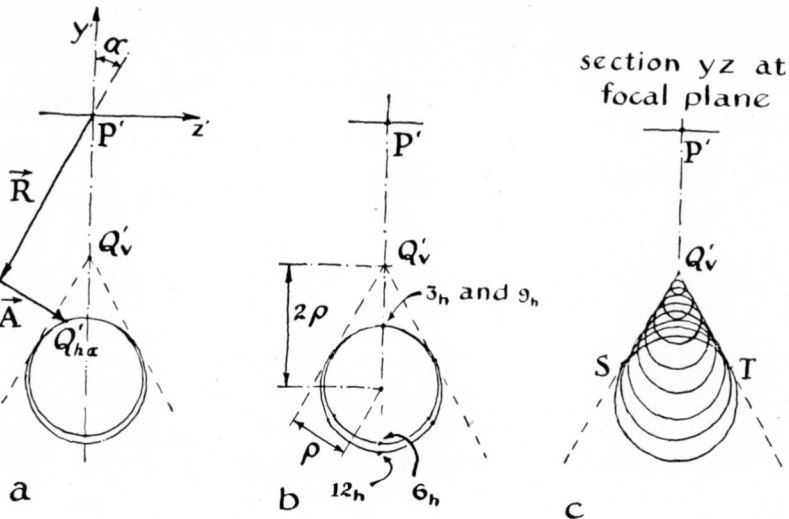

FIG. 16-4 Sections at the focal plane (to go with Fig. 16-3).

zone h are indicated at b. We shall deduce below that the centers of such double circle loci are located 2ρ below Q_V', so that all the double comatic circles together give the comet-shaped pattern illustrated in this figure at c. This pattern has a bright head at Q_V' and its shape is characterized by 60° angular boundaries.

We deduce this coma pattern by calculating the double comatic circles in the focal plane; that is, we determine the locus of points where rays reflected from the mirror at various facets on the zone h penetrate the focal plane. When the source point in Fig. 16-3a moves from P to Q through the distance y, in the plane of that figure, the angle of incidence at M will increase from i to $(i + \varphi)$, where $\varphi = \dfrac{y}{u}$. In contrast, the angle of incidence at M' will decrease from i to $(i - \varphi)$. On the other hand, if the source point is moved perpendicular to the plane of that figure by the distance z, this movement will not change the angles of incidence at either M or M', located at 12^h or 6^h. Such a perpendicular motion will simply rotate the azimuth of incidence at M about the facet normal—the dashed line MCN. Although this leaves the angles of incidence substantially unchanged at M and M' these angles are increased and decreased for facets at 3^h and 6^h. This rotation of azimuth at M is given by the angle $\Delta\psi = \dfrac{z}{u \sin i}$, where $u \sin i$ is the perpendicular distance from P to the facet normal. At intermediate angles, changes of both incidence and azimuth occur. For example, the motion y in the plane of

Fig. 16-3a will produce the following changes at the facet (h, α), resolving y into its orthogonal components parallel and perpendicular to the plane of incidence:

$$\Delta i = \frac{y}{u} \cos \alpha = \varphi \cos \alpha$$

$$\Delta \psi = \frac{y}{u \sin i} \sin \alpha = \varphi \frac{\sin \alpha}{\sin i}$$

A change of incidence Δi moves Q' in the focal plane in the opposite direction to α, by the vector amount

$$|\vec{R}| = \frac{v \, \Delta i}{\cos 2i}$$

Here v is the distance from M to P'. The factor $\cos 2i$ enters because v is inclined to the focal plane. A change of azimuth $\Delta \psi$ moves Q' in the focal plane in a perpendicular direction by the vector amount

$$|\vec{A}| = v \sin i \, \Delta \psi$$

Here $v \sin i$ is the lever arm, so to speak, on which Q' swings about the facet normal, MC. Taking projections of $\vec{R}$ and $\vec{A}$ parallel to the axes, z'- and y'-axes, we get the Cartesian coordinates of $Q_{h\alpha}'$ in the focal plane. As shown in Fig. 16-4a, these coordinates are

$$-y' = \frac{v\varphi \cos \alpha}{\cos 2i} \cos \alpha + v\varphi \sin \alpha \sin \alpha$$

$$z' = \frac{v\varphi \cos \alpha}{\cos 2i} \sin \alpha - v\varphi \sin \alpha \cos \alpha$$

Now, if we write $\cos^2 \alpha = \frac{1}{2}(1 + \cos 2\alpha)$ and $\dfrac{1}{\cos 2i} \cong 1 + 2i^2$, we get

$$-y' = v\varphi + v\varphi i^2 + v\varphi i^2 \cos 2\alpha$$

From Fig. 16-3 we see that the slant distance v is related to the focal distance $v_0 = \dfrac{r}{2}$ by the relation $v \cos 2i = v_0 - \dfrac{h^2}{2r}$. Here $\dfrac{h^2}{2r}$ is the sagitta of the parabola. But we make the approximation $i = \dfrac{h}{r}$, so that $v_0 - \dfrac{h^2}{2r} = v_0(1 - i^2)$.

Using our expression for $\cos 2i$, we get $v = v_0(1 + i^2)$. Setting $v_0\varphi = PQ_{V}'$, and $\rho = v_0\varphi_i^2$ we get, finally,

$$-y' \cong PQ_{V}' + 2\rho + \rho \cos 2\alpha$$

And, on writing $\sin \alpha \cos \alpha = \frac{1}{2} \sin 2\alpha$, we get

$$z' = v\varphi i^2 \sin 2\alpha \cong \rho \sin 2\alpha$$

These expressions for y' and z' are recognized as the parametric representation of a double circle shown resolved at b in Fig. 16-4. In that figure at c we show all the unresolved double circles superimposed, one for each of the different mirror zones. Obviously the envelope lines in this figure, at c, $Q_V'S$ and $Q_V'T$, make angles of $\sin^{-1}\left(\dfrac{\rho}{2\rho}\right) = 30°$ with the y'-axis, and give the $60°$ included angle, which is typical of the coma pattern. In the derivation

---

FIG. 16-5 Elby's measurement of the coma of an elliptical mirror.

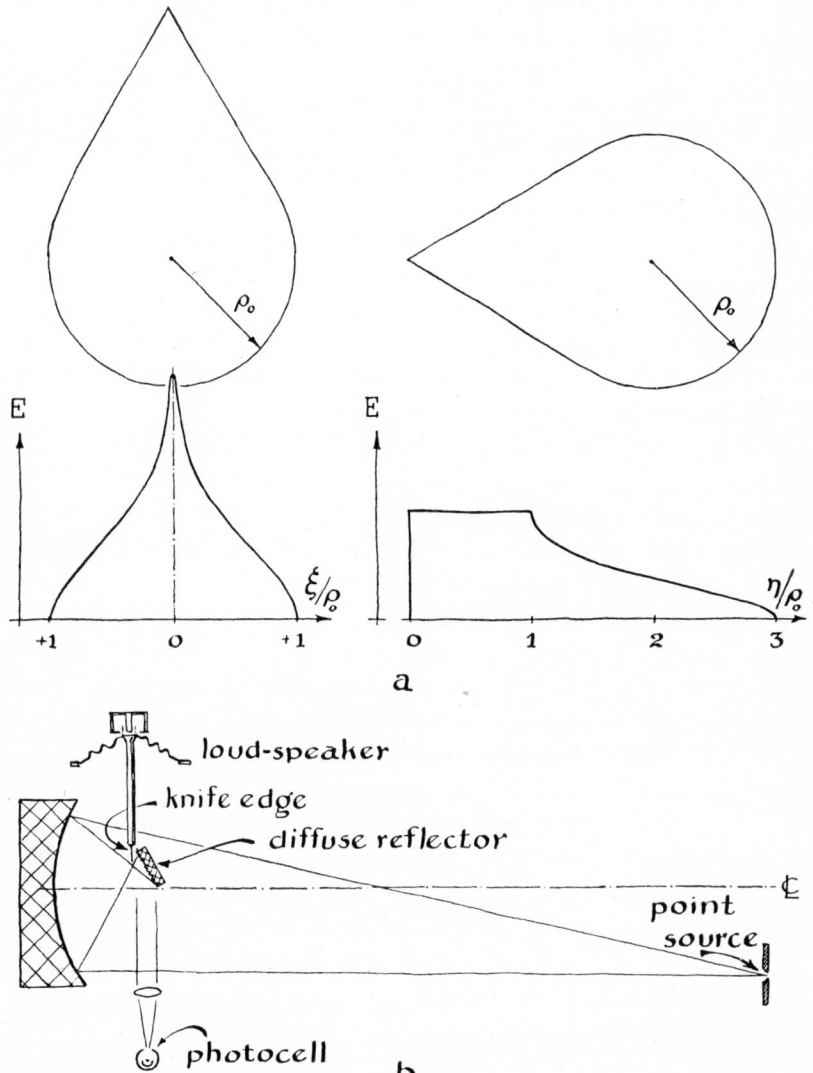

above, our approximations obscure discrimination between the resolved double circles at b.

Fig. 16-5 shows Frank Elby's half-resolved measures of distributions of energy in the comatic pattern that is characteristic of an elliptical mirror. The pattern was measured with an oscillating knife, as shown. This coma pattern is very much like that of a parabolic mirror, when the ellipticity approaches unity. The angle of projection in the case of the elliptical mirror was so great that all the light transmitted past the knife edge could not be managed with lenses. Actually, the light transmitted past the focal knife was received on a scattering surface and only the part of it scattered into the photocell was used as a sample for measurement.

16-5. *Astigmatism of a Single Surface*

Fig. 16-6 illustrates how astigmatism is produced on refraction at a facet on a spherical refracting surface. This figure at b shows the sections AB and CD. The fan of rays refracted along A and B exhibit a union at P_m', whereas those refracted along C and D exhibit a union at P_s'. These two foci, separated in depth, are characteristic of astigmatism. P_m' is the meridional line focus, and P_s' is the sagittal line focus. Sections of the incident and refracted wave fronts are shown in the figure at c and d. In order to clarify the meaning of the aberration of astigmatism, we shall consider the transformations which the refracting surface imposes on the incident spherical wave, following the manner of § 13-4 and the illustrations of Fig. 13-6. The wave front transformations can be described as the combination of a prism-like deviation, as if the facet $ABCD$ represented a prism face, and, in addition, a lens-like transformation of the curvature of the incident spherical wave front giving line focusings, similar to that mentioned in § 13-4. The different line focusings, for the AB and CD sections, are accounted for in terms of two transformations of wave fronts. For these transformations we use appropriate projections of the sagittae and chords of the arcs AB and CD (projections along the directions of incidence and refraction). The object distance for both line images is specified by the common slant distance u; the two image distances are measured, in the direction of prism refraction, by the slant distances v_s and v_m.

First, we consider the refraction in a plane perpendicular to the plane of incidence, represented by the arc CD. We determine the differences of optical path, both before and after refraction, between center and edge rays; and then, using the fact that wave surfaces are surfaces of equal phase, we invoke the implication of Fermat's principle; that all optical path distances from

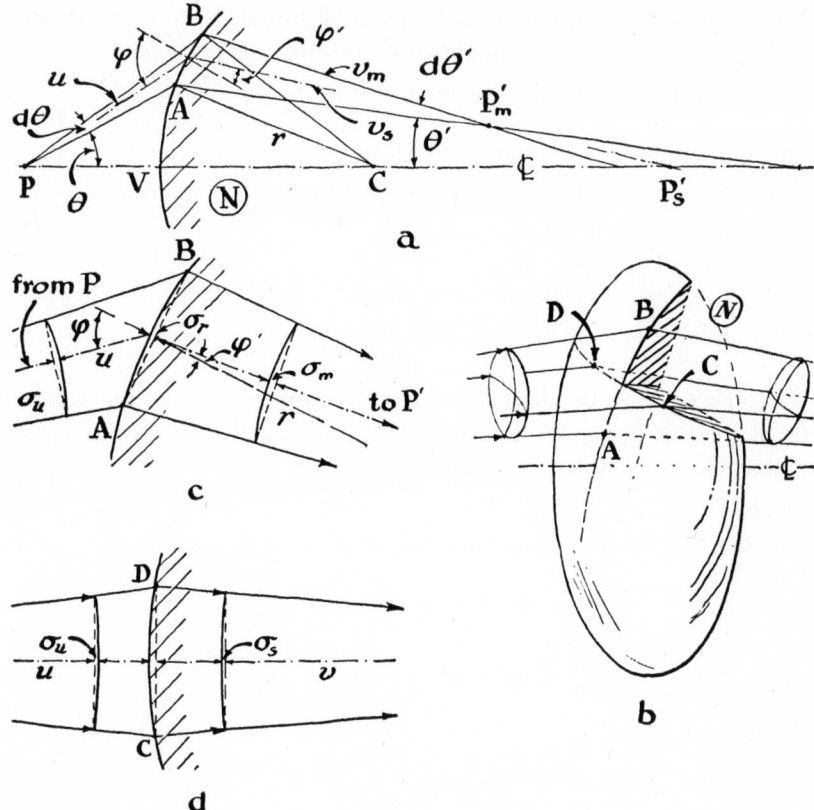

FIG. 16-6   Ray and wave calculation of astigmatism produced by a single refracting surface.

points on an incident wave to corresponding points on a refracted wave front are equal. Taking projections of the sagitta of the arc CD, and applying Fermat's principle,

$$\sigma_r' = \sigma_r \cos \varphi \qquad \sigma_r'' = \sigma_r \cos \varphi'$$

$$(\sigma_r' + \sigma_u) - N(\sigma_r'' - \sigma_{v_s}) = 0$$

$$\left(\frac{(CD)^2}{2r} \cos \varphi + \frac{(CD)^2}{2u}\right) - N\left(\frac{(CD)^2}{2r} \cos \varphi' - \frac{(CD)^2}{2v_s}\right) = 0$$

This relationship of path differences gives the same slant distance as Eq. 13-5a, namely

$$\frac{1}{u} + \frac{N}{v_s} = \frac{N \cos \varphi' - \cos \varphi}{r} \tag{a}$$

Next, considering refraction in the plane of incidence, represented by the arc AB, we again invoke the implications of Fermat's principle:

$$(\sigma_r \cos \varphi + \sigma_u) - N(\sigma_r \cos \varphi' - \sigma_{v_m}) = 0$$

And, taking proper projections of (AB), we get

$$\left(\frac{(AB)^2}{2r} \cos \varphi + \frac{[(AB) \cos \varphi]^2}{2u}\right) - N\left(\frac{(AB)^2}{2r} \cos \varphi' - \frac{[(AB) \cos \varphi']^2}{2v_m}\right) = 0$$

or

$$\frac{1}{u} + \frac{N \dfrac{\cos^2 \varphi'}{\cos^2 \varphi}}{v_m} = \frac{N \dfrac{\cos \varphi'}{\cos^2 \varphi} - \dfrac{1}{\cos \varphi}}{r} \tag{b}$$

This expression may be checked separately by means of the following geometrical analysis, applied to Fig. 16-6a. We take A and B to be infinitesimally separated, and take α as the inclination of r.

$$(AB) = r\, d\alpha$$
$$u\, d\theta = r\, d\alpha \cos \varphi \qquad v_m\, d\theta' = r\, d\alpha \cos \varphi'$$

$$\varphi = \alpha + \theta \qquad d\varphi = d\alpha + d\theta = d\alpha\left(1 + \frac{r \cos \varphi}{u}\right)$$

$$\varphi' = \alpha - \theta' \qquad d\varphi' = d\alpha - d\theta' = d\alpha\left(1 - \frac{r \cos \varphi'}{v_m}\right)$$

Substituting these values for $d\varphi$ and $d\varphi'$ in Snell's law differentiated,

$$\cos \varphi\, d\varphi = N \cos \varphi'\, d\varphi'$$

and on dividing by $r \cos^2 \varphi$, we get expression (b) above.

Both of these equations, (a) and (b), apply equally to mirrors. Fig. 16-7 shows reflection from a presumed flat mirror at 45°. Here the flat mirror is

FIG. 16-7 Astigmatic test for the deviation of a mirror from true flatness.

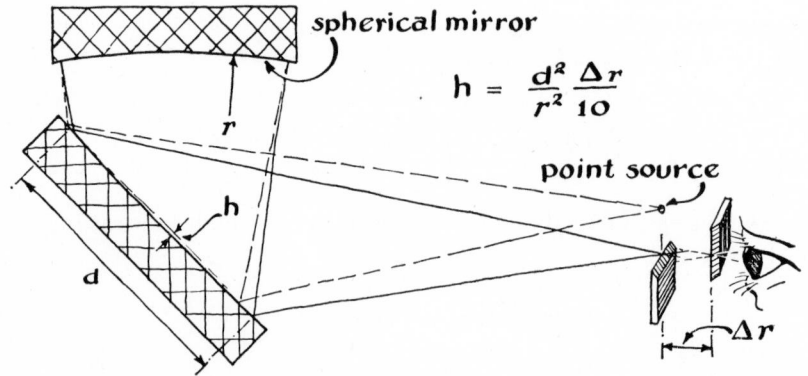

set at 45° in front of a spherical mirror of known true quality. If now the presumed flat actually has a very slight concavity or convexity, the effect of its radius of curvature will be to produce astigmatism at the focus of the true testing sphere. This manifest astigmatism may be evaluated qualitatively with the eyepiece test; or it may be determined quantitatively by means of horizontal and vertical knife-edge cut-offs, in the manner indicated in the figure. A student exercise shows that even the slightest convexity or concavity introduces easily measurable separations of the horizontal and vertical knife-edge cut-offs.

For a mirror, either at 45° as above, or at grazing incidence, we rewrite the expressions (a) and (b) with both terms in F positive. In the case of grazing incidence we write γ for the grazing angle, so that $\gamma = \frac{\pi}{2} - \varphi$ and $\cos \varphi = \sin \gamma \cong \gamma$. Thus for $u = \infty$, and $N = +1$,

$$\frac{1}{v_s} = \frac{\cos \varphi' + \cos \varphi}{r} \cong \frac{2\gamma}{r} \tag{a'}$$

$$\frac{1}{v_m} = \frac{2}{r \cos \varphi} \cong \frac{2}{\gamma r} \tag{b'}$$

It is evident that $(v_m - v_s)$ will become very large when γ gets small. In fact, this effect is so marked that the convexity of a quiet mercury surface of only 1 foot diameter, its surface conforming to the curvature of the earth with a sagitta of $\frac{\lambda}{40}$ for green light, introduces noticeable astigmatism in a parallel beam reflected at grazing incidence from it.

16-6. *Coddington's Equations for a Thin Lens*

Fig. 16-8 shows a symmetrical thin lens stopped centrally. The chief ray is inclined at angle φ, relative to the lens ⊄. A combination of expressions for the astigmatism at each lens surface, as follows, gives formulas for oblique penetration of the simple lens as a whole. We get one formula for the section of the wave front perpendicular to the plane of incidence of the chief rays, giving v_s, and another for the section of the wave front parallel to the plane of incidence, giving v_m. Taking the lens thickness as negligible, we write

$$\frac{1}{u} + \frac{N}{v_s'} = \frac{N \cos \varphi_1' - \cos \varphi_1}{r_1} \quad \text{and} \quad \frac{N}{v_s'} + \frac{1}{v_s} = \frac{\cos \varphi_2' - N \cos \varphi_2}{r_2}$$

for the two lens surfaces. On combining these expressions and eliminating v_s', we get the first of Coddington's equations:

$$\frac{1}{u} + \frac{1}{v_s} = \cos \varphi \left(\frac{N \cos \varphi'}{\cos \varphi} - 1 \right) \left(\frac{1}{r_1} + \frac{1}{r_2} \right) \tag{a}$$

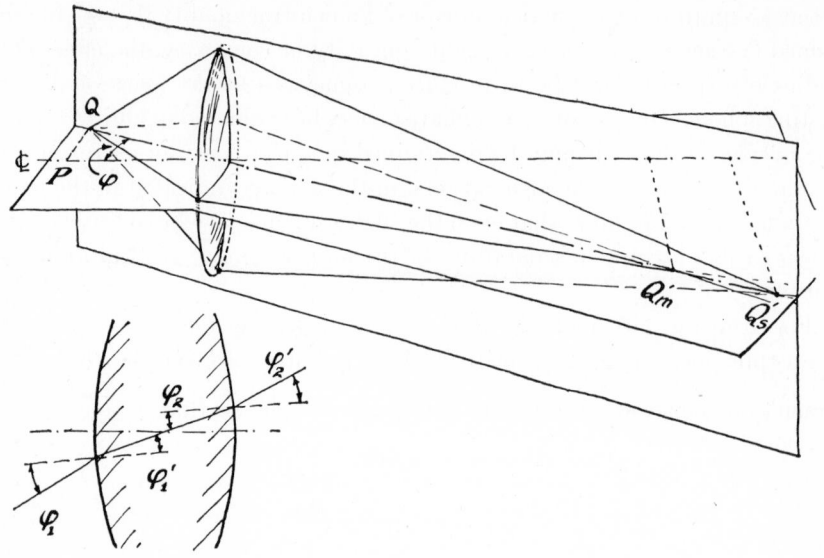

FIG. 16-8 Astigmatic foci for a lens, where the chief ray is incident at the angle φ with the $\mathcal{C}$. In Eqs. 16-6 a and b, $N \sin \varphi' = \sin \varphi$.

And elimination of v_m', similarly, gives the second Coddington equation:

$$\frac{1}{u} + \frac{1}{v_m} = \frac{1}{\cos \varphi} \left(\frac{N \cos \varphi'}{\cos \varphi} - 1 \right) \left(\frac{1}{r_1} + \frac{1}{r_2} \right) \tag{b}$$

16-7. *Curvature of Field*

The curvature of field of a thin lens is closely related to astigmatism. We may predict the curvature of field by the variation of $\left(\dfrac{v_m + v_s}{2} \right)$ with φ, whereas astigmatism is measured by the variation of the difference, $(v_m - v_s)$, with φ. The astigmatic images at focal distances v_m and v_s, for incident parallel light, are line images—one a tangential line at the slant distance v_m, and the other a radial line at the slant distance v_s. As Fig. 16-9 shows, between these line images the light is converged within a roughly circular *blur circle*, of diameter b. From Eqs. 16-6a and b, the blur circle diameter for incident parallel light is approximately

$$b = -(v_m - v_s) \tan \theta_0' \cong \varphi^2 f_0 \tan \theta_0'$$

where f_0 is the focal length for $\varphi = 0$. From the same equations, the oblique focal length, f_φ, can be similarly shown to be

$$f_\varphi = \frac{1}{2}(v_m + v_s) \cong f_0$$

This expression predicts that the focal surface of least confusion for a thin lens is a sphere centered at the center of the lens—showing that curvature of field and astigmatism are closely connected.

Whereas we previously have found that spherical aberration can be controlled by bending the lens at constant f_0, the equation above indicates that curvature of field is independent of bending. The curvature of field, however, can be reduced by a proper positioning of a stop so that oblique chief rays do not go through the center of the lens. Unfortunately, when curvature is thus reduced, astigmatism is increased. Fig. 16-10 shows how such a reduction of curvature may be effected. In inexpensive cameras such a reduction is sometimes used when curvature is more intolerable than a loss of resolving power at the edges of the field, due to increased astigmatism. Such use of a stop is called *artificial field flattening*.

A compound lens with components of equal and opposite power, $\mathfrak{F}_1 = -\mathfrak{F}_2$, having the aberrations of astigmatism and curvature of field in equal and opposite measure, will have the residual power $\Phi = \mathfrak{F}^2\tau$ (applying equation c of § 14-2). With a properly positioned stop such a system can be given reduced astigmatism or curvature of field; and, if the condition

$$\frac{N_1}{\mathfrak{F}_1} + \frac{N_2}{\mathfrak{F}_2} = 0 \tag{a}$$

FIG. 16-9   Geometry for location of circle of least confusion and for prediction of its diameter.

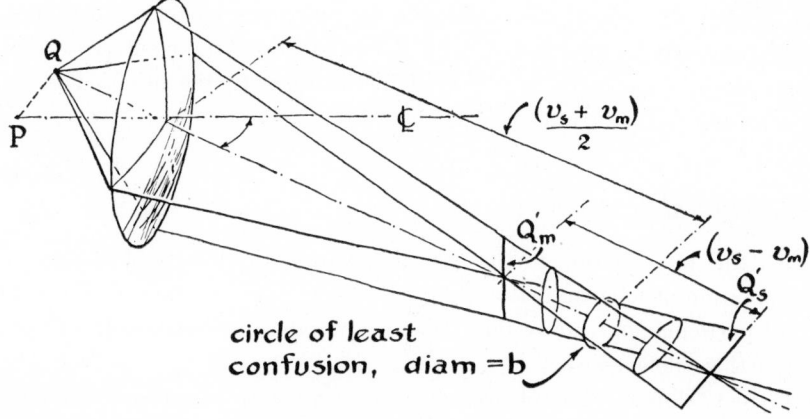

P Q $\mathfrak{C}$ $\frac{(v_s + v_m)}{2}$ Q'_m $\frac{(v_s - v_m)}{}$ Q'_s

circle of least confusion, diam = b

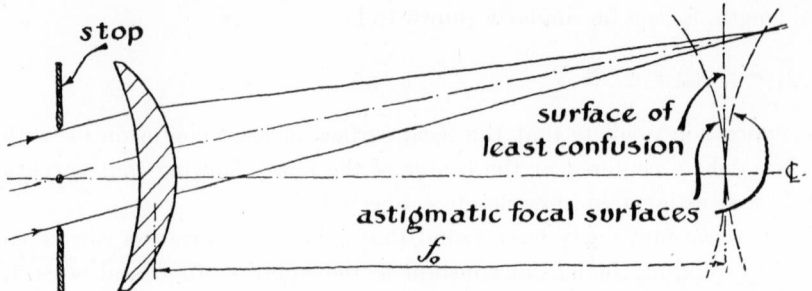

FIG. 16-10 Artificial field flattening by means of an aperture stop.

is also fulfilled, the system can be freed of both aberrations simultaneously. The above condition is called Petzval's condition, and it has been referred to previously in § 14-8.

16-8. *Distortion*

Finally we come to the fifth monochromatic aberration, distortion. It must be apparent by now that the simultaneous satisfaction of all the conditions for removal of aberrations is difficult to accomplish. Distortion may be controlled by the stop position; and compound lenses that are arranged with components symmetrically on both sides of a stop have reduced distortion.

Distortion is a manifestation of the increase, or decrease, of magnification as the obliquity of the chief ray increases. If the magnification increases with obliquity, we get so-called *pin-cushion* distortion; if it decreases, we get *barrel* distortion. Fig. 16-11 shows a simple lens with a preceding stop and the same lens with a following stop. The combination of these to form a symmetrical compound lens, with the stop between the components, is free from distortion. The first lens and stop combination manifests barrel distortion; the second combination manifests pin-cushion distortion; the combination with central stop compensates one type of distortion with the other. The compound system is distortion free, but $m = 1.0$.

16-9. *Optical Systems*

With this introduction to the five monochromatic defects of images, and some mention of longitudinal and lateral chromatism, we may now consider several particular optical systems that are notable for freedom from one or more of these defects.

Fig. 16-12 illustrates a spherical mirror with a stop at its center of curva-

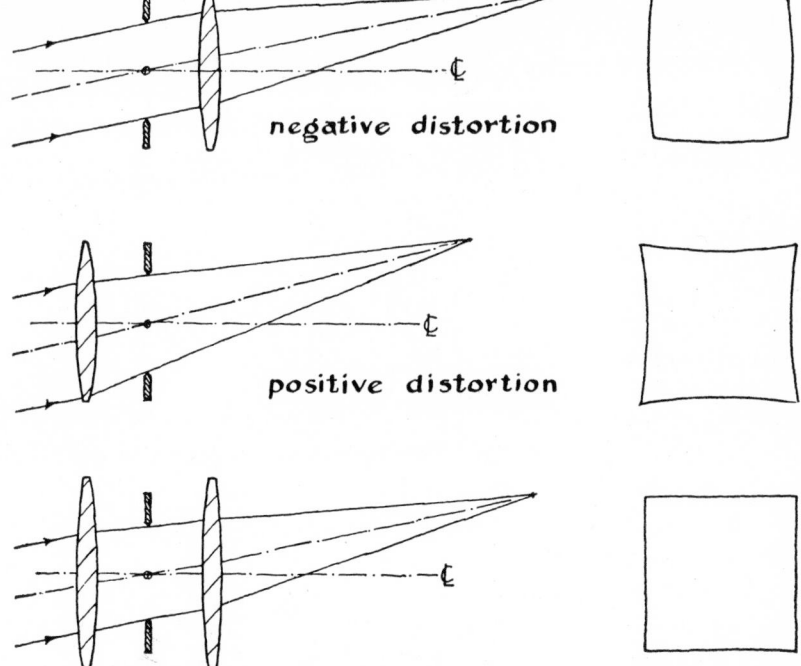

negative distortion

positive distortion

FIG. 16-11   Distortion, and its control by stops.

ture—a system from which the Schmidt camera derives. The Schmidt system
was not introduced until 1931.† To understand this system, consider a spherical mirror with a stop at its center of curvature pointed toward the stars of
the heavens. It is apparent that chief rays from each star will strike the
spherical mirror normally, for this position of the stop at the center of curvature of the mirror. And the images of all the stars will be free from coma as
well as astigmatism. But although the field of this combination is free of
the off-axis aberrations, it is strongly afflicted with curvature. When the aperture stop is opened, although the field aberrations remain absent, spherical
aberration becomes offensive. But Bernard Schmidt coped with this offensive
aberration by means of a corrector plate located at the stop. This plate did
not introduce offensive field aberrations because it had substantially no overall
focusing power, but it corrected out spherical aberration of the mirror over
the whole field. Although, for oblique parallel rays—a ray displaced a dis-

† A translation of Bernard Schmidt's original article by N. U. Mayall appears in Pubns.
of Astron. Soc. Pacific, *58*, 282 (1946). See also D. O. Hendrix and Wm. H. Christie,
Scientific American, August 1939; Albert G. Ingalls (ed.), *Amateur Telescope Making*,
Book III (1953, Scientific American).

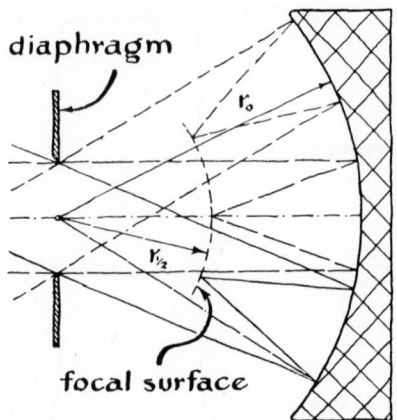

diaphragm

r_o

$r_{1/2}$

focal surface

FIG. 16-12 Geometric illustration of the Schmidt camera to show why it avoids coma, astigmatism, and distortion.

tance h from the chief ray strikes Schmidt's corrector lens at a greater distance h_{lens} from its center and so it gets an improper correction. But the distance $h - h_{\text{lens}}$, which is proportional to $h (1 - \cos \varphi)$, is small for chief rays of modest obliquity.

We may think of a facet on the corrector plate as a small-angle prism that produces the deviation $(N - 1)$ multiplied by a local prism angle. When the prism angle is small, as it is here, the deviation is independent of angle of incidence on the prism. Thus the corrector plate at the center of curvature of the spherical mirror is exceptionally free of fault for both a large aperture and a large field. This optical combination of Schmidt's has enjoyed extensive use in astronomy, culminating in the 48″ Schmidt telescope of Palomar Mountain. That telescope has a focal length of 10 feet; and, as its name implies, the corrector plate is 48″ in diameter. The primary spherical mirror of that telescope has a radius of curvature of 20 feet and its diameter is 72″. The diagonal of the rectangular photographic plate used is 19 inches. This plate is bent over a mandril in the plate holder to the 10-foot radius of curvature of the focal plane.

We have discussed Bouwers' somewhat similar correction of spherical aberration by a spherical mirror. In contrast to Schmidt's correction, Bouwers

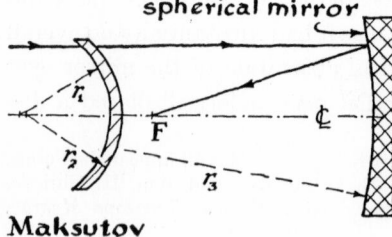

spherical mirror

r_1

F

$\math3{C}$

r_2

r_3

Maksutov

FIG. 16-13 Maksutov method of correcting spherical aberration.

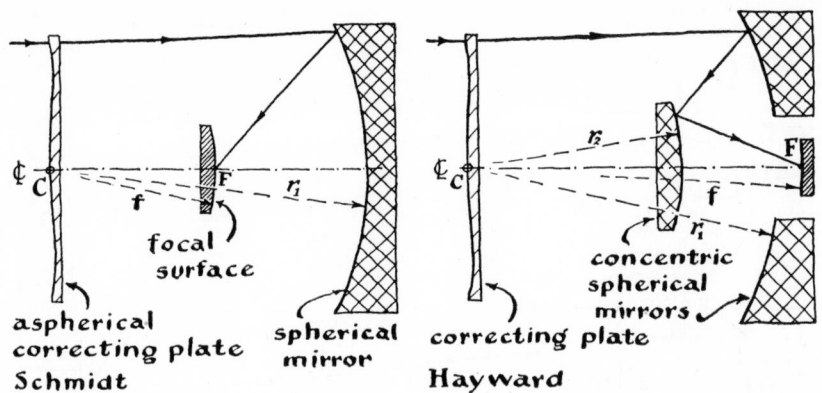

FIG. 16-14 Schmidt's method of correcting spherical aberration, and Hayward's modification of it.

uses a negative lens. A similar combination of a mirror and an $\mathfrak{F} \cong 0$ meniscus lens, described by Maksutov, is shown in Fig. 16-13. Both Bouwers and Maksutov, however, followed the principle of Schmidt's camera which removes coma and astigmatism easily by locating the corrector with the stop where it "wants to be"—at the center of curvature of the spherical mirror. The correctors themselves introduce but little coma or astigmatism and very little chromatic aberration. Two modifications of the Schmidt are shown in Fig. 16-14.

Fig. 16-15a shows the mirror optical system described by H. Ebert. This system has only recently found wide use in grating monochromators; the aberrations which the collimator reflection introduces at one side of a sphere are corrected in the telescope reflection at the other side of the same sphere. This type of correction was recognized and described by Professor M. Czerny and Dr. A. F. Turner,† as illustrated in the figure, at b. A central section of the collimated wave front in the Ebert or Czerny-Turner system is S-shaped, as shown. This, however, is just the type of incident wave front which the telescope reflection "likes" to correct, and it concentrates the light into a line image. The performance details of the Ebert optical system have recently been described by William G. Fastie‡ in the *Journal of the Optical Society of America*. Fastie's use of a single large sphere as originally proposed by Ebert, instead of two spheres as described by Czerny and Turner, simplifies the system for gratings and makes exact adjustments easier (especially, adjustments of the circular slits).

---

† M. Czerny and A. F. Turner, Z. Physik, *61*, 792 (1930).
‡ Wm. G. Fastie, J. Opt. Soc. Am., *42*, 641 (1952), and *43*, 1174 (1953).

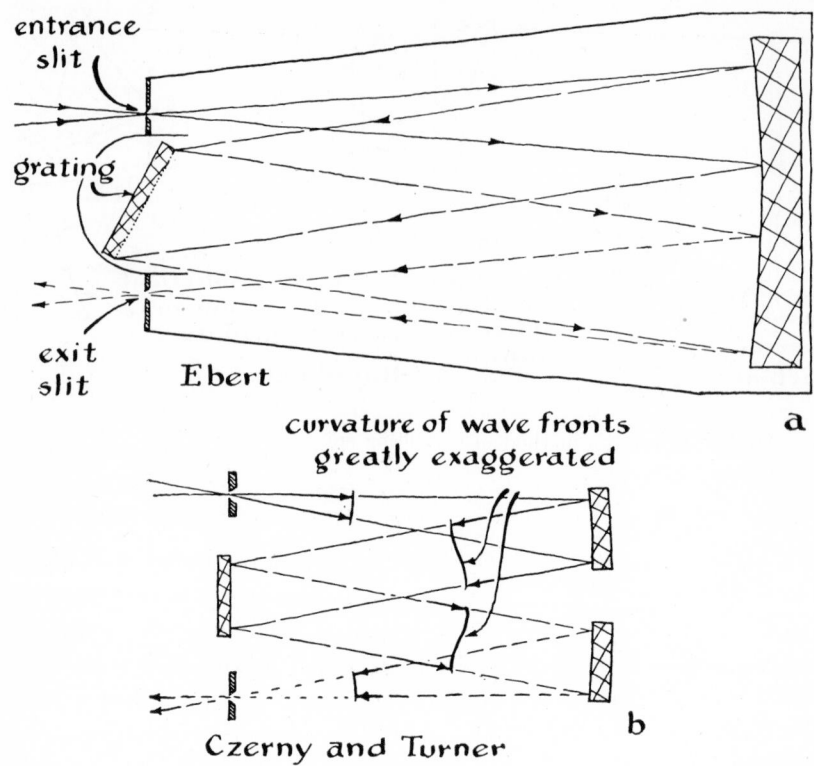

entrance slit

grating

exit slit

Ebert

curvature of wave fronts
greatly exaggerated

a

Czerny and Turner

b

FIG. 16-15 Corrections by the telescope mirror of the aberrations that the collimator mirror introduces in a spectrometer.

There are many recent developments in the mirror-lens combinations and in systems using aspheric surfaces. The descriptions of these fall beyond the scope of our work. These systems have proved particularly useful for microscopes used in spectral regions where suitably transparent glasses are not available.

Appendix **A**

Applications of Interferometry[†]

by W. Ewart Williams[‡]

The fundamental basis of the phenomenon of light interference is the *principle of superposition* due to Thomas Young. Suppose that due to a single wave train we have a displacement X in a given direction at a certain point and time and that another train acting by itself have a corresponding displacement Y, the principle states that the *instantaneous* resultant displacement of the two waves acting together is the algebraic sum of the separate displacements.

$$R = X + Y$$

This additive principle, which is only approximately valid for vibrations in matter, must hold exactly for light vibrations since practically the whole of optics is based on its correctness. Although not understood at the time, it is tacitly included in Huygens' method of constructing the new position of a wave front as the envelope of secondary wavelets. In one sense the term "interference" is a misnomer, for, as Young himself realized, this principle of superposition implies the absolute independence of the individual components of the resultant displacement. This is in accordance with the experiments of Ebert, which showed that one light beam, however intense, had no effect on another beam crossing its path.

The field of interference of light is so wide that it has become customary to divide it into two parts. The study of the interference effects due to particular shapes of wave fronts is termed *diffraction*, and the term *interference* is

---

[†] Introductory chapter from one of Methuen's Monographs on Physical Subjects, published by Methuen & Co. Ltd., London, and John Wiley & Sons, Inc., New York. Reprinted by permission of author and publishers.

[‡] 155 South Orange Grove, Pasadena, California.

limited to mean the effect of combining two or more separate beams that originally must have come from the same source.

When two or more beams "interfere" to form the familiar fringe effect, either we have a redistribution of the light, the bright parts collecting the energy from the darker portions, or a complementary pattern is formed elsewhere, the bright parts of which correspond to the dark portions of the observed pattern. Fresnel's biprism, the diffraction grating, and the Lummer plate are examples of cases giving an actual redistribution of the light in the field of view, while in Newton's rings and the Michelson and Fabry-Perot interferometers complementary patterns are formed.

Laws Governing the Interference of Light

Interference effects are only obtained if the interfering beams originate at a common source.

This law is a practical consequence of the enormously high frequency (of the order of 10^{15} per second) of visible light and the lack of homogeneity of known sources. Experiments indicate that the most nearly monochromatic light sources only radiate an unbroken train of waves for about 10^{-8} second. After that interval another train will be started, bearing no definite phase relationship with the first. If it were possible to photograph with an exposure time of this order or less, in all probability fringes could be obtained from separate sources. Again with a sufficiently short time of exposure we should be able to obtain interference effects between beams of different wavelengths or frequencies, as these effects are obtained with the longer electromagnetic or wireless waves, which are inherently similar. Practically, in consequence of the comparatively long period required to register the effects, the interfering sources must be *identical* in all respects, and this is only possible when the beams originate from a common source.

When the original beam has been plane polarized and is passed through a doubly refracting medium, we have the additional limitation (first given by Fresnel and Arago) that the vibrations must be analyzed on a common azimuth before the interference effects are observable. This follows since the transverse vibrations of the two beams emerging from the doubly refracting medium are mutually perpendicular.

Classification of Interference Phenomena

We can divide the methods of obtaining interference effects into two broad classes: (A) methods which require a point source or, if the interference effects are only required in one direction, a line source (division of wavefront);

(B) methods in which the beam is divided by partial reflection into two or more beams (division of amplitude).

In the first class, provided the point or line source is sufficiently fine, it is possible to have wave fronts with similar phases emerging in slightly different directions from the source. These can be further separated by mirrors, prisms, and lenses and eventually brought together again to produce interference bands. The greater the area over which the wave front must be of the same phase, or coherent, the smaller must be the angle the source subtends at the wave front. Examples of this class are the Fresnel biprism and mirrors, Lloyd's mirror, the Billet split lens, and the Rayleigh interferometer. When the slit is too wide, the fringes disappear. This disappearance of the fringes was actually used by Michelson in his method of measuring the angular diameter of stars and by Gerhardt in his adaptation of Michelson's method to obtain the diameters of ultra-microscopic particles. All diffraction gratings, including echelons, also belong to this class.

In the second class, where the beam is divided by partial reflection at a half silvered mirror, there is a point-to-point correspondence between the wave fronts of the transmitted and reflected beams. Any peculiarities in the one are also present in the other. It therefore follows that, however complex the original wave front may be, the clearness of the interference effects is not impaired. This means that, with division of amplitude, extended light sources may be used, so that in general the effects are much brighter. Examples of this class are the interference effects of thin films and the interferometer systems of Jamin, Mach, Michelson, Fabry-Perot, and Lummer-Gehrcke.

An interferometer of this second class can in special instances be modified and used as an instrument of the former class. It thereby gains many additional qualities and behaves in an entirely different way. A notable example of this is the modification by Twyman and Green of the Michelson interferometer.

Michelson, in his book *Light Waves and Their Uses*, restricted the term "interferometer" to denote any arrangement which separates a beam of light into two parts and allows them to reunite under conditions to produce interference. Common usage has, however, extended this definition to include all methods that divide the beam into any number of parts that are subsequently brought together to cause interference. For this reason it is convenient to subdivide each of the above classes according to whether we have two or more interfering beams. All the arrangements in Class A, with the exception of diffraction (including echelon) gratings, belong to the double-beam type, while the Fabry-Perot and Lummer-Gehrcke interferometers of Class B are examples of the multiple-beam type.

BIBLIOGRAPHY

Houston, *Treatise on Light* (Longmans), Chap. IX.

Preston, *Theory of Light*, 5th ed. (Macmillan), Chap. I, Sect. III, and Chap. II.

Schuster, *Theory of Optics* (Arnold), Chaps. I–IV.

Wood, *Physical Optics* (Macmillan), Chap. VI.

Bouasse & Carrière, *Interférences* (Delagrave, Paris).

Fabry, *Applications des interférences lumineuses* (Paris).

Michelson, *Light Waves and Their Uses* (out of print).

Michelson, *Studies in Optics* (Univ. of Chicago Press).

Gehrcke, *Die Anwendung der Interferenzen* (out of print).

Appendix **B**

Interferometers[†]

by J. Dyson[‡]

The essence of interferometry lies in the combination of two beams of light in such a way that reinforcement or cancellation of the illumination takes place at points depending on the phase differences of the two beams. This necessarily implies that there is a statistical correlation between the electro-magnetic vibrations in the two beams in the plane where the fringes are formed, and this in turn implies that the two beams have been derived from the same primary beam of light. Another condition is implied also, viz: that the light emitted by every elementary point of the source shall form an identical fringe pattern in the same place; otherwise the various patterns will overlap and reduce the contrast, perhaps to zero. When both these conditions are satisfied, the two beams are said to be "coherent." Thus the basic system of an interferometer is as shown in Fig. B-1, where the contents of the "black boxes" may be of many different forms.

The manner in which the two beams are split off from the primary beam constitutes a possible basis for the classification of interferometers. The principal ways in which the splitting can be effected are (a) aperture splitting, (b) diffraction splitting, (c) amplitude splitting.

A good example of the first method of splitting is given by the Rayleigh interferometer, shown in its essentials in Fig. B-2. Light passes through a slit, S_1, which is so narrow that, whatever the convergence of the primary beam might have been, it is spread by diffraction into a wide fan of light. This falls on a screen pierced by two slits, S_2 and S_3, which allow two narrow

† Thanks are due to Dr. T. E. Allibone, F.R.S., Director of the Research Laboratory, Associated Electrical Industries Ltd., for permission to publish this appendix.

‡ Research Laboratory, Associated Electrical Industries Ltd., Aldermaston Court, Aldermaston, Berkshire, England.

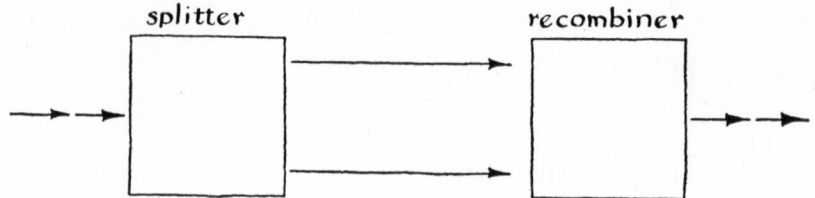

FIG. B-1   Basic system of an interferometer.

---

portions, well separated, of the wave front to pass. The light is rendered parallel by the lens, L_1, and focused again by another lens, L_2. The image, as shown, consists of a band of width determined by the width of S_2 and S_3 and by the magnification of the system. The two beams combine at this image and form interference bands parallel to the slits. A tube closed at the ends by transparent windows is inserted in the parallel portion of each beam, and the contents (liquid or gaseous) of the tubes determine the positions of the fringes by the path difference introduced between the two beams.

It can easily be seen that each point across the width of S_1 will give rise to its own set of interference fringes, and that, if S_1 is too wide, these sets of fringes will overlap and reduce the contrast. It can be shown that the condition for this loss of contrast not to be serious is that the angular width of the central maximum of the diffraction pattern due to S_1 should be large compared with the angle subtended by S_1 at the distance separating S_2 or S_3 from it.

This condition is common to aperture-splitting systems, and shows that, in general, the illumination in such systems cannot be very high because of the necessarily small width of S_1. It can be generalized into a theorem of great value in interferometry, as follows:[1] Consider Fig. B-3, where the

---

FIG. B-2   Rayleigh interferometer.

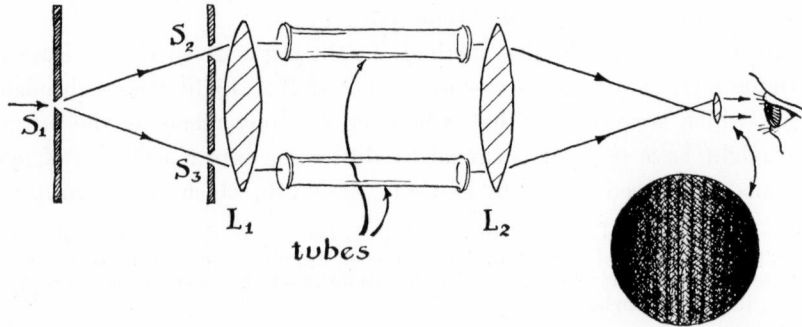

illumination of two points, P_1 and P_2, is produced by light received from a source, S, of uniform brightness. If the light passing through these points be combined by some means to form interference fringes, the intensity of the two interfering beams having been made equal (say by introducing an absorbing filter into one of them), the maximum and minimum illumination in these fringes will be $\mathfrak{E}_{max}$ and $\mathfrak{E}_{min}$. The contrast may be defined as $\mathcal{V} = (\mathfrak{E}_{max} - \mathfrak{E}_{min})/(\mathfrak{E}_{max} + \mathfrak{E}_{min})$.

The theorem then states that the value of $\mathcal{V}$ is given by the ratio of the *amplitudes* at P_1 and P_2 in the diffraction pattern produced by an aperture of the same shape and position as the boundary of S when the central maximum of the pattern is situated at either P_1 or P_2.

The value of this theorem lies in the fact that such diffraction patterns have been calculated for apertures of most shapes likely to be met with in practice. It remains true even if P_1 and P_2 are not at equal distances from the source (provided the diffraction pattern is calculated to take account of the "out of focus" effect).

The quantity $\mathcal{V}$ can also be defined as the "coherence" between the light at P_1 and at P_2.

Diffraction splitting is not often used, and is best exemplified in the case of the phase contrast microscope. The full treatment of this case lies outside the field of this appendix, but the state of affairs can be described roughly as follows. A small transparent refracting object is placed on the stage of the microscope (Fig. B-4) and illuminated by parallel light. If the optical thickness of the object is small, the light vectors are rotated through a small angle after passage through it. The resultant vector field can be regarded as consisting of a uniform field together with a small "difference vector" approximately in quadrature, and occurring only behind the object. The uniform field or "background" will be converged by the objective to its rear principal focus, where a very small disk of refracting material is placed. This disk is of

FIG. B-3   Illustration of Dyson's visibility theorem.

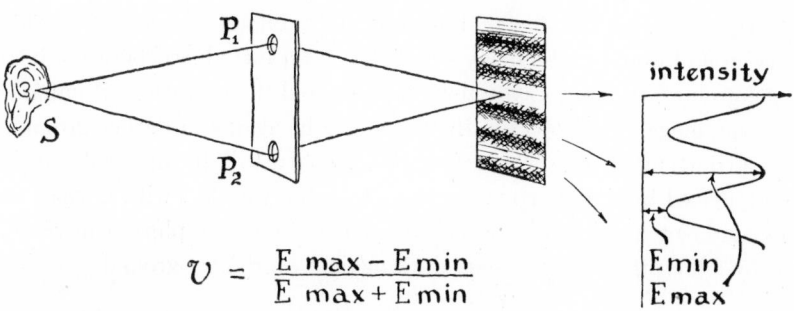

$$\mathcal{V} = \frac{E\,max - E\,min}{E\,max + E\,min}$$

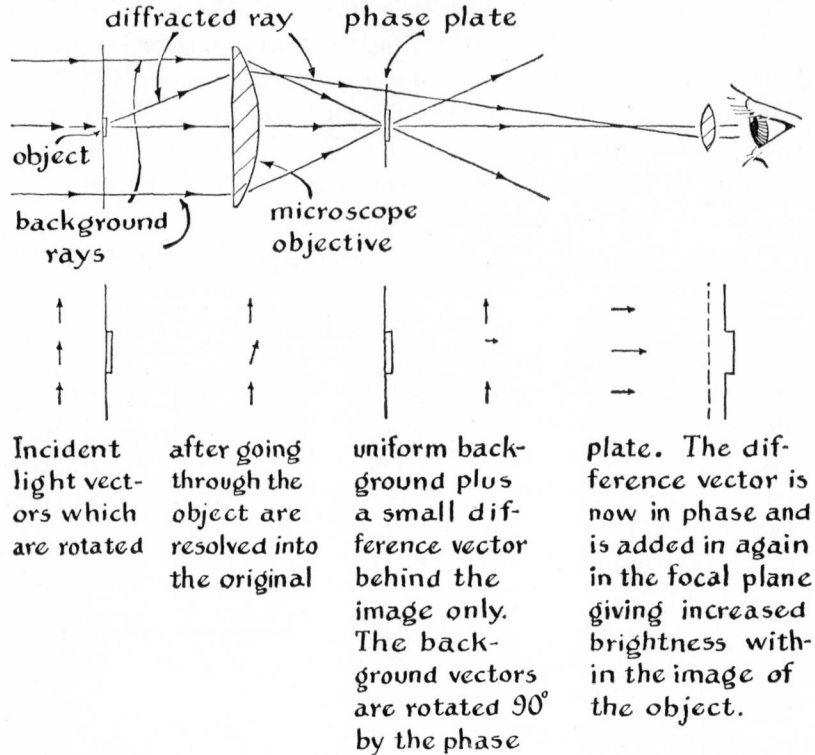

FIG. B-4 Diffraction splitting of a primary beam, as illustrated by the phase contrast microscope.

thickness sufficient to introduce a phase delay of 90° in the light passing through it. The background wave passes entirely through this disk, but the other wave front, characterized by the difference vectors, being restricted in lateral extent by the dimensions of the object, will spread by diffraction outside the disk and will largely avoid this phase delay.

The two waves are combined again in the image plane, and the illumination outside the image of the object will be that due to the background wave. Inside the image of the object, however, the "difference" vector is added, but, as an additional phase delay of 90° has been introduced between it and the background, the vectors are now in phase and the brightness is enhanced within the image. The object is then seen bright against a darker ground.

This arrangement gives "negative phase contrast." If the small circular disk is replaced by a plate covering the whole aperture but with a small area in the center etched away to a suitable depth, "positive phase contrast" is given, by which the object is seen dark against a brighter ground.

The foregoing analysis is approximate only, of course, as it does not explain where the extra energy for the brightening of the image comes from (in fact, it comes from a darkening of the rest of the field due to that portion of the diffracted beam which passes through the phase disk), nor is the well-known "halo effect" explained. For a more complete discussion, reference should be made to standard texts.[2]

Amplitude splitting covers a very wide field and can be achieved in a number of ways. The beam may fall on a partially reflecting surface, part of the light being reflected and part transmitted. This mechanism is responsible for the formation of Newton's rings and for the colors of soap bubbles, but a more modern use is in Michelson's interferometer. The arrangement is shown in Fig. B-5. Light is incident at 45° on a glass plate which is half silvered on one face. Part is transmitted and falls on a mirror, M_1, being reflected back along its path and partially reflected at the half-silvered surface, coming out at 90° to its original path. Part of the primary beam is also reflected at the half-silvered layer, is reflected back by another mirror, M_2, and is partially transmitted, emerging along the same path as the first beam.

A second glass plate of the same thickness as the half-silvered plate is placed in the path of this beam; this ensures that the same thickness of glass is traversed by both beams. If this were not so, a path difference varying with wavelength would be introduced because of the dispersion of the glass, in addition to a path difference varying with angle in the field of view.

The optical conditions are thus the same as would arise if the light were reflected by M_1 and a mirror in the position of the virtual image of M_2 produced by reflection in the half-silvered layer. As the two beams are in fact well separated, however, an object such as a plane-parallel plate or a tube

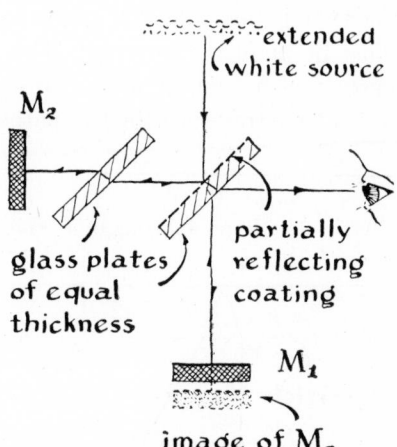

FIG. B-5 Amplitude splitting of a primary beam, as illustrated by Michelson's interferometer.

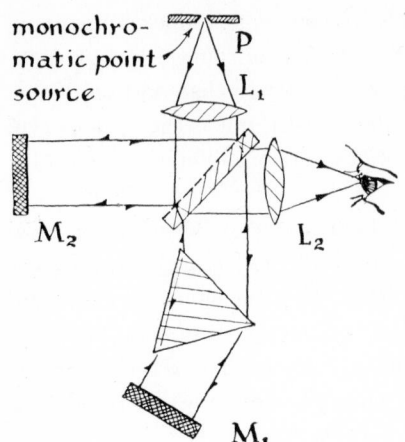

monochro-
matic point
source

P

L_1

M_2

L_2

M_1

FIG. B-6   Twyman-Green modification of Michelson's interferometer.

filled with gas may be placed in either beam in order to measure its optical thickness by the displacement of the fringes produced thereby.

It is easily seen that, unless the separation between M_1 and the virtual image of M_2 is very small, the fringes appear to be at infinity. To enable them to be seen at finite distances, a modification was introduced by Twyman and Green (Fig. B-6). The light now comes from a pinhole, P, at the principal focus of a lens, L_1, and another lens, L_2, is provided to focus the emerging light into the pupil of the eye. This form is used largely for testing of optical parts (here exemplified by a prism). The function of the use of collimated light is twofold. First, the fringes at infinity can now be seen at a finite distance (such as on one of the surfaces of the prism) because of the greatly increased depth of focus. Secondly, the light is made to traverse the optical part under test in a definite manner, thus making the result of the test explicit. It is not true that, as is sometimes stated in textbooks, the lens L_1 must be perfectly corrected. The fringe pattern obtained shows, after the manner of a contour map, the *difference* in shape of the two wave fronts which have traversed the two paths, so a small deviation of the primary wave

FIG. B-7   Scatter splitting of a primary beam.

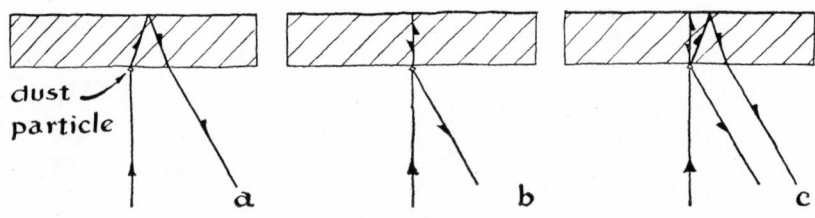

dust
particle

a

b

c

front from flatness will not affect the pattern, as this deviation will be present equally in both waves.

A second method of amplitude splitting is by scattering. The resulting phenomena were mentioned by Newton in his *Optics,* but have not been put to practical use till recently. The principle is illustrated in Fig. B-7. Consider a plane parallel back-silvered mirror with a thin layer of dust on the front. Light is scattered by the dust, and can find its way to the eye by two alternative routes, as shown in Fig. B-7a or B-7b, scattering taking place before reflection in the first case and after reflection in the second. The two paths are shown on the same diagram in Fig. B-7c. The path difference between the two emergent rays depends on the angle of scattering and is zero when this angle is zero. As the emergent rays are parallel, the resulting interference fringes are seen at infinity.

These fringes may easily be demonstrated by grinding one face of a piece of plate glass very lightly so that some of the original surface is left between the pits and silvering the other side. Light from a small bright source is reflected onto the ground side by a plate of unsilvered glass, and the fringes are seen through this plate, the mirror being adjusted so that the line of sight to the reflected image of the light source is normal to its surface.

The phenomenon has been used by Burch[3] to make a remarkably simple interferometer, which can be used for testing optical systems. A suitable surface is prepared as described above, and two replicas are taken from it by casting a film of collodion or other plastic onto the surface, or by pressing while hot against a sheet of thermoplastic material. The set-up in Fig. B-8 shows this device used to test a concave mirror, M. Light from a pinhole, P, is focused to a small image, I, on the surface of M by a lens, L, and one replica, R_1, is placed at the center of curvature of M. A partially reflecting surface is placed at 45° with the axis a short distance inside the center of curvature, and the other replica, R_2, is placed at the virtual image of R_1. R_2 is inverted

---

FIG. B-8   Scatter splitting of a primary beam, as employed in Burch's interferometer.

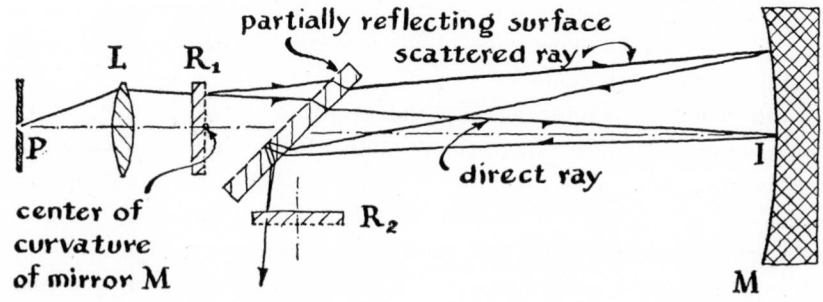

and turned front for back, as shown. Under these conditions, each point of R_1 is imaged into the corresponding point of R_2.

Some of the incident light is scattered by R_1 into a wide cone filling the aperture of the mirror. This light is focused back onto R_2, and some of it passes straight through without further scattering. Some of the light forming the image I is scattered by R_2, and, because of the point-to-point correspondence of the two replicas, the two beams emerge in coincidence.

An observer looking through R_2 will therefore see two superimposed fields of view. One, formed by light scattered by R_2 but not by R_1, will give a large diffuse illuminated patch; the other, by light scattered by R_1 but not by R_2, gives a sharp image of the mirror surface. These fields will be coherent where they overlap and will be crossed by interference fringes.

The first field (the "reference" field) is formed by light which has only struck the mirror at a small area, I, and the aberrations can be taken to be sensibly zero over this area. The position of this small area can be chosen so that this is true. The other field, however (the "test" field), is formed by light filling the aperture of M, and its wave front is therefore deformed by the errors of the mirror surface. Consequently, the interference fringes, constituting a contour map of the *difference* between the two wave surfaces, indicate directly the imperfections of the mirror.

The small reference area, in the field of view, is excessively bright because of light not scattered by either R_1 or R_2. Because of its concentration and brightness, it may be necessary to use a patch-stop at an appropriate place in the system after R_2 in order to avoid dazzle from this source.

This type of interferometer presents notable advantages in operation over the Twyman-Green type for testing purposes. In the latter case, the reference beam is widely separated from the test beam and tends to be affected differently by vibration and thermal drift. Furthermore, it has to have a linear aperture comparable with the system under test, which involves the use of large, very precise optical parts. In addition, if the system under test includes refracting elements, highly monochromatic light must be used because of the dispersion of these parts. All these disadvantages are avoided by the fact that, in the scatter-fringe interferometer, both test and reference beams traverse the same optics. This type of instrument may therefore be called a "common-path interferometer."

Another method of splitting, amplitude splitting by polarization, is rapidly coming into prominence. This may be illustrated by another common-path interferometer due to Dyson,[4] in many respects similar to that of Burch. Splitting is accomplished by means of a lens incorporating an element of birefringent material. In Fig. B-9 this lens is shown as a symmetrical triplet with a central negative component of Iceland spar and two positive components of glass. The crystal is cut with its optic axis in the plane of the lens. The optical behavior of the lens can be described in first approximation by assuming that it has two refractive indices equal to the ordinary and extraordinary indices for the crystal, and otherwise ignoring the effects of anisotropy, which, in fact, ultimately cancel out. The triplet is so designed that its overall power for the ordinary ray is zero, and positive for the extraordinary ray. Light polarized at 45° to the optic axis of the crystal is incident from the left, and is focused by a lens, L_1, to an image, I_1, on the mirror under test. This light is split into two beams polarized at right angles to each other in the crystal. The ordinary beam goes through without deflection to form I_1; the extraordinary beam is converged to a focus, I_2, and thereafter expands to fill the mirror aperture. A birefringent quarter-wave plate is placed after the triplet with its principal directions at 45° to the planes of polarization. As the light traverses it twice, it acts as a half-wave plate and rotates the planes of polarization through 90°. Thus the ordinary ray becomes the extraordinary ray on its return.

The initially ordinary ray is returned from I_1 to a point symmetrically opposite the point from which it left the triplet. Now being the extraordinary ray, it is converged to a focus, I_3. The initially extraordinary ray passes through the triplet undeflected and comes to a focus which, if the radius of the mirror is very large, coincides with I_3. If the mirror radius is in fact not very large, the same effect can be achieved by using a collimator lens, L_2, to image the mirror surface to infinity.

The two beams diverging from I_3 are deflected out of the system to a position accessible for viewing by an oblique, partially reflecting surface, and pass through an analyzer. It is convenient to use an interference filter placed at Brewster's angle[5] for the partially reflecting surface, as this will act as both polarizer and analyzer in addition, and give the maximum economy of light.

The interpretation of the fringe patterns is exactly the same as for Burch's interferometer. It has the advantages, however, of fewer adjustments, greater

FIG. B-9   Amplitude splitting by polarization, as employed in Dyson's interferometer.

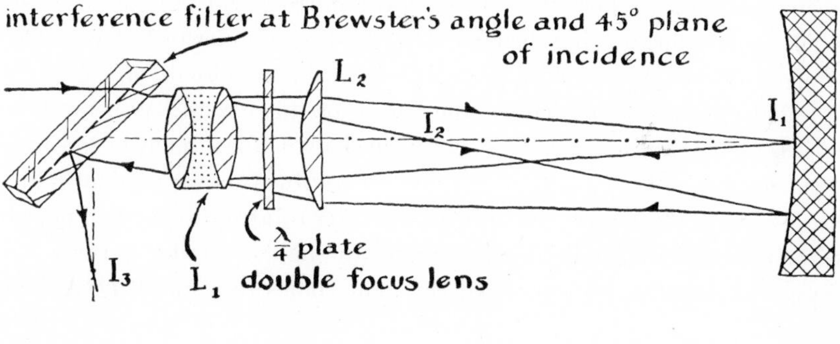

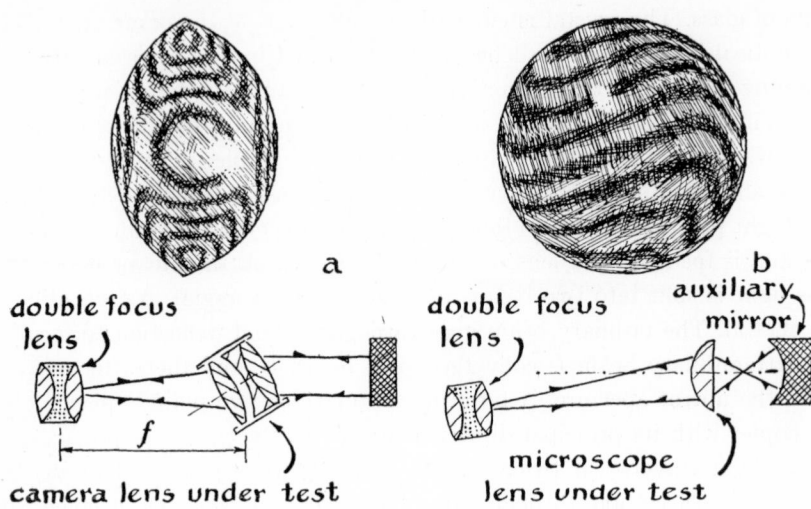

FIG. B-10   Interferograms taken with Dyson's interferometer.

light efficiency, and no dazzle due to light from I_1, as this image is only visible as a result of incomplete polarization and can be made quite inconspicuous. The fringe contrast is also better because the parasitic light is smaller.

Fig. B-10 gives two examples of interferograms taken with this instrument. Fig. B-10a is of a wide-angle photographic lens taken at 30° off axis, with the set-up shown below the picture. The lens errors are doubled by this means. The fringe distortions show the simultaneous presence of astigmatism and spherical aberration, but the perfect symmetry of the pattern indicates that coma is absent.

Fig. B-10b is of a 4 mm microscope objective, 0.15 mm from the center of the field. The asymmetry of the fringes is typical of the presence of coma.

A second type of interferometer avoids the use of the birefringent lens, splitting taking place at a polarizing interference filter. The arrangement is shown in Fig. B-11. Light polarized at 45° to the plane of incidence falls on a polarizing interference filter, half being reflected and half transmitted. The transmitted half is reflected by three plane mirrors and falls once more on the polarizing filter, passing through it to the system under test. A converging lens of symmetrical construction is placed within the circuit of the mirrors, its focal length being one half of the length of this circuit. As a result, the light is converged to a focus, I_1. This is placed at the focal point of the system under test, with the result that the light is returned on its original path. A quarter-wave plate beyond the interference filter rotates the plane of polarization through 90° by the double passage of the light, so reflection now takes place at the interference filter, and the light comes to a focus, I_2. A second

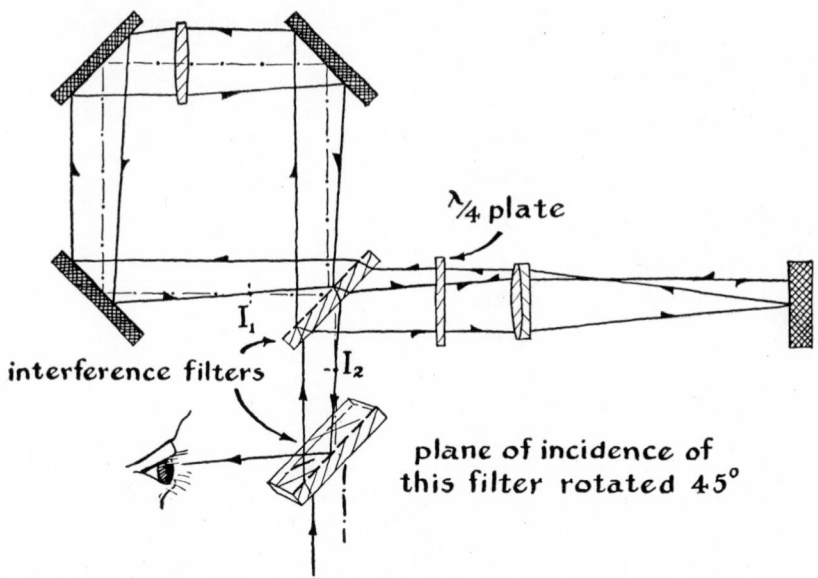

FIG. B-11 Splitting of a primary beam by means of a polarizing interference filter.

interference filter (which, as before, also acts as polarizer and analyzer) throws the light out of the system to a point convenient for viewing. The plane of incidence on the second interference filter should, of course, make an angle of 45° with the first, as Fig. B-11 indicates.

The reflected part of the primary beam comes to a focus in the aperture of the system under test and is returned to a symmetrically opposite point on the interference filter, through which it now passes and is converged by the lens to a focus coinciding with I_2. The two beams then pass out of the interferometer in coincidence. The interpretation of the fringe patterns is then identical with that previously described.

This system has a number of advantages over that previously described. As the beam splitting does not depend on the difference between the ordinary and extraordinary refractive indices of a crystalline material, the lens can be designed with surfaces of smaller curvature, and so may have a larger angular aperture. Also, as the light diverging from focus I_1 is returned exactly along its own path (save for aberrations) by the system under test, the latter can be tested under exactly the design conditions. The instrument previously described gives the sum of the aberrations for two focal positions, one on each side of the design position.

Polarization splitting has also been applied to the interference microscope. The phase contrast microscope described above is not ideally suited for the measurement of path differences in the specimen, a function of particular

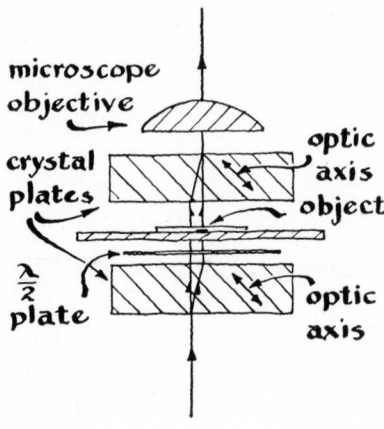

FIG. B-12   Lebedev's shearing type of interference microscope.

interest to biologists, and several workers[1,6,7,8,9] have developed microscopes in which interference is obtained between the field of view containing the specimen and another field (the "reference field") from which the specimen is absent. The optical thickness of the specimen can then be measured by the resulting displacements of the interference fringes.

An instrument developed early by Lebedev[6] is shown in principle in Fig. B-12. The object is placed between two thick birefringent crystal plates cut at 45° with the optic axis. The light from the condenser is polarized at 45° with the plane containing the optic axis, and is split into two beams on entering the lower plate. The ordinary beam passes through undeviated, whereas the extraordinary beam traverses the plate obliquely, emerging parallel with the ordinary beam but displaced laterally. The object is taken to be so small that it lies entirely in the extraordinary beam.

A half-wave plate is placed between the crystal plates so oriented as to rotate the planes of polarization through 90°. Consequently, in the upper crystal plate, the previously extraordinary beam becomes the ordinary one, and passes through without deviation. The previously ordinary beam, however, now suffers a lateral displacement, and the two beams emerge parallel and coincident. One beam now carries an image of the light source with the object superimposed, the other an image of the light source only. The paths are symmetrical, and so conditions for interference are satisfied. The resulting pattern of interference fringes is examined by means of a low-power microscope. As two images of the object, laterally displaced from each other, are seen in the field of view, this is known as a "shearing" type of interference microscope.

Another type of polarizing interference microscope was devised by Smith[9] (Fig. B-13). In the lower principal focal plane of the microscope condenser

FIG. B-13 Smith's interference microscope.

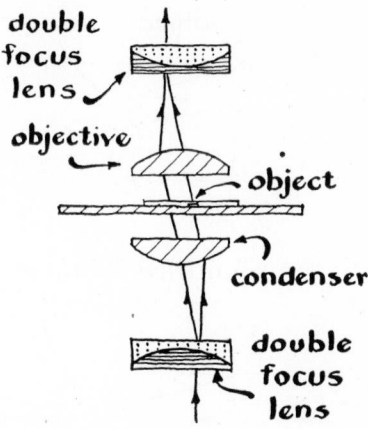

is placed a birefringent lens consisting of two components of quartz, one of positive and one of negative power, cut with the optic axis in the plane of the lens and with the optic axes of the two components at right angles. The combination behaves as a double focus lens of equal positive and negative powers. A similar lens is placed in the upper principal focal plane of the objective, but rotated through 90° with respect to the lower one. A light ray polarized at 45° with the optic axis of the lower lens is split into two rays polarized mutually at right angles, one converging towards the axis and the other diverging from it. After passing through the condenser, these are parallel, but cross the axis at different heights. Accordingly, two images of the light source are formed, separated from each other in an axial direction. The object (assumed small) is placed in one of these foci. After passage through the objective and upper birefringent lens, the two beams are recombined and emerge in coincidence. The observer then sees two images of the object, one considerably out of focus, and, if the object be small enough, only one image is noticeable. The paths once more are symmetrical, and, if an analyzer is used, interference contrast is observed.

Another type of polarizing interferometer has been devised for engineering purposes by Dyson.[10] A frequently arising problem is that of checking the alignment of the bearings of a turbo alternator or of a marine propeller shaft to close limits, or of measuring the straightness of path of an object sliding along machined ways, such as the saddle of a lathe. A closely analogous problem is that of checking the flatness of a large surface plate.

The interferometer is based on the system shown in Fig. B-14a. Two identical concave mirrors are placed facing each other with a separation equal to their common radius of curvature. This system has the property (easily verified by elementary means) that an object anywhere on the line joining

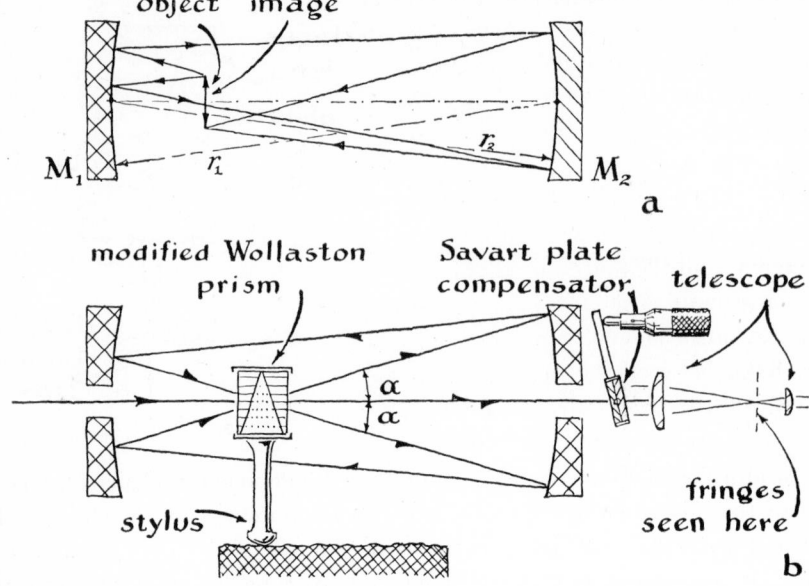

FIG. B-14 Dyson's straight-line interferometer.

the centers of curvature is imaged after successive reflection at the two mirrors into an image the same size as the object, coinciding with it in axial position but inverted. This means that the line of centers can be used as a reference "straight." To do this, the object consists of a small graticule, which, in the case of the problem of checking a surface plate, is mounted on a stand which is slid along the surface. So long as the motion is rectilinear and parallel with the line of centers, the image of the graticule will coincide with the graticule itself, but a small motion up or down will cause a mutual displacement of graticule and image. By measuring this displacement, the deviation from straightness of the path followed by the graticule can be measured.

However, the system as shown is not free from aberrations, so the image will not be sharp. This will reduce the accuracy with which the measurement can be made. To obviate this difficulty, the system shown in Fig. B-14b was devised. The mirrors are each pierced by a small central hole, and the graticule is replaced by a Wollaston prism modified by splitting one of the wedges into two and cementing one of these on each side of the other wedge in order to make the prism symmetrical. The prism is mounted in conjunction with a stylus in such a way that, as the stylus is run over the surface to be examined, the prism partakes of its vertical movements.

A beam of light polarized at 45° with the principal directions of the prism enters through the hole in one mirror and is split into two by the prism. These two beams each make an angle, α, with the axis, and, after reflection at both mirrors, re-enter the prism and are deflected again to emerge parallel with the axis, passing out of the mirror system via the hole in the other mirror. The two beams do not entirely coincide, but, because of the finite thickness of the prism, are separated by a few tenths of a millimeter. An observer looking into the prism will therefore see a system of interference fringes at infinity, and these are conveniently observed by means of a telescope fitted with cross-hairs in the focal plane of the eyepiece.

If the prism should move vertically by distance d, a path difference is introduced between the upper and lower light paths equal to $4d \sin \alpha$. The fringes will therefore move by a distance equal to $4d \sin \alpha/\lambda$ times the width of one fringe, where λ is the wavelength of the light being used. Assuming that a movement of 0.1 fringe width can be distinguished, the reading accuracy of the instrument is equal to $\lambda/(40 \sin \alpha)$. A model of this system made with $\sin \alpha = 0.062$ therefore had a reading accuracy of 0.2 μ, or 8 microinches. By the use of more sophisticated methods for detecting fringe movements, this accuracy could be considerably improved if the mechanical stability of the instrument justified it. It is usually more convenient to introduce a compensating path difference into the two beams, and so to return the colorless central fringe of the system to the cross-hairs for every measurement rather than to count the fringe displacement. A suitable compensator for this purpose is a tilting Savart plate placed before the telescope objective, the tilt being measured by a micrometer screw.

In the foregoing survey, many large and important classes of instruments have been omitted altogether for lack of space. Modern interferometry is such a lively and rapidly expanding topic that it is not possible in an appendix of reasonable length even to cover completely the principles of all the classes of instruments already existing, let alone to attempt a description of every individual. Consequently, a rather arbitrary selection has been made, with a bias to those types of instruments using polarization beam-splitting in view of its inherent interest and rapidly increasing importance.

The peculiar advantages of interferometric methods are already well known among scientists, and they are currently becoming recognized by technologists also, especially in engineering. It is in this field that the most rapid advance may be expected in the near future.

Students wishing to investigate the subject further may be referred to the writing of, among others, M. Françon, G. Nomarski, W. Krug, E. Lau, A. C. S. van Heel, and W. J. Bates.

REFERENCES

1. J. Dyson, Proc. Roy. Soc., *A204*, 170–187 (1950).

2. A. H. Bennett, M. O. Osterburg, H. Jupnik, & O. W. Richards, *Phase Microscopy*.

3. J. M. Burch, Nature, *171*, 889 (1953).

4. J. Dyson (in the press).

5. M. Banning, J. Opt. Soc. Am., *37*, 792 (1947).

6. A. A. Lebedev, Rev. d'Optique, *9*, 385 (1930).

7. J. St. L. Philpot, *Contraste de phase et contraste par interférences* (1952), p. 42.

8. J. Dyson, Nature, *171*, 743 (1953).

9. F. H. Smith, British Patent No. *639,014* (1950).

10. J. Dyson, Nature, *175*, 559 (1955).

The Kösters Double-image Prism

by J. B. Saunders†

The Kösters[1] double-image prism is an interferometer element that is relatively free from vibration effects, easy to adjust, and of a compactness that facilitates temperature control. The compound prism is made from a pair of nearly identical prisms. A 30°–60°–90° prism that is slightly more than twice as long as the desired prism is optically finished and cut into two equal parts. A semi-reflecting film of aluminum or silver is applied to the face opposite the 60° angle of one prism. This face is then cemented to the corresponding face of the other prism, the combination forming an equilateral prism (Fig. C-1a) with the semi-reflecting film bisecting one of the angles.

The adjustments of the prisms in relation to each other are critical and must be performed with interferometric precision. Fortunately, the surfaces of the prisms form several interferometers[2] that readily lend themselves to these adjustments. When the prism is correctly adjusted, it produces straight fringes that may be seen with white light; the central dark fringe appears in the center of the aperture, normal to the dividing plane; and the space between successive fringes has a value chosen for ease of reading fractions

† National Bureau of Standards, Washington, D.C.

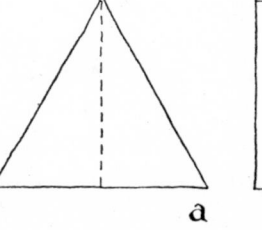 

FIG. C-1   Kösters double-image prism. A small angle is formed by the two 30° edges shown exaggerated in b.

a   b

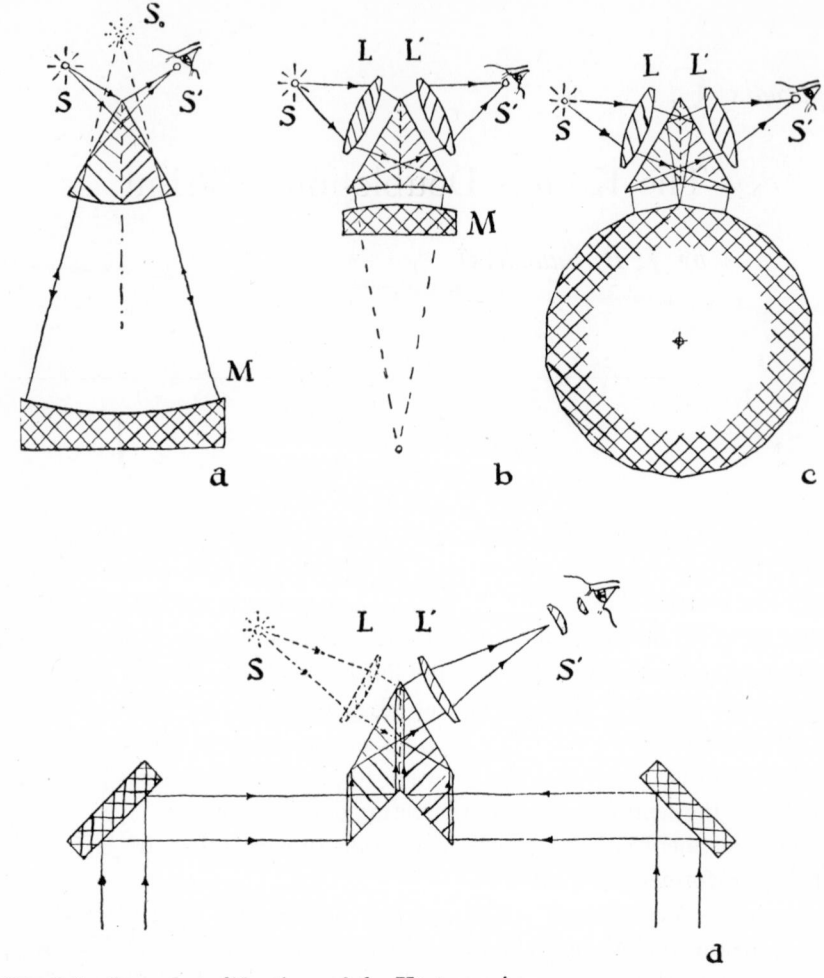

FIG. C-2 Several modifications of the Kösters prism.

of orders. This condition is determined by the relative position and orientation of the two component prisms.

If the surfaces of the film of cement are parallel, the spacing of the fringes is due to a built-in glass wedge. This wedge is introduced by rotating one prism relative to the other, as shown exaggerated in Fig. C-1b, about an axis normal to the dividing plane of the prism. The ends and base of the compound prism are reground to form common planes. A wedge may also be formed in the cement by adjustments of the prisms. By using a glass whose index of refraction differs from that of the cement, the glass and cement wedges may be combined to form an optical wedge that yields fringes for any one color equal in width to those for any other chosen color. By making the fringes for the C and F spectral lines equal in width, and coincident at one point in the aperture, the fringes will all appear black and white, with white light illumination, for one adjustment of the interferometer. This approximately fixed fringe width, for all colors, has advantages when micrometric measurements are used to evaluate fractions of orders.

The Kösters prism finds many applications, and several modifications of it have advantages for special purposes. Fig. C-2 shows a few modifications that have important applications. Fig. C-2a represents the form designated as the inverting interferometer.[3] It may be used for testing telescope objectives with starlight, positive lenses by autocollimation, and concave mirrors having any radius that is larger than that of the prism base. Fig. C-2b shows a form that may be used for testing negative lenses, convex mirrors, ball bearings, and cylindrical surfaces. The form shown in Fig. C-2c is for comparing angles, and the form shown in Fig. C-2d is for measuring the turbidity of air, or testing the quality of seeing conditions for telescopic observations. The last-mentioned form also constitutes a simple and relatively compact stellar interferometer for measuring the diameter of stars.

Fig. C-3 illustrates an application of the Kösters prism in metrology. This instrument is designed to measure the parallelism of gage block surfaces without wringing them to optical flats. The light from each lamp is collimated

---

FIG. C-3   Interferometer for testing the parallelism of gage blocks without wringing to an optical flat.

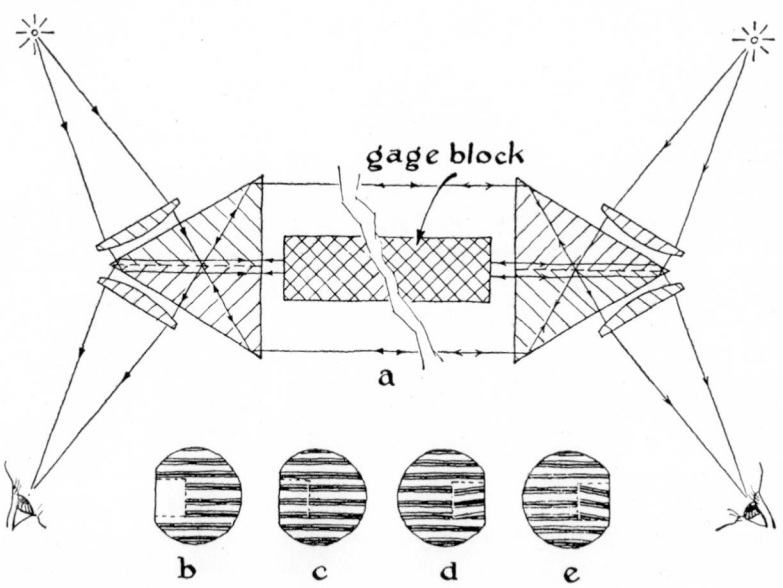

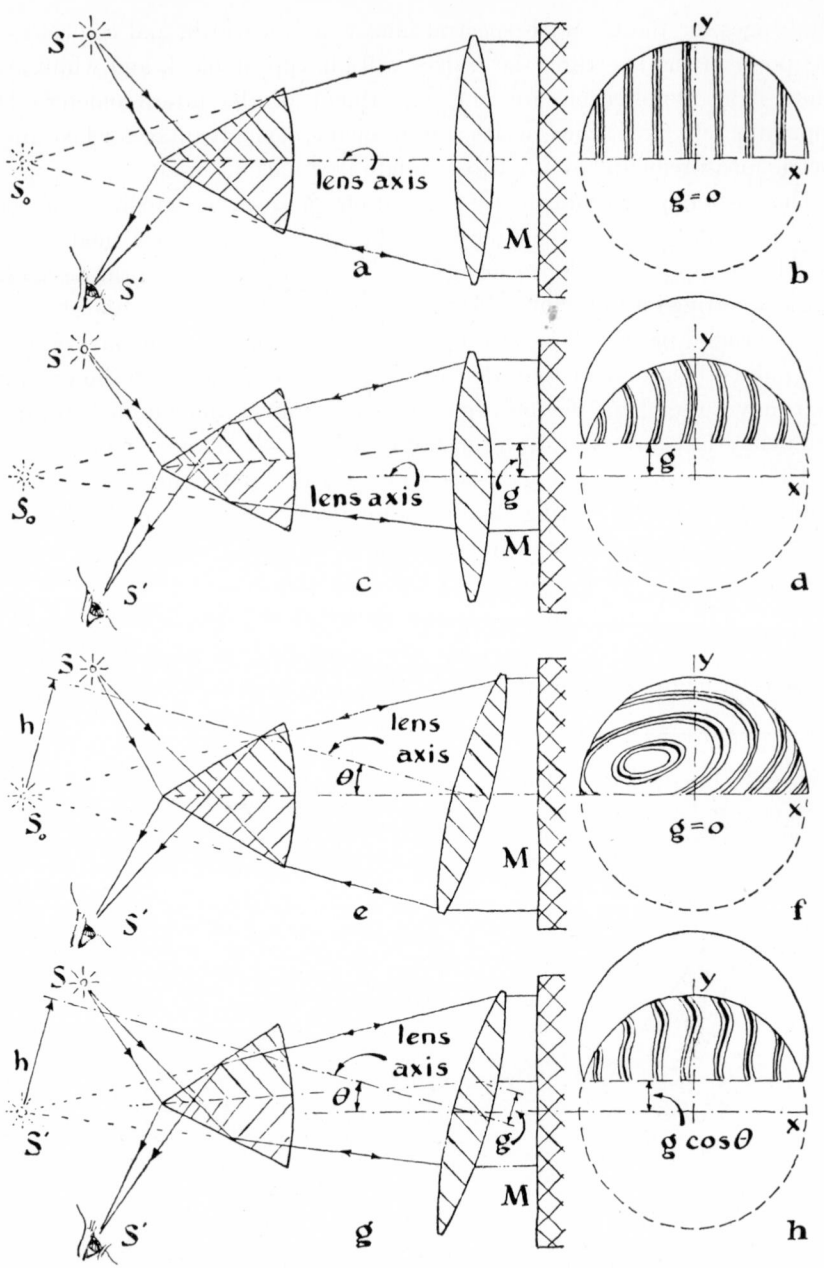

FIG. C-4 Arrangement of the inverting interferometer when testing lenses for (a) asymmetry, (c) shape of wave fronts and spherical aberration, (e) coma, and (g) astigmatism in combination with known coma and spherical aberration. The fringe patterns b, d, f, and h correspond, respectively, to adjustments a, c, e, and g.

before it enters a prism. Each beam is divided by a semi-reflecting plane (illustrated in Fig. C-1a) into two equal components. After total internal reflection each pair of components emerges normally from the base of a prism. The gage block intercepts and reflects parts of each component beam back through the prism. One half of each reflected component returns into the source and the other half is received by the eye, appearing as fringes of interference on the image of the gage block. The parts of the component beams not intercepted by the gage block are transmitted through a second prism to the other eye position. This light forms the background fringes. The background fringes that are formed by light from the right-hand side are seen at the left-hand eye position, and vice versa. Also, the light from the right-hand source that is reflected from the gage block is seen at the right-hand eye position, and vice versa.

Fig. C-3b represents the image observed at the left eye position before the gage block has been adjusted. When it is adjusted to reflect normally the light from the left source, fringes will appear on the rectangular image of the gage block, as illustrated in Fig. C-3c. The eye is then moved to the right eye position, where it sees a pair of fringe patterns similar to that shown in Fig. C-3d or C-3e, depending upon the direction of the wedge formed by the surfaces of the gage block. If Fig. C-3d represents the initial fringe pattern, Fig. C-3e is obtained by turning the gage block upside down, thus reversing the direction of the wedge. If the prisms are accurately aligned, the angle between the two sets of fringes in either d or e represents the angle between the end faces of the gage block. The angle between the fringes on the gage block in positions d and e represents twice the angle between the gage block faces, regardless of the background fringes.

An important application of a modified Kösters prism is the testing of lenses and mirrors. Fig. C-4 shows a sketch of the arrangements for testing lenses with one conjugate at infinity. Light from a source at S, after division into two component beams, diverges from two virtual but coincident images at S_0, where both the center of curvature of the prism base and the image point at which the lens is to be tested are located. The mirror returns one half of the light into the source, and one half of each component beam forms an image of S at S'. An observer's eye located at S' sees a set of interference fringes.

The flexibility of this instrument permits adjustments that are most favorable for the results sought. If the lens surfaces are not figures of revolution, or if the glass is not homogeneous, the arrangement of Fig. C-4a yields the corresponding asymmetrical aberrations only—excluding the effects of all symmetrical aberration types, such as zonal effects, and spherical aberration. A symmetrical lens will yield straight fringes, as indicated in Fig. C-4b.

Aberrations resulting from aspheric surfaces with zonal symmetry cannot be separated from true spherical aberration. The best and most useful test for a lens having zonal aberration is to measure the deviation of the resultant wave front from a best-fitting sphere. This is performed with the arrangement of Fig. C-4c, in which the prism has been rotated about an axis normal to the plane of the figure and through S_0. Data from the corresponding fringe pattern (Fig. C-4d), together with the results of asymmetry discussed above, yield the absolute shape of the wave front relative to a sphere whose center coincides with the chosen image point, S_0.

If the lens is symmetrical and free from zonal effects, and if secondary or higher-order aberrations can be ignored, the test for primary aberration types is simplified by testing for one type at a time. The equation for optical path difference, based on Conrady's equations and using Kingslake's[4] terminology, is

$$P = 2(y - g)\left\{ a_1 g[x^2 + y^2 - 2g(y - g)] + a_2 \left(\frac{h}{f}\right)[x^2 + y^2 - 2g(y - 2g)] \right.$$
$$\left. + 3a_3 g \left(\frac{h}{f}\right)^2 + g \frac{\delta l}{f^2} + \frac{\delta h}{f} \right\} \quad (1)$$

where P is the order of interference, x and y are coordinates of a reference point in the plane of the lens, g is the distance from the center of the lens to the dividing plane of the prism, f is the focal length of the lens, h is the image height or distance from the optic axis to the ideal image point, and δl and δh are displacements of the chosen image point from the corresponding ideal image point and are controlled by the adjustments of the interferometer. The quantities a_1, a_2, and a_3 are aberration constants of the lens, a_1 being a measure of the spherical aberration, a_2 of coma, and a_3 of astigmatism.

If we apply equation 1 to the conditions of Fig. C-4c, in which h is zero, we get

$$P = 2(y - g)\left\{ a_1 g[x^2 + y^2 - 2g(y - g)] + g \frac{\delta l}{f^2} + \frac{\delta h}{f} \right\}$$

The aberration constants, a_2 and a_3, have been eliminated and spherical aberration, a_1, isolated. An analysis of the fringe pattern, Fig. C-4d, using either Gates'[5] or Kingslake's[4] procedure, yields pure primary spherical aberration.

If we leave the prism of Fig. C-4a as it is and rotate the lens about an axis through its center and normal to the plane of the figure, we have the conditions shown in Fig. C-4e. The value of g is zero, and equation 1 becomes

$$P = 2a_2 \left(\frac{h}{f}\right) y(x^2 + y^2) + 2y \frac{\delta h}{f}$$

The effects of astigmatism and spherical aberration are absent, and the corresponding fringe pattern in Fig. C-4f yields pure coma.

The effects of astigmatism cannot, like those of coma and spherical aberration, be isolated. However, the arrangement of Fig. C-4g yields a mixture of all three aberration types. Since coma and spherical aberration are known from the tests described above, astigmatism remains the only unknown constant and is readily evaluated from the corresponding fringe pattern, Fig. C-4h.

REFERENCES

1. W. Kösters, *Interferenzdoppelprisma für Messzwecke*, Reichspatentamt Patentschrift Nr. *595*, 211 (1931).

2. J. B. Saunders, J. Research NBS, *58*, 21 (1957), RP2729.

3. J. B. Saunders, J. Research NBS, *58*, 30 (1957), RP2730.

4. R. Kingslake, Trans. Opt. Soc. (London), *27*, 95 (1925–26).

5. J. W. Gates, Proc. Phys. Soc. (London), *B68*, 1065 (1955).

Interferometry with Savart's Plate

by A. C. S. van Heel[†]

In interferometers the path difference in most cases is strongly affected by external influences. In a Michelson or Mach-Zehnder interferometer, for instance, even a very slight change in the relative position of the mirrors, produced, for example, by vibration or change of temperature, can alter the fringe pattern so much that exact measurement is made impossible. With other types these influences are sometimes smaller. If, however, an inter-ference pattern is produced by the two pencils in an anisotropic crystal, external influences affect both pencils (which pass through the crystal in practically the same way) to the same amount, and the change in the pattern is often negligible. The effect of temperature, for instance, is usually inap-preciable. There is consequently a trend to revert to the use of crystal-optic parts in instruments intended to determine small optical path differences or changes of path difference. Especially where the path differences are small, or at least smaller than ten or twenty times the wavelength, the use of chromatic polarization effects in crystals appears to be useful.

It is often advantageous to be able to measure in the neighborhood of path difference zero. White light can then be used. In this case, furthermore, the phenomenon is often a "linear" one, *i.e.* the dark (and light) fringes are equidistant. With a quartz wedge (cut with the optical axis parallel to the line of intersection of the two surfaces and placed between crossed Nicols) tapering out to thickness zero, the dark line of order zero is at the utmost tip. Such a wedge is not a practical instrument. Another device is therefore resorted to: compensation. Two crystal plates are combined. The light passes through one after the other, and they are made and mounted in such a way

---

[†] Laboratory for Technical Physics, Delft, Holland.

that one of the mutually perpendicular vibrations gains as much in optical path over the other in the first plate as it loses in the second plate. This is usually attained for a given angle of incidence, for which there is then complete compensation.

Of the many possible constructions we will describe only one, which has proved to be extremely useful and handy for many purposes. That is the polariscope named after Félix Savart (1791–1841). The name "polariscope" arose from the fact that it enables one to ascertain the presence of linearly polarized light in a pencil consisting mainly of natural (non-polarized) light. It also is useful for the study of elliptically polarized light, giving the proportions of the axes of the ellipse of vibration and their orientation. It is often mentioned in optical literature, and, because of its adaptability and usefulness for precision work, it has been revived by Lyot[1] and Françon.[2] It therefore seems worth while to describe briefly the working and some uses of this beautiful contrivance.

The construction of the Savart plate is as follows: From a uni-axial crystal, in practice either quartz or calcite, two plane-parallel plates of equal thickness are cut in such a way that the normal of the plates makes the same angle with the optical axis of the crystal (see Fig. D-1). The value of this angle is not critical but can best be chosen in the neighborhood of 45°. For calcite the natural cleavage surface, making an angle of 45.4° with the axis, has a suitable orientation. The two plates, shaped into squares, are mounted after one is

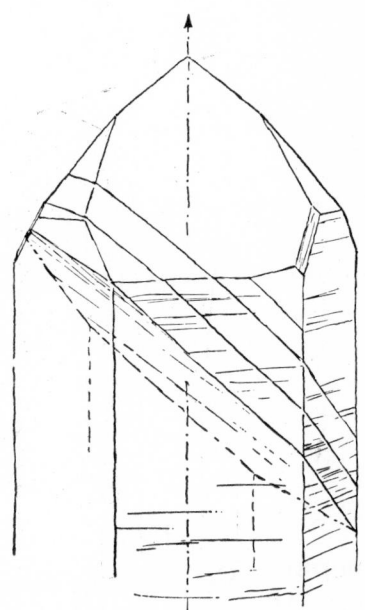

FIG. D-1   Orientation of Savart plates.

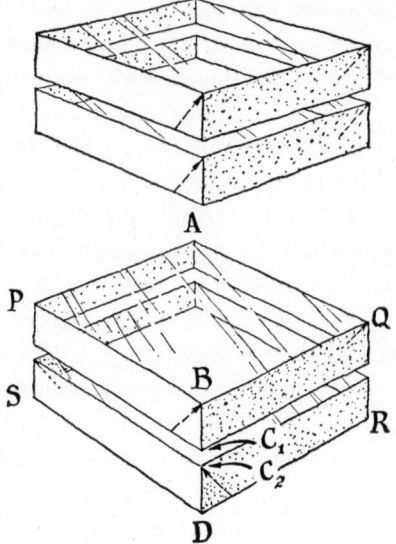

FIG. D-2 Savart plates.

turned 90° with respect to the other (see Fig. D-2, where the direction of the optical axes is indicated by – – –).

It is important to note that after transposition the position of the optical axis in the plates is symmetrical with respect to the vertical plane *ABD*, but not symmetrical with respect to the vertical plane *PQRS*. Let us assume now that a vertical plane-polarized pencil of light is incident from above on

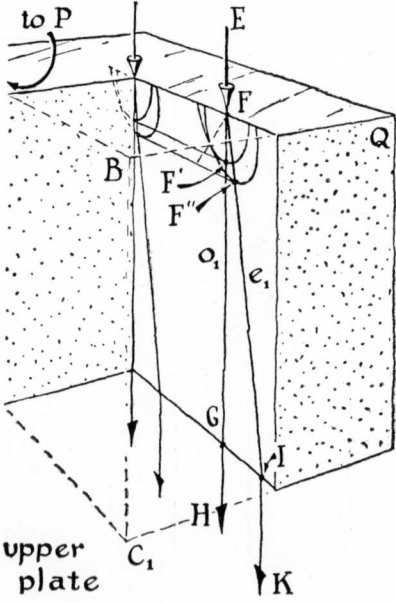

FIG. D-3 Rays through upper Savart plate.

the first surface of the upper plate, with the direction of the vibrations in the plane containing AB. The incident vibration will be split in the crystal along the direction PB and BQ. One "light ray," MN (see Fig. D-3, where a negative crystal, say calcite, is represented), will be split into the two rays FGH and FIK. Their direction follows from Huygens' construction, as indicated in the figure. After leaving the upper plate the emergent rays are parallel to the incident ray EF. The ray FGH is the ordinary ray and vibrates parallel to BQ; the ray FIK is the extraordinary ray and vibrates in the plane PBC_1.

Proceeding now to the lower plate, we imagine the rays GH and IK, emerging from the upper plate, to be transferred to another position to enable us to make the figure clearer. This is allowed if we preserve their directions. In Fig. D-4 the ray through K vibrates perpendicular to the plane C_2DR and will be transmitted as an ordinary ray, KLL', by the second plate. The ray through H, vibrating parallel to the plane C_2DR, is transmitted as an extraordinary ray, HMM'.

Reverting to Fig. D-3, we see that the time for o_1 to reach F' is the same as that for e_1 to reach F''. As $FI : FG$ is smaller than $FF'' : FF'$, we understand that the light at I gains over that at G. Correspondingly, the o_1e_2 ray at M in Fig. D-4 gains over the e_1o_2 ray at L. The two plates being of equal thickness, the total light path through both plates is the same for both rays: there is complete compensation. The pencil is completely extinguished when we put below the second plate an analyzer in the crossed position.

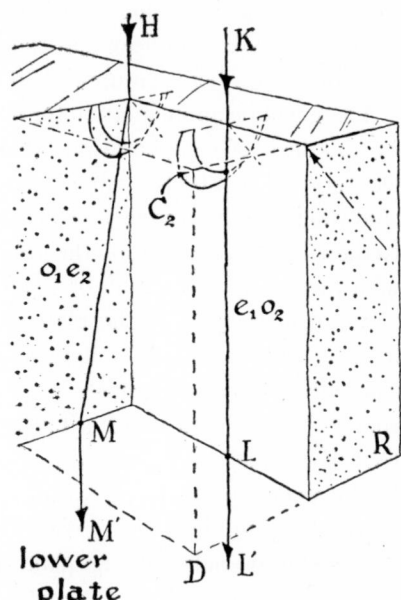

FIG. D-4   Rays through lower Savart plate.

The boundary of the pencil for the o_1e_2 ray is more to the left than for the e_1o_2 ray, and the distance ML is equal to $GI\sqrt{2}$. Assuming that the plates are cut at 45° to the axis, the distance $ML = g$ can be calculated from the refractive indices, N_o and N_e, and the thickness, t, of each of the plates. One finds that $g = t\dfrac{N_o^2 - N_e^2}{N_o^2 + N_e^2}\sqrt{2}$. For calcite $g = 0.154\ t$. For quartz $g = -0.0084\ t$.

So far we have assumed that the wave incident on the upper plate is vertical. We shall now look into the more general case of a parallel pencil of light coming from another direction. Let us first take the case that the normal to the incident wave front is lying in the plane ABD of Fig. D-2. The wave front thus, for different inclinations of the normal, revolves or swings about an axis parallel to PQ. The spheres and ellipsoids of Figs. D-3 and D-4 are then touched at other points than in the original case. In the upper plate another path difference between the two rays will result. But the same change in the path difference will result in the lower plate, and with the same sign, as the Huygens constructions in both the upper and the lower plate are symmetrical. This will hold for not too large inclinations. For large deviations from the perpendicular, higher-order effects will enter into play, which, however, we shall neglect.

Quite another effect will be reached by changing the directions of the incident pencil in the plane PQR of Fig. D-2. The plane waves will again touch the spheres and ellipsoids in the two plates at different points. In this case the symmetry is lacking, for if, for instance, the normal is inclined to the right, the tangent point in the upper plate will creep up in the direction of the optical axis. It is clear that there will be no complete compensation of the path difference between upper and lower plate.

There will be an inclination for which the residual path difference, produced by this lack of compensation, amounts to one wavelength. For that wavelength there will be the same intensity as for the vertical incident pencil, *i.e.* zero when the analyzer is crossed and maximum when the analyzer is parallel to the vibration incident on the upper plate. There will be maxima and minima with increasing inclination of the incident pencil.

As an inclination in the plane ABD of Fig. D-2 has no effect, the interference pattern will show light and dark fringes parallel to AB when light from all directions is incident on the upper plate. In white light there will be colored fringes with either a black center (Nicols crossed) or a white center (Nicols parallel).

The phenomenon being symmetrical to the plane ABD, the angular distance of the minima for a given wavelength must be an uneven function of the incidence angle i in the plane PQR. The first approximation is a linear function.

From this simple reasoning we find that a Savart plate between two polarizers in convergent light will show equidistant fringes of which the direction is AB in Fig. D-2. This is exactly what one observes.

A complete investigation of the phenomenon is not easily given, as it entails a three-dimensional analysis.[3] It can be shown that for wavelength λ the angular distance between minima equals λ/g. For $\lambda = 0.560$ micron the first minimum between crossed Nicols corresponds to the sensitive color when white light is used. This gives an angle of $0.0035/t$ for calcite and of $0.065/t$ for quartz. The angular distance of the fringes is, for instance, 0.0035 radian $= 0.20$ degree for a calcite Savart plate of which the components have a thickness of 1 mm and 0.0065 radian $= 0.37$ degree for a quartz Savart of which the components have a thickness of 10 mm.

When a parallel beam of white light is incident on the upper plate with an inclination i (in the plane PQR of Fig. D-2), the pencil emergent from the lower plate is parallel and in the same direction. If a lens is placed below the second plate, the light is concentrated in its focus. By putting the eye at the focus one sees the whole field in one interference color. The parallel incident pencil is usually produced by a collimator. In the frontal focal plane of the collimating lens a slit of suitable width is put. The slit must be parallel to the plane ABD in Fig. D-2. In order to produce the best contrast in the image, its angular width (as seen from the collimating lens) has to be about $\frac{1}{20}$ of the fringe distance.[4]

The Savart plate, in conjunction with an analyzer, is a sensitive instrument for detecting the presence of a direction of preference in the vibrations of a pencil of otherwise natural light. The direction in which the analyzer transmits vibrations is chosen as that of either AB or PQ in Fig. D-2. In the first case, the presence of a portion of linearly polarized light in the incident pencil of which the vibrations are in the direction PQ, the colored interference fringes with black center line will be observed when looking through the analyzer. This interference pattern will be flooded by the presence of natural light. The contrast is diminished by the latter, but will be best when the combination Savart plate plus analyzer is turned about the line of vision until PQ coincides with the direction of vibrations of the incident pencil. Again, the preferential direction of the incident vibrations being that of AB in Fig. D-2, the interference pattern will have a white central line.

This combination, called a Savart polariscope, enables one not only to detect the presence of polarized light in a pencil of natural light, but also to determine the direction of the vibrations.

A further refinement is described by Lyot,[1] who introduces two plane parallel glass plates before the polariscope in order to make the interference pattern visible in all orientations of the plate. Turning the plate about the

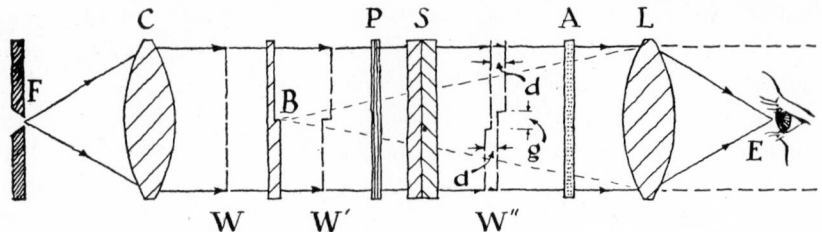

FIG. D-5 Arrangement for using Savart plates.

line of vision then results in a maximum and minimum visibility. In this way a better precision is obtained. Furthermore, the analyzer is a birefringent prism of the Rochon or Wollaston type by which both interference patterns (crossed Nicols and parallel Nicols) are shown simultaneously. The detectable amount of polarized light is about $\frac{1}{1000}$.

The Savart plate is eminently useful for measuring phase differences in a wave front. Assume that a transparent object has a step B (Fig. D-5), unobservable unless some special device is used to detect it (dark field illumination, phase contrast observation); it can be made visible and even accurately measurable by means of a set-up described by Françon.[4] A plane wave, W, is transformed into a wave, W', with a step. At S a Savart plate is put, preceded by a polarizer. The orientation of the directions of vibration in the Savart plate is 45° with reference to the direction of the polarizer. If W and S are not exactly parallel, two waves will emerge from S, displaced with respect to one another, having a path difference d and vibrating perpendicularly to one another. The displacement g has been discussed above. The value of d depends on the incident angle between W and the plane of S. The plate S ought to be turned together with P until the displacement is in a direction perpendicular to the boundary line of the step. In the region where the steps in the two wave fronts overlap, the phase difference is not d, but has a value, q, equal to the sum (or the difference) of d and s, where s is the path difference produced by the step in the object. The light then passes through an analyzer, A, and a lens, L. We assume A and P to be crossed. The lens forms at infinity an image of B, which can be observed by putting the eye behind it.

The illumination is provided by a slit, F, and collimating lens, C. By either a lateral shift of F or a rotation of S about an axis perpendicular to the plane of the drawing, the incident angle on S can be changed, with a corresponding change of the path difference, d.

The eye placed at E sees a field in one color (corresponding to the value of d), transversed by a strip in another color (corresponding to the value of $d + s$ or $d - s$); see Fig. D-6. In monochromatic light the strip has, in general, another intensity than the main field.

FIG. D-6   Appearance of interference in Savart interferometer.

Here now we have a contrivance for ascertaining the presence of phase differences in a wave front and for measuring them, even if the differences are large (in that case the measurement must be made in monochromatic light); there are methods of finding out how many whole wavelengths are comprised in the path differences.[5]

It is interesting to note that no diffraction phenomena occur as in the method of observation called phase contrast. Further advantages are that the interferometric arrangement can be mounted in the eyepiece (thus allowing the use of the usual object glasses in microscopy and in telescopes), and that measurements are rendered possible even when the phase differences are gradual. For the observation of such gradients we refer to Fig. D-7. The two wave fronts have suffered a mutual translation and show a range of colors (or of intensities in monochromatic light) in the region AB.

A special feature of Françon's set-up is that interposition of plane parallel plates of glass of rather bad quality in the light path has no serious effect on the interference pattern, provided they are put in such places that they are quite far out of focus. The irregularities produced by the faults in the glass plates are then spread out so widely over the field of view that they give a general contrast-diminishing flare. This fact is made use of for observations in a vacuum enclosed by plate glass windows. A possible set-up is sketched in Fig. D-8. This is intended to measure the thickness of coatings *in vacuo*. A screen, N, is placed near the surface on which the coating is being deposited, while the illuminating pencil just can pass a clean and an adjacent coated part of the surface.

Lastly, we point to the possibility of measuring at the same time the path difference and the light loss (produced by absorption or reflection) at a step. This can be illustrated by Fig. D-9. In the region AB the amplitudes of the two waves are equal, say a. In the strip BC the amplitude of the first wave is a, of the second wave $a(1 - \rho)$, the factor $1 - \rho$ being produced by the

FIG. D-7   Wave fronts in Savart interferometer.

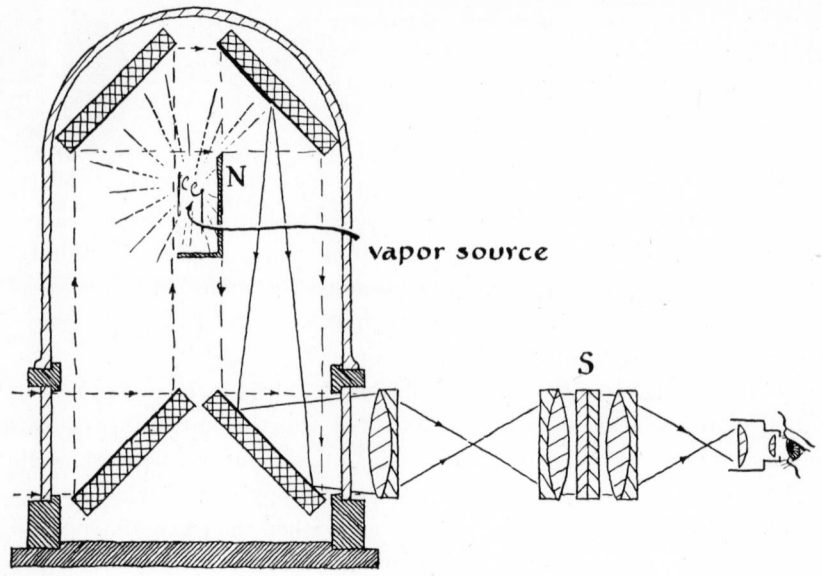

FIG. D-8 Application of Savart interferometer to measurement of thermally evaporated films.

light loss at the step. In the region CD the amplitudes again are equal, having the value $a(1 - \rho)$. It is impossible in the cases where $\rho \neq 0$ to equalize at the same time AB and BC and also BC and CD. In Fig. D-9b $\rho = 0$, the second wave bisects the step in B and C, and the strip BC is equalized with both sides of the field. In Fig. D-6c for $\rho \neq 0$ the strip is equalized with the half AB of the field, while in Fig. D-6d the strip is equalized with the half CD of the field.

It proved possible to measure the step with a precision of at least $\lambda/300$ and the transmission with a precision of 2%. For further refinements and for a discussion of sources of error the reader is referred to reference 5.

FIG. D-9 Wave fronts obtained with the arrangement of Fig. D-8.

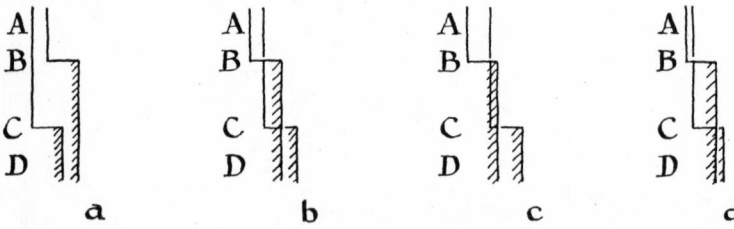

REFERENCES

1. B. Lyot, Rev. d'Optique, *5*, 108 (1926).

2. M. Françon, Rev. d'Optique, *31*, 65 (1952) and *32*, 349 (1953); Optica Acta, *1*, 50 (1954).

3. E. Mascart, *Traité d'optique*, II, p. 132 (1891, Gauthier-Villars, Paris); A. Schuster, *Introduction to the Theory of Optics*, Chap. IX, p. 122 and p. 212 (2nd ed., 1909, Arnold, London).

4. M. Françon, Optica Acta, *1*, 55 (1954–55).

5. A. C. S. van Heel & A. Walther, Optica Acta (to be published in 1957).

Apodization

by Pierre Jacquinot[†]

It has been explained in § 10-2 that the light diffracted far from the center of the diffraction pattern which constitutes the physical image of a point may be, in some cases, very troublesome, although its relative intensity is very small.

Only the central part of the diffraction curve needs to be considered when two points or two objects of equal strength have to be resolved, but a very faint object close to a strong one may be masked by the light corresponding to the subsidiary maxima of the curve. That is the reason why it may be useful in some cases to suppress, at least partially, this light, or, in other words, to cut down the "feet" of the curve. This operation may be called "apodization."

Apodization by a Suitably Shaped Aperture

If we need only to resolve two points (or two stars) with very different relative strengths, and if, in addition, we know the direction of the line joining them, we can use the diagonal square aperture; the example of Sirius and its dwarf companion has been given in § 10-2. The same is true for two spectral lines, provided the slit is reduced to a pinhole and the two spectral lines become two spectral "spots" or "points." In the case of the diagonal square aperture the relative illumination falls off very rapidly, as is shown in Table 10-1.

But many other aperture shapes are also able to give a very rapid falling off along a given direction on the observing screen, even more rapid than the diagonal square. For example, with an aperture shaped as in Fig. E-1a, the

---

† Université de Paris, Laboratoire A. Cotton, 1 Place Briand, Bellevue (S. & O.), France.

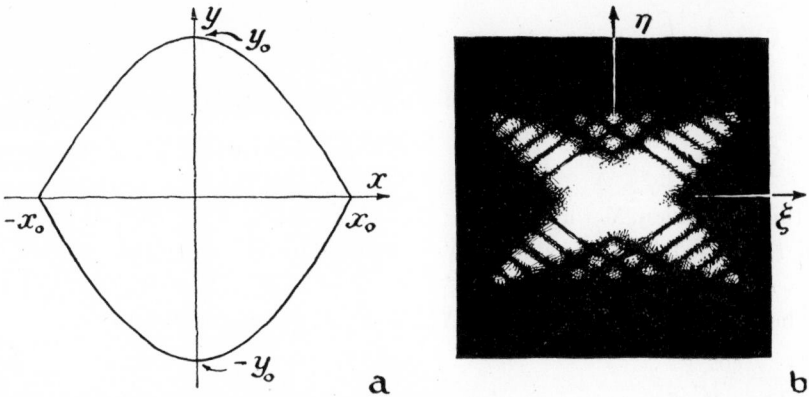

FIG. E-1 (a) Shape of a simple apodizing aperture. The two curves limiting the aperture are $y = y_0 \cos (\pi x/2x_0)$. (b) Diffraction pattern (heavily over-exposed) corresponding to a point source imaged through this aperture.

diffraction pattern is that of Fig. E-1b; it can be seen that the illumination falls off very rapidly along the axis. The two curves limiting the aperture shown in Fig. E-1a are simply the curves

$$y = y_0 \cos (\pi x/2x_0)$$

Of course, the complete calculation of the pattern shown in Fig. E-1b would be very difficult, but it is not too hard to compute the illumination along the ξ axis. The Kirchhoff integral is, with the notation of § 10-1,

$$\tilde{E}_p = 4x_0 y_0 \frac{j\mathcal{E}_0}{\lambda r_m} e^{j\omega\left(t-\frac{r_m}{c}\right)} \times \frac{1}{4x_0 y_0} \iint e^{j\frac{2\pi}{\lambda}(x \sin \alpha + y \sin \beta)} \, dx \, dy \tag{1}$$

The integration must be extended to the area of the aperture limited by the two curves.

Here the integral is no longer separable, but if we are interested only in the field along the ξ axis, we take $\sin \beta = 0$, and we have to compute the integral:

$$\tilde{X} = \frac{1}{4x_0 y_0} \int_{-x_0}^{+x_0} e^{j\frac{2\pi}{\lambda} x \sin \alpha} \left(\int_{-y}^{+y} dy \right) dx$$

and, since

$$\int_{-y}^{+y} dy = 2y_0 \cos (\pi x/2x_0)$$

$$\tilde{X} = \frac{1}{2x_0} \int_{-x_0}^{+x_0} \cos (\pi x/2x_0) e^{j\frac{2\pi x \sin \alpha}{\lambda}} \, dx$$

setting, as in § 10-1, $\xi = \dfrac{2\pi x_0 \sin \alpha}{\lambda}$, our integral over x becomes

$$\tilde{X} = \frac{1}{2x_0} \int_{-x_0}^{+x_0} \cos(\pi x/2x_0)e^{j\xi x/x_0}\, dx \tag{2}$$

The integration, giving the following result, can be easily carried out:

$$\tilde{X}_{\text{rel}} = \frac{\cos \xi}{1 - 4\xi^2/\pi^2}$$

In this expression $\tilde{X}_{\text{rel}}$ means the value of $\tilde{X}$ at any point defined by $\eta = 0$ and ξ, divided by the value of $\tilde{X}$ at the center of the diffraction pattern $(\xi = \eta = 0)$.

The relative illumination, $\tilde{X}_{\text{rel}} \cdot \tilde{X}_{\text{rel}}^* = E_p/E_0$, is then equal to

$$\frac{\cos^2 \xi}{(1 - 4\xi^2/\pi^2)^2}$$

and its maximal values, occurring approximately at $\xi = n\pi$, are given in Table E-1 and compared with corresponding values for the diagonal square aperture.

TABLE E-1

$y_0 \cos(\pi x/2x_0)$		*Square Diagonally*	
ρ/x_p	$\mathfrak{C}_p/\mathfrak{C}_c$	ρ/x_p	$\mathfrak{C}_p/\mathfrak{C}_c$
0	1.00	0	1.00
1.5	0	1.41	0
1.9	0.005	2.12	0.002
2.5	0	2.82	0
2.9	0.00085	3.53	0.00026
3.5	0	4.24	0
3.9	0.00025	4.95	0.000069
4.5	0	5.64	0
5	0.00010	6.36	0.000025
5.5	0	7.06	0
6	0.000049	7.77	0.000011

Instead of an aperture limited by the two curves $y = \pm y_0 \cos(\pi x/2x_0)$, different types of curves $y = \pm f(x)$ could be used, giving more or less apodization along the ξ axis. These methods of apodizing are very simple and efficient but are valid only for points and only along a given direction. In the case of stars this direction has to be known in advance or found by trial and error. In the case of spectral "lines" the lines have to be reduced to points, and this is not very convenient.

Apodization by an Aperture with Non-uniform Absorption

More general results are obtained if, instead of changing the shape of the aperture, the amplitudes of the wave trainlets emerging from the points of the aperture are made different. This can be done easily by introducing an absorption distributed non-uniformly over the aperture. If, for instance, the absorption is stronger near the edges of the aperture, one can guess that the light diffracted far from the center is weaker because this light is related to the edges.

With such "apodizing screens" apodization may be obtained for any kind of source: point, line, or extended source. The most general apodizing screen must obviously have rotational symmetry, but the complete theory of a screen with rotational symmetry is rather difficult and involves more difficult mathematics.

The One-dimensional Apodizing Screen

The theory is much simpler in the case of the "one-dimensional" apodizing screen: in this type of screen the absorption is a function only of x; that is, it is the same along a line parallel to the y axis, but variable along a line parallel to the x axis. This one-dimensional screen may be used in spectrographs because the source is a slit parallel to the y axis; and, although it is much more difficult to make than the shaped screen, it is much better because it is no longer necessary to reduce the slit to a point, and the spectrum has the conventional appearance of a spectrum with true spectrum lines.

Let us now see how such a screen behaves. The general equation is almost the same as equation 1 at the beginning of this appendix, and must be extended to the rectangular aperture, but we have now to take account of the variable absorption. This is done by introducing a factor, $t(x,y)$, describing the amplitude of the wavelet emerging from every point (x,y) of the aperture; in the one-dimensional case under examination this factor is a function of x only and is written $t(x)$. In this simple case the integral of the Kirchhoff differential is still separable into three parts as in § 10-1.

$$\tilde{E}_p = \tilde{X}\tilde{Y}\tilde{Z}$$

$\tilde{Y}$ and $\tilde{Z}$ have exactly the same form as in § 10-1, but now $\tilde{X}$ takes account of $t(x)$,

$$\tilde{X} = \frac{1}{2x_0} \int_{-x_0}^{+x_0} t(x)e^{j\frac{2\pi x \sin\alpha}{\lambda}} \, dx$$

or, introducing the variable, $\xi = (2\pi x_0 \sin\alpha)/\lambda$,

$$\tilde{X} = \frac{1}{2x_0} \int_{-x_0}^{+x_0} t(x)e^{j\xi x/x_0} \, dx \tag{3}$$

The student acquainted with Fourier transforms will recognize here that $\tilde{X}(\xi)$ is proportional to the Fourier transform of a function equal to $t(x)$ for $-x_0 < x < +x_0$ and equal to zero for $x > x_0$ and $x < -x_0$. But it is, of course, not necessary to know the Fourier transformation in order to solve our present problem.

It is obvious that equation 3 is a generalization of equation 2; we have not specified the form of the function $t(x)$ which was taken, as an example, equal to $y_0 \cos (\pi x/2x_0)$ in equation 2. The choice of the "best" function $t(x)$, that is, the choice of the distribution of absorption over the aperture giving a falling off of the illumination as rapid as desired and at the same time not spoiling unduly other qualities such as luminosity and resolving power, is not a simple matter.

The simplest method would be to try a number of functions such as $\cos (\pi x/2x_0)$ or $\cos^2 (\pi x/2x_0)$ or $(1 \pm x/x_0)$ or $(1 \pm x/x_0)^2$ or $e^{-k(x/x_0^2}$ (or some linear combination of these functions), computing the diffraction pattern for each of these cases and choosing the most satisfactory. But it is obvious that, although good screens may be found by this method, the "best" screen cannot be discovered except by very improbable good luck, for the number of functions to be tried is infinite!

So the problem has to be attacked in another way. One could believe that, according to equation 3, it would be possible to obtain *a priori* the desired distribution of the diffracted light by choosing a function $\tilde{X}(\xi)$ and then finding the corresponding distribution of the transmission over the apodizing screen by merely solving equation 3.

This equation can, in fact, be written in the inverse form:

$$t(x) \propto \int_{-\infty}^{+\infty} \tilde{X}(\xi) e^{-j\xi x/x_0} \, d\xi$$

But, in order to be acceptable, the function $t(x)$, the solution of this equation, should be equal to zero outside the aperture, *i.e.* for $|x| \geqslant x_0$, and this happens only for a very limited class of functions $\tilde{X}(\xi)$, which cannot be known in advance. Physically, this is not surprising and means only that diffraction does exist and that, with a limited aperture, the diffracted light has to fulfill certain conditions although its distribution may be changed. So the form of the distribution function $\tilde{X}(\xi)$ cannot be given *a priori*, but certain reasonable conditions can be asked for: for instance, that the ratio R of the energy falling outside a given domain of ξ to the total transmitted energy be as low as possible (without, of course, asking it to be zero, which is impossible).

The mode of computation is then the following: The function $t(x)$ is written in the general form of a trigonometric (or some other type) expansion with a limited number of terms and undetermined coefficients such as

$$t(x) = a_0 + a_1 \cos (\pi x/2x_0) + a_2 \cos (2\pi x/2x_0)$$

and the ratio R is calculated as a function of the coefficients a_0, a_1, a_2. The trigonometric form of the terms is advantageous because the diffraction pattern corresponding to each one (and hence the value of R) can be easily computed, as has been shown previously.

Minimization of R gives the values of the coefficients which define the best $t(x)$ among all the possible functions of the type considered and involving n coefficients. This minimum value of R sometimes turns out to be lower than is really useful, and in this case the mean transparency of the apodizing screen is generally too small for any practical use. One should then take $R > R_{min}$, and try to take advantage of this to improve the total transparency of the screen. Indeed, each value of $R > R_{min}$ corresponds to an infinity of screens, all satisfying a given relation between the coefficients $a_0 \cdots a_n$, instead of the unique screen which corresponds to the minimum value of R. These screens therefore form a particular group in the general family of functions $t(x)$, among which a new choice becomes possible which will take account of some other desirable quality of the screen, for example, its mean transparency. One can in this way find functions $t(x)$ which provide good apodization as well as good transparency.

The Apodizing Screen with Rotational Symmetry

This method can also be used for apodizing a circular aperture provided appropriate new developments of the function t, involving basic functions of $u = \sqrt{x^2 + y^2}$ instead of x, be introduced in the calculations. These appear easier if the basic functions are taken to be certain Bessel functions which have pronounced analogies with the cosine functions which were found to be so efficient in the one-dimensional case.

In all cases, apodization does not take place without a slight broadening of the "body" of the diffraction pattern, which reduces somewhat the resolving power for points or lines of equal strength. But, as a rule, the relative loss can be made less than 25 or 30%.

Extensive application of the method described above has been made and has permitted the computation of a number of apodizing screens with two, three, or four terms. Let us cite, for ordinary use, a circular screen with an apodization coefficient K:

$$K = \frac{R \text{ (uniform screen)}}{R \text{ (apodizing screen)}} = 125$$

The apodizing region goes from the second ring of the diffraction pattern on to infinity, and the total transmission of the screen is still as great as 0.4.

It is also possible to improve the diffraction pattern given by other types

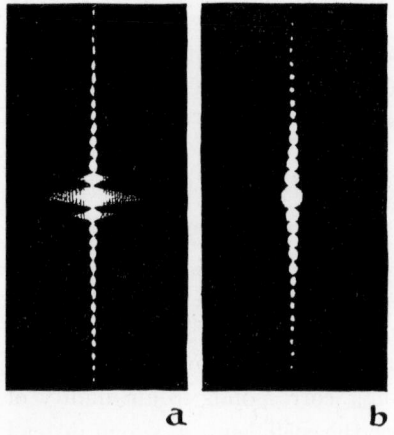

a   b

FIG. E-2 (a) Diffraction pattern (point source) of a rectangular aperture (the two sides of which are in the ratio 5 to 1). (b) Diffraction pattern of the same aperture, with a coated one-dimensional screen. The zones of equal absorption on the screen are vertical. The diffraction fringes along the horizontal axis have disappeared.

of apertures, such as, for instance, circular apertures with a central stop, which occur in many telescopes and reflection microscopes. The calculations are much more involved, but the results are still fairly good, although perhaps less striking than in the pure circular case.

These apodizing screens with non-uniform absorption are made by evaporating chromium on glass. The glass plate is placed behind a suitable mask and moved during the evaporation according to the type of screen; for one-dimensional screens the plate is moved uniformly back and forth, and for screens with rotational symmetry the plate is rotated.

Results

Fig. E-2 shows an example of the effect of a one-dimensional apodizing screen. Fig. E-2a shows the conventional diffraction pattern, strongly over-exposed, given by a *point source* and a rectangular aperture. In this case the lengths of the two sides of the rectangular aperture are in the ratio 5 to 1 so that the scales on the two arms of the diffraction "cross" are in the same ratio but rotated by 90°. Fig. E-2b shows the diffraction pattern given under the same conditions but with the aperture covered by an apodizing screen having its lines of equal absorption parallel to Oy. The light diffracted along the

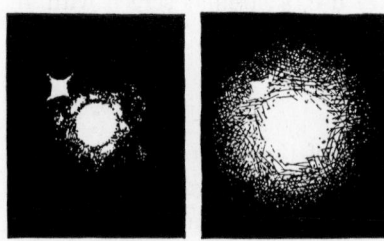

FIG. E-3 Point source and weak satellite (1000 times weaker) separated by 5 times the limit of resolution, photographed through a circular aperture with (left) and without (right) apodizing screen.

FIG. E-4 Intensity in the diffraction pattern of a circular aperture without (dotted lines) and with (solid lines) an apodizing screen. The scale for the ordinates is linear for the central part of the pattern $\rho/x_p < 2$ and logarithmic for the outer part.

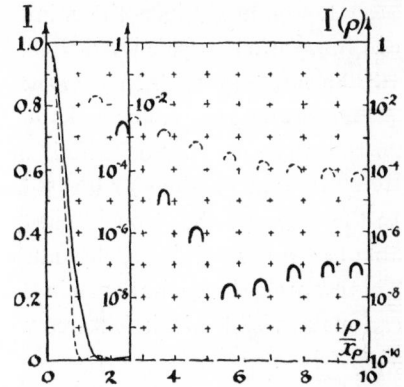

axis is so strongly reduced that the arm of the cross parallel to the ξ axis has completely disappeared; however, the central part of the diffraction pattern is substantially widened. If the source had been a slit parallel to Oy, the image would have been a line very well "apodized." Some examples of the effect of an apodizing screen with rotational symmetry are given in Figs. E-3 to E-5. In Fig. E-3 are shown two photographs of a point with a satellite (like Sirius and its companion) 1000 times weaker, the distance between them being about 5 times the conventional limit of resolution. Fig. E-3b is made without an apodizing screen and Fig. E-3a with an apodizing screen; in both cases the exposure time was chosen in order to get a normal exposure for the weak satellite. The gain in separation is very striking. The calculated variation of the relative intensity in the diffraction pattern of the screen used for this picture is given in Fig. E-4. The abscissae are given in the same units as in Table 10-1; the scale for the ordinates is linear for the central part of the pattern ($\rho/x_p < 2$) and logarithmic (in order to represent the very faint intensities) for the outer part. The dotted lines correspond to the uniform circular aperture and the solid lines to the same aperture with the apodizing screen; in the logarithmic part of the curve, only the maxima have

10^3 **times overexposed**

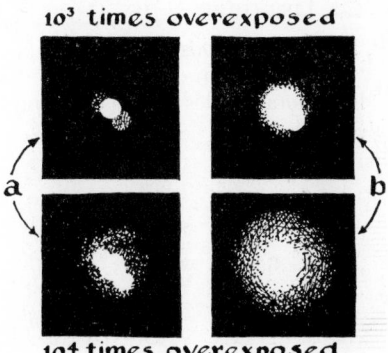

FIG. E-5 "Extended" object (diameter equal to 25 times the limit of resolution) and weak satellites photographed with different exposures times through a circular aperture with (a) and without (b) apodizing screen.

10^4 **times overexposed**

been represented since the minima are zero and their logarithms $-\infty$. It can be seen on these curves that the fifth maximum, for instance, is at least 10^4 times weaker with the screen.

Fig. E-5 shows photographs of an "extended" object (diameter equal to 25 times the conventional limit of resolution) strongly overexposed (respectively 10^3 and 10^4 times) with (E-5a) and without (E-5b) an apodizing screen. In the first case the strong halation due to diffraction is absent so that one is able to see several disks, the fainter of which are due to spurious reflections. This example shows that even in the case of an extended object the apodization can make visible weak features which would escape ordinary observation.

BIBLIOGRAPHY

R. Straubel, *Über Bildgüte* (1935, Pieter Zeeman Verhandelingen, Martinus Nijhoff), p. 302.

A. Couder & P. Jacquinot, "Méthode pour l'observation de radiations de faible intensité au voisinage d'une raie brillante," Comptes Rendus de l'Académie des Sciences (Paris), *208*, 1639 (1939).

G. Lansraux, "Sur la modification des figures de diffraction par les filtres d'intensité," Comptes Rendus de l'Académie des Sciences (Paris), *222*, 1434 (1946).

P. M. Duffieux, "Nouvelle méthode pour le calcul des diaphragmes de transparence non uniforme réduisant les effets de la diffraction," Comptes Rendus de l'Académie des Sciences (Paris), *222*, 1482 (1946).

P. Boughon, B. Dossier, & P. Jacquinot, "Détermination de fonctions pour l'amélioration des figures de diffraction dans le spectroscope," Comptes Rendus de l'Académie des Sciences (Paris), *223*, 661 (1946).

H. Osterberg & J. E. Wilkins, "The Resolving Power of a Coated Objective," J. Opt. Soc. Am., *39*, 553 (1949).

P. Jacquinot, "Quelques recherches sur les raies faibles dans les spectres optiques," Fifth Holweck Discourse (17 May 1950), Proc. Phys. Soc., B *63*, 969 (1950).

B. Dossier, "Recherches sur l'apodisation des images optiques" (a complete treatment of the question), Thèse, Paris (1953), Ed. de la Revue d'Optique Théorique et Instrumentale.

B. Roizen-Dossier, "L'apodisation des images optiques: cas particulier de l'obturation centrale," Proceedings of a Symposium on Astronomical Optics Held in Manchester (April 1955), edited by Zdenek Kopal, p. 163.

Appendix **F**

Application of Fourier Transformations in Optics: Interferometric Spectroscopy

by George A. Vanasse† and John Strong†

In 1911 Rubens and Wood[1] isolated the long wavelength radiation emitted by a Welsbach gas light mantle. These wavelengths, in the spectral region around $\lambda = 175\mu$, were isolated by the selective-refraction method of *focal isolation*—a method that depends on the peculiar optical properties of quartz. Fig. F-1 shows the wavelength dependence of the index

† Laboratory of Astrophysics and Physical Meteorology, The Johns Hopkins University.

FIG. F-1 Refractive index of quartz (ordinary ray) as a function of wavelength. An opaque region is indicated by the shaded area.

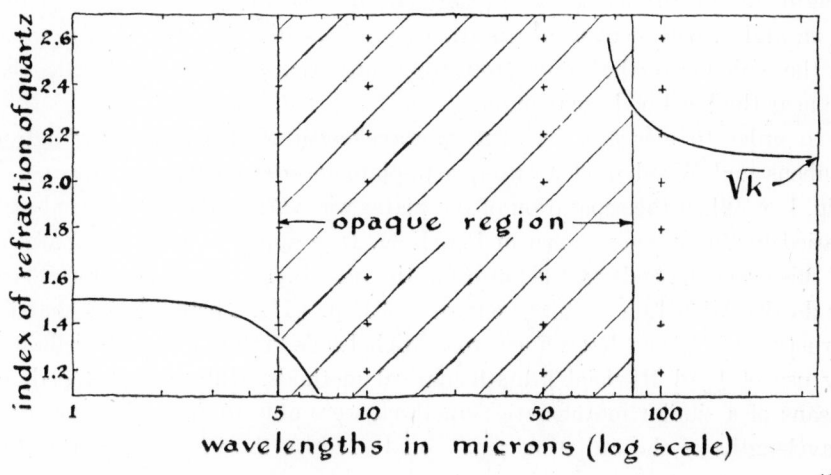

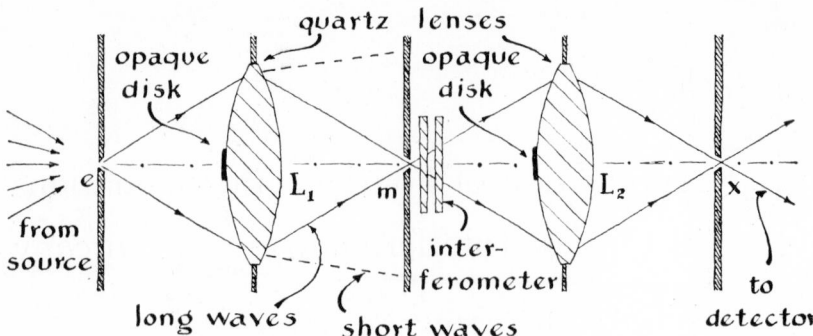

FIG. F-2 Focal isolation apparatus (after Rubens and Wood).

of refraction of crystalline quartz; the shaded area represents the spectral region where the quartz is opaque in substantial thickness. As the dispersion theory of Chap. IV requires, at very long wavelengths, where the reaction of material to electric light fields approaches its reaction to static fields because of their low frequency, the optical index approaches $\sqrt{\kappa}$. For quartz $\kappa = 4.44$, and Fig. F-1 shows that N^2 approaches this value rather closely around 300μ, where Rubens measured it.

The focal isolation method,[2] illustrated in Fig. F-2, depends on the fact that the index is approximately 1.5 for the short wavelengths that are to be disposed of ($\lambda < 5\mu$), while the index is greater than 2.0 for the very long wavelengths that are to be saved (beyond the opaque region of wavelengths $> 80\mu$). Thus, as Fig. F-2 shows, the short wavelengths from such a source passing through the entrance aperture e either are obstructed, or are not converged so they can go through the middle aperture m, while the long wavelengths do go through m. This combination of two lenses and the apertures e, m, and x, with central obscurations, thus isolates the longest heat waves in the emission of the source from the energy-rich shorter wavelengths that lie near the blackbody maximum.

In order to measure the average wavelength of this isolated radiation, Rubens and Wood used a two-quartz-plate interferometer, as indicated in Fig. F-2. When the separation of the plates was varied, the transmitted light varied in the manner shown in Fig. F-3a. It is apparent that the character of this curve depends on the nature of the radiations that are in interference, as do the Michelson visibility curves of Chap. VIII. There we saw that the structure of H_α was determined from Michelson's "interferometric visibility" by use of Lord Rayleigh's mathematical methods. Rubens and Wood, by means of a similar mathematical method, were able to determine the mean wavelengths of the radiation that would produce the interferogram of Fig.

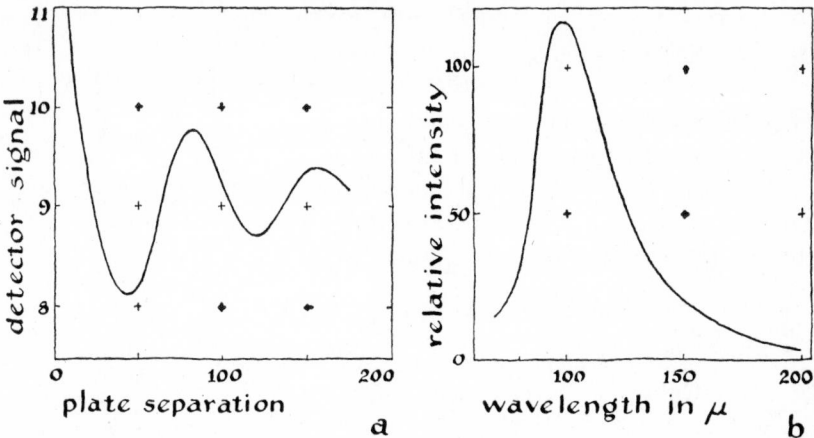

FIG. F-3 (a) Two-beam interferogram obtained by Rubens and Wood. (b) Result of Fourier analysis of curve a where abscissae are in Na-D wavelengths.

F-3a. Fig. F-3b shows their result. Furthermore, by dividing this spectral curve into narrow strips and calculating the interferogram that would result from the superposition of the interference patterns of all these narrow strips, they checked this analysis—arriving at a curve very similar to that of Fig. F-3a. From this they concluded: "it is apparent that we have experimental evidence of the presence of heat waves certainly 150μ and probably 200μ in length." Using a similar technique, Rubens and Von Baeyer[3] found that the high-pressure Hg source was richer in long wavelengths than the Welsbach mantle, and they determined that the Hg source emitted two bands of wavelengths having mean values of 218μ and 343μ.

The mathematical methods of transforming such an interferogram as Fig. F-3a into a spectrum, Fig. F-3b, have very general application in optics; this appendix is devoted to the description of these methods. As first examples of these methods, we shall apply them to problems that have been previously solved in the text. Following a description of the general methods, we shall then describe a spectroscopic procedure that involves transforming an interferogram into a spectrum.

The general mathematical methods referred to above involve the reciprocal Fourier transform relations:

$$g(v) = \frac{1}{\sqrt{2\pi}} \int_{-\infty}^{\infty} f(u)e^{-juv}\, du \tag{1}$$

$$f(u) = \frac{1}{\sqrt{2\pi}} \int_{-\infty}^{\infty} g(v)e^{+juv}\, dv \tag{2}$$

Here $g(v)$ and $f(u)$, called *Fourier pairs*, are the *Fourier transforms* of each other. The functions, $g(v)$ and $f(u)$, may be complex functions, or real. However, in order for the above relations to be valid, the function $g(v)$, or $f(u)$, must satisfy certain conditions. The following set of conditions is sufficient to allow the integral expansions of Eqs. 1 and 2:

(a) $g(v)$ must be a single-valued function of v in the interval $-\infty < v < \infty$, although a finite number of finite discontinuities are allowed;

(b) at a point of discontinuity v_0, the function will be given the mean value

$$g(v_0) = \lim_{\epsilon \to 0} \tfrac{1}{2}[g(v_0 + \epsilon) + g(v_0 - \epsilon)]$$

(c) the integral $\displaystyle\int_{-\infty}^{\infty} |g(v)|\, dv$ must exist.

In order to illustrate the uses of Fourier pairs in optics, we shall use the above relations in several familiar and simple applications.

Case I

First, we shall show that these relationships apply to the following functions, related as in Chap. X: one represents the distribution of field transmitted by a rectangular aperture when a plane wave falls on it; the other function represents the field of diffracted light falling on a distant screen. Here, in the y direction, we let

$$v = \frac{2\pi y_0 \sin \beta}{\lambda} = \eta \qquad\qquad u = y/y_0$$

$$f(u) = \sqrt{2\pi y_0}\,\mathcal{E}(y) \qquad\qquad E = \mathcal{E}(y)e^{j\omega(t - z/c)}$$

$$\qquad\qquad\qquad\qquad\qquad\qquad = \mathcal{E}(y) \text{ at the aperture,}$$
$$\qquad\qquad\qquad\qquad\qquad\qquad\qquad\text{at } t = 0,$$

making $uv = 2\pi y \sin \beta/\lambda$. As in Chap. X, y_0 is half the width of the diffraction aperture, and $\mathcal{E}(y)$ is the amplitude distribution of the light field transmitted by the diffraction aperture. $\mathcal{E}(y) = 0$ for $|y| > y_0$. $g(v)$ represents the light on the diffraction screen at the point defined by the angle β.

$g(v)$ has the dimensions $\left[\dfrac{\text{volt}}{\text{meter}} \times \text{meter}\right]$. Actually, as in § 9-5, this is but half the problem. A similar treatment along the X-axis would give a second function $g'(v)$, and the product of these is involved in the whole problem. The product of two half solutions, as shown in § 9-5, yields the dimensions of an illumination proportional to $\mathcal{E}^2$ multiplied by an area; *i.e.* it has the dimensions of a flux. This indicates what we mean by "$g(v)$ represents the light on the diffraction screen."

In order to determine $g(v)$, we apply Kirchhoff's differential and integrate over the aperture width, following Chap. X; the result is

$$g(v) = \int_{-y_0}^{y_0} \mathcal{E}(y) e^{-j\left(\frac{2\pi y \sin\beta}{\lambda}\right)} dy \tag{3}$$

Since the mathematical conditions imposed on the Fourier pairs are satisfied here, we may immediately write, from Eq. 2,

$$f(u) = \sqrt{2\pi} y_0 \mathcal{E}(y) = \frac{1}{\sqrt{2\pi}} \int_{-\infty}^{\infty} g(v) e^{j\left(\frac{2\pi y \sin\beta}{\lambda}\right)} dv$$

or

$$\mathcal{E}(y) = \frac{1}{2\pi y_0} \int_{-\infty}^{\infty} g(v) e^{j\left(\frac{2\pi y \sin\beta}{\lambda}\right)} dv \tag{4}$$

We now apply this interrelation, between the amplitude distribution in the diffraction pattern $g(v)$ and the amplitude distribution of the light field at an aperture, to the case of a single narrow aperture (single slit) where the function $\mathcal{E}(y)$ is equal to $\mathcal{E}_0$ for $-y_0 \le y \le y_0$, and equal to zero for $|y| > y_0$, as shown in Fig. F-4a.

As in Chap. X,

$$g(v) = \mathcal{E}_0 \int_{-y_0}^{y_0} e^{-j\left(\frac{2\pi y \sin\beta}{\lambda}\right)} dy = 2y_0 \mathcal{E}_0 \left(\frac{\sin\eta}{\eta} \quad \text{or} \quad \frac{\sin v}{v}\right)$$

Reciprocally, the Fourier transform of $g(v)$, whose square is represented by the curve of Fig. F-4b, should give a step function representing the amplitude distribution: $f(u) = \sqrt{2\pi} y_0 \mathcal{E}_0$ for $-y_0 \le y \le y_0$ and $f(u) = 0$ for $|y| > y_0$. We now show that this is true by integrating $g(v)$ as follows:

$$f(u) = \frac{2y_0 \mathcal{E}_0}{\sqrt{2\pi}} \int_{-\infty}^{\infty} \frac{\sin\eta}{\eta} e^{j\pi u/y_0} d\eta$$

FIG. F-4 (a) Amplitude distribution of light field at a narrow slit. (b) Fraunhofer diffraction pattern of the single slit.

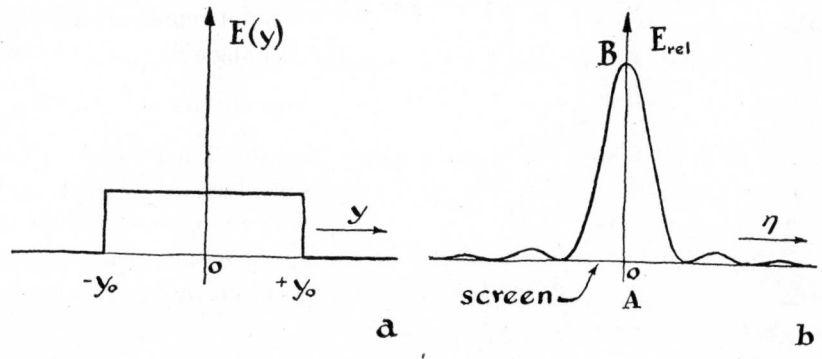

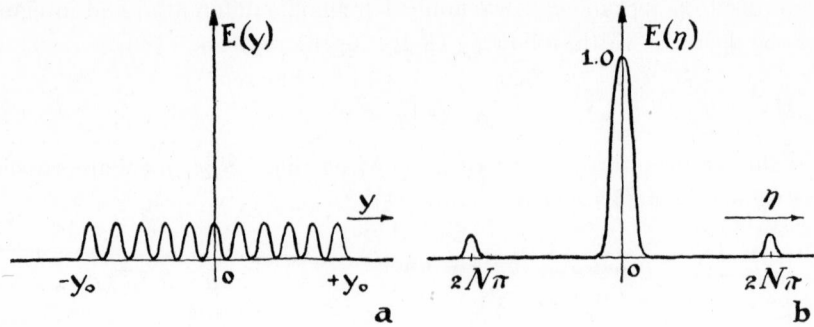

FIG. F-5 (a) Amplitude distribution at an aperture. (b) Fraunhofer diffraction pattern due to the aperture a.

This gives†

$$f(u) = 0 \qquad \text{for} \qquad y^2/y_0^2 > 1$$
$$= \sqrt{\pi/2}\, \mathcal{E}_0 y_0 \qquad \text{for} \qquad y/y_0 = \pm 1$$
$$= \sqrt{2\pi}\, \mathcal{E}_0 y_0 \qquad \text{for} \qquad y^2/y_0^2 < 1$$

which is the step function that was expected. Since this function has a discontinuity at $y/y_0 = \pm 1$, $f(u) = \sqrt{\pi/2}\mathcal{E}_0 y_0$ at these points (as is to be expected from condition b imposed on the Fourier pairs).

As a second example of application of Fourier pairs we take a wave of constant phase but now with the amplitude varying from $-y_0$ to $+y_0$ over $\mathfrak{N}$ cycles according to a cosine function. We shall assume that the amplitude is zero beyond these limits. Furthermore, as before, we take the illumination constant along the x direction. Fig. F-5a shows this wave front of variable amplitude and F-5b shows the principal maxima in its grating pattern. Here again we let $u = y/y_0$, and $v = \eta$, with $f(u)$ representing the amplitude in the modified wave front as function of y.

$$f(y/y_0) = 2 \cos^2 \pi \mathfrak{N} y/y_0 = 1 + \cos 2\pi \mathfrak{N} y/y_0.$$

The relative illumination along the η direction on the observing screen is obtained by evaluating

$$g(v) = \tilde{Y} = \frac{1}{2y_0} \int_{-y_0}^{y_0} \cos(2\pi\mathfrak{N}y/y_0)e^{-j\eta u/y_0}\, dy + \frac{1}{2y_0} \int_{-y_0}^{y_0} e^{-j\eta u/y_0}\, dy$$

which becomes, on integration,

†Since
$$\frac{2}{\pi}\int_0^\infty \sin\omega\cos\omega t\, \frac{d\omega}{\omega} = \begin{cases} 1 \text{ for } |t| < 1 \\ \tfrac{1}{2} \text{ for } |t| = 1 \\ 0 \text{ for } |t| > 1 \end{cases}$$

$$\bar{Y} = \frac{1}{2}\left[\frac{\sin{(2\mathfrak{N}\pi - \eta)}}{(2\mathfrak{N}\pi - \eta)} + \frac{\sin{(2\mathfrak{N}\pi + \eta)}}{(2\mathfrak{N}\pi + \eta)}\right] + \frac{\sin{\eta}}{\eta}$$

$$\bar{Y} = \left[\frac{4\mathfrak{N}^2\pi^2}{4\mathfrak{N}^2\pi^2 - \eta^2}\right]\frac{\sin{\eta}}{\eta}$$

This real quantity yields the illumination $\mathfrak{E}_p$ at a point p of the screen, relative to that in the center of the diffraction pattern $\mathfrak{E}_c$. This illumination, represented as $\bar{Y}\bar{Y}^*$, is (in part) illustrated in Fig. F-5b;

$$\bar{Y}\bar{Y}^* = \left[\frac{4\mathfrak{N}^2\pi^2}{4\mathfrak{N}^2\pi^2 - \eta^2}\right]^2\frac{\sin^2{\eta}}{\eta^2}.$$

This function has three principal peaks. This fact is apparent from consideration of the first integrated expression for $\bar{Y}$. The values where η is equal to zero, or $\pm 2\mathfrak{N}\pi\left(\text{with }\eta = \frac{2\pi y_0\sin{\beta}}{\lambda}\right)$, correspond to the maxima on the observing screen at $\sin{\beta} = 0$, and $\pm\frac{\mathfrak{N}\lambda}{y_0}$, or in the $k = 0$ and $k = \pm1$ orders of the grating spectrum, as described by

$$k\lambda = a\sin{\beta}$$

where a is equal to $y_0/\mathfrak{N}$. If we had taken simply the amplitude function $f(y/y_0) = \cos{2\pi\mathfrak{N}y/y_0}$, then the $\frac{\sin{\eta}}{\eta}$ term in $\bar{Y}$ would be absent. This means that we would have no zeroth order. In a fascinating article on the diffraction theory of microscopic vision A. B. Porter[4] has shown that a coarse grating, when illuminated with monochromatic radiation and observed in a microscope, will appear as a cosine modulated pattern if all of its orders but the $k = 0$ and ±1 are blocked out. Also, if only the $k = \pm2$ orders are not blocked out, it will appear as a similar pattern but with twice the actual number of lines per inch. Porter points out that a "simple grating," *i.e.* a grating producing a cosine function for the amplitude distribution of radiation transmitted at its surface, will show light in only two orders $k = \pm1$, with a microscope under monochromatic illumination, as determined mathematically above. Any complex grating which produces an even function for the amplitude distribution of radiation can be considered as made up of many such "simple gratings" with different spacings of lines. Upon scanning the spectrum produced by such a complex grating, each set of orders ($+$ and $-$) of the monochromatic diffracted light corresponds to the monochromatic spectrum that would be produced by the corresponding "simple grating." In principle the Fourier transform of any function can be obtained with a screen that gives an amplitude distribution function at the aperture which is identical with the function whose Fourier transform is desired. The Fourier transform of this function is then obtained by illuminating the aperture with plane mono-chromatic radiation and scanning the resultant diffraction pattern.[5]

Case II

We have shown that Eqs. 1 and 2 represent the interrelationship of the functions that describe the spatial distribution (perpendicular to the direction of propagation) of the amplitude over the unobstructed wave front and the amplitude distribution in the diffraction pattern that it produces.

Now we discuss the degradation of monochromaticity that is introduced by chopping (cf. Fig. 5-6) an infinitely long wave train into pulses. (See Chap. V.) We shall show that the same Fourier transform equations describe the relation between the amplitude spectrum of such degraded monochroma-ticity and the spatial distribution (now parallel to the direction of propaga-tion) of the amplitude in the pulse.

The spatial profile of amplitude in the pulse, consisting of infinitely many waves of different frequencies ω, traveling together superimposed in the z direction, can be expressed, at the instant $t = 0$, as

$$\tilde{f}\left(\frac{z}{c}\right) = \int_{-\infty}^{\infty} \mathcal{E}(\omega) e^{j\omega \frac{z}{c}} \, d\omega.$$

Here $\mathcal{E}(\omega)$ will be the spectral distribution of amplitudes to be calculated for a given pulse. For this calculation, using our transform relationships, we take the following for our variables and functions:

$$u = \frac{z}{c} \quad \text{and} \quad v = \omega, \quad \text{with} \quad uv = \frac{\omega z}{c}$$

$$g(v) = \sqrt{2\pi} \mathcal{E}(\omega)$$

The second of our reciprocal relationships gives

$$\mathcal{E}(\omega) = \frac{1}{2\pi} \int_{-\infty}^{\infty} \tilde{f}\left(\frac{z}{c}\right) e^{-j\omega \frac{z}{c}} \frac{dz}{c} \tag{5}$$

We may use this equation to determine the desired spectrum for a pulse chopped from a wave train of infinite length, of frequency ω_0. We take the field after chopping to be zero, at $t = 0$, for $|z| > l$; and we represent it by the equation $\mathcal{E} = \mathcal{E}_0 \cos \frac{\omega_0 z}{c}$ for $-l \leq z \leq l$. $\mathcal{E}_0$ is the amplitude of the chopped plane wave, traveling in the z direction. If the pulse length is taken to be $2l$, its spatial profile is given by

$$f\left(\frac{z}{c}\right) = \mathcal{E}_0 \cos \frac{\omega_0 z}{c} = R_l \mathcal{E}_0 e^{j \frac{\omega_0 z}{c}}$$

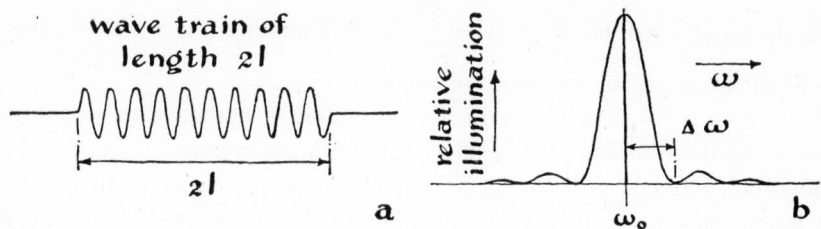

FIG. F-6 (a) Profile of wave train of length $2l$. (b) Spectral distribution of the wave train.

---

over the interval of length $2l$, and $f\left(\dfrac{z}{c}\right)$ is zero elsewhere. It was pointed out in Chap. V that such a pulse as this, shown at the left in Fig. F-6a, when analyzed by a spectrometer, contains frequencies other than ω_0—*i.e.* it contains side bands, such as appear at the right in Fig. F-6b. We could have a spectrum characterized by only one frequency, represented by the vertical line (b) at ω_0, only if the pulse length were infinite. The Fourier transform of the pulse at (a), as described by Eq. 5, has the following spectral amplitude distribution:

$$\mathcal{E}(\omega) = \frac{1}{2\pi c} \int_{-\infty}^{\infty} \mathcal{E}_0 e^{j(\omega_0 - \omega)\frac{z}{c}}\, dz = \frac{\mathcal{E}_0}{2\pi c} \int_{-l}^{l} e^{j(\omega_0 - \omega)\frac{z}{c}}\, dz$$

$$= \frac{\mathcal{E}_0}{2\pi c} \left[\frac{e^{j(\omega_0 - \omega)\frac{l}{c}} - e^{-j(\omega_0 - \omega)\frac{l}{c}}}{j(\omega_0 - \omega)\dfrac{1}{c}} \right]$$

or

$$\mathcal{E}(\omega) = \frac{\mathcal{E}_0 l}{\pi c} \left[\frac{\sin (\omega_0 - \omega)\dfrac{l}{c}}{(\omega_0 - \omega)\dfrac{l}{c}} \right],$$

and the illumination spectrum, as revealed by a spectrometer of sufficient resolving power, would be

$$\mathfrak{E}(\omega) = \frac{1}{2}\, c\kappa_0 \mathcal{E}^2(\omega) = \frac{\mathcal{E}_0^2 l^2 \kappa_0}{2\pi^2 c} \left[\frac{\sin (\omega_0 - \omega)\dfrac{l}{c}}{(\omega_0 - \omega)\dfrac{l}{c}} \right]^2$$

As Fig. F-6b shows, the first minimum on either side of ω_0 occurs when $(\omega_0 - \omega)\dfrac{l}{c} = \pi$ or $\Delta\omega = \dfrac{\pi c}{l}$. If we call this $\Delta\omega$ the band width of the pulse, then the band width becomes $\Delta\nu = \dfrac{1}{2l}$, in wave numbers, where $2l$ is a measure of the longitudinal coherence of the pulse.

It is not surprising that this spectral illumination has the same functional form as the distribution of illumination in a diffraction pattern, for the longitudinal coherence ($\mathcal{E} = 0$ for $|z| > l$) is functionally similar in this case to the lateral coherence ($\mathcal{E} = 0$ for $|y| > y_0$) of our former problem, and the same Fourier transform relations apply. And further, if the successive amplitudes of our pulse were enveloped by a function like an error function, say, we would expect to get a spectral function similar to the diffraction pattern function such as would be produced by an aperture apodized after the same error function.

Case III

We shall now treat[6] the case of interference between two coherent beams, one with the spectral amplitude distribution $\mathcal{E}(\omega)$ and the other with the distribution $\rho\mathcal{E}(\omega)$. We shall relate the superposition of these two beams with the path difference δ by which they are separated in depth. We describe the spatial profile of the first wave train by $\tilde{f}_1\left(\dfrac{z}{c}\right)$, and that of the second coherent wave train, in the same coordinate system, by $\tilde{f}_2\left(\dfrac{z}{c}\right)$, and write the two profiles as follows:

$$\tilde{f}_1\left(\frac{z}{c}\right) = \int_{-\infty}^{\infty} \mathcal{E}(\omega)e^{j\omega\frac{z}{c}}\, d\omega \qquad \text{and} \qquad \tilde{f}_2\left(\frac{z}{c}\right) = \int_{-\infty}^{\infty} \rho\mathcal{E}(\omega)e^{j\omega\left(\frac{z-\delta}{c}\right)}\, d\omega$$

By the principle of superposition the resultant profile is

$$\tilde{F}\left(\frac{z}{c}\right) = \tilde{f}_1\left(\frac{z}{c}\right) + \tilde{f}_2\left(\frac{z}{c}\right)$$

From Eq. 5 the Fourier transform of $\tilde{F}\left(\dfrac{z}{c}\right)$ gives the spectrum capable of producing such a profile; namely,

$$\tilde{E}_R(\omega) = \frac{1}{2\pi}\int_{-\infty}^{\infty} \tilde{F}\left(\frac{z}{c}\right)e^{-j\omega\frac{z}{c}}\frac{dz}{c} = \frac{1}{2\pi}\int_{-\infty}^{\infty} \tilde{f}_1\left(\frac{z}{c}\right)e^{-j\frac{\omega z}{c}}\frac{dz}{c}$$

$$+ \frac{1}{2\pi}\int_{-\infty}^{\infty} \tilde{f}_2\left(\frac{z}{c}\right)e^{-j\frac{\omega z}{c}}\frac{dz}{c}$$

If we take ρ and δ as constants with respect to ω, we have

$$\tilde{f}_2\left(\frac{z}{c}\right) = \rho e^{-j\frac{\omega\delta}{c}}\int_{-\infty}^{\infty} \mathcal{E}(\omega)e^{j\frac{\omega z}{c}}\, d\omega = \rho e^{-j\frac{\omega\delta}{c}}\tilde{f}_1\left(\frac{z}{c}\right)$$

and

$$\tilde{E}_R(\omega) = \mathcal{E}(\omega)\left[1 + \rho e^{-j\frac{\omega\delta}{c}}\right]$$

For a special case as in the Michelson interferometer when $\rho = 1$, with equal beams, we have the resultant illumination, for monochromatic irradiation,

$$\mathfrak{E}(\omega) = c\kappa_0\mathcal{E}^2(\omega)\left[1 + \cos\frac{\omega\delta}{c}\right] \tag{6}$$

This expression describes the illumination as a function of the path difference in two-beam interferometry, such as would be realized with a thin, dielectric plate if the plate thickness were variable. And this expression is the equivalent of Eq. 8-3d as applied in § 11-1. And below, to anticipate, we shall apply an extension of this treatment to the calculation of the superposition illumination for two coherent beams each characterized by the presence of many different frequencies.

We can determine the spectrum of the equal, interfering beams from the illumination observed in interference, as a function of δ. From Eq. 6, the total illumination, as a function of δ, is

$$\mathfrak{E}_T = c\kappa_0\int_0^\infty \mathcal{E}^2(\omega)\,d\omega + c\kappa_0\int_0^\infty \mathcal{E}^2(\omega)\cos\frac{\omega\delta}{c}\,d\omega = E(\delta)$$

The variation of total illumination, as δ is varied, is illustrated in Fig. F-7, where we assume that the interferometer transmission is ideal for all wavelengths.

The value of $\mathfrak{E}(\delta)$, at $\delta = 0$, is

FIG. F-7   Variation of total illumination as a function of the path difference δ between two interfering beams of non-monochromatic radiation.

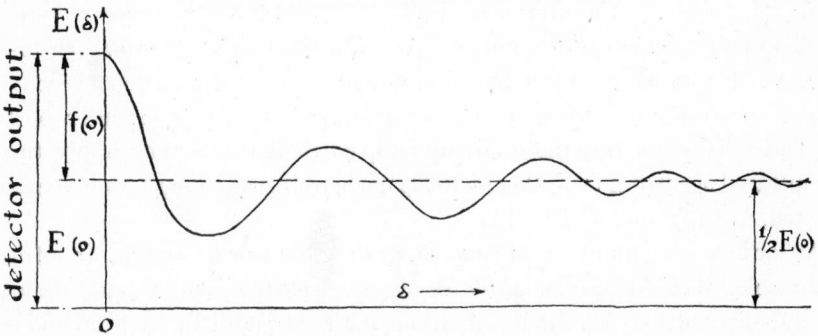

$$\mathfrak{E}(0) = 2c\kappa_0 \int_0^\infty \mathcal{E}^2(\omega)\, d\omega$$

and for a sufficiently large δ, with a non-monochromatic spectrum,

$$\mathfrak{E}(\delta) = \frac{1}{2}\,\mathfrak{E}(0) = c\kappa_0 \int_0^\infty \mathcal{E}^2(\omega)\, d\omega$$

Thus

$$\mathfrak{E}(\delta) - \frac{1}{2}\,\mathfrak{E}(0) = f(\delta) = c\kappa_0 \int_0^\infty \mathcal{E}^2(\omega) \cos\frac{\omega\delta}{c}\, d\omega \tag{7}$$

This function $f(\delta)$, taken alone, appears to represent unrealistic negative illuminations. Nevertheless, it is a convenient function for us here because it is easily determined from a record of illumination versus δ (which we call here an *interferogram*), and because it will yield the complex spectrum that produced it when we apply Fourier transformation.

Since $\mathcal{E}^2(\omega)$ in Eq. 7 is an even function as well as $f(\delta)$, we shall rewrite Eqs. 1 and 2 as follows:

$$g(v) = \sqrt{\frac{2}{\pi}} \int_0^\infty f(u) \cos uv\, du \tag{1a}$$

$$f(u) = \sqrt{\frac{2}{\pi}} \int_0^\infty g(v) \cos uv\, dv \tag{1b}$$

We use the second even function $f(u)$ to describe the character of the measured illumination. Here $u = \delta$. From Eq. 1a, where $v = \dfrac{\omega}{c}$,

$$g(v) = \sqrt{\frac{\pi}{2}}\, c^2\kappa_0\mathcal{E}^2(\omega) = \sqrt{\frac{2}{\pi}} \int_0^\infty f(\delta) \cos\frac{\omega\delta}{c}\, d\delta$$

or

$$\mathfrak{E}(\omega) = \frac{1}{2}\, c\kappa_0\mathcal{E}^2(\omega) = \frac{1}{\pi c} \int_0^\infty f(\delta) \cos\frac{\omega\delta}{c}\, d\delta \tag{8}$$

The energy distribution $\mathfrak{E}(\omega)$ can be determined from Eq. 8 when $f(\delta)$ is known for a two-beam interferogram.

The determination of a complex spectrum $\mathfrak{E}(\omega)$ by interferometry (using $f(\delta)$, and its transformation) has advantages over determination by conventional spectroscopy (using prisms or gratings). In the remainder of this appendix we point out these advantages, particularly as they apply to the far infrared spectrum ($\lambda = 25\mu$ to 1000μ), where the resolving power is energy limited, as explained in § 12-10.

An ordinary scanning spectrometer, grating, or prism for the far infrared embraces a band of wavelengths $\Delta\lambda$ at wavelength λ (which band depends on the slit widths). The determination of the strength of each component requires a time τ. If the scanning range of the instrument covers $\mathfrak{N}$ slit widths $\Delta\lambda$, then the scanning time for the complete spectrum takes $\mathfrak{N}\tau$ seconds. By contrast, in interferometry,[7] all components fall on the detector at greater or less strength for the total time $\mathfrak{N}\tau$. In this case we have, essentially, $\mathfrak{N}$ times as many measurements of each component. As pointed out in Appendix I, the noise (information-wise) should be lower by the factor $\dfrac{1}{\sqrt{\mathfrak{N}}}$. However, as component intensities are modulated at different frequencies, for identification, the average signal from each component is halved by this modulation, and the gain in signal-to-noise by interferometric modulation over conventional spectroscopy is $\frac{1}{2}\sqrt{\mathfrak{N}}$.

We shall now show how interferometric modulation is practiced.

Fig. F-8 shows a two-beam interferometer.[8] It is a lamellar grating (see

---

FIG. F-8   Two-beam interferometer designed by Strong and McCubbin.

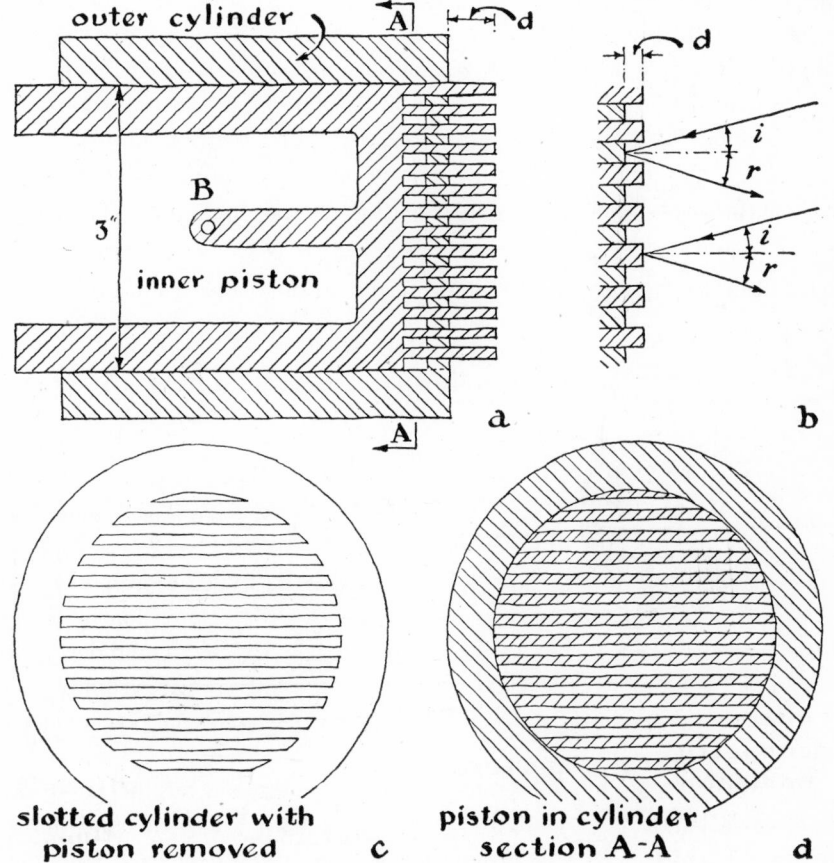

outer cylinder

A

d

B

3″

inner piston

d

i

r

i

r

A

a

b

slotted cylinder with
piston removed

c

piston in cylinder
section A-A

d

Fig. F-8b) used at approximately normal incidence, i. If the groove depth is d, the phase difference is $4\pi \dfrac{d}{\lambda} \cos i$ between rays, for the grating's central order, $k = 0$. The illumination in the various orders, after the results of Problem 10-4, is

$$\mathfrak{E} = \cos^2\left(\frac{k\pi}{2} + \frac{2\pi d}{\lambda}\right) \qquad \text{or} \qquad \cos^2\frac{2\pi d}{\lambda} \quad \text{for} \quad k = 0.$$

In terms of the optical path difference δ, the illumination in the $k = 0$ order is

$$\mathfrak{E} = \frac{1}{2}\left(1 + \cos\frac{\omega\delta}{c}\right) \tag{6'}$$

We see that Eq. 6′ is equivalent to Eq. 6. Thus everything that we said of Eq. 6 applies here. The interferogram $f(\delta)$ is obtained when δ, in Eq. 6′, is varied uniformly by means of a micrometer screw. This screw drives the piston to which the odd grating facets are attached (see Fig. F-9). Thus the energy entering this device and passing through the exit slit is modulated. This modulated energy is focused on a Golay radiation detector; and its output, as δ varies linearly with time, is written out by a pen recorder. Fig. F-10a,[9] for example, is the interferogram obtained when a globar is the black-body source, and the average wavelength, isolated by two reflections from CsBr, is about 120μ. Fig. F-10b shows its calculated Fourier transform. This transform is the complex spectrum of these source radiations after they have penetrated water vapor in the optical path.

FIG. F-9   Ebert mounting of the interferometric modulator shown in Fig. F-8.

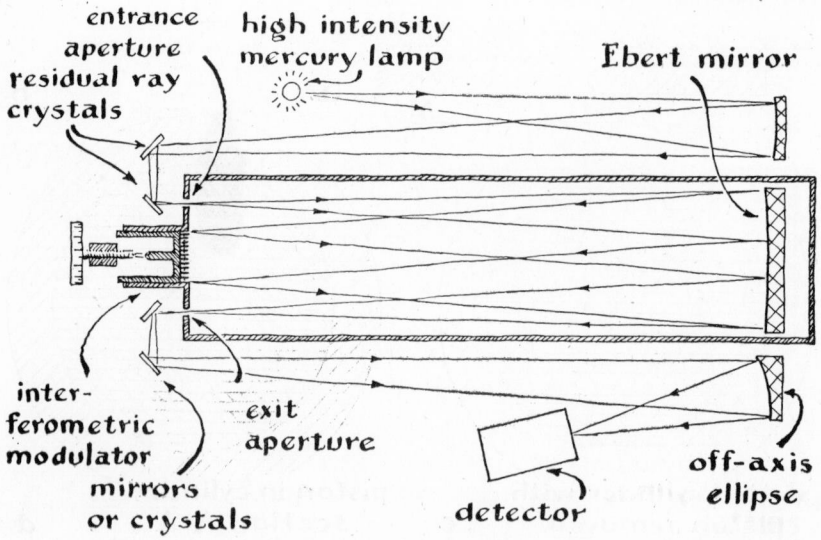

The resulting spectrum shows the absorption lines due to residual water vapor absorption. Even though the entire optical system was flushed with dry nitrogen while the interferogram was being recorded, there was enough residual vapor to show the water absorption lines.

The spectral resolving power can be calculated by using the Eq. $\Delta \nu = \dfrac{1}{2l}$, derived in Case II. Here, for the modulator,

---

FIG. F-10   (a) Interferogram obtained with the experimental set-up of Fig. F-9. (b) Spectral distribution of the radiation obtained by taking the Fourier transform of the interferogram.

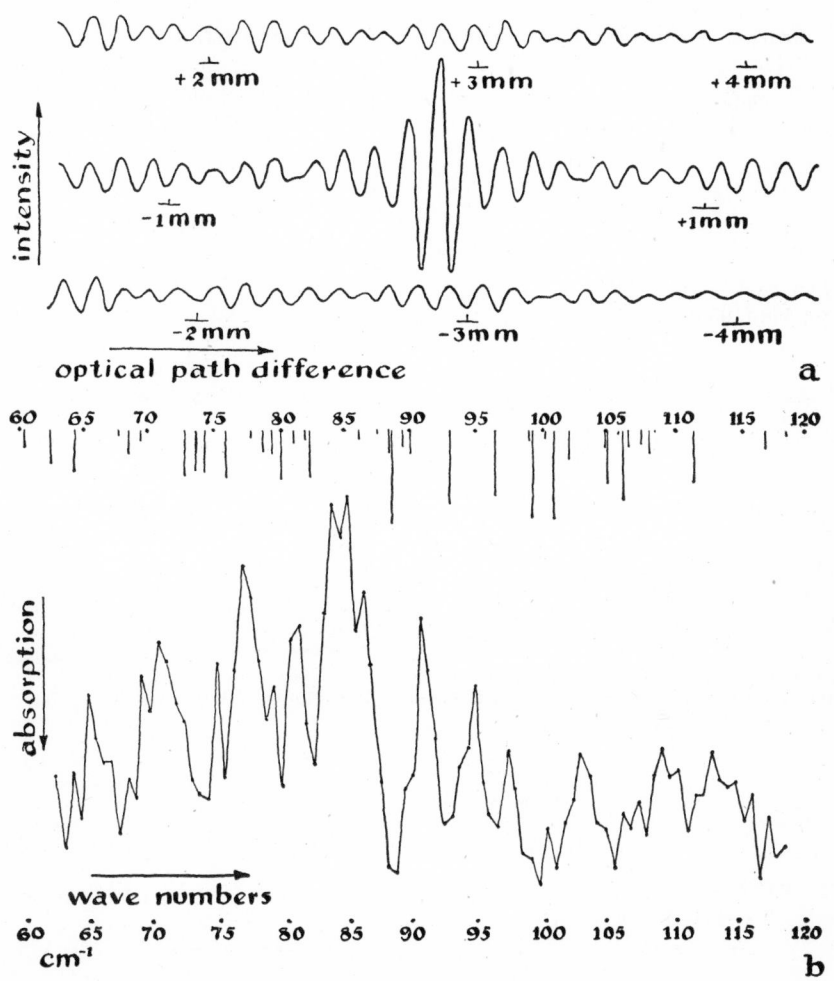

$$\Delta\nu = \frac{1}{4D} \qquad (9)$$

when the groove depth is varied from $0 \rightarrow D$ cm.†

The slit widths in the Ebert mounting of the modulator shown in Fig. F-9 were $\frac{1}{4}$ in. The only condition imposed on them is that they be so narrow that the first order spectrum of the shortest wavelength that is to be studied shall not pass through the exit slit onto the detector, for Eq. 6' is valid only for the zeroth grating order.

REFERENCES

1. Rubens & Wood, Phil. Mag., *21*, 249 (1911).

2. J. Strong, *Procedures in Experimental Physics* (1938, Prentice-Hall).

3. Rubens & Von Baeyer, Phil. Mag., *21*, 689 (1911).

4. A. B. Porter, Phil. Mag., *11*, 154 (1906).

5. Lawrence Mertz, J. Opt. Soc. Am., *46*, 548 (1956).

6. J. J. Hunzinger, Rev. d'Optique, *6* (1954).

7. P. Fellgett, in seminar given at Johns Hopkins in 1954.

8. J. Strong, J. Opt. Soc. Am., *47*, 354–357 (1957).

9. Gebbie & Vanasse, Nature, *178*, 432 (1956).

BIBLIOGRAPHY

R. W. Ditchburn, *Light* (1954, Blackie).

Jean Lecomte, *Le Spectre infrarouge* (1928, Les Presses Universitaires de France).

Louis A. Pipes, *Applied Mathematics for Engineers and Physicists* (1946, McGraw-Hill).

R. W. Wood, *Physical Optics*, 3rd ed. (1936, Macmillan, New York).

---

† The modulator that produced the interferogram of Fig. F-10a was driven a distance D equal to 4 mm, which gives a theoretical resolution of approximately $\frac{1}{2}$ cm^{-1}.

Appendix **G**

Some Modern Concepts of Light

by L. Witten†

The Ether

The main text of this book, in dealing with classical concepts, largely ignores two cardinal discoveries of theoretical physics made during the twentieth century. These are the theory of relativity and the quantum theory of particles and fields. The text refers only briefly to the old and historically important classical concept of the ether, with which many scientists were occupied at the turn of the century and until 1905. In this year Einstein introduced the special theory of relativity and showed the essential meaninglessness of the ether. Since the time of Fresnel, in order to explain polarization and interference, light has been generally conceived as a transverse wave motion; the medium which partook of the wave motion was called the ether and was believed to exist everywhere because it was considered inconceivable to have a wave motion without an underlying material medium which supported the wave and through which it propagated. It was said with some seriousness that the light waves were undulations and that the ether must exist in order to be the subject of the verb "to undulate."

All other wave motions known at that time were motions of parts of a material medium. Water waves traveling along the surface of a body of water provide us with an example of a traverse wave propagating through a medium. The water particles along the surface move up and down but remain essentially stationary as far as the wave propagation along the surface is concerned. Tne up and down motion of the water particles induces a similar up and down motion of the adjacent particles, and by the repetition of this process the

---

† RIAS, Inc., 7212 Bellona Avenue, Baltimore 12, Maryland.

wave disturbance propagates along the surface of the water. There is thus a horizontal propagation of an immaterial phenomenon taking place by virtue of a succession of vertical motions of material objects. The material objects permitting the light disturbance to propagate were conceived in an analogous way and were called the ether.

This invented material, the ether, must be endowed with rather remarkable properties to serve its purpose. The ether must pervade all space and permeate all matter because electromagnetic influences travel everywhere. The ether must be very thin and tenuous because the planets move through space with no apparent diminution in speed. If these waves are to be construed as transverse and not longitudinal elastic waves, the ether must possess a very high rigidity. These properties should be separately detectable and demonstrable by appropriate experimental means. The finest physical measurements and the most ingenious schemes have been used in an attempt to demonstrate the existence of the ether and to further reveal its character. All of them have failed.

Perhaps the most famous experiment along these lines was conducted by Michelson and Morley to detect the motion of the earth through the ether— the ether being considered at absolute rest as the earth rotates and travels around the sun. With a very delicate and clever optical experiment using Michelson's interferometer, Michelson and Morley attempted to measure the earth's rotation and its velocity around the sun with respect to the ether. The measurement failed to reveal any relative velocity—it appeared to be zero at all times of the day and of the year. As the ether thus resisted attempts to be experimentally detected, its properties became more and more mysterious; and the explanations of the failure to detect it became more and more cumbersome. Finally, and with great intellectual boldness, Einstein formulated his special theory of relativity, wherein he advanced the view that there is no necessity for an ether as a carrier of light. Einstein's views required the velocity of light to be identical in all directions for all uniformly moving frames of reference. At the surface of the earth (a planet moving through space) the velocity of light was experimentally found to be equally great in all directions in conformity with Einstein's doctrine. As a consequence of this doctrine, it is predicted that the observation of the ether by optical experiments is impossible. According to the modern view, in which concepts which in principle are not subject to experimental verification are scientifically meaningless, it becomes unnecessary and useless to discuss the ether and its properties.

Einstein's special theory of relativity, of course, has a lot more to say about physics and about the theory of knowledge, but we shall not pursue these matters here, except to say that the predictions of the theory are all in conformity with experimental observations. Also, Maxwell's equations continue to remain valid in the special theory; thus classical optics has been very little changed by the theory.

Light as Waves; Light as Particles

The wave-nature of light affords explanations of the main topics of physical optics: namely, reflection and refraction; velocity and dispersion; polarization and crystals; and, finally, interference and diffraction. "Light produces many effects in its interaction with matter. It ejects photoelectrons; it induces the property of electrical conduction in photoconductors; it produces fluorescence and phosphorescence in crystals . . ." (See § 4-10). The effects mentioned in the quotation, and not elaborated in the text, were discovered in the late nineteenth and the early twentieth century, and serve, in part, as the experimental basis of the quantum theory of light and of matter. These phenomena give convincing evidence of the particle nature of light, a concept which had seemed to perish in 1850 when Foucault showed that the velocity is greater in air than in water. The twentieth century thus not only raised the very old question of the nature of light again, but, as we shall see, answered it in an unexpected manner.

The photoelectric effect, briefly treated below, clearly requires for its interpretation that we consider light rays to behave like particles. In 1887 Heinrich Hertz succeeded in producing electromagnetic waves by means of an oscillatory current and thus proved convincingly the correctness of Maxwell's theory of light as electromagnetic waves. In these very experiments, which confirmed the existence of Maxwell's waves by producing them electromagnetically, Hertz noticed a curious effect that was to lead to the resurrection of the particle concept of light. The curious effect he noticed was that when ultraviolet light fell upon an air gap the air in the gap appeared to become a better conductor of electricity. This observation was the beginning of a long series of investigations of the photoelectric effect. The effect is quite simply summarized: if light of sufficiently short wavelength falls on matter (gas, liquid, or solid), the light knocks electrons out of the matter.

A typical set-up for studying this photoelectric effect is shown in Fig. G-1. When light of short wavelength, say ultraviolet light, falls upon a cathode, K, through the quartz window, W, emission of electrons from K is induced, and these electrons are gathered by the plate, P, thus completing the circuit.

Careful investigations have established the relation between the properties of light and the number and velocity of the electrons emitted. If the intensity of the light falling on a cathode is increased, the current flowing through the circuit is increased. The increase in current is due to the emission of a greater number of electrons from the surface, the electrons having the same velocity

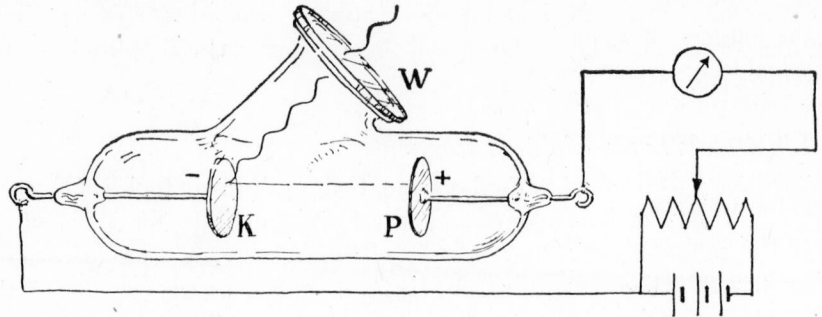

FIG. G-1 Typical set-up for studying the photoelectric effect.

distribution as before. Increasing the light intensity does not increase the electron velocity at all. In order to increase the electron velocity, the frequency of the incident light must be increased. The range of photoelectron velocities emitted from the cathode extends from zero up to a maximum velocity, v_{max}. As the frequency increases, the maximum electron velocity increases. As the frequency decreases, the maximum electron velocity decreases, until finally, at some frequency v_0, zero velocity is reached. Below v_0, a frequency which is characteristic for each particular surface, no photoelectric emission of electrons is observed.

To be quantitative, the maximum kinetic energy of the ejected electrons is proportional to the expression $(v - v_0)$, where v is the incident light frequency. Thus

$$\tfrac{1}{2}mv_{max}^2 = h(v - v_0) \tag{1}$$

In this expression m represents the mass of the electron and h is the proportionality constant, called Planck's constant. Accurate experiments show that this expression is extremely reliable. Further, the proportionality constant is the same as the constant introduced in Max Planck's equation for black body radiation.

These facts are all quite incomprehensible from the point of view of classical wave theory. Why should there be such a close connection between the frequency of the wave and the kinetic energy of the electrons? When the photoelectron ejection experiment is performed with very small particles, we should expect, according to classical wave theory, that an electron would not be emitted instantaneously, but only after the particle had absorbed energy from the incident wave train equal in amount to the electron ejection energy— the energy necessary to remove the electron from the metal plus the kinetic energy it acquires. This time can be appreciable if the particles are sufficiently minute. However, observation shows that this expectation is not realized; for the most minute particles electrons are occasionally emitted practically the instant of exposure to light.

In 1905 Einstein asserted that all of this peculiar photoelectric behavior of electrons could be explained by considering light itself as a shower of particles, each particle having an energy $h\nu$. Today we call these particles photons. A photon which strikes a metal surface can transfer its energy completely to one of the electrons in the surface. If the incident photon frequency is ν_0, the photon has the energy $h\nu_0$, which is just sufficient to remove the electron from the metal with zero velocity, as predicted by Eq. 1. And if the frequency is increased to ν, the energy which can appear as kinetic energy of the electron is $h(\nu - \nu_0)$.

Suppose now that the intensity of incident light is increased. This only means that there are more photons in the beam than previously, but the energy of each photon in the beam is still $h\nu$. Consequently, a large number of photons can eject a larger number of electrons than previously, but the maximum energy or velocity given to such electrons cannot be increased unless $h\nu$ is increased.

In 1905 the classical wave theory of light was on so strong a basis, both experimental and theoretical, that physicists were reluctant to accept Einstein's interpretation. Soon, however, a number of experimental facts accumulated which were immediately understandable on the basis of Einstein's picture, and not at all otherwise. These facts were concerned with the conversion of light into other forms of energy, and with the converse production of light. The new experimental facts culminated in the so-called Compton effect. In 1923 A. H. Compton observed the results of collisions between a photon and an electron. He could interpret his observations very neatly by assuming the photon to be a particle with a definite energy and momentum. With such a particle, and assuming that his collisions were, as regards interchange of energy and momentum, the same as elastic collisions between any two particles, he could readily explain the results of his experiments.

Thus, in the early twenties, the old argument as to the wave or corpuscular nature of light was again in the center of the scientific arena. Each side could bring forth a large and convincing body of supporting interpreted experimental data. But, very remarkably, by the late twenties the physicists had made such gigantic strides forward that they were able to resolve the conflict in a curious and delightful way, as we shall see.

Electrons as Both Particles and Waves

It is appropriate, when considering the particle nature of light, to consider also the wave nature of the particles of electricity, called electrons. The electron, discovered by J. J. Thomson in the late nineties, appeared to have all of the ordinary properties of a particle; it had a mass and an electric charge; it was localized in space; and it obeyed Newton's laws of mechanics for particles. Until the early twenties, electrons were never considered in any way other than as particles. But suddenly the picture changed. First, a theoretical physicist, Louis de Broglie, suggested that the wave and corpuscle concepts should apply simultaneously to electrons as they were already found to apply simultaneously to light. He supported this suggestion by considerations derived from the special theory of relativity and from the century-old work of Hamilton in mechanics. De Broglie's suggestion was, of course, later confirmed experimentally. According to it, every particle (such as an electron, atom, or molecule) has an associated wavelength, λ, which is expressed by

$$\lambda = \frac{h}{mv} = \frac{h}{p} \tag{2}$$

Here, again, h is Planck's constant, m is the mass of the particle, and v is the velocity associated with its momentum, p. This λ is called the de Broglie wavelength of the particle.

If an electron of charge e is accelerated through a difference of potential V, its kinetic energy is

$$\tfrac{1}{2}mv^2 = eV \tag{3}$$

and its de Broglie wavelength is

$$\lambda = \frac{h}{\sqrt{2mVe}} \tag{4}$$

For a potential difference of 100 volts the de Broglie wavelength is $\lambda = 1.22$ Å. This wavelength is of the order of magnitude of the wavelength of x-rays. These wavelengths of x-rays are measured by using crystals as diffraction gratings; the distances between atomic planes in crystals are of the same order of magnitude. Electrons of the same de Broglie wavelength as x-rays, on striking a crystal lattice, are diffracted in much the same pattern as the x-rays.

Further investigations have established similar wave-like properties for all types of particles, electrons, protons, ions, atoms, and molecules, thus confirming de Broglie's suggestion.[1]

The wave-particle situation for both electrons and light can be dramatically discussed by way of an experiment illustrated by Fig. G-2. Coherent light from a source S passed and diffracted by two slits finally forms the familiar diffraction-interference pattern on a photosensitive plate, P. Here photoelectrons are ejected and the number of ejected electrons from any point off P is proportional to the strength of the light falling on that point, more electrons being ejected in the regions of intensity peaks than in those of intensity minima, and the maximum velocity of the ejected electrons is determined by ν_0. Here we have, in a single experiment, light behaving both as a wave undergoing diffraction and interference and as particles, ejecting photoelectrons. Of particular importance in this experiment is that the interference pattern persists even if the light intensity is reduced to the level where only one photon at a time falls upon the screen—provided the photoelectrons are counted over a long enough period of time.

The analogue of this experiment could have been performed with electrons. In this instance the diffraction slits would be replaced by a crystal and the photosensitive plate replaced by a particle detector such as a Wilson cloud chamber. Again, in such an experiment, electrons behave both as waves and as particles.

The Uncertainty Principle

"The concept of light rays," to quote Paul Drude again, "is . . . introduced for convenience. It is altogether impossible to isolate a single ray and prove its physical existence. For [owing to diffraction] the more one tries to attain this end by narrowing with restricting apertures, the less does light proceed in straight lines, and the more does the concept of light rays lose its physical significance." (See § 13-1.)

Today an exactly parallel statement can be made about electrons. We can paraphrase the quotation and say, "The concept of electron rays is introduced for convenience. It is altogether impossible to isolate a single (infinitesimally narrow) ray and prove its physical existence. For (owing to diffraction) the more one tries to attain this end by narrowing with restricting apertures, the less do electrons proceed in straight lines, and the more does the concept of electron rays lose its physical significance."

The interpretations of the wave-particle duality of nature followed the theory of quantum mechanics introduced during the middle 1920s, and by

FIG. G-2 Experimental arrangement to demonstrate the wave-particle nature of light.

the end of that decade the interpretations had completely resolved the wave particle paradoxes. Although the exposition of quantum mechanics is altogether beyond our present scope, we shall give some of its results below.

The concepts of waves and of particles arise from our experiences with large-scale phenomena. When we carry over these concepts to small-scale phenomena—on the scale of atomic dimensions—we must realize that we are extrapolating experiences to a different realm, and they may be expected, at most, to serve only as guides. Our classical concepts of particle velocity and energy, developed by observing the mechanical behavior of large, macroscopic bodies, do not have universal applicability in experiments on the motion of microscopic particles such as electrons; and it is often more convenient to apply undulatory concepts to such particles.

To understand these limitations and restrictions on the dynamics of atomic and sub-atomic particles, we must first recognize that the precise position and velocity of electrons are not simultaneously known. Whereas in ordinary mechanics both the position and the velocity of microscopic bodies are assumed to be precisely observable, for atomic and sub-atomic particles this assumption, as we shall see, cannot be made.

Consider the behavior of a system of, say, electrons. To predict future behavior we must know the initial position and momentum of each electron. For a particular electron let the uncertainty in the measurement of the x component of position be Δx, and let the uncertainty in the measurement of the corresponding component of momentum be Δp. In 1927 Werner Heisenberg first pointed out a relationship between Δx and Δp. This relationship is based on the principle that it is impossible to measure precisely and simultaneously both the position and the momentum of a particle. This principle is called Heisenberg's uncertainty principle. The uncertainty, Δx, involved in the measurement of the coordinate of the particle and the uncertainty, Δp_x, involved in the simultaneous measurement of its momentum are governed by the relationship

$$\Delta x \cdot \Delta v \geq \frac{h}{m} \tag{5}$$

Here, again, h is Planck's constant. The value of h has been experimentally determined and is very small by ordinary standards, $h = 6.6 \times 10^{-27}$ erg sec. The uncertainty relationship holds in principle for ordinary large objects, also, but when the mass is large, as it is in macroscopic experiments, $\dfrac{h}{m}$, for all practical purposes, is vanishingly small.

Illustrations of the Uncertainty Principle

The uncertainty principle originates in the interaction between the atom or electron being observed and the mechanism used to observe it. Ordinarily a macroscopic object is observed by means of a light ray, and the object is so massive that we overlook the possibility that the light ray itself will appreciably affect the position or momentum of the object. However, in observing an electron, the action of the light ray on the electron can be demonstrated to produce changes of both position and momentum.

Suppose that the position of the electron being observed is to be determined by an imaginary microscope of very high resolving power given by (see Fig. G-3)

$$\Delta x \sim \frac{\lambda}{\sin \alpha} \tag{6}$$

where Δx is the distance between two points which can just be resolved by the microscope. Here λ is the wavelength of the light used to see the electron; and $\sin \alpha$ is the numerical aperture of the microscope objective, assumed to be filled by a cone of light coming from the illuminated electron. Δx becomes the uncertainty in the measurement of the x component of the position of the electron. Applying the principles of microscopy, to make Δx as small as possible, we must use light of very small wavelength. The minimum amount of light that could be used for such a determination would, of course, be a single photon having a momentum given by the de Broglie relationship of

$$p = \frac{h}{\lambda} \tag{7}$$

When the electron scatters the photon into the microscope, the electron itself will receive some momentum from the photon by particle-particle collision. Since the scattered photon can enter the microscope anywhere within the

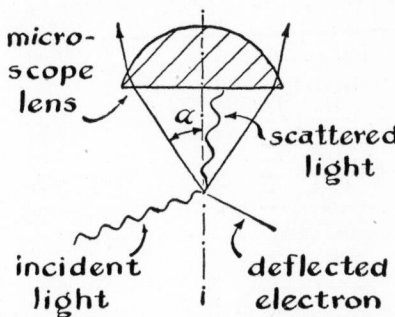

FIG. G-3 The uncertainty principle: a microscope is used to measure the precise position of an electron.

micro-scope lens

α

scattered light

incident light

deflected electron

angle α, the x component of the momentum of the electron will be unknown. The amount by which this angle of scattering is unknown yields

$$\Delta p_x \sim p \sin \alpha = \frac{h}{\lambda} \sin \alpha \tag{8}$$

Combining Eqs. 7 and 8, the product of the position and momentum uncertainties in this microscope observation of the electron is that given by Heisenberg's uncertainty principle:

$$\Delta x \cdot \Delta p_x \sim \frac{\lambda}{\sin \alpha} \cdot \frac{h}{\lambda} \sin \alpha = h \tag{9}$$

Objections might be raised to this conclusion. For example, the indeterminateness of the momentum, which is due to the uncertainty of the direction of the scattered photon, might be determined by measuring the recoil which the microscope receives from the photon. But the microscope itself cannot be observed within closer limits than those imposed by the uncertainty principle. Such a cascading of uncertainties foils the attempt to circumvent the implication of the analysis.

In the foregoing discussion we invoked the simultaneous use of the wave and corpuscular concepts of light. The resolving power of our microscope was given by an expression based on the wave nature of light; the momentum uncertainty was estimated on the basis of the photon or corpuscular nature of light.

Fig. G-4 illustrates another application of the uncertainty principle. Consider a parallel beam of electrons made to pass through a slit of width d. After passing through the slit the y component of the coordinate will be known with an accuracy approximately $\Delta y = d$. And, in passing through the

---

FIG. G-4 The uncertainty principle: a slit is used to measure the y component of momentum of the electrons.

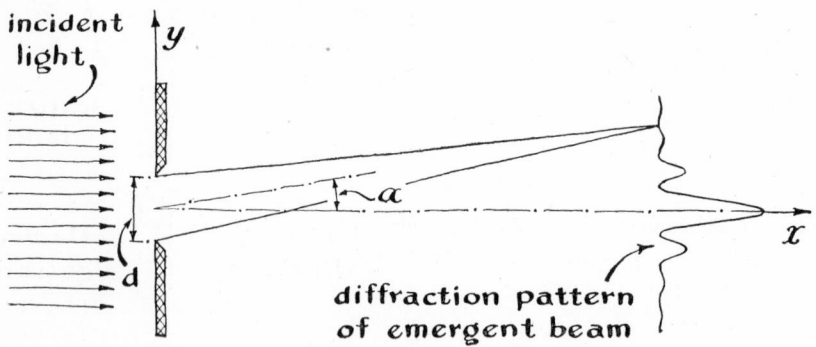

slit, the electrons are diffracted like waves with de Broglie wavelength, $\lambda = h/p_y$, and yield a diffraction pattern of angular width, α:

$$\sin \alpha \sim \lambda/d \tag{10}$$

where λ is the de Broglie wavelength of the electrons. The y component of the momentum of the electron after passing through the slit, from this diffraction pattern, is uncertain by the amount

$$\Delta p = p \sin \alpha = \frac{h}{\lambda} \sin \alpha \tag{11}$$

where p is the momentum of the electron in the direction of the beam and Δp is perpendicular thereto. Multiplying the two uncertainties again yields the product h:

$$\Delta x \cdot \Delta p = d \cdot \frac{h}{\lambda} \sin \alpha = d \cdot \frac{h}{\lambda} \cdot \frac{\lambda}{d} = h \tag{12}$$

The Uncertainty Relations for Waves

Our concepts of waves originally derived from macroscopic experience, just as our concepts of particles did. The concepts of wave amplitude, energy density, wavelength, etc., originated mainly from observations of water waves and from studies of the vibrations of elastic bodies. Just as our mechanical concepts for particles are limited by the uncertainty principle when applied to microscopic particles, so our wave concepts are limited in application by the uncertainty principle. We shall not go very deeply into this subject but merely illustrate the uncertainty principle by one particular case.

What is meant by an exact knowledge of the amplitude of a wave, such as an electric or a magnetic field strength? These quantities are conceived as defined at every mathematical point in space and for every mathematical instant of time. However, every actual physical measurement we can make involves a small but finite region of space of volume $\delta v = (\delta l)^3$, and also a certain definite span of time, $\delta t = \delta l/\mathsf{c}$. Although it may be a legitimate abstraction to allow the finite region of space and finite extension of time to shrink to a point in space and to an instant of time, it is usual not to make this abstraction in the considerations of measurements of field strengths. It is of course obvious that to measure a wave of wavelength λ, our space limitation, δl, must be smaller than λ (spatial volume $\delta v < \lambda^3$); and to observe a wave with period ν^{-1} our time limitation δt must be smaller than $\delta t = \delta l/\mathsf{c} < \nu^{-1}$.

The measurements of electric and magnetic field strengths are thus averages over a small region of space, $\delta v = (\delta l)^3$, and a small interval of time, $\delta t = \delta l/\mathsf{c}$, determined by the detailed method of measurement. Since electric and magnetic fields have, at least sometimes, another characterization as particles (photons), it is reasonable to expect that the uncertainty relations for the fields must be consistent with those for the particles. Thus the uncertainty principle, if applicable, would preclude a simultaneous knowledge within the space-time domain δv, δt of the exact knowledge of all components of the electric field strength, $\overline{E}$, and of all components of the magnetic field strength, $\overline{H}$. If these were exactly known, the energy and momentum of the small volume would be

$$E = \delta v \, \frac{1}{8\pi} \, (\overline{E}^2 + \overline{H}^2); \qquad G = \delta v \, \frac{1}{4\pi c} \, \overline{E} \times \overline{H} \tag{13}$$

Since δv could be made as small as possible, the energy and momentum could both be arbitrarily small (which would not be consistent with the particle theory). Now, from the quantum theory, the energy and momentum content of a small volume is permitted to have only discrete and finite values, $h\nu$ and $h\nu/c$ respectively. Accordingly there must be uncertainty relations involving $\overline{E}$ and $\overline{H}$ which prohibit the exact knowledge of both E and G within the volume δv at the same instant δt.

The uncertainty relation for particles, as was made obvious in the examples above, arises physically from the disturbance of the system by measurement; measuring a position disturbs the momentum of the system, and measuring a momentum disturbs the position. The uncertainty relation for waves has a similar origin and takes on a similar form for the electric and magnetic fields which constitute the wave motion. For such related fields the uncertainty relation is

$$\Delta E_x \Delta H_y \geq \frac{h}{\delta v \delta t} = \frac{hc}{(\delta l)^4} \tag{14}$$

with cyclic permutation for the other components. ΔE_x is the uncertainty in the measurement of the x component of E, ΔH_y is the uncertainty in the y component of H; relation 14 is an example of Heisenberg's uncertainty principle applied to fields. These uncertainty relations refer to a knowledge of E and H in the same volume element at the same interval of time; in principle, in different volume elements E and H may be simultaneously known to any degree of accuracy. The relations refer also to measurements in the same volume element which are simultaneous; a physical measurement of $\overline{E}$ disturbs the measurement of $\overline{H}$ made at the same time, but not if the measurement of $\overline{H}$ is made at some time later. Heisenberg, in a book[2] which discusses clearly the general problem of the physical picture of the quantum theory, gives a description of an actual measurement of the electromagnetic field and how the uncertainty principle applies to the measurement.

Complementarity

The uncertainty principle was first derived by Heisenberg from the mathematical formulation of the quantum theory. In order to understand more clearly the physical implications of this principle, Niels Bohr introduced the complementarity principle.[3] This principle states that the processes of atomic physics can be equally well understood in terms of waves or in terms of particles. Either the wave conception or the particle conception can be used for any atomic phenomenon. A necessary result of this is that atomic phenomena cannot be described with the completeness demanded by classical dynamics, but rather with the limitations set by the uncertainty principle.

The statement that the position of an electron is known to within an accuracy Δx means in the wave picture that a wave packet exists in the proper position with an approximate extension Δx. A "wave packet" is a wavelike disturbance which is appreciably different from zero only within a bounded region. This region is in motion and also changes its size and shape. The velocity of the wave packet is the same as the velocity of the electron but cannot be exactly defined because of the changing size and shape of the packet. This indefiniteness is considered as an essential feature of the electron's behavior; it is not evidence of the inapplicability of the wave picture. If Δv represents the velocity uncertainty of the electron, its momentum uncertainty is $\Delta p = m\Delta v$. We shall make an approximate derivation, from the laws of optics and by use of the de Broglie wavelength, $\lambda = h/p$, of the uncertainty relation $\Delta x \cdot \Delta p \geq h$.

Suppose the wave packet is made up of a superposition of plane sinusoidal waves, all having wavelengths near λ_1. Then there are about $n = \Delta x/\lambda_1$ wave maxima within the packet. Outside the packet, the plane waves must cancel by interference; this is possible only if the packet contains waves of wavelength $(\lambda_1 - \Delta\lambda)$ sufficiently short so that there are at least $n + 1$ maxima in the packet.

$$\frac{\Delta x}{\lambda_1 - \Delta\lambda} \geq n + 1 \tag{15}$$

Consequently

$$\frac{\Delta x \Delta\lambda}{\lambda_1{}^2} \geq 1 \tag{16}$$

The group velocity of a wave packet can be calculated to be

$$v = \frac{h}{m\lambda_1} \tag{17}$$

This is the velocity which de Broglie uses in defining the wavelength for the electron. The spreading of the packet can be represented by the range of velocities,

$$\Delta v = \frac{h}{m\lambda_1^2} \Delta\lambda \tag{18}$$

Since $\Delta p = m\Delta v$, it follows from 16 and 18 that

$$\Delta x \cdot \Delta p \geq \frac{\lambda_1^2}{\Delta\lambda} \cdot m \cdot \frac{h}{m\lambda_1^2} \cdot \Delta\lambda = h \tag{19}$$

This is the uncertainty principle.

The Probability Concept

By using the concept of wave packet it is evident that the particle picture has become indistinguishable from the wave picture. This duality must also extend to waves as well as particles. The illumination of light at a point, according to the wave theory, is proportional to E^2, the square of the amplitude of the electric field vector at the point. From the photon picture, the illumination of the light is determined by the rate, n, at which photons cross a unit area oriented perpendicular to their direction of motion. Consequently $n \sim E^2$. But this relationship can only hold for intense beams; if n is a small number, E^2 can be thought of as being proportional to the probability that a photon will cross a unit area perpendicular to the direction of motion in unit time.

For example, if light passes through a slit, it forms a diffraction pattern on a photographic plate with regions of great illumination alternating with regions of weak illumination. From the statistical point of view, the probability of a photon striking the plate is very great where the intensity is great and very small where the intensity is small. If the beam is very weak, so that only a few photons come through the slit per second, it is impossible to say where they will strike the plate. They will make a pattern which is apparently random. All one can say is that the probability of striking a certain portion of the plate is large when the wave theory predicts a great intensity and small when the wave theory predicts a small intensity. If, however, a sufficient time is allowed to elapse to permit a large number of photons to strike the plate, the resultant pattern will be the ordinary diffraction pattern expected from wave theory.

This type of description can also be applied to the diffraction of electrons by a crystal lattice or by a slit. The wave in this case is the de Broglie wave, and the probability of the electron striking a particular point of a photographic plate is proportional to the square of the amplitude of the de Broglie wave at that point.

The phenomena of transmission and reflection at a plane surface can also be explained from the particle point of view with a statistical interpretation. According to this, the particle associated with the incident wave has a certain probability of being reflected and a certain probability of being transmitted. The probabilities are proportional to the squares of the amplitudes of the corresponding waves.

From this discussion it is apparent that from the corpuscular point of view there is an indeterminacy or lack of causality in the description of phenomena. From the wave point of view, the wave functions necessary to describe the phenomena are continuous functions of space and time and obey definite differential equations. For light waves, the appropriate equations are Maxwell's equations; for waves associated with material particles one must use a wave equation first formulated by Schrödinger and since developed into that branch of quantum theory called wave mechanics.

REFERENCES

1. Louis de Broglie, *Physics and Microphysics* (1955, Pantheon Books). In this book the author's very interesting personal memories of the beginnings of wave mechanics are presented.

2. W. Heisenberg, *The Physical Principles of the Quantum Theory* (1930, Dover Publications).

3. N. Bohr, Nature, *121*, 580 (1928).

BIBLIOGRAPHY

F. K. Richtmyer, E. H. Kennard, & T. Lauritson, *Introduction to Modern Physics*, 5th ed. (1955, McGraw-Hill).

H. Semat, *Introduction to Atomic Physics*, 3d ed. (1954, Rinehart).

P. G. Bergmann, *Introduction to the Theory of Relativity* (1947, Prentice-Hall).

H. Reichenbach, *Philosophic Foundations of Quantum Mechanics* (1944, Univ. of California Press).

L. I. Schiff, *Quantum Mechanics*, 2nd ed. (1955, McGraw-Hill).

P. A. M. Dirac, *Quantum Mechanics*, 3rd ed. (1947, Oxford Univ. Press).

Appendix **H**

The Speed of Light

by C. Harvey Palmer, Jr.[†]

Introduction

The constant **c**, which represents the speed of light in vacuum, has played a truly spectacular role in the development of modern science. In electromagnetic theory it represents both the speed of propagation of plane electromagnetic waves of any wavelength through empty space and the speed of radio waves along a lossless transmission line or coaxial cable (in vacuum). In quantum mechanics **c** appears in combination with other constants in such places as the fine structure constant and the Rydberg constant. The special theory of relativity is based on the assumption of the constancy of the speed of light regardless of the motion of either the source or the observer, and it yields equations in which **c** is the reference standard—the fastest possible speed for material particles, as well as the conversion factor by which to calculate the energy of a particle at rest in the familiar relation $E = m\mathbf{c}^2$.

The value of **c** is now so precisely known[1] [$\mathbf{c} = (2.997928 \pm 0.000004) \times 10^{10}$ cm/sec] that it has been treated as an auxiliary constant, *i.e.* as one whose value is exactly known, in the evaluation of most other fundamental atomic constants. The accuracy with which it is known also permits making distance measurements, based on the travel time of light flashes, by such devices as Bergstrand's geodimeter,[2] which compete favorably in reliability with the best precision surveying techniques.

[†] Laboratory of Astrophysics and Physical Meteorology, The Johns Hopkins University.

A Short History of c

The extraordinary importance of **c** in present-day physics has come about as a result of the most exacting research by a number of brilliant minds during the last three centuries. Probably the earliest attempt to measure the speed of light was made by Galileo, early in the 17th century, as described in Chapter V. Several decades later Roemer studied the eclipses of the satellites of Jupiter and attributed the variations in the predicted times of the eclipses to the finite time required for light to traverse the varying distance between the earth and Jupiter. Roemer's explanation, presented to the French Academy in 1666, was not generally accepted until Bradley in 1727 discovered the aberration of light and obtained a speed in good agreement with Roemer's value.

The first successful terrestrial measurement of **c** was made by Fizeau in 1849. Fizeau used a rotating toothed wheel to interrupt a light beam directed toward a mirror 5.36 miles away, and he obtained a speed equivalent to 3.14×10^{10} cm/sec. The next year Foucault and Fizeau, independently, following a suggestion made by Wheatstone years earlier, measured **c** with a rotating mirror. This method gave improved accuracy. The same year, Foucault determined that light traveled more slowly through water than through air, a result which was in agreement with the wave theory of light and not in agreement with Newton's corpuscular theory. In 1859 Fizeau performed an experiment which showed that the speed of light in a medium (water) was affected by the motion of the medium. His result was in fair agreement with a prediction made by Fresnel in 1818 and based on the elastic-solid theory of the ether. (The factor by which the speed of light is changed in the moving medium is known as the Fresnel dragging coefficient.) In 1872 Airy, using a telescope filled with water, measured the angle of aberration of light, discovered by Bradley a century and a half earlier. Although the water slowed down the light traveling in the telescope, Airy found that the angle of aberration remained unchanged (because of the Fresnel dragging coefficient).

In 1865 James Clerk Maxwell published his classic paper entitled "A Dynamical Theory of the Electromagnetic Field," in which he pointed out the agreement between the calculated speed of propagation of an electromagnetic disturbance through a non-conducting field and the measured speed of light. The values of Table H-1 are taken from his book of 1873, *A Treatise on Electricity and Magnetism*, which contains a more finished treatment than the 1865 paper. From the agreement between the observed and calculated speeds Maxwell concluded that light was an electromagnetic disturbance, and that the speed of light in vacuum should be exactly equal to the ratio of the electrostatic unit of charge to the electromagnetic unit of charge.

TABLE H-1

Velocity of Light (Meters per second)	Ratio of Electric Units (Meters per second)
Fizeau: 3.14×10^8	Weber: 3.1074×10^8
Aberration etc., and	Maxwell: 2.88
sun's parallax: 3.08	Thomson: 2.82
Foucault: 2.9836	

To test the validity of Maxwell's theory, more accurate determinations of both quantities were made. Before 1900 measurements of the ratio of electric units had been made by a number of experimenters, including Weber and Kohlrausch (1857), Maxwell (1868), Rowland (1879), J. J. Thomson (1883), and Rosa (1889), to mention only a few. The determinations of this ratio reached their peak of perfection in the thorough and painstaking efforts of Rosa and Dorsey at the National Bureau of Standards. Their final results, published in 1906, are the earliest ones listed in a recent (1956) evaluation of the most probable value of **c** by Bearden and Thomsen.[1]

Meanwhile Cornu, in 1876, had measured **c** by Fizeau's toothed-wheel method, using a path length of 14.3 miles and an improved electrical method of measuring the speed of rotation of the wheel. A modification of Fizeau's method was also used in 1882 by Young and Forbes, who added an auxiliary distant mirror. Newcomb in 1885 used a rotating mirror method to determine **c** with high accuracy. His value was **c** = 2.99860×10^{10} cm/sec.

While at the U.S. Naval Academy in 1879, A. A. Michelson began his experiments on the speed of light, which represent, perhaps, the pinnacle of what might be termed the mechanical methods of determining **c**, *i.e.* those using a rotating disk or mirror to interrupt the light. Michelson made other determinations of **c** at Cleveland (1882) and at Mount Wilson (1927), where he used a path 22 miles long which had been measured by the U.S. Coast and Geodetic Survey to within $\frac{1}{5}$ inch! This was followed by an experiment in collaboration with Pease and Pearson in which the light traveled back and forth ten times within an evacuated pipe 3 feet in diameter and a mile long. This experiment, completed after Michelson's death in 1931, gave the value **c** = $(2.9977 \pm .0001) \times 10^{10}$ cm/sec.

In 1887 Michelson and Morley published the results of their famous experiment† to observe the ether drift. Their apparatus was a special Michelson interferometer with such long path lengths that the expected ether drift should have been readily measured. No ether drift (or at least very little) was detected. This astonishing result eventually led to the theory of relativity in 1905. Since then, other observers have used even more sensitive apparatus, but the once expected ether drift has never been found.

A new approach to the direct optical method of measuring **c** was made by Karolus and Mittelstaedt in 1928. They used a Kerr cell to interrupt the light electrically—at rates of the order of 20 million times per second—instead of mechanically with a rotating disk or mirror. This type of chopper was used later by Huttel and by Anderson. More recently Bergstrand[2] (1951) has made further refinements on the Kerr cell method by modulating both the Kerr cell itself and also the photomultiplier detector. Houstoun (1950) has invented an even faster chopper using a crystal grating. A direct determination of **c** from microwaves instead of visible light has been made by Aslakson (1951) with a radar system known as Shoran.

A very different kind of experiment for determining **c**, based on molecular spectroscopy, was carried out by D. H. Rank and associates in 1954 and by E. K. Plyler and associates in 1955.

It seems likely that the ultimate method of measuring **c** will be one using microwaves. As long ago as 1923 Mercier made a precision measurement of **c** by using electromagnetic oscillations ($\sim$100 Mc) to set up standing waves on a transmission line. He measured the wavelength and, with considerable ingenuity, the frequency and could thus compute the phase velocity. Essen and Gordon-Smith (1948) set up standing waves in a microwave cavity and determined **c** by a method analogous to that of Mercier. Hansen and Bol performed a similar experiment about 1950. Another type of experiment, based on a microwave interferometer, was carried out by Froome (1952, 1954) and at a much lower frequency by Florman (1955).

Principal Methods of Measuring c

The many methods of measuring the speed of light may conveniently be divided into four basic types: (1) group velocity experiments, (2) permittivity experiments, (3) phase velocity experiments, and (4) spectroscopic experiments.

Group Velocity Experiments

Group velocity experiments include those for which the travel time of the signal over a measured course is determined. All direct optical determinations

†See Ditchburn, *op. cit.* in the Bibliography, for details of the experiment.

and Aslakson's microwave experiment are of this type. A basic characteristic of this kind of experiment is the modulation or chopping of the light or microwave beam to "mark" it. This "marking" is necessary since in the case of electromagnetic waves, unlike water waves, one cannot observe the progress of a particular wave crest. On the other hand, one can determine the speed of a flash of light. The process of modulating or chopping the beam, however, introduces new frequencies (sidebands) so that the beam could no longer be purely monochromatic even if the source itself were purely monochromatic. In any medium having dispersion the speed of propagation of waves depends on wavelength, and the (non-monochromatic) signal travels at the group velocity rather than at the phase or wave velocity which is characteristic of monochromatic waves. If a stone is thrown into a quiet pond, the different velocities are readily observable. The individual wavelets, which travel at the phase velocity, can be seen rising up back of the main group, proceeding through it, and dying out ahead of the group. The group velocity is less than the phase velocity, a relation true for any medium in which the dispersion is normal. If the dispersion is anomalous, the group velocity exceeds the phase velocity and also the free space speed (with light), but the signal now travels at a still different speed, the signal velocity† (less than **c** for light). Michelson measured the speed of propagation of light in CS_2 (which has high dispersion) and found that the ratio of the free space speed to that in the medium is 1.758 whereas the ratio of the phase velocities, the index of refraction, is 1.64. The higher ratio corresponds to the group velocity.

Chopping a light beam may be done mechanically either by a rotating toothed wheel such as Fizeau used in 1849 or with a rotating mirror as in Foucault's or Michelson's experiments. Much higher chopping speeds and thus shorter flashes are produced with a Kerr electro-optic cell or with Houstoun's vibrating crystal grating. With microwaves the chopping is readily achieved by simply turning the microwave source on and off, *i.e.* by square wave modulation of some voltage.

Once the beam has been chopped, it is directed over a measured path and returned to the detector—the eye, a photomultiplier, or a radio antenna. In the case of electromagnetic waves, which travel at fantastic speed, one must use great finesse to measure the travel time accurately.

In Fizeau's experiment a beam of light entered through a hole in the side of a telescope and was reflected by a plane piece of glass toward the objective. This light was brought to a focus at the principal focus of the telescope objective, and thus emerged from the telescope as a parallel beam. It was at this principal focus that the toothed wheel chopped the light. The light flash

---

† See, for example, Stratton. Signal velocity equals group velocity except in an absorption band.

then traveled to the distant mirror (a lens with a mirror at its focus) and back to the telescope, passing through the toothed wheel and the diagonal glass and to the observing eyepiece. With the wheel at rest Fizeau observed either a steady light or darkness, depending upon whether or not a tooth blocked the light at the focal point of the telescope. When the wheel was rotated slowly, the light passing between the teeth had time to travel to the distant mirror and back through the same notch and be seen. But if the wheel rotated fast enough, the light transmitted through a notch returned just as a tooth had been turned into the focal point, and an eclipse was observed. At still faster speed another notch was in the position occupied by the notch through which the light had been transmitted, and the returning light could then pass through to the eyepiece again. Fizeau was able to compute **c** from the distance traveled by the light, the speed of rotation of the toothed wheel needed for an eclipse (or second reappearance of the light), and the number of teeth on the wheel.

The use of a rotating mirror has important advantages over the toothed-wheel method. At relatively slow speeds, as in Foucault's experiment, one can measure the displacement of an image caused by rotation of the mirror faces while the light is en route. At higher rotational speeds, if the distant mirror is far enough away, one can use a fixed eyepiece and crosshair. The eyepiece is placed so that with the mirror at rest in the proper position one can see the image of the slit source centered on the crosshair. If now the speed of rotation of the mirror is correctly adjusted, a new mirror face will occupy the position required to reflect the image to the crosshair. Calculation of the speed of propagation is made in much the same way as in Fizeau's method. The rotating mirror gives brighter images, and it is easier to judge the position of a slit image with respect to a crosshair than to judge the middle of an eclipse or the maximum brightness as is done in Fizeau's method.

The Kerr cell method used by Anderson[3] to measure **c** is rather different in principle from that of Fizeau or Foucault. Basically, Anderson's apparatus was a sort of Michelson interferometer. Anderson measured the r.f. pulse length and the r.f. chopping frequency. However, since it was really the speed of the light flashes that was measured, his was still a group velocity experiment and not a phase velocity one.

The apparatus is shown schematically in Fig. H-1. Light from a mercury arc was chopped by the Kerr cell into pulses of length L, from crest to crest, equal to the ratio of the speed of light to the r.f. frequency of interruption of the shutter, f. These light pulses, represented for simplicity by heavy arrows (bright) and dots (dark), were divided by the half silvered mirror $M_{1/2}$ into two beams. One beam traveled the path $M_{1/2}$, M_2, $M_{1/2}$, D while the other beam traveled the path $M_{1/2}$, M_1, $M_{1/2}$, D. Actually the return beams

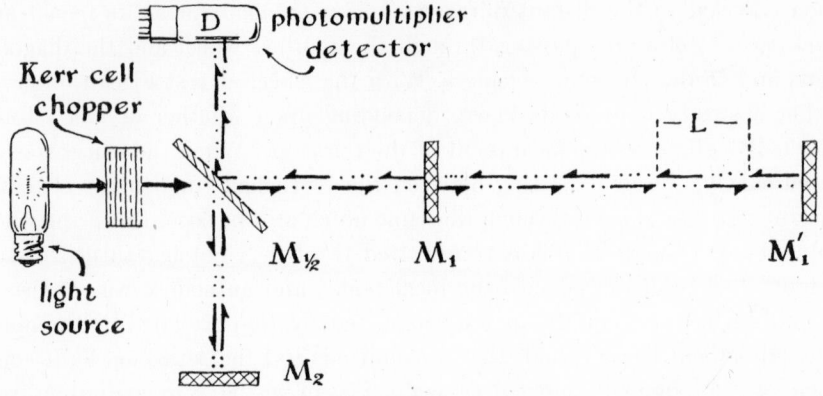

FIG. H-1 Anderson's method of measuring **c**.

were also divided by the beam splitter, and the light which was returned to the source was wasted. If the mirrors M_1 and M_2 were equidistant from $M_{1/2}$, the two bright crests of the light flashes reached the photomultiplier detector simultaneously, and a moment later the two dark troughs arrived together. Thus the photomultiplier output current fluctuated strongly at the r.f. chopping frequency. This signal was amplified and detected by a sensitive short wave receiver, which, for this position of the mirrors, gave a large output signal. On the other hand, if the mirrors M_1 and M_2 were not equidistant, but placed so that the round trip path to and from each of them differed by $L/2$ (or any odd multiple of $L/2$), the photocell was illuminated alternately by light coming from mirror M_1 and from M_2. Thus the photocell saw a steady light intensity, and no r.f. signal resulted.

The position of the movable mirror to give a null can be made extremely critical if the conditions of the experiment are optimum. Anderson fixed M_1 and adjusted M_2 to give the minimum r.f. signal. Then M_1 was removed and the light allowed to travel an additional distance to M_1' and back. The new position, M_1', was chosen so that the increased path (M_1 to M_1'), which was measured with a precision Invar tape, was an integral number of lengths L as nearly as possible. Since the chopping frequency was 19.2 Mc, the light path increase of 2×172 meters was very nearly 11 L each way. A small measured shift of M_2 restored the null again. The speed of light was given by $c = sf/\Re$, in which s is the length ($\sim 344/m$) by which the path has been increased with M_1 removed (including the small displacement of M_2 to restore the null), f the chopping frequency, and $\Re$ the number of lengths L in the distance s ($\Re = 22$). The integer $\Re$ was evaluated from a rough value for **c**. Measurement of the chopping frequency was made by electrical comparisons with a known frequency, that of radio station WWV at the National Bureau

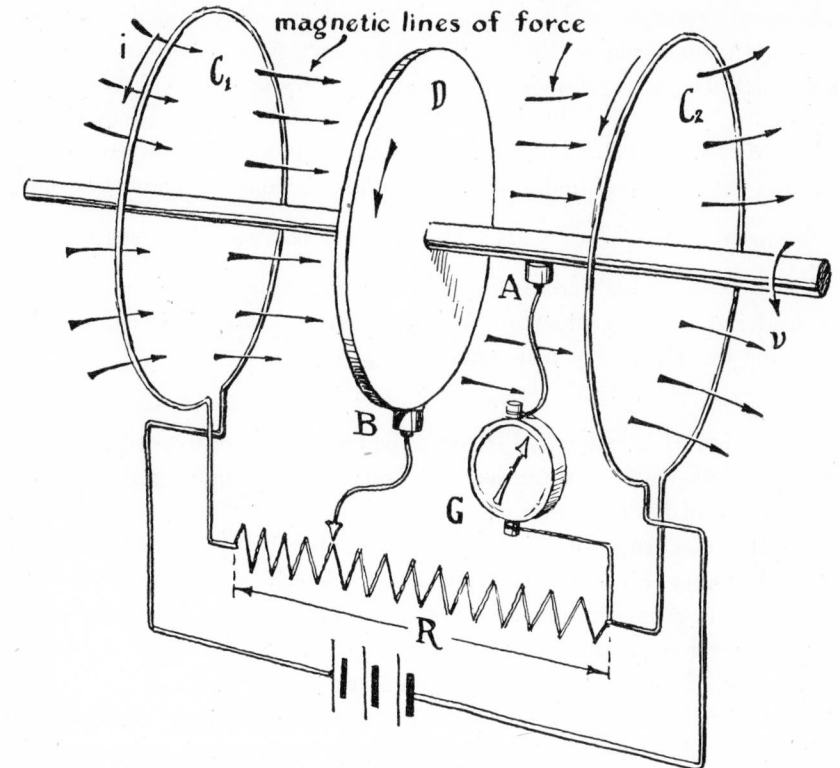

magnetic lines of force

FIG. H-2 Lorenz apparatus for measuring resistance.

of Standards. Anderson's final value for the speed of light was $c = (2.99776 \pm .00006) \times 10^{10}$ cm/sec.

Permittivity Method

According to Maxwell's equations, electromagnetic waves are propagated through empty space at a speed $c = 1/\sqrt{\kappa_0\mu_0}$, in which μ_0 is the permeability of space and κ_0 the permittivity of space. It is important to notice that it is the product of these two constants which determines c and not either one alone. One of the two constants is assigned a value (arbitrary) depending upon the system of electrical units chosen; the other constant is fixed by nature. In the MKS or absolute practical system of units, the system considered here, μ_0 is assigned the value $4\pi \times 10^{-7}$ henry/meter, and κ_0 is determined by electrical experiments.

The experiment to determine κ_0 depends upon a standard of resistance, so that it is necessary to fix this standard first. Lorenz devised an experiment based on the Faraday disk type of homopolar generator for this purpose.

The accurate resistance of a precision standard was determined in terms of mass, length, time, and μ_0. The apparatus is shown schematically in Fig. H-2. A brass disk, D, mounted on an axle rotated ν times per second in a magnetic field parallel to the axle. The flux, Φ, passing through the disk is equal to the product of the mutual inductance, M, of the coils and the current i. From Faraday's law, the e.m.f. induced between the axle contact A and the sliding rim contact B is $V = -\dfrac{\partial \Phi}{\partial t} = M i \nu$. This e.m.f. is balanced against a fraction, α, of the voltage drop across the resistance, R, which may, for simplicity, be considered to be a slidewire. When no current flows through the galvanometer, one has $\alpha i R = M i \nu$. Dividing by αi gives

$R = M\nu/\alpha$

In this relation M is calculated from μ_0, the dimensions of the coils, and their geometry. Both ν and α are measured. Thus the resistance standard can be fixed in terms of M, K, S, and μ_0.

The value of κ_0 can now be determined by an experiment such as that of Rosa and Dorsey. Since the capacity of a condenser is equal to the product of κ_0 and some geometric factor (*e.g.* the ratio of the area to the separation for an idealized plane parallel plate condenser), Rosa and Dorsey measured the capacity of, and calculated the geometric factor for, various specially constructed precision air condensers, including spherical, cylindrical, and plane parallel plate types.

The Maxwell bridge circuit used by Rosa and Dorsey to measure capacity is shown in Fig. H-3. A condenser of capacity C and a mechanically driven switch, S, take the place of the fourth arm of a Wheatstone bridge. Switch S causes the condenser to be alternately charged and discharged. If V is the potential of the junction of R_3 and the galvanometer with respect to ground, the charge flowing into the condenser (and out) is CV each time the switch operates. If the switch operates $\mathfrak{N}$ times per second, the total charge per second, *i.e.* the current, passing through the fourth arm of the bridge is

FIG. H-3   Maxwell commutator bridge for measuring capacity.

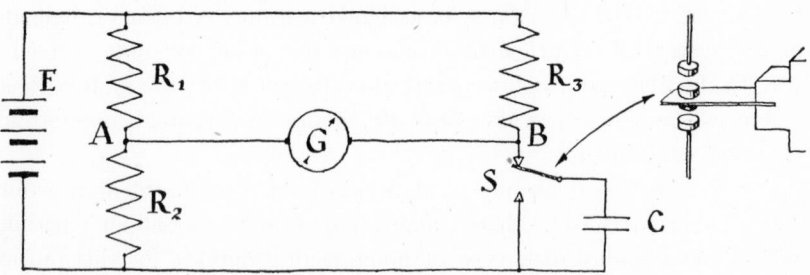

$\mathfrak{N}CV$. The condenser and switch are equivalent, so far as the sensitive, sluggish galvanometer, G, is concerned, to a resistance of magnitude $R' = V/\mathfrak{N}CV = 1/\mathfrak{N}C$. Thus, if the bridge is balanced, one has $R_1R' = R_2R_3$ or $R_1/\mathfrak{N}C = R_2R_3$, and the measured capacity is given by

$$C = R_1/\mathfrak{N}R_2R_3$$

A rigorous derivation of the balance condition† shows that both the galvanometer resistance and the battery and lead resistance must be included as a small correction.

Thus with the Maxwell bridge the capacity of a condenser is measured in terms of frequency and resistances which are in turn compared with the standard fixed by the Lorenz experiment. Rosa and Dorsey's final value for permittivity was given as $\kappa_0 = 8.851 \times 10^{-12}$ farad/meter. When their value is corrected for the present standard of resistance, their value for the speed of propagation becomes

$$c = 2.99800 \times 10^{10} \text{ cm/sec}$$

Phase Velocity Method

To determine c by a phase velocity method, both the frequency and the wavelength are measured; c is the product of the two. Perhaps the earliest accurate measurement of phase velocity was Mercier's experiment of 1923. Mercier used an r.f. oscillator having a frequency of the order of 100 Mc and set up electric standing waves on a transmission line. The wavelength was accurately measured by finding the positions of voltage nodes with a sensitive detector. The frequency was ultimately determined by comparison with a standard clock. More recent experiments of this type employ a resonant microwave cavity. A cylindrical container, for example, can be set into electrical oscillation in a number of modes depending on the frequency. Using particular modes of oscillation, one can accurately measure the wavelength from the geometry of the evacuated resonant cavity, and one can also measure the frequency at resonance by standard methods and so compute c.

An interesting variation on the phase velocity method is that of K. D. Froome,[5] who used two types of microwave interferometers. The earlier model is quite similar to a Michelson interferometer. Fig. H-4 shows a highly simplified diagram of this instrument. The source of energy is a 24,000 Mc microwave oscillator with a special frequency-stabilizing circuit. Energy from this source is directed through a wave guide to a hybrid junction, or "magic T," which divides the energy in such a way that half of it travels out one side arm and the other half out the other side arm, but none straight through to the detector. The fixed arm of the interferometer is completely

† See, for example, Glazebrook.

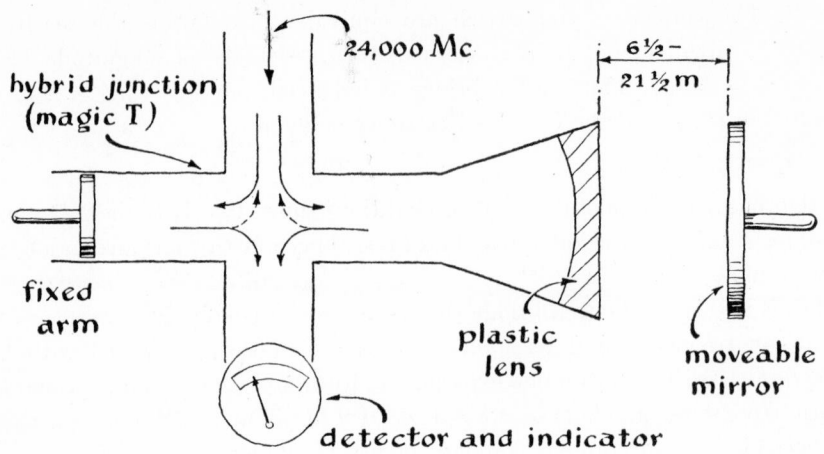

FIG. H-4   Highly simplified diagram of Froome's earlier microwave interferometer.

within the wave guide. The energy traveling down this fixed arm is reflected back to the magic T, and half the energy goes to the detector and indicator, the other half returning to the source. The energy traversing the variable arm of the interferometer toward the right passes through a horn with a plastic lens at its end and then through an air path to a movable mirror, whence it is reflected back through the horn to the magic T, which again directs half of this energy to the detector. As the movable mirror is brought toward or away from the horn, the detector signal passes through maxima and minima. Thus one can determine the wavelength by measuring the distance the mirror must be moved to pass between successive minima. (In this simplified explanation the several attenuators and the matching section needed in the wave guide circuit have been omitted. The components used to measure the oscillator frequency have been omitted from the diagram too.)

A most important correction to the measured wavelength is required. This correction arises from the diffraction effects associated with the relatively long wavelength (1.25 cm) used. Froome pointed out that in an optical interferometer the optical components are of the order of 10^4 wavelengths or more in size and the diffraction effects are small. For microwaves, on the other hand, since the components are much smaller compared to the wavelength, the diffraction effects are very large.

The lack of symmetry in this earlier interferometer, which causes uncertainties in the corrections needed, led Froome to design a symmetrical model using two transmitting horns with plastic lenses and two receiving horns of the same sort, which were mounted on a carriage rolling on a lathe bed. Signals from the two horns were combined in a magic T to give the detector signal much as before. The value reported by Froome is $c = 299{,}793.0 \pm 0.3$ km/sec.

It is worth mentioning that one of the chief advantages in using microwaves rather than light waves for the determination of c results from a peculiarity in frequency measurements. With electric circuits it is possible to obtain exact frequency multiples or submultiples by the use of nonlinear circuit elements such as crystals, class C amplifiers, etc. For this reason one can determine a high frequency with the same absolute error as a low frequency. Accordingly, the higher the frequency used, the smaller the percentage error may be made. With microwaves one can measure the frequency of the source, whereas with light one cannot; rather one must be content to measure a chopping frequency, which, at best, is far lower than microwave frequencies.

Molecular Spectra

The spectroscopic method of determining c depends upon measurements of molecular spectra in both the microwave and the infrared region. For diatomic and linear polyatomic molecules the rotation and rotation-vibration states may be expressed with great accuracy in terms of two constants of the molecule and the rotational quantum number. In the microwave region the possible frequencies of rotation of the molecules are given by the formula

$$\nu = 2B_0(J + 1) - 4D_0(J + 1)^3 \text{ Mc}$$

where ν is the frequency measured in Mc, B_0 is proportional to the reciprocal of the moment of inertia, D_0 is the centrifugal stretching constant, and J is the rotational quantum number, which is integral and can change only by ± 1. In the infrared, where the molecule executes vibration as well as rotation, the formula is a little more involved:

$$\tilde{\nu} = \tilde{\nu}_0 + (B_1 + B_0)m + (B_1 - B_0 - D_1 + D_0)m^2$$
$$-2(D_1 + D_0)m^3 - (D_1 - D_0)m^4$$

Here $\tilde{\nu}$ is the wave number corresponding to the spectral line (the reciprocal of the wavelength in centimeters), $\tilde{\nu}_0$ is the wave number of the band center, B_0 and D_0 the same as before but in units of wave number (cm^{-1}) instead of frequency (Mc), B_1 and D_1 are similar to B_0 and D_0 but apply to the upper vibrational state of the molecule, and, finally, m is the quantum number, which takes on the values $J + 1$ for the R branch and J for the P branch of the band.

If the microwave frequencies of the rotation lines for a simple molecule are measured, it is clear that one can evaluate B_0 (and D_0) in terms of frequency (Mc). On the other hand, by measuring the wavelengths of a number of lines in an infrared rotation-vibration band for the same molecule and converting these wavelengths to wave numbers, one can evaluate B_0 (and the other constants) in terms of wave number units. Thus the same constant can be expressed both in frequency units and in wave number units. From the definition of wave number we get

$$\frac{B_0 \text{ micro}}{B_0 \text{ IR}} = \frac{B_0 \text{ cycles/sec}}{B_0 \text{ waves/cm}} = \frac{\text{cm}}{\text{sec}} = c$$

An experiment of this type was carried out at the Pennsylvania State College in 1954 by D. H. Rank, J. N. Shearer, and T. A. Wiggins,[6] who studied the HCN molecule, and at the National Bureau of Standards in 1955 by E. K. Plyler, L. K. Blaine, and W. S. Connor,[7] who studied the CO molecule. The two final values are in close agreement with each other and with the results of other methods.

Two Simplified Experiments on the Speed of Light

Although extreme care and patience are required to make a truly precise determination of the speed of light, it is not difficult to measure c to about 1%, and such an experiment can show clearly the principles involved as well as the directions in which refinements must be made for high accuracy. Two types of experiment are outlined, a group velocity experiment and a permittivity experiment, both of which will yield results better than 1%.

Kerr Cell Experiment

An experiment patterned after Anderson's Kerr cell measurements was set up at Bucknell University.[8] Because of space limitations, the long path of Anderson was eliminated (and with it the precision). In the simplified experiment only one null position was determined, and the corresponding path difference was measured.

Fig. H-5 shows one arrangement of the apparatus. Light from a 500-watt projection lamp is polarized by P_1 so that its electric vector is at 45° with respect to the electric field between the Kerr cell plates. Lens L_1 converges the light to pass between the Kerr cell plates. Lens L_2 reconverges it to pass through a second Polaroid (crossed with respect to the first) and a mechanical chopper whose purpose is described later. The light passing through the auxiliary chopper is made parallel by lens L_3 (about 4″ in diameter) and strikes the beam splitter mirror, $M_{1/2}$. At this point half of the light is directed toward the fixed thick mirror,† L_4 and M_1, and half toward M_2, the movable mirror which slides on the optical bench—preferably several meters long (though a short one could be used if necessary). The return beams from M_1 and M_2 are again divided by the beam splitter mirror, and half the flux from each beam is converged by L_5 onto the photomultiplier cathode housed in a

---

† A thick mirror is a combination of a lens and a mirror.

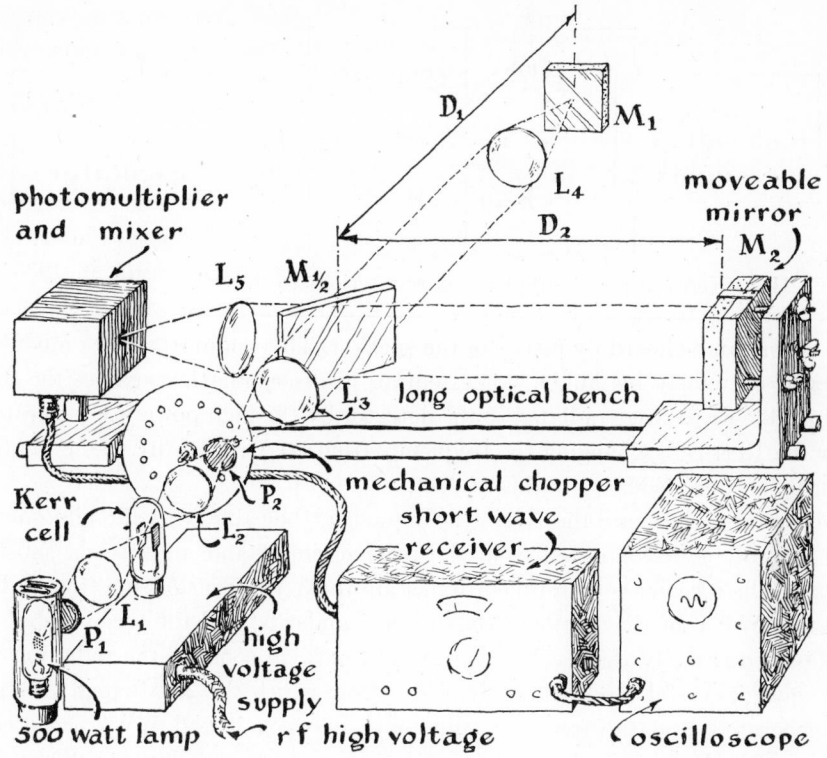

FIG. H-5 Arrangement of apparatus to measure c by a Kerr cell method.

carefully shielded box with a preamplifier stage. The distance, D_1, between M_1 and $M_{1/2}$, should be such that, when the movable mirror, M_2, is about halfway along the optical bench, one obtains a minimum output signal. The minimum signal occurs when the path difference between D_1 and D_2 is $L/4$ or about $3\frac{3}{4}$ meters for an r.f. frequency of 20 Mc.

A block diagram of the electrical components is shown in Figs. H-6 and H-7. Fig. H-6 shows the Kerr cell power supplies. An adjustable high d.c. voltage is impressed on the Kerr cell to polarize it, and a high r.f. voltage is superimposed to produce the light modulation. The r.f. choke keeps the r.f. voltage out of the d.c. supply, and the condenser keeps the d.c. voltage isolated from the ground. Fig. H-7 shows a block diagram of the detector circuit. The light flashes falling on the photomultiplier give rise to a 20 Mc voltage which is heterodyned to 6 Mc by a local oscillator not shown in the figure. This lower frequency signal passes through a coaxial cable to the sensitive short wave receiver. The 1000-cycle modulation impressed on the light beam by the auxiliary chopper described above appears on the oscillo-

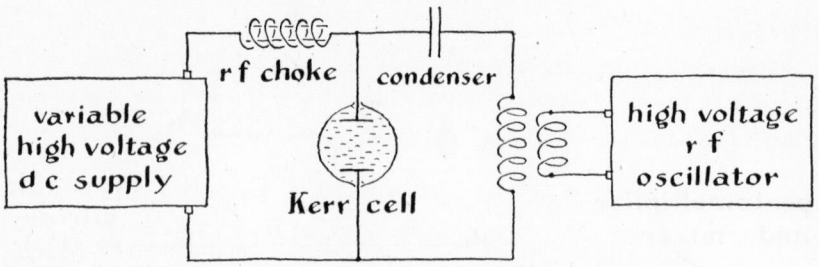

FIG. H-6   Kerr cell power supplies and connections to the cell.

scope or can be heard by means of the speaker. This audio frequency modulation is extremely helpful, if not essential, in discriminating against the unavoidable r.f. voltage radiated to the detector by the high power r.f. generator for the Kerr cell. Changing the frequency from 20 Mc to 6 Mc in the photomultiplier box also helps in this respect.

In order to measure the Kerr cell frequency, the cable from the photomultiplier unit is disconnected from the receiver input and an antenna substituted. The receiver is then tuned to pick up the 5 Mc transmission from radio station WWV of the National Bureau of Standards, and the receiver dial is calibrated. Finally the receiver is tuned to pick up some of the radiation of the quartz crystal (approximately 5 Mc) from which the 20 Mc high voltage is generated, and the frequency difference between it and WWV is noted. This method of measuring frequency is not very precise, but is sufficiently good.

In the Bucknell arrangement, the details of which are described in reference 8, it was possible to set the movable mirror for a null within about a millimeter on successive trials. In a typical good measurement, the r.f. frequency was found to be 19.88 Mc (*i.e.* 4×4.97 Mc) and the measured path difference 374.2 cm, which gave a value for c of 4×374.2 cm $\times 19.88 \times 10^6$ c.p.s. $= 2.98 \times 10^{10}$ cm/sec. The factor 4 comes from the fact that the path difference $D_1 - D_2$ is $L/4$.

FIG. H-7   Block diagram of electronic components for the Kerr cell experiment.

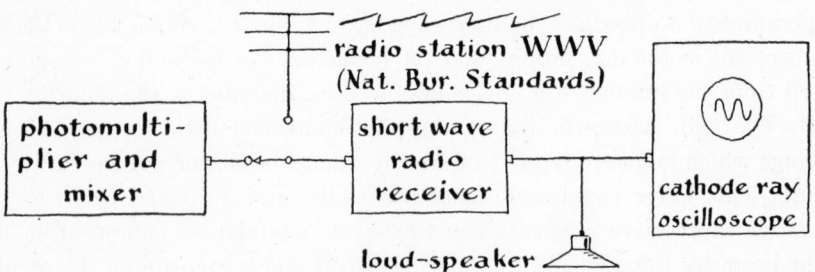

Permittivity Experiment

A simplified version of the experiment of Rosa and Dorsey was set up both at Johns Hopkins and at Bucknell University.[9] Two components had to be constructed for the experiment: a condenser and a switch. The condenser was constructed from two brass disks 8″ in diameter and $\frac{1}{2}$″ thick. One face of each plate was turned flat to about a thousandth of an inch over the surface on a lathe. The plates were mounted in a wooden frame as shown in Fig. H-8. The spacing of the plates was fixed by small glass squares cut from a microscope slide. Two separations of the plates were obtained by using either a single set of three spacers or by using three superposed pairs of spacers. With a parallel plate condenser of this kind a correction is required both for the edge effect in the electric field between the condenser plates and also for the glass spacers. Details of the construction of the condenser and the formulas needed for the two corrections are given in reference 9.

The single-pole double-throw switch for the experiment operates at about 1000 c.p.s. The center terminal of the switch must make low resistance contact with each outside terminal long enough to give the condenser sufficient

---

FIG. H-8   Construction of the parallel plate condenser used in the permittivity experiment

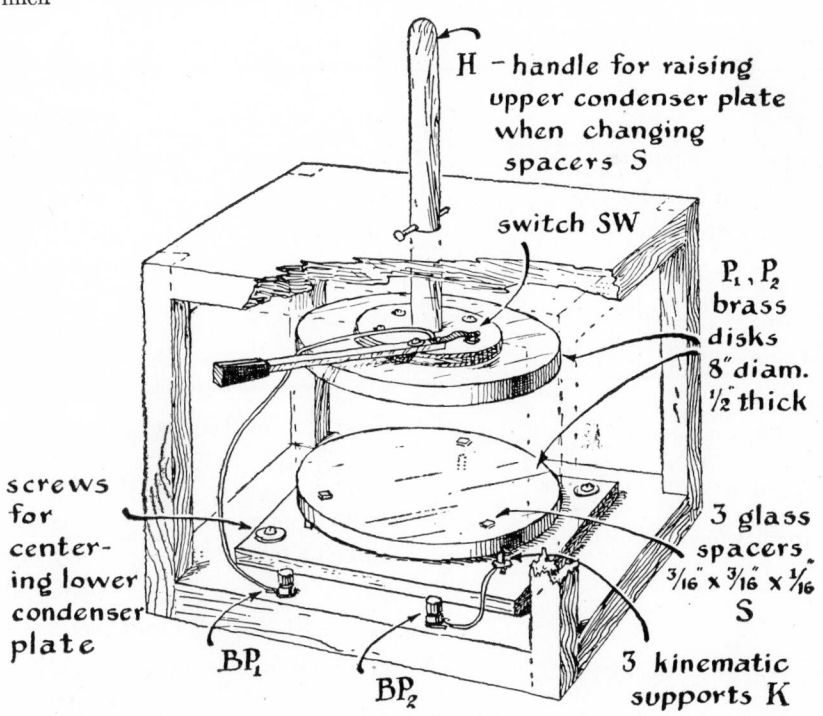

H – handle for raising upper condenser plate when changing spacers S

switch SW

P_1, P_2 brass disks 8″ diam. $\frac{1}{2}$″ thick

screws for centering lower condenser plate

3 glass spacers $\frac{3}{16}$″ × $\frac{3}{16}$″ × $\frac{1}{16}$″ S

BP_1

BP_2

3 kinematic supports K

time to charge and discharge completely in every cycle; at the same time the outside contacts must never short. Such a switch was constructed from an old General Radio tuning fork oscillator by mounting an insulated silver contact on one prong of the fork and a compensating mass on the other prong. The tuning fork was driven by a simple electronic circuit described in the reference. Proper adjustment of the switch to get optimum performance required some care and patience, but the final performance was very satisfactory. The contact time of the moving silver contact with each of the fixed silver buttons was just over 25% of the cycle; *i.e.* the switch allowed the condenser to charge or discharge about $\frac{1}{4}$ millisecond. For the Johns Hopkins arrangement a tuning fork switch was purchased from Riverbank Laboratories (Geneva, Ill.). This switch, which also operated at about 1000 c.p.s., required some modification for this experiment in that the center contact had to be insulated from ground.

The switch frequency is measured by comparison with a calibrated audio oscillator. The oscillator itself is calibrated against the 440 and 600 c.p.s. modulation of station WWV by using various Lissajous figures on a cathode ray oscilloscope. The details are given in the reference.

Using apparatus of this kind, it is possible to determine κ_0 to 1% and thus the speed of light to $\frac{1}{2}$%. A set of measurements is included in Table H-2.

T A B L E H-2 *Permittivity Experiment*

	G_1 (single spacers)	G_2 (double spacers)
Ratio area/separation	27.98 meters	14.00 meters
Correction for edge effect	1.32	1.16
Spacer correction	0.99	0.50
Total geometric factor	30.29 meters	15.66 meters
Measured capacities		
Condenser + "stray"	$C_1 = 392.2\ \mu\mu f$	$C_2 = 264.7\ \mu\mu f$
"Stray" (includes an added		
mica capacitor)	$C_0 = 126.7$	$C_0 = 126.7$
Net capacity of condenser	265.5 $\mu\mu f$	138.0 $\mu\mu f$
Capacity/geometric factor	$\kappa_0 = \dfrac{265.5\ \mu\mu f}{30.29\ \text{meters}}$	$= \dfrac{138.0\ \mu\mu f}{15.66\ \text{meters}}$
	8.78×10^{-12}	8.83×10^{-12}

$$\text{Av. } \kappa_0 = 8.81 \times 10^{-12} \text{ farad/meter} \pm 1\%$$
$$\text{Av. } c = 3.00 \times 10^{+8} \text{ meter/sec} \pm \tfrac{1}{2}\%$$

REFERENCES

1. J. A. Bearden & J. S. Thomsen, *A Survey of Atomic Constants* (1956, Johns Hopkins Univ.).

2. L. E. Bergstrand, Nature, *165*, 405 (1950); Ark. Fys., *2*, 119 (1950), and *3*, 479 (1951).

3. W. C. Anderson, J. Opt. Soc. Am., *31*, 187 (1941).

4. E. B. Rosa & N. E. Dorsey, Bull. U.S. Bur. Stand., *3*, 433 (1907).

5. K. D. Froome, Proc. Roy. Soc., *A213*, 123 (1952), and *A223*, 195 (1954).

6. D. H. Rank, J. N. Shearer, & T. A. Wiggins, Phys. Rev., *94*, 575 (1954).

7. E. K. Plyler, L. R. Blaine, & W. S. Connor, J. Opt. Soc. Am., *45*, 102 (1955).

8. C. H. Palmer, Jr., & G. S. Spratt, Am. J. Phys., *22*, 481 (1954).

9. C. H. Palmer, Jr., Am. J. Phys., *23*, 40 (1955).

BIBLIOGRAPHY

T. Preston, *The Theory of Light*, 5th ed. (1928, Macmillan, London). Contains good detailed history of earlier measurements of c.

R. W. Ditchburn, *Light* (1953, Interscience Publishers, New York).

W. F. Magie, *A Source Book in Physics* (1935, McGraw-Hill, New York).

R. Glazebrook, *Dictionary of Applied Physics* (1922, Macmillan, London), II, pp. 125–128, 955–958.

J. A. Stratton, *Electromagnetic Theory*, 1st ed. (1941, McGraw-Hill, N.Y.), pp. 338–340. Discusses signal velocity.

J. F. Mulligan, "Some Recent Determinations of the Velocity of Light," Am. J. Phys., *20*, 165 (1952).

J. P. Mulligan & D. F. McDonald, Am. J. Phys., *25*, 180 (1957).

Appendix I

Radiation Detectors and Measuring Devices

by Harold W. Yates[†]

Introduction

Electromagnetic radiation is detected by the effects produced when it interacts with a material body. By detection we mean the manifestation, either qualitative or quantitative, of the presence of the radiation, such manifestation being an impression on the senses of an observer. The detector, a material body with which the radiation is made to interact, can be considered as a coupling device linking the electromagnetic energy in the radiation field to some kind of display apparatus. The detector converts the electromagnetic energy into a different form of energy, usually electrical or mechanical, for display and measurement.

In many practical detectors in use today, the energy is converted from electromagnetic flux into electrical current, which is subsequently made observable to the senses. This is a result of the natural physical properties of the detectors themselves—most of them just work that way—and of the high state of development of the science of electronics today, which makes electrical energy the easiest form to work with accurately. There are, however, as we shall see, some detectors which do not depend on a conversion to electrical energy.

Any radiation, from the extremely short wavelength radiation present in cosmic rays to the longest radio waves, can be detected by one or more of the many detectors available today. The working ranges of a number of the more important detectors are shown in Fig. I-1.

[†] Barnes Engineering Co., Stamford, Connecticut.

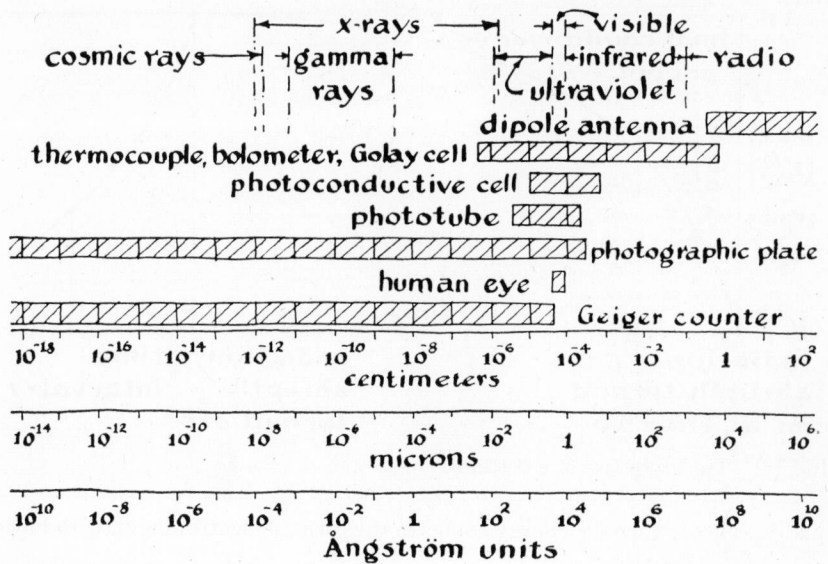

FIG. I-1 Wavelength of radiation.

Electromagnetic radiation, following the hypothesis of Max Planck, may be considered as discrete bundles, called quanta, each quantum carrying the amount of energy

$$E = \frac{hc}{\lambda}$$

where E is the energy of the quantum in joules, λ is the wavelength of the radiation in centimeters, c is the velocity of light in meters per second ($c = 2.99 \times 10^8$ m/sec), and h, Planck's constant of proportionality, has the value 6.624×10^{-34} joule seconds. The interaction of a quantum with a material body usually involves the transformation of the full amount, E, of the energy of the quantum into another form.

The quantum nature of electromagnetic radiation leads to the conception of the perfect detector as one which would respond to and count each quantum which strikes it and would yield no response when no quanta strike it. In other words, it is a quantum counter with 100% efficiency and no spurious counts. Such a detector is, of course, unknown, although several can closely approach this ideal. More commonly, the detector is a device which registers the number of quanta striking it with considerably less than 100% efficiency and is apt to give a certain number of spurious counts which are indistinguishable from real ones.

Although most detectors are basically quantum counters, some of them function purely on the amount of energy incident upon them and will respond

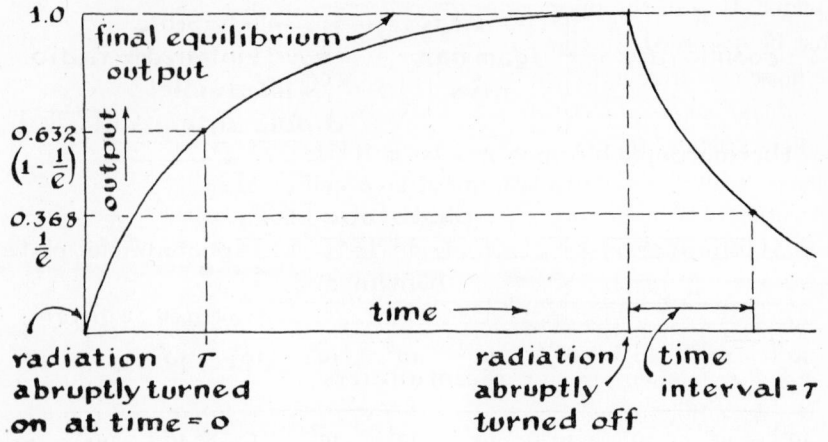

FIG. I-2   Time response of a detector.

equally well whether the energy comes in the form of electromagnetic radiation or in some other form.

Several terms involved in discussing the general properties of the many detectors will be defined here. The first of these is sensitivity, S, which is defined as the response of the detector to unit energy input. In the case of a pure quantum counter S is the quantum efficiency of the detector. In the case of an energy detector S has some form, such as volts/watt or amperes/lumen, depending on the significant output parameter of the detector.

The second general property of detectors is the time constant, τ. For a particular detector the time constant is defined as the time required for the output signal of the detector to rise to $\left(1 - \dfrac{1}{e}\right) = 0.632$ of its final value, when the detector is abruptly exposed to a continuing field of electromagnetic radiation; or to fall to $\dfrac{1}{e}$ of its peak value, when the radiation is abruptly removed (see Fig. I-2).

The third quantity we define is inherent noise, N. We shall discuss noise first, before we discuss sensitivity, because it is inherent in all detectors.

Noise[1]

The symbol $\mathfrak{N}$ expresses the spurious signal† response or output of the

---

·† Actually this spurious signal is usually made up of two components: (1) the random fluctuations which are inherent in the detector and essentially independent of the radiation focused on the detector for study and (2) the systematic fluctuations arising from some source in the field of view other than the source being studied. It is the first, or random, component that is usually meant by the word "noise," while the other is referred to as "background."

detector. It is in the presence of this spurious response that the true signal must be discerned. In the case of detectors having electrical output signal, the noise can originate from the following three sources:

1. *Johnson noise*, or voltage appearing at the terminals of a resistor because of the statistical thermal motion of the electrons within the resistor. This noise is "white," *i.e.* its spectrum contains all frequencies with equal relative intensities. The Johnson voltage (root mean square), within a frequency band Δf, is predicted by Nyquist's famous formula for thermal noise:

$$V_{rms} = 2\sqrt{kTR\Delta f}$$
$$= 1.29 \times 10^{-10} \sqrt{R\Delta f} \text{ volts at room temperature}$$

where T = temperature of resistor (°K)

k = Boltzmann constant (1.38×10^{-16} ergs/degree)

R = resistance (ohms)

Δf = frequency band over which the noise is measured, in cycles/second

2. *Current noise*, which can be a result (a) of the fact that an electric current is not continuous but made up of individual electrons (it is the statistical fluctuation of the flow of these electrons that gives rise to "shot" noise) or (b) of variations in contacts or in the resistance of a detector which carries a current. In the case of shot noise the fluctuations in current are given by

$$i_{rms} = \sqrt{2ei\Delta f}$$

where e = the electronic charge of the electron (1.6×10^{-19} coulomb) and i = the average value of current. The shot noise through a resistor, R, in which the current is detected as a voltage drop, is then $V_{rms} = i_{rms} \cdot R$.

The current noise due to fluctuations in the resistance of a resistor depends entirely on the material of the resistor and cannot be generalized in the form of a law. A high quality wire-wound resistor has very low current noise, much less than that due to Johnson noise, but a carbon resistor usually has current noise comparable to its Johnson noise. The current noise of a resistor of resistance R can be expressed in terms of an equivalent resistance r_{rms}. Such a resistor carrying a current i will thus exhibit statistical voltage fluctuations

$$V_{rms} = ir_{rms}$$

where r_{rms} is experimentally determined for each material. Semiconductor materials usually are characterized by high values of r_{rms}, which, to a first approximation, are independent of i until i becomes large enough for appreciable heating. This current noise, while still random, is not white but in general decreases as frequency increases.

3. *Background noise*, which may be due to statistical fluctuations in the number of quanta striking the detector and arising from its surroundings or may arise from unwanted radiation sources in the field of view rather

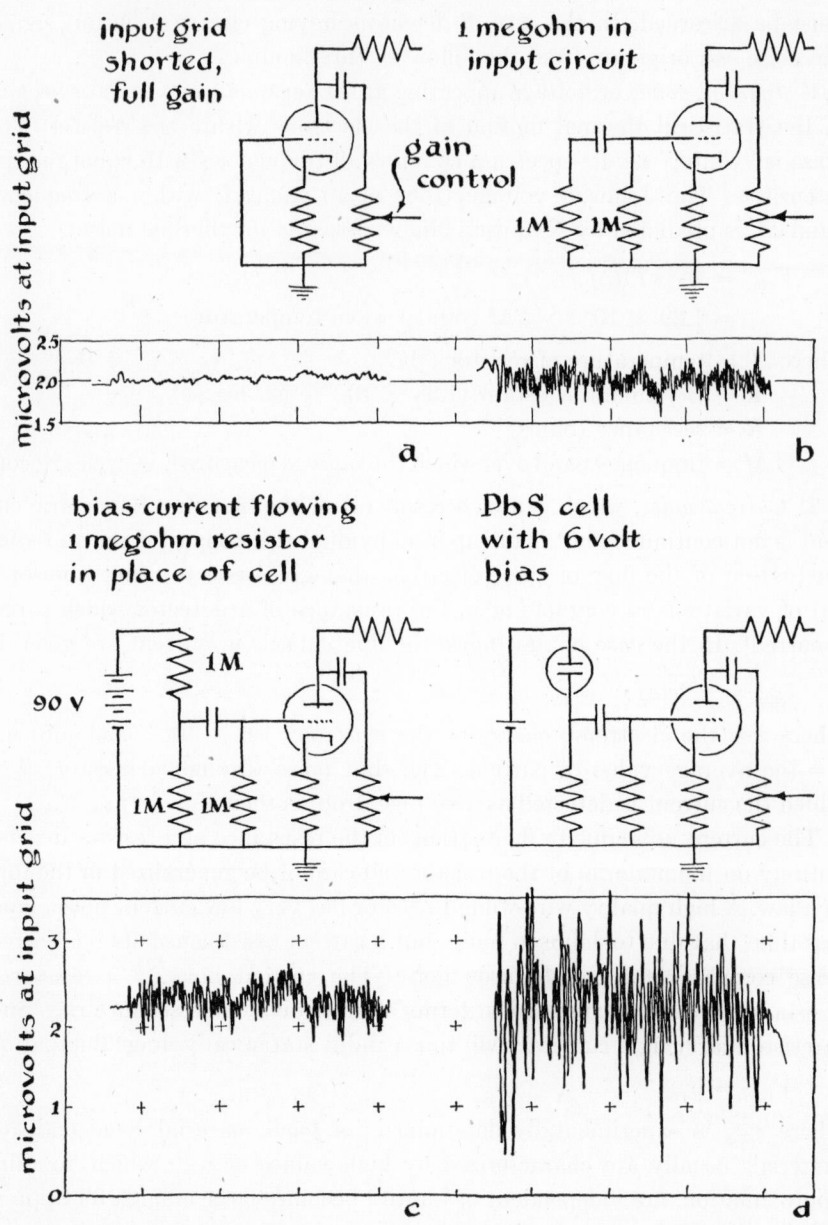

FIG. I-3   Noise appearing in typical lead sulfide (Kodak Ektron) cell detector circuit. Dark resistance of cell is 0.5 megohm.

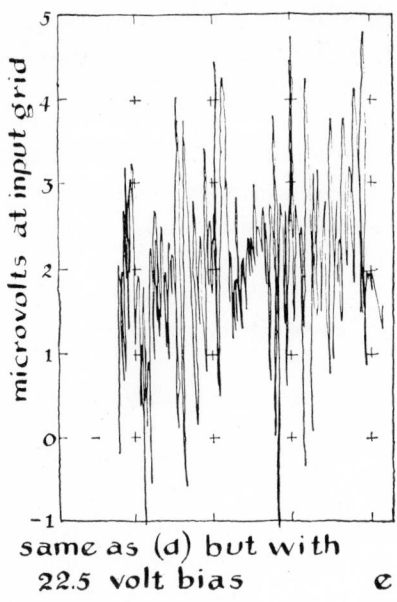

FIG. I-3 (*continued*)

same as (d) but with
22.5 volt bias   e

than from the source being measured. This noise can be considered a sort of quantum "shot" noise in the detector's environmental radiation, and it becomes appreciable in the infrared at around 5–15 microns, where the black-body peak, $\lambda_M = \dfrac{2880}{T}$, of most detector environments lies. Background noise represents a fundamental limitation to detectability.

In the case of a mechancial display system, such as we shall encounter in the Golay cell, background noise may come from the Brownian motion of the parts of the mechanical system and, being proportional to $\frac{1}{2}kT$, will represent limits in the same order of magnitude as the electrical ones just discussed.

Noise appearing in an actual detector application, for some typical cases, is illustrated in Fig. I-3. Here we consider a typical circuit such as that used with a lead sulfide photoconductor. Below the circuit involved we display the amplified and rectified electrical voltage due to noise alone. In Fig. I-3a the input to the amplifier is shorted, and the noise level is the "shot" noise of the tube, together with the current noise in the plate resistor. This noise can be held very low by proper design of the amplifier. In Fig. I-3b the grid circuit contains two 1-megohm resistors in parallel for an effective resistance of 0.5 meg. The displayed noise is Johnson noise, which for a bandpass of 6 cycles (supplied by a filter farther on in the amplifier) is predicted to be

$$V = 1.29 \times 10^{-10} \sqrt{5 \times 10^5 \times 6} = 0.25 \text{ microvolt}$$

In Fig. I-3c a third 1-megohm resistor is added where the detector will later be connected. Here a battery is connected to cause a current of about 50 μamp to flow. The effective impedance of the grid circuit (battery resistance negligible) is that of the three resistors in parallel or about 0.3 megohm. The displayed noise, which should be less than that in Fig. I-3b when no current is flowing, is seen to be slightly larger. Also, it is seen to contain a higher proportion of low frequencies due to current noise in the resistors, all of which are wire-wound, low-noise resistors. In Fig. I-3d a lead sulfide cell ($R = 0.5$ meg) is put in place of the wire-wound resistor and a current of 4 μamp passed through it. A large current noise is evident; increasing the current increases the noise linearly, as Fig. I-3e shows.

Detectors in General[2]

In discussing the wide variety of electromagnetic radiation detectors it will perhaps be helpful to classify them according to the nature of the process involved in their interaction with the radiant energy. This classification will be seen, in due time, not to be rigorous, for some detectors may be reasonably fitted into two or more categories; but such a classification does serve as a convenient starting point for comparisons.

When electromagnetic radiation interacts with a material body, and its energy is absorbed, it usually produces one of the following effects: (1) The material body is heated. (2) The chemical state of the material body is changed. (3) The physical state of the material body is changed (in some manner other than by simple heating). These several phenomena are, in general, characteristic of the energy of the quanta involved, and hence of specific regions of the spectrum. Heating effects are dominant for low energy quanta, chemical change is associated with medium and high energy quanta, and a change of physical state is characteristic of the high energy quanta.

Starting with the lowest quantum energy, *i.e.* the longest wavelength radiations, and proceeding to the shortest wavelengths, the most important detectors and their fundamental mechanisms of operation are:

1. Dipole antenna: electric field generated within a metal by the action of the electromagnetic radiation field.
2. Thermocouple: Seebeck effect, whereby an e.m.f. results from heating of a junction between two metals.
3. Bolometer: change in resistance of a solid with change in temperature produced by irradiation.
4. Golay pneumatic cell: thermal expansion of a gas.
5. Photoconductive cell: excitation of electrons within a crystal lattice.

6. Phototube: ionization of a solid.
7. Photographic plate: chemical decomposition.
8. Human retina: chemical decomposition.
9. Geiger counter: ionization of a gas.
10. Cloud or bubble chamber: ionization in a thermodynamically instable medium.

These detectors, except the first and the last,† will be discussed briefly with particular reference to the parameters which control their ultimate sensitivity, inherent noise, and time constant. Further discussion will be found in the literature listed in the bibliography at the end of this appendix.

Thermopile Detectors

The thermal detectors respond to a change in their temperature. Their response is independent of the nature of the energy producing this change of temperature. Since they cannot differentiate between being heated by absorbed electromagnetic radiation and being heated by any other means, it is necessary to isolate and identify that portion of a thermal detector's output that is due to the incident radiation being measured. The method of doing this is described in a later section (on a.c. modulation of radiation).

The first thermal detector‡ was simply a mercury-in-glass thermometer that absorbed radiation. With this device William Herschel[3] discovered the infrared region of the solar spectrum. Such liquid expansion devices are not, however, sufficiently sensitive for serious use today.

A thermocouple[4,5,6] used for radiation detection is the same in principle as one used in thermometry; but because of the peculiar requirements of radiation measurement its embodiment differs from the thermometry embodiment: a radiation thermocouple is made almost as small as possible, and the materials are chosen primarily for a high thermoelectric power rather than for such properties as mechanical strength or high melting point, which may be important in ordinary thermoelectric thermometry.

† The dipole antenna will be omitted from this discussion because it belongs appropriately to the study of electromagnetic theory and Maxwell's equations. It does not convert the electromagnetic radiation energy into another form and, except for resistance losses, is a completely reversible transducer. Its behavior is not associated with the quantum nature of light, and the photons it detects come, in general, from continuous wave generators and are therefore very long because of the large uncertainty in the time of their emission. Also, the cloud and bubble chambers are not treated here.

‡ The human skin, of course, was the first such detector but we are referring above to the first instrument devised by man. The skin of the forehead has a sensitivity such that a sensation of warmth can be just felt when 630 microwatts/cm² of radiant energy are incident on it. About 80,000 microwatts/cm² are required for the skin of the rest of the body. If less than several cm² of skin are exposed, these numbers are increased.

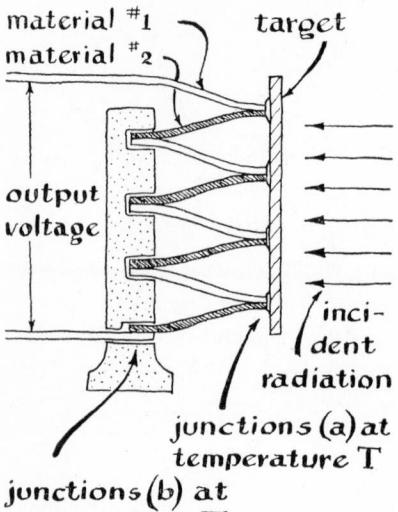

material #1   target

material #2

output
voltage

inci-
dent
radiation

junctions (a) at
temperature T

junctions (b) at
temperature T_o

FIG. I-4   Structure of a radiation thermopile composed of four couples. Junctions (a) are thermally connected to the target but are electrically insulated from it. Junctions (b) are thermally connected to the housing but electrically insulated from it.

The thermopile is made by connecting several, say $\mathfrak{N}$, identical thermocouples in series as shown schematically in Fig. I-4. The thermoelectric wires, 1 and 2, are joined to form two sets of junctions, (a) and (b). Junctions (a) have a target fastened to them to receive the measured radiation. The junctions themselves could be the target but for high sensitivity they are best constructed too small for practical use as a target. The flux of incident absorbed radiation, $\mathfrak{F}$, heats the target and, in turn, the thermoelectric wire junctions (a) attached to it. Junctions (b) are kept shielded from the radiation, and they are usually connected thermally to, but insulated electrically from, the case surrounding the target. Thus the junctions (b) are kept at ambient temperature

We derive the expression for the sensitivity of the thermopile by considering the balance of heat received and lost by the target. This heat balance gives the temperature rise as a function of the radiation input. On multiplying this temperature rise by the thermoelectric power of the materials used, we get the electrical signal voltage generated.

The total heat loss from the thermopile is the radiation loss from the target together with the heat conduction loss down the wires (which here is not negligible). The radiation loss, $\mathfrak{F}_R$, is given by the derivative† of the Stefan-Boltzmann law

---

† The radiation heat loss is actually $\mathfrak{F}_R = 2A\sigma(T^4 - T_0^4)$ where A is the area of the target and the factor 2 enters because both sides of the target are taken to be radiating. For simplicity, the emissivity of both sides is taken to be unity, although only one side is blackened. If $T \gg (T - T_0)$, the above expression is given to sufficient accuracy by the derivative or $\mathfrak{F}_R = 8A\sigma T^3 \Delta T$.

$$\mathfrak{F}_R = 8A\sigma T^3(T - T_0) \tag{1}$$

where A = target area (cm^2), σ = Stefan-Boltzmann constant (5.67×10^{-12} watts $\times$ cm^{-2} $\times$ deg. K^{-4}), T = temperature (°K) of hot junction (a), and T_0 = temperature (°K) of reference junction (b) and of environment. The loss, $\mathfrak{F}_c$, by conduction down the wires will be, for a thermopile of $\mathfrak{N}$ junctions,

$$\mathfrak{F}_c = \frac{\mathfrak{N}k_1\pi r_1^2}{l_1}(T - T_0) + \frac{\mathfrak{N}k_2\pi r_2^2}{l_2}(T - T_0) \tag{2}$$

where k_1 and k_2 = heat conductivities of the two thermocouple materials, r_1 and r_2 = radii of wires of the two thermocouple materials, l_1 and l_2 = lengths of the wires, and $(T - T_0) = (\Delta T)$ = temperature difference between the junctions (a) and (b). The total heat loss is the sum $\mathfrak{F}_R + \mathfrak{F}_c$, and this must equal the heat supplied in the form of radiation, $\mathfrak{F}$.

$$\mathfrak{F} = \mathfrak{F}_R + \mathfrak{F}_c = \left[8A\sigma T^3 + \mathfrak{N}\left(\frac{k_1\pi r_1^2}{l_1} + \frac{k_2\pi r_2^2}{l_2}\right)\right](T - T_0) \tag{3}$$

The temperature rise $(T - T_0)$ generates an electromotive force, V, given by

$$V = \mathfrak{N}P(T - T_0) \tag{4}$$

where P = thermoelectric power coefficient.

Eliminating $(T - T_0)$ between equations 3 and 4 gives

$$\frac{V}{\mathfrak{F}} = \frac{\mathfrak{N}P}{\left[8A\sigma T^3 + \mathfrak{N}\left(\dfrac{k_1\pi r_1^2}{l_1} + \dfrac{k_2\pi r_2^2}{l_2}\right)\right]} \tag{5}$$

The length, radius, and thermal conductivity of the thermocouple wires are conveniently combined in a more significant parameter, the resistance of the thermopile. If the specific resistivities of the two wire materials are ρ_1 and ρ_2, the resistance, R, of the thermopile is

$$R = z_1 + z_2 = \mathfrak{N}\left(\frac{\rho_1 l_1}{\pi r_1^2} + \frac{\rho_2 l_2}{\pi r_2^2}\right) \tag{6}$$

where z_1 and z_2 are the resistances of the $\mathfrak{N}$ lengths of wires 1 and 2 respectively. If the materials obey the Wiedemann-Franz law,[†] which they do fairly closely,

$$\rho_1 k_1 = \rho_2 k_2 = \rho k \tag{7}$$

and equation 5 becomes

$$\frac{V}{\mathfrak{F}} = \frac{\mathfrak{N}P}{\left[8A\sigma T^3 + \mathfrak{N}^2\rho k\left(\dfrac{1}{z_1} + \dfrac{1}{z_2}\right)\right]} \tag{8}$$

The simplifying approximation $z_1 = z_2$ will be made here, so that

---

† The Wiedemann-Franz law states that the ratio of the thermal to the electrical conductivity for a metal has the same value at any given temperature and that the value of the ratio for any given metal varies directly as the absolute temperature.

$$\frac{V}{\mathfrak{F}} = S = \frac{\mathfrak{N}P}{\left[8A\sigma T^3 + 4\mathfrak{N}^2 \dfrac{\rho k}{R}\right]} \tag{9}$$

or, since $R = \mathfrak{N}r$, where r is the resistance of one couple, equation 9 can be written as

$$S = \frac{\mathfrak{N}P}{8A\sigma T^3 + 4\dfrac{\mathfrak{N}\rho k}{r}} \tag{10}$$

For a given receiver area, A, the sensitivity in volts/watt does not increase linearly with $\mathfrak{N}$.

The time constant of the thermocouple is given by $\tau = \dfrac{C}{K}$ where C is the heat capacity (joules/°C) of the target and hot junctions and K is a Newton cooling coefficient (joules/°C sec) defined by the relationship

$\mathfrak{F}$ = heat loss (joules/sec) = $K(T - T_0)$.

The time constant for a practical thermocouple is of the order of 0.1 sec.

The signal from the thermocouple must be seen in the presence of the inherent noise of the thermocouple, which is almost entirely Johnson noise, although in some cases radiation background noise can become appreciable. Johnson noise is proportional to $\sqrt{R}$; and it is helpful to include the resistance, $R = \mathfrak{N}r$, in equation 10 and to define a parameter, $\dfrac{S}{N}$, called the signal-to-noise ratio, D. Dividing equation 10 by $\sqrt{\mathfrak{N}r}$, we have

$$D = \frac{S}{N} = \frac{\sqrt{\mathfrak{N}r}\,P}{8A\sigma T^3 r + 4\mathfrak{N}\rho k} \tag{11}$$

This expression is maximum for some given value of $\mathfrak{N}$, which can be determined by setting $\dfrac{dD}{dn} = 0$. Thus the optimum $\mathfrak{N}$ is found to be

$$\mathfrak{N}_{\text{opt}} = \frac{4A\sigma T^3 r}{\rho k}$$

Substituting this value of $\mathfrak{N}$ into equation 10, we find that the sensitivity of the optimized thermopile is

$$S = \frac{rP}{6\rho k} \tag{12}$$

For a typical example, let us consider a thermopile made of the Hutchins alloys (wire #1 being 95% Bi and 5% Sn and wire #2 being 97% Bi and 3% Sb). These alloys are characterized by $P = 1.05 \times 10^{-4}$ volt/°C, $\rho_1 k_1 = 12.4 \times 10^{-6}$, $\rho_2 k_2 = 12.2 \times 10^{-6}$. If we take $A = 1$ mm² or .01 cm², $T = 300°K$ (room temperature), and $r = 5\Omega$ (a practical value), we find that

$$\mathfrak{N}_{opt} = \frac{4A\sigma T^3 r}{\rho k} = 2.5$$

Since $\mathfrak{N}$ must be an integer, we take $\mathfrak{N} = 3$ and $R = 15$, getting

$$S = \frac{rP}{6\rho k} = \frac{5 \times 1.05 \times 10^{-4}}{6 \times 12.3 \times 10^{-6}} = 7.1 \text{ volts/watt}$$

If we place an RC low-pass filter on the output of the detector so that all frequencies above 1 c.p.s. are eliminated, we will have a bandpass characteristic of $\Delta f = 1$ c.p.s. The Johnson noise from a 15Ω thermocouple, as measured at the output of this filter, is then

$$V_{rms} = 2\sqrt{kTR\Delta f} = 5 \times 10^{-10} \text{ volt}$$

Our sensitivity of 7 volts/watt indicates that 7.2×10^{-11} watt of incident radiation would be required to give an output signal just equal to the r.m.s. noise voltage fluctuations. The 1-c.p.s. bandpass corresponds to a time constant for the filter of $\frac{1}{2\pi\Delta f} = 0.16$ sec, and this, rather than the time constant of the detector itself (usually ≈ 0.1 sec), would control the fidelity with which transient phenomena were measured. The figure, 7×10^{-11} watt, is in this case the "equivalent noise input"—a frequently encountered parameter—and is defined as the input power required, for a 1-c.p.s. bandwidth, to give an output signal voltage just equal to the r.m.s. noise voltage.

Bolometer Detectors

The bolometer—the word is compounded from the Greek words *bole* (throw or ray) and *metron* (measure)—was invented in 1880 by Langley[7] and since that time has appeared in a greater variety of forms than perhaps any other thermal detector. The bolometer function is based on the change in the resistance of a sensitive strip of material when its temperature changes. The radiation to be detected is absorbed by the strip and heats it, and this heating causes the change in its resistance. This change in resistance is manifest as a change in the potential drop across the bolometer when a biasing current, $\mathfrak{I}$, flows through it. (See Fig. I-5.)

The sensitive strip, if naturally highly reflecting, is covered on the irradiated side with a blackening material to cause it to absorb the radiation incident upon it. The mass of this black, usually finely divided particles of gold or platinum deposited on the strip by thermal evaporation in a partial vacuum, is usually small compared with that of the element itself, and its heat capacity can be neglected.

The following is a simple derivation of the sensitivity of a metal strip

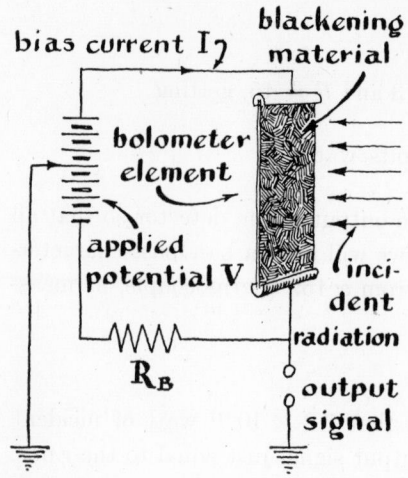

bias current I

blackening material

bolometer element

applied potential V

R_B

incident radiation

output signal

FIG. I-5 Bolometer and associated bridge circuit.

bolometer. Metals have a positive temperature coefficient of resistance; that is, their resistance increases with an increase of temperature.

The bolometer element (see Fig. I-5) is heated by the biasing current. Being warmer than its environment, it loses energy by thermal radiation and by conduction through its mounting supports and electrical lead wires. If $\mathfrak{F}$ is the amount of energy thus lost (watts), T is the temperature of the bolometer element, T_0 is the ambient temperature of the element's environment, then a cooling coefficient (watts/°C), K, is defined by

$$\mathfrak{F} = K(T - T_0) \tag{13}$$

When no radiation falls on the element, the heat loss must exactly equal the electrical current heating, for an equilibrium condition:

$$\mathfrak{F} = K(T - T_0) = \mathfrak{I}^2 R \tag{14}$$

where $\mathfrak{I}$ = the bias current flowing in the element as a result of the applied bias voltage (V) and R = resistance of the element at temperature T.

The temperature difference, $T - T_0$, and the resistance are related by

$$R = R_0[1 + \alpha(T - T_0)] \tag{15}$$

where R_0 = resistance of bolometer element at temperature T_0 and α = temperature coefficient of resistance for the material. On substituting equation 15 into 14, to eliminate the temperature, we get

$$K\left(\frac{R - R_0}{\alpha R_0}\right) = \mathfrak{I}^2 R \tag{16}$$

Now if the bolometer element is further heated by $\mathfrak{F}$ watts of electromagnetic radiations, its resistance will be increased by an amount, r, which leads to a reduction in the bias current of i. The new heat balance for this condition will be

$$K\left(\frac{R + r - R_0}{\alpha R_0}\right) = (\mathfrak{J} - i)^2(R + r) + \mathfrak{F} \tag{17}$$

Subtracting equation 16 from equation 17 yields

$$\left[\frac{K}{\alpha R_0} - (\mathfrak{J} - i)^2\right] r = [(\mathfrak{J} - i)^2 - \mathfrak{J}^2]R + \mathfrak{F} \tag{18}$$

We seek iR, which is the potential drop produced by $\mathfrak{F}$. From Fig. I-5 we have, for no input radiation equilibrium,

$$\mathfrak{J}(R + R_B) = V \tag{19}$$

and, for equilibrium with $\mathfrak{F}$ watts of electromagnetic radiation incident on the element,

$$(\mathfrak{J} - i)(R + r + R_B) = V \tag{20}$$

The condition for maximum signal voltage, iR, is found from setting $\dfrac{d(iR)}{dR_B} = 0$ to yield the condition $R_B = R$. Combining equations 19 and 20, we have

$$r = \frac{2i}{\mathfrak{J} - i} R \tag{21}$$

and substituting 21 into 18 gives

$$\frac{2RKi}{\alpha R_0(\mathfrak{J} - i)} + i^2 R = \mathfrak{F} \tag{22}$$

Now i is negligible, compared with $\mathfrak{J}$; and neglecting the term $i^2 R$, we have

$$\frac{iR}{\mathfrak{F}} = \frac{\alpha R_0 \mathfrak{J}}{2K} \tag{23}$$

as the sensitivity in volts/watt.

It is now more convenient to return to the temperature difference between the element and its environment, instead of the biasing current, $\mathfrak{J}$. Using equation 14, we write equation 23 as

$$\frac{iR}{\mathfrak{F}} = S = \frac{1}{2} \alpha R_0 \left(\frac{T - T_0}{RK}\right)^{1/2} \tag{24}$$

For a realistic bolometer strip, with an area of 1 mm², made of platinum of $\alpha = 3.8 \times 10^{-3}$, the temperature rise, $(T - T_0)$, might be of the order of 25°C, and R might be 50Ω. If we neglect heat losses by conduction, the radiation losses alone are found from the derivative of the Stefan-Boltzmann law, as in the case of the thermocouple.

$$\mathfrak{F} = 8A\sigma T^3(T - T_0) \qquad \text{(or 3.06} \times 10^{-4} \text{ watt in the present example)} \tag{25}$$

$$K = \frac{\mathfrak{F}}{(T - T_0)} \qquad \text{(or 1.2} \times 10^{-5} \text{ watt/°C in the present example)}$$

Substitution of all values into equation 24 gives

$$S = \frac{1}{2} \times 3.8 \times 10^{-3} \times 50 \left(\frac{25}{55 \times 1.2 \times 10^{-5}}\right)^{1/2} = 18.5 \text{ volts/watt}$$

This simple derivation is only an approximation. We have neglected all heat losses except radiation losses, and we have considered the bolometer to be a linear resistance element, whereas in reality it frequently is not.[8]

The inherent noise in the metal bolometer is primarily Johnson noise, although background noise from its environment can be appreciable.

If the bolometer is to be used as a detector of a d.c. radiation field, the ballast resistor, (R_B) in Fig. I-5, is replaced by an identical bolometer element, to eliminate drift. This added element is kept entirely away from the radiation to be measured and is intended to compensate for changes in the temperature of the environment, which otherwise would cause drift.

The parameters in the equation for the sensitivity show that S is linearly proportional to α and approximately linearly proportional to $\sqrt{R}$, $\sqrt{T - T_0}$, and $\frac{1}{\sqrt{K}}$. It is desirable therefore to have a large α, a large $(T - T_0)$, which is a manifestation of a large biasing current, and a small cooling constant. There is no theoretical limit on the magnitude of α. Both α and R are characteristic of the material. The biasing current, however, has a practical limit beyond which the above theory falls down—and the element will be damaged. K can be made small by operating the bolometer in a low ambient temperature environment, and perhaps by using blacking of selective emissivity, but this can be undesirable because of its effect on the time constant, which is given by

$$\tau = \frac{C}{K} \approx 0.1 \text{ sec}$$

for an ordinary metal bolometer where C is the heat capacity of the bolometer element (joules/°C).

The thermistor bolometer and the superconducting bolometer are interesting variations of the simple bolometer. These bolometers are characterized by a larger α than the metal bolometer. Values of α for several bolometer materials are given in Table I-1.

The thermistor bolometers are characterized by a higher resistance than metal bolometers, usually in the order of a megohm. Extensive development of these detectors is still in progress.[10] By increasing the cooling constant, K, time constants down to around 10^{-3} second have been achieved, while still maintaining fairly high sensitivity.

TABLE I-1 *Composition and Values of Thermal Coefficients of Resistance for Bolometer Materials*

Material	Bolometer Coefficient of Resistance, α	
Platinum	$+3.8 \times 10^{-3}$	Metal bolometers at room temperature
Nickel	$+6.0 \times 10^{-3}$	
Mixture of oxides of manganese and nickel or of oxides of manganese, nickel, and cobalt	-4.0×10^{-2}	Thermistor bolometers as originally developed by Bell Telephone Laboratories[9]
Columbium nitride	$+25$	Values of α at the transition between normal and superconducting state. This occurs at about 14.2°K for CbN and 1.4°K for Ta.
Tantalum	$+370$	

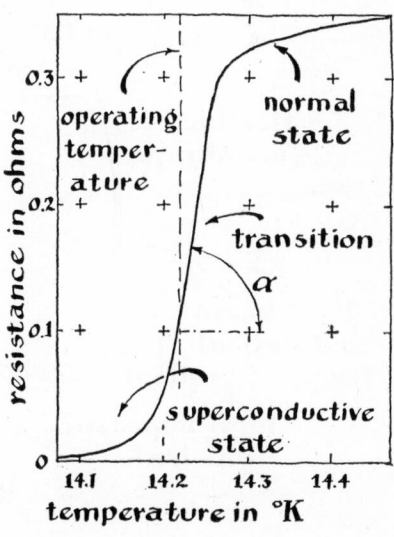

FIG. I-6 Resistance of typical CbN strip in the region of transition into the superconducting state. α is defined by the slope at the transition between normal ($T > 14.3°$K) and superconducting ($T < 14.2°$K) states.

The simplicity, ruggedness, reliability, and fairly good sensitivity of the metal bolometer—and now of the thermistor bolometer, in some applications —make them popular detectors.

The superconducting bolometer[11] has great theoretical promise in view of its large α (see Fig. I-6), the reduced heat capacity at low temperatures, the low ambient environment, and the small resistance of the element itself, which leads to a low Johnson noise. In practice, however, it has never realized its capabilities and remains more of a laboratory curiosity than a practical detector.

The Golay Cell[12]

The Golay pneumatic detector is the thermal detector that was developed by Dr. Marcel Golay. The Golay cell contains a small confined volume of gas that is heated by the energy to be detected. Expansion of this gas, produced by absorbed energy, is observed as the motion of a flexible membrane. Fig. I-7 illustrates the simplified essentials of the detector.

The radiation enters the cell through a transparent window. This radiation is absorbed by and heats a target, which, in turn, heats the surrounding working gas by contact. Expansion of the working gas pushes out the flexible membrane, and the motion of the membrane is detected by the equivalent of an optical lever. A small hole in the chamber wall supplies a high impedance

FIG. I-7   Simplified elements of Golay pneumatic radiation detector.

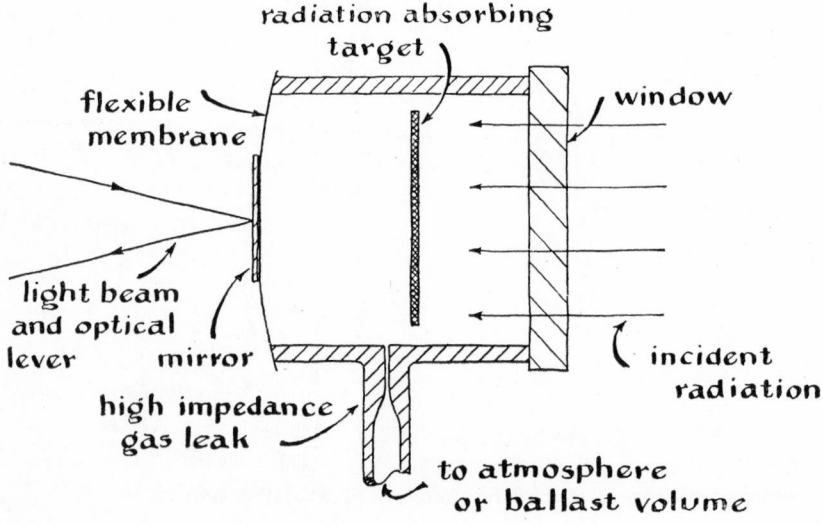

leak to the atmosphere, or to a reservoir, and the time constant of this leak is made long. Slow variations of ambient pressure are equalized by this leak to prevent drift or cell rupture.

Although the final output of the cell is an electrical one (motion of the flexible membrane varies the intensity of a light beam striking a phototube whose current represents the output), the smallest energy observable is determined by the Brownian motion (noise) of the flexible membrane. This limit is equal to or better than that of other thermal detectors. And the time constant can be made less than 10^{-3} second.

The Eye

The two most important detectors whose fundamental process is a change of chemical state are the eye and the photographic plate.

FIG. I-8   Horizontal section of right human eye. The limit of peripheral vision is 104° from the optic axis of the lens system. The image is in sharpest focus near the point where the optic axis intersects the retina and deteriorates as the edges of the retina are approached. The fovea centralis, near the optic axis intercept on the retina, is the area of greatest sensitivity and resolution. About 15° away from the optic axis is the blind spot of the retina. This is the point of entry for the optic nerve, and it is devoid of any light-sensitive receptors. The iris diaphragm attempts to maintain an image of uniform brightness on the retina by opening and closing as the brightness of the field of view varies. It operates only at fairly high brightness and is then only moderately successful.

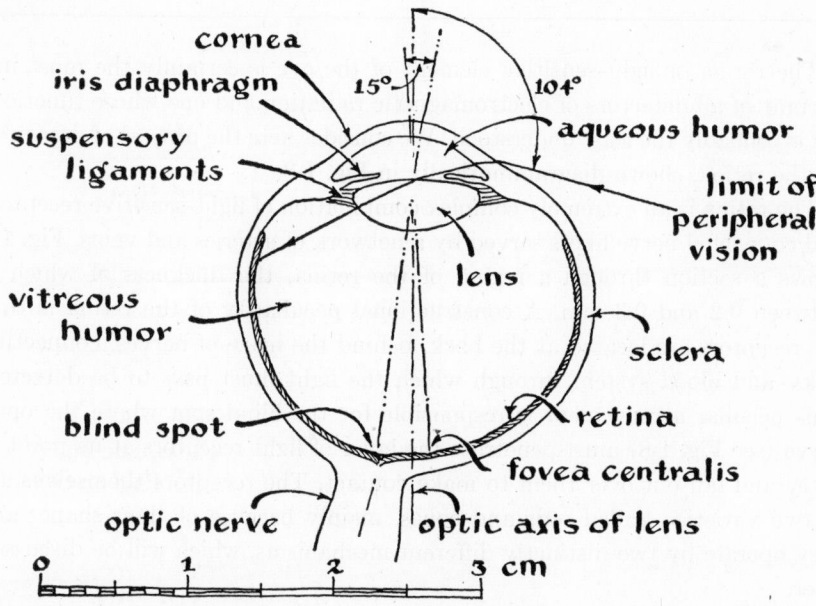

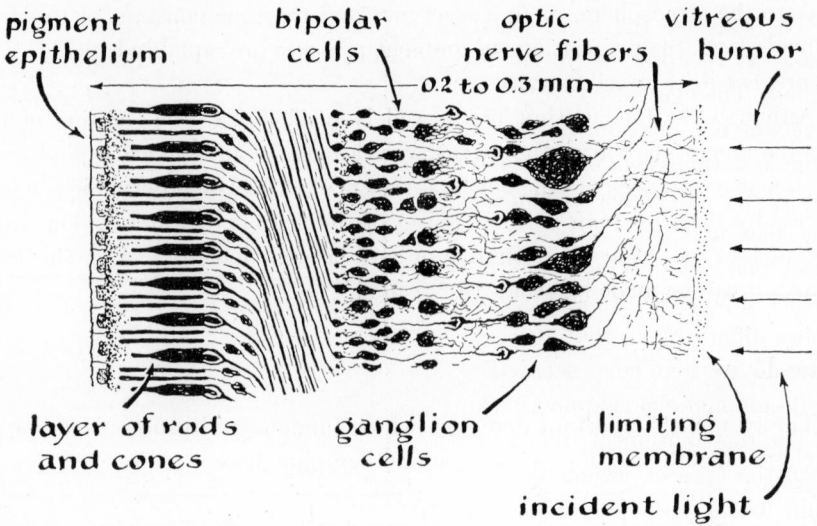

pigment epithelium   bipolar cells   optic nerve fibers   vitreous humor

0.2 to 0.3 mm

layer of rods and cones   ganglion cells   limiting membrane

incident light

FIG. I-9 Cross-section of a human retina. The section shown is near the yellow spot (so named because it contains a yellow pigment), which is about 1.5 mm in diameter and at the center of which lies the fovea centralis. The retina is normally transparent but has been stained to enable the observer to see the structure. Notice the crowding to the right of the connections between rods and cones and the bipolar cells. This crowding, not observed in the peripheral regions, is due to the fact that there are no cells or nerve fibers overlaying the fovea and that the foveal rods and cones must connect to cells in the surrounding area.

The retina or light-sensitive element of the eye is certainly the most important of all detectors of electromagnetic radiation and one whose functioning is probably the least understood. We consider here the process of detection by the retina, shown diagrammatically in Fig. I-8.

The retina is an extremely complex combination of light-sensitive receptors and connected nerve fibers served by a network of arteries and veins. Fig. I-9 shows a section through a region of the retina, the thickness of which is between 0.2 and 0.3 mm. A constructional peculiarity of the retina is that the receptors are located at the back, behind the maze of nerves, connecting links, and blood system through which the light must pass to be detected. This peculiar arrangement is responsible for the blind spot where the optic nerve (see Fig. I-8) must penetrate the layer of light receptors at its point of entry and fan out over them to make contact. The receptors themselves are of two varieties, called rods and cones, mainly because of their shape; and they operate by two distinctly different mechanisms, which will be discussed later.

The rods are the more sensitive of the receptors and are responsible for vision at very low brightness levels, of the order of 2×10^{-3} candle/meter^2 or less. The eye is said to be "dark-adapted" when functioning at these low levels of illumination. After leaving a brightly lighted area, an adaptation time in darkness of 20 minutes to an hour or more is required to achieve this sensitivity. At higher brightness levels the rod mechanism is saturated, and the cones take over as the active detectors. This transition from dark-adapted (scotopic) vision to high light level (photopic) vision is not abrupt, but occurs rather gradually. In addition to their different sensitivities, the rods and cones differ also in that the cones are responsible for color vision while the rods do not give color sensation. This accounts for the observation that the dark-adapted eye is almost completely unable to distinguish colors.

The distribution of rods and cones in the retina is singularly associated with the type of seeing for which they are used. The cones, which are far more useful to the human than the rods, are concentrated most heavily in the area of the fovea centralis, and the eye automatically brings the image of the most important object in its field of view to this point. Receding from the fovea, the population density of cones falls rapidly and then levels off, while the rods, which are completely absent in the center of the fovea, rise in concentration to a maximum at about 20° on either side of the fovea and then fall off as the edges of the retina are approached. Fig. I-10 shows the population in the retina of rods and cones as a function of the apparent angular separation from the fovea centralis. In all there are over 100 million rods and about 6½ million cones in the average retina. There are, however, only about 1 million nerve fibers in the optic nerve; so it is obvious that many rods and/or cones are connected to a single nerve fiber, this phenomenon of

FIG. I-10 Distribution of rods and cones in the human retina.[19]

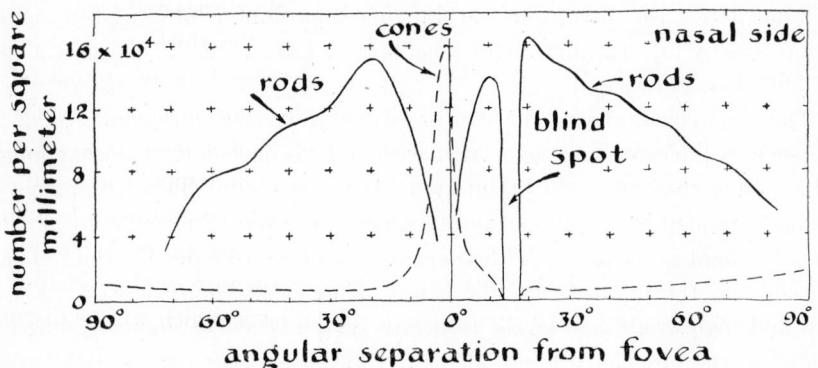

number per square millimeter

16 x 10⁴

cones

rods

nasal side

rods

blind spot

angular separation from fovea

90° 60° 30° 0° 30° 60° 90°

multiple connection being far more pronounced at the edges of the retina than in the vicinity of the fovea centralis. The superior acuity of foveal vision is due to the fact that the foveal cones are closely packed and do not have multiple or cross-connections; and also to the fact that the fovea is not over-laid by cells, blood vessels, and nerve fibers, as is the remainder of the retina. The increased population of rods away from the fovea is illustrated by the fact that a very dim object, such as a faint star, can only be seen if an observer does not look directly at the object, and is best seen by looking about 20° to one side of it. This is called averted vision.

Rod vision, at its wavelength of peak sensitivity, about 5100 Å, comes very close to the limiting detector sensitivity of an ideal photon counter. It has been shown[13] that the average retina will respond with a 60% probability to from 5 to 14 quanta, which, at this wavelength, corresponds to from 2.0 to 5.6×10^{-11} erg. These numbers, representing the quanta actually absorbed by the rods, are obtained by multiplying the energy incident on the cornea by the optical efficiency of the eye, which is here shown as approximately 1%. This low efficiency is due primarily to the fact that only a small part, 10 or 20%, of the light striking the retina is effectively absorbed by the rods, absorption and reflection in the optical system of the eye accounting for the additional loss.

The stimuli used in these experiments were in the form of 1-millisecond pulses of light since the eye's ability to integrate a total signal does not extend beyond about $\frac{1}{10}$ second. If the same total amount of energy had been presented to the eye over a long period, say $\frac{1}{2}$ second, it would not have been seen. Experiments of this nature indicate a "time constant" for the eye of about $\frac{1}{10}$ second. The sensation value depends on the amount of light in the stimulus—and the rise and decay curves are not the same. It is the fact that the eye has a "time constant," and persistence of vision, that makes motion pictures possible. As long as the frame rate of the camera is kept above about 16 per second—and 24 frames per second is standard—the eye will not detect the discontinuities. Further smoothing is produced in modern projectors by interrupting each frame once to present to the observer a total of 48 flickers per second.

The time constants for the two processes, rod vision and cone vision, are somewhat different, that associated with rod vision being somewhat longer. These time constants can be inferred from the critical flicker frequency or highest frequency of modulation at which the eye is just able to perceive that the light is flickering. This frequency is about 10 c.p.s. for the rods and depends on the intensity for the cones, being between 10 and 50 c.p.s. for low and high brightness levels, respectively. These values indicate time constants in the order of 0.1 second for rod vision and from 0.1 to 0.02 second

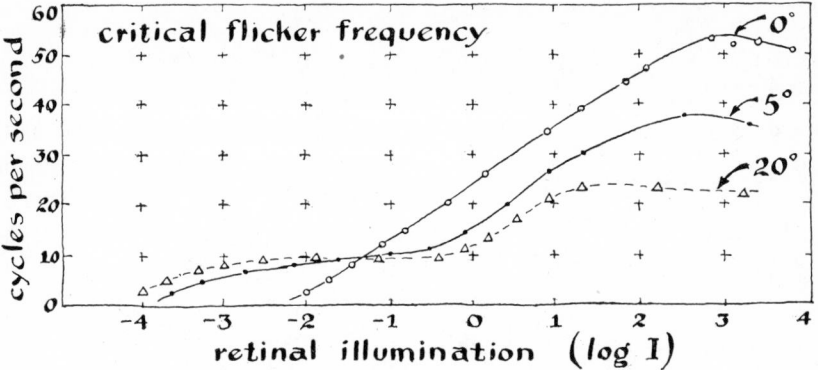

FIG. I-11 Curve of critical flicker frequency[14] as a function of retinal illumination for white light at the fovea, and at 5° and 20° above the fovea. The fovea contains only cones. The population is about even for rods and cones at 5°, and at 20° is the peak population of rods.

for cone vision. Fig. I-11 shows the dependence of critical flicker frequency on brightness level.

Some extremely interesting experiments have recently shown that the eye is essentially an a.c. detector functioning by virtue of the fact that the constant tiny motions of the eyeball keep the retinal image in constant motion over the retina itself. If this motion of the eye is frustrated, as has been done by an ingenious optical system,[15] the result is that the stabilized image on the retina (first observed more sharply) fades after about a minute almost to invisibility.

The actual mechanism by which the rods function is fairly well established to be a chemical decomposition and regeneration. A purple pigment, known as rhodopsin or "visual purple," is present in the rods, and this is decomposed into a yellow pigment, known as retinene or "visual yellow," and a protein,[16] by the action of light. Reconstruction of the decomposed rhodopsin is continuously carried out in the rods, and the kinetics of these reactions determine in part the time constant of the eye for rod vision. The exact nature of the process by which this reaction is converted into an electrical impulse in the optic nerve is still not clearly understood. Rhodopsin has been isolated from frog retinas and studied under laboratory conditions. Its spectral absorption curve shows a peak at 5100 A and matches almost perfectly the relative visibility curve for rod vision (Fig. I-12).

Cone vision is not so well understood as rod vision. The photosensitive chemicals associated with cone vision have been more difficult to identify and isolate than rhodopsin. At least one, known as iodopsin, has been isolated

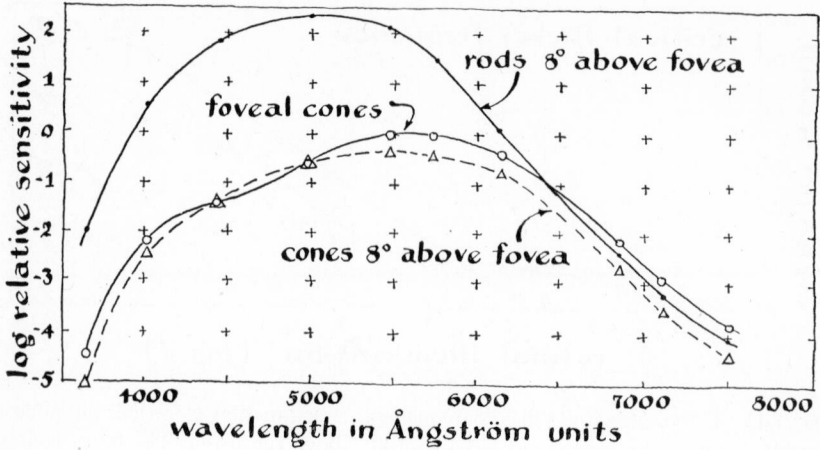

FIG. I-12 Spectral sensitivities (1/threshold) of dark-adapted foveal cones, peripheral (8°) rods, and peripheral cones.[35] All sensitivities are relative to the maximum sensitivity of the fovea.

and studied. Since the cones are also able to distinguish color, they must be more complex than the rods.

The curve for cone vision sensitivity, as a function of wavelength, shows (see Fig. I-12) a peak sensitivity at 5600 Å. The cones are therefore more red-sensitive than the rods, a phenomenon that is demonstrated by the dark-adapted eye's difficulty in seeing red. Of the theories of color perception, that originally advanced by Thomas Young in 1801 is the most generally accepted. According to this theory, elaborated by Helmholtz, there are at least three distinctly different types of cones[†] with peak sensitivities at or near the wavelength of the three primary colors, red, yellow, and blue. The reader is referred for a more complete discussion of this fascinating subject to the many excellent books on the retina and its processes.[19,20,21] A thorough discussion of the theories of retinal color mechanism can be found in a paper by S. A. Talbot.[18]

[†] These three types of cones have never been distinguished in an actual retina. An interesting hypothesis containing all the elements of the Young-Helmholtz theory except the existence of three distinct cone types has been recently published by Ségal.[17] He proposes that rhodopsin and its photolytic products are the only photosensitive substances in the retina, and that they act in addition as filters. Detection of the primary colors would take place in different sections of the retina and by different photosensitive substances. Red would be detected by microcrystals of rhodopsin in the pigment epithelium (see Fig. I-9) behind the retina, which then acts as a filter passing red light only; blue detected by a yellow photolytic product of rhodopsin observed in the bulbous end of the fibers in front of the layer of rods and cones; and yellow detected by the rhodopsin in solution in the cones themselves.

Photographic Plate for Short Wavelengths

The photographic plate† differs from all the other detectors discussed here in that it is a totally integrating device and hence has no time constant associated with its process. In its idealized form it would provide one grain of developable silver bromide for each photon incident upon it and would have no silver bromide grains made developable by any other means (*i.e.* it would have no fog).

A photon of visible or near ultraviolet light will usually produce one sensitivity speck (a free silver atom) on the silver bromide grain which absorbs it, but on the average from 10 to 100 such sensitivity specks are required to make one grain developable. Therefore, the quantum efficiency of the plate is from 0.1 to 0.01 for these wavelengths. Quantum efficiency rises as the wavelength decreases, and below 1 Å a single incident quantum may produce many developable silver halide grains. In terms of energy, of course, the sensitivity of the photographic plate becomes lower as the wavelength of the incident radiation is reduced. This is characteristic of a detector which is a quantum counter. Actually, the photographic plate lies somewhere between the quantum counter and the energy detector in its spectral sensitivity characteristics, as is shown in Fig. I-13. The fact that the higher energy quanta are less efficient on a pure energy basis is probably due to the fact that many

† It will be assumed that the student is acquainted with the elements of the photographic process. For a detailed treatment he may refer to the texts on the subject.[22]

FIG. I-13   Sensitivity of two typical X-ray emulsions as a function of wavelength.[23]

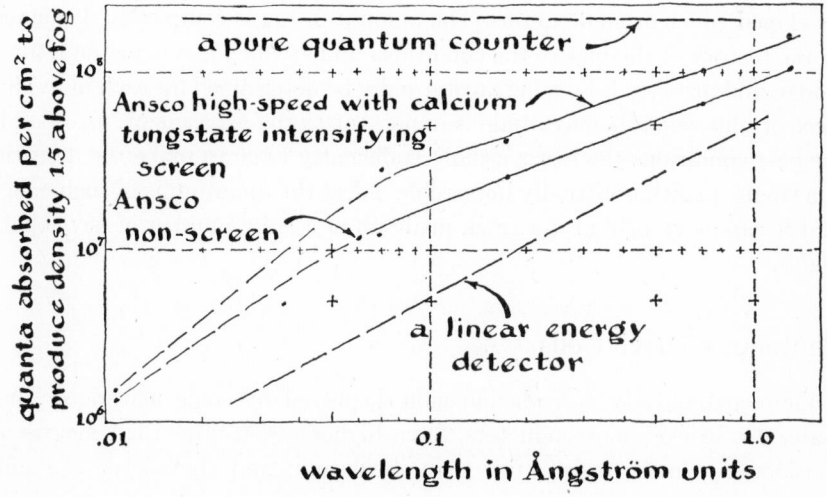

of the development centers produced by them are inside the silver bromide grains where they will not contribute to the grain's developability. This internal latent image effect is somewhat offset by the fact that, as the wavelength of the incident radiation is reduced, photo or recoil electrons are ejected from the emulsion by the primary radiation, and these too produce sensitivity specks and hence add to the latent image.

Silver bromide is insensitive to radiation of wavelength greater than 4500 A. In order to extend the sensitivity to longer wavelengths, certain dyes—usually of the carbocyanine group—are usually added to be adsorbed on the silver bromide grains in about monomolecular layers. In this manner the sensitivity can be extended to an extreme of 13,000 Å (1.3μ). The gelatine matrix in which the silver bromide crystals are embedded is opaque to radiation between 2300 Å and about 10 Å, and special techniques must be employed to use the photographic plate as a detector in this region. The most common and effective technique is to coat the emulsion with a fluorescent material which is excited by the incident radiation and then emits longer wavelength quanta to which the emulsion is transparent. At much shorter wavelengths, below about 10 Å, the gelatine becomes transparent again, and the photographic plate is very useful in the X-ray or gamma ray region. The emulsion has a lower absorbing power for these high energy quanta, however; so, to increase their absorbing power, X-ray films are made with thick emulsions rich in silver bromide on both sides of the substrate and are further supplied with calcium tungstate ($CaWO_4$) fluorescent intensifying screens.

For extremely high energy quanta (or particles) very thick emulsions have been employed in which the incident quantum produces so many latent images it leaves a track in the emulsion.

The equivalent of noise for the photographic plate is fog or spuriously developed silver bromide grains. The amount of fog is completely dependent on the history of the plate—the conditions under which it was manufactured, stored, and processed. Fogging must usually be determined for each individual piece of film when its magnitude is important to the experiment. Even under the best conditions the fog is usually sufficiently large to make the detection of a single quantum virtually impossible unless the quantum is of such energy that it can leave behind it a track made up of a large number of developable grains.

Photoconductive Cells

Photoconductivity is a phenomenon displayed by some materials. These materials, called semi-conductors, have higher resistivities than metals but considerably lower resistivities than insulators; and they have a smaller resistance when they are irradiated with electromagnetic radiation. The wavelength of this irradiation, to be thus effective, must be shorter than some critical wavelength which is a characteristic of each specific material. This long-wavelength cut-off is usually in the infrared region between 1 and 8 μ. Most photoconductive materials have a fairly uniform quantum efficiency down through the visible and ultraviolet, but they are generally inferior to the phototube for radiations to which the latter is responsive. Photoconductivity therefore finds use primarily for infrared detectors.

Photoconductivity was first discovered in selenium by Smith[24] in 1873. It attracted little attention as an infrared detector until just before and during World War II. The tremendous effort in military applications of infrared radiation resulted in the development of a number of photosensitive materials, notably thallium sulfide, lead sulfide, lead telluride, and lead selenide. Many other materials are known to be photosensitive, and research in this field is very active today. However, the three lead compounds (PbS, PbTe, and PbSe) remain at the moment the most important detector materials.

The photoconductive cell consists of a thin, flat wafer of a photosensitive semi-conductor mounted and provided with electrical connections, very much as is the conventional bolometer (see Fig. I-5 and also Fig. I-3). The photosensitive material is usually deposited on a glass substrate by condensation from the vapor state or by deposition from chemical solution. The change in resistance, on irradiation, is detected by the resulting change of an electrical bias current flowing through the semi-conductor, just as in the case of the conventional bolometer. The photoconductor differs from the metal bolometer, however, in that its output is not primarily a manifestation of a change in its temperature, but results from the increased number of conduction electrons that are freed from bound states by the absorption of incident irradiation. The process can be considered as a sort of photoelectric effect in which the absorbed quantum raises an electron from a bound state to the conduction state. The electron eventually returns to a bound state, giving up its excitation energy in the form of heat. The speed with which this relaxation takes place largely determines the time constant of the detector. The smallest energy required to excite a bound electron determines the long-wavelength limit of the photoconductor's sensitivity. One of the aims of modern research is to find new materials, and new means to decrease this minimum energy for existing materials, in order to extend the detector's sensitivity to longer and longer wavelengths. An explanation of the photoconductor's properties (*i.e.* absolute sensitivity, long-wavelength cut-off, and time constant) in terms of the physical variables of the crystal itself cannot be pursued here, and the student is referred to the literature for this interesting branch of solid state physics.[25]

It is difficult to give firm values for the sensitivities of photoconductive cells because they depend strikingly on the method of manufacture, impurities present, temperature, and humidity. Cells produced under research, or on development projects, are usually unique in all their properties. There are only a few examples of reproducible photoconductive cells being commercially manufactured by a controlled process. One of these is the Ektron lead sulfide cell[26] manufactured by the Eastman Kodak Co., whose properties will be briefly reviewed here.

The Ektron cell is chemically deposited on glass and can be made in almost any size or configuration. Considering a 1×1 mm sensitive area, the dark resistance of the cell is about 0.5 megohm at 20°C. The limiting noise is current noise (see Fig. I-3), and the signal-to-noise ratio, $\frac{S}{\Re}$, exhibits a maximum at some particular bias current. Currents much above this value should not be used, or the sensitive area will be damaged by current heating. The time constant is around 10^{-3} sec. Cooling the cell to -40°C increases the resistance to 5 megohms and increases the time constant to about 4×10^{-3} sec but improves the ratio $\frac{S}{\Re}$ by a factor of about 25. The relative response of the Ektron cell as a function of wavelength is shown in Fig. I-14 for two cell

FIG. I-14   Relative spectral response of Eastman Ektron (PbS) cell.[25]

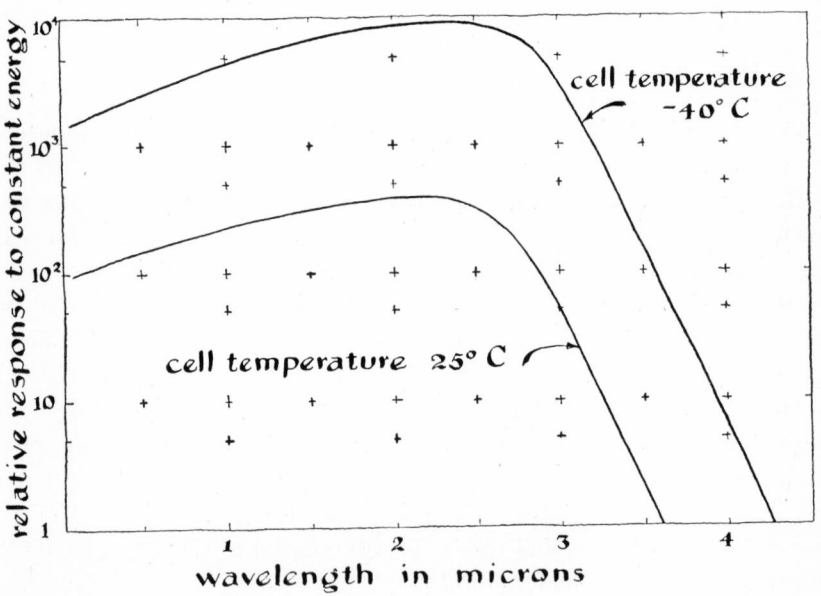

temperatures, 25°C and −40°C. The apparent decline in sensitivity as the wavelength is reduced below 2μ is due to the fact that the PbS cell's mechanism is a quantum phenomenon—one quantum excites one electron into the conduction band. Actually, the quantum efficiency is fairly uniform down through the ultraviolet.

The minimum detectable power measurable with the lead sulfide cell lies in the range from 10^{-12} to 10^{-13} watt. Such a sensitivity makes it from 100 to 1000 times more sensitive than the thermocouple or bolometer.

PbTe and PbSe cells are sensitive and usable out to longer wavelengths— to about 5μ and 8μ respectively (see Fig. I-15). Both of these materials must be cooled to around 100°K before they have sufficient sensitivity to be useful. The dark resistance of these cells may be a few thousand ohms at room temperature, increasing to between 10 and 100 megohms for a square sensitive area at 100°K. The minimum detectable power is from 10^{-11} to 10^{-12} watt for the PbTe cell, making it from 10 to 100 times more sensitive than a thermocouple or bolometer. The time constant of the PbTe cell is from 10^{-5} to 10^{-4} second. The lead selenide cell is actually of comparable or only slightly greater sensitivity than the thermocouple, but its time constant—usually less than 10^{-5} second—makes it very useful for the study of rapid phenomena.

FIG. I-15   Relative spectral response of typical PbSe and PbTe cells at 100°K. (The curves are not normalized to each other. Actually PbSe is much less sensitive than PbTe and on an absolute response plot would be shifted down approximately two cycles from the PbTe curve.)

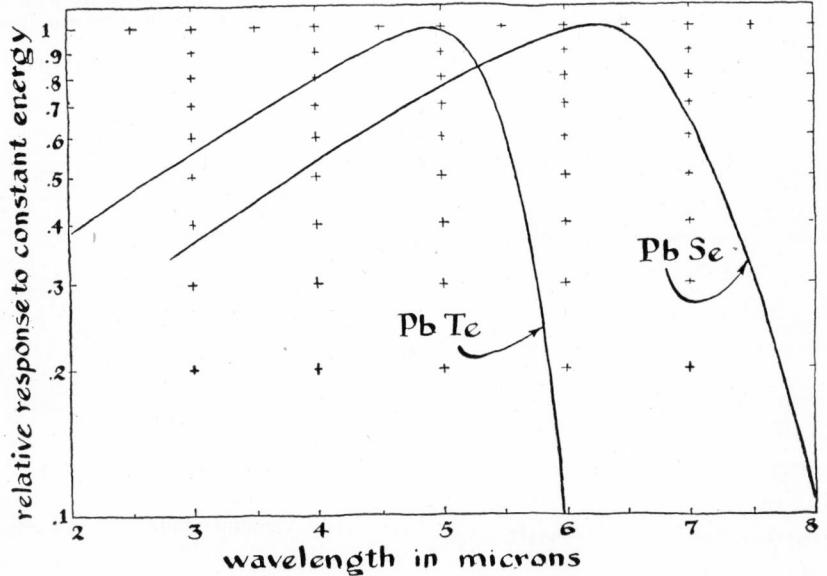

Of the many materials currently being studied and developed, indium antimonide[27] seems to be one of the most promising photoconductor materials. The sensitivity is so variously given that it is difficult to select a value to quote here, but it is evidently more sensitive than the PbSe cell, and its sensitive range extends to 8μ. The time constant is remarkably small, of the order of 10^{-6} to 10^{-5} second, and the resistance of most samples is found to be of the order of a few hundred ohms, thus considerably reducing current noise as well as Johnson noise. Of other materials, selenium and germanium show promise, provided certain impurities have been added to enhance their sensitivities and extend the long-wavelength cut-off farther out into the infrared.

The Phototube[28]

The phototube is a very simple device (Fig. I-16) composed of a cathode, usually a fairly large area, and an anode, which is frequently a thin wire or small button placed to one side of the cathode so as not to obstruct the incident radiation. Phototubes are housed in a glass envelope, which must be transparent† to the radiations to be studied. A high vacuum sufficient to give the mean free path of an electron appreciably larger than the distance between anode and cathode must exist inside the envelope. The potential difference applied to the tube is not critical as long as it is sufficiently high

---

† In the ultraviolet region, either special window materials are used, or the envelope of an ordinary phototube is coated with a fluorescent material which, on irradiation, emits radiations to which the envelope is transparent. The relative quantum efficiency of this latter arrangement is low, however (5% or so), because most of the fluorescence quanta do not strike the photocathode.[29]

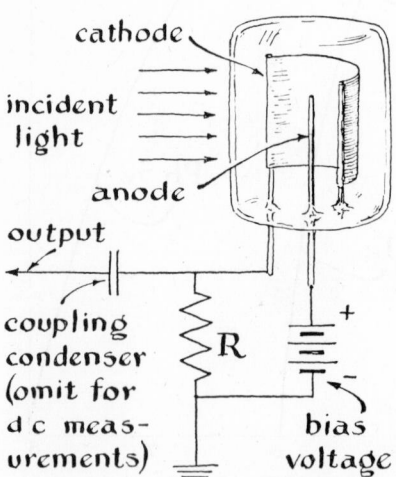

FIG. I-16 Vacuum phototube and associated circuit.

to remove quickly all photoelectrons from the vicinity of the cathode and prevent their building up a space charge. For potential differences between this value, which is about 90 volts for most commercial tubes, and those high enough to cause auto-emission from the cathode (> 1000 volts or so), the photocurrent is nearly independent of the applied voltage. Typical cathode materials for the ultraviolet region are platinum, thorium, and tantalum; for the visible or near infrared various composite surfaces, such as silver–rubidium-oxide–rubidium (called S-3 surface in commercial classification system) and silver–cesium-oxide–cesium (S-1), are employed. These cathodes will withstand high photoemission current densities, thus giving the vacuum phototube a remarkable dynamic range.

The output signal and the noise from a vacuum phototube circuit are both usually determined by a ballast resistor (see Fig. I-16). The phototube can be considered as a current valve allowing (ideally) one electron to pass for each photon† that strikes its cathode; the output signal (voltage) is the product of this current multiplied by the ballast resistance (R). If we can neglect current noise and shot noise, the Johnson noise is predominant, and this noise (voltage) is proportional to $\sqrt{R}$.

Since the signal is proportional to R and the noise is proportional to $\sqrt{R}$, the signal-to-noise ratio is seen to be proportional to $\sqrt{R}$. Thus the largest practical ballast resistor should be used. Assuming $R = 10^7$ and $\Delta f = 1$ c.p.s., Johnson noise would be 4×10^{-7} volt r.m.s. To have a d.c. signal voltage of the same magnitude would require a current of 4×10^{-14} ampere, or 2.5×10^5 photoelectrons/second. At a quantum efficiency of 0.1 this means that 2.5×10^6 quanta per second are required to give a d.c. signal current just equal to the r.m.s. noise fluctuations. If the incident radiation has a wavelength of 4000 Å, this limit corresponds to an incident energy flux of about 1.25×10^{-12} watt.

The photomultiplier tube is an important variation of the vacuum phototube. It incorporates a compact, high-gain amplifier inside the same envelope with the photocathode. The mechanism of this amplifier depends on so-called secondary emission. The photoelectrons from the cathode are accelerated and focused onto an electrode. By secondary emission each photoelectron ejects from 5 to 10 secondary electrons. These secondary electrons are then accelerated and focused onto a following identical electrode from which each of them ejects $\mathfrak{N}$ = from 5 to 10 secondary electrons. This process, repeated 9 times in a typical, popular modern photomultiplier, results in an overall amplification of $\mathfrak{N}^9 \cong 10^6$. The secondary emission surfaces are usually com-

---

† Actually the quantum efficiencies of most photoemissive surfaces lie in the region from 0.1 to .001, so that from 10 to 1000 quanta are required for each photoelectron.

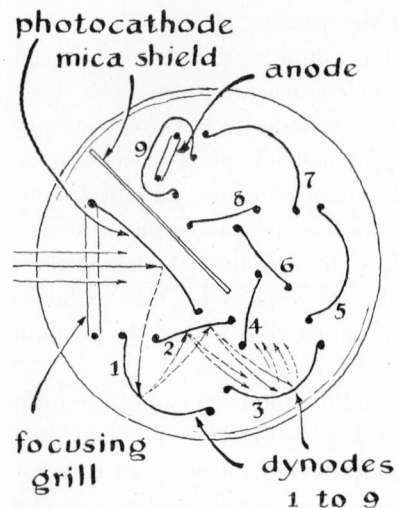

photocathode
mica shield
anode
9
8
7
6
5
4
2
1
3
focusing
grill
dynodes
1 to 9

FIG. I-17 Electrostatic photomultiplier tube[29] (horizontal section). Dynode 1 is made about 90 volts positive with respect to the cathode, dynode 2 is 90 volts positive with respect to dynode 1, and so on, each dynode being 90 volts positive with respect to its predecessor. The anode is made positive with respect to dynode 9, and the current flowing between the anode and dynode 9 is the output signal.

posed of some typical photocathode material, and the electron beam arising from one electrode (frequently called a dynode) is focused onto the next by magnetic or electrostatic deflection. Fig. I-17 shows the very compact dynode arrangement in a photomultiplier where the focusing is electrostatic, and it is achieved with nothing more than the accelerating potentials themselves.

Under most conditions the photomultiplier has a lower noise level than can be achieved with a simple phototube circuit. Whereas the Johnson noise in the large ballast resistor determines the noise level in the simple circuit, noise in the photomultiplier is shot noise in the dark current† and shot noise in the signal current itself. For an example, consider the dark current from an antimony-cesium cathode,[30] which is about 10^{-13} ampere at room temperature. The shot noise in this current is

$$i_{rms} = \sqrt{2ei\Delta f} = 1.7 \times 10^{-16} \text{ amp } (\Delta f = 1 \text{ c.p.s.})$$

which represents the flow of about 10^3 electrons per second. In order to have a d.c. signal equal to this r.m.s. a.c. noise current (neglecting shot noise in the signal current itself), we would need, assuming a quantum efficiency of 0.1, 10^4 photons per second. At a wavelength of 4000 Å this represents an incident energy of 5×10^{-15} watt.

By refrigerating the tube to very low temperatures (say 70°K), the thermionic emission from the cathode can be reduced to around 10^{-19} ampere, representing about 2 electrons per second dark current. At such current levels the detection of single photoelectrons can be treated in a statistical

---

† The dark current (current flowing in the absence of any incident radiation) results from leakage current through the materials of the tube and thermionic emission from the cathode. The thermionic emission usually determines the limitations of the tube.

manner, and the phototube can be considered to approach very closely the perfect quantum counter.

The time constants of the phototube and the photomultiplier are very small, and with special construction can be made as low as 10^{-7} or 10^{-8} second.

The Geiger Counter[31]

The Geiger, or Geiger-Müller, counter[32] is a particle counter, and thus a useful device in nuclear physics. It is also used extensively as a detector of high energy quanta such as X-rays, gamma rays, and cosmic rays, and of ultraviolet light.

The counter was first described by Geiger and Müller in 1928. It consists of a hollow metal cylinder, a thin wire suspended on the axis of this cylinder, and a surrounding envelope, usually glass, to contain some selected gas at a particular pressure (see Fig. I-18). A potential, usually in the order of 1000 volts, is applied between the cylinder and the wire. The cylinder is at a negative potential (cathode) with respect to the wire (anode). The wire is connected to ground through a ballast resistor.

There is an electrical breakdown of insulation in the gas between the anode and the cathode when a high energy charged particle or a photon produces one or more free electrons in this space. A free electron can be produced by the ionization of a gas molecule, by photoemission from the inner cathode surface, or by Compton recoil in the metal cathode shell. In the absence of any ions or electrons in this gas, the resistance of the counter is very high, but a single electron freed by ionization between the electrodes is sufficient to break down this insulation of the gas and permit a large instantaneous current to flow. Electrons freed by ionization are accelerated in the electric field, and inelastic collisions with neutral gas molecules will ionize or excite

---

FIG. I-18   Typical Geiger counter and circuit arrangement.

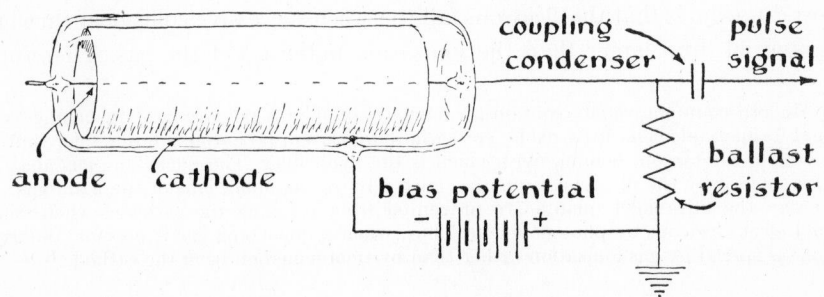

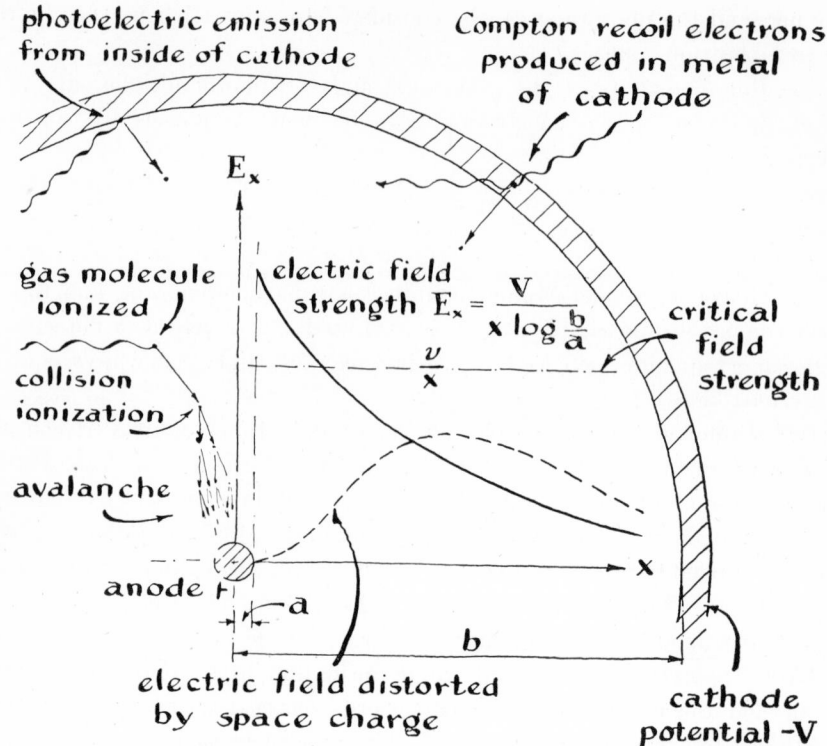

FIG. I-19   Electric field strength in the interelectrode space of a Geiger counter.

them if the electrons' energy is sufficiently high. Thus one electron may excite or ionize many molecules on its way to the anode. Furthermore, recombinations of electrons and ionized molecules can produce ultraviolet quanta capable of photoelectric ionization elsewhere in the counter.† Because of these effects a chain-reaction production of ion-electron pairs, which is called a Townsend avalanche, can be set off in the Geiger counter. The necessary condition for the occurrence of this avalanche to produce a high current of short duration is that the electrons fall (between successive collisions) through a potential drop larger than the ionization potential of the gas. Therefore,

---

† Modern counters usually contain a certain amount of a polyatomic "quenching" gas (ethyl formate, alcohol, nitric oxide, etc.) whose ionization potential is lower than that of the simple gas (argon, helium, etc.) which is the main filler. The quenching gas absorbs and is ionized by the ultraviolet quanta emitted by recombining ions of the main gas. In this way the ultraviolet quanta are prevented from reaching the cathode, where they would eject electrons by photoemission. Thus, when a quenching gas is present, the avalanche is spread by gas ionization rather than by photoemission from the cathode.

the mean free path in the gas, X, which is a function of the pressure, must be such that

$$\overline{E} \cdot \overline{X} > Vi$$

where E = electric field strength (see Fig. I-19), Vi = ionization potential, and X = mean free path of electron (its value is the same everywhere within the counter).

Too low a field in the counter, or too high a gas pressure, will fail to produce the avalanche. Too high a potential, on the other hand, will initiate a sustained corona discharge. The region between these extremes, called the counter plateau, is that required for the counter to operate as intended.

Once initiated, the discharge dies out because of the geometrical configuration of the counter. From the point of initiation in the gas the freed electron proceeds toward the wire anode (see Fig. I-19), producing more ion pairs on the way. For this reason and also because the electric field strength is greater near the anode, the greatest density of produced ions is in its vicinity. The massive positive ions move sluggishly in the electric field compared with the fleet electrons, and as the electrons evacuate themselves from the interelectrode space and arrive at the anode, there is left a sheath of positive ions surrounding the anode. This space charge lowers the field strength near the anode below that required to maintain multiplication of ionization, so that the discharge automatically stops. After these positive ions move far enough from the anode to restore the field strength near it (see Fig. I-19), the counter will be operative again. The time for the positive ion to reduce sufficiently is called the "dead time." It depends on the geometry of the counter, the potential applied, and the nature and pressure of the filling gas. For a typical counter it is in the order of 10^{-4} to 10^{-3} second.

When the positive ion sheath reaches the cathode, the electric field in the counter is restored, and the counter is operative again. A positive ion in the sheath draws an electron from the cathode and becomes neutralized, emitting in the process a quantum of radiation whose energy is the difference between the ionization energy of the gas molecule and the work function of the cathode material. If this energy difference exceeds the cathode work function sufficiently, *i.e.* if gas ionization energy is larger than twice the cathode work function, the quanta so produced will emit photoelectrons from the cathode and rekindle the avalanche. To prevent this, a small amount of "quenching" gas is added. Its ionization potential is lower than that of the main gas filler, and ionization is transferred by collision from the main gas molecules to the quenching gas molecules. The energy difference between the ionization potentials will then appear in the form of low energy quanta that are unable to produce photoemission. The positive ion sheath that ultimately reaches the cathode is then made up of quenching gas ions that, on being neutralized, emit quanta which have too little energy to cause photoemissions; and the avalanche is not rekindled.

For a description of the variations employed to suit specific counter applications the student is referred to more complete discussions of counter theory.[31]

In some cases, by the proper selection of configuration, materials, and gas pressure, with reference to the wavelength of the quanta to be counted, counting efficiency can be made as high as 99.8%. In addition, spurious counts can be reduced to a negligible level by using the proper tube voltage and by carefully eliminating any radioactive contaminants from the materials used in the counter. With care, the Geiger counter can be made, over a range of wavelengths from 1.0 to 50 Å, to approach the perfect detector very closely. The efficiency will be appreciably less for wavelengths below 1 Å or above 50 Å. Fig. I-20 shows the spectral sensitivity of a typical ultraviolet photon counter.[33] In those cases where efficiency cannot, for one reason or another, be made near 100%, or where spurious counts are high, the reliability of the counter's information is a matter of statistical evaluation.[34]

FIG. I-20 Spectral sensitivity of typical far ultraviolet photon counter.[33] The counter has a chrome iron cathode and a lithium fluoride entrance window, and contains 630 mm partial pressure of helium and 7 mm of ethyl formate quenching gas. Between 3000 and 1500 Å the response is attributed to photoelectrons from the surface of the cathode. Ethyl formate absorbs strongly from 1800 to 1500 Å, causing a reduced response in this region, and the increased response below 1500 Å is due to photoelectrons from deeper within the cathode material—the "internal photoelectric effect" which is characteristic of metals and is more efficient than the surface phenomenon. At 1180 Å photoionization of the ethyl formate causes another rise in the sensitivity curve (photoionization of the helium would require light of wavelength < 50 Å), and at 1050 Å the entrance window becomes opaque.

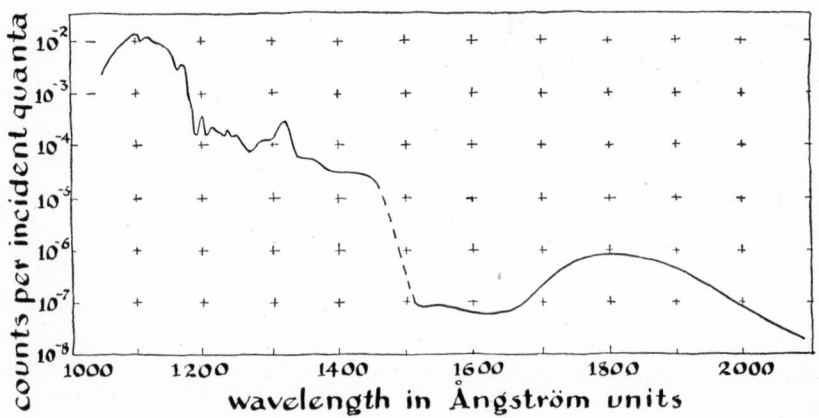

A.C. Modulation of Radiation

For several reasons it is frequently advantageous to modulate the radiation striking a detector so as to produce an a.c. output signal instead of a d.c. signal. The chief virtues of such a procedure are:

D.c. amplifiers for weak currents or potentials are less stable and more difficult to construct and operate than a.c. amplifiers. It is possible, of course, to create an a.c. signal from a d.c. electrical response by means of a commutator. Then the a.c. signal, after amplification, can be rectified for final display. However, beginning with an a.c. electrical response, produced by chopped irradiation, obviates this process.

When modulated or chopped radiation is used, there is no drift as with a d.c. system.

By modulating only the radiation that is to be measured, it is possible to discriminate against unmodulated spurious radiation within the field of view of the detector. Scattered light in a monochromator is an example of such spurious radiation. The radiation striking the detector in a monochromator includes both that which has properly traversed the dispersing system and that which has arrived by being reflected without dispersion from the optical surfaces, or scattered. If a chopper is so arranged that only the properly dispersed radiation is modulated, only its contribution will be measured. The unmodulated reflected or scattered radiation will produce a d.c. electrical response that is ignored by the a.c. amplifier.

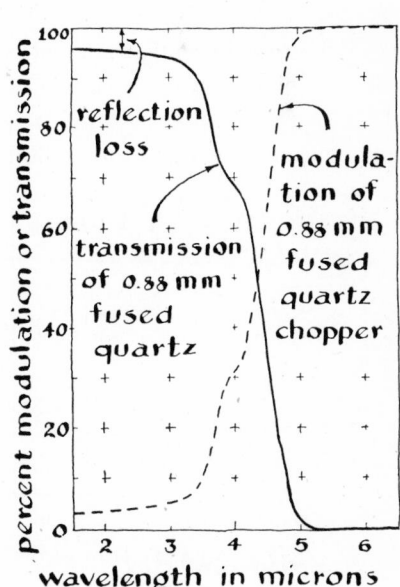

FIG. I-21 Filtering effect obtained by using a chopper blade made of Pyrex rather than a uniformly opaque material.

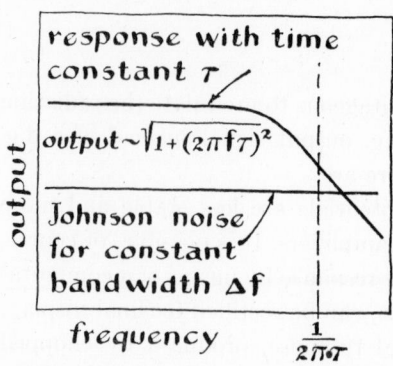

FIG. I-22 Detector response in the presence of white noise.

By using a chopping (modulating) blade which is transparent to certain wavelengths, it is possible to obtain an ingenious optical filtering characteristic (see Fig. I-21). Radiations to which the chopper is transparent are not modulated. The electrical detector response resulting from them will therefore be a d.c. signal that is ignored by an a.c. amplifier. Thus only those radiations to which the chopper is opaque will be modulated and detected.

By using a narrow bandwidth a.c. amplifier that is tuned to a modulation frequency, it is possible to discriminate against a considerable portion of the inherent detector noise, and thus, for a given $\frac{S}{\mathfrak{N}}$, a smaller radiation signal can be measured. In the case of a detector whose limiting noise is Johnson noise, any frequency, f, is satisfactory that is appreciably less than $\frac{1}{2\pi\tau}$, where τ is the time constant of detector, as before. Since Johnson noise is white, the noise discrimination depends only on the bandwidth, Δf; it is independent of the chopping frequency chosen. Fig. I-22 shows the manner in which the detector response varies with frequency. Here we consider time constant τ and sinusoidal modulation of the radiation.

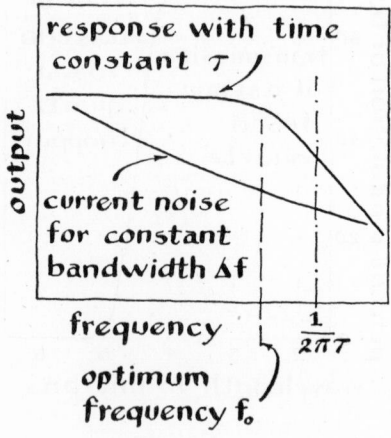

FIG. I-23 Detector response in the presence of noise which is a function of the frequency.

In cases where the noise amplitude is frequency-dependent, as current noise in a photoconductive detector, there are preferred frequencies at which to work. Fig. I-23 illustrates the choice of frequency in such a case, in order to obtain the best signal-to-noise ratio.

When a detector of time constant τ is used with modulated incident radiation, modulated at frequency f, the output of the detector will be multiplied by factor $\mathfrak{f}$.

$$\mathfrak{f} = \frac{1}{\sqrt{1 + (2\pi f\tau)^2}}$$

for sine wave modulation of incident radiation and

$$\mathfrak{f} = \frac{1}{\tanh\left(\dfrac{1}{4f\tau}\right)}$$

for square wave modulation.

The time constant of an a.c. amplifier having bandwidth Δf is $\dfrac{1}{2\pi\Delta f}$. This time constant affects the amplification factor in the manner described above. It is common practice to rectify the output of the amplifier for display by a d.c. meter together with a low-pass RC filter. Such a combination of amplifier and meter provides a variable overall time constant that is easily controlled by switching condensers in and out of the circuit.

REFERENCES

1. For a more complete discussion of the sources of noise and the mathematical methods of working with noise see Goldman, *Frequency Analysis, Modulation and Noise* (1948, McGraw-Hill), Chaps. VI–IX; Van der Ziel, *Noise* (1954, Prentice-Hall Electrical Engineering Series), Chap. 8.

2. A classification of detectors in terms of their sensitivity and speed of response has been published by R. Clark Jones, who has pursued the subject of ultimate sensitivity with rigor:
   R. C. Jones, "A New Classification System for Radiation Detectors," JOSA, *39*, 327 (1949);
   R. C. Jones, "Factors of Merit for Radiation Detectors," JOSA, *39*, 344 (1949);
   R. C. Jones, "Performance of Visible and Infrared Detectors" in *Advances in Electronics* (1953, Academic Press, New York), Vol. V.

3. William Herschel, "Experiments on the Refrangibility of the Invisible Rays of the Sun," Phil. Trans. Roy. Soc. (London), *90*, 284 (1800). In the 1809 abridgment this paper appears on p. 688 of Vol. 18.

4. R. J. Havens, "Theoretical Comparison of Heat Detectors" (abstract), JOSA, *36*, 355 (1946).

5. Lord Rayleigh, "On the Thermodynamic Efficiency of the Thermopile," Phil. Mag., *20*, 361 (1885).

6. D. F. Horning & B. J. O'Keefe, "The Design of Fast Thermopiles and the Ultimate Sensitivity of Thermal Detection," R.S.I., *18*, 474 (1947).

7. S. P. Langley, Nature, *25*, 14 (1881); and Proc. Am. Acad. Arts and Sciences, *16*, 342 (1881).

8. R. C. Jones, JOSA, *43*, 1 (1953).

9. W. H. Brattain & J. A. Becker, JOSA, *36*, 354 (1946).

10. E. M. Wormser, JOSA, *43*, 15 (1953).

11. D. H. Andrews, W. F. Brucksch, W. T. Zeigler, & E. R. Blanchard, R.S.I., *13*, 281 (1942); N. Fuson, Jr., JOSA, *38*, 845 (1948).

12. M. J. E. Golay, R.S.I., *18*, 347, 357 (1947), and *30*, 816 (1949). The Golay cell is manufactured by the Eppley Laboratories, Newport, R.I.

13. S. Hecht, S. Shlaer, & M. H. Pirenne, "Energy, Quanta and Vision," J. Gen. Physiol., *25*, 819 (1942).

14. Hecht & Verrup, J. Gen. Physiol., *17*, 263 (1933).

15. L. A. Riggs, F. Ratliff, J. Cornsweet, & T. N. Cornsweet, JOSA, *43*, 495 (1953).

16. A. Wald & A. B. Clark, J. Gen. Physiol., *21*, 93 (1937).

17. J. Ségal, *Le Mécanisme de la vision des couleurs* (1953, G. Doin, Paris), reviewed by D. L. MacAdam, JOSA, *46*, 1002 (1956).

18. S. H. Talbot, "Recent Concepts of Retinal Color Mechanism," JOSA, *41*, 895 (1951).

19. M. H. Pirenne, *Vision and the Eye* (1948, Chapman Hall, now Pilot Press, London).

20. R. Granit, *Sensory Mechanisms of the Retina* (1947, Oxford Univ. Press).

21. E. N. Willmer, *Retinal Structure and Color Vision* (1946, Cambridge Univ. Press).

22. C. E. K. Mees, *The Theory of the Photographic Process* (1952, Macmillan).

23. H. Hoerlin, JOSA, *39*, 891 (1949).

24. W. Smith, Am. J. Science, *5*, 301 (1873).

25. R. A. Smith, "Recent Developments in the Detection and Measurement of Infrared Radiation," Scientific Monthly, Jan. 1956; E. Burstein & P. H. Egli, "The Physics of Semiconductor Materials" in *Advances in Electronics and Electron Physics* (1955, Academic Press, New York); K. Lark-Horovitz, "The New Electronics" in *The Present State of Physics* (1954, American Association for the Advancement of Science).

26. Eastman Kodak Company, *Kodak Ektron Detector* (Pamphlet U-2).

27. D. G. Avery, D. W. Goodwin, W. D. Lawson, & T. S. Moss, Proc. Phys. Soc. (London), *B67*, 761 (1954); T. S. Moss, Proc. Phys. Soc. (London), *B67*, 775 (1954).

28. V. K. Zworykin & E. G. Ramberg, *Photoelectricity* (1949, Wiley).

29. F. Johnson, K. Watanabe, & R. Tousey, JOSA, *41*, 702 (1951).

30. R. W. Engstrom, JOSA, *37*, 420 (1947).

31. H. Friedman, Proc. I.R.E., *37*, No. 7, p. 791 (1949).

32. H. Geiger & W. Muller, Phys. Zeit., *29*, 839 (1928), and *30*, 489 (1929).

33. T. A. Chubb & H. Friedman, R.S.I., *26*, 493 (1955).

34. John Strong, H. V. Neher, A. E. Whitford, C. Hawley Cartwright, and Roger Hayward, *Procedures in Experimental Physics* (1938, Prentice-Hall, N.Y.), Chap. VII by H. V. Neher.

35. G. Wald, Science, *101*, 653 (1945).

Microwave Experiments and Their Optical Analogues

by Gordon Ferrie Hull, Jr.[†]

The advent of vacuum tubes for the production of microwaves and of fixed crystals for detecting them has made it possible to demonstrate readily all the optical properties of electromagnetic radiation. The apparatus which is described makes use of radiation of wavelength 3.2 cm, and the experiments which will be carried out are essentially those for free space microwaves, demonstrating the analogous phenomena of geometrical and physical optics. These experiments have been partially described in a previous paper.[1]

Because a microwave generator produces plane polarized, coherent, monochromatic radiation, one would expect some differences to occur between the optical properties of free space microwaves and those of light. Such differences do occur, for a light source is seldom strictly monochromatic; its wavelength is less by a factor of 10^5, and its radiation is not coherent. Consequently, the optical analogues of microwaves must be considered as analogues and not as identities.

Generator and Receiver for 3.2 cm Microwaves

The generator and receiver for 3.2 cm microwaves are shown in Fig. J-1. A Western Electric 2 K 24 or 723 A/B reflex klystron is used as the microwave source.[2] This tube requires two power supplies, one regulated at 300 volts and about 30 ma., to accelerate the electron stream through the cavity, and another to apply a variable negative voltage from 0 to -300 volts between the cavity and the repeller. A means of modulating the repeller voltage with

[†] Dartmouth College.

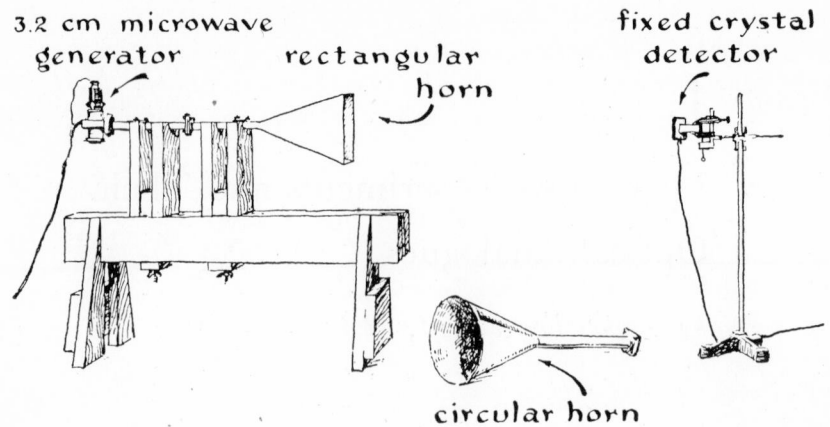

3.2 cm microwave generator rectangular horn fixed crystal detector circular horn

FIG. J-1 Transmitter and receiver with horn radiators for 3 cm microwave experiments.

an audio oscillator should also be provided. This tube will deliver about 30 mw. power, in a frequency range from 8500 to 9700 mc. (from 3.5 to 3.1 cm). The coaxial output of the 723 A/B tube is coupled to a standard $0.5'' \times 1''$ (outside dimensions) rectangular wave guide having a wall $0.056''$ thick. The TE_{01} mode[3] is excited in the rectangular guide, and the microwaves are propagated along it and radiated from the 20 db horn, which is shown in Fig. J-1 mounted together with the 723 A/B tube and its section of wave guide on a wooden bench. Wave guide mounts for the 723 A/B tube and also power supplies can be bought from suppliers of microwave apparatus or can be made. The details of construction of the wave guide mount are given in the Western Electric circular describing the operation and performance of the 723 A/B tube.

The receiver, which is shown clamped to a rod stand in Fig. J-1, consists of a short section of wave guide into which a Western Electric 1N23B fixed crystal detector is appropriately mounted. This piece of wave guide is shorted at one end and has a plane rectangular coupling flange at its open end. The crystal output is connected by coaxial microphone cable to either an audio amplifier and loudspeaker for demonstration or to a vacuum tube voltmeter if precision measurements are required. If the crystal is "square law," as it usually is, the vacuum tube voltmeter will measure directly the relative microwave power received. Although such a receiver can be built, it is simpler to buy a wave guide crystal mount from one of the suppliers of microwave apparatus. Since the receiver is small, it can be used as a probe for exploring radiation coming from different directions. It can be held in the hand for this purpose and moved about. Since the TE_{01} mode has the electric vector across the short dimension of the wave guide, the receiver is also an analyzer for the polarization of the microwave radiation. Finally, the receiver can be mounted on a wooden bench similar to the one on which the transmitter is mounted. Other sections of wave guide and horns can be attached to the receiver by means of the coupling flange.

In Fig. J-1 two horn radiators are shown: a rectangular one attached to the generator on the wooden bench and a circular one in the foreground.[4] Each horn has an absolute gain of 20 db. The rectangular horn, which is attached to a standard 0.5″ × 1″ rectangular guide, has the dimension of $3.6\lambda \times 4.45\lambda$ for its open end and an axial length of 6λ where λ is the free space wavelength. The circular horn has a diameter of 4.4λ for its open end and an axial length of 6λ, and is attached to a standard 1″ (outside diameter) circular wave guide having a wall 0.032″ thick. Each of these horns has a total beam width of about 8° at the half-power points. The circular horn can be substituted for the rectangular horn on the transmitter, and the TE_{11} mode[5] will be excited in the circular wave guide. If this substitution is made, a standing wave will be produced in the section of rectangular guide from the generator because of the sharp discontinuity at the rectangular-circular wave guide junction. For demonstration purposes the discontinuity is not troublesome. For many measurements, however, it is desirable to eliminate the standing wave. This can be done by the insertion of a transition section of guide, which changes gradually from rectangular to circular wave guide. Such a transition section is shown in Fig. J-2.

A number of other wave guide components, also shown in Fig. J-2, are useful in wave guide measurements. At the bottom of Fig. J-2 is a 0.5″ (internal diameter) circular guide loaded with a polystyrene rod. When the polystyrene rod is removed, the wave guide diameter is below cutoff, and hence the microwaves are not transmitted through it. Next is the transition section from rectangular to circular guide, followed by a twist section to change the polarization through a right angle, and, last, two rectangular wave guide bends. With the apparatus shown in Figs. J-1 and J-2, the experiments with wave guides previously described by the author for 10 and 20 cm microwaves can be performed with 3 cm microwaves.[6]

Transmission and Reflection

Besides the transmission of 3 cm microwaves through the various wave guide components shown in Fig. J-2, the transmission of free space microwaves through various dielectrics such as sheets of glass and plywood can be demonstrated by inserting the dielectrics between the transmitter and the receiver. These dielectrics are also partial reflectors. For these demonstra-

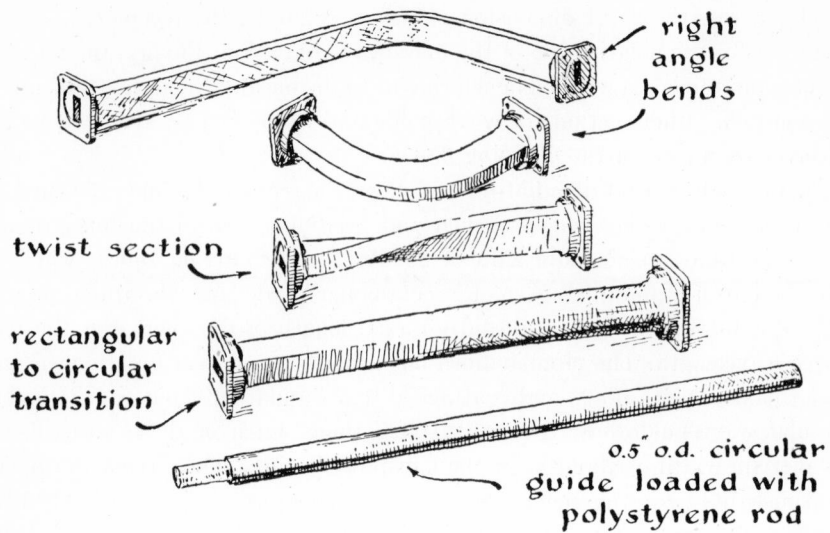

FIG. J-2 Wave guide components for demonstrating transmission of 3 cm microwaves through wave guides.

tions the receiver should be at some distance from the horn radiator. A sheet of copper and $\frac{1}{8}''$ mesh copper screen are excellent reflectors and prevent the transmission of the 3 cm microwaves. Standing waves in air can be produced by reflection from a copper sheet and an approximate wavelength measurement made. A plywood sheet which has been wound with wire spaced $\frac{1}{8}''$ apart completely stops and reflects 3 cm microwaves when the wires are parallel to the electric vector, but transmits the radiation when the wires are at right angles. If the wire is spaced 0.5'' on the plywood sheet, it will be found to be about half reflecting and half transmitting when the wires are oriented parallel to the electric vector.

Instead of using a sheet of brass as a reflector, one can also use an ordinary plane mirror of silvered glass. It is easy to show for microwaves as for light that the angles of incidence and reflection are equal. Two silvered mirrors or two brass sheets at right angles will reflect the microwave radiation in the direction from which it came. Finally, a concave spherical mirror, if large in aperture, will focus the microwave radiation sharply. The author uses a concave mirror, of silvered glass, 12'' in diameter and 20'' in focal length, for this purpose. For short focal lengths and large diameters, parabolic reflectors of metal are used.

Interference and Diffraction

Because of the long wavelength of microwaves, compared with light, interference from double or multiple slits can be demonstrated with large slits spaced only a few wavelengths apart. In Fig. J-3 two brass plates 10″ square are shown with two and four slits. The slits are $\lambda/2 \times \lambda$ in size and are 2λ apart. By making the slits narrow, the diffraction pattern due to a single slit covers a total angle of more than 180° and has little effect on the interference pattern produced by the slits. To show the interference from two slits, the brass plate is placed in a holder clamped to the wooden bench supporting the transmitter, a few inches from the open end of the horn. The receiver can then be moved about in front of the double slit screen to locate the maxima. Because the slit spacing is 2λ, there are maxima at 30° and 90° on each side of the central maximum. A brass plate with two slits spaced 4λ apart will give four maxima on each side of the central maximum. Finally, the brass plate shown with four slits spaced 2λ apart gives an interference pattern which is the combination of the two double slit patterns in which the slit spacings are 2λ and 4λ.

To measure the interference pattern from double slits quantitatively, the apparatus is arranged as shown in Fig. J-4. The receiver, clamped to a rod stand, is fastened with a wire 5–6′ long to the base of the wooden holder

---

FIG. J-3   Double and quadruple slits and zone plates for demonstrating interference and diffraction of 3 cm microwaves.

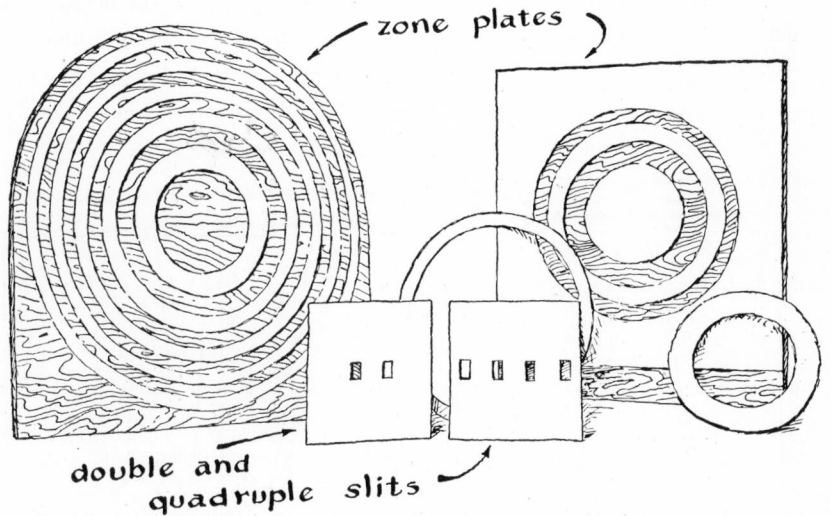

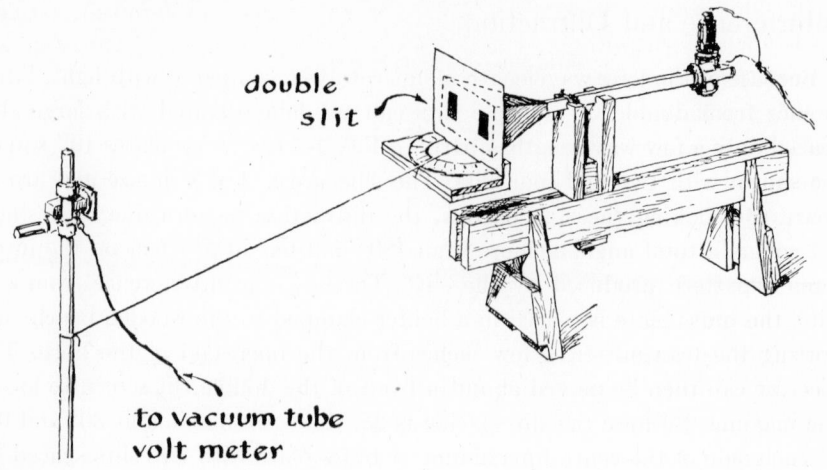

FIG. J-4 Apparatus for measuring interference patterns from double slits.

supporting the brass plate containing the double slit. A large protractor is also fastened to the wooden holder just below the wire, so that, as the receiver is moved along the arc of a circle, whose radius is the length of the wire, the angle which the receiver makes with the axis of the double slit can be measured with the protractor. The output of the receiver is connected to a vacuum tube voltmeter. In Fig. J-5 are shown quantitative measurements of interference maxima and minima from double slits spaced 2λ, 3λ, and 4λ apart. The solid curves are the theoretical interference patterns. It will be noted that the experimental measurements of maxima and minima occur at the correct angles but that the intensities of the maxima decrease with increasing

FIG. J-5 Quantitative measurements of interference patterns from double slits spaced 2λ, 3λ, and 4λ apart.

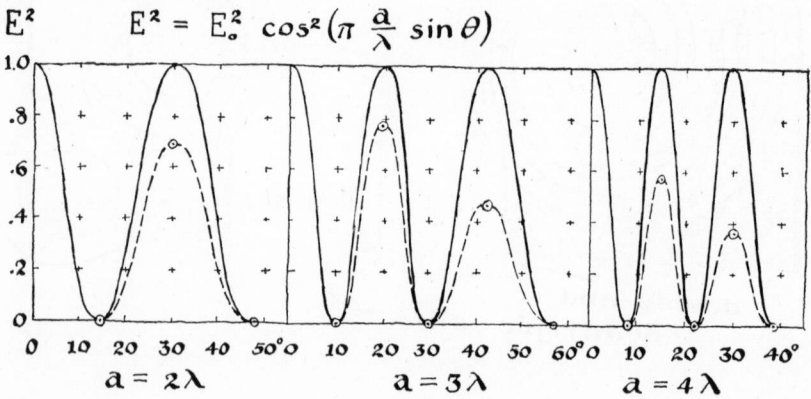

angle. This is due in part to the fact that the diffraction pattern from a single slit is not uniform with angle and decreases in intensity slowly with increase in angle from the central axis.

As a corollary to Young's double slit experiment, interference can be produced by means of reflection as in Lloyd's single mirror experiment. All that is required is to place a brass plate just to one side and in front of the horn radiator and then investigate the interference pattern with the receiver.

In Fig. J-3 two zone plates are shown for demonstration of Fresnel diffraction. The fixed zone plate on the left has even-numbered half-period zones cut from galvanized sheet iron and tacked to plywood 0.5″ thick, thus exposing the odd-numbered, half-period zones. The zone plate on the right has four zones, cut from galvanized sheet iron, which are supported on a plywood board with two pins at the top of each zone. These zones are made removable so that the effect of removing successive zones one after another can be demonstrated when the receiver is placed at the focal point of the zone plate. Each zone plate has a focal length of 10λ, and the radii of the zones are given by the usual equation $r_n = [nf\lambda + (n\lambda/2)^2]^{1/2}$, where n is the number of the zone, f the focal length, and λ the free space wavelength. The zone plates should

FIG. J-6   Removable zone plate for measurement of Fresnel diffraction.

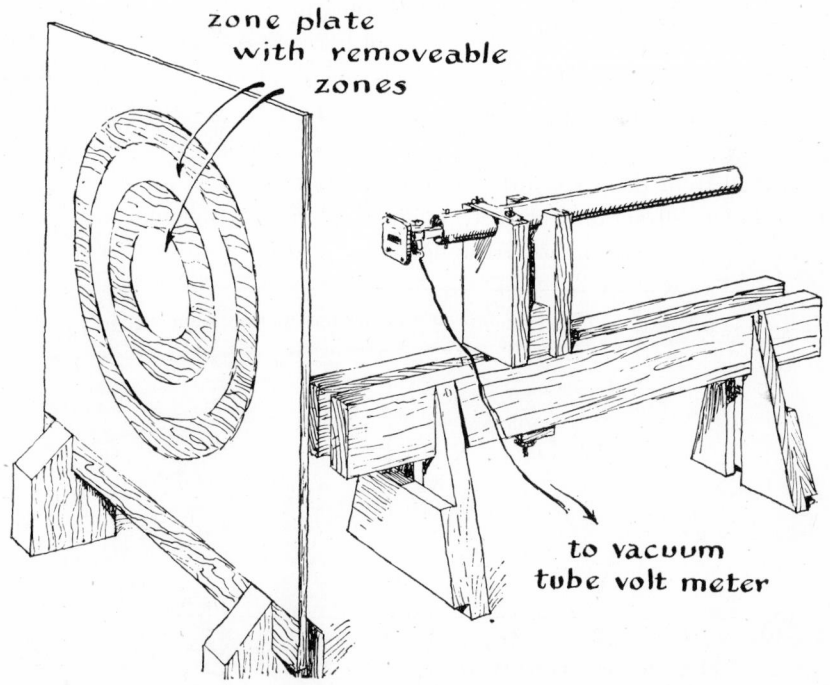

zone plate
with removeable
zones

to vacuum
tube volt meter

be placed at some distance from the transmitter to ensure that a plane wave will strike the zone plate. When this is done, the focal point is very sharp and can be easily located within $\pm\lambda/2$.

To measure the removable zone plate quantitatively, the apparatus is arranged as shown in Fig. J-6. The removable zone plate is supported in front of a wooden bench, and the receiver, connected to a vacuum tube voltmeter, is mounted on the wooden bench in an appropriate holder at a distance from the zone plate equal to the plate's focal length, 10λ. First the illumination of the signal $\frac{1}{2}c\kappa_0 E_0^2$ from the transmitter is measured in the absence of the zone plate, and this value is used for reference. After the zone plate is placed in front of the bench, successive zones are removed one at a time, and the illumination $\frac{1}{2}c\kappa_0 E^2$ of the received signal is measured. In Fig. J-7a the ratio in db, $10\log_{10}\dfrac{E^2}{E_0^2} = 20\log_{10}\dfrac{E}{E_0}$, is plotted against the number of zones removed. When the first half-period zone is removed, the received amplitude, compared with the received signal in the absence of the zone plate, is doubled (increased 6 db). Removal of the second zone decreases the received signal by 27 db (a factor of 1 to 22.5 in amplitude), signifying almost complete cancellation of the amplitude from the first by that from the second. Removal of the third zone increases the signal again, and removal of the fourth zone decreases the signal, as expected, but does not produce complete cancellation. The reason for this is shown in Fig. J-7b, in which successive zones are removed. The fact that the straight line is not horizontal shows that the amplitude from each zone is not constant. This is due to the fact that the amplitude of the wave front over the whole aperture of the zone plate is not constant, being largest at the center and slowly decreasing radially outward, as would be expected from the transmitter horn diffraction pattern.

Fraunhofer diffraction from a rectangular or circular opening is exhibited by the radiation patterns of a rectangular or circular horn or of a parabolic

---

FIG. J-7 Quantitative measurements of Fresnel half period zones.

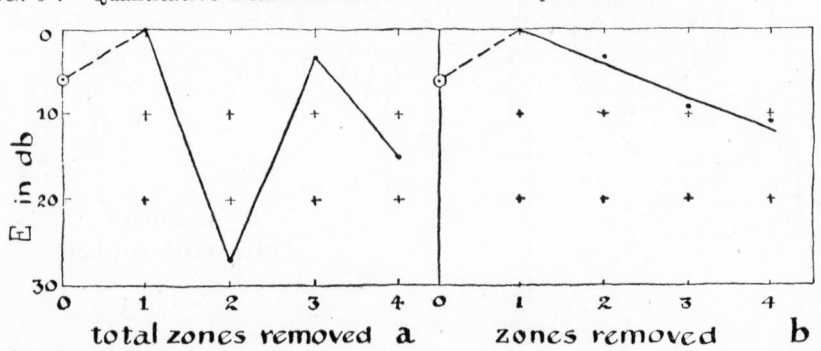

reflector.[7] To demonstrate or measure a Fraunhofer diffraction pattern, a horn with receiver attached should be mounted on a rotating stand at a large distance from the microwave transmitter. The transmitter and the horn whose pattern is to be measured should first be lined up, and then, as the horn is rotated, the received microwave power as a function of angle is measured. The vacuum tube voltmeter connected to the crystal detector of the receiver will measure relative microwave power directly, provided the crystal obeys the square law. To obtain the diffraction pattern of a parabolic reflector, the receiver should be mounted at the focus of the parabola and the entire assembly rotated about a vertical axis, as is done with the horn.

Another interesting demonstration, as well as an instrument for precision measurement, is the microwave Michelson interferometer. This instrument, shown in Fig. J-8, consists of the microwave transmitter; two totally reflecting mirrors of brass 10″ square, mounted on movable supports on the wooden benches; a half-reflecting mirror made by winding wires 0.5″ apart on a 12″ plywood board, which is mounted on a rotating support on a wooden bench; and the receiver, which is shown clamped in a rod stand in the foreground. The transmitter, mirrors, and receiver must be carefully lined up. When one of the totally reflecting mirrors is moved slowly along the wooden bench, the receiver will indicate the passage of maxima and minima corresponding to bright and dark fringes in the optical case, as shown in Fig. J-9. From precise measurement of the distance between minima or maxima the

FIG. J-8   Michelson interferometer for 3 cm microwaves.

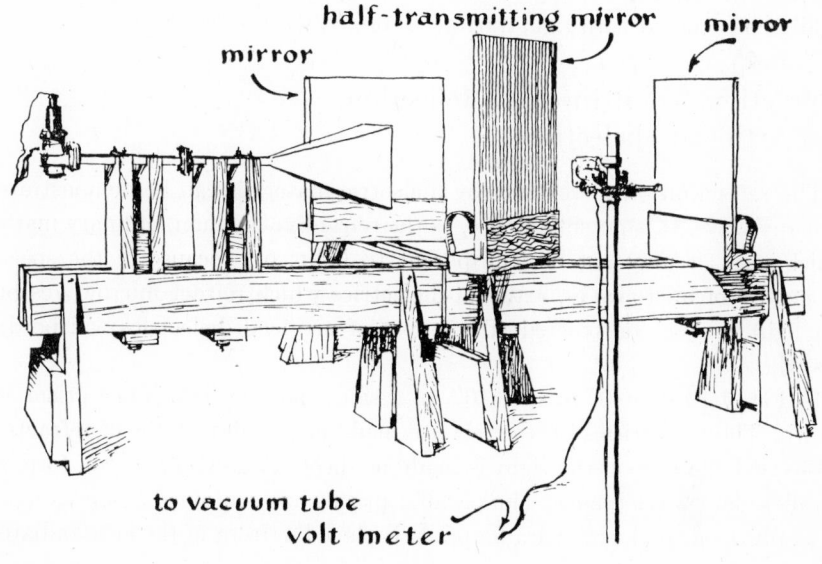

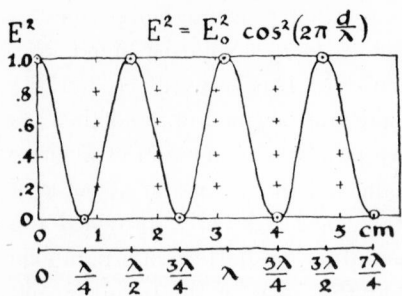

FIG. J-9 Measured and calculated Michelson interferometer fringes.

wavelength of microwaves can be obtained, as in the optical case. In fact, when the interferometer is properly adjusted, the minima are extremely sharp and are from 30 to 40 db below the maxima in intensity. If several sheets of dielectric such as glass or plywood are placed in one arm of the interferometer, the receiver will indicate maxima and minima as the sheets are removed one after another. The index of refraction of a dielectric can be measured by noting the fringe shift produced by the insertion of a known thickness of the dielectric in one of the interferometer arms, as is done in the optical case. In the microwave region the index of refraction of ordinary window glass is about 2.0 and that of plywood about 1.3. Hence, four sheets of single weight window glass or four sheets of $\frac{1}{4}''$ plywood inserted in one arm of the interferometer will produce a shift of about one fringe. The microwave Michelson interferometer is capable of high precision and is especially useful in measuring the dielectric constants of artificial microwave dielectrics, which will be discussed in the next section and which cannot be placed inside a 3 cm wave guide. As a precision instrument, the interferometer must be rigidly constructed with rigid mirrors equipped with screw drives.

Refraction, Total Internal Reflection, and Artificial Dielectrics

The refraction of microwaves by dielectric materials can be demonstrated in many ways. It is possible to use, as the refractive medium, ordinary matter which may or may not be transparent to light; or, because of the special properties of microwaves, artificial dielectrics which refract microwaves but not light can be constructed. Two types of artificial dielectrics will be discussed.

In Fig. J-10 two 60° prisms, 10'' on a side, and one right angle prism are shown. The two prisms on the left are made of paraffin (index of refraction 1.47), and the one on the right is made of sheets of galvanized iron forming parallel-plate wave guides. The paraffin prisms are contained in forms made of $\frac{1}{4}''$ plywood. If the 60° paraffin prism is placed in front of the horn radiator,

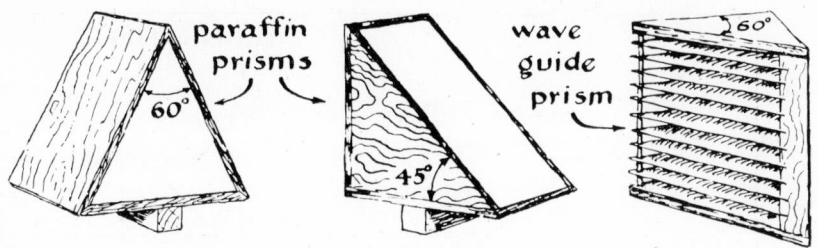

FIG. J-10   Paraffin and wave guide prisms for 3 cm microwaves.

it is found that the microwave beam is bent in the same way that a light beam is bent by a glass prism. Holding the microwave receiver in the hand and rotating the prism, one can locate the angle of minimum deviation, which for a 60° paraffin prism is 33°. With a right angle prism made of paraffin, total internal reflection of the 3 cm microwave beam can be demonstrated. If the microwave receiver is brought close to the totally reflecting surface of this prism, the presence of surface waves, the distance by which these waves emerge from the paraffin surface, and their polarization can be readily measured.

The expression for the amplitude of the surface wave emerging from the totally reflecting surface of the paraffin right angle prism[8] is

$$E_y = E_{oy} \exp 2\pi j \left(ft - \frac{x \sin r}{\lambda} \right) \exp \left(-\frac{2\pi z}{\lambda} \sqrt{N^2 \sin^2 i - 1} \right)$$

where $j = \sqrt{-1}$, f is the frequency, t the time, λ the wavelength, i the angle of incidence, r the angle of refraction, and N the index of refraction; the coordinate x and y axes are those indicated in Fig. J-11. The first exponent represents the propagation constant and the second the attenuation constant. This wave is traveling along the surface in the x direction. Its wave front, or surfaces of constant phase (yz plane), is at right angles to the surfaces of constant amplitude (yx plane). According to the second exponential term, the amplitude of the surface wave is damped out exponentially as z increases. As indicated in the upper right-hand part of Fig. J-11, the rate at which the surface wave is damped out as z increases can be measured. The measurements are shown in Fig. J-11 for a paraffin right angle prism which gives a slope of 5 db/cm (or 1.79/cm). Using the second exponential term in the equation, the expression $20 \log_{10} \frac{E_y}{E_{oy}}$ is calculated with $\lambda = 3.2$ cm, $N = 1.47$, and $i = 45°$. This gives a slope of 5.2 db/cm (or 1.82/cm), which is to be compared with the experimental measurement of 5 db/cm.

The wave guide prism in Fig. J-10 operates only as a prism when the electric vector is parallel to the metal plates of the prism. For this case the TE_{01}

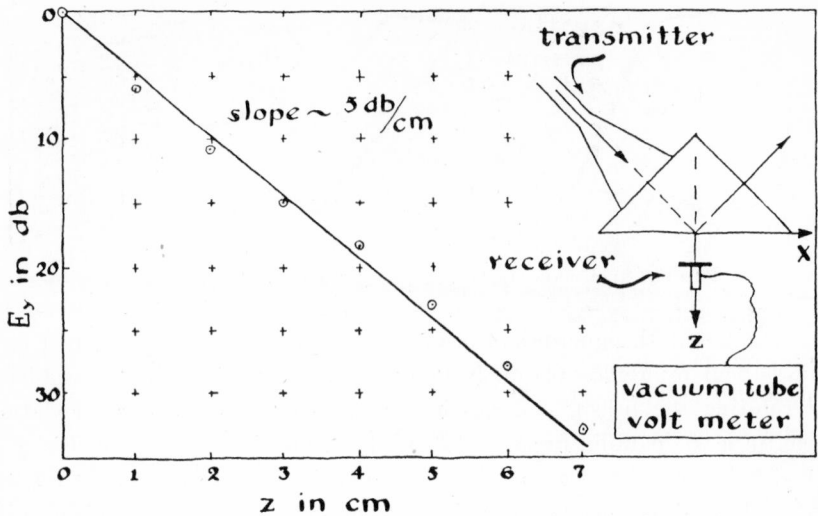

FIG. J-11 Measured damping of the emerging wave at the surface of a totally reflecting right angle paraffin prism.

mode for a parallel plate wave guide is excited, and the wave velocity in the parallel plate guide is greater than the free space velocity. The wave velocity is given by $v = c[1 - (\lambda/2b)^2]^{-1/2}$, where c is the velocity of light, λ the free space wavelength, and b the spacing between the metal plates.[3] The index of refraction of the parallel plate wave guide is then $N = [1 - (\lambda/2b)^2]^{1/2}$. The prism shown has an index of refraction of 0.6, corresponding to a plate spacing of $b = 0.79''$. The plate spacing is critical and should be maintained to within ±2 percent. Because the index of refraction is less than unity, the microwave beam is bent in the opposite direction from what it is for the paraffin prism. By rotating the wave guide prism, an angle of minimum deviation can be located with the aid of the receiver; it is $-24°$. If the wave guide prism is oriented with its plates perpendicular to the electric vector, the TM_{00} mode, whose wave velocity is the same as in free space,[3] is excited, and consequently the microwave beam is not deviated. The artificial wave guide dielectric prism therefore operates only for plane polarized microwaves with the electric vector parallel to the wave guide plates. Evidently this dielectric will exhibit a type of double refraction for unpolarized microwaves, and later we shall describe experiments in which this property of the metal plate dielectric is used.

Besides prisms, lenses also can be constructed. Plano-convex lenses of paraffin can easily be made by filling watch glasses of 8″ or 10″ diameter with paraffin. Glass lenses and lenses made of artificial dielectrics can also

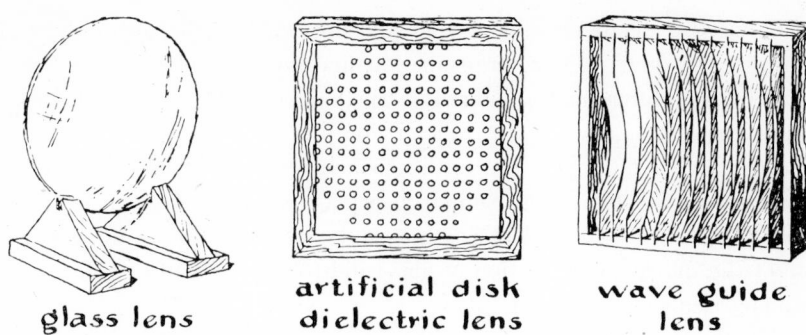

glass lens

artificial disk
dielectric lens

wave guide
lens

FIG. J-12 Three lenses for 3 cm microwaves. From left to right are a glass lens, an artificial disk dielectric lens, and a wave guide lens.

be used. In Fig. J-12 three lenses are shown. The lens at the left is a 10″ glass, plano-convex, condensing lens; the lens in the center is an artificial dielectric lens made up of an array of thumb tacks; and the one on the right is a lens made up of parallel-plate wave guides. All of these lenses exhibit the usual properties expected of lenses. Their focal length for 3 cm microwaves can be determined experimentally within $\pm\lambda$, which, considering that the lens diameters are about 8λ, is reasonable precision.

The parallel-plate wave guide lens has the same refractive properties as the wave guide prism previously described. The galvanized iron lens plates are supported in a wooden frame 12″ × 12″ with a plate spacing of $b = 0.79″$ ± 2 percent, which gives an index of refraction of 0.6. Since the index of refraction is less than unity, a converging lens is plano-concave. Such a lens has a focal length of 12λ, a diameter of 9λ, and a radius of curvature of 4.8λ. With these particular dimensions, it is not necessary to zone or step the lens, and the departure of the spherical surface from the true ellipsoidal surface is not greater than 0.2″, or a phase difference of $\lambda/16$ at the extreme, which is within the tolerance limits for this type of lens.[9] To obtain the correct radius for each plate, it is simpler to draw the entire lens to scale and take off the radii with dividers rather than calculate each radius individually. Like the wave guide prism, the wave guide lens operates as a lens only when the electric vector is oriented parallel with the plates.

It is interesting to note at this point that we can define an index of refraction for a parallel-plate wave guide from which we can make the same type of calculations as for an ordinary dielectric. For example, we can calculate the reflection coefficient. Also, we can define a characteristic wave impedance for the parallel-plate wave guide and for free space, and from these quantities calculate the reflection coefficient. Both methods must yield the same value for the reflection coefficient, and consequently one would expect analogies to exist between transmission line theory and optics.[10] Another example of this analogy is the quarter-wave transmission line and coated lens. Two transmission lines of different characteristic impedances can be connected, without producing reflection, by a third transmission line, a quarter wavelength long, whose characteristic impedance is the geometric mean of the characteristic impedances of the two lines. Similarly, a glass lens coated with a quarter-wavelength thickness of dielectric whose index of refraction is the geometric mean of the indices of refraction of glass and free space will be reflectionless for one particular wavelength.

Another interesting analogue is the microwave equivalent of molecular arrays, which is exemplified by the lens made of thumb tacks shown in the center of Fig. J-12. Since a piece of metal whose dimensions are small compared with a wavelength can be driven in forced oscillation by an electromagnetic radiation field, an array of identical metal pieces such as spheres, disks, or rods should behave in the same way that a dielectric made up of a molecular array behaves when exposed to light. In other words, an array of disks have a dielectric constant $x_e = 1 + \mathfrak{N}\mathfrak{p}/\kappa_0$, where $\mathfrak{p}$ is the polarizability of the disk, $\mathfrak{N}$ the number of disks per unit volume, and κ_0 the electric inductive capacity of free space. This is the same as the classical expression for a dielectric in which $\mathfrak{p}$ is the polarizability of a molecule and $\mathfrak{N}$ the number of molecules per unit volume.

The expression for x_e is applicable only when the disks are far from resonance and there is no interaction between the fields of adjacent disks. In general, we would expect an artificial disk dielectric to obey the Clausius-Mossotti equation at long wavelengths and to exhibit the phenomenon of anomalous dispersion at short wavelengths when the microwave frequency approaches the resonant frequency of the metal disks.

The general criterion for the design of an artificial dielectric made up of an array of identical metal elements is that the dimensions of the elements should be less than $\lambda/4$ and the spacing of the elements less than λ. If the spacing is greater than λ, diffraction, similar to X-ray diffraction by crystals, occurs. Furthermore, the metal elements should be thin in the direction of propagation of the microwaves.[11] On the basis of this criterion, the artificial disk dielectric lens shown in the center of Fig. J-12 was constructed. Since the polarizability of a metal disk is $\frac{2}{3}\kappa_0 d^3$, where d is the disk diameter, the dielectric constant or square of the index of refraction of a disk dielectric is $N^2 = x_e = 1 + \frac{2}{3}\mathfrak{N}d^3$. The actual array used is shown in Fig. J-13, and the dimensions shown are $d = 1$ cm, $s_1 = 1.3$ cm, and $s_2 = 1$ cm. These dimensions give $\mathfrak{N} = 1.18$ disks per cm³ and a calculated index of refraction $N = 1.33$. The metal disks used were thumb tacks $\frac{3}{8}''$ (0.95 cm) in diameter, which were stuck into sheets of polystyrene foam 1 cm thick and 8″ square. Poly-

FIG. J-13   Arrangements of disks in artificial disk dielectric lens for 3 cm microwaves.

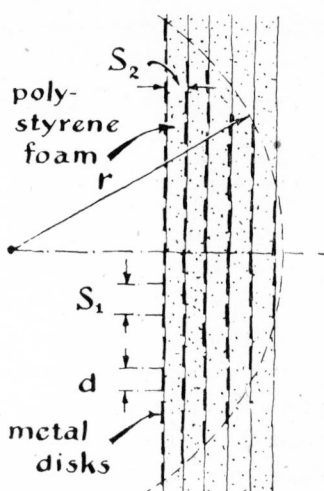

styrene foam has a density of about 1.5 pounds per cubic foot and an index of refraction of 1.01. It has practically no refractive effect on microwaves. The lens is plano-convex with a radius of curvature of 5″ and a calculated focal length of 15″. Five sheets of polystyrene foam are used. As shown in Fig. J-13, the disks in alternate layers are staggered. This is done in order to increase the number of disks per cm³. To determine the radius of the circular area to be covered by thumb tacks on each sheet of foam, it is simpler to draw the lens to scale and take off the radii with dividers rather than calculate each radius individually. The positioning of the thumb tacks is best accomplished by marking out the circular area and dotting in the thumb tack centers on thin paper. This paper is then placed on the foam sheet and the thumb tacks pushed through the paper into the foam. After all the sheets have been filled with the required number of tacks, the sheets are put together and supported in a wooden frame. The lens, when completed and measured, is found to have a focal length of 10″ instead of the calculated value. This means that the index of refraction of the disk dielectric is 1.5 instead of the calculated value of 1.33. The discrepancy between calculated and measured index of refraction is to be expected because of the effect of the Clausius-Mossotti equation and because the diameter of the thumb tacks is slightly larger than $\lambda/4$. Unlike the wave guide dielectric lens, the disk dielectric lens operates independently of the polarization of the microwave radiation. Also, the index of refraction of the disk dielectric—but not of the wave guide dielectric—remains essentially constant for longer wavelength microwaves.

The index of refraction of artificial dielectrics can be measured with high precision with the microwave Michelson interferometer discussed in the previous section. The procedure is the same as in optics: one inserts a sheet of artificial dielectric about 1' square and of known thickness in one arm of the interferometer and measures the fringe shift, from which the index of refraction can be calculated.

Polarization

As has been pointed out, the microwave radiation from the transmitter is plane polarized. For polarization experiments it is often desirable to have elliptically or circularly polarized radiation. Elliptically polarized radiation is easily obtained by placing a sheet of glass, polystyrene, or other dielectric in front of the horn radiator, with the plane of the sheet at 45° to the electric vector and parallel to the direction of propagation. Elliptically and circularly polarized microwaves can also be obtained by use of the artificial wave guide dielectric discussed in the previous section. If this dielectric is made with a plate spacing to give an index of refraction of 0.6 when the electric vector is oriented parallel to the plates, it will also have an index of refraction of unity when the electric vector is at right angles to the plates. Consequently, if an appropriate thickness of wave guide dielectric is placed in front of the horn radiator with the plates oriented at 45° to the electric vector, elliptically, circularly, or plane polarized radiation will result. The thicknesses for a quarter- or half-wave plate of wave guide dielectric are calculated in the same way as in optics, using 0.6 for the extraordinary and unity for the ordinary index of refraction. These thicknesses are 0.79″ and 1.58″ for the quarter- and half-wave plates, respectively, and the thickness tolerance is $\pm 2\%$. The plates are supported in a wooden frame 12″ square and have a plate spacing of $0.79″ \pm 2$ percent, the same spacing as the wave guide prism and lens. Although the quarter- and half-wave plates behave in a manner similar to that of those used in optics, the wave guide dielectric is not exactly similar in its double refracting properties to a uniaxial crystal, for the wave guide dielectric does not have an optic axis. In general, microwaves pass through the wave guide dielectric with two components, one of which travels faster than the other, and whose amplitudes depend upon the orientation of the electric vector of the incident microwave radiation with respect to the plates forming the wave guides.

One can extend the principle of the quarter-wave plate to a circular wave guide so that the radiation from a circular horn will be circularly polarized. All that is necessary is to split up the microwave radiation in a circular wave guide operating in the TE_{11} mode into two components of equal amplitudes and at right angles and to delay the phase of one component by $\lambda/4$ with respect to the other. This can be achieved by inserting in a circular piece of wave guide a sheet of dielectric along the diameter and oriented at 45° to

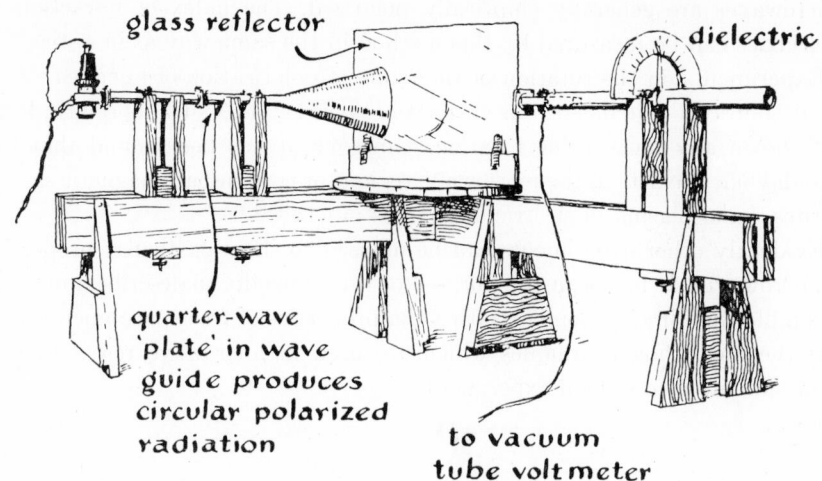

glass reflector

dielectric

quarter-wave 'plate' in wave guide produces circular polarized radiation

to vacuum tube voltmeter

FIG. J-14 Transmitter and receiver arranged for producing plane-polarized 3 cm microwaves by reflection from glass at Brewster's angle.

---

the electric vector of the incident microwaves. The proper thickness and length of such a dielectric are not easily calculated but can be found experimentally. The horn radiator shown attached to the transmitter in Fig. J-14 has a polystyrene strip $\frac{1}{16}''$ thick and 2.5″ long across the diameter of the 1″ standard circular wave guide leading to the horn. The strip is oriented at 45° to the electric vector of the incident microwaves, and the horn radiates circularly polarized radiation. If this polystyrene strip is replaced by one of the same thickness and 5″ long, the result is a half-wave plate, and the radiation coming from the horn is plane polarized with the electric vector rotated 90°.

A number of interesting experiments can be performed with various types of polarized microwaves. Circularly polarized microwaves can be plane polarized by reflection from a dielectric at the Brewster angle. In Fig. J-14 the circular horn equipped with a quarter-wave plate radiates circularly polarized microwaves which are incident upon several sheets of window glass held in a rotating support equipped with a protractor on the wooden bench. Microwave radiation reflected and transmitted by the glass sheets is investigated with the receiver, which is mounted in a rotating holder on another wooden bench. A protractor is fastened to the rotating holder so that the polarization of the reflected and transmitted microwaves can be measured. When the glass plates are adjusted at the Brewster angle, the reflected microwaves are plane polarized, with the electric vector vertical, while the transmitted microwaves are generally elliptically polarized. The index of refraction of dielectrics can be measured by this method in the same way as in optics.

Experiments on the rotation of the plane of polarization of microwaves by sugar solutions, liquids such as turpentine, and crystals such as quartz have not shown any measurable rotation. However, it has been found that the Faraday effect exists in the microwave region for certain paramagnetic salts,[12] ferrites,[13] and plasma in electric discharges in gases.[14]

Evidently other experiments can be devised to show particular properties of microwaves. It has been the purpose of this appendix to describe a number of simple experiments demonstrating the properties of free space microwaves and the measuring techniques which are used in microwave research, and their similarity to optical experiments

REFERENCES

1. G. F. Hull, Jr., Am. J. Phys., *17*, 559 (1949). The author is indebted to the American Journal of Physics for permission to use the paper in this appendix.

2. J. R. Pierce, Bell Lab. Rec., *23*, 287 (1945).

3. J. C. Slater & N. H. Frank, *Electromagnetism* (1947, McGraw-Hill, New York), Chap. 11.

4. G. C. Southworth and A. P. King, Proc. I.R.E., *27*, 95 (1939).

5. Slater & Frank, *loc. cit.*

6. G. F. Hull, Jr., Am. J. Phys., *13*, 384 (1945).

7. G. F. Hull, Jr., Am. J. Phys., *15*, 111 (1947).

8. Cf. p. 125.

9. W. E. Kock, Proc. I.R.E., *34*, 828 (1946).

10. Slater & Frank, *op. cit.*, Chap. 10.

11. W. E. Kock, Bell Sys. Tech. J., *27*, 58 (1948).

12. M. C. Wilson and G. F. Hull, Jr., Phys. Rev., *74*, 711 (1948).

13. C. L. Hogan, Bell Sys. Tech. J., *31*, 1 (1952).

14. F. R. Arams, Electronics, *27*, 168 (1954).

Appendix **K**

The Wave Theory of Microscopic Image Formation

by F. Zernike†

The Microscopic Specimen

The trained microscopist uses various, often complicated techniques to prepare his specimen for the microscope. A few indications must suffice here.

Suppose, for instance, that he has to study a piece of paper in order to find out what kind of fibers it has been made of. It is of no use simply to put it under the microscope, for its smooth surface will hardly show the structure. A much better suggestion is to impregnate the paper with oil so that it becomes translucent. It will then be put on a glass slide so it can be illuminated from below. To avoid irregular refraction at the upper surface, the oiled paper is covered with a coverglass (thickness $\frac{1}{6}$ mm), the space between being filled with oil. The only remaining difficulty may be that the paper is too thick; there will be some six layers of fibers on top of each other. Now this is not as troublesome as it may seem. The microscope, it should be remembered, has a very small depth of focus, which means that it will focus only a very thin slice of the specimen. Even so, it may well be that the microscopist prefers to see the fibers separately. He then has to treat the specimen so as to loosen its structure, the fibers being spread out into a single layer between slide and coverglass.

This example illustrates the general principles according to which the vast majority of specimens are prepared. They are cut—or sometimes ground— into thin slices, of thickness comparable to the dimensions of the details to be studied. Small loose objects—such as bacteria or blood cells—are spread

---

† Groningen University, Netherlands.

out so as not to overlap. The specimen is further mounted in a suitable liquid or resin, of refractive index close to that of the structural parts. This does away with most of the refraction of the light transmitted; that is, it makes the specimen translucent.

The Need for Special Treatment

The optics of the microscope calls for a special treatment for two reasons: because the object is observed in transmitted light which changes only slightly by passing through it, and because the details of the object are of a size comparable to the wavelength. Since these characteristics cause diffraction effects to be very prominent, the wave nature of light is best taken into account from the outset.

Let us start with the simple case of a parallel beam of incident light—that is, a plane wave—falling on an object consisting of very small black particles. Each particle obliterates a small part of the wave; it can be said to make a hole in the wave front. The diffraction effect of these holes is found as follows. Instead of subtracting a small patch of wave front, each particle may be said to add a negative piece—that is, a wave of opposite phase. In this way the transmitted light is seen to consist of two parts that behave differently: (1) the unchanged incident wave, which will be called the *direct light;* (2) the wavelets starting from the black particles, to be called the

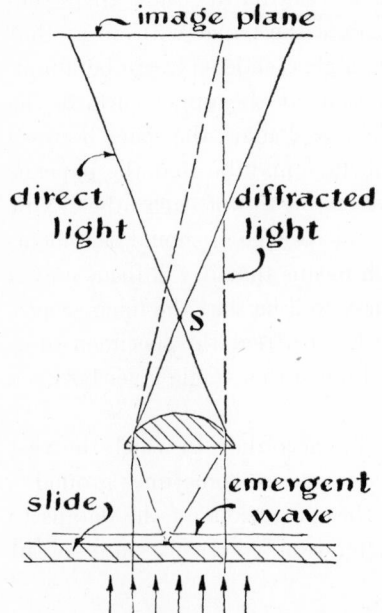

FIG. K-1 Schematic representation of image formation by an objective. (Here we have an advancing particle, as Fig. K-4 shows.)

diffracted light. As shown in Fig. K-1, the direct light is just a parallel beam, the same as if no particles were there. The microscope objective brings this beam to a focus S a short distance above the lenses or even inside the lens system of the objective. From there on it spreads out again and gives an evenly illuminated field in the image plane. The diffracted light, on the contrary, diverges from each particle and is concentrated by the objective in the image plane. In a word, this light behaves exactly as we should expect from the familiar ray-optic treatment of the microscope; it enables the objective to form an enlarged image of the object, to be further enlarged by the eyepiece.

Experiment

In order to demonstrate by experiment the existence and behavior of the direct and the diffracted light, an objective with accessible back focal plane should be used. Objectives of 8–10 ✕, of focal distance 20–16 mm, will do; see Fig. K-2. Practically parallel light may be obtained from a frosted bulb at a few meters distance. Of course, any substage condenser must be removed and the plane mirror used. It should be realized that the focus at S is, in reality, a small image of the light source. As an object a glass slide finely sprayed with India ink will serve, or finely powdered galena (PbS) embedded in Canada balsam. The first gives round dots, the second squares and broken squares. Now adjust the microscope and observe the image, with magnifica-

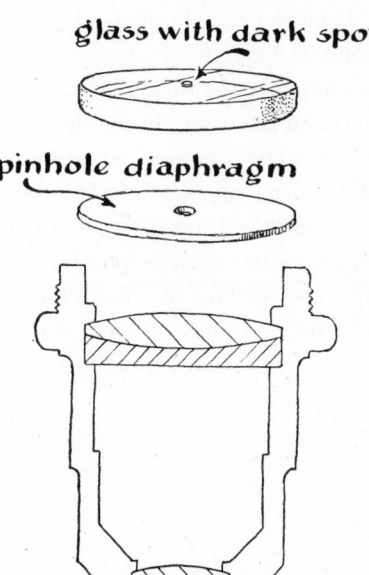

glass with dark spot

pinhole diaphragm

FIG. K-2 Low-power objective with diaphragms for alternately obliterating diffracted and direct light.

tion about 100 ×, in the ordinary way designated as "bright field." Then cut out a circular piece of metal foil so as to fit on the upper lens of the objective, and make a pinhole of about 0.5 mm in its center. Put this in place, remove the eyepiece, and adjust the mirror so that the maximum amount of light comes through the hole. Now replace the eyepiece and observe the image. The field of view should now be evenly illuminated, and it will be if the black particles in your object are all quite small. Ordinarily there will also be some larger ones; and these will show diffuse disks because they diffract the light, mostly over small angles, so that a good part of it falls through the pinhole. In other words, the intended separation of direct and diffracted light does not quite succeed. For the complementary part of the experiment, replace the pinhole diaphragm by a glass disk with a black spot of 1 mm in the center, and adjust the microscope mirror so that the image of the light source is covered by this spot. In the eyepiece the dark particles will then appear bright on a dark ground. This demonstrates the reality of the diffracted light.

After these experiments it may seem surprising that the direct light and the diffracted light together give the ordinary bright field image—that is, *black* particles on a bright background. Clearly this must be a case of destructive interference, for which some further explanation is needed.

Equality of Optical Paths

Consider the simple case of a convex lens making an image O' of an object point O (Fig. K-3). According to ray-optics the action of the lens consists in changing the divergent rays from O into convergent rays coming together in O'. In the wave theory the equivalent of this is as follows. O emits spherical waves, the surfaces of equal phase being spheres with O as center. The action of the lens consists in retarding these waves in proportion to the thickness of glass they traverse, so as to change the incident convex wave front S into an emerging concave spherical wave S', which contracts towards its center O'. The new surface of equal phase S' may be constructed by setting out equal lengths along the various rays from S, the retarding action of the glass being taken into account by multiplying each distance l by the corresponding index

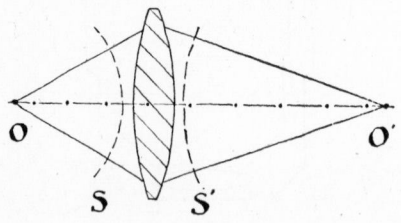

FIG. K-3 Equality of optical paths along all rays from O to O'.

FIG. K-4 Embossed wave front directly above transparent object.

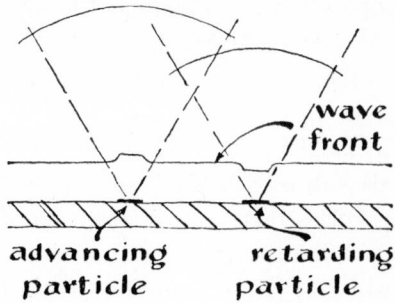

N. The so-called optical path ΣNl is therefore equal along all rays between S and S', and hence also between O and O'. This leads to the general theorem: the optical path ΣNl from an object point through an aberration-free optical system to the corresponding image point is equal along all rays.

Returning now to the microscopic image of a black particle, we found that the incident plane wave gives rise to two waves, the direct and the diffracted wave, which on starting from the particle were in opposite phase. These waves follow different ways through the objective and unite in the image plane. According to the general theorem, they must show the same phase relation in the image as in the object, and will therefore give destructive interference in the image.

Transparent Objects

Now take the case of small objects which do not change the intensity of the light, *i.e.* which are absolutely transparent. As a test object, finely powdered common salt embedded in Canada balsam will do very well. Consider the wave front passing through the transparent object. There is no change of amplitude in this case, the only effect consisting in the various retarding by the different thickness and different refraction of the various details. The wave front directly above the transparent object may therefore be visualized as carrying an embossed impress of the object (Fig. K-4). In order to find the further course of this wave, we use the same method as before.

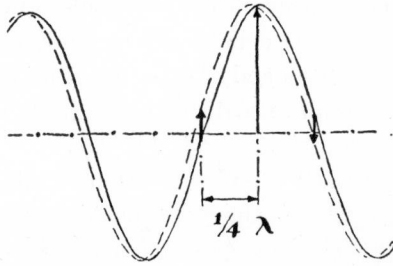

FIG. K-5 Original wave (full curve) and slightly retarded wave (broken curve). The difference "retarded *minus* original" is a wave motion, $\frac{1}{4}\lambda$ behind.

Suppose that a small particle has slightly retarded the wave, compared with the surrounding background. The retarded wave can be obtained by adding to the original one a wave of much smaller amplitude, which is one quarter wave behind (Fig. K-5). This will again spread out by diffraction and be reunited in the image plane. Its existence is demonstrated by inserting the glass disk with black spot which cuts off the direct light. The transparent particles then appear bright on a dark ground. It is in the ordinary bright field that the difference from the black particles appears. The smaller particles will be seen to disappear altogether as soon as the microscope is exactly focused. This is as it should be: the direct light and the diffracted light are reunited in the image plane with the same relative phases as they started with from the object, thus forming again a wave of equal amplitude everywhere and therefore of equal intensity, *i.e.* no image at all.

For two reasons the exact equality of amplitude is not realized in practice. First of all, the diffracted light is partly spread out from each detail of the object in such inclined directions as are not taken in by the objective. This causes a deficiency in the image plane which upsets the exact balance of amplitudes. Especially sharp edges of larger particles will show as darker lines for this reason. Secondly, the trained microscopist tries hard to see something and therefore changes the focus, even without noticing it, until he sees some image of the transparent details. To understand the effect of defocusing, remember that the general theorem of equal optical paths applies only to distances from an object point to its conjugate image point. If now the microscope is lowered by a small distance ϵ from the exact focus, the path along the axis, which is that of the direct light, will be shortened by ϵ, whereas a ray inclined by ϑ will be shortened by $\epsilon \cos \vartheta$. This then gives a relative retardation for the inclined ray of $(1 - \cos \vartheta)\epsilon$. The result of this will be discussed presently.

The Phase-contrast Method

To recapitulate the case of transparent details, their visibility is bad because their diffracted light is $\frac{1}{4}\lambda$ behind the direct light. If this could artificially be changed into $\frac{1}{2}\lambda$, we would bring it into the condition shown ordinarily by absorbing details. Now the diffracted light and the direct light overlap near the object and near the image plane, but are practically separated in the upper focal plane of the objective (Fig. K-1). It is there that they can be influenced separately. Indeed, this was made use of above, where they were alternately screened off. We now want to retard the diffracted light by an additional $\frac{1}{4}\lambda$ or, what is the same thing, advance the direct light by this amount. Instead of the black spot in the center of a glass disk, we need a phase-advancing spot. This is obtained in the simplest way by etching. Rayleigh described, some fifty years ago, how to make very shallow etchings in glass surfaces, without spoiling their optical quality, by the slow action of very dilute hydrofluoric acid. If a small spot in the center of the glass disk is etched out to a depth of δ, the direct light passing through will cover this distance in air, whereas the surrounding diffracted light goes through δ in glass. The etched plate, called a *phase plate*, thus causes a difference in path of $(N - 1)\delta$, and δ must be about $\frac{1}{2}\lambda$.

The result of inserting the phase plate is quite remarkable. Thin transparent details which in the ordinary bright field are hardly visible appear darker, in good contrast with the bright background, just as if they were absorbing (Fig. K-6). This phase-plate method of observing phase-changing objects in good contrast is called, for short, the *phase-contrast method*. In the case here discussed the direct light is advanced, with the result that thicker or higher-

FIG. K-6   Two specimens, in bright field (left) and in phase contrast (right). Upper figures: living tissue culture; lower figures: diatom. For halftone reproductions of these see *Science*, *121*, 345–349 (11 March 1955).

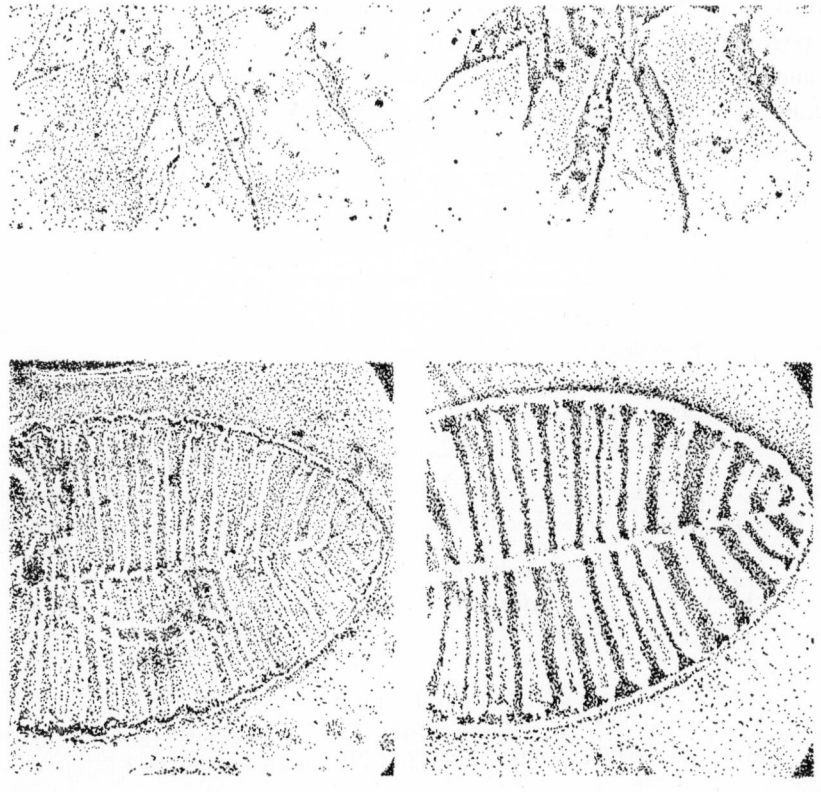

refracting particles appear darker. This is called *positive* phase contrast. Sometimes it may be more useful to make such particles appear brighter, which is evidently obtained by using a phase plate which retards the direct light—*negative* phase contrast.

Returning for a moment to the out-of-focus bright field image discussed a little earlier, we can now express the result of defocusing in the newly introduced terms. The retardation of the inclined ray *inside* focus causes a kind of positive phase contrast, a poor improvised kind because the amount $(1 - \cos \vartheta)\epsilon$ is variable over the aperture of the objective. Outside focus the inclined rays are advanced, giving a negative phase contrast. Of course, the variability of the phases with which the diffracted light arrives in the image plane makes for a blurred image, and the observer chooses a position which gives a compromise between an image that is bad because of blurring and one that is without contrast because it is too near focus.

Two further points are of importance in the practical realization of phase contrast. The first is the size and shape of the phase-advancing spot on the phase plate. Remember that this phase spot must cover the image of the light source. Now it is only for the sake of simplicity that we started with a point source. In practical use this is undesirable as it causes spurious shadows from all dust particles anywhere in the light path. To avoid these, the light source should be rather broad. On the other hand, phase contrast depends on the possibility of separating the direct and the diffracted light. Now the diffracted light fills the whole aperture of the objective, and therefore a part of it falls on the phase spot and is lost for the contrast image. The phase spot should

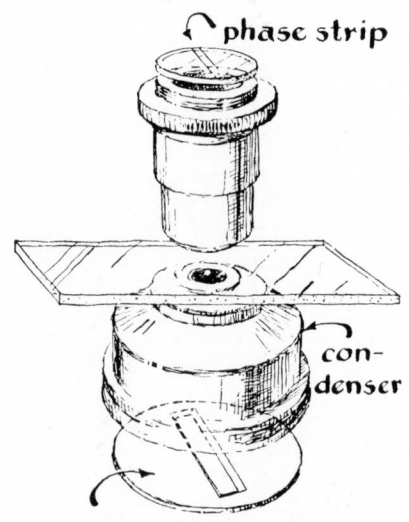

phase strip

con-denser

slit diaphragm

FIG. K-7 Arrangement of condenser and objective for phase contrast.

for this reason be as small as possible. The best compromise is obtained with a long and narrow shape. A straight phase strip 0.4 mm wide will do for laboratory experiments. A narrow ring shape is still better and is now in general use. In order to adapt the light source to these special shapes, we use a substage condenser having in its lower focal plane an opaque disk from which the desired form is cut out. This acts as a secondary light source, which is imaged by the condenser and objective on the phase plate (Fig. K-7).

In the second place, the phase-changing spot, strip, or ring is expressly made absorbing—for instance, by depositing a thin metal layer on it. It is only in this way that it becomes visible so that the image of the source can be exactly adjusted to be covered by it. Another advantage is thus obtained at the same time. Suppose that the ring absorbs 75% of the light. This means that the direct light arriving in the image plane is reduced to a fraction of 0.25. As it acts by interference with the diffracted light, we need the square root, which gives the amplitude ratio as 0.50. The relative importance of the diffracted light is therefore doubled; *i.e.* the contrast is doubled, which has a very marked effect.

The Abbe Theory

Some special features of microscope optics are more readily explained by a different approach, first given by Abbe in 1873. It must suffice here to give the main lines of his arguments. The theory aims primarily at finding the circumstances under which a fine structure can be seen as such under the microscope—that is, can be *resolved*. It therefore starts with a regularly striated object having equidistant parallel lines—in other words, a grating. The action of such a transparent grating on a parallel beam of light is well known. Let the grating be horizontal on the microscope stage and the incident beam vertical, in the direction of the optical axis. The transmitted light will be split into a number of beams, making angles ϑ_m for which

$$\sin \vartheta_m = m\lambda/p \tag{1}$$

where p is the distance between adjacent lines and m is any whole number between $-p/\lambda$ and $+p/\lambda$. These beams, so far as they enter the objective, are each brought to a focus S_m in the upper focal plane of the objective. With white light each S, except S_0, will represent a spectrum, and we shall give it the usual name of spectrum of the mth order, although a single wavelength only need be considered for our purpose. Abbe calls the row of spectra the *primary image;* it is a multiple image of the light source. It may be seen on removing the eyepiece and looking down the microscope tube (Fig. K-8).

Above the points S_m the beams will spread out and overlap in the eyepiece.

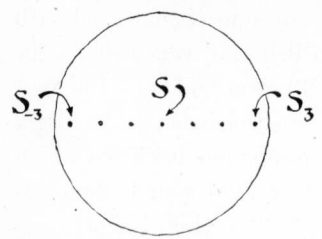

FIG. K-8 Multiple image of point source in back focal plane of objective. Note: The central S corresponds to S_0 in the text.

According to the Huygens principle each point S may be considered as a new source, all these sources being coherent as images of the one original source. They will therefore give interference fringes in the eyepiece. Take, for instance, S_0 and S_1; let these be a distance d_1 apart. If the image plane is a distance a above the focal plane (a is the optical tube length), the spacing of the fringes q caused by S_0 and S_1 is found to be

$$q = \lambda a/d_1$$

If the angle ϑ_1 is small, there is no doubt about d_1, which will simply be

$$d_1 = f\vartheta_1 = f\lambda/p \qquad \text{and} \qquad q = a\lambda \cdot p/f\lambda = pa/f$$

Here the wavelength has dropped out, and the fringe spacing is equal to the spacing in the object multiplied by a/f, which is the magnification according to geometrical optics. This is the result of S_0 and S_1, and a further condition is, evidently, that all other adjacent pairs give the same spacing.

They will if the whole row of S's is equidistant. Or the distance d_m of S_m from the optical axis must satisfy

$$d_m = md_1 = mf\lambda/p = f \sin \vartheta_m$$

Clearly this is a general condition to be satisfied by any microscope objective: the distances of all emergent rays from the axis must be proportional to the sines of the angles of the corresponding incident rays. This is known as the *sine condition*. It has been derived in a number of different ways. When Abbe discovered it, he wondered how the then existing objectives, especially the high powers, could have been constructed without taking the sine condition into account. He found that they satisfied the condition very well indeed. The manufacturers had evidently arrived at the right design by many years of practical trial and error.

Now suppose that one puts finer and finer gratings under the same microscope. The spectra S get farther and farther apart. As long as S_1 still appears in the objective, there will be interference fringes revealing the striated structure of the object. But as soon as ϑ_1 exceeds the aperture angle α of the objective, only S_0 comes through, and an evenly illuminated field results. For the limiting p_{lim} we find hence

$$\sin \alpha = \sin \vartheta_1 = \lambda/p_{\text{lim}} \qquad \text{or} \qquad p_{\text{lim}} = \lambda/\sin \alpha \qquad (2)$$

This is for axial illumination. If oblique illumination is used, the incident beam making an angle β with the axis, the grating formula (1) changes into

$$\sin \vartheta_m - \sin \beta = m\lambda/p$$

Here β can be nearly equal to $-\alpha$, so that the point S_0 lies at the edge of the objective aperture. The limit is then reached when S_1 is at the opposite edge (Fig. K-9), and this gives

$$2 \sin \alpha = \lambda/p_{\text{lim}} \qquad \text{or} \qquad p_{\text{lim}} = \lambda/2 \sin \alpha$$

Having found this formula, Abbe introduced the name *numerical aperture* for $\sin \alpha$ as a measure of the resolving power of an objective. The advantage of a large aperture angle had been recognized before, but the exact dependence of the resolving power on α was unknown.

If the object is embedded in a medium of index N, the above reasoning will hold with the difference that the wavelength λ will be replaced by λ', the wavelength in the medium, and α by α', the aperture angle in the medium. With $\lambda' = \lambda/N$ formula (1) becomes

$$p_{\text{lim}} = \lambda/N \sin \alpha' \qquad (3)$$

In the ordinary case, when there is air between the cover glass and the front lens of the objective, the inclined beam will be refracted according to $\sin \alpha = N \sin \alpha'$, and (3) reduces to (2). In order to obtain a smaller limit from (3), the air is replaced by a special oil of index 1.52, equal to that of the cover glass and of the front lens of the objective. Of course, the objective must be

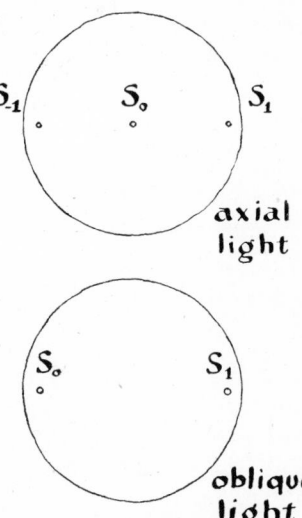

FIG. K-9 Multiple back focus image with grating near limit of resolving power, with axial and with oblique light.

specially designed for this *homogeneous immersion* method. It is only logical to extend the term "numerical aperture" to the denominator in (3):

numerical aperture $= N \sin \alpha'$

With $\alpha' = 60°$, this attains 1.32, the usual value for immersion objectives.

Appendix L

Modern Trends in Methods of
Lens Design

by M. Herzberger[†]

The earliest lenses were made by trial and error, and the excellence of the lens depended upon the skill and experience of the artisan. With the discovery of the refraction law, it became possible to trace rays through a proposed lens and thus predict its performance before the blank was put on the grinding block. The modern era of lens design may be said to have begun in the 1880's when Ernst Abbe insisted that every lens made by the firm of Zeiss should be perfected on the drawing board. This has become the procedure followed by all optical shops, but it has been found to be not without its drawbacks. The simplest rays to trace are those lying in the meridional plane—the plane containing the object point and the optical axis. However, these account for only a small fraction of the light in the image, and thus they give a very imperfect picture of the performance of the lens. Many attempts have been made to render the tracing of skew rays practicable, but these were all defeated by the lack of a suitable mathematical language and the slowness of the methods of calculation that designers were forced to employ.

Some years ago, the writer developed by vector methods a set of formulas that are suitable for modern high-speed computers.[1] A complete picture of the performance of a lens can now be given[2] by imagining the exit pupil to be divided into elementary areas of equal size and by tracing a ray from the selected object point through each of these areas to the image plane. Fig. L-1 shows how this is done for a certain lens designed for aerial photography. The right-hand column of sketches shows the pattern made by the intersections of the selected rays with the exit pupil for a series of field angles, the meridio-

---

[†] Research Laboratories, Eastman Kodak Company, Rochester 4, New York.

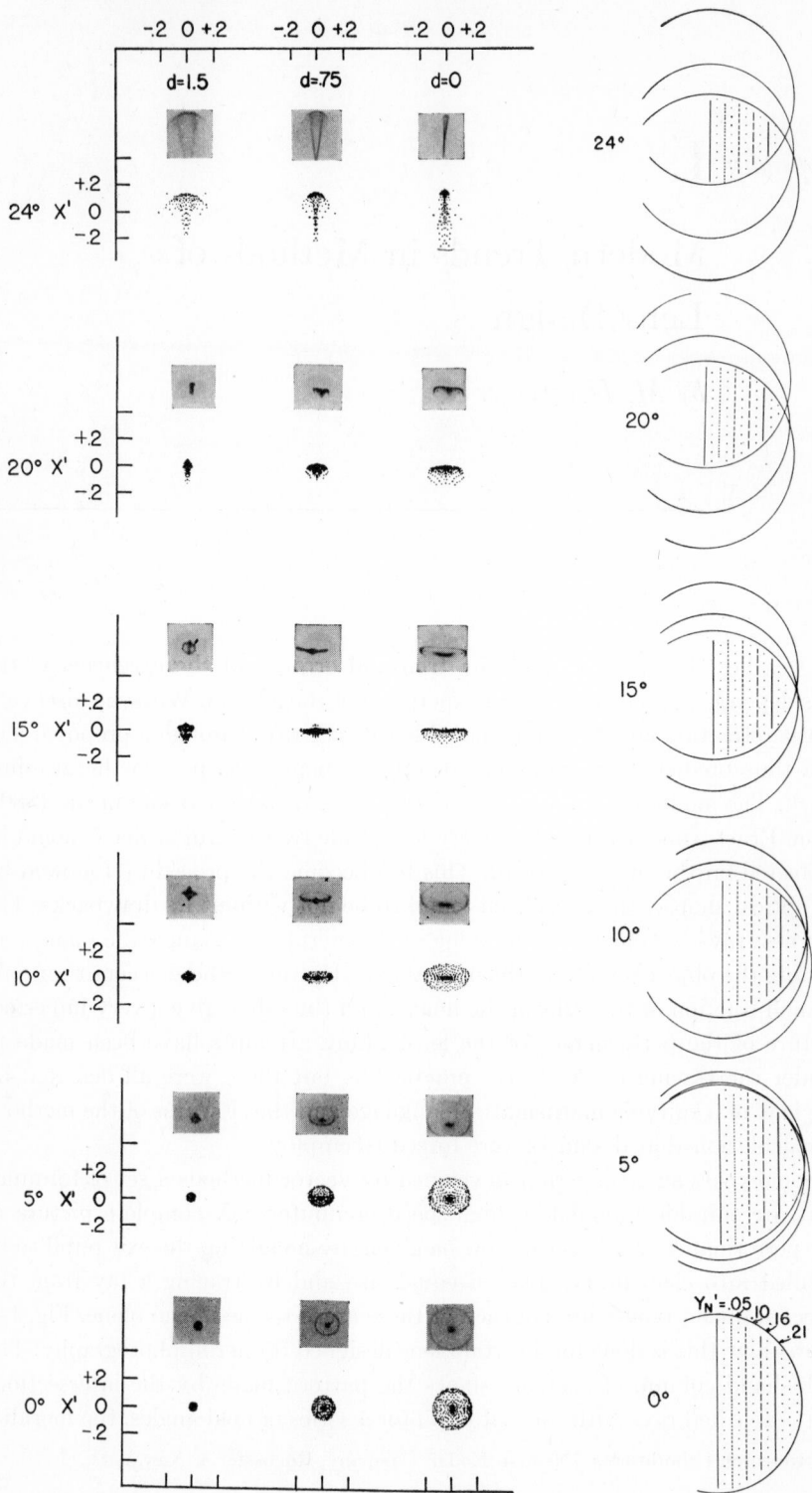

nal plane being a vertical plane normal to the paper. The x-direction is assumed to be parallel to the meridional plane, the y-direction normal to it, and the z-direction along the optical axis. The intersecting circles in the figure represent the vignetting produced by the diaphragm and the mounts of the lenses. It is clear that the vignetting at large field angles must profoundly modify the image pattern, and indeed the designer can often turn it to good advantage.

To determine the character of the image, an image plane must be assumed, and in Fig. L-1 three planes have been considered. The plane at $d = 0$ contains the Gaussian focus while the other two planes are respectively 0.75 and 1.50 mm nearer the lens. The intersection points (x', y') have been computed for the three image planes and six field angles, and when they are plotted, they give the patterns shown in the figure. These patterns are called spot diagrams, and they give a qualitative indication of the distribution of light in the images. Their accuracy has been shown by Dr. R. N. Wolfe, who made photomicrographs of the corresponding images after the lens was constructed.[3] Photomicrographs are shown in Fig. L-1 just above the corresponding spot diagrams.

This procedure, although informative, calls for the tracing of a large number of rays. Since the function representing the path of the rays is continuous, it should be possible to trace only a few rays and determine the intersection points of the others by interpolation, and the writer has developed formulas for doing this.[4] Now it is necessary to trace only from three to five meridional rays and three or four skew rays (most of which do double duty because of the bilateral symmetry of the system about the meridional plane), and then the intersection points of any number of rays can be rapidly computed by simple interpolation.

But the advantage of the interpolation method does not stop here. Ever since W. R. Hamilton in 1828 enunciated the general principles of image formation, attempts have been made to derive analytical formulas for indicating the behavior of individual lenses. The critical difficulty is that trigonometrical functions enter into the exact formulas, and these must be expanded into power series for use in algebraic formulas. Since the series must be truncated

---

FIG. L-1 Comparison of the performance of a certain $f/5.6$, 300-mm Tessar lens as indicated by spot diagrams and by bench tests. The column of sketches at the right shows the vignetted aperture for field angles of from 0° to 24° and the intersections of the rays from the object point with the exit pupil. The three columns of spot diagrams show the theoretical distribution of light on three image planes 0.75 mm apart. The corresponding photographs above the spot diagrams show the distribution as determined by testing the finished lens on a lens bench. Scales are in millimeters.

at some term, the algebraic formulas hold only approximately. Gaussian or first-order formulas take into account the linear terms alone. It was Ludwig Seidel[5] who in 1856 clearly enunciated the geometrical meaning of the third-order terms. Studies of the fifth-order terms have appeared sporadically in the literature, but these terms have rarely been used systematically.

Although the third-order theory, which is accurate for only a vanishingly small region near the optical axis, has often been extended to finite fields and apertures, it becomes increasingly inaccurate as the terms of higher order become increasingly significant. The new interpolation formula does not suffer from this limitation. The difference between it and the Seidel procedure may be summarized by saying that the Seidel procedure is based upon an *approximation* that fits perfectly in the vicinity of the axis, while the new procedure is based on a formula that fits very closely over the entire aperture and field and which is evaluated by *interpolation*.

The interpolation formula consists of a fifth-order power series for the *x*-coordinate of the image and a similar power series for the *y*-coordinate. By combining the terms of the same order for the two coordinates, a mathematical model of the optical system can be made that can be divided for analysis into five separate aberrations. These aberrations have comparatively simple geometrical forms although they bear no simple relation to the familiar Seidel aberrations. The new analysis has the great computational convenience that changes in the design data affect the aberrations differently and hence, if one is found to be of controlling importance, it can be reduced without greatly affecting the others. This is illustrated by Fig. L-2. The spot diagrams in the columns labeled from I to V represent the aberrations of the orders from the first to the fifth of the actual image of a certain lens, while the last column, marked *T*, represents the image itself. This image is usually smaller than diagrams I and III because the sign of the third-order aberration shown at III is opposite to that of the first-order aberration shown at I, while the total aberration *T* is the vector sum of these individual aberrations.

The figure is divided into three sections, *A*, *B*, and *C*, the first row of each section representing the image on the axis and the other two rows representing the images at field angles of 7° and 13°. Section *A* is for the lens as originally designed. The image *T* at 13° is extremely large and unsymmetrical, and a study of the figure shows that this poor quality is caused by the second-order aberration II.

Fig. L-3 shows at *A* the Staeble-Lihotzky or isoplanasie† condition for this lens, and it is clear that isoplanasie is overcorrected. The design data were

---

† This word was coined by E. Lihotzky to describe the correction of asymmetry in the presence of spherical aberration. It is equivalent to the offense against the sine condition (OSC) of Conrady's *Applied Optics and Optical Design* (1929, Oxford Univ. Press, London).

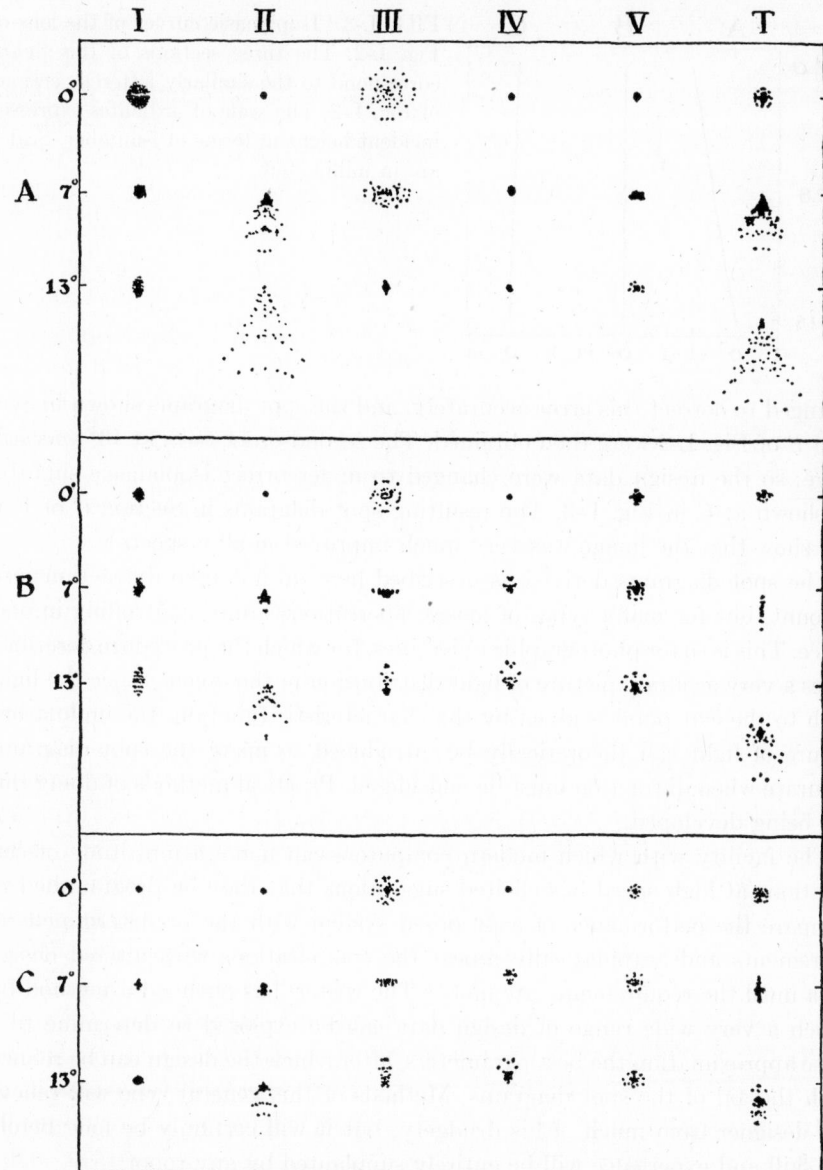

FIG. L-2 Diagram showing how the performance of a certain lens was improved by analyzing the spot diagrams. Columns I–V in section A represent the terms of orders 1–5 at each of three field angles of the original lens; column T represents their sum, the actual image. Section B is for the lens when isoplanasie is corrected, and section C shows the great over-all improvement that results when the isoplanasie is slightly undercorrected.

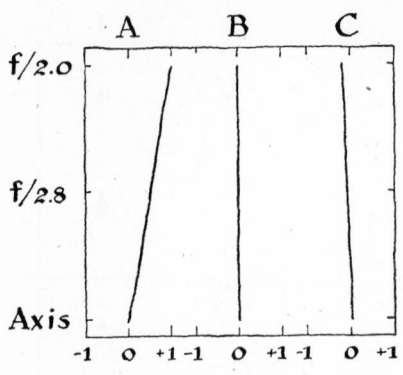

FIG. L-3   Isoplanasie curves of the lens of Fig. L-2. The three sections of this figure correspond to the similarly lettered sections of Fig. L-2. The scale of ordinates expresses incident height in terms of *f*-number. Scales are in millimeters.

changed to correct this error accurately, and the spot diagrams shown in Section *B* of Fig. L-2 were then obtained. The second-order term at 13° was still large; so the design data were changed to undercorrect isoplanasie slightly, as shown at *C* in Fig. L-3. The resulting spot diagrams in Section *C* of Fig. L-2 show that the image was very much improved in all respects.[6]

The spot diagrams derived as described here do not take diffraction into account, but for many types of lenses, aberrations are of controlling importance. This is so for photographic objectives, for which the procedure described gives a very accurate picture of light distribution in the image. Since the light path to the exit pupil is given by the characteristic function, the undulatory nature of light can theoretically be introduced to make the spot diagrams accurate when diffraction must be considered. Practical methods of doing this are being developed.

The facility with which modern computers can make a multitude of calculations at high speed has elicited suggestions that they be programmed to compare the performance of a proposed system with the predetermined requirements and automatically repeat the computations with altered design data until the requirements are met.[7,8] The writer[9] has outlined a method by which a very wide range of design data can be explored to determine to a close approximation the best parameters, after which the design can be refined with the aid of the spot diagrams. Methods of this general type will relieve the designer from much of his drudgery, but it will certainly be long before his skill and experience will be entirely supplanted by any robot.

REFERENCES

1. M. Herzberger, J. Opt. Soc. Am., *47* (in press).

2. M. Herzberger, Chap. 6 in *Optical Image Evaluation*, National Bureau of Standards Circular No. 526 (1954, Government Printing Office, Washington).

3. M. Herzberger, J. Opt. Soc. Am., *37*, 485 (1947).

4. M. Herzberger, J. Opt. Soc. Am., *47* (in press).

5. L. Seidel, Astronomische Nachrichten, *43*, 289 (1856).

6. See also M. Herzberger, *Modern Geometrical Optics* (1957, Interscience Publishers, New York).

7. G. Black, Proc. Phys. Soc. (London), *68*, 729 (1955).

8. D. P. Feder & L. E. Sutton, J. Opt. Soc. Am., *46*, 368 (1956) (abstract of oral paper).

9. M. Herzberger, J. Opt. Soc. Am., *47*, 345 (1957) (abstract of oral paper).

Appendix **M**

Graphical Ray Tracing

by E. W. Silvertooth[†]

The scientific worker who is only occasionally concerned with geometric optical design may spend a conspicuous amount of time becoming familiar with and adept in the practice of trigonometric ray tracing. Any device which can aid in the reduction of "look-ups" in mathematical tables and of numerical calculation should be a welcome adjunct to the designer's facilities.

Several graphical methods of ray tracing have been described in the literature, each of which is a variation on Fig. M-1a,[1] where ray DC in a medium of refractive index N_1 is refracted at the interface separating mediums N_1 and N_2 ($FG \parallel EC$). In the case where N_1 is unity (1.00029), the arc of radius FC is known as the "air circle" and the more general arc of radius CD as the "wavelength circle." For convenience it may be preferable to employ the

[†] 3681 Chevy Chase Drive, Pasadena, California.

FIG. M-1   Construction of a ray.

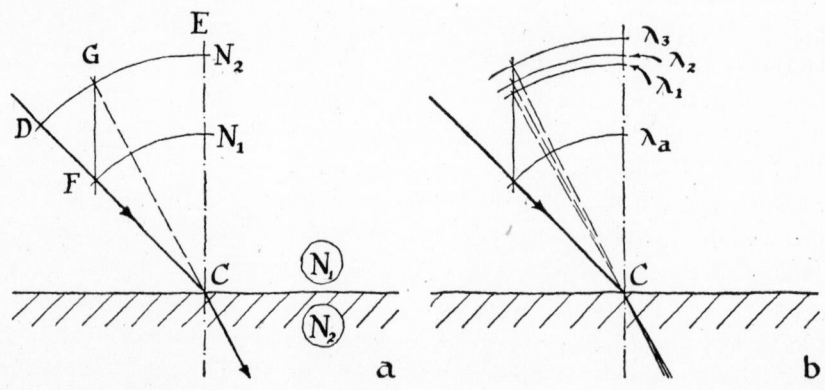

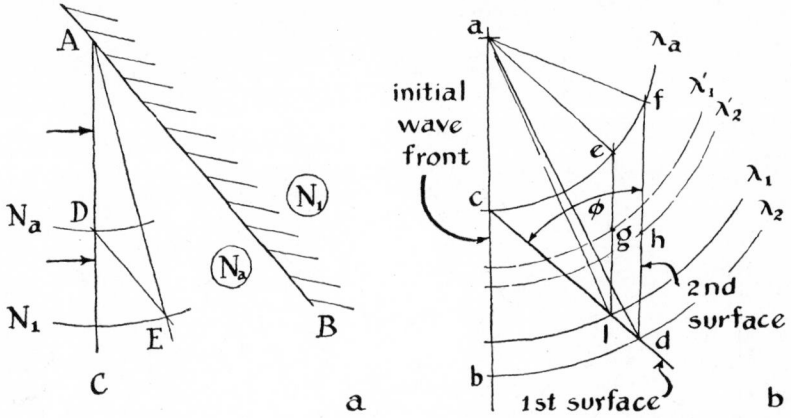

FIG. M-2   Construction of a wave front.

labels λ_a and λ_{5890} (typ.) for each wavelength traced, with supplementary notation for the glass type. Thus rays for several wavelengths may be traced as in Fig. M-1b.

Alternatively, and particularly for compound prism systems, it is usually preferable to construct the ray normals, or wave fronts, as in Fig. M-2a. In this case the wave front AC is rotated into AE after refraction at the interface $AB(DE \parallel AB)$. An example of systemization of the method suitable to the design of a prism train is shown in Fig. M-2b. Here the initial wave front ab is at $< bcd$ with the first interface cd. After refraction the wave fronts become al and ad respectively for the wavelengths λ_1 and λ_2. Where the second interface df is in contact with air, the respective wave fronts become ae and af after the second refraction ($le \parallel df$). Where the second interface is in contact with another medium having wavelength circles λ_1' and λ_2', the wave fronts would then be ag and ah respectively. In cases where a specified amount of dispersion for a prism or prism train is required, a few trials will yield sufficient information to make a graphical interpolation plot for the final prism angles within the accuracy of the method.

One very simple device useful for ray tracing is Smith's Ray Plotter,[2] shown in Fig. M-3a. A sheet of thin Plexiglass about $2'' \times 6''$ may be scribed and filled with the normals XY and OC, after which fine holes A, B, and C are drilled as indicated. To use the plotter (Fig. M-3b), place XY tangent to the boundary of the two media with O coincident with the intercept at the boundary of the ray to be traced. Insert a pin at A, and rotate the plotter until C lies on the incident ray. Pricking the paper at B and extending a line through this point and the intercept will define the refracted ray. Conversely,

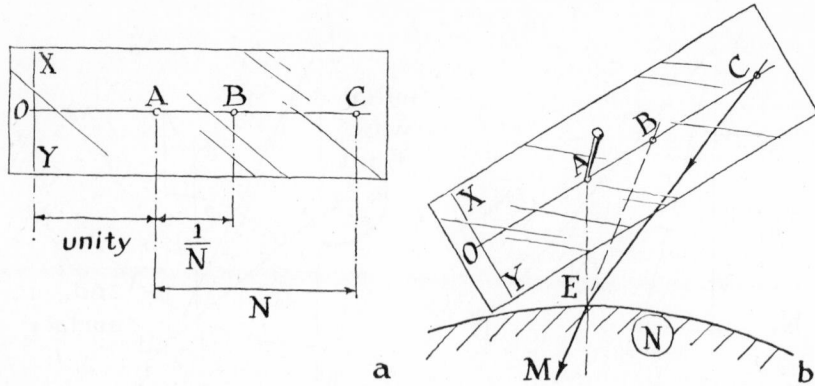

FIG. M-3 Smith's ray plotter.

starting with the ray in the denser medium, the plotter may be rotated until B is aligned with an extension of that ray which locates C and hence defines the ray in the rarer medium. While this device is useful for ray plotting at a single wavelength of a particular glass for which the holes have been drilled, it is not conveniently adjustable for other indexes. Similar plotters which may be adjusted have been described, such as Wray's Optical Ray Plotter[3] and the BK Ray Plotter.[4]

Another method, described by van Albada,[3] makes use of a proportional divider with an adjustable pivot such that the spans between the pairs of points on opposite ends are in proportion to the relative index of the two media being traced. Fig. M-4a illustrates the method of operation. The incident ray is extended, and the longer pair of the divider legs are adjusted to a span Mm, where OM is the radius of the refracting surface. The dividers are reversed and the point of tangency of the arc Mm' is determined. The refracted ray is then defined by D and the point of tangency with the arc m'. The use of proportional dividers may be obviated by constructing an auxiliary nomograph as in Fig. M-4c.

Perhaps the most convenient and practical method of graphical ray tracing is that described by J. H. Dowell.[1] Fig. M-5 is taken from his paper. o', with its associated set of concentric circles, is generally prepared on a separate sheet and located at any point convenient to the layout of the optical system. $o'a'$ is drawn parallel to the chosen incident ray and $a'b'$ is drawn parallel to ar_1, from which the refracted ray ab may be constructed parallel to $o'b'$. "When as many rays as possible have been traced without confusion of the direction line and surface normals, the figure at the left is simply rotated to a fresh position and so on until it is filled up, when it can be replaced by another.

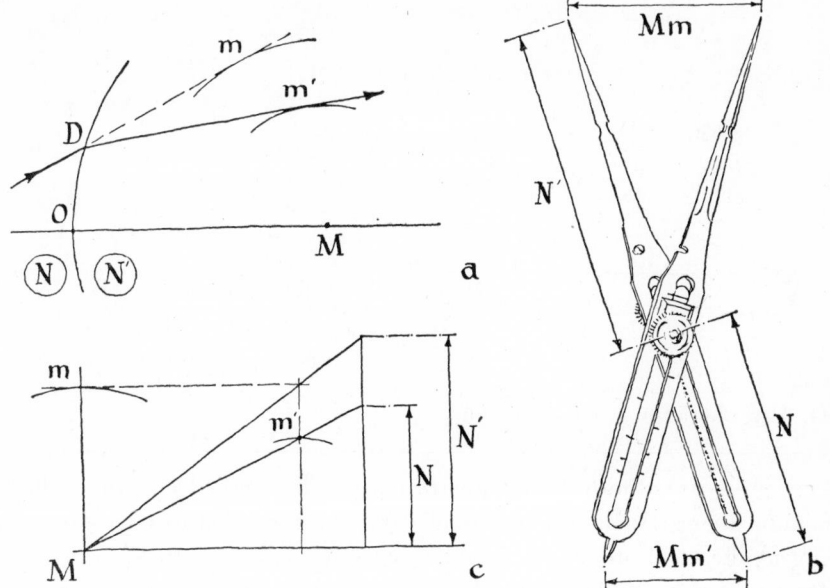

FIG. M-4   Method of proportional dividers.

In this way, no matter how many rays may be traced through the system, there will be a complete absence of such workings on the lenses themselves as would result from use of the ordinary graphical methods. I would particularly like to draw attention to the value of this method for the use of students as it provides a means of studying the form of lenses in an attractive way. For instance, if a bundle of equally spaced parallel rays are traced through lenses

FIG. M-5   Ray tracing with a drafting machine.

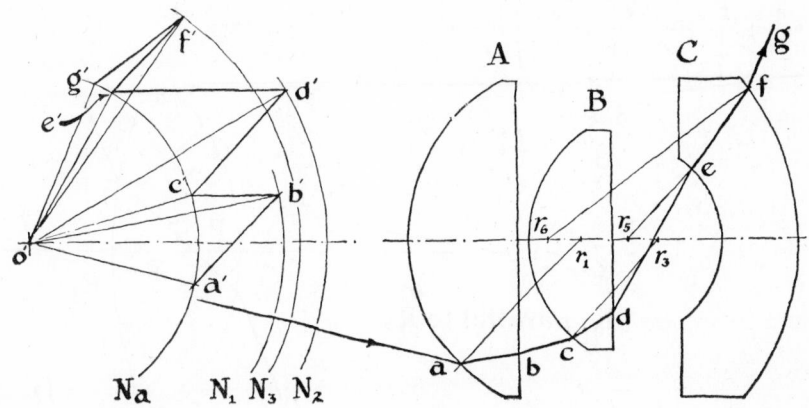

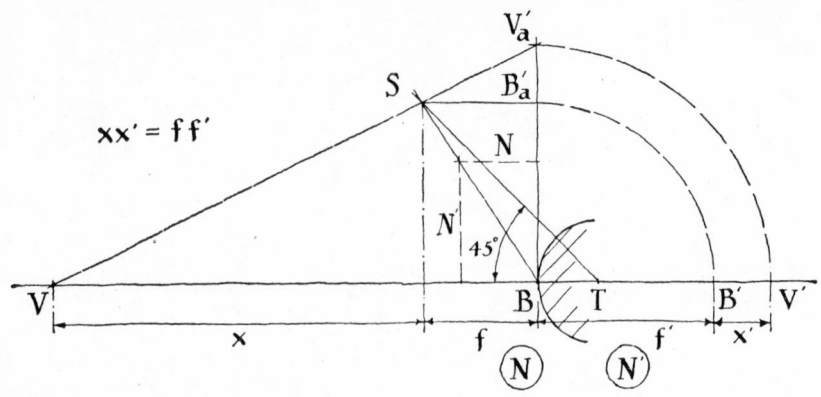

FIG. M-6 Location of the pivot point.

---

of various forms, the resultant aberration can be studied and the radii for minimum aberration determined. Similarly, achromatic lenses, effects of stops, doublets, etc., will prove very fascinating exercises."

Prior to undertaking a detailed study of the ray behavior in a given system, it is desirable to establish paraxial benchmarks from which may be referenced the principal and image surfaces, curves of spherical aberration, etc. Since the paraxial rays lie in the immediate vicinity of the optical axis, accurate location of nodal and focal points by the method described would be impractical. Paraxial focal and image points for a refracting surface may be located graphically by an alternate method shown in Fig. M-6.[3] The point S is called

---

FIG. M-7 Location of nodal points.

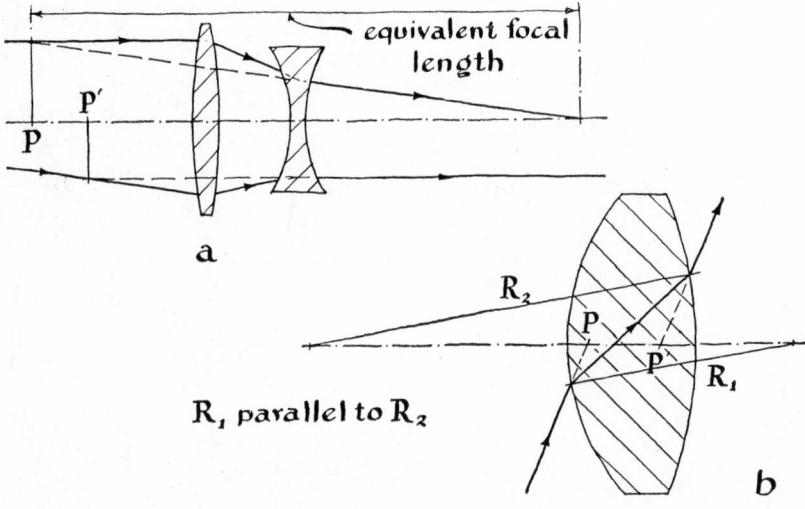

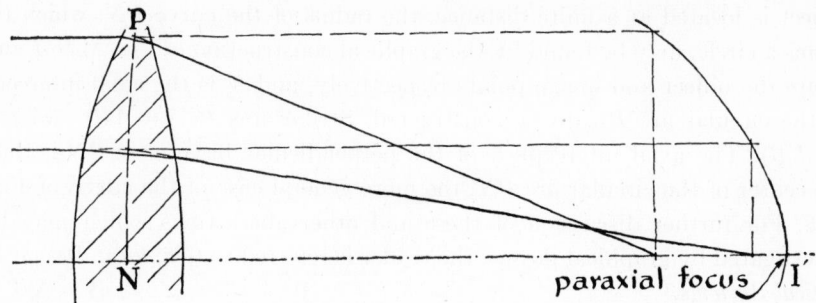

FIG. M-8 Curve of spherical aberration.

the pivot point and is unique for each refracting surface. By locating S for each refracting surface it is possible to locate the overall object, image, and focal points for a complete optical system. It is probably more convenient, however, to run a numerically calculated paraxial ray trace through the system. This is quite direct since functions of the angles are not involved. Equations for a paraxial trace may be found in any full text on geometric optics. For approximate purposes, nodal points and the equivalent focal length may be located as in Fig. M-7a, or, alternatively, the nodal points as in Fig. M-7b.

Having obtained the nodal points and paraxial focal points, one may proceed to examination of the various defects of the system. For example, the curve of spherical aberration may be constructed as in Fig. M-8. Offense against the sine condition, or measure of coma, may be assessed from the curve PN of Fig. M-8. In the case shown, where the object is at infinity, the curve PN for a coma-free system is a circle with center I'. In the case where the

FIG. M-9 Construction of a coma-free wave front.

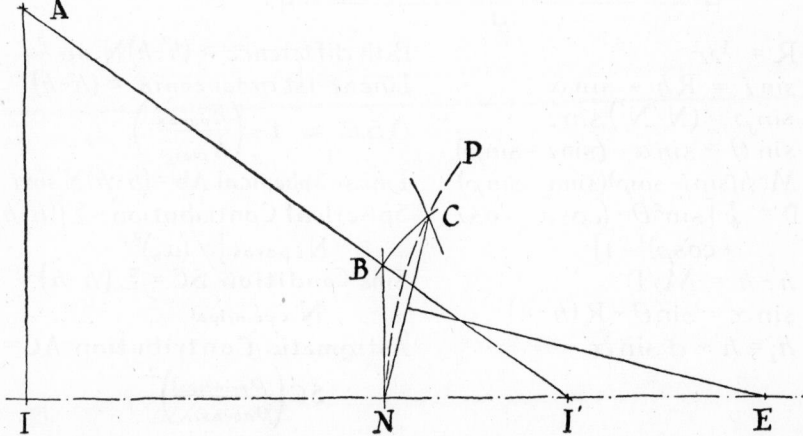

object is located at a finite distance, the radius of the curve PN, which remains a circle, may be found by the graphical construction of Fig. M-9. I and I' are the object and image points respectively, and N is the axial intercept of the circular arc PN to be constructed. Strike arcs IC ($= AB$) and $I'C$ ($= I'B$). The axial intercept E of the perpendicular bisector of CN is then the center of the circular arc PN, the more general case of the curve of Fig. M-8. For further discussion of these and other aberrations which may be investigated by graphical means, the reader is referred to *Graphical Design of Optical Systems*.[3]

In the design of high aperture illumination systems it is frequently desirable to employ aspheric surfaces to reduce spherical aberration and Fresnel reflection losses at high angles of incidence. Second degree parabolic surfaces are frequently employed in such applications. In this instance the slope of the radius for a given zone, required for the Dowell method, may be obtained from the bisector of the angle formed by a line parallel to the optical axis displaced

---

FIG. M-10 Ray-tracing method of T. Smith. In regard to the last three equations see reference 7.

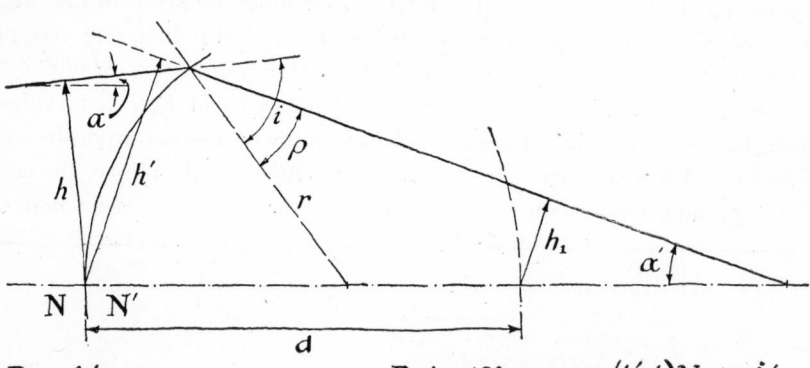

$R = 1/r$

$\sin i = Rh + \sin \alpha$

$\sin \rho = (N/N')\sin i$

$\sin \theta = \sin \alpha - (\sin i - \sin \rho)$

$M = h(\sin i - \sin \rho)(\sin \alpha + \sin \rho)$

$D = \frac{1}{4}\left[\sin^2\theta + (\cos \alpha + \cos i + \cos \rho)^2 - 1\right]$

$h' - h = M/D$

$\sin \alpha' = \sin \theta - R(h'-h)$

$h_1 = h' + d \sin \alpha'$

Path difference $\approx (h'-h)N \sin \frac{i}{4}$

Linear 1st order coma $\approx (h'-h)$

O.S.C. $\approx 1 - \left(\dfrac{h'_{parax.}}{h_{marg.}}\right)$

Linear Spherical Ab $= (h'-h)N\sin i$

Spherical Contribution $= \Sigma[(h'-h) \, Ni_{parax.}]/(\alpha'_k)^2$

Sine Condition $SC = \Sigma(h'-h) \, Ni_{principal}$

Astigmatic Contribution $AC = SC\left(\dfrac{\rho_{principal}}{\rho_{parax.}}\right)^2$

to the desired zonal radius, and a line connecting the focus of the parabola and the constructed parallel intercept with the parabola.

Finally, for completeness, it is possible by methods of descriptive geometry to graphically trace skew rays, or rays which are generally not in a plane containing the optical axis. The complexity of the construction is, however, sufficiently great to discourage practical use.

The early stages of graphical lens system design are generally carried on somewhere near unity scale. As the design progresses, it is desirable to increase the scale in the interest of increased accuracy. A 50 mm high aperture photographic objective can usually be determined within 1% when laid out 5 × scale. For some applications, notably illumination systems, this degree of accuracy may well terminate the design. However, in image-forming systems the final design must normally be established by recourse to numerical calculations.

A convenient set of equations (Fig. M-10) for computing rays in a plane which includes the optical axis are described by T. Smith.[5,6] In this method the normal refraction terms are separated from those representing aberrations. By expressing the latter as a fraction with the first order aberration as the numerator, and a correcting factor, which may take various forms, as the denominator, rays may be traced exactly through the system by use of a short table of cosines in terms of sines. A considerable saving in time is thus effected

---

FIG. M-11   Analogue computer for sine-cosine conversion.

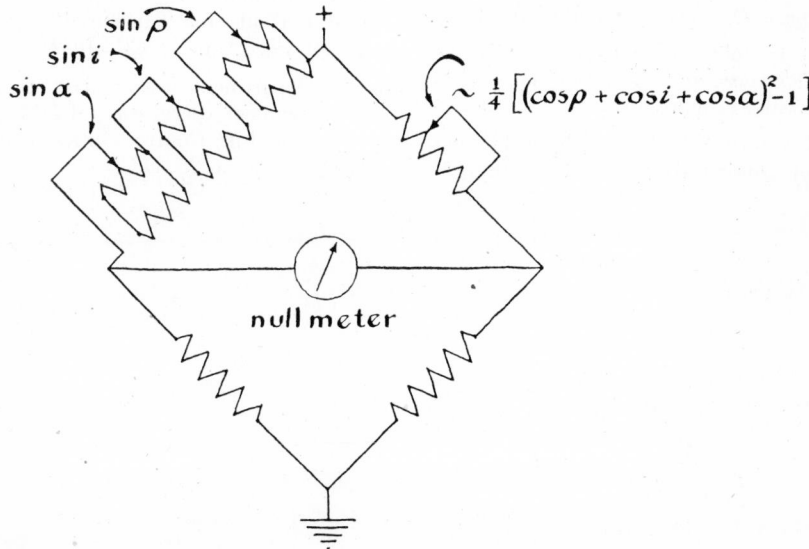

$$\sim \tfrac{1}{4}\left[(\cos\rho + \cos i + \cos\alpha)^2 - 1\right]$$

in the calculations, and the estimation, without calculation, of the aberrations of other rays is facilitated. Mr. Donald Hendrix, of the Mount Wilson Observatory, has eliminated entirely the need for consulting tables through the use of a small analogue computer which performs the essential transform $y = \sqrt{1 - x^2}$. Such a computer, shown in Fig. M-11, plus a desk calculator to perform multiplication and division, can provide accuracy to one part in the sixth place where the computer is accurate to three places. After familiarity with the method is established, computing can proceed at the rate of about two minutes per surface.

An excellent source of information on the subject of optical systems may be found in *Design of Fire Control Optics:* Volume 1, *Application of Fundamental Laws of Geometric Optics to Individual Optical Components;* and Volume 2, *Complete Computations for Fire Control Optical Components and Systems.*[8]

REFERENCES

1. J. H. Dowell, "Graphical Methods Applied to the Design of Optical Systems," Proc. Optical Convention of 1926, II, 965–981 (University Press, Aberdeen).

2. T. Smith, Trans. Opt. Soc., XXI, No. 3 (1919–20).

3. L. E. W. van Albada, *Graphical Design of Optical Systems* (Pitman, Bath).

4. B. K. Johnson, *Practical Optics for the Laboratory and Workshop* (Turnbull & Spears, Edinburgh).

5. T. Smith, "On Tracing Rays Through an Optical System" (Third Paper), Proc. Phys. Soc. (London), *32*, 252 (1920).

6. "Notes on Mr. T. Smith's Method of Tracing Rays Through an Optical System," J. Opt. Soc. Am., *5*, 14, 334 (Jan. 1921). Smith's table is reproduced in this paper.

7. Erwin Delano, "A General Contribution Formula for Tangential Rays," J. Opt. Soc. Am., *42*, 631 (1952).

8. U.S. Department of Commerce, Office of Technical Services, ORDM2-1: Design of Fire Control Optics, I and II (Washington 25, D.C.).

Appendix **N**

Fiber Optics

by Narinder S. Kapany†

Introduction

The conduction of light along a transparent dielectric cylinder due to multiple internal reflections is well known. Until recently the application of this phenomenon has been limited to transporting light from one point to another. In recent years, however, applications of this phenomenon have received close attention, and a new practical branch of optics is being developed. It appears to offer many fascinating possibilities and requires the investigation of a number of new techniques. A search in the literature has revealed that Baird[1] and Hansell[2] commented on one of these possibilities in connection with the development of television. In 1952 work on a fiberscope was initiated independently by Hopkins and Kapany[3,5] at the Imperial College, London, and by van Heel[4] in Holland. Since then the field of fiber optics has found various other uses in different parts of optics, and work is now in progress on many different types of applications.

Light is conducted from one point to another along a transparent fiber without much loss of energy, and geometrical optics is found to hold for most calculations on the conduction of light along transparent rods or fibers if their diameter is many times the wavelength of light. When the diameter is smaller, diffraction effects play a role, and the fibers act as wave guides.[6,7] Experiments have been conducted, so far, with glass fibers 25 microns and more in diameter, and work using 12-micron fibers is also in progress. The fiber length is obviously limited by the light transmission characteristic of the material. The physical properties, surface structure, inhomogeneities, and variation in diam-

---

† Physics Research Department, Armour Research Foundation of Illinois Institute of Technology, Chicago, Illinois.

eter play an important role in the light conduction property of a fiber. A large number of available fibers were therefore tested in a manner to be described later, and it was found that glass fibers hold promise for most uses in fiber optics.

Fibers are capable of isolating an element of an image surface and conveying it to another point along flexible axes. This phenomenon has been termed "static scanning." A well-aligned bundle of fibers can thus be used for the transmission of optical images. Because of the flux-integrating property of the individual fibers, any detail finer than the fiber diameter will not be transmitted by static scanning. The resolving power of a bundle of fibers used to convey an image is therefore limited by the diameter of the component fibers. It has also been demonstrated that the resolving power of a bundle of fibers can be increased substantially by employing the principle of dynamic scanning. An experiment on these lines demonstrated a gain of 100 percent in resolving power. Appropriate assemblies of transparent dielectric cylinders may thus be employed for transmission or modification of optical images, and such applications of fiber optics are called static and dynamic scanning devices. On the other hand, single dielectric cylinders may be used for image scanning and photorefractometry. Finally, transparent fibers have been employed as light funnels, changing the shape of an illuminated surface but keeping the image area constant. A detailed discussion of these techniques will follow at a later stage in this appendix.

Optical Properties of Single Dielectric Cylinders

The light transmission and angular relationship of the entrance to the emergent cone of light for a straight dielectric cylinder are dependent on its constancy in diameter, surface structure, and isotropy. The mechanism of light conduction in an ideal transparent cylinder is clear, but a slight departure from cylindricity and any surface structure tend to cause light losses. In a curved cylinder, it is found that the skew rays contribute a great deal towards the modification of flux distribution in the emergent light cone. It is observed that the light transmission of a bent glass rod falls when the bending radius is equal to or smaller than 20 times its diameter. This loss is accounted for by the behavior of skew rays, which tend to escape through the fiber walls at a radius many times greater than is required for the meridional rays to escape.

Fig. N-1a shows a meridional ray incident at an angle i to the axis of a straight fiber whose axial length is equal to L. The angle of the refracted ray at the fiber wall $\phi = \dfrac{\pi}{2} - i'$. Any ray which enters a fiber having a flat end

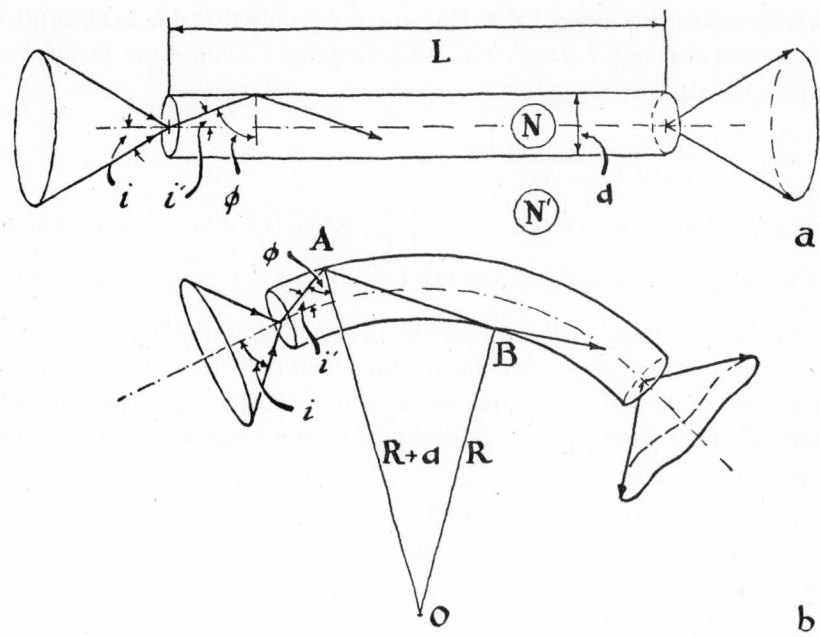

FIG. N-1 Passage of light along cylindrical fibers. (a) Ray path in a straight fiber. (b) Fiber bent on a circular arc and the modification of its radiation pattern.

will therefore meet the walls of the fiber at an angle greater than critical incidence provided the critical angle $\left(\phi_c = \sin^{-1} \dfrac{N'}{N} \right)$ condition is satisfied by the two media. It is easy to see that any ray after refraction at the flat end will be trapped in a glass fiber ($N \geq 1.5$) placed in air. Furthermore, the length of path $P(i)$ of a multiply reflected ray in a straight fiber is independent of the diameter d and is given by

$$P(i) = \frac{L}{\sqrt{1 - \left(\dfrac{\sin i}{N} \right)^2}} = L \sec i' \tag{1}$$

where N is the refractive index of the fiber material and i' the angle of the refracted ray. The length of optical path between any two successive reflections in a straight fiber is the same, and the direction of the emergent ray is the same as or opposite to that of the entrance ray as the number of reflections is even or odd. For a ray inclined at 40° to the fiber axis ($N = 1.5$) the length of path is only 10 percent greater than the axial length of the fiber. Consequently, the absorption by the glass path is not substantially greater for oblique rays than for the axial ray.

In a clean fiber the loss of light along any ray is only that due to absorption along a path of length L sec i'. The transmissivity $\mathfrak{T}(i)$ for a ray inclined at an angle i is given by

$$\mathfrak{T}(i) = \exp\left[\frac{-\alpha L N}{\sqrt{N^2 - \sin^2 i}}\right] \tag{2}$$

where α is the absorption coefficient of glass. When the value of i goes from $0°$ to $90°$, the transmission goes from $\exp(-\alpha L)$ to $\exp\left[\frac{-N\alpha L}{\sqrt{N^2 - 1}}\right]$. This expression assumes total reflection within the walls of the fiber, and freshly drawn glass fibers closely approximate this condition. However, if films of dirt, grease, or a coating material are present upon the fiber, a certain percentage α' may be lost at every reflection. The transmission expression then becomes

$$\mathfrak{T}(i) = (1 - \alpha')^{\mathfrak{N}} \cdot \exp\left[\frac{-\alpha L N}{\sqrt{N^2 - \sin^2 i}}\right] \tag{3}$$

where

$$\mathfrak{N} = \frac{L \sin i}{d\sqrt{N^2 - \sin^2 i}}$$

Here $\mathfrak{N}$ is the number of reflections for a given ray in the fiber. The transmission of a uniform cone of light incident on a fiber may thus be calculated by numerical integration. It can be shown that, for a $10°$ incident cone of light in a fiber 50μ in diameter and 50 cm long ($N = 1.5$ and $\alpha = 1\%/\text{cm}$), the transmission is 60% when the fiber surface is clean ($\alpha' = 0$). However, when $\alpha' = 1\%$, the light transmission falls to only 4%. This indicates the critical necessity of keeping the fiber surface clean.

Fig. N-2 is a plot of the percentage of emergent light E for a fiber 30 inches long ($\alpha = 2\%/\text{inch}$), P being the length of optical path and $\mathfrak{N}$ the number of reflections as a function of the incident angle I. It is to be noted that, whereas the light transmission for the axial ray is 55%, a ray inclined at $40°$ to the axis transmits nearly 51%. Also the length of the optical path is $1.1L$ for a ray inclined at $40°$ to the axis. The number of reflections $\mathfrak{N}$ for a given ray in a fiber increases as the axial length increases or the diameter decreases. For example, in a fiber 25μ in diameter and $1''$ long ($N = 1.5$), when $i = 10°$, $\mathfrak{N} = 116$; when $i = 5°$, $\mathfrak{N} = 58$.

Fibers drawn from molten glass may have large variations in diameter, and such fibers act essentially as conical reflectors. It will be shown that the photometric efficiency of a light guide with constrictions of radial symmetry is reduced for rays exceeding a given angle to the axis. Fig. N-3a shows a conical channel in which two rays enter at different inclinations to the axis. The solid

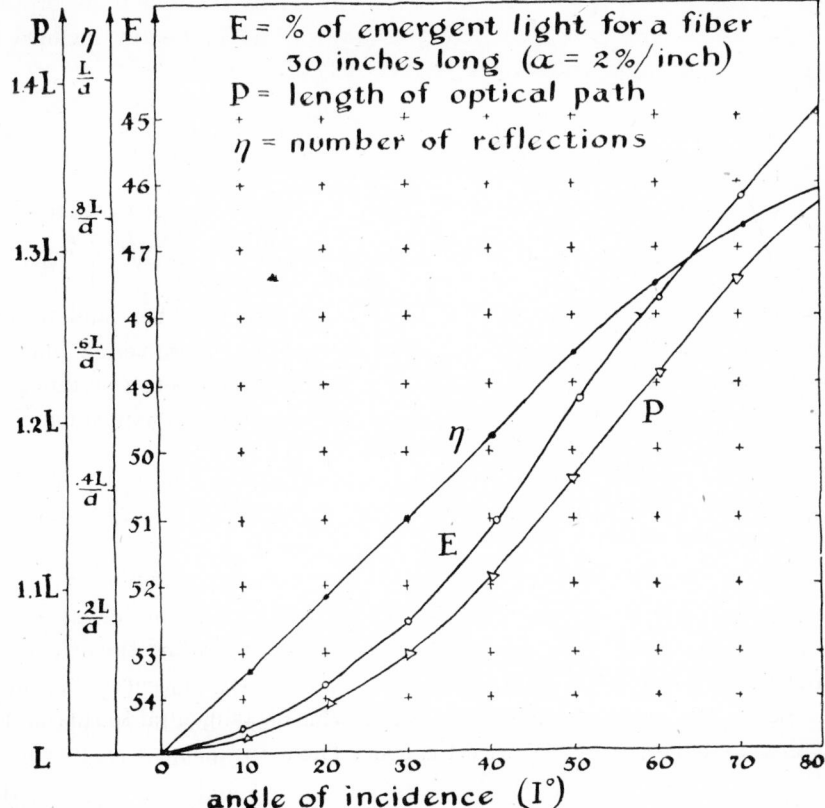

FIG. N-2 Transmission, path length, and number of reflections in a straight fiber as a function of angle of incidence.

ray, inclined at an angle less than the "returning angle," passes through the cone after multiple internal reflections, whereas the dotted ray, inclined at a greater angle, is "returned." This is true only for a metallic cone; for a glass cone, however, this ray will escape through the wall when the angle $\phi_\Re$ subtended by the ray at any reflection is less than critical incidence $\Big($ *i.e.* when

$\phi_\Re \leq \sin^{-1} \dfrac{N'}{N} \Big)$. The condition for a ray in the meridian plane to be transmitted through a hollow reflecting cone is simply found by the geometrical construction in Fig. N-3a. C_1D_1, A_1B_1, C_2D_2 . . . are the successive images of the sides AB and CD formed by repeated reflections. If the incident ray intersects the polygon $DBD_1B_1D_2B_2$ . . . , formed by the successive positions of the exit end, it will be transmitted, otherwise not. Fig. N-3b shows $ABCD$ the

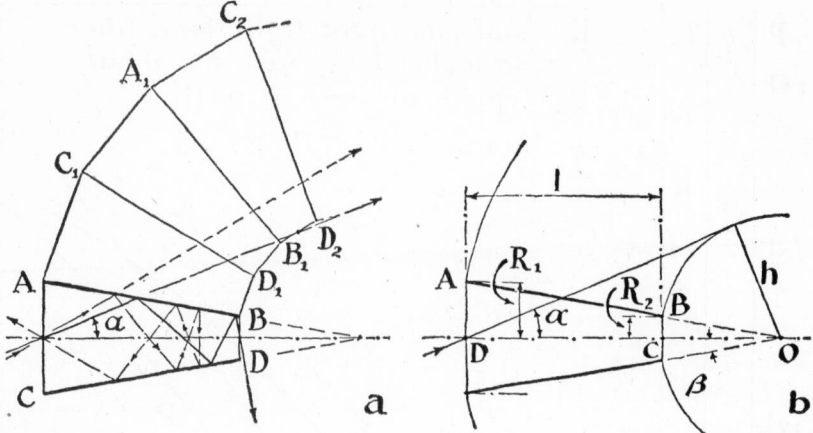

FIG. N-3 Ray path in a conical reflecting channel. (a) The solid ray, inclined at angle α, passes the exit end of the cone, and the broken ray "returns." (b) Approximate method of calculating the "returning angle."

half section of a conical channel. A circle with center at the projected vertex of the cone O and radius equal to the distance h between O and the exit pupil of the cone is a near approximation to the polygon formed by successive positions of the exit end. The limiting ray is that which is tangential to this circle. Let this ray make an angle α with the cone. Then

$$\sin \alpha = \frac{h}{h+l} = \frac{R_2}{R_1} \tag{4}$$

FIG. N-4 Polar diagram of light emerging from a straight fiber illuminated with a point source on the axis.

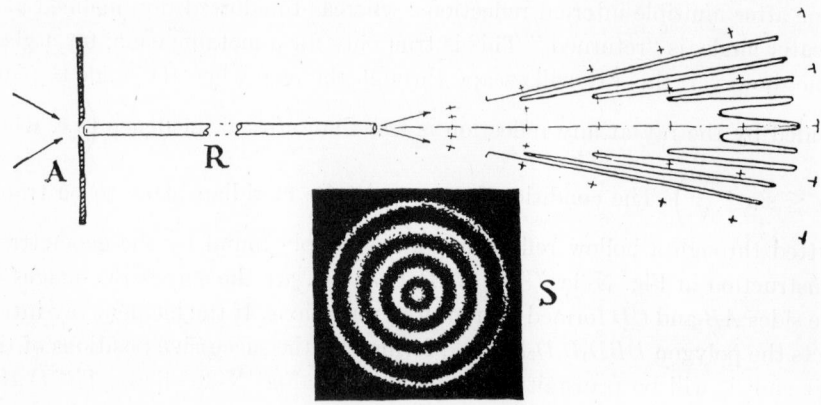

where R_1 and R_2 are the radii of the entrance and exit ends of the channel. The tolerance in variation of diameter therefore depends on the radii of the two ends and is independent of the axial length. For a ray inclined at 30° to the axis a 50 percent variation in diameter is tolerable. With adequate laboratory control, however, glass fibers with ±5 percent variation in diameter can be produced.

Suppose that light is condensed on a small aperture placed on the axis at one end of a straight glass rod, whereby a cone of light is passed axially through the cylinder (Fig. N-4). Now, if a screen is placed at a distance away from the exit end, annular dark and bright rings are observable. Each ring corresponds to a reflection that a part of the light cone suffers at the wall of the cylinder. Fig. N-4 illustrates this polar distribution of light at the exit end of the cylinder. It is found that, in spite of the ring formation, the extreme angle of the exit cone is equal to that of the entrance cone. The number of bright rings is equal to the number of reflections of the extreme ray. No such ring pattern is observable when the entire entrance end of the rod is illuminated uniformly and the angles of the entrance and emergent light cones remain nearly equal.

On the other hand, because of the complicated behavior of the skew rays, the radiation pattern of curved fibers is not easily predictable. The variables are (1) the direction cosines and the incident height h of the rays, (2) the ratio

FIG. N-5   Radiation pattern of various bent fibers.

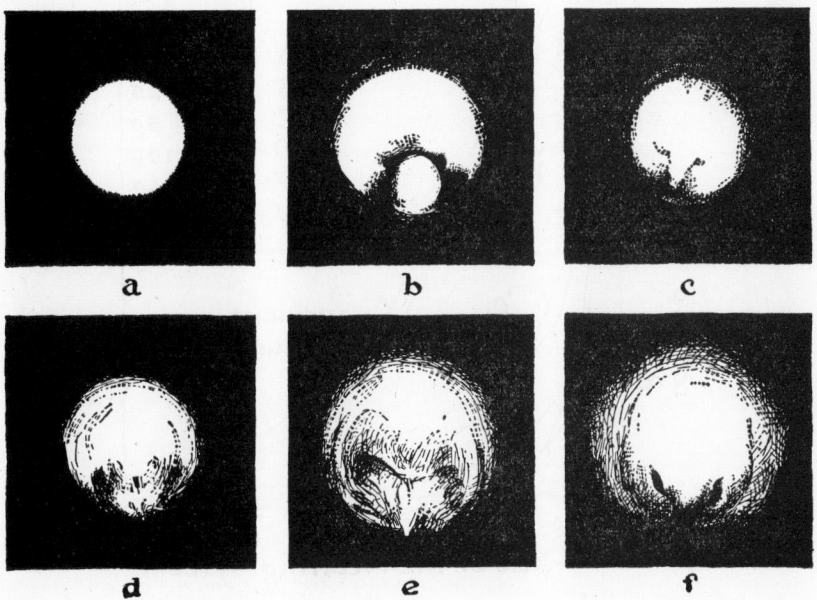

a   b   c

d   e   f

of the bending radius R to the fiber diameter d, and (3) the ratio of the axial length L of the fiber to its diameter. Fig. N-1b shows a glass fiber bent into a circular arc of radius R. For a meridional ray incident at an angle $i = 40°$ to the fiber axis the minimum permissible bending radius of the fiber is found when $\sin \theta \geq 0.71$. With reference to Fig. N-1b the condition becomes

$$(R + h) \cos 25° \geq 0.71 (R + d) \tag{5}$$

so that the bending radius must satisfy $R \geq 3.5d$. In practice, however, it is found that the rays in the meridional plane contribute little to the total conduction of a bent cylinder. The condition for minimum permissible bending radius is violated because rays in other planes escape from the walls when the cylinder is bent on a radius 20 times its diameter.

The radiation pattern of bent fibers was studied experimentally.[8] Fibers of various diameters and lengths were bent on circular arcs of different radii. Light was condensed at one end of the bent fibers, and the radiation pattern from the emergent end was studied. A general broadening of the emergent cone was observed, and photometric measurements were made to study the redistribution of light flux in the emergent cone. Fig. N-5 shows the radiation patterns of six fibers, whose axial lengths are 25d, 50d, and 100d and whose

---

FIG. N-6 Percentage of light flux as a function of f/ratio of emergent light cone for various bent fibers.

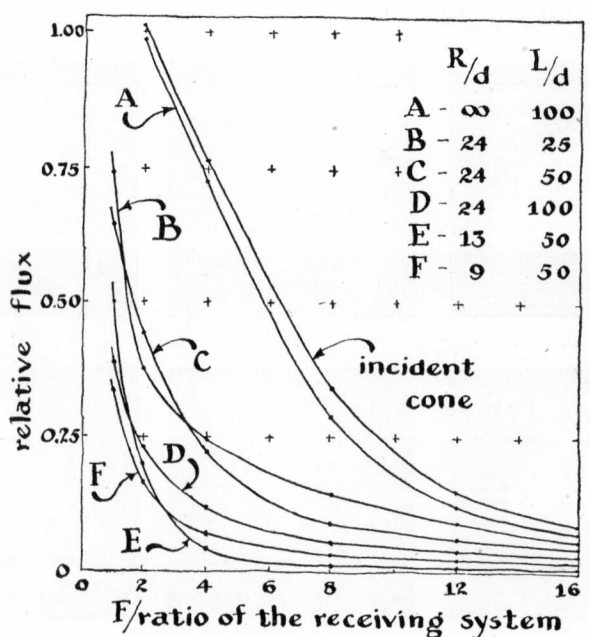

	R/d	L/d
A	∞	100
B	24	25
C	24	50
D	24	100
E	13	50
F	9	50

incident cone

relative flux

F/ratio of the receiving system

bending radii are 9.4d, 13.5d, 24d, and ∞. The radiation patterns of straight rods are symmetrical, and a gradual broadening and asymmetry occur for shorter lengths of fibers bent on steeper radii. However, a general symmetry along the *y*-axis is observable in these patterns. Fig. N-6 shows the photometric measurement of light distribution in the emergent cone from six bent fibers. A lens with iris diaphragm was used to collect the light from the emergent end of the fiber, and the *f*-ratio of the lens is plotted as a function of the photocell response. A knowledge of the photometric behavior of bent fibers is relevant to the design of light funnels, etc.

Very fine fibers are being hot-drawn of materials such as glass and quartz, and extruded fibers of synthetic materials are available. A large number of available fibers were tested for surface structure, homogeneity, birefringence,

---

FIG. N-7 Interference patterns of various fiber surfaces. Coplanar wave fronts (a) and tilted wave fronts (b); interferograms for a perfect cylindrical fiber, nylon, polyacrylonitrile, and glass fiber.

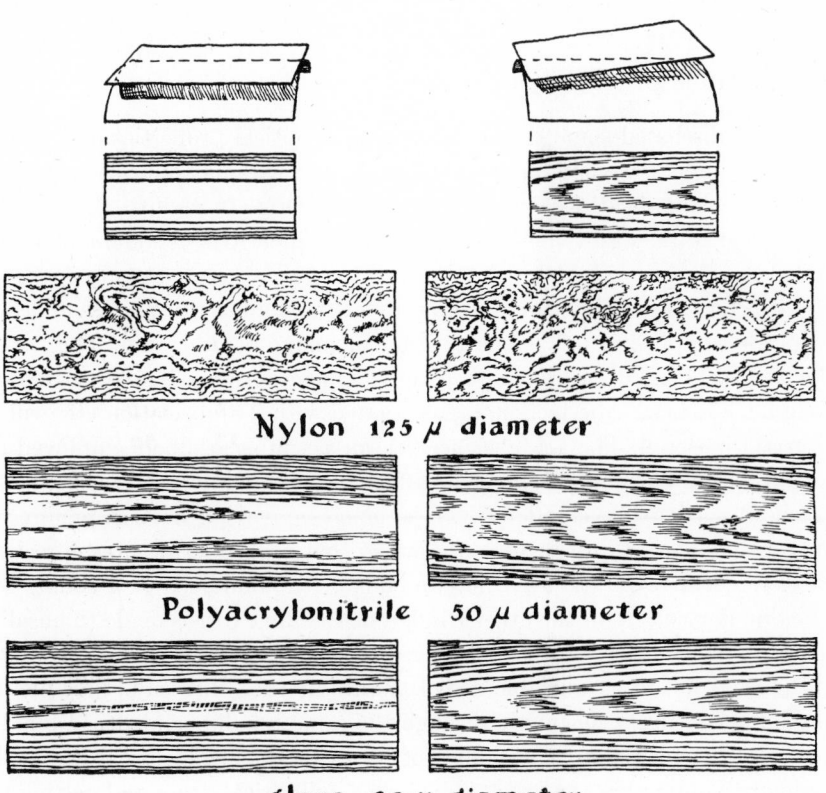

Nylon 125 μ diameter

Polyacrylonitrile 50 μ diameter

glass 60 μ diameter

a   b

and transmissivity with applications in fiber optics in mind.[8] Glass and quartz fibers have shown most promise for optical uses, the former being easily procurable. The surface structure and cylindricity are measured on the Linnik type of microinterferometer, and continuous measurements of their variation in diameter are made for longer lengths. Fig. N-7 shows the microinterferometric pattern of three types of fiber surfaces. The expected interference pattern of a perfect cylindrical wave front with a plane reference wave front is shown at the top when the two wave fronts are coplanar (a) and when they are tilted (b). It is seen that, whereas nylon and polyacrylic fibers show marked surface structure, glass fiber has good cylindricity and surface quality. Electron micrographs of drawn glass fibers have revealed small isolated hills nearly 400 Å wide and 150 Å high, but this structure should not appreciably effect the conduction property of glass fibers. It is also observed that a difference exists between the refractive index of a glass fiber and that of the parent bulk glass. In a typical case the index of the fiber was found to be lower than that of the bulk glass index by 0.0008. This difference is characteristic of the mode of production.

Applications of Fiber Optics

Having discussed some factors affecting the optical properties of single dielectric cylinders, we now proceed to outline the various applications of fiber optics. Such applications have been grouped (Fig. N-8) into three major categories: static and dynamic scanners, single dielectric cylinders, and light funnels.

Static and dynamic scanning devices employ appropriate assemblies of fibers for the purposes of image transfer or image modification. The *flexible fiberscope* (a) is used for transporting optical images from one point to another along flexible axes. The techniques associated with its fabrication etc. will be discussed in detail. The *field flattener* (b), an assembly of fibers employed for the purpose of flattening the image surface of a curved field lens system, will also be discussed in some detail. These image-conveying devices require the fibers to be aligned in a regular fashion so that each fiber occupies precisely the same relative position at the two ends of the bundle. For a coding and decoding device (c), on the other hand, a deliberate effort is made to misalign the component fibers.[9] An object viewed through such a random fiber bundle will present an unrecognizable coded picture at the other end. This picture may be decoded by viewing it through the same bundle or identical bundle. It is possible to make two bundles of fibers producing identical coding by randomly winding long lengths of fibers on a cylindrical drum and cutting the

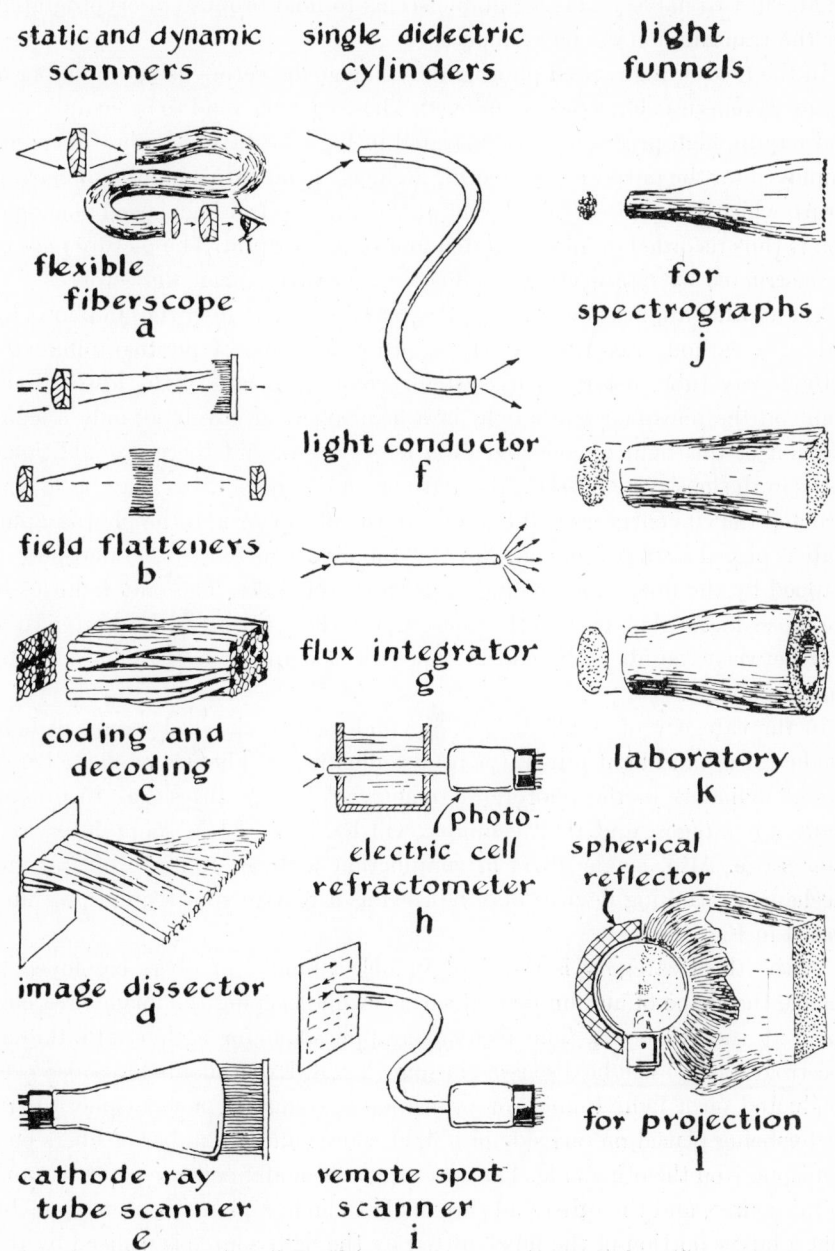

static and dynamic scanners

flexible fiberscope
a

field flatteners
b

coding and decoding
c

image dissector
d

cathode ray tube scanner
e

single dielectric cylinders

light conductor
f

flux integrator
g

photo-electric cell refractometer
h

remote spot scanner
i

light funnels

for spectrographs
j

laboratory
k

spherical reflector

for projection
l

FIG. N-8 Applications of fiber optics fall into three categories: static and dynamic scanners, single dielectric cylinders, and light funnels.

bundle in two halves. This technique seems to hold promise in cryptography for the transmission of coded images.

In the field of high speed photography, image dissectors using a number of prism systems etc. have been employed. These systems tend to be complicated and require high precision. As illustrated in Fig. N-8d, transparent fibers are employed for the purpose of dissecting an image plane into a number of narrow strips, which are made to lie in a row at the other end. A high speed phenomenon is thus recorded on film mounted in a rotating drum. The picture may be reconstructed by reprojecting the image through the image dissector.

Yet another application of an aligned bundle of short fibers lies in photography of a cathode ray tube (e). Conventional means of photographing the cathode ray tube, using a lens system, suffer from large light losses. Each point on the phosphor emits light in a hemispherical envelope; only a small fraction of this light is collected by a lens system, and there are additional losses in the lens system itself. Alternatively, a bundle of fibers may be so constructed that its entrance end lies next to the phosphor and the photographic plate is placed at the other end. Most of the light emitted by the phosphor is trapped by the fibers and conducted towards the plate. The only light losses are those due to glass path in the fibers and to the 9.4 percent dead interstitial area between the fibers. A gain in exposure of many times is achievable by this method.

In the category of single dielectric cylinders, two classical usages of light conductor (f) and light integrator (g) are illustrated. The use of single transparent cylinders in the photorefractometry of liquids (h) seems to present many advantages, and this technique will be discussed in some detail at a later stage. Also, single fibers in conjunction with appropriate optical and mechanical scanning devices may be used as a remote spot scanner, as illustrated in Fig. N-8i.

In the third category is the light funnel, a bundle of fibers employed to change the shape of an illuminated surface while keeping the image area constant at the two ends. The particular application for stellar and Raman spectrographs (j) will be discussed in some detail. Experiments have also been conducted on a light funnel for projection systems (l), a hemispherical reflector being placed on one side of a light source and a number of fibers on a hemisphere on the other side. The entrance ends of all fibers are placed normal to the source, and the other ends are made to lie in a plane. It seems possible that a larger portion of the flux emitted by the light source is utilized by this system.

Static and Dynamic Scanners

Flexible Fiberscope

A well-aligned bundle of fibers, with good optical insulation of individual fibers, is particularly useful for viewing objects lying at places remote from direct observation. The flexibility of this system allows it to be navigated along curved channels. Many applications of this unit for remote viewing in mechanics and other fields are possible. In particular, the endoscopic examination of the internal portions of the body is borne in mind. The existing instruments, which consist of periscopic doublets and field lenses, are found to have only limited flexibility, and their light transmission and image quality are unsatisfactory. Because of the existing gastroscope's lack of flexibility, four blind regions inside the stomach are unobservable.[10] The use of flexible fiber bundles will ease the passing of the instrument inside the patient and will also improve the image quality and light transmission. Color photography of remote areas is made possible by this system.

A typical optical assembly of a flexible fiberscope system is illustrated in Fig. N-9. A two element lens system is employed to form the image of the object on the entrance end of the fiber bundle, and an eyepiece or a camera lens may be used for receiving the image. Whereas conventional instruments make use of a small electric bulb at the distal end for illuminating the object, coarser fibers conduct light from an outside source to the object. Because of the comparatively high light transmission of a fiber bundle, a strong light source may not be necessary even for the purposes of color photography. The mechanical design of such an instrument is of interest. Flexible tubings have been developed which allow required flexures to the instrument with appropriate mechanical controls. One such instrument is being constructed with the addi-

---

FIG. N-9   Optical arrangement of a typical flexible fiberscope.

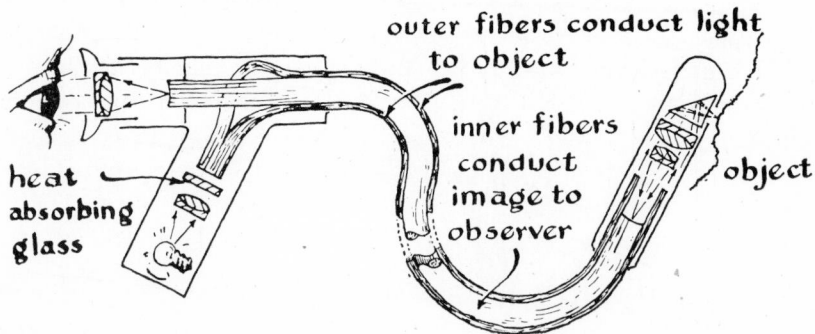

tional feature of a mechanism for rotating the 90° prism at the object end. Thus mosaic color pictures of the entire inside of the stomach seem possible for the purpose of permanent records and reference.

For alignment of the fibers in a bundle, use is made of nominally infinite lengths of constant-diameter transparent glass fibers. Long lengths of fibers are wound on a former having a peripheral groove of square cross section, layer by layer, so that each layer is made to lie in the groove of the previous layer. It is to be noted that for perfect alignment the direction of winding of successive layers of fibers on the cylindrical former has to be the same. Loss of detail and resolving power occurs if there is any error in winding. After winding the required number of layers, the bundle of fibers may be gripped at different points by mechanical clamps, the spacing of which is determined by the length of fiber bundles required. The fiber stack is sawn at different points between the clamps by means of a very fine diamond saw, and the two ends are polished. There are also other mechanical methods of aligning the fiber bundles.

The mechanism used for winding continuous lengths of glass fibers on the aligning former is illustrated in Fig. N-10. Long lengths of glass fiber may

FIG. N-10   Fiber-aligning mechanism.

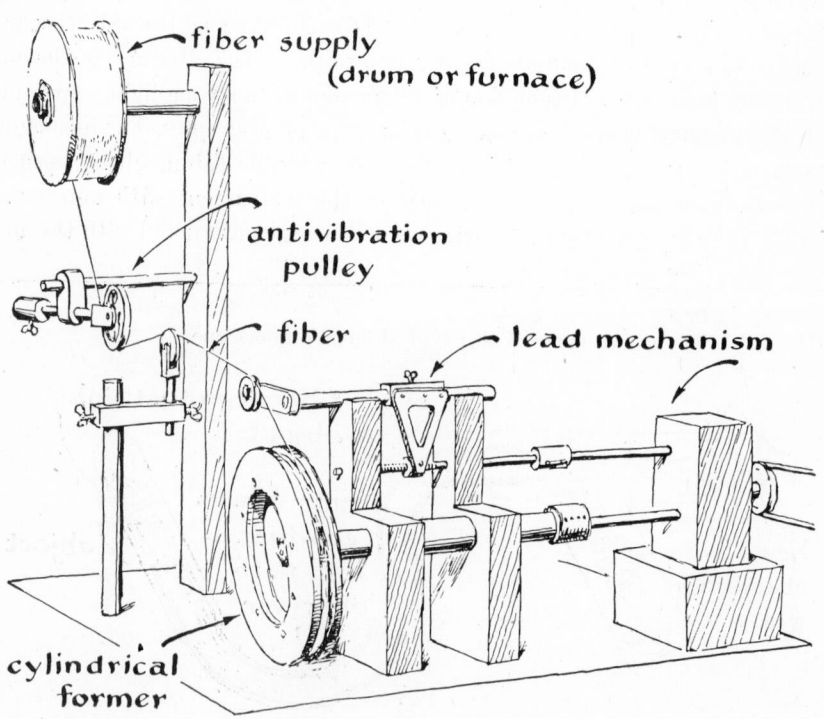

be wound on a cylindrical supply drum or used directly to feed this aligning mechanism. Continuous lengths of fiber are thus fed, under constant tension, to the cylindrical former by way of anti-vibration pulleys and constant-tension devices. A precision lead mechanism is used, and a smooth running pulley guides the fiber on the rotating former. The former itself is provided with a peripheral groove of square cross section, and the clamps are made an integral part of the former. A number of fiber bundles have been aligned, using fibers 50μ and 25μ in diameter and of various lengths. One such fiber bundle consisting of approximately 50,000 fibers is shown in Fig. N-11. These bundles are found to be quite strong and flexible, and the image quality and light transmission are found to be satisfactory.

Field Flattener

A well-aligned bundle of fibers can also be used to flatten the curved field of an anastigmatic lens system. The entrance end of a bundle of fibers is given the shape determined by the Petzval sum of the preceding lens system. The other end of this bundle may then be made plane if the system is used

FIG. N-11   A 50 μ fiber bundle (consisting of 50,000 fibers) bent over a circular arc of 2 inches radius.

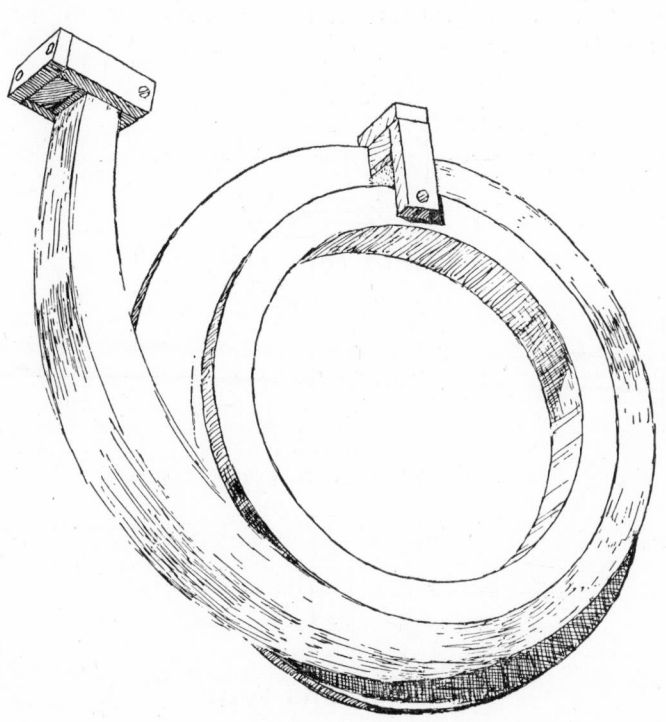

for direct observation or photography. On the other hand, if the field flattener
is followed by some lens system, it may be necessary to curve also that end
of the fiber bundle to a shape determined by the image surface of the lens
system. The obvious application of a field flattener for a photographic system
is illustrated in Fig. N-8b. The entrance end of the fiber bundle is curved to
the shape of the image surface of the lens system, and the flat emergent end
is placed directly in contact with the photographic plate. In this case, however,
the limitations in resolution are clear. The techniques of fabrication of large
bundles consisting of fibers less than 25μ in diameter have not yet been perfected, and work on these lines is still in progress.

The use of field flatteners in visual instruments presents a completely different set of conditions. For example, large field lenses in a periscopic system
cause additional field curvature and chromatic aberration of the exit pupil. It
seems possible to eliminate these large field lenses and also achieve a gain in
image quality by the use of a field flattener.[11] One point, however, needs some
consideration. In order to preserve the entrance and exit pupil relation of the
complete optical system, the principal ray must pass directly down the axis
of the fiber. This necessitates the introduction of field lenses in front of and
behind the fiber bundle. Three systems of field flatteners which allow the

FIG. N-12 Three types of field flatteners: (a) field lens type, (b) flat end type,
and (c) Fresnel lens type.

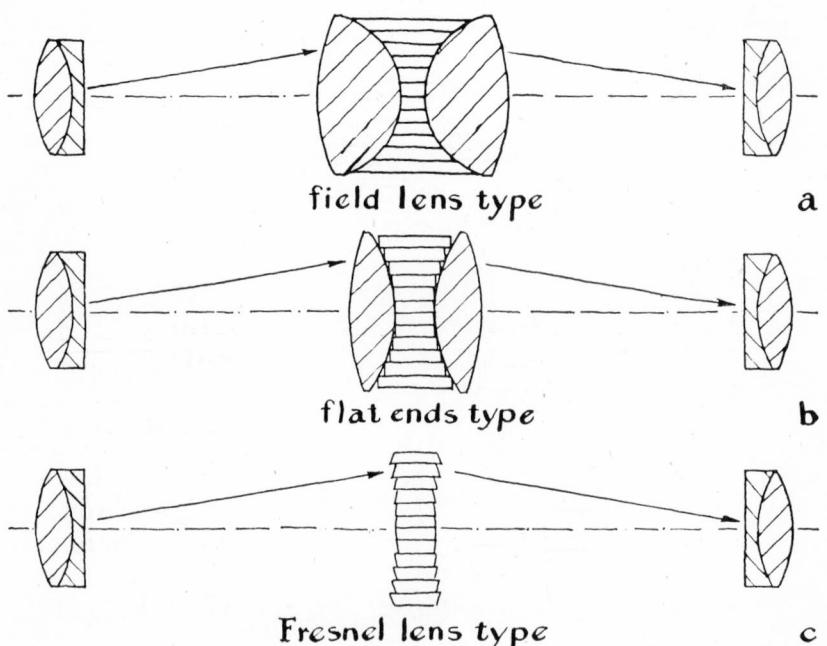

field lens type   a

flat ends type   b

Fresnel lens type   c

principal ray to pass down the axis are illustrated in Fig. N-12. The field lens type (a) is so constructed that the fiber bundle, with the required curves at the two ends, is placed between the two field lenses. The first surface of the field lens is strong enough to bend the principal ray straight down the axis. However, the radii of the curves on the fiber elements are steeper, and the large field lenses are undesirable because of their size and weight. Fig. N-12b shows the flat ends type, where both surfaces of the field lens contribute towards bending the principal ray down the fiber axis. Consequently, the lens elements are much weaker, with comparative reduction in size and weight. Finally, Fig. N-12c illustrates the Fresnel lens type field flattener, so called because the power required of the field lens in order to bend the principal rays is given to the entrance and exit ends of the component fiber. The similarity of this system to the well-known Fresnel lens is evident. Although the ends of the fiber bundle lie on the required curves, the individual fiber ends are prism-shaped so as to bend the rays. This system offers the advantage of weight reduction over the other types of field flatteners.

The constructional details of a Fresnel lens type field flattener are understood with reference to Fig. N-13a. The lens L_1 forms the image on the

FIG. N-13 Procedure adopted for the fabrication of the Fresnel lens type field flattener.

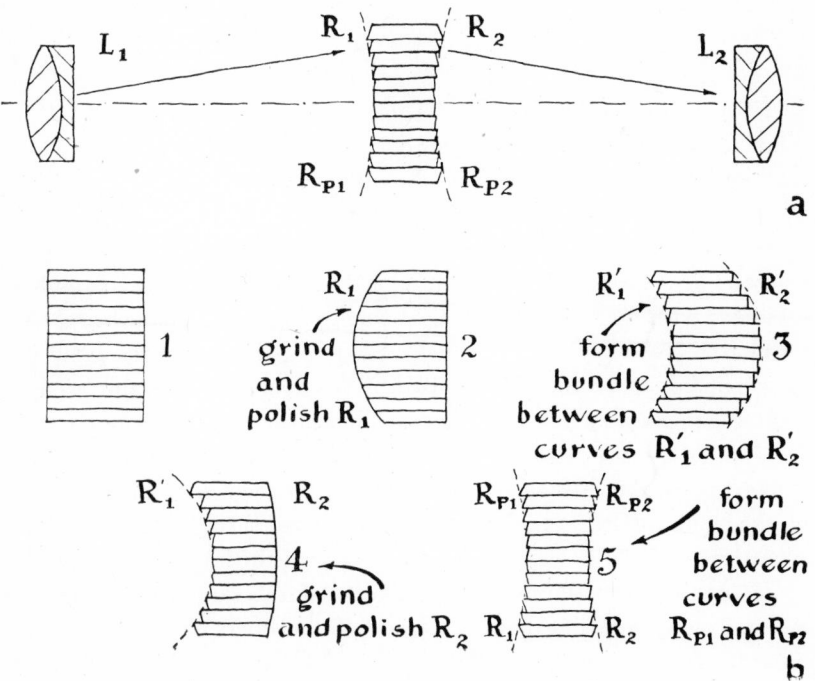

entrance end of the fiber bundle, and lens L_2 is used to pick up the image from the other end. Two ends of the fiber bundle are required to lie on curves R_{P_1} and R_{P_2}, where R_{P_1} and R_{P_2} are the radii of the image surfaces of lenses L_1 and L_2. In order to maintain the entrance and exit pupil relationship, the entrance and emergent ends of the component fibers must have radii R_1 and R_2 formed on them. The procedure adopted for the construction of such a field flattener is illustrated in Fig. N-13b. A well-aligned bundle of fibers is embedded in a low-index cement, and the two ends are ground flat. Then the curve R_1 required at the entrance ends of all fibers is ground and polished. The fiber bundle is now heated to the softening temperature of the cement and pushed in between radii R_1' and R_2', where

$$R_1' = \frac{R_{P_1}R_{P_2}R_2}{R_{P_2}R_2 - R_{P_1}R_2 + R_{P_1}R_{P_2}} \quad \text{and} \quad R_2' = \frac{R_1'R_1}{R_1 - R_1'} \tag{6}$$

This operation requires great care to avoid axial rotation of the component fibers. A curve of radius R_2 is now ground and polished on the other end of the fiber bundle. Finally, the fiber bundle is again heated to the softening temperature of the embedding material and pushed between curves R_{P_1} and R_{P_2}. Working models of Fresnel lens type field flatteners have shown encouraging results. It also seems possible to give the two surfaces of the field flattener any desired aspheric shape to completely correct the off-axis spherical aberration.

Image Transfer on Static and Dynamic Scanning with Fiber Bundle

When an image is formed upon one end of an aligned bundle of transparent fibers, the relayed image viewed at the other end is limited in resolution by the fiber size. The multiple reflections occurring within an individual fiber tend to integrate any variations in illumination across its entrance aperture so that the exit aperture of the fiber appears uniformly bright. If the bundle is held fixed with respect to the image being scanned, the image brightness is, in effect, sampled at an evenly spaced array of points. The resolving power for a line test object is therefore limited by the fiber diameter, the orientation of the test object, and the state of optical insulation of the component fibers. The sampling interval is determined for a closely packed bundle mainly by the diameter of the fibers, and this sampling interval imposes a restriction upon the band width of signals which can be passed through the static bundle. The pattern of the fiber ends becomes obtrusive if the image is so viewed that they are resolved; furthermore, any defects of the bundle, such as broken fibers, and any variation in transmission of the fibers are apparent in this method of static scanning.

A substantial gain in resolution and image quality is achieved by the technique of dynamic scanning.[12] The band width of signals which may be transmitted through the bundle is increased, and all effects associated with the pattern of the fiber ends are integrated when the bundle rapidly and randomly scans the image to be transmitted. It has been experimentally determined that a scanning amplitude of four or five fiber diameters is sufficient to blur out all signs of bundle structure in the final image, and the frequency of the random scanning is not critical. If the motion sweeps out all portions of the image format equally, a bundle composed of identical fibers can be thought of as a filter whose frequency response is that of a uniformly illuminated disk equal in diameter to the component fibers.

In measuring the resolving power of a static bundle,[13] it has been observed that, when the test object is just above the limit of resolution and revolved in its own plane, lines are not resolved in any other position but three. These three positions are found when the lines joining the centers of the fibers coincide with the test line. This "greater resolution" is very critical and is observed only on those parts of the field where alignment is perfect. Fiber bundles as ordinarily fabricated have a nearly hexagonal close-packed arrangement, so that each fiber is surrounded by six other fibers. This close packing increases the number of sampling points within a given area of the image. The maximum spatial frequency which the bundle can transmit will, therefore, lie between $1/2d$ and $1/1.74d$ when it is used as a static scanner. Dynamic scanning, on the other hand, can be thought of in terms of a single fiber, which is made to scan the image format uniformly. If all portions of the format are swept out equally, the frequency of the transmitted image is obtained by multiplying the frequency of the incident image and the frequency response of a uniform disk equal to the fiber in diameter. Since this is true for each fiber in the bundle, it is true for the complete bundle as well. Clearly, a bundle is a more efficient scanner than a single fiber. It has been shown[14] that the frequency response of a uniform disk is given by

$$Y(W) = 2J_1(\pi WD)/(\pi WD) \tag{7}$$

where W is the spatial frequency and D the disk diameter. The influence of dynamic scanning on image quality and resolution has been experimentally investigated[12] with a well-aligned fiber bundle mounted on an appropriate scanning mechanism. It is to be noted that an essential criterion of dynamic scanning is that the two ends of the fiber bundle be provided synchronous motion. It is found that random scanning motion with an amplitude of four or more fiber diameters produces a substantial improvement in image quality. Fig. N-14 shows a three bar test object, a static image, and a dynamic image. The line width in the test object is given in terms of the fiber diameter. It is apparent that dynamic scanning blurs out the structure of the fiber bundle and produces thereby a much more acceptable and useful image. The increase in resolving power is quite marked. The images of the three bar test chart were microdensitometered, and the contrast transmission on dynamic scanning is calculated from these data. Fig. N-15 shows the resulting values plotted upon the theoretical response of a uniform disk. In general, the agreement is marked, especially near the limit of resolution.

The above discussion on resolution of a fiber bundle is valid only when the fibers are optically insulated. A marked deterioration in image contrast and resolution may occur if there is any light leakage between neighboring fibers at the line contact. The penetration of light incident at an angle greater than the critical angle through a thin film due to the phenomenon of frustrated total internal reflection is well known. The magnitude of the flux leakage into the second medium is dependent on the thickness and refractive index of the

FIG. N-14 Images of three bar test charts on static and dynamic scanning with a fiber bundle.

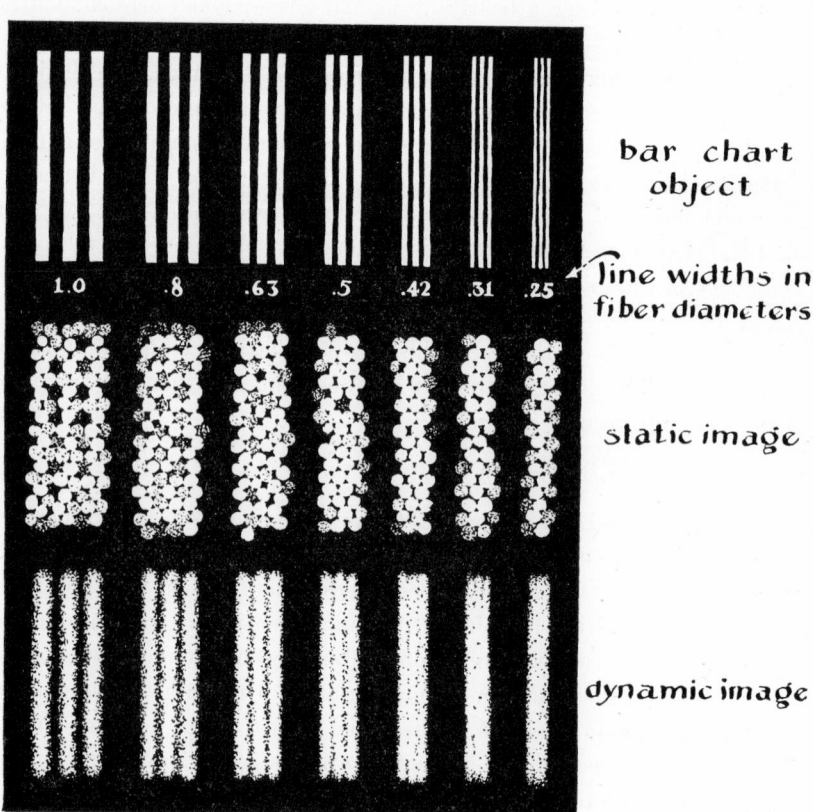

FIG. N-15 Frequency response of a fiber bundle on dynamic scanning. The solid curve shows the frequency response of a uniform disk, and the points are obtained experimentally with a fiber bundle.

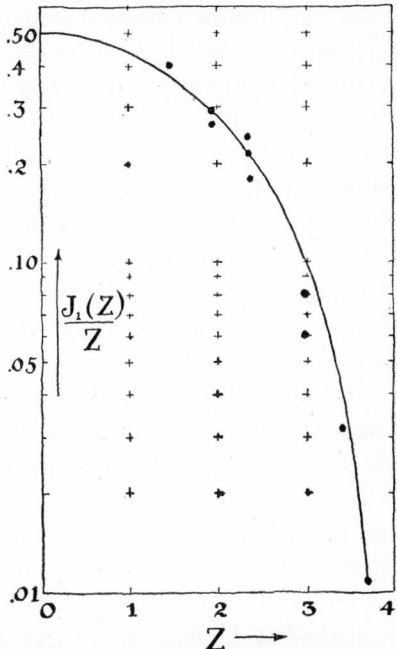

thin film, the angle of the incident beam, the wavelength of the light, and the degree of polarization. An "order of magnitude" calculation of light leakage from a light-conveying fiber may be made from the computation of transmission arising from frustrated total reflection at a thin film having plane parallel interfaces as a function of air film thickness h. This information pertaining to plane parallel interfaces is then utilized in calculating the magnitude of light leakage in neighboring dielectric cylinders. Fig. N-16 is such a plot for

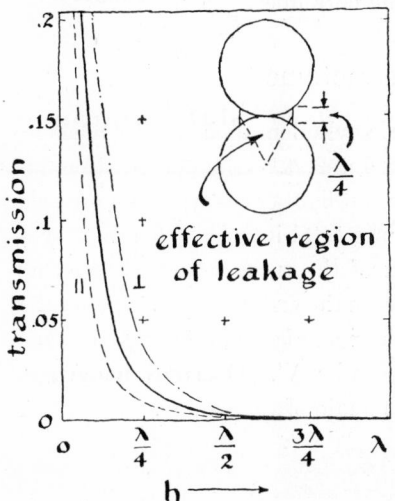

FIG. N-16 Light leakage at adjacent fibers due to frustrated total internal reflections.

a wave incident at a medium of index 1.5. The percentage of light transmission for the plane surfaces is plotted as a function of the air film thickness h, for parallel ‖ and perpendicular ⊥ planes of polarization. The curve shows that light leakage for unpolarized light is appreciable only at air film thickness below $\lambda/4$. In fact, of course, the boundary conditions for the film formed by contact between two cylinders would be quite different. However, we assume that light is lost only from the region of contact for which $h \leq \lambda/4$ but that within this region the loss is complete. This simplifying assumption allows the problem to be treated merely by calculating the fraction of the area of a given fiber wall for which $h < \lambda/4$ between adjacent fibers, as illustrated in Fig. N-16.

These considerations and practical experience have shown that the optical insulation problem plays a role for fibers below 50μ in diameter. For the purpose of optical insulation of smaller fibers a thin coating of a low-index material is formed on the fiber during the aligning process. This low-index coating separates the fibers by a small amount and prevents any light leakage between them. A very high degree of cleanliness is required during the coating process; otherwise considerable light losses may occur at every reflection in the fiber. A method of drawing glass fibers down to 12μ diameter with a glass core of high refractive index and a glass coating of low refractive index, of the order of $1–2\mu$ thickness, has been developed. This method consists of inserting a glass rod in a glass tube of appropriate refractive index and drawing this assembly into a "concentric insulated fiber" through a hollow cylindrical furnace. Nominally infinite lengths of such fibers have been drawn on a rapidly rotating drum. Fibers drawn in this manner yield the desired optical insulation with the additional advantage of higher light transmissivity, due to the protected interface of the wall of the core fiber.

Single Dielectric Cylinders: Photorefractometer

Among the applications of single dielectric cylinders their use for measurement of the refractive index of liquids is of interest.[15] It is evident that in a straight rod the angles of entering and emergent light cones are equal provided the index difference across the rod wall allows total internal reflection for all rays in the rod. However, when the index of the surrounding material approaches the rod index, some loss of light occurs through refraction of rays that are inclined at an angle less than critical incidence. Fig. N-17 shows a transparent rod of index N immersed in a surround of index N'. The critical angle ϕ_c for a given value of N and N' is related to the angle of the emergent cone α_c as follows:

$$\alpha_c = \sin^{-1} \sqrt{N^2 - N'^2} \tag{8}$$

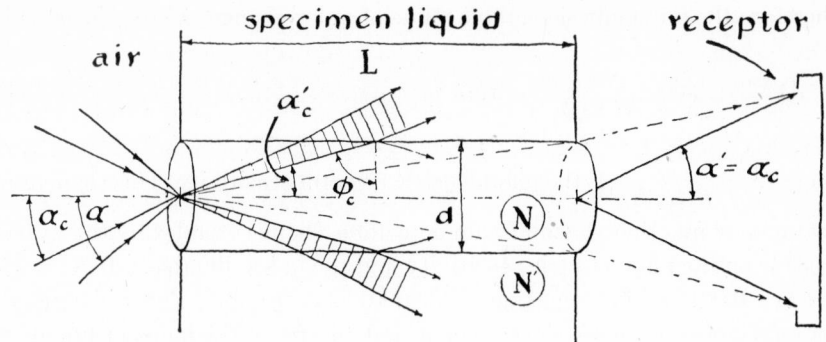

FIG. N-17   Principle of a dielectric cylinder photorefractometer.

It can be seen that for small values of α_c the quantity $\Delta\alpha_c/\Delta N'$ is large. Thus large changes in α_c result in small changes in the index of the surround. For $\alpha \geq \alpha_c$ and a uniform incident light cone of $I(0)$ watts/steradian the emergent energy $E(N, N')$, which is proportional to the solid angle of the emergent cone, is given by

$$E(N, N') = 2\pi I(0)[1 - \sqrt{1 - N^2 + N'^2}] \tag{9}$$

When $N' \to N$, the function $E(N, N')$ becomes proportional to $(N - N')$. Change of E is given by the expression

$$\mathcal{S} = -\frac{dE}{dN'} = \frac{N'}{\sqrt{1 - N^2 + N'^2}}$$

It is to be seen that as $N' \to N$ and $N' \to N^2 - 1$, $\mathcal{S} \to \alpha$. If δ is the index difference between the rod and the surround, then

$$\alpha_c \approx \sqrt{2N\delta} \tag{10}$$

and

$$E(N, N') = N\delta + \frac{N^2 - 1}{2}\delta^2 + \ldots \qquad \cong N\delta \tag{11}$$

and

$$\mathcal{S} = N + (N^2 - 1)\delta + \ldots \qquad \cong N \tag{12}$$

Therefore

$$\frac{\mathcal{S}}{E(N, N')} = \frac{1}{\delta}\left[1 + \frac{N^2 - 1}{2N} \cdot \delta + \ldots\right] \cong \frac{1}{\delta}$$

This indicates that, for small values of δ,

$$\frac{\mathcal{S}}{E(N, N')} \approx \frac{1}{\delta}$$

Therefore the minimum detectable value of index change $|\Delta N'|_{min}$ is given by

$$|\Delta N'|_{min} = \frac{K}{S/E(N, N')} \approx K \cdot \delta \tag{13}$$

where $K = \dfrac{\Delta E}{E(N, N')}$ is the photoelectric receptor sensitivity. This elementary treatment shows that the minimum detectable change of index by this system depends on the photoelectric sensitivity and the index difference. If $K = 0.01$ and $\delta = 0.001$, an index change of 0.00001 is detectable. This indicates the possibility of continuous measurement and control of the index of liquids by photoelectric means.

Experimental investigations have been conducted on a rod refractometer of the type illustrated in Fig. N-18. Glass rods of various known refractive indices are mechanically mounted on a turret type immersion chamber through which the liquid is passed. The difference of index between the rods is so arranged as to cover a wider range and maintain higher sensitivity. An appropriate rod can be brought into position for measurement on a given liquid. A monochromator is used to condense light on the axis of the glass rod, and a precision photomultiplier is placed to measure the emergent light flux. It has been found that the theoretically expected sensitivity is achievable, and the milliammeter is calibrated in terms of refractive index, each rod corresponding to a different scale. However, it seems desirable to use only straight rods; curved rods[16] necessitate repeated calibration. Instruments have been designed, on this principle, to measure the refractive index of liquids automatically and also actuate appropriate control mechanisms.

FIG. N-18   Multiple rod recording photorefractometer.

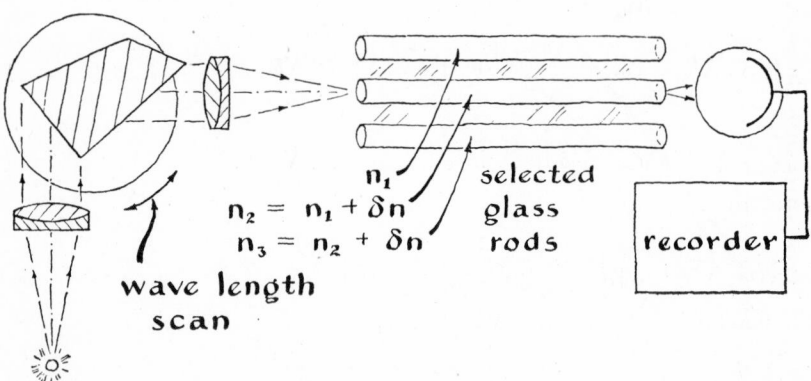

Light Funnel for Stellar Spectrograph

The fractional amount of light flux in a circular star image that is passed by the slit of a stellar spectrograph is often very low. The magnitude is highly dependent on the *f*-ratio of the optical elements and the atmospheric seeing conditions. Fig. N-19 shows the percentage transmission of the light flux as a function of slit width for four seeing conditions on the 100-inch *f*-30 Mount Wilson telescope, measured by Dunham.[17] Under more usual seeing conditions (Seeing 3) the image apparently subtends nearly four seconds of arc at the telescope, and a 60µ slit passes just 10 percent of the flux contained in the star image.

Use has been made of an image slicer[18] and an image transformer,[19] which consist of parallel mirror assemblies to slice the circular image into strips and direct them through the slit. Substantial gain is reported by the use of these principles, but the optical assembly requires considerable accuracy. On the other hand, an appropriate assembly of fibers has been suggested[20] to transform the shape of a beam of light. For stellar spectrographs a circular bundle of fibers is placed at the star image, and the other ends of the fibers are made to lie in a row, thereby forming the entrance slit of the spectrograph. Although the arrangement of fibers has been changed, the image area at the two ends of a light funnel is the same. Light losses occur because of the length of path in glass and the 9.4% dead interstitial area between the fibers. A slight broadening of the light cone, due to bent fibers, was discussed earlier.

FIG. N-19   Light transmission through the slit of a 100-inch telescope (Mount Wilson) for various seeing grades.

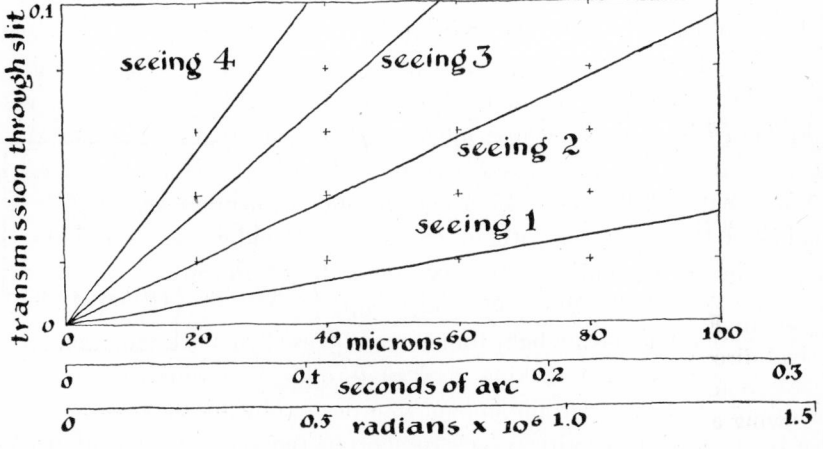

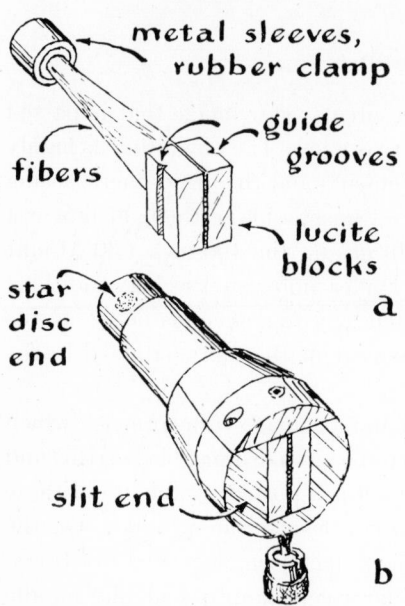

metal sleeves,
rubber clamp

guide
grooves

fibers

lucite
blocks

star
disc
end

a

slit end

b

FIG. N-20 Construction of a light funnel for a stellar spectrograph.

Fig. N-20 illustrates the assembly of a light funnel, and such light funnels have been constructed of optical glass fibers 200μ, 100μ, and 50μ in diameter. In a typical case the diameter of the entrance end of the fiber bundle was about 1 mm, and 250 fibers of 50μ diameter were employed. Consequently, the slit length was about 12.5 mm. In order to minimize light losses due to absorption and scatter, the length of the funnel should be a minimum, within limits prescribed by the allowable bending radius. In this case the total length of the light funnel was chosen to be 60 mm. The construction of such a light funnel requires a layer of fibers which is wound regularly on the fiber-aligning machine described earlier. This layer is clamped and thence provides the slit end. The other ends of the fibers are then mechanically clamped in a circular form. In order to facilitate grinding and polishing of the two ends, and also from other considerations, it has been found desirable to embed the entire light funnel in a plastic whose index is lower than that of the glass. With reference to Eq. 8 it may be seen that on embedding fibers of refractive index 1.5 in a material of index 1.45, a light cone of 22 degrees semi-angle can be conducted by the fibers without any losses. Grinding and polishing then proceed in the conventional manner. Laboratory performance tests of these light funnels have shown encouraging results.

The manner in which a light funnel may be used on a spectrograph is illustrated in Fig. N-21. A rocking glass plate optically oscillates the slit in a lateral direction by a certain amount in order to obviate the fiber graininess. Use is made of a cylindrical lens to shorten the spectrum height, thereby achieving a gain in exposure.[18] However, no cylindrical lens is required for a

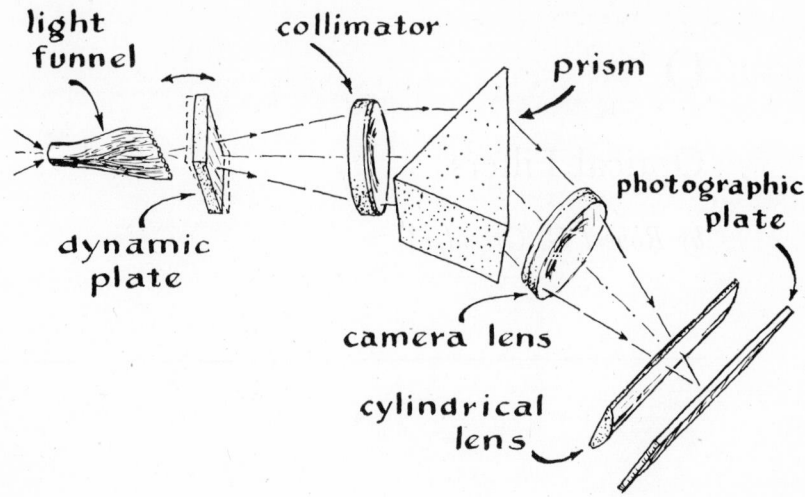

FIG. N-21   Use of a light funnel and a cylindrical lens in a stellar spectrograph.

photoelectric spectrograph. An appropriate aligning device is also incorporated with the light funnel in order to guide the star image.

REFERENCES

1. J. L. Baird, British Patent Specification No. *285,738* (1927).

2. C. W. Hansell, U.S. Patent Specification No. *1,751,584* (1927).

3. H. H. Hopkins & N. S. Kapany, Nature, *173*, 39 (1954).

4. A. C. S. van Heel, Nature, *173*, 39 (1954).

5. H. H. Hopkins & N. S. Kapany, Optica Acta, *4*, 164 (1955).

6. D. G. Kiely, *Dielectric Aerials* (1953, Methuen, London).

7. O. Schriever, Ann. Phys., *63*, 645 (1920).

8. N. S. Kapany, J. Opt. Soc. Am., *47*, 413 (1957).

9. W. Brouwer & A. C. S. van Heel, Optica Acta, *2*, 49 (1955).

10. R. Schindler, *Gastroscopy* (1950, Univ. of Chicago Press).

11. N. S. Kapany & R. E. Hopkins, J. Opt. Soc. Am., *47*, 594 (1957).

12. N. S. Kapany, J. A. Eyer, & R. E. Keim, J. Opt. Soc. Am., *47*, 423 (1957).

13. N. S. Kapany, Thesis, University of London (1955).

14. P. Lindberg, Optica Acta, *1*, 80 (1954).

15. N. S. Kapany & J. N. Pike, J. Opt. Soc. Am., *47*, 1109 (1957).

16. E. Karrer & R. S. Orr, J. Opt. Soc. Am., *36*, 42 (1946).

17. T. Dunham, Jr., *Vistas in Astronomy* (1955, Pergamon, New York), II.

18. I. S. Bowen, Astrophys. J., *88*, 113 (1938).

19. J. Strong & W. Benesch, J. Opt. Soc. Am., *41*, 252 (1951).

20. N. S. Kapany, Proc. Symposium on Astronomical Optics and Related Subjects, Manchester University, p. 288 (1955).

Appendix O

Optical Filters

by Robert G. Greenler†

As long as we have an interest in the interaction of different wavelengths of radiation with matter, we must have an accompanying concern with filters which serve to isolate various spectral regions for study.

Logically, optical filters can be classified according to the physical mechanism by which they isolate wavelengths. Most optical filters classified in this way will fall into the following categories:

1. Selective absorption.
2. Selective reflection.
3. Scattering.
4. Interference.
5. Polarization.
6. Refraction.

Many filters will fall into two or more of these categories, utilizing more than one of the processes in their operation.

A different description which can be applied to filters refers to band pass, long wavelength pass, and short wavelength pass filters. Such a designation is likely to reflect the frame of reference of the classifier. For example, a filter which transmits all radiation between 3000 Å and 1μ and absorbs all other radiation could be considered to be a long wavelength pass filter by the ultraviolet spectroscopist, no filter at all by an investigator whose perception is visual, and a short wavelength pass filter by the provincial infrared spectroscopist. To a person with a more cosmopolitan view of the radiation spectrum, it could be either a wide or narrow band pass filter, depending on the nature of his interest in the spectrum.

† Research Laboratories, Allis-Chalmers Manufacturing Company, Milwaukee 1, Wisconsin.

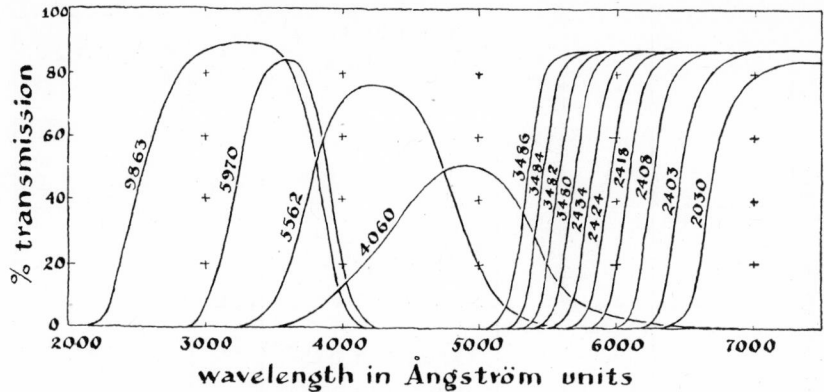

FIG. O-1 Transmission data from *Glass Color Filters* (Corning Glass Works).

In practice, the questions which arise usually concern what filters are available for a certain wavelength region. Therefore, a general description will be given of the principal types of optical filters that are most commonly employed in each wavelength region.

Visible

A large variety of glass color filters is produced by dissolving or suspending various coloring matter in glass. Fig. O-1 illustrates a few of the many such filters that are produced by the Corning Glass Works. These filters† are durable and have stable transmission characteristics. The Eastman Kodak Company produces Wratten filters in the form of thin films of gelatin to which various organic dyes have been added. More than a hundred filter characteristics are currently available, and combinations of these filters offer great flexibility in solving particular optical problems. Fig. O-2 illustrates a few Wratten and Corning filters which are useful in isolating strong lines of a mercury arc.

Both of these types of filters function by selective absorption, and in general the absorption bands exhibited by solids are not sharp. It is therefore difficult to achieve a sharp transmission cut-off or a narrow pass band with such filters.

Interference filters can be designed to give good transmission at a predetermined wavelength with a sharp reduction in transmission on each side. The simplest form of interference band pass filter is just a solid Fabry-Perot inter-

---

† Glass filters are available from the Corning Glass Works. British Chance glass filters are available through the Alpha American Corporation. German Jena glass filters are available through the Fish-Schurman Corporation.

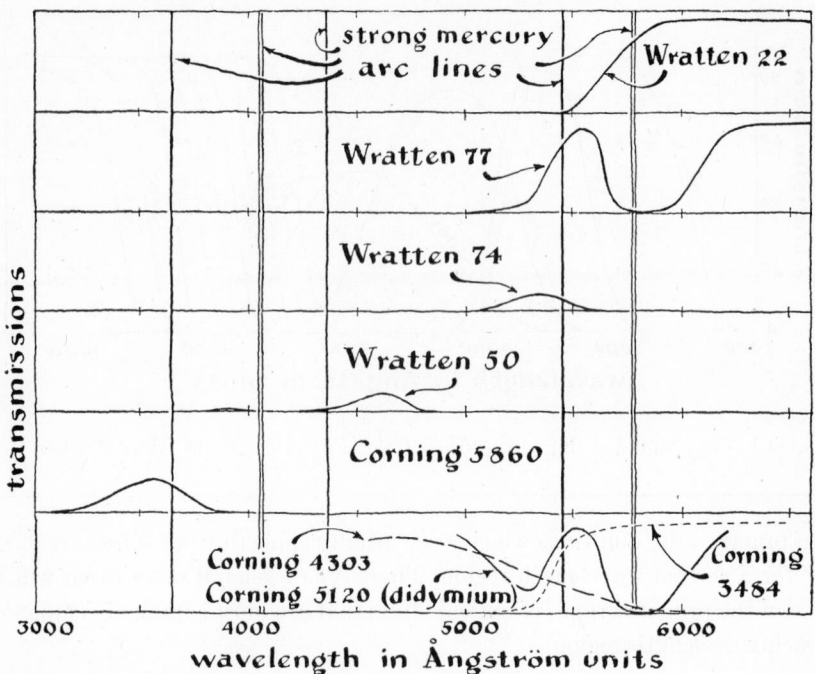

FIG. O-2   Gelatin and glass filters useful for isolating strong mercury arc lines. From *Kodak Wratten Filters* (Eastman Kodak Co.) and *Glass Color Filters* (Corning Glass Works).

ferometer used in a low order of interference. It is made by depositing a semi-transparent film of metal on a glass support by thermal evaporation, followed by a spacer film of transparent dielectric material and by another semitransparent metallic film. As in the Fabry-Perot interferometer, a high reflectivity of the metallic films yields a narrow pass band. The reflectivity can be increased by increasing the metallic film thickness, but the accompanying increase in absorption reduces the peak transmission of the filter. A practical compromise between the desire for a narrow pass band and a high peak transmission results in commercial filters for the visible region which have transmission peaks 150 Å wide with peak transmissions of about 30%.

The substitution of high-reflecting stacks of non-absorbing dielectric films for the semireflecting metal films produces both narrower band passes and higher transmission peaks.[1] The high-reflecting stacks are composed of layers (which are a quarter wavelength thick for a specified wavelength) of alternate high and low indexes of refraction. Fig. O-3 illustrates the characteristics of such a filter composed of fifteen layers of NaF and ZnS. The maximum trans-

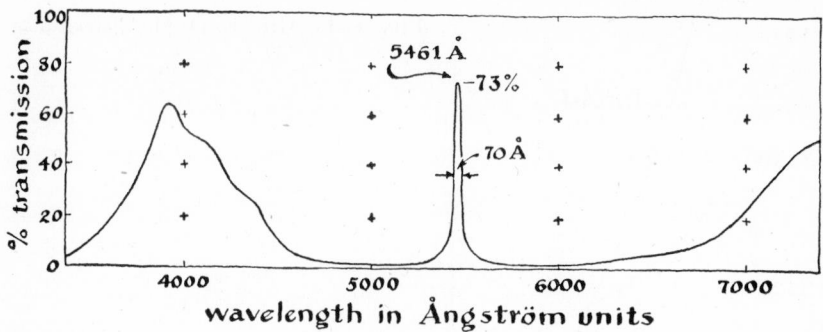

FIG. O-3   Fifteen-layer interference filter (Baird Associates).

mission is 73% with a band width of about 70 Å. Similar filters are available†
with peaks located anywhere between 3900 Å and 7800 Å. In the case of the
all-dielectric interference filter, the penalty which is levied for the reduced
band width is the reduced free spectral range of the filter. While the solid
Fabry-Perot filter may be constructed so that a peak at 5000 Å has no accompanying peaks in the visible region, the filter illustrated in Fig. O-3 has
considerable transmission at 1000 Å on either side of the main peak. This
transmission may be effectively disposed of with glass or gelatin filters. The
properties of such filters are treated in § 12-3. The fact that the wavelength
of peak transmission varies with the angle of incidence is both an advantage
and a disadvantage: an advantage in that it provides an easy way to adjust
the position of the pass band; and a disadvantage in that it requires the filter
to be used in approximately parallel light if the narrow pass band is to be
preserved.

A great number of specialized problems may be solved by specially designed
vacuum-deposited multilayer films. For example, a filter might be designed
to reflect very strongly one particular emission line while having a high transmission for another line in the immediate neighborhood. Filter mirrors consisting of thin layers of metals and dielectrics have been developed[2] to reflect
strongly certain spectral regions while absorbing others.

An interesting type of filter which has application in the ultraviolet and
infrared as well as the visible is the Christiansen filter, which was first described by Christiansen[3] in 1884. McAlister[4] has investigated the practical
application of these filters and suggested some interesting improvements in
design. One form of the filter is a cell packed with small particles of a transparent optical material and filled with an appropriate liquid. Fig. O-4 illus-

† Solid Fabry-Perot filters and multilayer filters are available from Baird Associates,
the Bausch and Lomb Optical Co., Axler Associates, the Farrand Optical Co., and others.

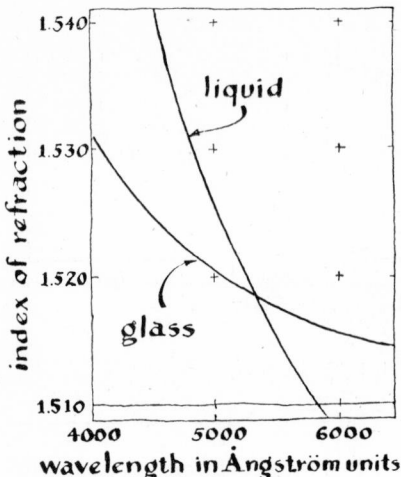

trates the dispersion curves for a suitable combination consisting of crushed borosilicate crown glass and a solution of carbon disulphide in benzene. At the point on the wavelength scale where these two dispersion curves cross, the mixture is optically homogeneous, and light of that wavelength will pass through the cell undeviated by reflection or refraction at the solid-liquid interfaces. However, the difference in index between the solid and the liquid becomes progressively greater for wavelengths farther from this crossover point, and light is scattered by reflection and refraction in traversing the cell. Fig. O-5 shows the spectral transmission of a set of five such filters made with borosilicate crown glass particles in a mixture of carbon disulphide and benzene of varying proportions. Such spectral purity is attained only if the cell

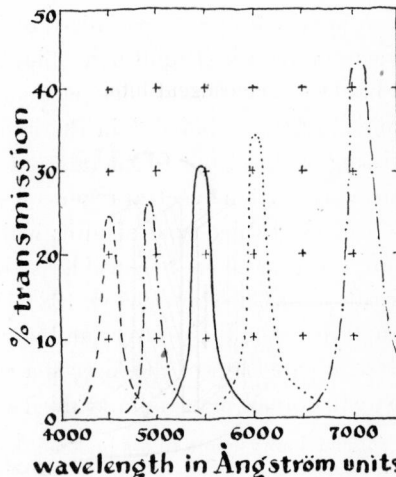

FIG. O-5   Christiansen filters (after E. D. McAlister[6]).

is placed in the collimated beam of a small light source and the scattered light is eliminated by focusing the beam onto an aperture which is just the size and shape of the image of the source. This requirement defines one of the serious limitations of this filter; it cannot be used in an image-forming system —*e.g.*, in front of a camera lens.

Since the refraction index of a liquid changes more rapidly with temperature than the index of a solid, the wavelength of peak transmission of such a filter varies with temperature. Weigert[5] used this fact to construct a tunable band pass filter of crown glass chips in methyl benzoate which transmits red light when the temperature of the cell is 18°C and blue light when the temperature is 50°C. This sensitivity to temperature requires careful thermostatic control if the transmission of the filter is to remain constant in intense beams of light. McAlister has added cooling vanes in the body of large cells to keep the temperature of the entire cell uniform and equal to that of the surrounding thermostated bath.

The birefringent filter (sometimes called a polarization interference filter, or the Lyot-Öhman filter) is an ingenious device for isolating a spectral band a few Ångstrom units wide. It was described in 1933 by Lyot[6] in France and described and built independently by Öhman[7] in Sweden in 1938. Fig. O-6 illustrates the construction of the filter. The birefringent crystal quartz plates are cut with their optical axes parallel to the large faces. The axes of the polarizers are oriented at 45° to the quartz optical axes. Linearly polarized light, incident on the first quartz plate at 45° to the optical axis, would have its plane of polarization rotated by 90° if the plate were a half wave plate (or if the optical path difference between the ordinary and extraordinary rays were any odd multiple of half waves). If the plane of polarization is rotated 90°, the radiation will not be transmitted by the second polarizer. Since the

---

FIG. O-6   Birefringent filter.

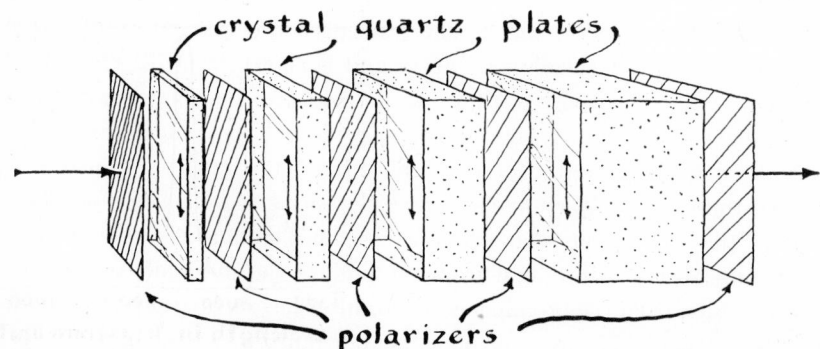

crystal quartz plates

polarizers

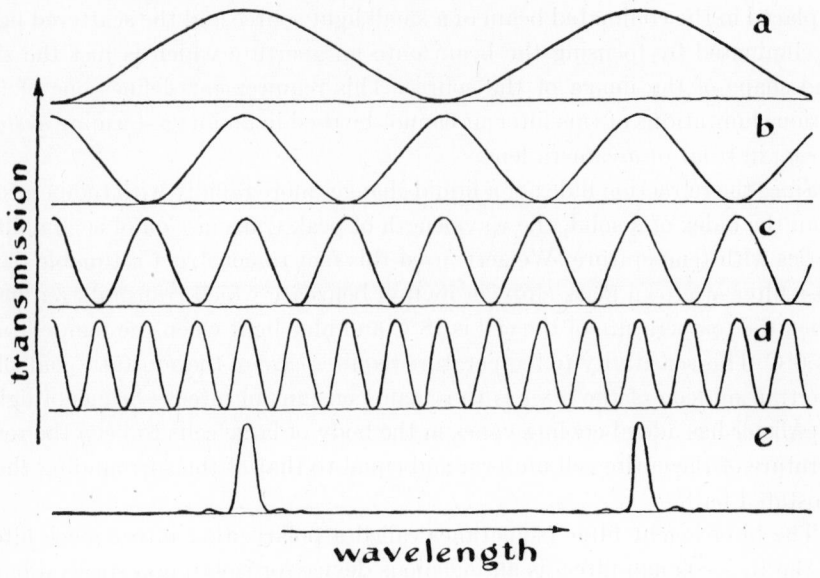

FIG. O-7 Schematic transmission of the birefringent filter.

phase difference introduced between the ordinary and extraordinary rays depends on wavelength as well as on the thickness of the quartz plate, the same plate may be a 5/2 wave plate for one wavelength, a 7/2 wave plate for a different wavelength, a 9/2 wave plate for still another wavelength, etc. Each of these wavelengths will be blocked by the polarizer following the plate, while those wavelengths for which the difference in optical path between the two polarizations is an even number of half waves will be completely transmitted. Fig. O-7a illustrates the transmission of the first quartz plate and its polarizers. If each quartz plate is made twice as thick as the preceding one, it will have twice as many transmission maxima and minima in a given wavelength interval. The transmission curves for the second, third, and fourth plates are illustrated in Fig. O-7b, c, and d. The transmission of the entire filter is the product of all these transmission curves and is shown in O-7e. These peaks now occur far enough apart so that one of them may be isolated with a conventional glass or interference filter. Evans[8] has reviewed and expanded the theory of such filters, and a filter which he has built has been in use at the High Altitude Observatory at Climax, Colorado, since 1943. Its purpose is to aid in the viewing of solar prominences by eliminating the tremendous amount of scattered white light from the limb of the sun while transmitting the H_α emission from the prominences. This filter has six quartz plates ranging in thickness from 1.677 mm to 53.658 mm and transmits a band of width 4.1 Å centered on the H_α line of hydrogen at 6563 Å. Baird Associates, Inc., now make similar birefringent filters out of synthetic ammonium dihydrogen phosphate crystals. These filters have band widths as small as 1 Å.

Ultraviolet

There is no shortage of long wavelength pass filters for the ultraviolet spectral region. Actually, this is just an optimist's statement of the fact that many solids and liquids which are transparent in the visible become non-transmitting in the ultraviolet because of absorption by electronic excitation. Ordinary window glass does not transmit wavelengths below about 3200 Å. Water vapor and oxygen absorption make air opaque below 2000 Å. Quartz is transparent down to 1800 Å; sapphire, to 1500 Å; barium fluoride, to 1400 Å; calcium fluoride (fluorite), to 1300 Å; and lithium fluoride has been found to transmit appreciably down to 1000 Å. Brode[9] has measured the ultraviolet transmission of many organic liquids, and Wood[10] lists a number of substances whose natural absorption make them useful for isolating various parts of the ultraviolet spectrum.

Thin films of silver show a narrow band of transmission between 3160 Å and 3260 Å which makes them useful for isolating this spectral region. The absorption coefficient for silver rises on either side of this region because of the absorption by free electrons of longer wavelengths and the absorption caused by bound electrons of shorter wavelengths. The alkali metals show similar, more extended regions of transparency in the ultraviolet. Potassium films have been used[11] to isolate the region below 3150 Å, but the problem of keeping the film (in vacuum) from breaking up into droplets of potassium at room temperature has continued to be a serious limitation to the practical use of these films as ultraviolet filters. O'Bryan[12] has devised a method of increasing the stability and permanence by sealing the potassium film between two flat quartz plates.

The problem of making satisfactory multilayer interference filters for the ultraviolet has been the problem of finding a suitable high-index material which is transparent at short wavelengths. The commonly used high-index material for the visible region is zinc sulphide ($N = 2.4$), but it is strongly absorbing, even in very thin layers, below 4000 Å. The low-index materials, magnesium fluoride ($N = 1.4$) and sodium aluminum fluoride (cryolite, $N = 1.4$), which are commonly used in the visible, are transparent as far down as 2000 Å. Recently films of lead chloride[13] ($N = 2.2$) and antimony trioxide[14] ($N = 2.1$) have been shown to be quite transparent above 3000 Å in the ultraviolet.

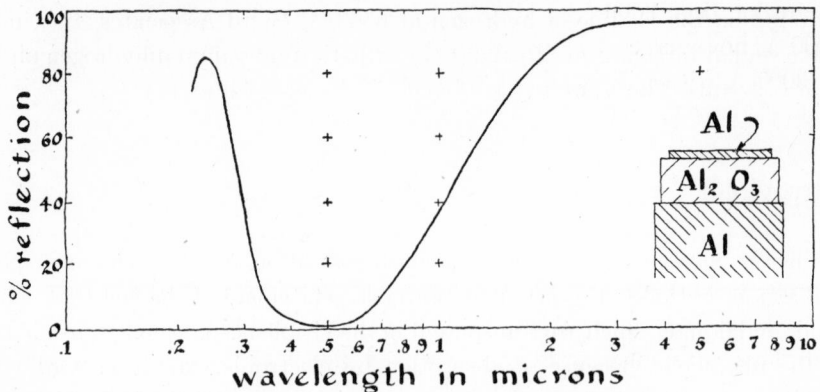

FIG. O-8   Ultraviolet reflection interference filter (after G. Hass[4]).

A simple reflection interference filter for the ultraviolet is shown in Fig. O-8. This is a filter of the type described by Hadley and Dennison,[2] consisting of a transparent layer deposited on a reflecting surface and covered by a semi-reflecting metal film. Hass has formed the transparent layer of Al_2O_3 by anodic oxidation of the aluminum surface and covered it with a thin semitransparent layer of aluminum. Such a filter can also be useful to reflect infrared radiation while absorbing visible radiation.

The Christiansen filter has some application in the ultraviolet. Kohn and Fragenstein[15] have used fused quartz chips in benzol-alcohol mixtures to span

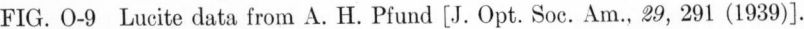

FIG. O-9   Lucite data from A. H. Pfund [J. Opt. Soc. Am., *29*, 291 (1939)].

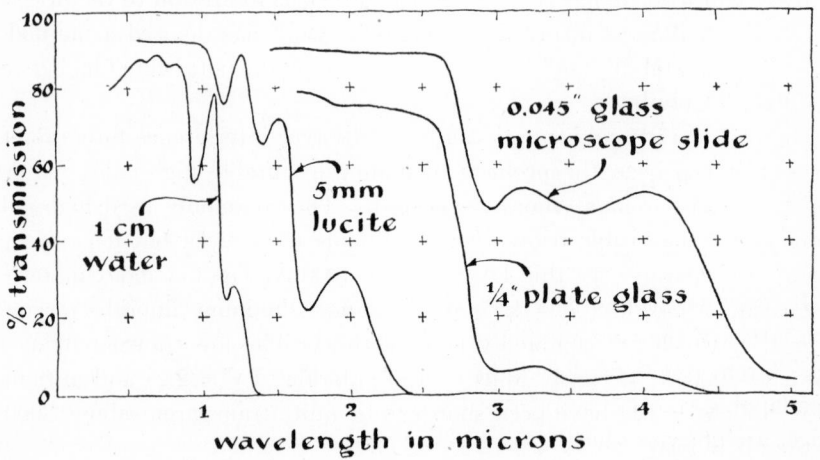

the region from 3000 Å to 4000 Å. Benzol absorbs wavelengths shorter than 3000 Å; however, Sinsheimer and Loofbourow[16] cover the region from 2350 Å to 3000 Å with quartz chips in mixtures of purified decahydronaphthalene and cyclohexane.

Infrared

Just as there are many long wavelength pass filters for the ultraviolet, there are many short wavelength pass filters for the infrared; *i.e.* there are many optical materials which are transparent in the visible but opaque in the infrared. This infrared absorption is due to vibrational excitation of either the atoms in the crystal ions or the ions in the crystal lattice. Fig. O-9 shows the transmission limits in the infrared for some common substances. Table O-1 lists the useful limit of infrared transmission for several substances. The limit is calculated as the point of about 30% transmission in a 3 mm sample.

TABLE O-1

Crystal	*Long Wavelength Transmission Limit (Microns)*
SiO_2	4
LiF	7
MgO	8
CaF_2	11
NaF	13
NaCl	20
KCl	25
AgCl	26
KBr	35
KI	37
KRS-5 $\left(\begin{array}{l}\text{synthetic thallium}\\\text{bromide-iodide}\end{array}\right)$	42

If there were an equal abundance of long wavelength pass filters covering the infrared, the combination of the appropriate short and long wavelength pass filters would produce a selection of band pass filters in the regions of overlapping transmission. Of course, since many of these natural absorption edges are not very sharp, the resulting pass band would have low peak trans-

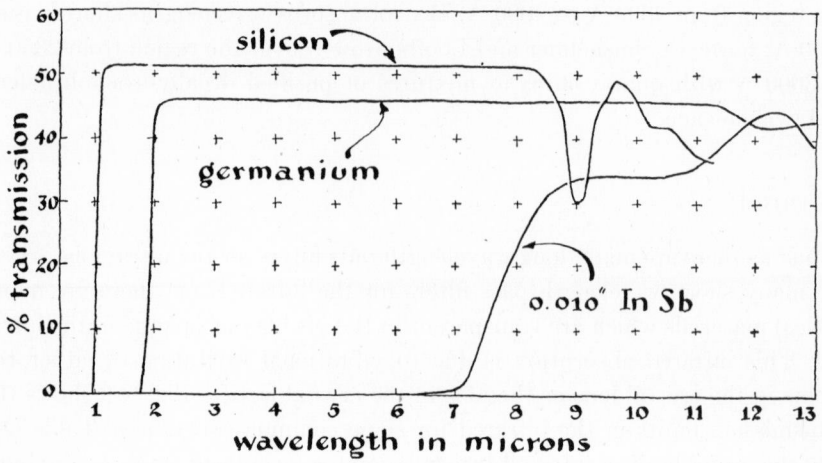

FIG. O-10 Silicon, germanium and indium antimonide transmission.

mission if it were narrow. High peak transmission could be achieved only by tolerating a wide pass band.

The only materials which act as long wave pass filters in the infrared at wavelengths shorter than 25μ are semiconductors. The short wavelength absorption arises from the transition of electrons from a valence band to a conduction band in the crystal. The quanta of short wavelength radiation have energies greater than the gap between the conduction and valence bands in the crystal, and this radiation is absorbed by electrons making the transition. Long wavelength radiation has quantum energies less than the energy gap in the semiconductor and so is not absorbed by the crystal. Since this energy gap is quite sharply defined, semiconductors can show sharp short wavelength cut-off characteristics. If impurity atoms are present which provide electron energy levels between the valence and conduction bands, the crystal will absorb beyond the cut-off wavelength. Impurity content must be kept below 1 part in 10^7 for good transmission of semiconductor crystals. Materials with such purity have recently become available as a result of investigations into the electrical properties of semiconductors. Fig. O-10 shows the transmission of silicon, germanium, and indium antimonide crystals.† The transmissions of about 50% can be attributed to the high surface reflection resulting from the remarkably high indexes of refraction of these materials; $N = 3.6$ for Si, 4.1 for Ge, and about 4 for InSb. This reflection loss may be largely eliminated

---

† Germanium and silicon filters (and infrared lenses) are commercially available from Baird Associates. Indium antimonide crystal plates are available from Ohio Semiconductors, Inc.

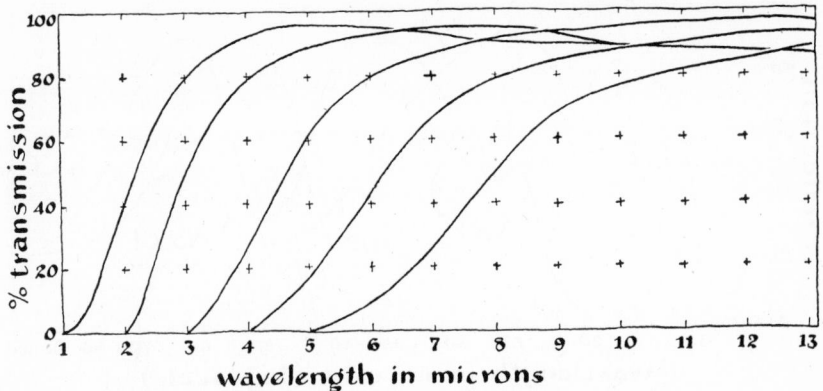

FIG. O-11   Data from *Kodak Far Infrared Filters* (Eastman Kodak Co.).

over a restricted spectral range by reflection-reducing overcoats (see § 12-1). Other materials which may be useful, in a pure state, as long wavelength pass filters are tellurium and indium arsenide, with absorption edges at 3.5μ and 3.8μ, respectively.

A series of long wavelength pass filters manufactured by the Eastman Kodak Company are illustrated in Fig. O-11. The cut-off wavelength may be specified in half-micron steps between 1 and 5 microns.

Selective reflections of crystals provide a standard way of isolating a part of the infrared spectrum. The reflectivity of an absorbing medium in air is given by the formula

$$\Re = \frac{(n-1)^2 + k^2}{(n+1)^2 + k^2}$$

FIG. O-12 n and k at an absorption band.

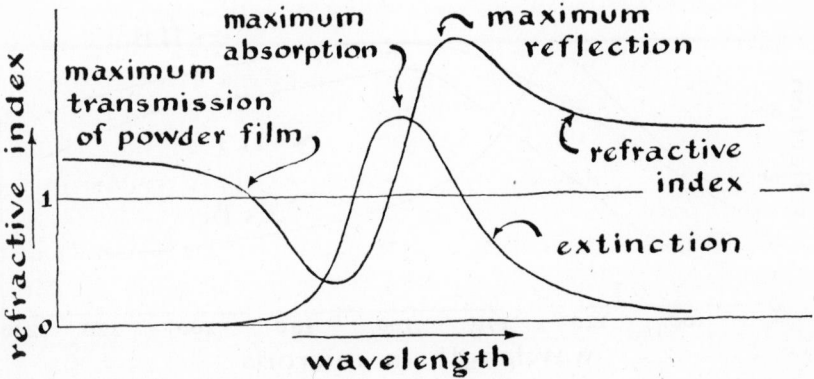

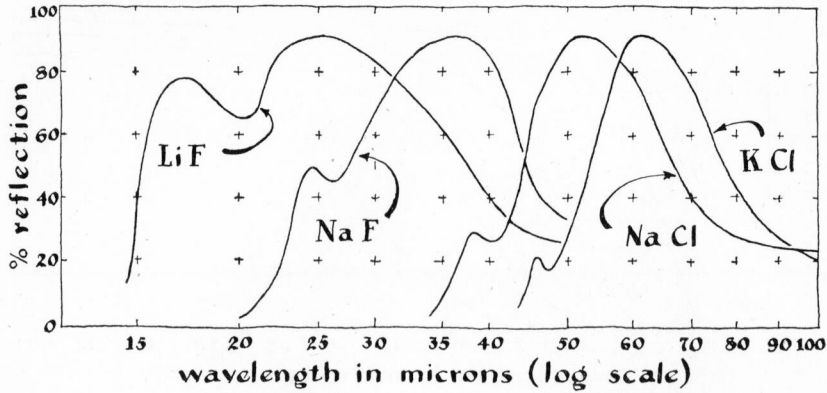

FIG. O-13 Residual ray crystals. LiF and NaF after H. W. Hohls [Ann. Phys., *29*, 433 (1937)]; NaCl and KCl after M. Czerny [Z. Physik, *65*, 600 (1930)].

where n and k are the index of refraction and the extinction coefficient of the crystal. The reflectivity will have a high value if either n or k is large. In the infrared, k becomes great for those wavelengths which excite atomic or crystal lattice vibrations, and crystals exhibit high reflectivity for these wavelengths. Fig. O-12 illustrates the classical relationship between n and k. It can be seen that the peak reflectivity will not occur at the absorption peak but at somewhat longer wavelengths where the rapidly increasing value of n also contributes to $\mathfrak{R}$. Figs. O-13 and O-14 show the reflectivities of some crystals, and Table O-2 gives the peak of the reflection curves (also called residual ray

FIG. O-14 Far infrared residual ray crystals [after W. M. Sinton and W. C. Davis, J. Opt. Soc. Am., *44*, 503 (1954)].

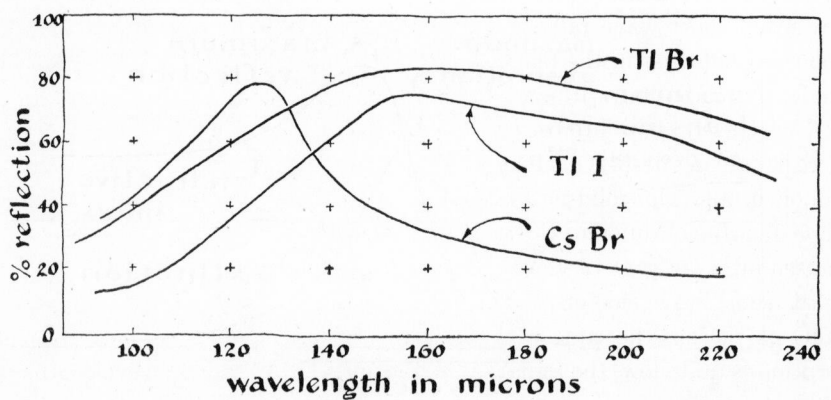

or reststrahlen curves) for various other crystals in the infrared and far infrared.

TABLE O-2

Crystal	Main Residual Ray Peak (*Microns*)
Quartz	9 and 21
Al_2O_3 (corundum)	15
CaF_2	23
LiF	26
NaF	36
NaCl	52
KCl	63
KBr	83
KI	94
CsBr	125
TlI	155
TlBr	160

It may be that one reflection from a crystal surface still gives too much
unwanted radiation. In such a case, successive reflections from more than one
crystal will rapidly sharpen the residual ray peak. For example, a reflectivity
of 90% after three reflections will still yield 73% of the initial energy, whereas
three reflections from a surface of 5% reflectivity will reduce the energy to
0.01% of its incident value. The efficiency of this process may be improved[17]
by letting the radiation strike the first crystal at the polarizing angle for the
unwanted short wave radiation. If the next crystal is so oriented that the
polarized light lies in the plane of incidence and is incident at the polarizing
angle, this radiation will be entirely transmitted.

A very useful and interesting filter can be easily made by holding a crystal
of rock salt in the MgO smoke produced by burning a length of magnesium
ribbon in air. The solid curves of Fig. O-15 show two filters resulting from
MgO deposits of different thicknesses. These powder filters are actually Christiansen filters. Fig. O-12 shows that the dispersion curve crosses the line $n = 1$
at the short wave side of the absorption peak. At this point the index of the
MgO particles is the same as the index of the surrounding air, and, since absorption is quite low, the radiation is transmitted. At shorter wavelengths the

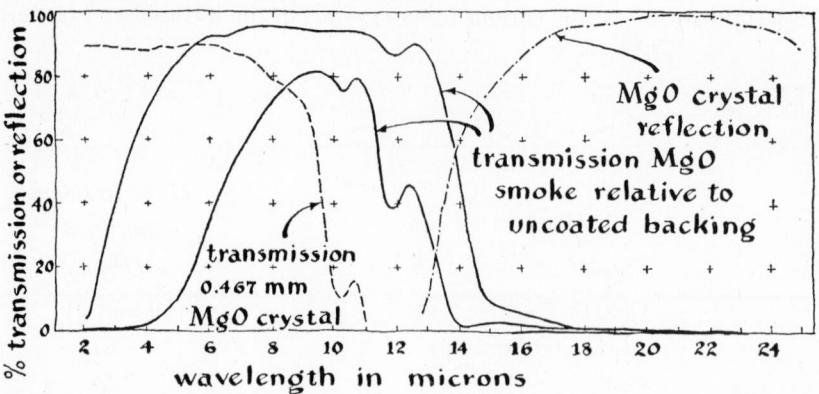

FIG. O-15   MgO crystal transmission [after J. C. Willmott, Nature, *162*, 996 (1948)]; MgO crystal reflection [after R. P. Madden, Johns Hopkins University, Dissertation (1956)].

index of MgO is sufficiently different from that of air so that the radiation is scattered by reflection and refraction. At longer wavelengths, the MgO particles are opaque, and they both scatter and absorb the incident radiation. An interesting verification of the mechanism is given by Barnes and Bonner[18] when they show that the transmission peak shifts to a shorter wavelength if the MgO film is immersed in CCl_4. In this case the peak transmission occurs where the index of MgO matches that of CCl_4. They also show the Christiansen transmission peaks for various alkali halide and quartz powders. Fig. O-15 also shows the transmission and reflection of MgO crystals. The relative positions of all these curves are explained by the discussion of Fig. O-12. It is of some interest to note that the absorption bands of the powder filter at 10.3μ and 11.9μ correspond to absorption bands in the MgO crystal.

Interference filters have the great advantage of a band pass or cut-off which may be controlled and established at precisely the desired wavelength. Fig. O-16 illustrates a simple three-layer filter[19] with a 1μ band pass located at 10μ. The peak can be located anywhere in the region from 5μ to 20μ. Tellurium, which is used here for the high-index layer, has an index of about 5. The use of such a high-index material permits narrow pass bands to be obtained with only a few layers. Additional filtering is required to cut out the secondary transmission peaks occurring at shorter wavelengths.

Fig. O-17 illustrates two types of interference filters now commercially available for isolating a pass band in the near infrared. They are composed of layers of germanium and cryolite and are currently available† in the region

† Filters are available from the Bausch and Lomb Optical Co., the Eastman Kodak Co., and Baird Associates.

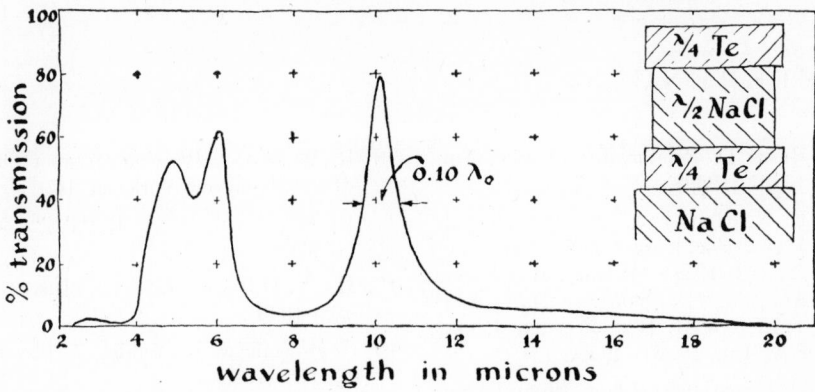

FIG. O-16 Infrared interference filter (after R. G. Greenler[22]).

from 1.4μ to 4.5μ. The long wave pass filter has a secondary transmission which comes up at about half the cut-off wavelength. In the specific case illustrated, this short wavelength transmission is absent because the germanium absorbs strongly below 1.7μ. In this wavelength range, interference filters are made with band widths as narrow as 1/50 of the wavelength of the peak transmission.

FIG. O-17 Infrared band pass and long wavelength pass filters for the infrared (Bausch and Lomb Optical Co.).

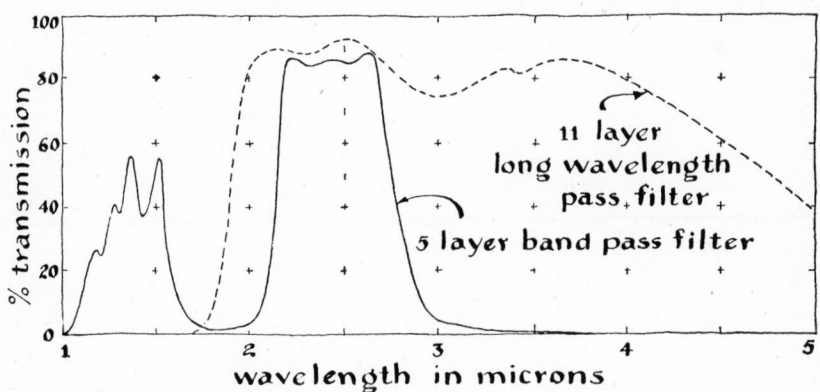

REFERENCES

1. Harry D. Polster, J. Opt. Soc. Am., *42*, 21 (1952).

2. L. N. Hadley & D. M. Dennison, J. Opt. Soc. Am., *37*, 451 (1947), and *38*, 483 (1948); G. Hass, J. Opt. Soc. Am., *45*, 945 (1955); G. Hass, A. F. Turner, & H. H. Schroeder, J. Opt. Soc. Am., *46*, 31 (1956).

3. C. Christiansen, Ann. Phys. Chem., *23*, 298 (1884), and *24*, 439 (1885).

4. E. D. McAlister, Smithsonian Misc. Coll., *93*, No. 7 (1935).

5. F. Weigert, H. Staude, & E. Elvegard, Z. Phys. Chem., *B2*, 149 (1929).

6. B. Lyot, Comptes Rendus, *197*, 1593 (1933).

7. Y. Öhman, Nature, *141*, 157 (1938).

8. J. W. Evans, J. Opt. Soc. Am. *39*, 229 (1949).

9. W. R. Brode, J. Phys. Chem. *30*, 56 (1926).

10. R. W. Wood, *Physical Optics* (1934, Macmillan, New York), p. 16.

11. R. W. Wood, Phys. Rev., *44*, 353 (1933).

12. H. M. O'Bryan, Rev. Sci. Instr., *6*, 328 (1935).

13. S. Penselin & A. Steudel, Z. Physik, *142*, 21 (1955).

14. W. L. Barr & F. A. Jenkins, J. Opt. Soc. Am., *46*, 141 (1956).

15. H. Kohn & K. von Fragenstein, Physik. Z., *33*, 929 (1933).

16. R. L. Sinsheimer & J. R. Loofbourow, Nature, *160*, 674 (1947).

17. M. Czerny, Z. Physik, *16*, 321 (1921).

18. R. B. Barnes & L. G. Bonner, Phys. Rev., *49*, 732 (1936).

19. R. G. Greenler, J. Opt. Soc. Am., *47*, 130 (1957).

Diffraction Gratings

by R. P. Madden[†] *and John Strong*[†]

This appendix develops equations for predicting the efficiencies of *plane* diffraction gratings in their various orders of spectra; and it evaluates the aberrations of the *concave* grating as used on the Rowland circle, or in parallel light.

Grating Blazes

First, let us consider the efficiency of a plane diffraction grating with triangular-shaped grooves. By efficiency we mean the fraction of incident monochromatic flux that is diffracted into a given order of the grating. For practical reasons it is desirable to know the efficiency as a function of wavelength for the various orders. This is referred to as the "blaze" of the grating.

Since Rowland's original attempt to calculate the efficiency of a blazed grating[1] (in 1893), several corrections and extensions of the theory have been made. These improvements are included here. The treatment to follow involves certain fundamental and significant assumptions which must be kept in mind when applying the results. Rigorous calculation of the diffraction from such a grating surface requires finding a solution to Maxwell's equations satisfying the boundary conditions on this grooved surface (using its complex index of refraction, $n - jk$). This is, as pointed out in Chap. III, a formidable task, and as yet no such calculation has been made for a grating with triangular grooves. In the following treatment, the reflecting facets are considered to be "simple" apertures, where the direction of propagation of the incident ray is reflected. The fundamental assumptions are that the incident field is unaffected by the presence of the diffraction surface, and that fringe

[†] Laboratory of Astrophysics and Physical Meteorology, The Johns Hopkins University.

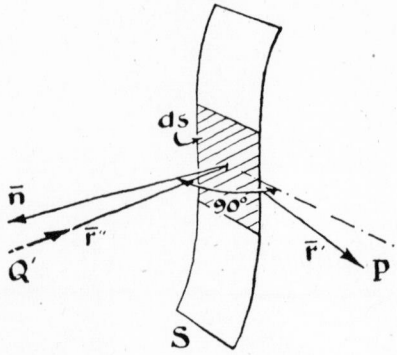

FIG. P-1 Vector definitions for the fundamental diffraction problem.

effects at the facet edges are negligible. These assumptions are reasonable when the dimensions of the grating grooves are large compared with the wavelength. In such cases we can use the scalar wave theory. However, it is well to keep in mind the fact that many diffraction gratings are ruled with a spacing of the same order of magnitude as the wavelength, and here the assumptions are not "good" ones. We shall later see that the experimentally determined efficiencies for such gratings depend upon the direction of the plane of polarization of the incident light—demonstrating the inadequacy of the scalar wave theory for describing this phenomenon in detail.

Further, it should be mentioned that the effects of multiple diffraction are not considered in the following treatment. A certain fraction of the light diffracted from one side of a grating groove will be intercepted by the other side, and rediffracted. This is neglected here, and it is apparent that this neglect will become more important as the grooves are made deeper with respect to the spacing.

Fig. P-1 shows the basic geometry of the diffraction problem. Let Q' be the source point, S the diffraction surface, and P the point at which we desire to know the field. $\bar{n}$ is the unit vector normal to the surface at dS and pointing out of the half-space containing P. At Q' we assume a point light source giving a (scalar) field $Ae^{j\omega t}$ at unit distance from Q'. The field at dS is therefore

$$\frac{Ae^{j\omega\left(t-\frac{r''}{c}\right)}}{r''}$$

where r'' is the distance from Q' to dS. The disturbance at point P, at distance r' from dS, due to the (scalar) Huygens wavelet emitted from dS, is given by Kirchhoff's differential (in a form somewhat more general than that given in § 9-1):

$$d\bar{E}_P = \frac{jA}{2\lambda r''r'} e^{j\omega\left(t-\frac{r'+r''}{c}\right)} \left[-\cos{(n, r'')} - \cos{(n, r')}\right] dS \tag{1}$$

FIG. P-2 Additional vector definitions for
the reflection-diffraction problem. Here $\bar{r}$ is
the image of $\bar{r}''$ reflected in dS.

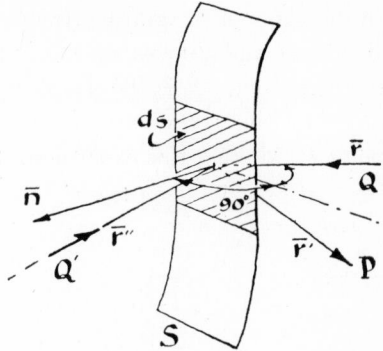

We now consider the situation for Fraunhofer diffraction from a reflection
grating. In adapting the Kirchhoff expression to our best use, we set $\dfrac{A}{r''} = \mathcal{E}_0$.
$\mathcal{E}_0$ then represents the amplitude of the field when parallel light is falling on
the grating. Also, we introduce the vector $\bar{r}$ as shown in Fig. P-2. Here $\bar{r}$ is
the mirror image of $\bar{r}''$—the mirror being the grating surface element. This
change alters the above equation, since

$$\cos(n, r'') = -\cos(n, r)$$

Hence we have

$$d\tilde{E}_P = \frac{j\mathcal{E}_0}{2\lambda r'}\, e^{j\omega\left(t - \frac{r' + r}{c}\right)} \left[\cos(n, r) - \cos(n, r')\right] dS \tag{2}$$

The integral of $d\tilde{E}_P$ over the entire diffraction surface, S, gives the total
field at P. To facilitate this integration, we express distances r and r' in terms
of a new coordinate system located on the grating surface and the distances
from Q and P to the origin of this new coordinate system. We choose a Cartesian coordinate system, as shown in Fig. P-3, with the x-axis perpendicular

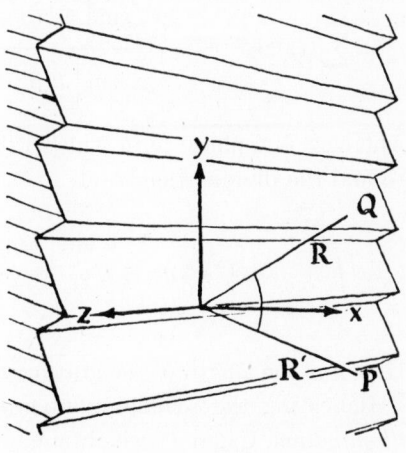

FIG. P-3 Grating surface coordinate system.

to the plane of the grating, the y-axis in the plane of the grating and perpendicular to the grooves, and the z-axis in the plane of the grating but parallel to the grooves. R is the distance from the source point to the origin of the x, y, z grating coordinate system, and R' the distance from this origin to the point P. Making this expansion, we get

$$\tilde{E}_P = \frac{j\mathcal{E}_0}{2\lambda r'}\, e^{j\omega\left(t - \frac{R+R'}{c}\right)}.$$

$$\int_S [\cos(n, r) - \cos(n, r')] e^{jk[(\alpha'-\alpha)x + (\beta'-\beta)y + (\gamma'-\gamma)z]}\, dS \quad (3)$$

where $k = \dfrac{2\pi}{\lambda}$; α, β, γ and α', β', γ' are the direction cosines of the incident and diffracted rays in the grating coordinate system.

The grating surface has a structure which is periodic in the y coordinate. For the case of Fraunhofer diffraction, therefore, the integral over the whole surface (Eq. 3) may be evaluated by integrating over one groove and summing over all ($\mathfrak{N}$) grooves. Also, we let $dS = dz \cdot dl$ where $dl = \sqrt{(dx)^2 + (dy)^2}$ is a line element in either surface of the groove and lies in the xy plane. Hence, Eq. 3 becomes:

$$\tilde{E}_P = \frac{j\mathcal{E}_0}{2\lambda r'}\, e^{j\omega\left(t - \frac{R+R'}{c}\right)} \sum_{n=0}^{\mathfrak{N}-1} e^{jkna(\beta'-\beta)} \cdot \int_{z=-\frac{h}{2}}^{\frac{h}{2}} e^{jk(\gamma'-\gamma)z}\, dz$$

$$\int_{\substack{\text{one}\\ \text{groove}}} [\cos(n, r) - \cos(n, r')] e^{jk[(\alpha'-\alpha)x + (\beta'-\beta)y]}\, dl \quad (4)$$

The summation and first integral are evaluated as follows:

$$\sum_{n=0}^{\mathfrak{N}-1} e^{jkna(\beta'-\beta)} = e^{j\Phi}\, \frac{\sin\left[\mathfrak{N}k(\beta'-\beta)\dfrac{a}{2}\right]}{\sin\left[k(\beta'-\beta)\dfrac{a}{2}\right]}$$

where $e^{j\Phi}$ is a phase term which will drop out when we multiply $\tilde{E}_P$ by $\tilde{E}_P{}^*$ to find the illumination. And

$$\int_{z=-\frac{h}{2}}^{\frac{h}{2}} e^{jk(\gamma'-\gamma)z}\, dz = h\, \frac{\sin\left[k(\gamma'-\gamma)\dfrac{h}{2}\right]}{k(\gamma'-\gamma)\dfrac{h}{2}}$$

Here h is the length of the grating grooves, measured in the y-direction.

Indicating the remaining integral in (4) by $\tilde{W}$, and calculating now the *illumination*, $\mathcal{E}_P$, at P, we obtain

$$\mathfrak{E}_P = \tfrac{1}{2}c\kappa_0 \tilde{E}_P \tilde{E}_P{}^*$$

$$= (\tfrac{1}{2}c\kappa_0)\frac{\mathcal{E}_0{}^2 h^2}{4\lambda^2 r'^2} \cdot \frac{\sin^2\left[\mathfrak{N}k(\beta' - \beta)\dfrac{a}{2}\right]}{\sin^2\left[k(\beta' - \beta)\dfrac{a}{2}\right]} \frac{\sin^2\left[k(\gamma' - \gamma)\dfrac{h}{2}\right]}{\left[k(\gamma' - \gamma)\dfrac{h}{2}\right]^2} \tilde{W}\tilde{W}^* \qquad (5)$$

To evaluate $\tilde{W}$, we first split the integration into two parts—one for each side, or facet, of the groove.

$$\tilde{W} = \int\limits_{\substack{\text{unprimed}\\\text{facet}}} [\cos(n, r) - \cos(n, r')]e^{jk[(\alpha' - \alpha)x + (\beta' - \beta)y]}\, dl$$

$$+ \int\limits_{\substack{\text{primed}\\\text{facet}}} [\cos(n', r) - \cos(n', r')]e^{jk[(\alpha' - \alpha)x + (\beta' - \beta)y]}\, dl' \qquad (6)$$

Case I: We shall first restrict the solution by the conditions $-(90 - \epsilon) \leq \phi \leq (90 - \epsilon')$ and $-(90 - \epsilon) \leq \psi \leq (90 - \epsilon')$. These conditions state that neither groove is in the shadow of the other for either the incident or the diffracted ray. We shall later consider the case where shadowing does occur.

From Fig. P-4 we can calculate the following relationships:

$$dl = \sqrt{1 + c^2}\, dy$$

$$dl' = \sqrt{1 + c'^2}\, dy$$

$$(\alpha' - \alpha) = \cos\psi + \cos\phi = \rho$$

$$(\beta' - \beta) = \sin\psi + \sin\phi = \mu$$

$$[\cos(n, r) - \cos(n, r')] = \frac{\rho + c\mu}{\sqrt{1 + c^2}}$$

$$[\cos(n', r) - \cos(n', r')] = \frac{\rho - c'\mu}{\sqrt{1 + c'^2}}$$

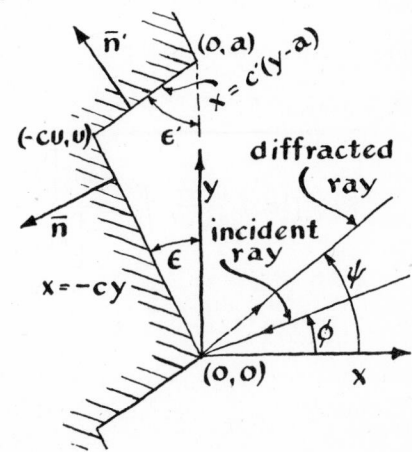

FIG. P-4 Geometry of one grating groove shown in grating coordinate system. Here,

ϵ, ϵ' are the facet angles
$c = \tan\epsilon = -$ slope of unprimed facet
$c' = \tan\epsilon' = $ slope of primed facet
$a = $ groove spacing
$u = $ value of y at the bottom of the groove
ϕ, ψ are the angles of incidence and diffraction
$\bar{n}, \bar{n}'$ are the facet normals

Making these substitutions in Eq. 6 and eliminating the variable x, we obtain

$$\tilde{W} = \int_{y=0}^{u} (\rho + c\mu)e^{jk(\mu - c\rho)}\, dy + \int_{y=u}^{a} (\rho - c'\mu)e^{jk(\mu + c'\rho)y - jkc'a\rho}\, dy \qquad (7)$$

Evaluating these integrals and multiplying by the complex conjugate, we have[2]

$$\tilde{W}\tilde{W}^* = \frac{\lambda^2}{\pi^2}\left[\frac{(\rho + \mu c)^2}{(\mu - c\rho)^2} \sin^2 \frac{\pi u(\mu - c\rho)}{\lambda}\right.$$

$$+ \frac{(\rho - \mu c')^2}{(\mu + c'\rho)^2} \sin^2 \frac{\pi(a - u)(\mu + c'\rho)}{\lambda}$$

$$\left.+ 2\frac{(\rho + \mu c)(\rho - \mu c')}{(\mu - c\rho)(\mu + c'\rho)} \sin \frac{\pi u(\mu - c\rho)}{\lambda} \sin \frac{\pi(a - u)(\mu + c'\rho)}{\lambda} \cos \frac{\pi a\mu}{\lambda}\right] \qquad (8)$$

Substituting Eq. 8 in Eq. 5, we have the illumination at the point P as a function of the constants of the grating and the angles of incidence and diffraction (subject to the condition of Case I).

We are now prepared to calculate the grating efficiency. As defined earlier, it is the fraction of incident monochromatic flux which is diffracted into a given order of the grating. We shall assume for this calculation that the exit slit (at the focus of a lens) is sufficiently wide to pass all but a negligible amount of the flux of wavelength λ that is diffracted into the order in question. Let this flux be $\mathfrak{F}$. Then

$$\mathfrak{F} = \int_{\substack{\text{exit}\\ \text{slit}}} \mathfrak{E}_P\, d\eta\, d\xi$$

where η and ξ are axes having their origin at P with η parallel to the grating y-axis, and ξ parallel to the grating z-axis. But $d\eta = \dfrac{r'd\psi}{\cos\psi}$ and $d\xi = r'd(\gamma' - \gamma)$. Hence (from Eq. 5)

$$\mathfrak{F} = (\tfrac{1}{2}\mathbf{c}\kappa_0)\frac{\mathfrak{E}_0^2 h^2}{4\lambda^2 r'^2} \int_{-\infty}^{\infty} \frac{\sin^2\left[k(\gamma' - \gamma)\dfrac{h}{2}\right]}{\left[k(\gamma' - \gamma)\dfrac{h}{2}\right]^2} r'd(\gamma' - \gamma)\cdot$$

$$\int_{\substack{\text{one order}}} \frac{\sin^2\left[\mathfrak{N}k\mu\dfrac{a}{2}\right]}{\sin^2\left[k\mu\dfrac{a}{2}\right]} \tilde{W}\tilde{W}^* \frac{r'd\psi}{\cos\psi} \qquad (9)$$

The first integral is readily evaluated:

$$\int_{-\infty}^{\infty} \frac{\sin^2\left[k(\gamma' - \gamma)\dfrac{h}{2}\right]}{\left[k(\gamma' - \gamma)\dfrac{h}{2}\right]^2} r'd(\gamma' - \gamma) = \frac{\lambda r'}{h}$$

In the remaining integral we notice that $\tilde{W}\tilde{W}^*$ is a slowly varying function (as was $\tilde{X}_s \tilde{X}_s'$ in § 10-3), which does not change appreciably in the range of ψ over which $\dfrac{\sin^2\left[\mathfrak{N}k\mu\dfrac{a}{2}\right]}{\sin^2\left[k\mu\dfrac{a}{2}\right]}$ has significant value. Evaluating $\tilde{W}\tilde{W}^*$ at the center of the diffraction pattern (given by the grating equation, $m\lambda = \mu a$), we then take it outside the integral. Thus:

$$[\tilde{W}\tilde{W}^*]_{\text{at } m\lambda = \mu a} \cdot \int_{\substack{\text{one} \\ \text{order}}} \frac{\sin^2\left[\mathfrak{N}k\mu\dfrac{a}{2}\right]}{\sin^2\left[k\mu\dfrac{a}{2}\right]} \frac{r'd\psi}{\cos\psi} = [\tilde{W}\tilde{W}^*]_{m\lambda = \mu a} \cdot \left(\frac{\lambda\mathfrak{N}r'}{a\cos\psi}\right)$$

Making these substitutions in Eq. 9 and evaluating $\tilde{W}\tilde{W}^*$ at $m\lambda = \mu a$,[3] we obtain

$$\mathfrak{F} = (\tfrac{1}{2}c\kappa_0) \frac{\mathcal{E}_0^2 S_g}{\cos\psi} \frac{1}{4\pi^2 m^2} \cdot$$

$$\left[\frac{(\rho + \mu c)}{\left(\dfrac{\mu - c\rho}{\mu}\right)} - \frac{(\rho - \mu c')}{\left(\dfrac{\mu + c'\rho}{\mu}\right)}\right]^2 \sin^2\left[\frac{m\pi c'}{(c + c')}\left(\frac{\mu - c\rho}{\mu}\right)\right] \quad (10)$$

for $-(90 - \epsilon) \le \phi \le (90 - \epsilon')$, $-(90 - \epsilon) \le \psi \le (90 - \epsilon')$, where $S_g = \mathfrak{N}ah$ is the area of the grating illuminated and m is the order number.

Case II: Let us now consider the angles of incidence or diffraction at which shadowing effects occur. It will be sufficient to consider a grating for which ϕ or $\psi > (90 - \epsilon')$. By a suitable redefinition of the variables we can obtain the solution for ϕ or $\psi < -(90 - \epsilon)$. When ϕ or $\psi > (90 - \epsilon')$, the groove facet designated by a prime is either no longer illuminated or no longer in view from the point P. In either case this facet cannot make a contribution to the field at point P. The contribution from the unprimed facet, meanwhile, is modified because part of this side of the groove is either in shadow or cannot be seen from P. We adjust our derivation to the blinding of one facet and foreshortening of the other by changing the integration limits in Eq. 7. Thus

$$\int_{y=0}^{u} U(y)\, dy + \int_{y=u}^{a} V(y)\, dy \text{ becomes } \int_{y=0}^{fu} U(y)\, dy$$

Here f is a geometrical foreshortening or shadow factor having the value

$$f = \frac{\sin (\epsilon + \epsilon')}{\sin \epsilon'} [\cos \epsilon - \sin \epsilon \tan (\theta - \epsilon)] \tag{11}$$

where θ is equal to the greater of ϕ or ψ (and necessarily $\theta > (90 - \epsilon')$).

Considering the final result for the case of shadowing, the flux of wavelength λ passed by the exit slit integrated over the diffraction pattern for one order, as before, becomes

$$\mathfrak{F} = (\tfrac{1}{2}c\kappa_0)\mathcal{E}_0{}^2 \frac{S_g}{\cos \psi} \frac{1}{4\pi^2 m^2} \cdot \frac{(\rho + \mu c)^2}{\left(\dfrac{\mu - c\rho}{\mu}\right)^2} \sin^2\left[\frac{m\pi c'f}{(c + c')}\left(\frac{\mu - c\rho}{\mu}\right)\right] \tag{12}$$

This solution blends smoothly with the "no shadow" solution (Eq. 10) at the point where $\theta = \phi$ or $\psi = (90 - \epsilon')$.

Comparison Between Theory and Observation

Several infrared diffraction gratings were made with different triangular groove forms to measure their blaze characteristics. To determine the efficiency, the flux of wavelength λ "transmitted" by the grating in one order was compared with the flux "transmitted" by a plane silver mirror using the

FIG. P-5 Experimental blaze measurements compared with theoretical predictions for a grating blazed at 19 microns with $a = 28$ microns, $\epsilon = 20°$.

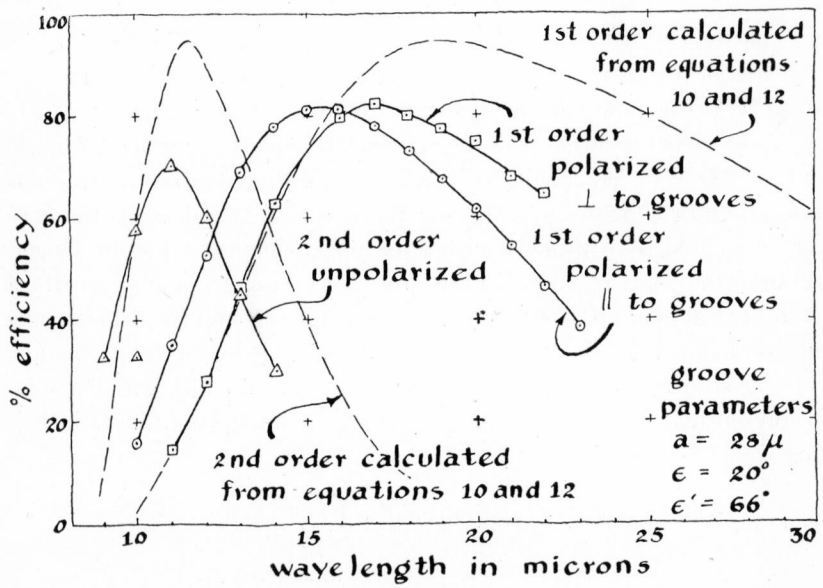

same incident beam of parallel light: An Ebert mounting was used with $\psi - \phi = 8.9°$. In the authors' experimental arrangement the optical "stop" in the system was not the grating, as in normal use with a spectrometer; rather, it was the width of the beam incident alternately on the grating and the mirror. The ratio of these measured fluxes, or the efficiency, was determined for incident light polarized both parallel and perpendicular to the grating grooves. These efficiencies are plotted as a function of λ for two different gratings in Figs. P-5 and P-6.

These two gratings were each blazed to have their peak theoretical efficiencies at the same wavelength (19 microns). However, one grating was blazed at this wavelength with a groove spacing of 28 microns and a facet angle of 20° while the other was ruled with a spacing of only 15.8 microns and a correspondingly greater facet angle, 37°. Comparing the experimental results with the theory will be particularly interesting for these two gratings since they both violate the fundamental assumption of the theory—namely that the groove dimensions be large compared with λ.

The efficiencies for these two gratings may be calculated from Eqs. 10 and 12, as indicated below. The groove spacings were determined by the ruling-engine gear ratios, and facet angles by reflecting visible monochromatic light

---

FIG. P-6 Experimental blaze measurements compared with theoretical predictions for a grating blazed at 19 microns with $a = 15.8$ microns, $\epsilon = 37°$.

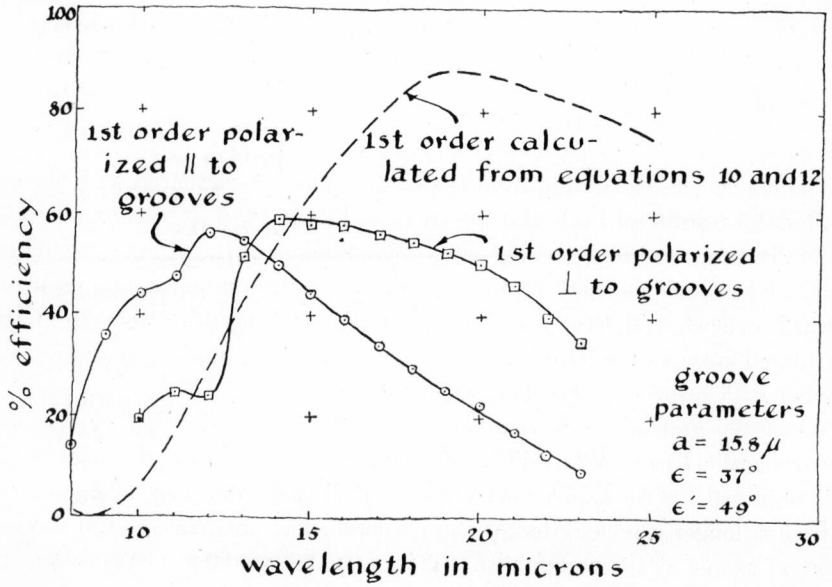

from the tipped grating face—the width of the facet face being from 20 to 50 times the visible wavelengths for these gratings.

Letting the reflectivity of the mirror be $\Re$ and the cross-sectional area of the incident beam be S_S, the flux "transmitted" by the silver mirror would be

$$\mathfrak{F}_m = (\tfrac{1}{2}c\kappa_0)\mathcal{E}_0{}^2 S_S \Re \tag{13}$$

The flux "transmitted" by the grating is given by Eqs. 10 and 12. For our experimental arrangement, the area of grating illuminated, S_g, may be expressed, in terms of the "stop" area, S_S, by

$$S_g = \frac{S_S}{\cos \phi}$$

The ratio of the grating flux to the mirror flux is then the expected efficiency —namely,

$$\frac{\mathfrak{F}_{\text{grating}}}{\mathfrak{F}_{\text{mirror}}} = \frac{1}{R \cos \phi \cos \psi} \frac{1}{4\pi^2 m^2} \cdot$$

$$\left[\frac{(\rho + \mu c)}{\left(\dfrac{\mu - c\rho}{\mu}\right)} - \frac{(\rho - \mu c')}{\left(\dfrac{\mu + c'\rho}{\mu}\right)}\right]^2 \sin^2\left[\frac{m\pi c'}{(c + c')}\left(\frac{\mu - c\rho}{\mu}\right)\right] \tag{14}$$

$$\text{for} \left\{ \begin{array}{c} -(90 - \epsilon) \le \phi \le (90 - \epsilon') \\ -(90 - \epsilon) \le \psi \le (90 - \epsilon') \end{array} \right\}$$

and

$$\frac{\mathfrak{F}_{\text{grating}}}{\mathfrak{F}_{\text{mirror}}} = \frac{1}{R \cos \phi \cos \psi} \frac{1}{4\pi^2 m^2} \cdot$$

$$\frac{(\rho + \mu c)^2}{\left(\dfrac{\mu - c\rho}{\mu}\right)^2} \sin^2\left[\frac{m\pi c'f}{(c + c')}\left(\frac{\mu - c\rho}{\mu}\right)\right] \tag{15}$$

$$\text{for} \quad \phi \text{ or } \psi > (90 - \epsilon')$$

The efficiencies calculated from these equations are compared with the experimental results for both gratings in Figs. P-5 and P-6. It was found to be generally true, for these gratings and others measured, that the experimental first order peak efficiencies were lower than the calculated ones, and occurred toward somewhat shorter wavelengths. Both of these deviations from the calculated curves were found to become larger as the groove spacing became smaller with respect to the blaze wavelength.

When the gratings were studied using polarized light, with the electric vector parallel to and perpendicular to the grooves, a different efficiency curve was obtained for the two polarizations. In all cases the peak efficiency occurred at longer wavelengths for the perpendicular polarization and did not drop off as fast as the wavelength was increased. The difference between the efficiency curves for the two polarizations increased as the groove spacing was made smaller with respect to the wavelength.

The concept of a merit factor is useful when selecting an infrared grating for a particular application. Three things are considered in the calculation of the merit factor for a grating: The first is the efficiency determined by the experimental procedure just described. The second is that in most practical applications the grating is the "stop" in the system, and therefore, as the grating rotates to higher angles, a smaller and smaller area of the incident illumination is intercepted. We may correct our measured efficiencies for this loss of beam area by a cos ϕ correction factor. And the third is that we consider that gratings with different groove spacings used for the same wavelength region have different dispersions. When resolution is energy limited, an increase in dispersion can be utilized by opening the slits, allowing a greater "transmission" of flux for the same resolution (as was pointed out in § 12-10). Thus we define the merit factor by

$$\text{merit factor} = \frac{\text{efficiency} \times \cos \phi}{\text{dispersion}} = \text{efficiency} \times \frac{\cos \phi}{a \cos \psi}$$

For a Littrow mounting, the merit factor is the efficiency ÷ a. The merit factors for the gratings of Figs. P-5 and P-6, using the experimentally determined efficiencies, are shown in Fig. P-7. This comparison demonstrates that

FIG. P-7 Comparison of the experimentally determined merit factors for the two gratings of Figs. P-5 and P-6.

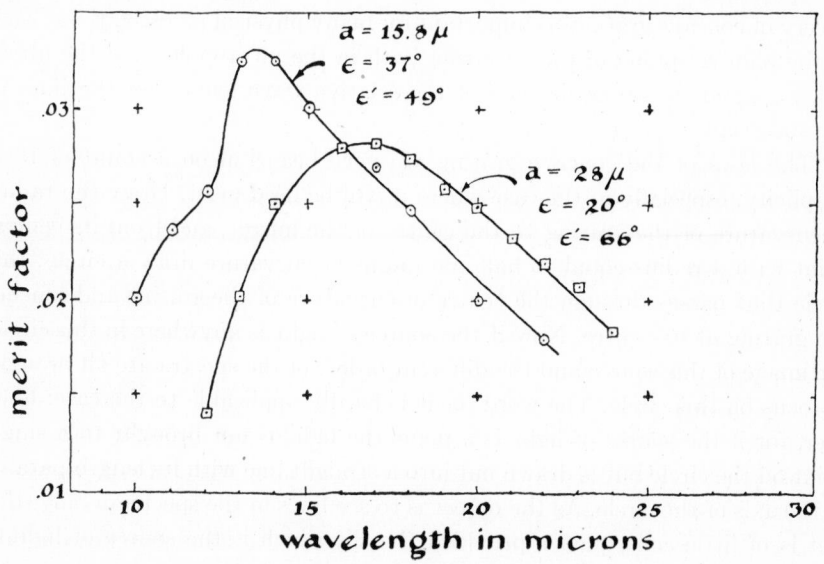

the higher dispersion available with the 15.8 micron spacing gives merit comparable with that of the grating with $a = 28$ microns, in spite of a much lower blaze efficiency.

In conclusion it should be reiterated that it is at present necessary to determine grating blaze efficiencies experimentally when the groove dimensions are not large compared to the wavelength. This fact is adequately demonstrated in Figs. P-6 and P-7. The scalar theory works well within its limitations but a more elegant theory is needed to explain the blaze characteristics of many gratings in use today.

Concave Gratings

Professor Rowland made the first large precision diffraction gratings. He invented the concave grating and subsequently used it to establish standard wavelengths (by the method of coincidences of § 12-6). We may quote from Professor Rowland's early writing[4] to define the characteristics of the concave grating:

. "All gratings hitherto made have been ruled on flat surfaces. Such gratings require a pair of telescopes for viewing the spectrum; these telescopes interfere with many experiments, absorbing the extremities of the spectrum strongly; besides, two telescopes of sufficient size to use with six inch gratings would be very expensive and clumsy affairs. In thinking over what would happen were the grating ruled on a surface not flat, I thought of a new method of attacking the problem, and soon found that if the lines were ruled on a spherical surface the spectrum would be brought to a focus without any telescope. This discovery of concave gratings is important for many physical investigations, such as the photographing of the spectrum both in the ultra-violet and the ultrared, . . . and the determination of the relative wave lengths of the lines of the spectrum.

"The laws of the concave grating are very beautiful on account of their simplicity, especially in the case where it will be used most. Draw the radius of curvature of the grating to the centre of the mirror, and from its central point with a radius equal to half the radius of curvature draw a circle; this circle thus passes through the centre of curvature of the grating and touches the grating at its centre. Now if the source of light is anywhere in this circle, the image of this source and the different orders of the spectra are all brought to focus on this circle. The word focus is hardly applicable to the case, however, for if the source of light is a point the light is not brought to a single point on the circle but is drawn out into a straight line with its length parallel to the axis of the circle. As the object is to see lines in the spectrum only, this fact is of little consequence provided the slit which is the source of light is parallel to the axis of the circle. Indeed it adds to the beauty of the spectra, as the horizontal lines due to dust in the slit are never present, as the dust has a different focal length from the lines of the spectrum."

The concave grating was an important invention because of its ingenuity, and because of great practicality in both its construction and its use. Since the advent of the process of thermal evaporation of highly reflecting aluminum films, the concave grating has lost some of its significance. Since the early 1930's, such films have been extensively used for the reflecting surfaces of astronomical telescopes. They find extensive use for the reflecting surfaces of collimator and telescope mirrors in spectrographs. And, indeed, most diffraction gratings are now ruled in aluminum films.[5]

To understand the importance of Professor Rowland's invention, we shall compare a conventional spectrograph (using a plane grating and two mirrors, for collimator and telescope) with the Wadsworth use of a concave grating (with only one auxiliary collimator mirror), and a concave grating worked on the Rowland circle (with no auxiliary mirrors). Taking $\Re = 70\%$ for the speculum surfaces that would have been used for these mirrors (before aluminum), the maximum efficiencies would have been in the ratio $0.7:0.5:0.35$. Now, with aluminum, these ratios are $0.9:0.8:0.7$.

Professor Rowland's concave grating is practical to construct because the spherical mirror it uses for the grating blank is the easiest contour by far to generate with high optical quality (as contrasted with a parabolic contour, *etc.*). Furthermore, the grating grooves, required to be equally spaced along the chord of the grating, rather than along its arc, are the only groove spacings that the ruling engine can accomplish.

One prominent physicist in 1883 did not think that Rowland's concave grating would give good spectra when the grating was significant in size. But this was because he did not appreciate that Rowland's grating lines, uniform along the chord, were unequally spaced along the arc to just the degree required. It is this point that we shall examine quantitatively below.

Rowland Circle

Fig. P-8 shows a spherical grating and its Rowland circle, of diameter equal to the radius of the spherical surface on which the grating is ruled. We shall show that the Rowland circle arrangement works well for a wide grating aperture. Consider that the entrance slit lies on the grating normal at N and is illuminated with light of wavelength λ. The ray diffracted by the center of the grating at V, toward P, satisfies the equation

$$m\lambda = \delta r_V = a \sin \alpha \tag{16}$$

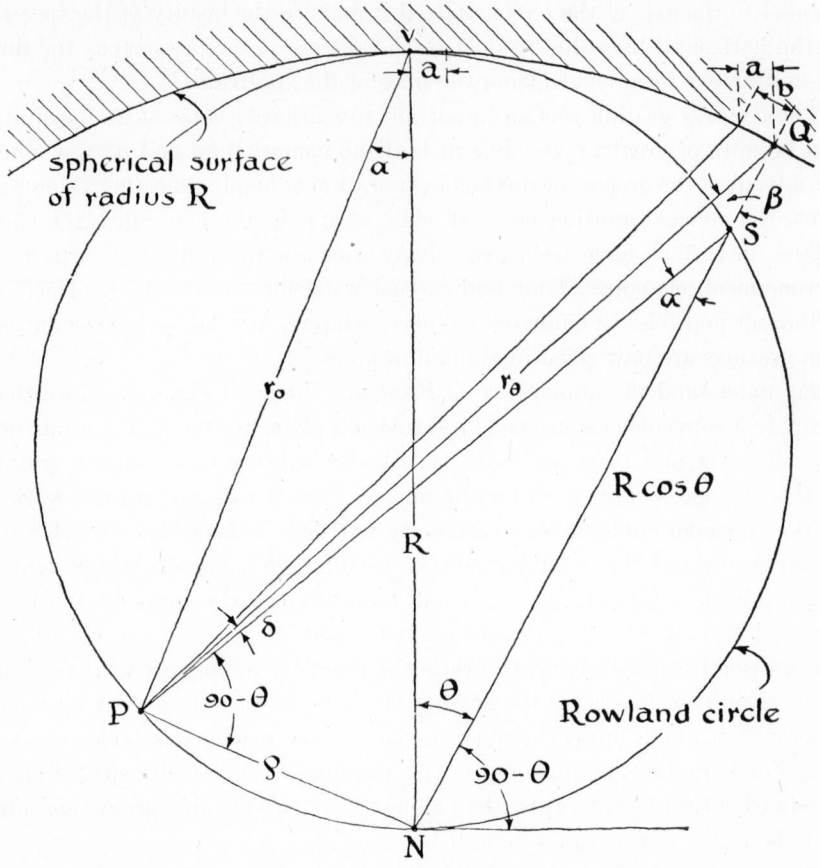

FIG. P-8 Grating focal properties on Rowland's circle.

And the ray diffracted by the grating at Q, toward P, where the grating spacing is $b = \dfrac{a}{\cos \theta}$, yields the path differential per groove of δr_Q. This δr_Q must be very nearly equal to $m\lambda$ if the grating is to give a sharp spectrum line over the aperture from V to Q.

$$m\lambda \overset{?}{=} \delta r_Q = b \sin \beta = a \frac{\sin \beta}{\cos \theta} \tag{17}$$

For purely pedagogic reasons our figure shows an exaggerated spacing and an exaggerated aperture of the grating. Actually, at the margin of a typical grating, $\theta = \theta_0$ will be only about .01 radian.

It is our primary purpose to show that ϵ_Q, in the equation below, is insignificant.

$$\epsilon_Q = \delta r_V - \delta r_Q = m\lambda \left(1 - \frac{1}{\cos \theta} \frac{\sin \beta}{\sin \alpha} \right) \tag{18}$$

To evaluate this quantity a brief digression occupied with determination of the trigonometric functions involved in the geometry of rays in Fig. P-8 is necessary.

First, we consider $\triangle PQN$ and apply the law of sines, remembering that δ, the difference between α and β, will be a very small angle.

$$\frac{\sin \beta}{\rho} = \frac{\sin \left(\frac{\pi}{2} - \theta + \delta \right)}{R} = \frac{\sin \theta \times \delta + \cos \theta}{R}$$

And, considering $\triangle PSN$, again applying the law of sines,

$$\frac{\sin \alpha}{\rho} = \frac{\sin \left(\frac{\pi}{2} - \theta \right)}{R \cos \theta} = \frac{1}{R}$$

Therefore, combining these expressions in Eq. 18,

$$\epsilon_Q = -m\lambda\delta \tan \theta \cong -m\lambda\theta\delta$$

From Fig. P-8, when θ is small,

$$\delta = \frac{R(1 - \cos \theta) \sin \beta}{r_\theta} \cong (1 - \cos \theta) \tan \beta \cong \frac{\theta^2}{2} \tan \beta$$

since $r_\theta \cong R \cos \beta$.

We take a path difference over $\Delta \mathfrak{N} = \dfrac{R d\theta}{b}$ grating grooves and then sum over all $\mathfrak{N}$ grooves: $\displaystyle\sum \epsilon_{\Delta\mathfrak{N}} = -\int \frac{m\lambda R \tan \beta}{2} \theta^3 \, d\theta$. On integrating from $\theta = 0$ out to θ_0, the total accumulation of path retardation between the welded Huygens wavelets wave front, and a spherical arc about P through V becomes

$$\Delta r = \frac{m\lambda R \tan \beta}{8a} \theta_0^4$$

For a typical case of a nominal six-inch grating, with five inches of ruling, using $R = 21$ feet, and $\tan \beta = \frac{1}{2}$ for the $m = 1.0$ order, we get $\theta_0 \cong 10^{-2}$, giving

$$\Delta r = \frac{\lambda}{400}$$

Although this deviation is completely insignificant, compared with a Rayleigh tolerance of $\Delta r = \dfrac{\lambda}{4}$, even this $\dfrac{\lambda}{400}$ can be reduced fourfold by a slight change of focus, as Problem 13-7 taught us. This reduction results because the path difference has the character of a spherical aberration

Wadsworth Mounting

In 1906 F. L. O. Wadsworth[6] described a use of the concave diffraction grating in parallel light. This Wadsworth arrangement gives stigmatic images, as contrasted to the astigmatic ones that are obtained with lines on Rowland's circle. The excellence of this arrangement was recognized and developed by W. F. Meggers and K. Burns.[7] They used a mirror collimator to make the light from an entrance slit parallel.

As a further exercise in geometrical optics, and one particularly suited for application of our "mathematics of modest rigor," we shall show that the Wadsworth arrangement, represented in Figs. P-9 and P-10, gives excellent stigmatic spectra for lines lying on the central normal to the grating.

We first determine where parallel rays incident at the top of the grating at angle α_0 are reflected down and diffracted onto the central normal from a point M that is on both the grating and the meridional plane (perpendicular to the plane of Fig. P-8 and containing V and N). Then, with this point as a focus, we show that the convergence error for the Huygens wavelets diffracted along the equatorial arc VQ (of Fig. P-10) onto this point is, indeed, small.

Fig. P-9 represents the above meridional plane, and the line with dots represents the projection of a ray incident at M. This incident ray is one that makes the angle α_0 to the plane of Fig. P-9. It is indicated, by a dashed line, in Fig. P-10. Consider the intersection of a plane perpendicular to Fig. P-10, that contains the dashed line to M, and the plane perpendicular to Fig. P-9, that contains the line MN. If $\alpha_0 = 0$, this intersection lies at the angle φ below a plane MT. On the other hand, if $\alpha_0 = \dfrac{\pi}{2}$, that intersection lies in a plane

---

FIG. P-9   Ray in the meridional plane from facet M to stigmatic focus F.

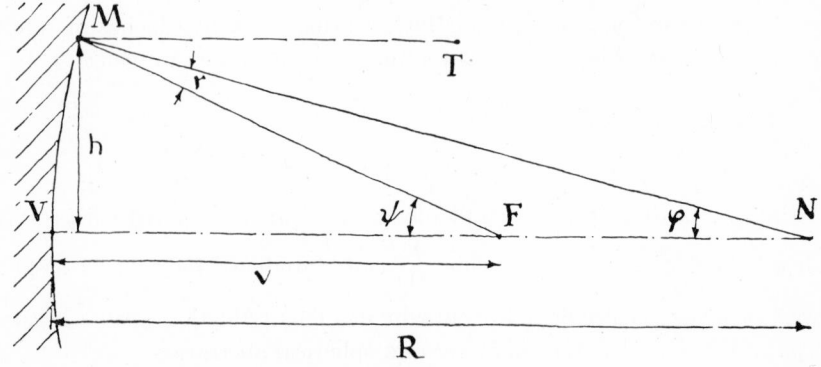

MT perpendicular to Fig. P-9. Intermediately, as the application of a little geometrical imagination will reveal, the intersection line makes an angle $\varphi' = \varphi \cos \alpha_0$ with the plane MT.

Now the ray diffracted at M into the meridional plane will be reflected down there, below the plane MN by this same angle $r = \varphi \cos \alpha_0$. Here we consider reflection in one plane and the diffraction in the other. With this fact we calculate v, the distance from M to the focus F. If all angles are small,

$$\psi = r + \varphi; \qquad \frac{h}{v} \cong \psi; \qquad \frac{h}{R} \cong \varphi$$

giving

$$v = \frac{R}{1 + \cos \alpha_0}$$

FIG. P-10   Grating focal properties with incident parallel light—Wadsworth mounting.

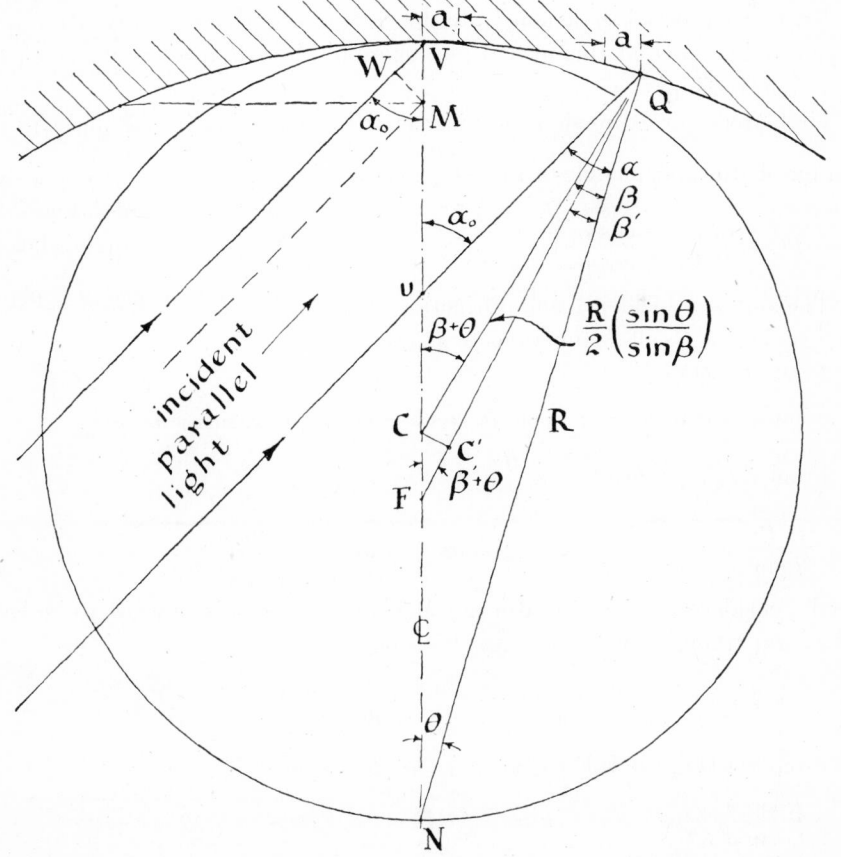

It is interesting to derive this formula also by the method of sagittae of § 13-4: Fig. P-10 shows the projection of the point M, for the ray striking the grating at the height h above the plane of the figure. It is diffracted and reflected into the meridional plane there. At the instant when this ray is diffracted and reflected, the ray in the wave front at W, in the plane of Fig. P-10, that is also diffracted into the meridional plane at V, has yet the distances $WV + VM$ to go before it will be under the point M. We set this distance (by the method of sagittae) equal to the sagitta of a meridional circle centered on the focal point F. The resulting equation gives the image distance v that we desire.

$$\frac{h^2}{2v} = WV + VM = \frac{h^2}{2R}(1 + \cos \alpha_0) \text{ since } VM = \frac{h^2}{2R}$$

and

$$v = VF = \frac{R}{1 + \cos \alpha_0}$$

Before we proceed to demonstrate the error of diffraction for rays from the equatorial arc VQ, at this focus F, let us establish several necessary geometrical relationships.

First, from our expression for v we get the distance between F and C, the center of Rowland's circle of radius $\frac{R}{2}$, as

$$FC = \frac{R}{2}\left(\frac{1 - \cos \alpha_0}{1 + \cos \alpha_0}\right)$$

The angles of incidence and diffraction at Q, α, and β', their difference, and the angle β at C, all as functions of θ, are:

From $\triangle UQN$:

$$\sin \alpha = \sin(\alpha_0 - \theta) = \sin \alpha_0 \cos \theta - \cos \alpha_0 \sin \theta$$

From $\triangle CNQ$, since $CN = \frac{R}{2}$:

$$\frac{R}{2 \sin \beta} = \frac{R}{\sin(\beta + \theta)} \qquad \text{giving} \qquad \tan \beta = \frac{\sin \theta}{2 - \cos \theta}$$

And, considering the numerator and denominator of this tangent to be legs of a right triangle, and solving for the sine,

$$\sin \beta = \frac{\sin \theta}{\sqrt{5 - 4 \cos \theta}} \simeq \frac{\sin \theta}{1 + \theta^2 - \frac{7}{12}\theta^4}$$

From $\triangle CQN$, the side CQ is

$$\frac{R}{2}\frac{\sin \theta}{\sin \beta}$$

From $\triangle CC'F$, using our expression for FC, we get $(\beta - \beta')$, in terms of θ, as follows:

$$(\beta - \beta') = \frac{FC \sin (\beta' + \theta)}{CQ} = \left(\frac{1 - \cos \alpha_0}{1 + \cos \alpha_0}\right)\frac{\sin (\beta' + \theta)}{\dfrac{\sin \theta}{\sin \beta}}$$

And, invoking the series expansions that are characteristic of our "mathematics of modest rigor," which has more than abundant quantitative precision here, we get

$$\beta' = (\cos \alpha_0) \times \theta$$

We are now ready to continue with our demonstration of the errors of diffraction along the arc VQ. We proceed exactly as before, except now

$$m\lambda = \delta r_V = a \sin \alpha_0 \tag{16'}$$

and

$$m\lambda \overset{?}{=} \delta r_Q = \frac{a}{\cos \theta} (\sin \alpha + \sin \beta') \tag{17'}$$

making

$$\epsilon_Q = \delta r_V - \delta r_Q = m\lambda \left[1 - \frac{1}{\cos \theta}\left(\frac{\sin \alpha + \sin \beta'}{\sin \alpha_0}\right)\right] \tag{18'}$$

Substitution for $\sin \alpha$ and $\sin \beta'$ gives

$$\epsilon_Q = \delta r_V - \delta r_Q = m\lambda \frac{\cot \alpha_0}{6} \theta^3$$

On taking the aggregate for $\dfrac{Rd\theta}{b}$ grooves and integrating, as before, we get

$$\Delta r = m\lambda R \frac{\cot \alpha_0 (\cos^2 \alpha_0 - 2)}{24} \theta_0^4$$

which yields the same type of inconsequential error that we got before on the Rowland circle.

REFERENCES

1. H. A. Rowland, Phil. Mag., *35*, 397–419 (1893).

2. R. D. Hatcher & J. H. Rohrbaugh, J. Opt. Soc. Am., *46*, 104 (1956), have calculated $\widetilde{W}\widetilde{W}^*$. Our result is identical with theirs, and we have adopted their definition of symbols.

3. R. D. Hatcher and J. H. Rohrbaugh have evaluated $\widetilde{W}\widetilde{W}^*$ at $m\lambda = \mu a$, although their result, as published in reference 2, is in error by a slight omission.

4. H. A. Rowland, Phil. Mag. [4], *13*, 469 (1882).

5. J. Strong, Publs. Astron. Soc. Pacific, *46*, 18 (1934).

6. F. L. O. Wadsworth, Astrophys. J., *3*, 54 (1906).

7. W. F. Meggers & K. Burns, Bureau of Standards Paper 441, *18*, 185 (1922).

Mathematical Review

by Trevor Williams†

Complex Numbers

It is in solving the general quadratic equation that complex numbers are first encountered. The quadratic formula states that the two roots of the equation

$$ax^2 + bx + c = 0$$

are given by

$$x_1 = \frac{1}{2a}\left(-b + \sqrt{b^2 - 4ac}\right) \qquad \text{and} \qquad x_2 = \frac{1}{2a}\left(-b - \sqrt{b^2 - 4ac}\right)$$

The quantity $b^2 - 4ac$ is called the discriminant of the quadratic, and when it is negative we are faced with the problem of attaching a meaning to the square root of a negative number. For example, the roots of the equation

$$x^2 - 2x + 5 = 0$$

are

$$x_1 = 1 + \sqrt{-4} \qquad \text{and} \qquad x_2 = 1 - \sqrt{-4}$$

But, since

$$(+2)\cdot(+2) = +4 \qquad \text{and} \qquad (-2)\cdot(-2) = +4$$

also, it looks as if -4 did not possess a square root. Indeed, the Rule of Signs of algebra states that the product of like signs is always positive, and hence the square of any number must be positive. Therefore the square root of a negative number cannot exist, for, if it did, it would be a number which when squared produced a negative number.

But, for that matter, negative numbers "cannot" exist either. Nobody has ever held in his hand $-1\frac{1}{2}$ apples. All numbers are mathematical abstractions

---

† The Johns Hopkins University.

of varying degrees of sophistication. When negative numbers were needed to facilitate calculations, they were *invented;* and so, in their turn, were negative squares when they were needed. A so-called imaginary unit was defined, privileged in that its square was equal to -1, but satisfying the rules of algebra in every other respect. This new number was called i, which is what it is still known as in the mathematical literature; but physicists generally refer to it as j, reserving i as the symbol for electric current. If, then, we write

$$j^2 = -1$$

we have immediately

$$\sqrt{-4} = \sqrt{(-1)\cdot(+4)} = \sqrt{-1}\cdot\sqrt{+4} = 2j$$

and the solutions of the above quadratic become simply

$$x = 1 + 2j \qquad \text{and} \qquad x = 1 - 2j$$

These are typical *complex numbers.* Here both have a *real part* of $+1$, and the *imaginary parts* are, respectively, $+2j$ and $-2j$. It is because these numbers consist of two parts that they are called complex, not because they are complicated. As for j, it is no more imaginary than any other kind of number; all numbers are equally products of the human imagination.

Exercise 1 What does $\sqrt{-x}$ equal when $x > 0$?

Exercise 2 Solve the equation $x^2 + x + 1 = 0$.

Exercise 3 It is readily seen that if the sum of two real numbers is 10, their product cannot be greater than 25. Find two numbers whose sum is 10 and product is 50.

Exercise 4 Show that $j^3 = -j$, $j^4 = +1$, $j^5 = j$, $j^6 = -1$, and so on.

Exercise 5 By first squaring and then cubing each of the solutions to Exercise 2, show that both these numbers are cube roots of unity. How many cube roots of unity are you now aware of the existence of?

It is a familiar fact that the real numbers may be represented as points on a straight line extending to infinity in both directions. Since we have seen that imaginary numbers, having negative squares, cannot be real, this means that they will not "fit" anywhere on this straight line, among the real numbers. Now multiplication of a real number by j has the property that when it is done twice in succession it has the same net effect as a single multiplication by -1. But, geometrically, multiplication by -1 is the same as a rotation of 180° about zero. Therefore we are led to interpret multiplication by j as a rotation of 90° about zero. Hence the real multiples of j all lie on a line perpendicular to the real axis and intersecting it at O. Such numbers are called *pure imaginaries,* and the line is called the *imaginary axis.* If these real and imaginary axes are used to define a Cartesian coordinate system, any complex number $\bar{w} = x + jy$ may be represented as the point (x, y), as indicated in

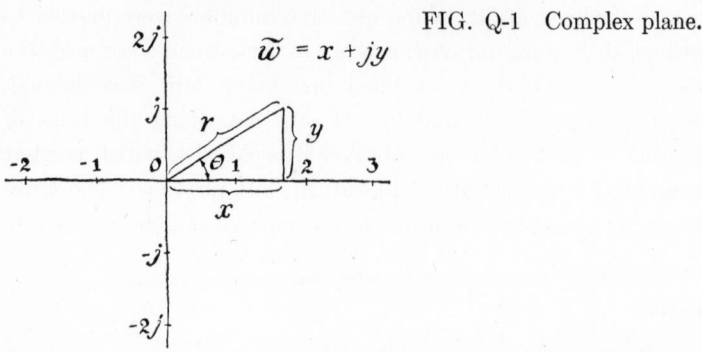

FIG. Q-1   Complex plane.

Fig. Q-1. The x,y-plane is most often referred to in this context as the *complex plane*. Zero, lying on both axes, is the only number which is both purely real and purely imaginary; it is usually spoken of as the *origin* when it is considered as a point in the complex plane.

The distance, r, from the origin to the point $\tilde{w}$ is called its *absolute value*, or *modulus*. The angle, θ, which the line from $\tilde{w}$ to the origin makes with the positive real axis is called the *argument* of $\tilde{w}$. These two quantities are written, respectively, $|\tilde{w}|$ and arg $\tilde{w}$. The number $x - jy$ is the reflection in the real axis of the number $x + jy$ and is called its *complex conjugate*. The negative of $w = x + jy$ is defined by the equation $-w = -x - jy$ and is the "reflection" of w through the origin.

Exercise 6   Bearing in mind the fact that the customary algebraic rules all carry over to calculations with complex numbers, with the single additional fact that $j^2 = -1$, show that, if $w_1 = x_1 + jy_1$ and $w_2 = x_2 + jy_2$, then

$$w_1 + w_2 = (x_1 + x_2) + (y_1 + y_2)j$$

and

$$w_1 w_2 = (x_1 x_2 - y_1 y_2) + (x_1 y_2 + x_2 y_1)j$$

Exercise 7   Interpret the results of Exercise 4 as rotations of the complex plane.

Exercise 8   Show, algebraically and geometrically, that if w_1 is the complex conjugate of w_2, then w_2 is the complex conjugate of w_1. (Since the relationship is reciprocal, such pairs are usually referred to simply as complex conjugates.)

Exercise 9   Show that the product of two complex conjugates is equal to the square of their (common) absolute value.

Exercise 10   Using the results of Exercises 6 and 9, show that the quotient of two complex numbers is a complex number:

$$\frac{x_1 + jy_1}{x_2 + jy_2} = \frac{x_1 x_2 + y_1 y_2}{x_2{}^2 + y_2{}^2} + \frac{x_2 y_1 - x_1 y_2}{x_2{}^2 + y_2{}^2} j$$

Exercise 11   Show that there are no new "complex-complex" numbers, *i.e.*, that

$$(x_1 + jy_1) + (x_2 + jy_2)j$$

is itself merely a complex number.

Exercise 12   Show, by squaring $\frac{1}{2}\sqrt{2}(1-j)$, that this number is a square root of $-j$. Hence there are no "hyper-complex" numbers based on $\sqrt{-j}$. What is the other value of $\sqrt{-j}$?

Exercise 13   Prove that the complex conjugate of a product is the product of the complex conjugates of the separate factors.

Exercise 14   From the results of Exercises 2 and 5, show that there are three cube roots of unity, all equal to 1 in absolute value, and having arguments of $0°$, $120°$, and $240°$, respectively. Interpret this geometrically in the complex plane.

It may well seem at this point that the only thing complex numbers are good for that real numbers are not is the solution of quadratic equations. They are, however, of much more basic importance than that. The so-called Fundamental Theorem of Algebra states that *every* algebraic equation of degree n has precisely n roots in the complex plane. This very important and difficult theorem was first proved rigorously in 1797 by the great German mathematician Gauss. He was eighteen at the time, and this proof was his doctoral dissertation!

What the mathematician terms *analysis*, the enormous discipline that stems from the differential and integral calculus, has its roots deep in the wonderfully fertile soil of the complex plane. It is only here that analysis finds its most natural expression, and the interplay between apparently unrelated phenomena is laid bare. Thus, if we recall the power-series expansion of e^w,

$$e^w = 1 + w + \frac{w^2}{2!} + \frac{w^3}{3!} + \frac{w^4}{4!} + \frac{w^5}{5!} + \cdots$$

and write $w = j\theta$, we find

$$e^{j\theta} = 1 + j\theta - \frac{\theta^2}{2!} - j\frac{\theta^3}{3!} + \frac{\theta^4}{4!} + j\frac{\theta^5}{5!} + \cdots$$

Now

$$\cos\theta = 1 + \frac{\theta^2}{2!} + \frac{\theta^4}{4!} + \cdots$$

and

$$\sin\theta = \theta + \frac{\theta^3}{3!} + \frac{\theta^5}{5!} + \cdots$$

Upon combining these results, we find that

$$e^{j\theta} = \cos\theta + j\sin\theta$$

In the complex plane the exponential and trigonometric functions merge together, so to speak. In this identity, θ is, of course, measured in radians, and the special case $\theta = \pi$ yields

$$e^{j\pi} = -1$$

Euler, the celebrated Swiss mathematician who discovered this relation,

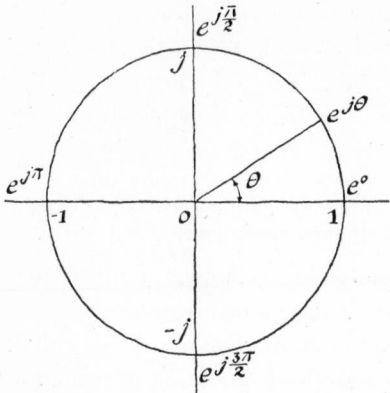

FIG. Q-2   Unit circle.

rightly considered it one of the most wonderful in all of analysis, connecting, as it does, the four most important numbers of mathematics: e, π, j, and 1.

Since the real part of $e^{j\theta}$ is $\cos \theta$ and the imaginary part is $\sin \theta$, its absolute value is $\sqrt{\cos^2 \theta + \sin^2 \theta} = 1$; and its argument is θ. Thus, $e^{j\theta}$ is a number making an angle θ with the positive real axis and lying on the circumference of the *unit circle*, *i.e.*, the circle centered at the origin and having unit radius (see Fig. Q-2). This leads to another representation of complex numbers, for the number

$$re^{j\theta} = r \cos \theta + jr \sin \theta$$

has an absolute value of r and an argument of θ, and hence is nothing but the number $w = x + jy$ of Fig. Q-1. The product of two complex numbers,

$$w_1 = r_1 e^{j\theta_1} \qquad \text{and} \qquad w_2 = r_2 e^{j\theta_2}$$

is merely

$$w_1 w_2 = r_1 r_2 e^{j(\theta_1 + \theta_2)}$$

I.e., to multiply two complex numbers, multiply their absolute values and add their arguments.

Exercise 15   By actually multiplying out the right-hand side, show that

$$x^4 - x^2 + 2x + 2 = [x - (1+j)][x - (1-j)](x+1)^2$$

Hence this quartic has four roots: the complex conjugate pair $1 \pm j$, and the "double" root -1. By forming a table of the first four powers of $1 + j$, show directly that $x = 1 + j$ satisfies the given equation. Do the same for -1 and $1 - j$.

Exercise 16   Prove that e^w is a periodic function of w, with period $2\pi j$.

Exercise 17   By writing the equation that the nth root of k must satisfy, and using the Fundamental Theorem of Algebra, show that every number has exactly n nth roots in the complex plane. Show in fact that the nth roots of unity are the numbers

$$\cos \frac{2\pi m}{n} + j \sin \frac{2\pi m}{n} \quad (m = 0, 1, \ldots, n-1)$$

Exercise 18   Prove the identities

$$\cos \theta = \frac{1}{2} \left(e^{j\theta} + e^{-j\theta} \right)$$

$$\sin \theta = \frac{1}{2j} \left(e^{j\theta} - e^{-j\theta} \right)$$

Exercise 19   Two complex numbers are equal if and only if their real parts are equal and their imaginary parts are equal. Hence, by multiplying the equations

$$e^{j\theta} = \cos \theta + j \sin \theta \qquad \text{and} \qquad e^{j\phi} = \cos \phi + j \sin \phi$$

together, prove (simultaneously!) the addition formulas for the sine and cosine.

Exercise 20   Derive, similarly, the "triple-angle" formulas,

$$\cos 3\theta = \cos^3 \theta - 3 \cos \theta \sin^2 \theta$$

$$\sin 3\theta = 3 \cos^2 \theta \sin \theta - \sin^3 \theta$$

Exercise 21   Recalling the formula for the sum of a geometric series,

$$a + a + \cdots + a^{n-1} = \frac{a^n - a}{a - 1}$$

show that

$$e^{j\theta} + e^{2j\theta} + \cdots + e^{(n-1)j\theta} = \frac{e^{(n-\frac{1}{2})j\theta} - e^{\frac{1}{2}j\theta}}{e^{\frac{1}{2}j\theta} - e^{-\frac{1}{2}j\theta}}$$

and hence that

$$\cos \theta + \cos 2\theta + \cdots + \cos (n-1)\theta = \frac{\sin \dfrac{n-1}{2} \theta \cos \dfrac{n}{2} \theta}{\sin \frac{1}{2}\theta}$$

and

$$\sin \theta + \sin 2\theta + \cdots + \sin (n-1)\theta = \frac{\sin \dfrac{n-1}{2} \theta \sin \dfrac{n}{2} \theta}{\sin \frac{1}{2}\theta}$$

Exercise 22   By breaking the left side down into its complex factors, and regrouping these four factors before multiplying out again, prove that

$$(a^2 + b^2)(c^2 + d^2) = (ac - bd)^2 + (ad + bc)^2$$

Check directly also. (The efficacy of complex notation here is clearly not in proving the result, but in discovering it. The above identity is important in the *theory of numbers*, the branch of mathematics that deals with properties of integers. It shows that if A is a sum of two squares and B is also, then so is AB.)

Fourier Series

Periodic functions are extremely common in physics and arise whenever any kind of wave motion is considered. A function $f(x)$ is said to be periodic with

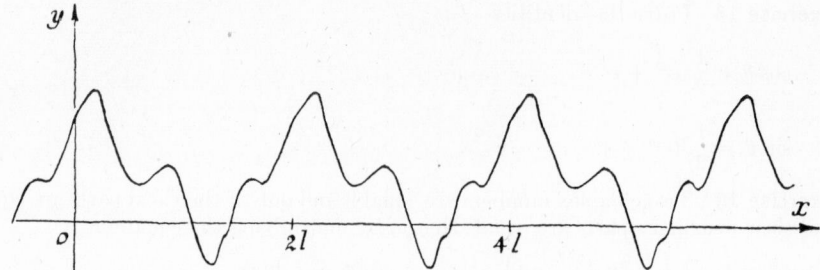

FIG. Q-3   Arbitrary periodic function.

---

period $2l$ if $f(x + 2l) = f(x)$ for every value of x. Fig. Q-3 is the graph of an arbitrary periodic function. The abscissa, x, might, *e.g.*, represent time, and the ordinate, y, stand for the displacement of some relatively complicated linear oscillator.

The simplest example of a periodic function is

$$f(x) = A_1 \cos\left(\frac{\pi x}{l} + \phi_1\right)$$

the mathematical equivalent of simple harmonic motion. Here A_1 is the amplitude, $2l$ the period, and ϕ_1 the phase angle. The curve shown in dashes in Fig. Q-4 is typical; one complete period has been graphed. The function

$$f(x) = A_2 \cos\left(\frac{2\pi x}{l} + \phi_2\right)$$

is similar to the former except that its period is l. It also repeats itself with a period $2l$, however, as can be seen from the fact that when the first curve

---

FIG. Q-4   Superposition of two sine waves.

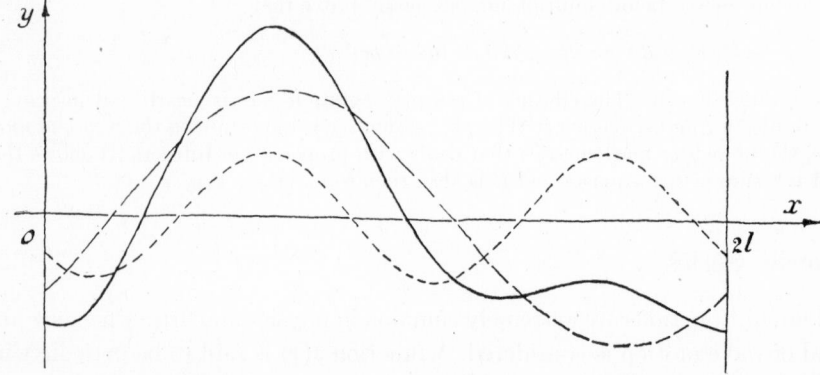

makes one complete oscillation the second makes two and is again back at its starting place (see the dotted curve in Fig. Q-4). Hence the sum of two such functions (the solid curve) is also a periodic function with period $2l$. It corresponds physically to the superposition of two vibrations, and musically to the sounding together, out of phase, of a fundamental tone and its octave, or first harmonic.

In general, then, the function

$$f(x) = A_n \cos \left(\frac{n\pi x}{l} + \phi_n \right) \quad (n = 1, 2, 3, \ldots)$$

has period $2l/n$ and therefore certainly the period $2l$ as well. By the addition formula for the cosine, this function may also be written as

$$f(x) = a_n \cos \frac{n\pi x}{l} + b_n \sin \frac{n\pi x}{l}$$

where

$$a_n = A_n \cos \phi_n \quad \text{and} \quad b_n = -A_n \sin \phi_n$$

(Notice that there were two degrees of freedom, A_n and ϕ_n, in the original representation, and there are also two degrees of freedom, a_n and b_n, in the new representation, so that nothing has been lost or gained by the transformation.) The superposition of terms of this sort,

$$\tfrac{1}{2}a_0 + a_1 \cos \frac{\pi x}{l} + b_1 \sin \frac{\pi x}{l} + a_2 \cos \frac{2\pi x}{l} + b_2 \sin \frac{2\pi x}{l} + \cdots$$

$$+ a_n \cos \frac{n\pi x}{l} + b_n \sin \frac{n\pi x}{l} + \cdots$$

will yield in general a very complicated function of x, but one which, nevertheless, is periodic with period $2l$. The constant $\tfrac{1}{2}a_0$ allows one to adjust the height of the graph and corresponds to the term $n = 0$; the reason for the factor $\tfrac{1}{2}$ will become apparent later.

Now the inverse operation to superposition, namely, analysis into component tones, is equally familiar in acoustics and optics. Indeed, the problem usually faced in practice is: given the curve of Fig. Q-3, or the solid curve of Fig. Q-4, what harmonics are present and with what amplitudes? In other words, given the function $f(x)$ which is known to have the period $2l$, what are the coefficients, a_0, a_1, b_1, a_2, b_2, etc., in the following trigonometric expansion?

$$f(x) = \tfrac{1}{2}a_0 + a_1 \cos \frac{\pi x}{l} + b_1 \sin \frac{\pi x}{l} + \cdots$$

$$+ a_n \cos \frac{n\pi x}{l} + b_n \sin \frac{n\pi x}{l} + \cdots$$

The task of determining these coefficients seems at first sight virtually hope-less, and yet it turns out to be surprisingly easy. For we have only to multiply both sides of the equation by $\cos \frac{m\pi x}{l}$ and integrate from $-l$ to l. Notice that, if $m \neq n$,

$$
\int_{-l}^{l} \cos \frac{m\pi x}{l} \cos \frac{n\pi x}{l} \, dx
$$

$$
= \frac{1}{2} \int_{-l}^{l} \left\{ \cos \frac{(m-n)\pi x}{l} + \cos \frac{(m+n)\pi x}{l} \right\} dx
$$

$$
= \frac{l}{2\pi} \left[\frac{1}{m-n} \sin \frac{(m-n)\pi x}{l} + \frac{1}{m+n} \sin \frac{(m+n)\pi x}{l} \right]_{-l}^{l} = 0
$$

whereas, if $m = n$, we have

$$
\int_{-l}^{l} \cos^2 \frac{n\pi x}{l} \, dx = \frac{1}{2} \int_{-l}^{l} \left\{ 1 + \cos \frac{2n\pi x}{l} \right\} dx = l
$$

It may similarly be shown that

$$
\int_{-l}^{l} \cos \frac{m\pi x}{l} \sin \frac{n\pi x}{l} \, dx = 0
$$

without any restriction at all on m. These are known as the *orthogonality relations* for the sine and cosine, and they show at once that

$$
a_m = \frac{1}{l} \int_{-l}^{l} \cos \frac{m\pi x}{l} f(x) \, dx
$$

since all the other terms in the expansion of $f(x)$ integrate out to zero. This formula holds good even for $m = 0$, as the student may readily check; this is why the first term was written with the factor of $\frac{1}{2}$. There is one more or-thogonality relation,

$$
\int_{-l}^{l} \sin \frac{m\pi x}{l} \sin \frac{n\pi x}{l} \, dx = 0 (m \neq n), \quad = l(m = n)
$$

and by virtue of it one finds that

$$
b_m = \frac{1}{l} \int_{-l}^{l} \sin \frac{m\pi x}{l} f(x) \, dx
$$

These formulas were derived and used extensively by Fourier in his *Analyti-cal Theory of Heat*. For this reason, such trigonometric expansions of periodic functions are called *Fourier series*. The possibility of such an expansion, and, indeed, even the formulas for the coefficients, were, however, known to Euler and the French mathematician Clairaut considerably before Fourier's time. The Fourier expansion, for all its apparent naturalness and simplicity, was an incredibly rich and far-reaching find. The mathematicians of the eighteenth century were accustomed to thinking of a function of x only as something that could be written down in a single equation. As we shall see in a moment, a Fourier series is capable of representing functions which result from "piecing together" several different equations. Thus the mathematicians had the necessity thrust upon them of re-evaluating and broadening their thinking, and the importance of this to the history of mathematics can scarcely be overestimated.

Before we proceed to examples of Fourier series, two special cases of Fourier's Theorem deserve attention. If $f(x)$ is an *even* function of x, *i.e.*, if

$$f(-x) = f(x)$$

for all x, then it is readily seen that the first half of the integral for b_m (the part from $-l$ to 0) exactly cancels the second half, so that $b_m = 0$; whereas the first half of the integral for a_m is equal to the second half. Therefore, we may expand an even function in the form

$$f(x) = \tfrac{1}{2}a_0 + a_1 \cos \frac{\pi x}{l} + a_2 \cos \frac{2\pi x}{l} + \cdots$$

where

$$a_m = \frac{2}{l} \int_0^l \cos \frac{m\pi x}{l} f(x)\, dx$$

Such an expansion is usually referred to simply as a "cosine series."

Likewise, if $f(x)$ is *odd*, *i.e.*, if

$$f(-x) = -f(-x)$$

then it may be expanded in a "sine series,"

$$f(x) = b_1 \sin \frac{\pi x}{l} + b_2 \sin \frac{2\pi x}{l} + \cdot$$

where

$$b_m = \frac{2}{l} \int_0^l \sin \frac{m\pi x}{l} f(x)\, dx$$

It is very easy to remember which functions can be expanded in a cosine series and which in a sine series. For example, since

$$\sin(-x) = -\sin x$$

the sine is itself an odd function. Thus a series of sines will be an odd, periodic function, and *vice versa*. Similarly, it is because the cosine is an even function that it may be used to expand even functions.

Let us find, now, the Fourier series for the following function, assuming that it has the period 2π:

$$f(x) = +1(0 < x < \pi), \quad = -1(-\pi < x < 0)$$

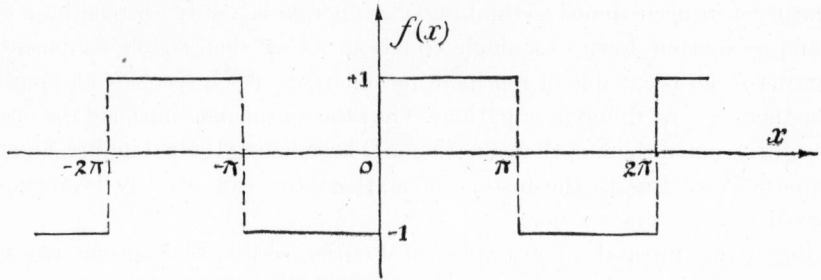

FIG. Q-5.

We find ourselves in a quandary as to what to call $f(0)$, but, as we shall see, the series settles this question for us automatically.

Clearly, $f(-x) = -f(x)$, and we may limit our attention to a sine series. Since $l = \pi$, and $f(x) = 1(0 < x < \pi)$, the formula yields at once

$$b_m = \frac{2}{\pi} \int_0^\pi \sin mx\, dx = -\frac{2}{m\pi} \cos mx \Big]_0^\pi$$

When m is odd, $\cos m\pi = \cos \pi = -1$; when m is even, $\cos m\pi = +1$; hence we find that

$$b_m = 0 \ (m \text{ even}), \quad = \frac{4}{m\pi} \ (m \text{ odd})$$

and the series for $f(x)$ is

$$f(x) = \frac{4}{\pi} \left(\sin x + \frac{1}{3} \sin 3x + \frac{1}{5} \sin 5x + \cdots \right)$$

When $x = 0$, every term in the series vanishes. Now, looking at Fig. Q-5, we see that, as x approaches 0 from the right, $f(x)$ always remains equal to $+1$; or, as it is usually written,

$$f(0+) = +1$$

Similarly, as x approaches 0 from the left, $f(x) = -1$:

$$f(0-) = -1$$

Thus we see that the value to which the series actually converges is the average,

$$\tfrac{1}{2}[f(0+) + f(0-)]$$

of the right and left limits at the discontinuity of the function being expanded. The same is true for any Fourier series.

Suppose we write $x = \tfrac{1}{2}\pi$ in the series just derived. It is readily seen that

$$\sin \tfrac{1}{2}\pi = +1, \quad \sin \tfrac{3}{2}\pi = -1, \quad \sin \tfrac{5}{2}\pi = +1, \quad \sin \tfrac{7}{2}\pi = -1$$

and so on. Also we clearly have, from Fig. Q-5,

$$f(\tfrac{1}{2}\pi) = +1$$

Therefore, the series yields

$$1 - \frac{1}{3} + \frac{1}{5} - \frac{1}{7} + - \cdots = \frac{\pi}{4}$$

This beautiful result is generally known as Gregory's Series. There are other ways of deriving it, but it is typical of Fourier series that they easily give rise to identities of this sort.

Exercise 23 Show that the function

$$f(x) = x(-\pi < x < \pi)$$

has the Fourier series

$$f(x) = 2[\sin x - \tfrac{1}{2} \sin 2x + \tfrac{1}{3} \sin 3x - + \ldots]$$

By setting $x = \tfrac{1}{3}\pi$, show that

$$1 - \frac{1}{2} + \frac{1}{4} - \frac{1}{5} + \frac{1}{7} - \frac{1}{8} + \frac{1}{10} - \frac{1}{11} + - \cdots = \frac{\pi}{3\sqrt{3}}$$

Exercise 24 The parabola

$$y = \frac{1}{\pi} x(\pi - x)$$

is tangent to the curve $y = \sin x$ at $x = 0$ and $x = \pi$. (Prove this.) How much do they differ at $x = \tfrac{1}{2}\pi$? Show that the Fourier sine series for the parabola is

$$y = \frac{8}{\pi^2}\left[\sin x + \frac{1}{3^3} \sin 3x + \frac{1}{5^3} \sin 5x + \ldots \right]$$

Setting $x = \tfrac{1}{2}\pi$, evaluate the infinite series

$$1 - \frac{1}{3^3} + \frac{1}{5^3} - \frac{1}{7^3} + - \cdots$$

Exercise 25 Letting $x = \tfrac{1}{6}\pi$ and $\tfrac{1}{3}\pi$ in the example worked out in the text, prove the identities

$$1 + \frac{1}{5} - \frac{1}{7} - \frac{1}{11} + \frac{1}{13} + \frac{1}{17} - \frac{1}{19} - + + - \cdots = \frac{\pi}{3}$$

$$1 - \frac{1}{5} + \frac{1}{7} - \frac{1}{11} + \frac{1}{13} - + \cdots = \frac{\pi}{2\sqrt{3}}$$

Exercise 26 A sine-wave with the negative lobes missing may be analyzed by expanding the function

$$f(x) = \cos x \left(0 < x < \frac{\pi}{2}\right), \quad = 0 \left(\frac{\pi}{2} < x < \pi\right)$$

in a cosine series. (Describe the resulting graph, to make sure.) Show that this expansion is

$$f(x) = \frac{1}{\pi} + \frac{1}{2}\cos x + \frac{2}{\pi}\left[\frac{1}{2^2-1}\cos 2x - \frac{1}{4^2-1}\cos 4x + \frac{1}{6^2-1}\cos 6x - +\ldots\right]$$

Exercise 27 A sine-wave with the negative lobes inverted may be analyzed by expanding the function

$$f(x) = \sin x \ (0 < x < \pi)$$

in a cosine series. (Why?) Show that the expansion is

$$f(x) = \frac{2}{\pi} - \frac{4}{\pi}\left[\frac{1}{2^2-1}\cos 2x + \frac{1}{4^2-1}\cos 4x + \frac{1}{6^2-1}\cos 6x + \ldots\right]$$

Exercise 28 Assuming that α is *not* an integer, expand the function

$$f(x) = \sin \alpha x \ (-\pi < x < \pi)$$

in a Fourier series and thereby show that

$$\frac{\pi}{2}\cdot\frac{\sin \alpha x}{\sin \alpha\pi} = \frac{\sin x}{1^2-\alpha^2} - \frac{2\sin 2x}{2^2-\alpha^2} + \frac{3\sin 3x}{3^2-\alpha^2} - +\ldots$$

Setting $x = \frac{1}{2}\alpha$, derive the "partial fractions expansion,"

$$\frac{\pi}{4}\sec\frac{\pi\alpha}{2} = \frac{1}{1^2-\alpha^2} - \frac{3}{3^2-\alpha^2} + \frac{5}{5^2-\alpha^2} - +\ldots$$

Exercise 29 Expand $f(x) = \cos \alpha x$ similarly and find the partial fractions expansion of cotan $\pi\alpha$.

Exercise 30 Prove the orthogonality relationships stated without proof in the text.

Exercise 31 Suppose that, when $-l < x < l$, two arbitrary functions, $f(x)$ and $F(x)$, have Fourier expansions

$$f(x) = \frac{1}{2}a_0 + a_1\cos\frac{\pi x}{l} + b_1\sin\frac{\pi x}{l} + \ldots$$

$$F(x) = \frac{1}{2}A_0 + A_1\cos\frac{\pi x}{l} + B_1\sin\frac{\pi x}{l} + \ldots$$

To multiply these two together, every term in the first must be multiplied against every term in the second, and all the products added. Show that, if the result is integrated between $-l$ and $+l$, "most" of the terms vanish on account of the orthogonality relationships, and those that are left yield Parseval's Theorem,

$$\frac{1}{l}\int_{-l}^{l} f(x)F(x)\,dx = \frac{1}{2}a_0A_0 + a_1A_1 + b_1B_1 + a_2A_2 + b_2B_2 + \ldots$$

Exercise 32 Deduce, as a special case of the previous exercise, that

$$\frac{1}{l}\int_{-l}^{l} [f(x)]^2\,dx = \frac{1}{2}a_0^2 + a_1^2 + b_1^2 + a_2^2 + b_2^2 + \ldots$$

Applying this to Exercise 22, prove that

$$1 + \frac{1}{2^2} + \frac{1}{3^2} + \frac{1}{4^2} + \ldots = \frac{\pi^2}{6}$$

Our discussion has been of a purely *formal* nature and has ignored the many subtle issues involved in the theory of Fourier series. There is, however, one point which should be cleared up. We have produced a formula for computing coefficients, a_m and b_m, in a certain infinite series, but we have no guarantee that the resulting series converges. We should like to be able to say that, owing to some property or other of the function $f(x)$ being expanded, the resulting Fourier series must converge for every value of x and, furthermore, converge to the corresponding value of $f(x)$. Fortunately, there *is* such a property, and most of the functions that arise in practice possess it.

A function is said to be *piecewise continuous* in an interval if it is continuous throughout the interval except for possibly a finite number of finite jumps. If its derivative is also piecewise continuous in the interval, the function is called "piecewise smooth" in the interval. This is the desired property. Piecewise smooth functions have convergent Fourier series, and the series is always equal to the functional value.

Although the above property is *sufficient*, it is by no means *necessary*, for functions can be exhibited which are not piecewise smooth, but still have decent Fourier series, *i.e.*, series which converge and yield the appropriate answer. On the other hand, necessary conditions are known, but they are also known not to be sufficient, for examples of functions satisfying them but not possessing decent Fourier series have been constructed. At present, mathematicians do not know (although they would greatly like to) any condition which is both necessary and sufficient for a function to possess a decent Fourier series. Such a condition would be completely equivalent to the property of possessing a decent Fourier series. Since no such condition is known, all we can say at present is that the property of possessing a decent Fourier series seems to be separate and distinct from any other known properties of functions.

Exercise 33 Show that, if a function $f(x)$ of period $2l$ is expanded in a series,

$$f(x) = \sum_{-\infty}^{\infty} a_n e^{n\pi jx/l}$$

(Σ is capital Greek sigma and stands for a summation taken over the indicated range of values of n), then the formula for the coefficients is

$$a_n = \frac{1}{2l} \int_{-l}^{+l} f(x) e^{-n\pi jx/l}\, dx \quad (n = 0, \pm 1, \pm 2, \ldots)$$

From this result deduce the familiar sine-cosine expansion by pairing off certain terms in the series.

Exercise 34 Let

$$s_n(x) = \sin x - \tfrac{1}{2} \sin 2x + -\ldots + (-1)^{n-1} \frac{1}{n} \sin nx$$

[*Cf.* Exercise 23; a sum like $s_n(x)$ is often called a *partial sum* of the infinite series in question.] Plot s_1, s_2, s_3, and s_4 in different colors on the same sheet of graph paper. (Since they are odd functions, it is enough to show only the portion between $0°$ and $180°$. Taking x in steps of $15°$ and working with three or four decimals is sufficient.) Also plot on the same graph the line $y = \frac{1}{2}x$. (Note that x is measured in radians here.) Using Exercise 21, show that

$$s_n'(x) = \frac{\cos \dfrac{n+1}{2}(x - \pi) \sin \dfrac{n}{2}(x - \pi)}{\cos \frac{1}{2}x}$$

Show that there are always n extrema of $s_n(x)$ in the interval $0 \leqslant x \leqslant \pi$. What is $s_n'(\pi)$? $s_n'(0)$? What is the difference between odd n and even n in the ways the curves leave the origin? How can increasing n make the partial sums approximate better to the straight line when they always start from the origin in the wrong direction?

Exercise 35   Show that at its maximum at $x = \dfrac{n}{n+1}\pi$

$$s_n(x) = \sin \frac{\pi}{n+1} + \frac{1}{2}\sin \frac{2\pi}{n+1} + \ldots + \frac{1}{n}\sin \frac{n\pi}{n+1}$$

and, invoking the definition of an integral, show that this approaches the limiting value

$$\int_0^\pi \frac{\sin x}{x}\,dx$$

as $n \to \infty$. Sketch the graph of the integrand and note its behavior at $x = 0$. The value of the integral is about 1.852, which is 18% larger than $\pi/2$. The fact that the partial sums overshoot the top of the jump and undershoot the bottom is known as Gibbs' Phenomenon, in honor of J. Willard Gibbs (1839–1903), the famous American scientist who was one of the first to notice it. It is typical (even to the factor of 18%) of the behavior of a Fourier series at a discontinuity.

There is a great body of ideas closely related to the concept of a Fourier series. One of the most important of these is the so-called *Fourier integral*. The Fourier series may be said to be *discrete*, and the Fourier integral its *continuous* analogue, in much the same way that a spectrum may consist either of a series of discrete lines or of a continuum. The meaning of this will become clear in the following discussion.

We saw, in Exercise 33, that a function $f(x)$, of period $2l$, may be expanded in a series of exponentials,

$$f(x) = \sum_{-\infty}^{\infty} a_n e^{n\pi jx/l}$$

where

$$a_n = \frac{1}{2l}\int_{-l}^{+l} f(x)e^{-n\pi jx/l}$$

The restriction of periodicity may be lifted by letting $l \to \infty$; any aperiodic function may be thought of as a periodic function of infinite period. The mathematics involved in the passage to the limit is relatively straightforward. The final result is the representation

$$f(x) = \int_{-\infty}^{\infty} F(u)e^{jxu} \, du$$

where

$$F(u) = \frac{1}{2\pi} \int_{-\infty}^{\infty} f(x)e^{-jxu} \, dx$$

Notice that the Fourier *series* tells how to break $f(x)$ down into its harmonics, $e^{n\pi jx/l}$, and the formula for a_n tells what the amplitude of the nth harmonic is. These harmonics go by steps of π/l and, when $l \to \infty$, the steps get smaller and smaller and the arguments of the harmonics go over into the continuous variable, u, of the Fourier *integral* representation. The Fourier integral is thus indeed the continuous analogue of the Fourier series. The Fourier integral tells how to break down any function into its harmonics, e^{jxu}, and the formula for $F(u)$ tells what the amplitude of the uth harmonic is.

The functions $f(x)$ and $F(u)$ are said to be *Fourier transforms* of one another. It is an unexpected and striking development that, aside from the factor of 2π and the differing signs in the exponential, the two equations are quite symmetric in u and x. Some authors strive for still more symmetry by inserting a factor of $1/\sqrt{2\pi}$ in front of *both* integrals. This amounts, of course, to a slight redefinition of what is meant by a Fourier transform.

Wave trains afford the most graphic means of picturing Fourier transforms. The actual physical amplitude is a function of time, $A(t)$. The so-called spectral amplitude is a function of angular frequency, $f(\omega)$. The intensity is given by

$$I(\omega) = |f(\omega)|^2 + |f(-\omega)|^2$$

FIG. Q-6   Gaussian pulse and its amplitude spectrum.

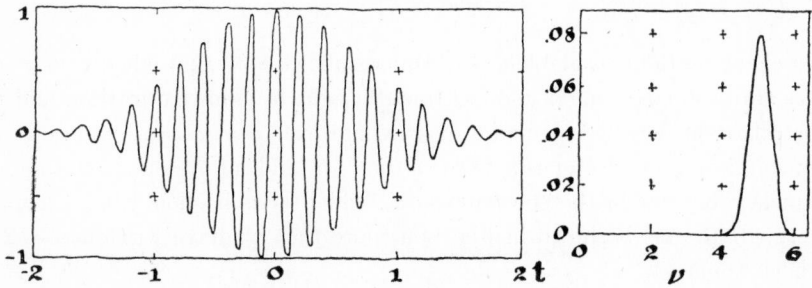

The physical and spectral amplitudes are Fourier transforms of each other:

$$A(t) = \int_{-\infty}^{\infty} f(\omega)e^{j\omega t}\, d\omega$$

$$f(\omega) = \frac{1}{2\pi} \int_{-\infty}^{\infty} A(t)e^{-j\omega t}\, dt$$

E.g., suppose we have a *Gaussian pulse*, with (see Fig. Q-6)

$$A(t) = e^{j\omega_0 t} e^{-t^2/\tau^2}$$

Here, small τ means a narrow pulse and large τ a broad pulse. The spectral amplitude is given by

$$2\pi f(\omega) = \int_{-\infty}^{\infty} e^{-\left[\frac{t^2}{\tau^2} + j(\omega - \omega_0)t\right]}\, dt$$

$$= \int_{-\infty}^{\infty} e^{-\left[\frac{t}{\tau} + \frac{1}{2}j\tau(\omega - \omega_0)\right]^2 - \frac{1}{4}\tau^2(\omega - \omega_0)^2}\, dt$$

upon making use of the familiar algebraic device of completing the square. Introducing a new variable,

$$y = \frac{t}{\tau} + \frac{1}{2}j\tau(\omega - \omega_0), \qquad dt = \tau\, dy$$

the integral becomes

$$\tau e^{-\frac{1}{4}\tau^2(\omega - \omega_0)^2} \int_{-\infty}^{\infty} e^{-y^2}\, dy$$

(Actually, y should run from $\frac{1}{2}j\tau(\omega - \omega_0) - \infty$ to $\frac{1}{2}j\tau(\omega - \omega_0) + \infty$, and a proper justification for replacing this by an integration along the real axis, as we have done, requires arguments from the theory of functions of a complex variable which are outside the scope of this discussion.) This last integral is familiar from statistics; its value is $\sqrt{\pi}$. Hence,

$$f(\omega) = \frac{\tau}{2\sqrt{\pi}} e^{-\frac{1}{4}(\omega - \omega_0)^2 \tau^2}$$

and

$$I(\omega) = \frac{\tau^2}{4\pi} \left[e^{-\frac{1}{2}(\omega - \omega_0)^2 \tau^2} + e^{-\frac{1}{2}(\omega + \omega_0)^2 \tau^2} \right]$$

As a check on the reasonableness of this result we note that when $\tau \to \infty$ the wave train becomes one of monochromatic light of infinite duration, and the spectral frequency becomes zero everywhere except at $\omega = \omega_0$, where it becomes infinite. The latter is a Dirac δ-function and informs us that the only frequency present in the spectrum is ω_0, which is as it should be. A typical intermediate case is shown in Fig. Q-6 (note that we have written $\omega = 2\pi\nu$, as is customary).

Exercise 36 Calculate the distribution of intensity of frequencies for a damped wave train where

$$A(t) = e^{j\omega_0 t}e^{-t/\tau} \quad (0 \leqslant t < \infty)$$

Discuss the behavior of the solution for large and small τ. Graph a typical example.

Exercise 37 Do the same for a monochromatic train of finite duration:

$$A(t) = e^{j\omega_0 t} \quad (0 \leqslant t \leqslant 1), \quad = 0 \text{ (otherwise)}$$

Note the interesting fact that even monochromatic light must be thought of as consisting of a distribution of frequencies unless it is of infinite duration.

Exercise 38 Assuming that $f(x)$ is an even function, show that its Fourier transform is given by

$$F(u) = \frac{1}{\pi}\int_0^\infty f(x) \cos xu \, dx$$

Hence show that $F(u)$ is also even and that therefore

$$f(x) = 2\int_0^\infty F(u) \cos xu \, du$$

Write $F_c(u) = 2F(u)$. This is called the *Fourier cosine transform* of $f(x)$. What are the two equations relating $f(x)$ and $F_c(u)$? Assuming that $f(x)$ is an odd function, define its *Fourier sine-transform* analogously, and find the corresponding pair of equations in this case.

Exercise 39 Find the Fourier sine- and cosine-transforms of the following functions: e^{-x}, xe^{-x}, and $f''(x)$ [assuming that the transform of $f(x)$ is already known].

Vector Analysis

Numbers answer the question "How much?" but when we ask "How much and in what direction?" the answer is given by a *vector*. The statement that I ate two apples provides a complete quantitative description; but the statement that I walked two miles becomes complete only when I specify that I was heading east, say. Vectors supply the geometric context of measurements, and this accounts for their special usefulness in physics.

Now, merely defining numbers is of no use until we have also defined the operations of arithmetic, which permit us to calculate with them. By the same token it is not enough merely to define vectors as "directed numbers"; we must also define appropriate *operations* on them. As we shall see, there is an unexpectedly rich and varied assortment of vector operations, making them a particularly powerful tool for the physicist.

A vector may be drawn as an arrow with a given length and direction, but no absolute location in space. Thus it tells me that I walked two miles east, but not where I started from. Hence any two vectors which are equal in length and parallel may be considered identical. We may remove the redundancy by

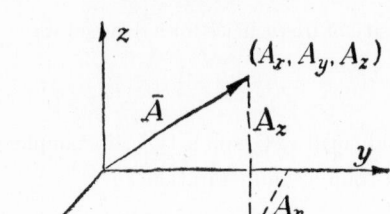

FIG. Q-7   Three-dimensional vector.

picturing our vectors as all having their tails tacked down at the origin; then there is only one which has length 2 and points to the east. (Technically, such vectors are called "bound" vectors, to differentiate them from "free" vectors, which may originate from any point in space; but the distinction is primarily of academic interest.) Actually, the vectors we shall be considering will all be three-dimensional—or, as a particular case of this, two-dimensional—and may therefore be described by three numbers, *e.g.*, length, azimuth, and elevation.

Now, since the tail is pinned down, we can equally well specify the vector $\bar{A}$ by giving the Cartesian coordinates (A_x, A_y, A_z) of the head (see Fig. Q-7). These are called the *components* of the vector. When $A_z = 0$, the vector lies in the (x,y)-plane; if we are thinking of it as a two-dimensional vector, we write it simply $\bar{A} = (A_x, A_y)$, but if we mean to consider it in three-space we write it $\bar{A} = (A_x, A_y, 0)$. Any vector equation may be reduced ultimately to three regular equations in the components, and any vector operation may be described in terms of arithmetic operations on the components.

A vector is like a "three-dimensional number," and an ordinary number is the same as a "one-dimensional vector." Ordinary numbers are referred to as scalars when vectors are under discussion. The simplest vector operation is multiplication by a scalar. If $\bar{A}$ is a vector and c is a positive scalar, we *define* $c\bar{A}$ as the vector with the same direction as $\bar{A}$ and c times as long. If c is negative, then, of course, the direction of the vector must be reversed. In particular, when $c = -1$, we have the vector $-\bar{A}$, which has the same length

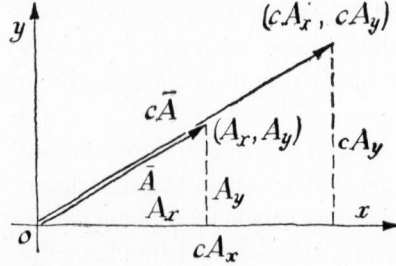

FIG. Q-8   Multiplication by a scalar.

FIG. Q-9 Vector addition.

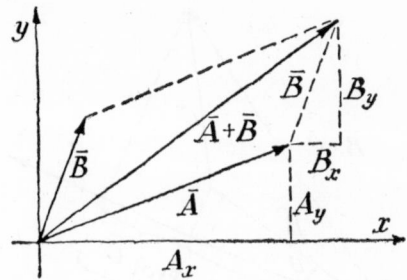

as $\bar{A}$, but points in the opposite direction. It is clear that the components of $c\bar{A}$ are (cA_x,cA_y,cA_z); see Fig. Q-8, where this is shown in two dimensions.

If I walk one mile east and then one mile north, I wind up 1.414 miles northeast of where I started. This suggests the following rule for the *addition* of vectors (see Fig. Q-9). To add $\bar{A}$ and $\bar{B}$, translate $\bar{B}$ parallel to itself so that its tail coincides with the head of $\bar{A}$; then $\bar{A} + \bar{B}$ is the vector with its tail at $\bar{A}$'s tail and its head at $\bar{B}$'s head. This is often called the *parallelogram law*, for if $\bar{A}$ and $\bar{B}$ are thought of as adjacent sides of a parallelogram, then $\bar{A} + \bar{B}$ is the diagonal from their common vertex. It is obvious (Fig. Q-9) that we may equally well add vectors by adding their corresponding components, i.e.,

$$(A_x,A_y) + (B_x,B_y) = (A_x + B_x, A_y + B_y)$$

A vector of unit length is called a *unit vector*. The unit vectors in the x-, y-, and z-directions are designated as **i**, **j**, and **k**, respectively. We thus have

$$\mathbf{i} = (1,0,0) \qquad \mathbf{j} = (0,1,0) \qquad \mathbf{k} = (0,0,1)$$

By application of the rule for multiplication of a vector by a scalar and the rule for addition of two vectors, we find at once that

$$\bar{A} = A_x\mathbf{i} + A_y\mathbf{j} + A_z\mathbf{k}$$

This is also clear geometrically.

Exercise 40 Prove the last two statements.

Exercise 41 Which of the following are vectors and which scalars? Velocity, acceleration, mass, force, momentum, energy, electric intensity, magnetic intensity, wavelength, time, density.

Exercise 42 What is the length of a vector in terms of its components? What is the length of the sum of two vectors in terms of their separate lengths and the angle between them?

Exercise 43 Show geometrically, and by examining the components, that $\bar{A} + \bar{B}$ is the same vector as $\bar{B} + \bar{A}$. (*I.e.*, vector addition is "commutative.") In the same two ways, show that if $\bar{C}$ is added to $\bar{A} + \bar{B}$, the result is the same vector as when $\bar{B} + \bar{C}$ is added to $\bar{A}$. (Vector addition is "associative.") Finally, in like fashion show that if the vector sum of $\bar{A}$ and $\bar{B}$ is multiplied by a scalar c, the result is $c\bar{A} + c\bar{B}$. (Multiplication by a scalar is "distributive.")

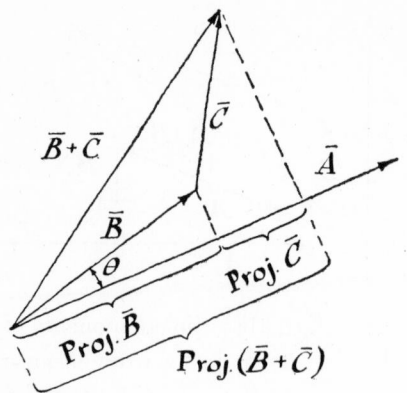

FIG. Q-10   Proof of distributive law for dot product.

Suppose that two vectors, $\bar{A}$ and $\bar{B}$, of length A and B respectively, make an angle θ with one another. Then, their *dot product*, $\bar{A} \cdot \bar{B}$, is defined as $AB \cos \theta$. Contrary to the operations discussed above, which resulted in a vector, the dot product gives rise to a scalar. Clearly, $\bar{A} \cdot \bar{B} = \bar{B} \cdot \bar{A}$, so that the operation is commutative. Furthermore, $B \cos \theta$ equals the projection of $\bar{B}$ upon $\bar{A}$. Now it is apparent that the projection of $\bar{B} + \bar{C}$ upon $\bar{A}$ is the sum of the projection of $\bar{B}$ on $\bar{A}$ and the projection of $\bar{C}$ on $\bar{A}$ (*cf.* Fig. Q-10). But A times the projection of $\bar{B}$ on $\bar{A}$ equals A times $B \cos \theta$, which equals $\bar{A} \cdot \bar{B}$, by definition. Similarly for A times the projection of $\bar{C}$ as well as A times the projection of $\bar{B} + \bar{C}$. Therefore

$$\bar{A} \cdot (\bar{B} + \bar{C}) = \bar{A} \cdot \bar{B} + \bar{A} \cdot \bar{C}$$

I.e., the dot product is distributive. The dot product is sometimes also called the "inner product" or "scalar product."

Exercise 44   Show that $\bar{A} \cdot \bar{A} = A^2$, where A is the length of $\bar{A}$.

Exercise 45   What does it mean if the dot product vanishes and neither $\bar{A}$ nor $\bar{B}$ is a *null vector* (vector of length zero)?

Exercise 46   Show that $i \cdot i = j \cdot j = k \cdot k = 1$ and $i \cdot j = j \cdot k = k \cdot i = 0$. By repeated application of the distributive law show that $(\bar{A} + \bar{B}) \cdot (\bar{C} + \bar{D}) = \bar{A} \cdot \bar{C} + \bar{A} \cdot \bar{D} + \bar{B} \cdot \bar{C} + \bar{B} \cdot \bar{D}$. How does this generalize to more terms? Using the results proved here and in Exercise 40, prove that $\bar{A} \cdot \bar{B} = A_x B_x + A_y B_y + A_z B_z$. Compare with Exercises 42 and 44.

Exercise 47   What is the dot product of force by the displacement through which it acts?

Exercise 48   Compare $\bar{A}(\bar{B} \cdot \bar{C})$ with $(\bar{A} \cdot \bar{B})\bar{C}$. What is the easiest way to see that they are different?

The *cross product*, $\bar{A} \times \bar{B}$, of two vectors, unlike the dot product, is a vector. Its magnitude is, using the same notation as before, $AB \sin \theta$, which is readily recognized as the area of the parallelogram which has $\bar{A}$ and $\bar{B}$ as concurrent sides. Its direction is perpendicular to the common plane of $\bar{A}$ and $\bar{B}$, and so

FIG. Q-11 Cross product of two vectors.

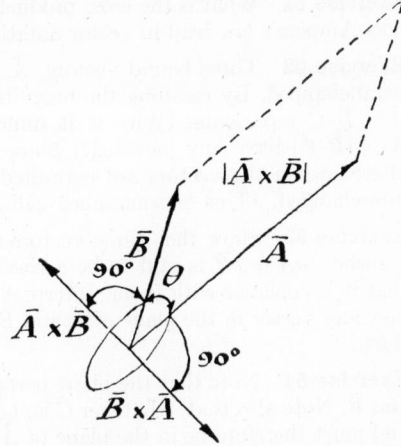

oriented that a right-hand screw would advance along it if the screw were rotated *from Ā to B̄* (Fig. Q-11). Hence interchanging $\bar{A}$ and $\bar{B}$ has the effect of reversing the direction of the cross product, and we have

$$\bar{A} \times \bar{B} = -\bar{B} \times \bar{A}$$

The cross product is also referred to as the "outer product" or "vector product."

The real wonder is not that the cross product fails to commute, but that people are as perennially surprised by it as they are. It is the rule rather than the exception for the outcome of several operations to be sensitive to the order in which they are performed; *e.g.*, your mother's father is surely not the same person as your father's mother.

It may be shown that, like the dot product and the scalar product, the cross product obeys the distributive laws, *i.e.*,

$$\bar{A} \times (\bar{B} + \bar{C}) = \bar{A} \times \bar{B} + \bar{A} \times \bar{C}$$

and

$$(\bar{B} + \bar{C}) \times \bar{A} = \bar{B} \times \bar{A} + \bar{C} \times \bar{A}$$

The proof is a little involved, and we omit it.

Exercise 49 What does it mean if the cross product is a null vector but neither $\bar{A}$ nor $\bar{B}$ is a null vector?

Exercise 50 Show that $\mathbf{i} \times \mathbf{i} = \mathbf{j} \times \mathbf{j} = \mathbf{k} \times \mathbf{k} = 0$ and that $\mathbf{i} \times \mathbf{j} = \mathbf{k}$, $\mathbf{j} \times \mathbf{k} = \mathbf{i}$, and $\mathbf{k} \times \mathbf{i} = \mathbf{j}$. (Note that the last three equations, *e.g.*, may be obtained from one another by *cyclic permutations* of $\mathbf{i}$, $\mathbf{j}$, and $\mathbf{k}$, *i.e.*, by replacing $\mathbf{i}$ by $\mathbf{j}$, $\mathbf{j}$ by $\mathbf{k}$, and $\mathbf{k}$ by $\mathbf{i}$. This is typical of all vector identities.) Proceed as in Exercise 46 and obtain the result

$$\bar{A} \times \bar{B} = (A_y B_z - A_z B_y)\mathbf{i} + (A_z B_x - A_x B_z)\mathbf{j} + (A_x B_y - A_y B_x)\mathbf{k}$$

which expresses the cross product specifically in terms of its components. (Note again how x, y, and z permute cyclically between the three components.)

Exercise 51 What is the cross product of distance (lever arm) by momentum? How does Ampère's law read in vector notation?

Exercise 52 Three bound vectors, $\bar{A}$, $\bar{B}$, and $\bar{C}$, may be considered the edges of a parallelepiped. By recalling the magnitude and direction of $\bar{A} \times \bar{B}$, determine what $\bar{A} \times \bar{B} \cdot \bar{C}$ represents. (Why is it unnecessary to write this as $(\bar{A} \times \bar{B}) \cdot \bar{C}$? Does $\bar{A} \times (\bar{B} \cdot \bar{C})$ have any meaning?) Show from this result that $A \times B \cdot C$ remains unaltered when the vectors are permuted cyclically, and when the dot and cross are interchanged. (This is sometimes called the *box product*, or *triple scalar product*.)

Exercise 53 Show that three vectors are coplanar if and only if their box product vanishes. $a\bar{A} + b\bar{B}$ is said to be a *linear combination* of the vectors $\bar{A}$ and $\bar{B}$. Show that it is coplanar with them. Interpret geometrically, and show (also geometrically) how any vector in the plane of $\bar{A}$ and $\bar{B}$ may be expressed as a linear combination of them.

Exercise 54 Note that the plane perpendicular to $\bar{A} \times \bar{B}$ is the plane containing $\bar{A}$ and $\bar{B}$. Note also that, whatever $\tilde{C}$ is, $(\bar{A} \times \bar{B}) \times \tilde{C}$ must be perpendicular to $\bar{A} \times \bar{B}$ and must therefore lie in the plane of $\bar{A}$ and $\bar{B}$. Hence show that $(\bar{A} \times \bar{B}) \times \tilde{C}$ must be of the form $a\bar{A} + b\bar{B}$, where a and b are scalars. In fact,

$$(\bar{A} \times \bar{B}) \times \tilde{C} = (\bar{A} \cdot \tilde{C})\bar{B} - (\bar{B} \cdot \tilde{C})\bar{A}$$

Prove this by computing the x-component on each side; if these are equal, so are the y- and z-components, for they are cyclic permutations of it. (Draw a diagram.)

The position of a mass point may be specified by a vector drawn to that point from some arbitrary origin. Such a vector is traditionally called a *radius vector* and denoted by $\bar{r}$ in mechanics. If the particle is moving along some trajectory, the radius vector is a function of the time in the usual sense that, once the time is known, the position is uniquely given in terms of it. We write

$$\bar{r} = \bar{r}(t)$$

This is a typical example of a vector ($\bar{r}$) which is a function of a scalar (t). Now we are accustomed to think of velocity as the time derivative of position. Can this definition be made to carry over to the present case, where the posi-

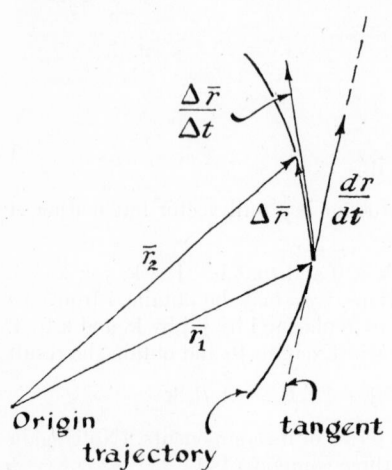

FIG. Q-12 Differentiating a vector.

tion is a vector? To do so requires being able to differentiate a vector with respect to a scalar. Does this make sense?

Fig. Q-12 shows two positions of the particle which are separated by a time interval, Δt. The derivative should be the limiting value of the change in $\bar{r}$, divided by the change in t:

$$\frac{d\bar{r}}{dt} = \lim_{t_2 \to t_1} \frac{\bar{r}_2 - \bar{r}_1}{t_2 - t_1} = \lim_{\Delta t \to 0} \frac{\Delta \bar{r}}{\Delta t}$$

But $\Delta \bar{r}$ is the difference of two vectors, which, as we have seen, is a vector; and Δt is a scalar. Also, dividing a vector by a scalar means multiplying by the reciprocal of the scalar, and we know that the result of such a scalar multiplication is a vector. Hence, in the limit, we find that the derivative of a vector is itself a vector. It is not hard to show (in fact, the geometry suggests so forcibly) that the derivative is always tangent to the trajectory and equal in magnitude to ds/dt, the speed at which the mass point is tracing out the arc length along the trajectory.

Having defined the derivative of a vector, it is natural to ask how the various vector operations we are acquainted with behave under differentiation. Suppose, *e.g.*, that we have two vectors, $\bar{A}(u)$ and $\bar{B}(u)$, which are functions of a scalar, u; can we write down a simple expression for the derivative of their cross product? Well, the change in $\bar{A} \times \bar{B}$ is given by

$$(\bar{A} + \Delta\bar{A}) \times (\bar{B} + \Delta\bar{B}) - \bar{A} \times \bar{B} = \bar{A} \times \bar{B} + \bar{A} \times (\Delta\bar{B})$$
$$+ (\Delta\bar{A}) \times \bar{B} + (\Delta\bar{A}) \times (\Delta\bar{B}) - \bar{A} \times \bar{B}$$
$$= \bar{A} \times (\Delta\bar{B}) + (\Delta\bar{A}) \times \bar{B} + \text{second-order differential}$$

This is a direct application of the distributive law for the cross product. Dividing this equation by Δu, we find that

$$\frac{d}{du}(\bar{A} \times \bar{B}) = \bar{A} \times \frac{d\bar{B}}{du} + \frac{d\bar{A}}{du} \times \bar{B}$$

In other words, the cross product differentiates in exactly the same way as the product of two scalar functions. It is important to observe that the *order* of the two factors must be maintained in performing the differentiation of the cross product; *i.e.*, in the above identity A stands before B in the expressions occurring on both sides of the equality sign. An interchange of order anywhere would, of course, introduce a minus sign. We may write down the same proof, word for word, merely replacing the cross by a dot, to show that the dot product also differentiates according to the familiar rule, and likewise for the scalar product.

The most elegant and satisfactory treatment of Newtonian mechanics is by way of vectors. Once the equations of motion are set down in vector notation, the important relationships drop out almost automatically. Unfortunately, there is no room in a brief introduction like this for a fuller discussion, and the student is referred to the standard texts for the details.

Exercise 55 Show from Fig. Q-12 that $\Delta\bar{r} = \Delta s \overline{T}$, where Δs is the change in arc length along the trajectory and $\overline{T}$ is a unit vector in the direction of the tangent. Hence prove that $\dfrac{d\bar{r}}{dt} = \dfrac{ds}{dt}\,\overline{T}$, as stated in the text.

Exercise 56 Carry through the proofs of the laws for differentiating the dot and scalar products alluded to above. Upon what, precisely, do they hinge?

Exercise 57 The radius vector of a particle lying always on a circle of radius r satisfies the vector equation

$$\bar{r}\cdot\bar{r} = r^2$$

Prove this. Show that therefore

$$\bar{r}\cdot\dot{\bar{r}} = 0$$

I.e., the velocity vector stands in what relationship to the position vector? Show that r may be written

$$\bar{r} = r\cos\theta\mathbf{i} + r\sin\theta\mathbf{j}$$

Calculate $\dot{\bar{r}}$ and show that it satisfies the last equation. Assuming that the angular velocity, $\dot{\theta}$, is constant, calculate $\ddot{\bar{r}}$ and show that the acceleration is *centripetal* (directed along the inward normal).

Exercise 58 A *helix* is a space curve having the shape of a spring (or screw thread). Show that the vector equation of a mass point describing a helix with constant velocity is

$$\bar{r} = r\cos(at)\mathbf{i} + r\sin(at)\mathbf{j} + bt\mathbf{k}$$

Calculate the velocity and acceleration.

Exercise 59 Under the influence of external forces, a particle describes a certain trajectory. A switch is thrown, cutting off the external field. How does the particle move? What is its velocity vector? What does the field supply to the particle when it is turned on?

Let us take a vacation from vectors for the moment, and consider a function of two independent variables, x and y. Call it $f(x, y)$. One way of depicting it would be to draw a locus in the x,y-plane of all those points for which $f(x,y)$

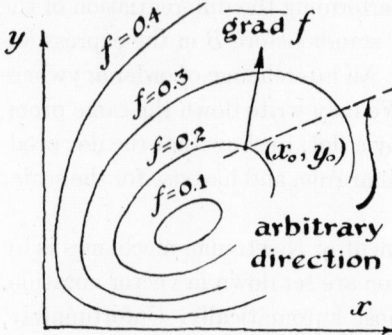

FIG. Q-13 Level lines and gradient of a function.

is equal to some constant. By taking different values of the constant, we would obtain a series of loci, and each locus might be labeled with the appropriate value of f. Such a series of loci is called a *relief map*; the term is borrowed from geography, where the various loci represent places of equal elevation.

Let us consider now an arbitrary point, (x_0, y_0). It will lie on one of these loci (see Fig. Q-13). Draw any straight line through the point. Along it the function, f, will assume a series of values. We could plot f as a function of position along the line, and therefore we could find the rate of increase (derivative) of f at the point (x_0, y_0), in the direction of this line. Consider now a "near-by" locus of constant f differing only slightly from the value of f at (x_0, y_0). If we look at a small region around the point (x_0, y_0), these two loci will be roughly parallel. Hence the normal to the locus through (x_0, y_0) will be the shortest distance between the two loci. Thus f increases most rapidly along the normal, where it accomplishes the increase in value from one locus to the next in the shortest possible distance. If we mark off along the normal a length equal to the derivative of f in that direction, we obtain a vector (our vacation was certainly brief!) which is called the *gradient* of f at the point (x_0, y_0). The gradient of a function of any number of variables is defined in a completely analogous fashion.

Now a derivative in a fixed direction is known in the calculus as a *directional derivative*. Let us consider the directional derivative in an arbitrary direction, making an angle θ with the gradient (Fig. Q-14, an enlargement of part of Fig. Q-13, shows this). Along the gradient, the distance between the two near-by loci is Δs, say; but along the other direction the distance is $\Delta s / \cos \theta$ (remember that the loci are almost parallel). Since in both cases we are moving from one locus of constant f to another, Δf is the same for both. Thus the derivative in an arbitrary direction is

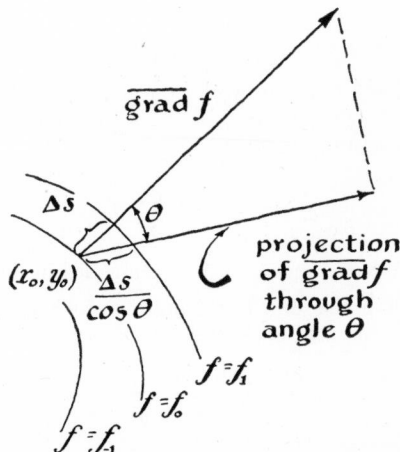

FIG. Q-14 Finding rate of increase in arbitrary direction.

$$\frac{\Delta f}{\Delta s/\cos\theta} = \cos\theta\,\frac{\Delta f}{\Delta s} = \cos\theta\,|\,\overline{\text{grad}}\,f\,|$$

In other words, it is the projection of the gradient in that direction; or, if **u** represents a unit vector in the given direction, the partial derivative in that direction equals $\mathbf{u}\cdot\overline{\text{grad}}\,f$.

The derivatives in the directions of the coordinate axes are customarily designated as $\dfrac{\partial f}{\partial x}$, $\dfrac{\partial f}{\partial y}$, etc. Hence a particular consequence of what we have just shown is that the projection of the gradient in the x-direction is $\dfrac{\partial f}{\partial x}$, its projection in the y-direction is $\dfrac{\partial f}{\partial y}$, etc. But these projections are, by definition, its components. Therefore,

$$\overline{\text{grad}}\,f = \frac{\partial f}{\partial x}\mathbf{i} + \frac{\partial f}{\partial y}\mathbf{j}$$

in two dimensions, while in three dimensions

$$\overline{\text{grad}}\,f = \frac{\partial f}{\partial x}\mathbf{i} + \frac{\partial f}{\partial y}\mathbf{j} + \frac{\partial f}{\partial z}\mathbf{k}$$

In one dimension the gradient reduces to the derivative.

Exercise 60   Consider $f(x, y) = \dfrac{x^2}{a^2} + \dfrac{y^2}{b^2}$. What are the loci of constant f? Sketch out a few (say, $f = 0, 1, 4,$ and 9). What are $\dfrac{\partial f}{\partial x}$, $\dfrac{\partial f}{\partial y}$, and $\overline{\text{grad}}\,f$? Draw the last to scale for a point on each of the loci.

Exercise 61   We say that we have a *force field* if at every point of the space a unit test body would experience a certain force if placed there. This force, $\overline{F}$, is said to be derivable from a potential, V (which is a scalar function of position), if

$$\overline{F} = \overline{\text{grad}}\,V$$

Note that the work done in moving this test body a distance dr is given by $\overline{F}\cdot\overline{dr}$. (Why?) But $\overline{dr} = dx\mathbf{i} + dy\mathbf{j} + dz\mathbf{k}$. (Explain.) Show therefore that the work done is equal to

$$\frac{\partial V}{\partial x}\,dx + \frac{\partial V}{\partial y}\,dy + \frac{\partial V}{\partial z}\,dz$$

What is such an expression called in the calculus? Interpret.

Exercise 62   In the notation we have been using, the potential function of a gravitational field is given by

$$V(r) = \frac{1}{r} = \frac{1}{\sqrt{x^2 + y^2}}$$

Show that $\dfrac{\partial V}{\partial x} = -\dfrac{x}{r^3}$ and similarly for the other partial derivatives. Prove therefore that

$$\overline{\text{grad}}\ V = -\vec{r}/r^3$$

Show that this is the inverse-square force law. (Pay attention to both the magnitude and the direction of the vector.) Because of the minus sign in the last expression, the force is often defined as the *negative* of the gradient of the potential.

Exercise 63 Show that if the various derivatives of $f(x,y)$ are plotted along all directions through (x_0,y_0), the locus is a circle with the gradient as diameter. Extend to 3-space.

Exercise 64 Show that

$$\overline{\text{grad}}\ (fg) = f\,\overline{\text{grad}}\ g + g\,\overline{\text{grad}}\ f$$

Exercise 65 A field is said to be *conservative* if no work is done in moving a body around a closed path (one which ends where it started). Using Exercise 61, prove that a force derivable from a potential has a conservative field.

Exercise 66 If a function of two variables is expressed in terms of polar coordinates, it is natural to set up a *local coordinate system* at the point (r,θ), with unit vectors, $\mathbf{u}_r$ and $\mathbf{u}_\theta$, pointing in the directions of increasing r and θ respectively. Since these vectors are perpendicular and make an angle θ with the x- and y-axes, show that

$$\mathbf{u}_r = \cos\theta\mathbf{i} + \sin\theta\mathbf{j}$$

and

$$\mathbf{u}_\theta = -\sin\theta\mathbf{i} + \cos\theta\mathbf{j}$$

(Draw a picture.) Noting that $r = \sqrt{x^2 + y^2}$ and $\theta = \tan^{-1}\left(\dfrac{y}{x}\right)$, prove that

$$\frac{\partial r}{\partial x} = \frac{x}{r} = \cos\theta \qquad \text{and} \qquad \frac{\partial\theta}{\partial x} = -\frac{y}{r^2} = -\frac{\sin\theta}{r}$$

and find $\partial r/\partial y$ and $\partial\theta/\partial y$ similarly. Recalling that

$$dV = \frac{\partial V}{\partial r}\,dr + \frac{\partial V}{\partial\theta}\,d\theta$$

show that

$$\frac{\partial V}{\partial x} = \frac{\partial V}{\partial r}\cos\theta - \frac{1}{r}\sin\theta\frac{\partial V}{\partial\theta}$$

and find the corresponding expression for $\partial V/\partial y$. Hence prove that

$$\overline{\text{grad}}\ V = \frac{\partial V}{\partial r}\mathbf{u}_r + \frac{1}{r}\frac{\partial V}{\partial\theta}\mathbf{u}_\theta$$

This is the representation of the gradient in polar coordinates.

Exercise 67 Clarify the reason why it is unimportant to distinguish between fixed and local coordinate systems in setting up the gradient in Cartesian coordinates.

We have just seen that the gradient of a scalar function is a vector function of position; *i.e.*, every point of space has a certain vector "attached" to it. Whenever the latter is the case, we say that we have a *vector field*. Thus a *force field* is a vector field, for, at every point of it, a test body is, by definition, acted upon by some definite force, and force is a vector. Now we are familiar from the calculus with the many ways in which a scalar function of position in space of several dimensions may be differentiated and integrated. It is reasonable to expect that a vector function of position will have an even richer variety of such properties, and this is indeed the case. We shall now sketch out the most important of these.

We take for our first illustration the motion of an ideal fluid. If we look at a fixed point of space at a given instant, we will find a certain fluid density and fluid velocity. We say that we have a *steady state* if the velocity and density of matter flowing past any point do not change with time. In other words, whenever we look at it, the motion of the fluid appears the same. The *flux* at any point is defined as the product of the velocity and density at that point. The flux is thus a vector (why?) and we have another realization of a vector field.

Consider a region of space bounded by a given surface. Matter is flowing through, and the *net efflux* is the difference between the total amount entering the region and the total amount leaving. Let us confine our attention to an infinitesimal rectangular parallelepiped of dimensions dx, dy, and dz, centered at (x,y,z); see Fig. Q-15. Its vertices are located at the points $(x \pm \frac{1}{2}dx, y \pm \frac{1}{2}dy, z \pm \frac{1}{2}dz)$. We define

$$\bar{u}(x,y,z) = \rho(x,y,z)\bar{v}(x,y,z)$$

where $\bar{u}$ is the flux, ρ the density, and $\bar{v}$ the velocity, and all are functions of position. The net efflux is the algebraic sum of the fluxes across the six faces; we consider them in opposite pairs. By an argument familiar from the kinetic theory of gases, the only particles flowing through a given face, say $ABCD$, in unit time, will be those which lie within the parallelepiped that has $ABCD$

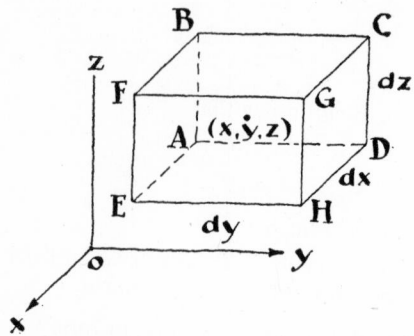

FIG. Q-15 Volume element in rectangular coordinates.

FIG. Q-16   Flux through one face.

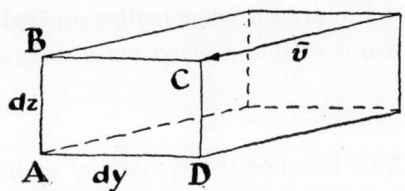

as its base and $\bar{v}$ as its slant edge (*cf.* Fig. Q-16). Plainly, the altitude of this parallelepiped is v_x, the x-component of the velocity; hence its volume is $v_x\,dy\,dz$. Thus the mass of fluid passing through $ABCD$ in unit time is

$$\rho v_x\,dy\,dz = u_x\,dy\,dz$$

Now $\bar{u}$, and hence also u_x, is not constant across $ABCD$, but varies slightly. We may, however, approximate to it by its value at the center of $ABCD$, the coordinates of which are $(x - \tfrac{1}{2}dx, y, z)$. Retaining the first two terms in the Taylor series expansion, we find for the flux across $ABCD$ the value

$$\left[u_x(x,y,z) - \frac{1}{2}\,dx\,\frac{\partial}{\partial x}\,u_x(x,y,z) \right] dy\,dz$$

Similarly, the flux across $EFGH$ is given by

$$\left[u_x(x,y,z) + \frac{1}{2}\,dx\,\frac{\partial}{\partial x}\,u_x(x,y,z) \right] dy\,dz$$

and the net contribution of the two faces is

$$\frac{\partial u_x}{\partial x}\,dx\,dy\,dz$$

[We suppress the argument (x,y,z) as superfluous from now on.] In like manner, the net efflux across $ABFE$ and $DCGH$ is

$$\frac{\partial u_y}{\partial y}\,dx\,dy\,dz$$

and the net efflux across $ADHE$ and $BCGF$ is

$$\frac{\partial u_z}{\partial z}\,dx\,dy\,dz$$

Thus the over-all net efflux, *per unit volume*, at (x,y,z), is

$$\frac{\partial u_x}{\partial x} + \frac{\partial u_y}{\partial y} + \frac{\partial u_z}{\partial z}$$

This quantity is called the *divergence* of the vector field at the point in question. It is obviously a scalar, and it measures the rate at which the field is "coming apart." It is written "div $\bar{u}$."

If matter is being neither created nor destroyed at (x,y,z), in other words, if it is neither a *source* nor a *sink*, then we must have

$$\text{div } \bar{u} = -\frac{\partial \rho}{\partial t}$$

This is called the *equation of continuity*. In particular, if the fluid is *incompressible*, then ρ is constant and the equation becomes simply

$$\text{div } \bar{u} = 0$$

Exercise 68 A rigid body is rotating about an axis with angular velocity ω; this is indicated by a vector, $\bar{\omega}$, of length ω, directed along the axis according to the right-hand rule. Take as the origin of position vectors $\bar{r}$ an arbitrary point on the axis. If we "freeze" the system at an instant of time, every point in the body will have a certain instantaneous linear velocity $\bar{v}$. Show that $\bar{v} = \bar{\omega} \times \bar{r}$. Which is the position vector? Which the field vector? What does the vector field look like? Show that it has zero divergence. (Use Exercise 50.)

Exercise 69 The inverse-square field was seen in Exercise 62 to be of the form $-\bar{r}/r^3$. Identify the position and field vectors. What is the appearance of the field? Show that, considered as a three-dimensional field, its divergence vanishes everywhere, but that this is not so when it is considered as a field in the plane. What is its divergence in the latter case? (What is the general expression for the divergence in two-space? Would the proof differ materially from the one given here for three-space?)

Exercise 70 Prove that, if $V(x,y,z)$ is a scalar function, then

$$\text{div } \overline{\text{grad}} . V = \frac{\partial^2 V}{\partial x^2} + \frac{\partial^2 V}{\partial y^2} + \frac{\partial^2 V}{\partial z^2}$$

This is referred to as the *Laplacian* of the function. (Before you begin calculating, examine the left-hand side of the equation to see whether the indicated order of operations is meaningful.) What is the Laplacian of the following function?

$$\cosh x \cosh y \cosh z$$

Exercise 71 Suppose we have a two-dimensional vector field whose field vectors, $\bar{u} = \bar{u}(\bar{r})$, have components u_x, u_y in a rectangular coordinate system, and u_r, u_θ in a polar coordinate system (*cf.* Exercise 66). Show that

$$u_x = u_r \cos \theta - u_\theta \sin \theta$$

$$u_y = u_r \sin \theta + u_\theta \cos \theta$$

(Why does r not enter explicitly in these equations?) Using the "chain rule" for partial differentiation, and the expressions for $\partial r/\partial x$ etc., already obtained in Exercise 66, prove that the divergence is given by

$$\frac{1}{r}\left[\frac{\partial}{\partial r}(ru_r) + \frac{\partial u_\theta}{\partial \theta}\right]$$

in polar coordinates.

Exercise 72 Consider $\bar{w}^n$, where $\bar{w}$ is a complex number and n an integer. Note that a complex number may be thought of as a vector by considering its real and imaginary parts as components. Thus $\bar{w}^n$ may be regarded as a vector associated with the point w in the complex plane. We therefore have a vector field. Show that its components in a local polar coordinate system are, say, z_r and z_θ, where

$$z_r = r^n \cos (n - 1)\theta$$

and

$$z_\theta = r^n \sin (n - 1)\theta$$

and that hence the divergence of the field is

$$2nr^{n-1} \cos (n - 1)\theta$$

Suppose now that we have a function, $f(\varpi)$, that may be expressed as a power-series in ϖ. Show, from the last result, that the vector field formed by it in the sense just discussed has a divergence equal to twice the real part of its derivative, $f'(\varpi)$.

Exercise 73 Prove, by an easy application of Exercise 71 to Exercise 66, that the Laplacian (introduced in Exercise 70) is given by the expression

$$\frac{1}{r} \frac{\partial}{\partial r} \left(r \frac{\partial V}{\partial r} \right) + \frac{1}{r^2} \frac{\partial^2 V}{\partial \theta^2}$$

in polar coordinates. Apply this to the two-dimensional case of the potential, V, of Exercise 62 and compare with Exercise 69.

We shall consider two important ways in which an integration may be performed over a vector field. The first results in a *surface integral* and arises most naturally from the idea of the divergence of a fluid discussed above. In fact, let S be a closed surface and dS an element of area (see Fig. Q-17), and let ρ, $\bar{v}$, and $\bar{u}$ have the same meaning as before. We again see that the only particles flowing through dS in unit time will be those lying within the "cylinder" which has dS as base and v as slant height. But the altitude of this cylinder is merely the component of v normal to dS. This suggests that we ought to represent the size and orientation of dS by a vector of length dS, normal to dS. Calling this vector $d\bar{S}$, the volume of the cylinder is then simply $\bar{v} \cdot d\bar{S}$ (why?), and the flux at the point in question is $\bar{u} \cdot d\bar{S}$ (why?). To obtain the total efflux through S, we sum over all such elements of area; in the limit, this becomes the surface integral

$$\iint_S \bar{u} \cdot d\bar{S}$$

A famous theorem of Gauss states that this is identically equal to

$$\iiint \operatorname{div} \bar{u} \, dx \, dy \, dz$$

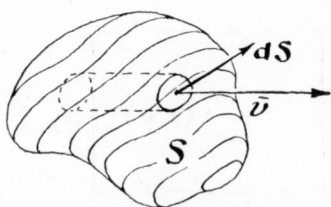

FIG. Q-17   Flux through area element of surface.

Clearly, both integrals express the same quantity, *viz.*, the total efflux through the surface, and their equality is thus obvious, although the theorem may be proved without recourse to physical intuition.

The second integral we want to consider is the so-called *line integral.* Suppose we have a force field and we move a test body through a closed loop inside the field. Now, if the field were conservative, the net work done in this manner would be zero, but not all fields are conservative, as anybody who has run round the block before breakfast knows. (What keeps the field at the earth's surface from being conservative?) The work in question is merely the work done in moving the test body along an infinitesimal arc length, dl, summed over all such arc lengths. If we let $d\bar{l}$ be a vector whose length and direction are those of the arc dl, the element of work is $\bar{F} \cdot d\bar{l}$, and the total work round the loop is written, for obvious reasons, as the line integral

$$\oint \bar{F} \cdot d\bar{l}$$

Let us examine this in Cartesian coordinates, as we did the divergence.

For ease of calculation we confine our attention to an infinitesimal rectangle in the x,y-plane, with vertices $(x \pm \frac{1}{2}dx, y \pm \frac{1}{2}dy)$; see Fig. Q-18. Along the lower edge we are only working against F_x, the x-component of the force, and the work done is

$$F_x(x, y - \tfrac{1}{2}dy)\, dx = \left[F_x(x,y) - \tfrac{1}{2}dy\, \frac{\partial}{\partial y}\, F_x(x,y) \right] dx$$

The expression for the work done along the upper edge is similar, except for a minus sign which enters because the path is described backwards:

$$-\left[F_x(x,y) + \tfrac{1}{2}dy\, \frac{\partial}{\partial y}\, F_x(x,y) \right] dx$$

The net contribution of the horizontal edges is therefore

$$-\frac{\partial F_x}{\partial y}\, dx\, dy$$

and, by the same reasoning, the net contribution of the vertical edges is

$$+\frac{\partial F_y}{\partial x}\, dx\, dy$$

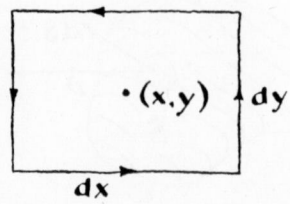

FIG. Q-18.

(Why is this quantity not negative like the first?) Thus the net work round the path is given by

$$\left[\frac{\partial F_y}{\partial x} - \frac{\partial F_x}{\partial y}\right] dx\, dy$$

The expression inside brackets is known as the *curl* of the two-dimensional vector field, $\bar{F}$.

The extension of this notion to three dimensions turns out to be rather interesting, for it develops that the right way to define the curl is as a vector. Its z-component is the quantity we have just derived, and its x- and y-components are, as usual, obtained by cyclic permutations of the z-component:

$$\overline{\text{curl}}\ \bar{F} = \left(\frac{\partial F_z}{\partial y} - \frac{\partial F_y}{\partial z}\right)\mathbf{i} + \left(\frac{\partial F_x}{\partial z} - \frac{\partial F_z}{\partial x}\right)\mathbf{j} + \left(\frac{\partial F_y}{\partial x} - \frac{\partial F_x}{\partial y}\right)\mathbf{k}$$

It may be shown, *e.g.*, that the work done in moving a test body round the perimeter of *any* element of surface area, dS, however oriented, is given by

$$d\bar{S}\cdot\overline{\text{curl}}\ \bar{F}$$

where $d\bar{S}$ is the "surface" vector defined above in connection with surface integrals. (Check that this gives the result we calculated for the two-dimensional curl.)

There is an integral theorem, due to Stokes, for the $\overline{\text{curl}}$, which is quite analogous to Gauss's theorem for the divergence. It states that

$$\iint\limits_{S} \overline{\text{curl}}\ \bar{F}\cdot d\bar{S} = \oint\limits_{C} \bar{F}\cdot d\bar{l}$$

Here S is a "cap" rather than a closed surface, and the line integral on the right side of the identity is taken along the boundary of S (*cf.* Fig. Q-19).

Exercise 74 Indicate how Stokes's theorem may be proved by breaking S up into a large number of small areas. Write down two expressions for the work done in moving round the perimeter of any one of these elements of area. Note that in the line-integral form for the work, the common boundary of any two adjacent elements of area is described twice, in opposite senses, and so leads to cancellation, except along C, which is a free edge to all areas of which it forms part of the boundary. Sketch out a similar

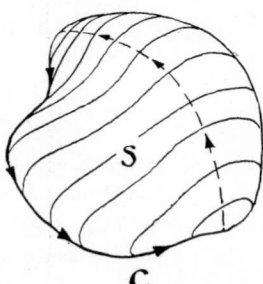

FIG. Q-19 Surface S with free boundary C.

proof of Gauss's theorem for the divergence, and state explicitly the cause of the corresponding cancellation when the efflux is summed over elements of volume.

Exercise 75 The *vector differential operator*, ∇ (pronounced "del"), is defined as

$$\nabla = \mathbf{i}\frac{\partial}{\partial x} + \mathbf{j}\frac{\partial}{\partial y} + \mathbf{k}\frac{\partial}{\partial z}$$

Show that

$$\overline{\text{grad}}\ V = \nabla V$$

$$\text{div}\ \bar{u} = \nabla \cdot \bar{u}$$

$$\overline{\text{curl}}\ \bar{u} = \nabla \times \bar{u}$$

and the Laplacian, defined in Exercise 70, $= \nabla \cdot \nabla V = \nabla^2 V$, as it is usually written. (The ∇-notation is systematic and compact and throws light on the various identities covered in the following exercises. Considerable caution, however, must be observed in using ∇ in any of the vector identities we know, involving repeated application of the dot and cross products. For ∇ must operate on everything that follows it, and hence it is not permissible to permute its position according to the familiar rules. The only sure procedure is always to go back to the fundamental definitions in terms of the components.)

Exercise 76 Using Stokes's theorem, prove that a field is conservative (Exercise 65) if and only if its curl vanishes. Write down the three scalar equations that this condition yields. Prove the identity

$$\overline{\text{curl}}\ \overline{\text{grad}}\ V = 0$$

and deduce the result of Exercise 65.

Exercise 77 Prove that

$$\text{div}\ (V\bar{u}) = V\ \text{div}\ \bar{u} + \bar{u}\cdot\overline{\text{grad}}\ V$$

and

$$\overline{\text{curl}}\ (V\bar{u}) = V\ \overline{\text{curl}}\ \bar{u} + \bar{u} \times \overline{\text{grad}}\ V$$

(The latter, a vector identity, may be proved by establishing the equality of the x-components on both sides, and permuting cyclically, as in Exercise 50.)

Exercise 78 Prove, similarly, that

$$\text{div}\ (\bar{A} \times \bar{B}) = \bar{B}\cdot\overline{\text{curl}}\ \bar{A} - \bar{A}\cdot\overline{\text{curl}}\ \bar{B}$$

and

$$\overline{\text{curl}}\ (\bar{A} \times \bar{B}) = \bar{A}\ \text{div}\ \bar{B} - \bar{B}\ \text{div}\ \bar{A} - \left(A_x\frac{\partial\bar{B}}{\partial x} + A_y\frac{\partial\bar{B}}{\partial y} + A_z\frac{\partial\bar{B}}{\partial z}\right)$$

$$+ \left(B_x\frac{\partial\bar{A}}{\partial x} + B_y\frac{\partial\bar{A}}{\partial y} + B_z\frac{\partial\bar{A}}{\partial z}\right)$$

(The latter identity cannot be expressed without explicit use of the vector components unless one has recourse to what are called "dyadics." We shall not consider them here.)

Exercise 79 Consider the vector field introduced in Exercise 68. Note that div $\bar{r} = 3$. (Why?) Using the last result of the previous exercise, show that the curl of the linear velocity at any point is equal to twice the angular velocity. (Hence the name "curl.")

Exercise 80 Prove that the curl of a two-dimensional vector field is given by

$$\frac{1}{r}\left[\frac{\partial}{\partial r}(ru_\theta) - \frac{\partial u_r}{\partial \theta}\right]$$

in polar coordinates. (Refer to Exercise 71.)

Exercise 81 Show that the curl of the vector field defined in Exercise 72 is equal to

$$2nr^{n-1}\sin(n-1)\theta$$

and generalize to the case of an arbitrary function, $f(\overline{w})$.

The Second-order Differential Equation

If we have a particle of mass λ, attached to a spring of compliance γ, moving against a resistance ρ, and acted upon by a driving force $f(t)$, its equation of motion will be

$$\lambda\frac{d^2x}{dt^2} + \rho\frac{dx}{dt} + \frac{1}{\gamma}x = f(t)$$

We have here a relationship connecting a dependent variable, x, and its derivatives with an independent variable, t. What is desired is a solution displaying x as a function of t, and involving, presumably, certain initial conditions. Such an equation is called, of course, a differential equation, and its *order* is defined as the order of the highest derivative involved. Thus the above is a second-order differential equation. It is further said to be *linear*, because x and its derivatives do not enter in powers higher than the first, and *ordinary* (rather than *partial*), because there is only one dependent variable. If, in addition, $f(t) = 0$, the equation is called *homogeneous*. This case is of fundamental importance both mathematically and physically, and we shall begin by restricting our attention to it alone.

Our equation is now simply

$$\lambda\frac{d^2x}{dt^2} + \rho\frac{dx}{dt} + \frac{1}{\gamma}x = 0$$

which, as we know, describes the motion of a particle vibrating under the influence of internal forces alone. We know that such a particle moves sinusoidally when $\rho = 0$; we have only to substitute $x = A\cos\omega t$ or $x = B\sin\omega t$ in order to see this. When $\rho \neq 0$, however, the presence of the first derivative rules out this simple solution. Still, the trigonometric functions must be hidden somewhere in the more general solution if it is to reduce to the familiar one when $\rho = 0$. But let us recall the beautiful identity

$$e^{j\omega t} = \cos\omega t + j\sin\omega t$$

connecting the trigonometric and exponential functions. This suggests that it might be a good idea to consider possible exponential solutions to our equation. Indeed, if we write

$$x = Ae^{j\omega t}$$

where A and ω are as yet undetermined constants, we find at once that

$$\frac{dx}{dt} = j\omega Ae^{j\omega t} \quad \text{and} \quad \frac{d^2x}{dt^2} = -\omega^2 Ae^{j\omega t}$$

and, substituting these back into the differential equation, it becomes

$$\left(-\lambda\omega^2 + j\rho\omega + \frac{1}{\gamma}\right) Ae^{j\omega t} = 0$$

Now one way of satisfying this equation is to set

$$Ae^{j\omega t} = 0$$

In other words, the solution we assumed turns out to be identically zero. This is just another way of saying that one physically possible situation is for the particle to remain eternally at rest in its equilibrium position. This is hardly what we were looking for, however, and we rule it out as trivial. The only other way, then, of satisfying the above equation is if we set

$$-\lambda\omega^2 + j\rho\omega + \frac{1}{\gamma} = 0$$

Since we have not yet put any restrictions on ω, this is plainly possible. But the interesting situation arises that, since this is a quadratic in ω, there are two values of ω which will satisfy it. We started out by looking for one solution and have wound up with two! This, however, is completely in agreement with the situation we had with $\rho = 0$, where the sine and cosine both turned out to be solutions.

We are not yet finished, however. It is true that we have the two solutions, say

$$x_+ = A_+e^{j\omega_+t} \quad \text{and} \quad x_- = A_-e^{j\omega_-t}$$

where

$$\omega_+ = \frac{-j\rho + \sqrt{\dfrac{\lambda}{\gamma} - \rho^2}}{-2\lambda} \quad \text{and} \quad \omega_- = \frac{-j\rho - \sqrt{\dfrac{\lambda}{\gamma} - \rho^2}}{-2\lambda}$$

(by the quadratic formula). In other words, x_+ and x_- each satisfy the original equation:

$$\lambda\frac{d^2x_+}{dt^2} + \rho\frac{dx_+}{dt} + \frac{1}{\gamma}x_+ = 0$$

$$\lambda\frac{d^2x_-}{dt^2} + \rho\frac{dx_-}{dt} + \frac{1}{\gamma}x_- = 0$$

But, if we add these two equations, we see immediately that

$$\lambda \frac{d^2}{dt^2}(x_+ + x_-) + \rho \frac{d}{dt}(x_+ + x_-) + \frac{1}{\gamma}(x_+ + x_-) = 0$$

I.e., $x_+ + x_-$ is also a solution. We have thus arrived at the most general solution of the equation we set out to solve, *viz.*,

$$x = A_+ e^{j\omega_+ t} + A_- e^{j\omega_- t}$$

Here ω_+ and ω_- are given explicitly in terms of λ, ρ, and γ by the equations we have already written down, and A_+ and A_- put two degrees of freedom at our disposal with which to meet the initial conditions. *E.g.*, we can start the particle off from any position we like with any preassigned velocity. This checks with physical intuition; mathematically, it is true in general that the number of degrees of freedom in the complete solution of a differential equation is always equal to its order.

It seems at first sight a little strange that the solution to our equation should come out in terms of complex numbers, when it must obviously be real. The paradox is easily explained, for A_+ and A_- turn out in practice to be complex, and their net effect is such as to make the imaginary part of the solution vanish.

Exercise 82   Assume the initial conditions $x = x_0$, $v = v_0$, when $t = 0$. Show that they require

$$A_+ + A_- = x_0 \qquad \text{and} \qquad \omega_+ A_+ + \omega_- A_- = -jv_0$$

and, solving for A_+ and A_-, derive the solution

$$x = \frac{1}{\omega_- - \omega_+} \{(\omega_- x_0 + jv_0)e^{j\omega_+ t} - (\omega_+ x_0 + jv_0)e^{j\omega_- t}\}$$

From this find the corresponding expression for v. Check that these two solutions do indeed satisfy the initial conditions.

Exercise 83   In the preceding exercise substitute for ω_+ and ω_- their values as given by the quadratic formula, and assume that $\rho < \sqrt{\lambda/\gamma}$ (small damping). Show that the solution becomes

$$x = e^{-\rho t/2\lambda}\left\{ x_0 \cos\left(\frac{t}{2\lambda}\sqrt{\frac{\lambda}{\gamma} - \rho^2}\right) + \frac{x_0 \rho + 2\lambda v_0}{\sqrt{\frac{\lambda}{\gamma} - \rho^2}} \sin\left(\frac{t}{2\lambda}\sqrt{\frac{\lambda}{\gamma} - \rho^2}\right) \right\}$$

and find the expression for v. Check the initial conditions. Do the same for the case where $\rho > \sqrt{\lambda/\gamma}$ (over-damping).

Exercise 84   When the modes, ω_+ and ω_-, are equal, the solution given in the text reduces to

$$A_+ e^{j\omega_+ t} + A_- e^{j\omega_+ t} = (A_+ + A_-)e^{j\omega_+ t}$$

I.e., the two constants, A_+ and A_-, coalesce into a single constant, $A = A_+ + A_-$. We have apparently lost a degree of freedom, which seems very strange. Exercise 82 shows, however, what is happening in this case, for, when $\omega_+ = \omega_-$, the solution obtained there reduces to the indeterminate form $0/0$. Assume that ω_+ is fixed, and let $\omega_- \to \omega_+$. Using l'Hôpital's Rule from the calculus, show that the solution of Exercise 82 approaches the limit

$$x = \{x_0 + (v_0 - j\omega_0 x_0)t\}\, e^{j\omega_0 t}$$

where we have written ω_0 in place of ω_+ to emphasize the fact that when $\omega_+ = \omega_-$ the discriminant of the quadratic is zero. Find the expression for v and check the initial conditions. (Another way of stating the above result is that, for equal modes, $\omega_+ = \omega_- = \omega_0$, the most general solution of the equation has the form $(A + Bt)e^{j\omega_0 t}$.)

Exercise 85   Note that in the previous exercise we must have $\rho = \sqrt{\lambda/\gamma}$ (critical damping) and that hence $\omega_0 = j/2\sqrt{\lambda\gamma}$. (Why?) Substituting, find the expressions for x and v in terms of x_0, v_0, λ, and γ.

Exercise 86   Consider the nth-order differential equation

$$a_n \frac{d^n x}{dt^n} + \ldots + a_1 \frac{dx}{dt} + a_0 x = 0$$

Try the solution $x = Ae^{\omega t}$ (the factor j would complicate the algebra unnecessarily in this case). What equation must ω satisfy? How many roots has this equation? What is the most general solution of the differential equation? Does this solution have the proper number of degrees of freedom? What do you conjecture happens to this solution when one of the modes is double? When one is triple?

We turn back now to the inhomogeneous equation with which we began our discussion. Suppose that, by some means or other, we happen to know a solution of it. Such a solution is called a *particular* solution and often written x_p. We thus have

$$\lambda \frac{d^2 x_p}{dt^2} + \rho \frac{dx_p}{dt} + \frac{1}{\gamma} x_p = f(t)$$

Suppose also that we know the complete solution of the homogeneous equation obtained by replacing $f(t)$ by zero. This is called the *complementary* solution, and written x_c:

$$\lambda \frac{d^2 x_c}{dt^2} + \rho \frac{dx_c}{dt} + \frac{1}{\gamma} x_c = 0$$

Adding these two equations, we find that

$$\lambda \frac{d^2}{dt^2}(x_p + x_c) + \rho \frac{d}{dt}(x_p + x_c) + \frac{1}{\gamma}(x_p + x_c) = f(t)$$

I.e., the sum of a particular solution and the complementary solution is itself a solution to the original equation. It is, in fact, the most general solution, for it contains in the complementary part two arbitrary constants, which is precisely the number of degrees of freedom that should be associated with the equation. We have not, it is true, said anything about how we are to find a particular solution, but, before we consider that problem, let us give some thought to the physical meaning in back of what we have been saying.

The particular solution is just the *steady-state* solution. It describes the motion of the particle after the driving force has been acting on it for a long while. Clearly, if the system is stable, the particle will ultimately move in the same manner, no matter what its position and velocity when the force was first applied may have been. The complementary solution, on the other hand, describes the *transient* response of the system. This will obviously depend on the initial conditions; hence we need our two degrees of freedom here. The over-all motion of the particle is the superposition, or sum, of the transient (complementary) and steady-state (particular) motions. Thus the physics and the mathematics each say the same thing in their own way.

Next let us suppose that we have a solution, x_1, corresponding to the function $f_1(t)$, and also a solution, x_2, corresponding to the function $f_2(t)$. Then

$$\lambda \frac{d^2 x_1}{dt^2} + \rho \frac{dx_1}{dt} + \frac{1}{\gamma} x_1 = f_1(t)$$

and

$$\lambda \frac{d^2 x_2}{dt^2} + \rho \frac{dx_2}{dt} + \frac{1}{\gamma} x_2 = f_2(t)$$

and, adding these two equations, we find

$$\lambda \frac{d^2}{dt^2} (x_1 + x_2) + \rho \frac{d}{dt} (x_1 + x_2) + \frac{1}{\gamma} (x_1 + x_2) = f_1(t) + f_2(t)$$

In other words, $x_1 + x_2$ is a solution corresponding to $f_1(t) + f_2(t)$. This is a very important property because, in effect, it allows us to "build up" solutions to the equation. Specifically, the driving force $f(t)$ is in practice always a periodic function of t. This means that we can expand it in a Fourier series:

$$f(t) = \sum_{-\infty}^{\infty} a_n e^{j\omega_n t}$$

where $\omega_n = n\pi/l$. (How do you know we can?) Thus the problem reduces to solving the equation

$$\lambda \frac{d^2 x}{dt^2} + \rho \frac{dx}{dt} + \frac{1}{\gamma} x = a_n e^{j\omega_n t}$$

Once we know the solution to this, we may sum over all values of n to obtain the solution to the original equation. (Why?) But physical intuition tells us that the *steady-state* solution to this, which is what we are looking for, must have the same frequency as the driving force, for ultimately the particle must have that frequency forced upon it. Hence we try a solution of the form $x = A e^{j\omega_n t}$ and find that the equation becomes

$$\left(-\lambda \omega_n{}^2 + j\rho \omega_n + \frac{1}{\gamma} \right) A e^{j\omega_n t} = a_n e^{j\omega_n t}$$

In order to satisfy this we must set

$$A = \frac{a_n}{-\lambda\omega_n{}^2 + j\rho\omega_n + \dfrac{1}{\gamma}}$$

Thus the particular solution of our original equation is given by

$$\sum_{-\infty}^{\infty} \frac{a_n}{-\lambda\omega_n{}^2 + j\rho\omega_n + \dfrac{1}{\gamma}} e^{j\omega_n t}$$

and the most general solution is, as before, obtained by adding to this the complementary solution already discussed.

Exercise 87 Assuming a driving force of $\mathcal{F} \cos \omega t$, obtain the results for x and v derived in Chapter I of the text. (What is the "Fourier series" of imaginary exponentials that produces $\cos \omega t$?)

Exercise 88 A ball with coefficient of restitution equal to 1 is dropped from height h onto a table. Show that the force it is subject to is similar to the graph shown in Fig. Q-5. Hence write down and solve the equation of motion of the ball.

Problems

CHAPTER I

1-1. Show that the velocity of water waves, of wavelengths longer than 12 cm or less than 2 mm, can be expressed with error no greater than 1% by one term or the other in the last equation of Chap. I. Take $g = 980$ cm sec^{-2}, $\sigma = 70$ dynes cm^{-1}, and $\rho = 1$ gm cm^{-3}.

1-2. Let a driving force of variable frequency, ω, but constant amplitude, $\mathfrak{F}$, be applied to produce simple harmonic motion of a mass λ, with moment of friction ρ. Show that the frequencies for maximum velocity and maximum displacement are, respectively,

$$\omega_{v \,=\, \max} = \sqrt{\frac{1}{\gamma\lambda}}$$

$$\omega_{x \,=\, \max} = \sqrt{\frac{1}{\gamma\lambda} - \frac{\rho^2}{2\lambda^2}}$$

1-3. When, in the problem above, ω goes to the limit zero, show that x becomes $\gamma\mathfrak{F}$ and that v vanishes.

1-4. In Problem 1-2, if ρ is small so that $\omega_+\omega_- \cong \dfrac{1}{\gamma\lambda}$, show that the frequencies ω_+ and ω_- on either side of the resonant frequency, $\omega_{v \,=\, \max}$, where $v_{\omega_\pm} = \dfrac{\mathfrak{F}}{\sqrt{2}\rho}$, are different by the amount

$$(\omega_+ - \omega_-) = \frac{\rho}{\lambda}$$

1-5. Show that, at the frequencies above, the phase of the motion becomes $\varphi = \pm\dfrac{\pi}{4}$ when ρ is small.

1-6. Consider tension in a spinning bicycle tire such as that produced by the centrifugal force. If the tire's peripheral velocity is v, and if its mass per unit length is μ, show that this tension, $\mathfrak{F}$, is such that

$$v = \sqrt{\frac{\mathfrak{F}}{\mu}}$$

One way to solve for v is to equate the sum of all vertical forces in half the tire to $2\mathfrak{F}$.

CHAPTER II

2-1. Imagine a drumhead provided with x- and y-coordinates, and let the drumhead be excited by the waves

$$\bar{w}_1 = A e^{j\omega\left(t-\frac{x}{v}\right)}$$

$$\bar{w}_2 = A e^{j\omega\left(t-\frac{y}{v}\right)}$$

Show that the resultant motion along a line through the origin of coordinates, and at 45° inclination to the x-axis, is described by

$$\bar{w} = 2A e^{j\omega\left(t-\frac{1}{v}\sqrt{\frac{x^2+y^2}{2}}\right)}$$

2-2. Consider three wave motions passing through some point P, either all transverse or all longitudinal. Furthermore, let the propagation directions (and polarizations, if the wave motion is transverse) be such that the three motions lie, respectively, along the three edges of a regular tetrahedron with its apex at P. If the three motions are

$$\tfrac{1}{2}e^{j\omega t}; \qquad e^{j(\omega t + \frac{2}{3}\pi)}; \qquad e^{j(\omega t - \frac{2}{3}\pi)}$$

show that the resultant motion, perpendicular to the tetrahedron side that lies opposite P, is $6^{-\frac{1}{2}} e^{j(\omega t + \pi)}$.

2-3. If two unequal but geometrically parallel fields of the same frequency, $\bar{E}_1$ and $\bar{E}_2$, are superimposed, and if only their phases vary with time, each varying randomly and without restraint, show that the time average amplitude of their sum is $\sqrt{\mathcal{E}_1{}^2 + \mathcal{E}_2{}^2}$.

2-4. We may express a circular wave motion that is propagating in the z-direction by the sum of two motions, one horizontal and parallel to the x-axis, and the other vertical and parallel to the y-axis. For example:

$$\bar{w}_R = e^{j\left(\omega t - \frac{\omega z}{v_R}\right)} \mathbf{i} + e^{j\left(\omega t - \frac{\omega z}{v_R} + \frac{\pi}{2}\right)} \mathbf{j}$$

or

$$\bar{w}_L = e^{j\left(\omega t - \frac{\omega z}{v_L}\right)} \mathbf{i} + e^{j\left(\omega t - \frac{\omega z}{v_L} - \frac{\pi}{2}\right)} \mathbf{j}$$

Here $\mathbf{i}$ and $\mathbf{j}$ are the unit vectors directed along the x- and y-axes. $\bar{w}_R$ and $\bar{w}_L$ are right and left circular motions. If $\dfrac{1}{v_L} - \dfrac{1}{v_R} = \delta$, and if $\delta \ll 1.0$, show that the total motion, if both waves are superimposed, is a linear oscillation oriented at the azimuth angle $\theta = \dfrac{\omega z}{2}\delta$.

2-5. Show that the superposition sum of two wave motions of equal ampli-
tude, $\mathfrak{a}$, but of slightly different frequencies, ω and $\omega - \Delta\omega$, can be
described as the product of $2A$ and two circular functions. And show,
further, that the frequency of one of these circular functions is the
mean frequency while that of the other is the difference frequency.

2-6. Is the helix referred to in § 2-4 a right- or left-handed screw; and in
which sense is it rotating?

CHAPTER III

3-1. The tangential boundary conditions on E and H for electromagnetic
waves that are incident in vacuum on a glass surface at angle i, for
the electric vector polarized in the plane of incidence, are

$$E_0 \cos i - E' \cos i = E_g \cos r$$

$$H_0 + H' = H_g$$

taking the vectors as shown in Fig. 3-4a. Eliminate the fields in the
glass, and solve for $r_\pi = \dfrac{E'}{E_0}$. Using the identities

$$\sin i \cos i - \sin r \cos r = \sin (i - r) \cos (i + r)$$

$$\sin i \cos i + \sin r \cos r = \sin (i + r) \cos (i - r)$$

show that

$$r_\pi = \frac{\tan (i - r)}{\tan (i + r)}$$

3-2. When the electric vector of incident light is polarized in a plane perpendicular to the plane of incidence, the tangential boundary conditions are

$$E_0 + E' = E_g$$

$$H_0 \cos i - H' \cos i = H_g \cos r$$

when we ignore the change of sign shown in Fig. 3-4b. Show that these
equations yield

$$r_\sigma = -\frac{\sin (i - r)}{\sin (i + r)}$$

Eliminate the fields in the glass, as above, in order to solve the boundary conditions for r_σ.

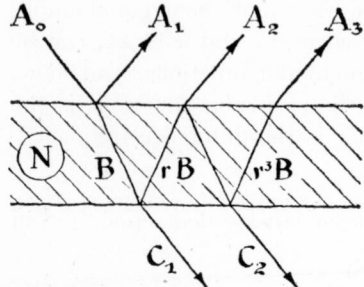

3-3. The accompanying figure illustrates notation used to describe the incident, reflected, and transmitted rays involved when light strikes a plane parallel glass plate. The letters here represent $\mathcal{E}$, the amplitudes of the oscillating electric fields. Illuminations are proportional to $\frac{1}{2}c\kappa_0 N\mathcal{E}^2$. By applying the principle of conservation of energy, or equality of power influx and efflux at each surface, show that

$$B_1 = \frac{A_0}{\sqrt{N}}\sqrt{1 - r^2}$$

$$C_1 = A_0(1 - r^2)$$

$$A_2 = rA_0(1 - r^2)$$

$$C_2 = r^2A_0(1 - r^2)$$

$$A_3 = r^3A_0(1 - r^2)$$

3-4. A hollow sphere coated inside with a diffusing, highly reflecting, white paint is much used in photometry. It is called an integrating sphere. Show that diffusely reflected light from any elementary facet of the inside surface casts a constant illumination on any other inside facet.

CHAPTER IV

4-1. The index of refraction of thermally evaporated dielectric films, such as those deposited on camera lens surfaces to reduce surface reflection, is less than the index of the massive dielectric material evaporated. For example, a cubic crystal with an index 1.39, when evaporated as a reflection-reducing film, may exhibit an index 1.38. Assuming the reduced index of the film to be due to its being an optically homogeneous mixture of crystal and voids ($N = 1.0$), show, from the formula of Clausius and Mossotti, that the fraction of the film that is occupied by the voids is .02.

4-2. Show that, when the oscillators that represent a dilute gas are characterized by a ρ that is small enough to make $\rho \ll 2\omega$, the maxima and minima of $(N - 1)$ occur at $\omega = \omega_0 \pm \frac{1}{2}\rho$.

4-3. If, after a sodium atom in vacuum emits a wave train ($\lambda = 5892$ Å) that is three centimeters long, the oscillator friction term will have caused the emitted light to decay to $\dfrac{1}{e}$ of its original illumination, show that $\rho = 10^{10}$. To get this result, solve for the constants in the particular solution that applies when the forcing function is set equal to zero:

$$x = x_0(Ae^{p_1 t} + Be^{p_2 t})$$

4-4. Show that the flux transmitted normally through a very thin absorbing prism of unit height, neglecting reflection losses and refraction, is $\dfrac{c}{2\omega k\varphi}$ $\mathfrak{E}$. Here jk is the imaginary part of the complex index of refraction of the prism material. Its prism angle is φ; and $\mathfrak{E}$ is the incident illumination.

4-5. Show, for normal reflection from silver and aluminum, that $\mathfrak{R}_{Ag} = .98$, $\mathfrak{R}_{Al} = .90$, $\varphi_{Ag} = -28°$, and $\varphi_{Al} = -18°$. Use the values of n and k given in Table 4-1. Here $\mathfrak{R} = \tilde{r}\tilde{r}^*$, and $\tilde{r} = re^{j\varphi} = \left(\dfrac{1 - \tilde{N}}{1 + \tilde{N}}\right)$.

4-6. Solve for reflections at a glass-metal interface, taking $N_g = 1.5$, and $\tilde{N}_{Al}$ and $\tilde{N}_{Ag}$ values from Table 4-1. Show that $\mathfrak{R}_{Ag\text{-}glass} = .96$, $\mathfrak{R}_{Al\text{-}glass} = .86$, and $\varphi = -41°$ and $-27°$ respectively. Use $\tilde{r} = \left(\dfrac{N_g - \tilde{N}_m}{N_g + \tilde{N}_m}\right)$.

4-7. Show that the A's and B's in Cauchy's formula are 1.602 and 1.577, and 0.89×10^6 and 1.78×10^6, for flint glass and liquid CS_2, using the following values for N:

	λ	Flint glass	CS_2
C	6563 Å	1.624	1.618
D	5892 Å	1.628	1.628
F	4861 Å	1.641	1.652

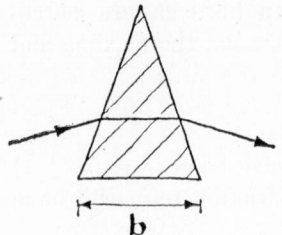

b

4-8. The prism of the accompanying figure is worked at minimum deviation; *i.e.*, it is penetrated symmetrically. If α is the extinction coefficient, it will exhibit an overall flux transmission of $\left(\dfrac{1 - e^{-2\alpha b}}{2\alpha b}\right)$. Derive this formula, assuming that reflection losses are negligible.

CHAPTER V

5-1. In the formula for water waves and ripples of § 1-6, show that the group and phase velocities are related as follows:

$$2\left(\frac{u - v}{v}\right) = \frac{\dfrac{a}{\lambda} - \dfrac{\lambda}{b}}{\dfrac{a}{\lambda} + \dfrac{\lambda}{b}}$$

Here a and b are the two terms under the radical sign of the formula for v.

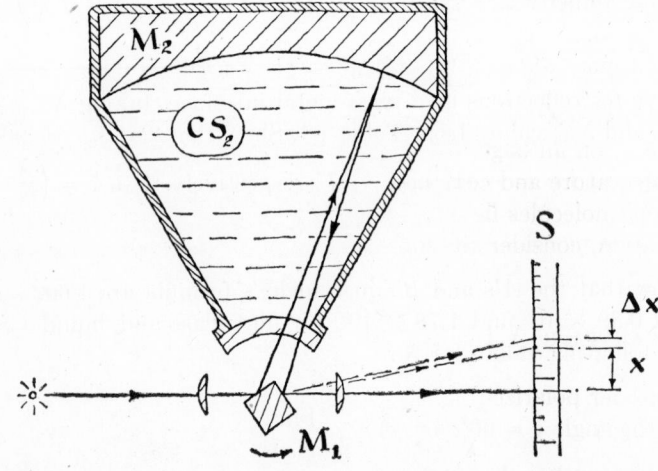

5-2. The accompanying figure shows an idealized rotating mirror apparatus for measuring the group velocity of light in CS_2. When M_1 rotates slowly, the light rays will strike the scale S at $x = 0$. But if M_1 rotates rapidly, the point where the light strikes S will be deflected, the red light, λ_R, being less deflected than the blue light, λ_B. Let the difference of wavelengths be $(\lambda_R - \lambda_B) = \Delta\lambda$; and let the spot separation, of deflections for red and blue, be Δx, with an average deflection x. Show that

$$\frac{\Delta x}{x} = \frac{\lambda \dfrac{d^2N}{d\lambda^2} - 2\dfrac{\lambda}{N}\left(\dfrac{dN}{d\lambda}\right)^2}{\left(N + \lambda\dfrac{dN}{d\lambda}\right)}\Delta\lambda$$

5-3. Show that the flux scattered per electron, when an oscillator is engulfed in a light flux of unit illumination, is approximated by $\dfrac{q^4}{6\pi m^2 \kappa_0^2 c^4}$, which is independent of wavelength. Here we may take $\omega \gg \omega_0$ for our oscillator, and neglect ρ. This is J. J. Thomson's approximation. It was the formula used by Barkla in the early days of atomic physics to determine the number of electrons in an atom from the measured scattering of X-rays.

5-4. Show that 15.1 seconds is the maximum apparent lengthening or shortening expected in the measured period of the one of Jupiter's moons that is eclipsed regularly (on an average of once every 1.75 days). It will be assumed, for our present purpose, that Jupiter is stationary and many times farther from the sun than the earth; and we shall consider that the plane of the earth's orbit contains Jupiter, and that that orbit is of such size that it takes 990 seconds for light to cross its diameter.

5-5. Using dimensional analysis, determine the relationship between the period of a simple pendulum, supposing it to depend on a suspended point mass, its length, and the acceleration of gravity.

5-6. Calculate the number of air molecules in a cube that is one wavelength $(0.5\ \mu)$ on an edge. Consider the air to be at 1 atm pressure and 0°C temperature and composed of molecules of molecular weight 29. How many molecules lie on an edge of this cube? For purposes of this calculation, consider the molecules to be in cubic array, as in a crystal.

CHAPTER VI

6-1. Consider polarized light incident on a glass plate of index $N = 1.732$ at the angle $i = 60°$. Let the incident illumination be taken as unity, and show that the illuminations within the glass are $\dfrac{N}{3}$ and $\dfrac{N}{4}$ for the parallel and perpendicular planes of polarization. Use the results of problems for Chap. III to obtain these results.

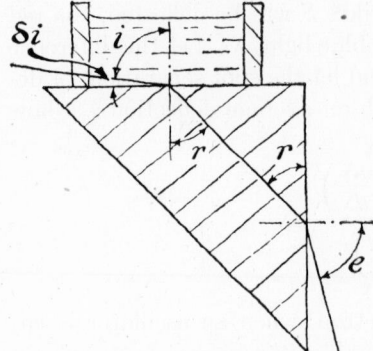

6-2. The Pulfrich refractometer has the main optical component shown in the accompanying figure. In addition, it has an illuminator, and a telescope and circle to determine the angle e. The Pulfrich refractometer may be used to measure N_1 (the index of the liquid on top of the 90° prism of index N_0). Here $N_0 > N_1$. If i is the angle of incidence at one facet of the 90° prism, and e is the angle of emergence at the other facet, taking $N = 1$ for air, show that

$$N_1 = \sqrt{\frac{N_0^2 - \sin^2 e}{1 - \cos^2 i}} \cong \sqrt{N_0^2 - \sin^2 e}$$

In the approximation above, if i is within $\delta i = .01$ of $\frac{\pi}{2}$, an error in N_1 of about one unit in the fourth decimal place is introduced (taking $N_1 \cong 1.5$ and $N_0 = 1.7$).

6-3. Referring to the accompanying figure, show by differentiation that errors, δi and δe, introduce the uncertainty δN_1 that is given below.

$$-\frac{\delta N_1}{N_1} = \left(\frac{\sin e \cos e}{N_0^2 - \sin^2 e} \right) \delta e + (\cot i) \delta i$$

Assuming $\delta i = 0$, show that $\delta e = 1^{\min}$ yields N_1 good to better than one unit in the fourth decimal place (taking $N_1 \cong 1.5$ and $N_0 = 1.7$).

6-4. Show that the ratios of incident to transmitted illuminations at the $N_1 - N_0$ interface above, for π- and σ-components, are

$$4 \frac{N_0}{N_1} \sec^2 r (\delta i)^2 \qquad \text{and} \qquad 4 \frac{N_1}{N_0} \sec^2 r (\delta i)^2$$

Furthermore, assuming $\delta e = 0$, calculate these transmissions for a value of δi that introduces an error of just one unit in the fourth decimal place (of N_1).

6-5. The degree of polarization of m parallel plates, of index N, penetrated at the polarizing angle, and when account is taken for multiple reflections, is given as follows:

$$\beta = \frac{m}{m + \left(\dfrac{2N}{1 - N^2}\right)^2}$$

Calculate β for $N = 2.4$ (selenium) and for $m = 2, 4$, and 6.

6-6. Show that, at Brewster's angle, the reflectivity of a selenium surface ($N = 2.4$) for the σ-component is approximately 50%. Neglect k for selenium.

6-7. Show that $r_\sigma{}^2 = r_\pi$ for any dielectric at $i = 45°$.

CHAPTER VII

7-1. Calculate the transmission of three ideal Polaroids laid one on another. Let the top and bottom have their azimuths of easy passage crossed, with the middle one lying with its azimuth oriented at 45° between them.

7-2. Show, from the following table (from Jenkins and White), that the rotary power of crystalline quartz along its optical axis is given by the following Cauchy-type formula:

$$\Re = 0.912 + \frac{6.7 \times 10^{-8}}{\lambda^2}$$

F	4861 Å	32.76
D	5892 Å	21.72
C	6438 Å	18.02

7-3. All the faces of a natural calcite rhomb are parallelograms with corner angles either 101° 55′ or 78° 05′. These faces meet to form the dihedral angles, 105° 05′ or 74° 55′. The optical axis is inclined symmetrically to all the faces that form the blunt corner, as shown in Fig. 7-4. Show, by spherical trigonometry, that the angle at which it penetrates those faces is $\psi = 45°.4$.

7-4. Consider two ideal analyzers forming a Lippich analyzer, as shown in Fig. 7-27, with the angle $2\Delta\theta$ between their azimuths of easy passage. Show that the azimuth of polarized light, that gives the same transmission through both parts of the field, does not quite bisect the two azimuths of easy passage of the Lippich, but makes an angle $\delta\theta$ with that bisector; and that

$$\delta\theta = \tan^3 \Delta\theta$$

Neglect transmission losses for electric fields parallel to azimuths of easy passage.

7-5. It is claimed that a trained observer in saccharimetry can match two brightness fields, such as above, to 1%; *i.e.*, $\Delta\mathfrak{B} = \dfrac{\mathfrak{B}}{100}$. Show that this leads to an angle uncertainty

$$\epsilon = \frac{1}{400}\tan\Delta\theta$$

and that, when $\Delta\theta$ (of the problem above) is 2°, then $\epsilon < 20$ seconds of arc. For a bright source $\Delta\theta$ can be set small, making ϵ only a few seconds of arc; whereas, without the Lippich device, the accuracy of setting would be a minute or more.

7-6. Consider Fresnel's compound prism of L- and R-quartz, for demonstration of the velocities of circular polarized light along the optical axis. Show that the angular separation of R- and L-circular polarized components of incident plane polarized sodium light, produced by it, is 4 minutes of arc. For calculation of the refractions at the 152° prism faces use

$$N_L \sin i = N_R \sin (i \pm \delta i)$$

Since δi is small, it is helpful to expand the right sine. Do not neglect to take account of the refractions at the final face.

7-7. Show that, for sodium light, the half deviations, δ_R and δ_L, produced by the 30° half-prisms of quartz differ by ± 11.5 seconds of arc. It is because of the canceling of one of these deviations by the other that a Cornu prism produces undoubled lines.

7-8. Consider the Rochon and Wollaston prisms of Fig. 7-21 with 10° prism components, both cut from calcite. Calculate the deviations at the inclined face.

CHAPTER VIII

8-1. Set up expressions for $(\mathfrak{E}_{5890} + \mathfrak{E}_{5896})$ for a Young experiment with yellow sodium light illumination. Let $\mathfrak{E}_{5890} = 2\mathfrak{E}_{5896}$ in the incident light, and in the order $k = 0$. Next, set up an expression for composite maximum and minimum total illumination as function of the variable, k_{5890}. Using the mathematics of modest rigor, with $\dfrac{\lambda_{5890}}{\lambda_{5896}} = 1 - \epsilon$, solve for $k_{5890}(1 + \kappa)$, the fractional order where composite maxima and minima occur. Here $\epsilon \ll 1.0$; and even for high orders, $\kappa \ll 1.0$. Show that the fractional order corresponding to maxima and minima is $\kappa = .074$ when $k = 250$.

8-2. Show that the visibility of fringes, when $k = 400$, in the situation depicted above, is $\mathcal{V} = 0.43$.

8-3. Consider a broad spectrum line at λ_0—its spectral illumination being

$$\mathcal{E}_x = \mathcal{E}_0 e^{-\left(\frac{x}{\xi}\right)^2}$$

Here ξ is a constant, small compared with unity; and $x = \dfrac{\lambda - \lambda_0}{\lambda_0}$.
Show that the illumination contributed by a narrow wavelength band, dx, in a Young pattern, is

$$d\mathcal{E} = 2\mathcal{E}_x \left[1 + \cos 2\,\frac{\pi a}{\lambda_0 \mathcal{D}}\, y(1 - x) \right] dx$$

Here, since $x \ll 1.0$, we write $(1 + x)^{-1} = 1 - x$.

Using this expression for $d\mathcal{E}$, show that the total illumination, expressed in terms of the phase difference on the observing screen appropriate to the point y_1 from the pattern center, *i.e.*

$$\Delta\varphi_1 = 2\,\frac{\pi a}{\lambda_0 \mathcal{D}}\, y_1$$

is given by

$$\left(2 \int_{-\infty}^{\infty} \mathcal{E}_x\, dx\right)\left[1 + \cos \Delta\varphi_1\, \frac{\int_{-\infty}^{\infty} \mathcal{E}_x \cos (x\Delta\varphi_1)\, dx}{\int_{-\infty}^{\infty} \mathcal{E}_x\, dx} + \right.$$

$$\left. \sin \Delta\varphi_1\, \frac{\int_{-\infty}^{\infty} \mathcal{E}_x \sin (x\Delta\varphi_1)\, dx}{\int_{-\infty}^{\infty} \mathcal{E}_x\, dx} \right]$$

Finally, integrating (use Peirce's *A Short Table of Integrals*), and using the fact that one of the functions is odd, show that the visibility of the interference bands at y_1 is given by

$$\mathcal{V} = e^{-\left(\frac{\pi a}{2\lambda_0 \mathcal{D}}\right)^2 \xi^2 y_1^2}$$

8-4. Using the above result, together with numbers that can be estimated from the visibility curve of Fig. 8-7, determine ξ and the approximate half width of the red cadmium line.

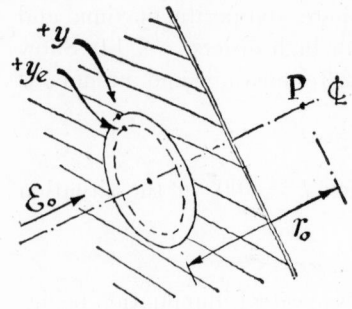

CHAPTER IX

9-1. The figure shows a circular aperture of radius y in an opaque screen. It is illuminated with plane waves of wavelength λ that are propagating normally to the screen. Show that the illumination at P, on the ₵, at the distance r_0 beyond the screen, is given by $\mathfrak{E}_{P}' = 2c\kappa_0\mathcal{E}_0^2 \sin^2 \frac{\varphi}{2}$ where $\mathcal{E}_0$ defines the incident illumination, and

$$\varphi = \pi \left(\frac{y^2 - y_e^2}{r_0^2} \right)$$

Here y_e is the largest radius that is contained by the aperture within which there is an even number of Fresnel half-period zones.

9-2. Let the screen with circular opening, above, be replaced by a circular obstruction of the same radius, y. Show that $\mathfrak{E}_{P}'' = \mathfrak{E}_0$.

Furthermore, considering that the superposition sum of these electric fields at P, $\tilde{E}_{P}'$, and $\tilde{E}_{P}''$ must give no diffraction at P and therefore that $\mathfrak{E}_P = \mathfrak{E}_{P}'' = \mathfrak{E}_0$, it would appear that we have run upon a violation of the principle of conservation of energy. Show that superposition of $\tilde{E}_{P}'$ and $\tilde{E}_{P}''$ gives the incident field, $\tilde{E}_0$, at P; and explain.

9-3. Fluctuations (see Fig. 9-10) of the electric field in the Fresnel diffraction fringes produced by a straight edge, as P moves from darkness into full light, become closely approximated by $\left(1 \pm \dfrac{1}{\pi u_2 \sqrt{2}} \right) \mathcal{E}_0$ as y_P moves into the light, and as the bands begin to fade out. Demonstrate this.

9-4. In the case set forth above, show that the visibility of fringes, just before the bands disappear, is given by $\mathcal{V} = \dfrac{\sqrt{2}}{\pi u_2}$.

9-5. Show that the zone plate of § 9-3 focuses parallel light at the distance $f = \dfrac{\Delta S}{\pi \lambda}$ beyond, where ΔS is the area of each zone. And further, show that a source point, at the distance u to the left of the plate, will be finally focused at the distance v beyond it, to the right; and that

$$\frac{1}{u} + \frac{1}{v} = \frac{1}{f}$$

This expression is the same as the thin lens formula. To obtain this expression, consider the zone plate as two superimposed congruent plates, and consider the rays from the source as diffracted parallel to the ¢, after the first plate, and as diffracted to converge onto the final focus beyond, by the second congruent plate. The constant optical path difference per zone, associated with the first diffraction, plus that per zone associated with the second diffraction, must give a total path difference per zone of $\dfrac{\lambda}{2}$ at the focus.

9-6. Predict qualitatively, by means of Cornu's spiral, the Fresnel diffraction pattern that is produced by the straight edge of a thin, partially transparent, metallic film. Consider that the metallic film absorbs 1/5 of the incident illumination and reflects none. Assume that the propagation constant in the metal is unity. To obtain the desired result, consider the unabsorbed wave front as resolved into two superimposed in-phase components. Make the resolution so that one of these components will have the same amplitude as the amplitude that is transmitted by the absorbing film. Thus, together, the film-transmitted wave front and this contiguous component, equally strong, form a coherent constant-phase plane background wave, of infinite extent. The other resolved component, falling on the observing screen, alone, would produce diffraction bands on the observing screen qualitatively the same as if produced by an opaque straight edge. Predict the Fresnel pattern for both components together on the observing screen from these considerations.

CHAPTER X

10-1. Whereas Rayleigh's criterion for angular separation gives a dip between the two maxima of two superimposed single-slit diffraction patterns, of two equal incoherent line sources, Sparrow's more realistic criterion for resolving power calls for an angular separation that just fails to give any such dip. Whereas the Rayleigh separation is $\delta \alpha_R = \dfrac{\lambda}{2x_0}$ (giving $\xi = 0$ for one pattern at the point where $\xi = \pm \pi$ for the other), the

Sparrow separation is slightly less. If $\xi = \pm(\pi - 0.4)$ for one pattern at the point where $\xi = 0$ for the other, then the superposition sum of illuminations midway between the equal patterns, $2\left[\dfrac{\sin^2\left(\dfrac{\pi}{2} - 0.2\right)}{\left(\dfrac{\pi}{2} - 0.2\right)^2}\right]$,

will be approximately equal to $\left[1 + \dfrac{\sin^2(\pi - 0.4)}{(\pi - 0.4)^2}\right]$. The purpose of this problem is to find out how much $\xi = \pm(\pi - 0.4)$ is in error. To accomplish this, set $\xi = (\pi - 0.4 + \epsilon)$; and, using the mathematics of modest rigor, determine the value of ϵ that satisfies Sparrow's criterion to the next degree of approximation, making the aggregate illumination at this ξ, and at $\xi = 0$, for one of the patterns, equal.

10-2. It is found by experiment that a recording monochromator, worked with equally wide slits, draws a response curve for a monochromatic line that is well represented by the Gaussian shape, $y = y_0 e^{-x^2}$ (using the meanings of x given in Problem 8-3). Consider the case where the wavelength separation of two equal recorded spectrum lines gives a record that just satisfies Sparrow's criterion. Taking $y_0 = 1.0$, show that the wavelength separation is $x = \sqrt{2}$, whereas the width of a line alone, at half maximum, is $x = \log_e 2 = 0.69$.

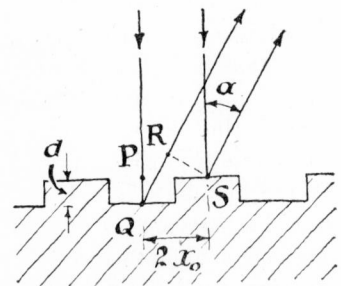

10-3. The figure shows a lamellar grating contour. Consider the equally wide facets to be perfectly reflecting, each of width $2x_0$, and with groove depth d. Show that the phase difference between beams diffracted from adjacent grooves in the direction α, when the grating is normally illuminated, is

$$\Delta\varphi = \frac{2\pi d}{\lambda}\left[(1 + \cos\alpha) + \frac{2x_0}{d}\sin\alpha\right]$$

10-4. In a manner similar to that employed for the amplitude grating, show that the fraction of the amplitude diffracted at angle α, for the lamellar grating above, is given by

$$\tilde{E}_\alpha = \tilde{E}_0 \tilde{X}_s \{\tilde{Y}\tilde{Z}\}\bar{B}\tilde{C}, \qquad \text{where} \qquad \tilde{C} = \left(1 + e^{j\frac{\Delta\varphi}{2}}\right)$$

(Here other terms have the meanings defined in § 10-3.)

10-5. From the above expression we get the illumination in the direction α as

$$\mathfrak{E}_\alpha = \tfrac{1}{2}c\kappa_0\mathcal{E}_0{}^2 X_s X_s{}^* \{\tilde{Y}\tilde{Z}\}\,\{\tilde{Y}\tilde{Z}\}^* \tilde{B}\tilde{B}^* \cos^2 \frac{\Delta\varphi}{2}$$

From these considerations show that $\mathfrak{E}_\alpha$ is proportional to $\dfrac{\sin^2 \xi}{\xi^2}$, as is characteristic of a rectangular aperture of width $4\mathfrak{N}x_0$, in the limiting case where ξ becomes $\dfrac{4\mathfrak{N}x_0\pi \sin \alpha}{\lambda}$, and where $\delta = 0$.

CHAPTER XI

11-1. Derive the formula for Newton's rings,

$$\rho = \sqrt{k\lambda R}$$

11-2. For an evaporated dielectric $\dfrac{\lambda}{4}$-film of index n onto glass of index N, show that, for normal incidence, the term

$$\left(\frac{n^2 - N}{n^2 + N}\right)\tilde{E}_0$$

expresses the approximate sum of the first two reflected beams and that it gives the exact sum when all the reflected beams are accounted for.

11-3. Consider a transparent high-index film on black glass that is found to produce maximum and minimum reflectivities of 4% and 30%, respectively, for normal incidence. Show that $\mathfrak{r}_e{}^2$ reduces to $\left(\dfrac{N_3 - 1}{N_3 + 1}\right)^2$ when $\Delta\varphi$ is an even number of π's; and, from this, determine the glass index, N_3. Then solve for the approximate film index, N_2, considering only the first two reflected beams. Use this approximate value for N_2 to fix the small term in the denominator of the exact expression for $\mathfrak{r}_o$, that takes account of multiple reflections, and solve for N_2 to the next approximation. Compare this result with that obtained by the formula of the problem above.

11-4. It seems proper that a transparent film on a perfectly reflecting surface should show interference bands in reflected light; but such is not the case. Show that $\tilde{\mathfrak{r}}\tilde{\mathfrak{r}}^* = \mathfrak{R} = 1.0$ from Eq. 11-8a, $\mathfrak{r} = \dfrac{\mathfrak{r}_{1,2} + e^{-j\Delta\varphi}}{1 - \mathfrak{r}_{2,1}e^{-j\Delta\varphi}}$, for any transparent film of index $(\mathfrak{r}_{1,2})$, and for any thickness $(\Delta\varphi)$ whatever.

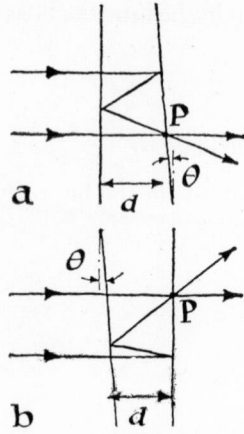

a

d

b

d

11-5. The accompanying figure shows (a) incident light normal to the *first* face of two slightly tipped partially transparent plane silver films immersed in a medium of index N, and (b) the light incident normal to the *second* face. Let the tip be small, $\theta \ll 1.0$. To second approximation, show that the path differences at P between the direct beam and the twice internally reflected beam are, respectively,

$$2Nd(1 - 2\theta^2) \quad \text{and} \quad 2Nd(1 - \theta^2)$$

In demonstrating the validity of these expressions, neglect phase shifts at reflection.

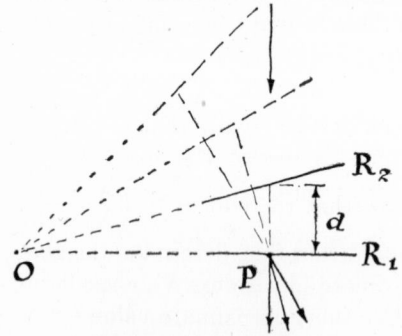

11-6. The accompanying figure shows Brosell's method of determining successive path differences at P for the arrangement shown at b above. The multiple reflection images of the partially silvered reflector R_2 in R_1 are shown. They all pivot about the point O where the silver films, extended, intersect. Normals to these images, dropped from P, represent the successive total internal paths. Show, for $\theta \ll 1.0$, that the path difference for successive beams through P is

$$md\left[1 - \left(\frac{m^2 + 2}{6}\right)\theta^2\right]$$

where m is 2, 4, 6, etc.

11-7. Consider the reflectivity to be averaged over the channeled-spectrum light reflected from a transparent thick plate ($d \gg \lambda$, and $r_{1,2} = -r_{2,1}$). Show by integration, using formula No. 860.1 in Dwight's *Tables of Integrals*, that the averaged intensity reflection coefficient is

$$\mathfrak{R}_1 = \frac{2r^2}{1 + r^2}$$

Consider N to be independent of wavelength; and use Eq. 11-8a for the amplitude reflection coefficient, r.

This result says, when r^2 is small, that the intensity reflection coefficient is the sum of the intensity reflection coefficients for the two surfaces of the plate. Thus the reflectivity of plate glass, $r^2 = .043$, to a precision of 4.3%, is $2 \times 4.3\%$ or exactly $\dfrac{2 \times .043}{1.043}$.

11-8. Show that the same result is obtained if we take account of multiple reflections of a thick plate, and if we use $\mathfrak{R}_0$, the intensity reflection coefficient for either surface of the plate rather than the amplitude reflection coefficients ($r_{1,2}$ and $r_{2,1}$, with the phase difference accounted for). That is, the overall reflection of the two surfaces, $\mathfrak{R}_1$, is

$$\mathfrak{R}_1 = \frac{2\mathfrak{R}_0}{1 + \mathfrak{R}_0}$$

We consider that the first surface reflects $\mathfrak{R}_0$ and transmits $(1 - \mathfrak{R}_0)$. The second beam emerging from the glass, accordingly, is $\mathfrak{R}_0(1 - \mathfrak{R}_0)$; the third is $\mathfrak{R}_0^2(1 - \mathfrak{R}_0)$, etc. In summing these beams, the expression

$$\frac{1}{1 - x} = \sum_0^\infty x^m$$

is used. The overall transmission becomes

$$\mathfrak{T}_1 = \left(\frac{1 - \mathfrak{R}_0}{1 + \mathfrak{R}_0} \right)$$

11-9. By a similar logic show that the overall reflectivity of two thick equal parallel transparent plates, separated by a distance that is large compared with λ, is

$$\mathfrak{R}_2 = \frac{2\mathfrak{R}_1}{1 + \mathfrak{R}_1} \quad \text{and} \quad \mathfrak{T}_2 = \frac{1 - \mathfrak{R}_1}{1 + \mathfrak{R}_1}$$

Now generalize this result to the cases of four, eight, sixteen, etc., plates. Show finally that the overall reflection of an infinite number of plates is $\mathfrak{R}_\infty = 1.0$.

The result of this problem suggests why layers of finely divided clear crystals (*e.g.* paint pigments) are white.

CHAPTER XII

12-1. Consider a train of transparent 60° prisms, their indices all being $N = 1.5$. Let them be arranged in series so that light of wavelength λ penetrates each one, successively, at minimum deviation. Show that $\dfrac{\mathfrak{T}_\pi}{\mathfrak{T}_\sigma}$ is approximately 1.5 for a train of two prisms and 2.5 for a train of four.

12-2. Calculate $\frac{1}{2}(\mathfrak{T}_\pi + \mathfrak{T}_\sigma)\dfrac{d\alpha}{d\lambda}$ to determine the energy-limited resolving power at $\lambda = 10\mu$ for one prism of prism angle 60° and another of angle 80°. Let these prisms be made of the same material, with each worked at minimum deviation. Take $N = 1.5$, and $\dfrac{dN}{d\lambda} = 0.007$ per micron (approximating the properties of the prism material NaCl).

12-3. Calculate $\dfrac{d\alpha}{d\lambda}$ for a blazed grating with groove spacing $a = 10\mu$. From this dispersion, and assuming $\mathfrak{T} = \frac{1}{2}$ for both π- and σ-polarizations, compare the energy-limited resolving power for this grating with that of the prisms above, taking the same projected working aperture, S, for both prism and grating.

12-4. Consider a transmission grating that is blazed in the first order for $\lambda = 0.5\mu$ when it is illuminated at normal incidence from the back. It is made of plastic with $N = 1.5$. It has 15,000 grooves per inch ($a \cong 1.7\mu$). Now consider that this grating is aluminized to become a blazed reflection grating for normally incident light. Which order (for $\lambda = 0.5\mu$) will lie nearest its reflection blaze?

12-5. Let the student show, by means of the integral of the product,

$$\int_{-\infty}^{\infty} \left(\frac{\sin \xi_1}{\xi_1}\right)\left[\frac{\sin (\xi_2 - \xi_1)}{(\xi_2 - \xi_1)}\right] d\xi_1 = \pi \frac{\sin \xi_2}{\xi_2}$$

that the final image in Fig. 10-15c has the same width as the intermediate image. Here ξ_1 is the phase parameter for the first image over which we integrate; and ξ_2 is that for the final image, here held constant. Both lenses, L_1 and L_2 of Fig. 10-15c, are assumed to be stopped by slits (of the same width, $2x_0$).

This equality of images may be established by means of the following hints, referring to tables for integrals.

$$\int_{-\infty}^{\infty} \frac{\sin x}{x} dx = \pi$$

$$\frac{1}{\xi_1(\xi_2 - \xi_1)} = \frac{1}{\xi_2}\left(\frac{1}{(\xi_2 - \xi_1)} + \frac{1}{\xi_1}\right)$$

Substituting $(\xi_2 - \xi_1) = x$ and $d\xi_1 = dx$ gives our answer.

12-6. Integrate the illumination in the Fraunhofer diffraction pattern over the area of the observing screen to get the total flux falling on it. Let the diffraction aperture be a square of area $(4x_0y_0)$ that is illuminated with a plane wave. Let this illumination be represented by $\frac{1}{2}c\kappa_0\varepsilon_0^2$. Show that the calculated flux, integrated over all observing screen area, is exactly equal to that passed by the square aperture.

12-7. A principle of spectral resolving power says the resolving power is $\delta\nu = \dfrac{1}{\tau c}$ cm^{-1}, where τ is the extreme time spread between the arrival of beams at the detector that were emitted simultaneously at the source.

Show that the zigzag path time is $\dfrac{l}{v \sin i'}$ in a Lummer-Gehrcke plate; and, using this principle, that

$$\delta\nu \cong \frac{1}{l(N^2 - 1)} \text{ cm}^{-1}$$

Similarly, from this principle, derive the spectral resolving power for a grating (a) illuminated normal to its surface; and (b) when it is worked in the Littrow arrangement, with $\alpha = \alpha'$, as shown in Fig. 10-14b.

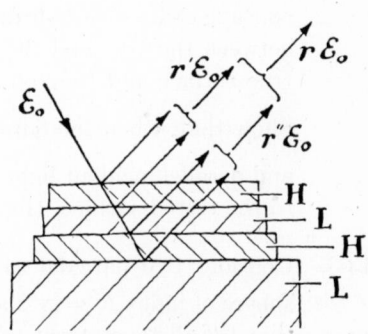

12-8. Referring to the accompanying figure, where a plate of index $N = 1.5$ is covered by two $\dfrac{\lambda}{4}$ films of index $N_H = 5.3$, with one of index $N_L = 1.5$ between, calculate r_e', the resultant of multiple reflections involving only the two top interfaces; then calculate r_e'' for the bottom two interfaces; finally calculate r_e for both r_e' and r_e'' together:

$$r_e = \frac{r_e' + r_e''}{1 - r_e'r_e''}$$

Finally, from $\mathfrak{R} = r_e^2$ show that two plates, each coated with three films as shown, will form a Fabry-Perot interferometer that is characterized by $\mathfrak{N} = 186$ beams.

12-9. Show that the expression

$$(N - \cos \alpha)d - a \sin \alpha = k\lambda$$

applies to the downward diffraction of a transmission etalon; and furthermore, show that, in the limit when $a \to 0$, where $\dfrac{d}{a} = \tan \varphi$, the expression for the $k = 0$ order of diffraction reduces to the expression for the minimum deviation for a half prism:

$$\frac{\sin (\varphi + \alpha)}{\sin \varphi} = N$$

CHAPTER XIII

13-1. By means of the sagitta formula determine the range at which a ship goes "hull down" because of the curvature of the sea's surface (that is, the range at which a man on the deck of one ship will see an equal ship hull down). Assume that eyes and hull top are both 26.4 feet above sea level. Take the radius of curvature of the sea's surface as 4000 miles.

13-2. This problem is concerned with Wight's test for mass-produced convex lens surfaces. Consider a surface of radius r to be tested by a true concave contact test surface, by reading the Fizeau interference bands between the true and the manufactured surfaces. Let the radius of the true surface and the nominal radius of the tested surface both be r. Show that, when the outside radius for the transparent test glass is $\dfrac{r}{3}$, and parallel incident light is used, the interfering beams will be normal to the tested surface. Here $N = 1.5$.

13-3. Consider two paraxial rays and a marginal ray that strikes a glass sphere of index $N = \sqrt{2}$ and radius r at $i = 45°$. Show by ray tracing that the distance from the paraxial focus to where the marginal ray crosses the ¢ is 0.48 r.

13-4. By the trigonometry and geometry of rays, show that the longitudinal spherical aberration of a spherical mirror that is used to focus parallel light is

$$\Delta x = h \left(\frac{1}{\sin 2i} - \frac{1}{2 \sin i} \right)$$

Furthermore, using the mathematics of modest rigor, show that this longitudinal aberration is approximately

$$\Delta x = \frac{h^2}{8f} + \frac{3}{128} \frac{h^4}{f^3}$$

13-5. If h_0 represents the marginal parallel ray that is focused by a spherical mirror, show that all the rays inside h_0 that are focused by the mirror are contained within a circle of radius $b = \dfrac{h_0^3}{16f^2}$, lying somewhere between the paraxial focus, at $\Delta x = 0$, and the marginal ₵ cross-over, at $\Delta x = \dfrac{h_0^2}{8f}$.

13-6. Whereas a non-uniform metallic reflecting film of proper thickness, $\delta x_1 = \dfrac{h_0^4 - h^4}{8r_0^3}$, deposited on a concave sphere of radius r_0, will parabolize it, show that a second non-uniform film that is at most only 1/4 as thick as the maximum thickness of the first, namely $\delta x_2 = \dfrac{h^2(h_0^2 - h^2)}{8r_0^3}$, will also parabolize it; but that such a second parabolization is accompanied by a fractional change of the central radius of curvature of $\dfrac{h_0^2}{4r_0^2}$.

13-7. Consider a spherical wave of radius r_0 that is converging onto a focus F. Let θ be the angle of obliquity of its rays with respect to a certain ₵. Show that the radial difference between the wave front and a sphere that is centered about a point on the ₵, but at a distance Δf from F, is $\dfrac{r_0}{64} \theta^4$. Here $\theta = \theta_0$ represents the marginal rays; and $\theta_0 \ll 1.0$.

13-8. Consider a rectangular wave front of width w that is focused to a point by a lens. If the wave front is plane, it will give a diffraction pattern of angular half width, $\alpha = \dfrac{\lambda}{w}$. Furthermore, it seems logical that an out-of-plane facet of this wave front cannot be inclined away by more than the angle $\dfrac{\alpha}{2}$ and give a good point image. Consider the wave front to be divided into two halves, each plane, and with a relative tip $\dfrac{\alpha}{2}$. Show how this consideration leads to Lord Rayleigh's approximate criterion that a lens must not distort the wave front more than by $\dfrac{\lambda}{4}$ if it is to yield a good point image.

CHAPTER XIV

14-1. If the plate in a spectrograph holder is photographically focused and then a glass filter of index N and thickness d is put in the plate holder in front of the emulsion of a second unexposed plate, pushing the emulsion back with respect to the plate holder's edge, show that the plate holder must be moved a distance $\dfrac{d}{N}$ toward the camera lens to reestablish focus.

14-2. Calculate the ratio of longitudinal spherical aberrations, $\Delta\left(\dfrac{1}{v}\right)$'s, for a plano-convex lens for two cases: one where parallel light is incident on the convex surface of the lens, and the other incident on the plane surface. Take $N = 1.5$ and $f = 1$ meter. Compare this with the aberrations when the lens of the same f is properly "bent" following the teaching of §14-6 and Fig. 14-11.

14-3. Calculate the thick lens $\mathfrak{F}$, and H and H', for a glass sphere of index $N = 1.5$ and radius r. This locates the principal foci. Now check this result by ray-tracing an initially parallel paraxial ray through the two spherical surfaces.

14-4. Show that the separation of the principal planes of a bi-convex lens of index $N = 1.5$, radius r, and thickness d is

$$d\left(\frac{2r - d}{6r - d}\right)$$

14-5. (From T. Smith) "Show that, as the thickness of *any* lens (of $N = 1.5$) diminishes, the distance between the unit points approximates to one-third of the thickness of the lens."

14-6. Calculate prism angles for a compound Amici prism. Let the outside prisms be made of flint glass, with CS_2 for the center prism. Consider prism angles small. Design two compound prisms that will (a) disperse the spectrum without deviation of the D-rays and (b) deviate the C- and F-rays equally. In case b calculate the angle between these rays and the D-rays, assuming that the small CS_2 prism angle is 5.74°.

14-7. (From T. Smith) "The axial distance between a fixed object and the image formed by a thin lens in air is X and this distance is unaltered when the lens is moved through an axial distance a. Show that the focal length of the lens is

$$f = \frac{X^2 - a^2}{4X}$$

14-8. For a prism of angle A, producing deviation D, we have $D - A = i_1 + r_2'$ and $A = r_1 + r_2'$. From Snell's law, and differentiation, show that $\dfrac{dD}{di_2} = 0$ only if we have symmetrical penetration of the prism— $i_1 = r_2'$ and $r_1 = i_2'$.

14-9. Some instruments use a thin spherical glass hood or dome with concentric spherical optical surfaces through which one may look radially. Show qualitatively that the optical power of such a hood is negative; and, quantitatively, that the power is approximately equal to the power of either surface multiplied by the ratio of thickness to radius, and divided by N.

14-10. Two equal, long, cylindrical lenses, of focal length f, are arranged with their axes of symmetry parallel in the x-z plane of a Cartesian coordinate system. The surfaces of one lens lie normal to this plane near the plane $z = +f$; those of the other lens lie near the plane $z = -f$. This combination acts as a stigmatic image inverter when put in front of a telescope, in parallel light. However, astigmatism enters when this combination is inserted between the objective and eyepiece of a telescope, in converging light. Set up equations for the meridional image position for this combination as function of the location of a point object.

CHAPTER XV

15-1. In the accompanying figure, at a, we see two paths from A to C, with reflection at a mirror. Show that the path *via* B' is greater than that

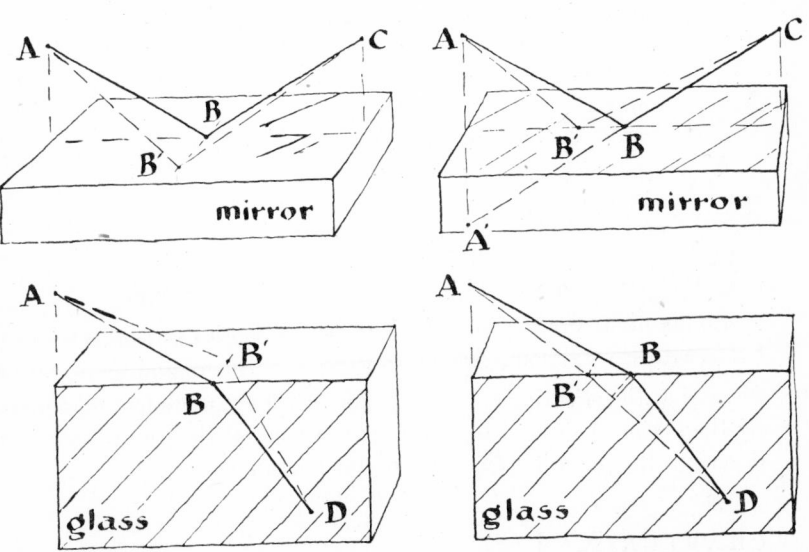

via B when B lies in the plane containing A and C that is perpendicular to the mirror, and if B' lies opposite B. And, furthermore, referring to b, show that the path is less for the path *via* B than for that *via* B' when both are in the perpendicular plane, if the angle of incidence at B

is equal to the angle of reflection there. It is of interest that Hero of Alexandria formulated the law of reflection (first stated by Euclid) as the "path of shortest distance."

15-2. Referring again to the figure of Problem 15-1, at c, show that, similarly, the path from A to D that lies in the plane that is perpendicular to a refracting surface and contains both A and D, with refraction at B, is shorter than the path with refraction at a point just outside this plane, at B' opposite B. And, finally, referring to d, show that the path ABD is shorter than an adjacent path in the plane $AB'D$ if Snell's law applies to the angles at B. To accomplish this, show that the Snell relationship between i, r, and N satisfies

$$\delta\left[\int n(s)\, ds\right] = [AB + N(BD)] - [AB' + N(B'D)] = 0$$

15-3. In § 15-9 we calculated the average transmission for a Pfund cell when the entrance window was a slit of length $2y_s$. Make a similar calculation for a circular entrance window of diameter $2y_s$.

15-4. Show that the linear magnification of a Galilean telescope for an object at great distance, but not at infinity, is $-\dfrac{f_e}{f_o}$, and not $-\dfrac{f_o}{f_e}$ as for angular magnification.

15-5. A Keplerian telescope and an equivalent Galilean telescope have objectives of diameter 2 and focal length 20, and eye lenses of diameter 1 and of focal lengths, respectively, $+10$ and -10.

When used as relay systems to transform parallel light to parallel light, show that they have magnifications -2 and $+2$; and further, that they have fields $1/30$ and $1/10$.

When used as viewing telescopes, show that the viewing fields, with eye rings of diameter 1 located a distance 10 behind the eye lenses, are, respectively, $1/10$ and $1/30$.

Finally, show, for the case above, that field lenses of power $F_P = 0.067$ and $F_V = 0.05$ are required, located at the common focus, to give maximum projection field (no eye ring) and maximum viewing field (with the eye ring specified).

15-6. Show that the effective reciprocal aperture of a camera lens is increased by the factor $(1 + m)$ when an object lies closer than infinity. Here m describes the magnification for the photographed image.

CHAPTER XVI

16-1. In Huygens' time telescopes made from a single thin lens were given long focal lengths (100 feet or more). The purpose was to increase the ratio of planetary image size to the blur circle arising from chromatic aberration. Show that $\dfrac{\alpha}{h_0}\dfrac{f_D}{\nu}$ gives this ratio—α being the planet's angular radius; f_D the focal length; h_i the half-diameter of the lens; and ν its reciprocal dispersion.

16-2. Check the results for Problem 13-3 with the Seidel correction term given in § 16-1.

16-3. In considering curvature of field, we have shown that the locus of circles of least confusion for a simple lens is a sphere. Determine the character of the locus of sagittal and meridional foci.

16-4. By means of the sagitta formula of Fig. 13-6a, show that the slant distances for a spherical mirror of radius r, say u and v of Fig. 13-7c, are related to ₵ distances, p and q, as follows:

$$u = p - \left(\frac{p-r}{pr}\right)\frac{h^2}{2} \qquad v = q - \left(\frac{q-r}{qr}\right)\frac{h^2}{2}$$

Furthermore, show that $\dfrac{p-r}{r-q} = \dfrac{u}{v}.$

Combine these expressions to get a measure of spherical aberration

$$\frac{1}{q_0} - \frac{1}{q} = \frac{h^2}{r^3}\left(1 - \frac{r^2}{pq}\right)$$

which yields $q = q_0$ for a sphere worked near its center of curvature, and confirms the value $\dfrac{y^2}{4r}$ of p. 294 when the sphere is worked in parallel light. Here q_0 is the value of the ₵ distance for $h = 0$ that is given by $\dfrac{1}{p} + \dfrac{1}{q_0} = \dfrac{2}{r}.$

16-5. Extend the considerations of the problem above to the single refraction shown in Fig. 14-5b:

First establish the following relationships between the ₵ distances, $p = P_1C_1$ and $q = C_1P_1'$, in terms of the angle of the radius to the point of refraction, $\alpha = \dfrac{h}{r}$:

Defining q_0 by $\dfrac{1}{p} + \dfrac{N}{q_0} = \dfrac{N-1}{r}$, show that

$$q_0 - q = \frac{h^2}{2r^2}\left(\frac{N}{p} + \frac{1}{q}\right)(q_0 - r)(q - r)$$

or

$$\frac{1}{q_0} - \frac{1}{q} \cong -\frac{h^2}{2r^2qq_0}(q_0 - r)^2\left(\frac{N}{p} + \frac{1}{q_0}\right)$$

16-6. By means of the formulas of the two problems above, show that spherical aberrations cancel for a Mangin mirror (*cf.* Fig. 14-12) when $N = 1.5$ for its glass, and when F and C_1 are coincident (with $r_1 = \frac{3}{4}r_2$).

Index

Abbe, 533, 537; deduction of sine condition, 333–336; on optical glass, 324; sine condition, 329–330, 534; theory of microscope, 533

Aberrations: astigmatism, 303, 362, 365, 367, 399; chromatic, 319; coma, 356–357; curvature of field, 366; distortion, 368; of light (manifest by starlight), 81–82; longitudinal and lateral chromatic, 356; measurement of longitudinal spherical, 297; spherical, 314, 315, 355; spherical, formula for single refracting surface, 681; spherical, formula for spherical mirror, 681

Absorption: cell, White's, 347; constant, 65–66; and emission, spectral lines in, 83; law of, 65–66; lines, water vapor, 433

Achromatic doublet, 319–320, 323, 356; origin of, 320

Achromatic lens, 320

Achromatized fringes, 394–395

Airy's disk, 203, 277, 293

Amici prism, 305

Amplifier time constant, 505

Amplitude: division of, 161, 223, 374, 381

Amplitudes: vector addition of, 25

Analog computer, 552

Analysis: dimensional, Lord Rayleigh's, 104; of polarized light, 144; spectral, with two-beam interference, 172; vector, 633 et seq.

Analyzer: error of, 665; Lippich, 148; for polarized light, 105, 112, 146–148, 665

Anderson: calcite achromat, 323

Angle, critical, 116; and plane of incidence, 50, 679–680

Angular aperture, 306; and angle of projection, 341–342

Angular and lateral magnification, 680

Anisotropic compliance (mechanical), 134

Anisotropic oscillator (electromechanical), 127

Anomalous dispersion, 97

Aperture: angular, 306; circular, diffraction pattern of, 417; numerical, 337, 342, 535; rectangular, diffraction by, 190, 201; stop, 341

Aplanatic points, 314, 332

Apodizing, 206; screens, 410, 413, 415

Arago: interference of polarized light, 178; rediscovery of Poisson's spot, 186

Artificial dielectrics, 520, 521

Astigmatism: of concave grating, 608 et seq.; and curvature of field, related, 367; of curved refracting surface, 362; of lens, 365; of mirror, 365; of plane-refracting surface, 303; testing for, 399

Averted vision, 488

Azimuth of polarized light, 146

Babinet, compensator of, 143

Babinet, principle of, 200

Bartholinus: description of Iceland spar, 111

Beam splitting, 379–381, 383–384

Bees, perception of light by, 109

Bell: origin of achromatic doublet, 320

Bentonite for Kerr cell, 155

Biaxial crystals, 138

Billet: split lens, 166–167, 375

Birefringent filter, 585

Blackbody radiation, 75–76

Blazed grating, 217–218

Blur circle, or circle of least confusion, 276, 367

Bohr, 447

Bolometer: detectors, 479; thermistor, 482; time constant, 482

Boundary conditions: Fresnel's reflection coefficients from, 50; at interface, 48, 124

Bouwers: catadioptric optical system, 318, 370

Boyle: discoverer of Newton's rings, 222

Boys' rainbow cup, 224

Bradley: velocity of light from stellar aberration, 81

Brewster's angle, 110, 385, 523

Brightness, 54; and flux, interrelation of, photometric magnitudes, 57; and illumination, interrelation of, photometric magnitudes, 59

Broca-Pellin prism, 305

Broglie, de, 440
Brownian motion, 473, 485
Bunsen and Kirchhoff on Fraunhofer lines, 84
Burch interferometer, 383
Butler-Edser interference bands, 225

c—the "velocity" or speed of light: early determinations, 81; history of, 451; by Kerr cell experiment, 462; by molecular spectra, 461; by permittivity method, 457, 465; by phase velocity, 459
Calcite: achromat, 323; extraordinary wave surface, 133; optical axis, direction in, 129; ordinary and extraordinary rays, 128; refraction for normal incidence, 130; refractions generally, 135; rhomb angles, 665; rhomb, Malus' polarization detector, 139; rhombs in series, 131; spheroidal wavelets in (Huygens), 129
Camera, 343, 369
Candela, 55
Cardioid reflector, 314
Carpenter's rule method of considering periodic film structures (Turner), 252
Cauchy: formula for dependence of N on λ, 68–69
Caustic, 298
Channeled spectrum, 225, 232, 673; thickness determination by, 234
Chief ray, 349
Chopping: optimum frequency, 504; selective, 503
Christiansen: anomalous dispersion, 97; filter, 584, 588
Chromatic aberration, 319; longitudinal and lateral, 356
Circle of least confusion, 276, 367
Circularly polarized light, 119 *et seq.*
Clausius and Mossotti formula, 68, 521
Coherence of light, 158
Coherent beams, 377
Collinear transformations, 327
Colors: complementary, 167; interference, 167
Coma, 332; of parabolic mirror, 357; pure, testing for, 399; of thin lens, 356
Compensator of Babinet and Soleil, 143–144
Complementarity, 447
Complex numbers, 10, 616 *et seq.*
Complex refractive index, 65
Compton effect, 439
Concave grating, 608 *et seq.*; astigmatism of, 608 *et seq.*
Cones: optical, 487
Conic sections, 281
Conjugate points and planes, 326
Conrady: equation, 398

Conservation of energy, 164
Constant: dynamic dielectric, 65; optical, at low frequencies (metal), 69; optical, for visible light (metal), 71; Planck's, 438; time: detector for, 470; of wave motion, 65
Conventions of sign: for electromagnetic fields, 52; for geometrical optics, 286, 288, 308
Cooke: photo-visual triplet, 323
Cornu prism, 151, 666
Cornu spiral, 193; Ditchburn's comments on approximation of, 196
Cornu-Jellett analyzer, 148
Corpuscular theory, 8
Cotton and Mouton effect, 155
Counter: Geiger-Müller, 499
Counting efficiency, 502
Critical angle, 116
Crystals: biaxial, 138; positive and negative, 149; right- and left-handed, 149 *et seq.*; uniaxial, 149
Curvature of field, 366–367
Curvature of surface, 282
Curve of shape, 299
Czerny-Turner optical arrangement, 371

D lines: Fraunhofer, 84–85; of sodium, 82
Dark current, 498
Davis and Sinton: double monochromator resolving power, 220
Dennison-Hadley filters, 588
Detectable power, minimum, of photoconductive cells, 495
Detector: bolometer, 479; classification, 474–475; Ektron, 472; Golay cell, 484; human skin as, 475; indium antimonide, 496; noise, 470; phototube, 496; thermopile, 475; time constant, 470
Deviation minimum, 214, 303–304, 678; of prism, 304
Dextrorotatory quartz, 150
Dielectric constant: dynamic, 65
Dielectrics: artificial, 520, 521
Diffraction, 374; beam splitting, 379; of diffraction pattern, 674; double slit, 160; Fraunhofer, *see* Fraunhofer diffraction; Fresnel, *see* Fresnel diffraction; fringes and interference bands contrasted, 226; grating, 207, 215, 216; and interference combined, 207; Kirchhoff's elaboration of Fresnel's theory, 4; limitation on resolving power, 210; in microscope, 425, 527; microwave, 511; pattern of circular aperture, 417; single slit, 29, 198, 203, 422; theory, Fresnel's essay on, 186; theory, Kirchhoff's, 180
Direct-vision prism, 678

Dispersion: angular, of prism, 304; anomalous, 97; of optical glass, 319; reciprocal (ν), 319
Displacement: electric, 42
Displacement (electric) current, 44
Distortion, 368
Ditchburn's comments on approximations of Cornu spiral, 196
Dolland: patentee for achromatic doublet lens, 320
Doppler line width, 86
Double refraction, 128 *et seq.*
Double slit, 160
Doublet: achromatic, 319–320, 323, 356; origin of, 320
Drude: remarks of on light rays, 276
Dyson: alignment interferometer, 389; interferometer, 385; visibility theorem, 379

Ebert, 373; optical system of, 371
Echelle grating, 267
Echelon, 264; reflection (Williams), 266
Edser-Butler interference bands, 225
Einstein, 436, 439
Ektron detector, 472
Ektron lead sulfide cell, 494
Elby: oscillator knife-edge test, 362
Electric displacement, 42
Electrical polarization and index of refraction, 64
Electromagnetic fields: conventions of sign for, 52
Electromagnetic spectrum: from gamma rays to radio, 39
Electromagnetic wave motion: history of, 40
Electro-optical shutter, 92, 156
"Elliptical" mirror, 362
Elliptically polarized light, 119 *et seq.*
Emission and absorption: spectral lines in, 83; of electrical dipole, 73–74
Energy: conservation of, 164
Energy-limited resolving power, 270, 674
Entrance and exit pupils, 341–342
Equations: differential, for light waves, 45; Fresnel, 659; lens, 306, 311; Maxwell, 40–42, 46; microscope, 290; of motion for harmonic oscillator, 64; second-order differential, 651
Etalon, 238, 265
Ether, 1, 435, 436
Evans, 586
Exit pupil, 341–342
Extraordinary ray, 128
Extraordinary wave surface in calcite, 133
Eye, 485; dark-adapted, 487; Gullstrand's, 313; optical efficiency of, 488; time constant of, 488

f-number, 342
Fabry-Perot interferometer, 231; film pack for, 675; resolving power of, 260
Fan of rays, 303
Faraday effect, 155; for microwaves, 524
Faraday's "Thoughts on Ray Vibrations," 41
Fastie: optical system, 371
FECO interference bands, 232
Fermat: law, 333, 679–680; principle, 363
Fibers: bent, 560; composite glass, 574; field flattener, 567; image transfer by, 570; loss of light in, 556; quartz, 562; refractive index of, 562; resolving power of, 571; scanning with, 571
Fiberscope, 565
Field: angle, 345; curvature of, 366–367; depth of, 343; flattener with fibers, 567; of Galilean and Keplerian (or astronomical) telescopes compared, 680; lens, 349; stop, 345; of view of optical system, 344
Figuring: of optical surfaces, 293; by thermal evaporation, 677
Film structures: periodic, carpenter's rule method of considering (Turner), 252
Films, evaporated: aluminum, for gratings, 609; $\frac{\lambda}{4}$, reflectance formula for, 671; metal, reflection of, 72; metallic, 245; multilayer, 252; multiple, 252; non-reflecting, 244, 248; non-uniform, 251; reflection-eliminating, 248; reflection-enhancing, 245; reflection-reducing, 248; silver, 72, 262, 587; stack, for Fabry-Perot interferometer, 675; tellurium, 594; thermal evaporation of, 408
Filters: alkali metal, 587; birefringent, 585; Christiansen, 584, 588; Dennison-Hadley, 588; germanium, 590; glass, 581–582; indium antimonide, 590; infrared, 591, 595; interference, 253, 385, 582–583; multilayer, 587; silicon, 590; silver film, 72, 587; Wratten, 581–582
Fizeau interference bands, 226
Flicker frequency, 489
Flux, 54; and brightness, interrelation of, photometric magnitudes, 57
Focal isolation, 419–420
Focal power, 287, 308, 309
Focal ratio, 342
Focus, sagittal and meridional (*i.e.* tangential), 366, 367
Foucault: knife-edge test, 294; velocity of light in air and water, 2, 10
Fourier: integral, 35, 630; series, 621 *et seq.*
Fourier transforms, 421–422, 631; grating pattern by, 424; monochromaticity of pulse by, 426; two-beam interference by, 428

Françon, 401, 406

Fraunhofer: D lines, 84–85

Fraunhofer diffraction: by circular aperture, 203; multiple- and single-slit patterns compared, 209; by rectangular aperture, 190, 201; by single split, 29; by square aperture, square-on, and along diagonal, 205; square and circular patterns contrasted, 203; transition from Fresnel to, 192

Free spectral range, 258–259, 583

Frequency, 18

Fresnel: biprism, 375; double mirror and double prism interference experiments, 166–167; elaboration of Huygens' ideas, 4; equations, 659; fringes, visibility of, 668; integrals, 192; interference of polarized light, 178; mirror, 375; prism for demonstrating theory of circular light in crystals, 150; reflection coefficients from boundary conditions, 50; rhombs, 118; theory of diffraction, Kirchhoff's elaboration of, 4; two-beam interference experiments, 161, 172

Fresnel diffraction: by circular aperture, 211; essay on theory of, 186; by slit and strip, 198; by straight edge, 197

Fringes: achromatized, 394–395; contrasted with interference bands, 226; of equal inclination, 231; of equal thickness, 227; localized, 228, 229; visibility of, 168, 668; white-light, 167, 171, 237

Froome interferometer, 460

Frustrated total reflection, 124; at fiber surfaces, 573

Full wave or tint plate, 143, 154

Galilean telescope, 338; field compared to telescope of Kepler, 680

Galileo: attempt to measure c, 81

Gases: refractive index of, 66

Gaussian formulas, 287 et seq.

Gaussian line: visibility of interference bands for, 667

Gaussian pulse, and spectrum, 631

Gaussian spectrum lines: resolving power with, 670

Gaussian theory, 352

Gaviola: caustic test, 298

Geiger-Müller counter, 499

Geometry of optical surfaces, 280

Gerhardt interferometer, 375

Germanium filters, 590

Gibson: thumbnail biography of Thomas Young, 164

Glare, 120

Glass: composite fibers, 574; dispersion of, 319; filters, 581–582; model for, 67; optical, 323 et seq.; strain in, device for observing, 155

Glasses: polaroid, 120

Golay cell, 484

Graphical ray tracing, 546

Gratings: aluminum films for, 609; blazed, 217–218; blazes, 597, 604; concave, 608 et seq.; diffraction, 207, 216; echelle, 267; echelon, 264; efficiency, 606; energy-limited resolving power, 674; etalon, 238, 265; lamellar, 431, 671; merit factor, 607; mounting, Wadsworth, 612 et seq.; pattern by Fourier transform, 424; production of, 218, 608; reflection, 216, 608 et seq.; replica, 216; resolving power, 215; Rowland circle, 609; types of, 216

Grimaldi, 22

Group velocity, 92, 93; experiments, 453

Gullstrand's eye, 313

Hagen and Rubens formula for metallic reflection, 71

Haidinger: brush, 106, 111; interference bands, 231

Half-wave plate, 142, 146, 388; for microwaves, 523

Hall: maker of first achromat, 320

Hamilton, 539

Harmonic oscillator: equations of motion for, 64; in light field, 63

Harrison: originator of echelle grating, 267

Hartmann test, 355

Heisenberg, 442

Helmholtz, 490

Hendrix, 552

Herschel, 475

Hertz, 437; discovery of electromagnetic waves, 47

Hg^{198}, 168

Hooke: early proponent of concept of light as wave motion, 1, 222

Hutchins: thermoelectric alloys, 478

Huygens: ideas about nature of light, 1–2; ideas, Fresnel's elaboration of, 4; spheroidal wavelets in calcite, 129; wave trainlets, 158

Hydrogen line spectrum, 172

Hyperfocal distance, 344

Iceland spar, 111

Illumination, 54; and brightness, photometric magnitudes, 59; and energy density, photometric magnitudes, 60

Images: of coherent sources, 219; real and virtual, 274; stabilized retinal, 487; testing of, 353; transfer with fibers, 570; transformer or slicer, 577

Impedance: electrical and mechanical, 13

Incidence: angle and plane of, 50, 679–680

Index of refraction, 65; for artificial dielectrics, 521; complex, 65; of fibers, 562; of gases, 66
Indium antimonide detector, 496
Indium antimonide filters, 590
Infraction, 222
Infrared: double reflection in, 135
Infrared filters, 591, 595
Infrared polarizers, 123, 124
Intensity, 54
Interference, 374; analysis for two-beam, 161, 172; colors, Young's and Lloyd's, 167; and diffraction combined, 207; diffraction pattern, multiple-slit, 207 *et seq.*; double-slit, experiment (Young), 160; experiments, Fresnel's double mirror and double prism, 166–167; filters, *see* Interference filters; microscope, 387–389; of microwaves, 511; multiple-beam, 255; order of, 163; of polarized light: Arago and Fresnel, 178; spectrometer, 259; by thin dielectric plate, 223; two-beam, by Fourier transform, 428; used for spectrum analysis, 172
Interference bands: Butler-Edser, 225; and diffraction fringes contrasted, 226; FECO, 232; Fizeau, Haidinger and FECO, contrasted, 226; for Gaussian line, visibility of, 667; of Haidinger, 231; localized, 228; *versus* fringes, 226; visibility of, 168, 233; with white light, 167, 171, 237
Interference filters, 253, 582–583; at Brewster's angle, 385; polarization by, 386
Interferogram, 430
Interferometer: Burch, 383; common-path, 384; Dyson, 385, 389; Fabry-Perot, 231; film stack for Fabry-Perot, 675; Froome, 460; Gerhardt, 375; inverting, 395; Lummer-Gehrcke, 268; Michelson, 235 *et seq.*, 381; for microwaves, Michelson, 515; quartz plate, 420; Rayleigh, 375, 377; stellar, 173, 395; Twyman and Green, 375, 382
Interferometric modulation, 431
Interferometry: low-order, multiple-beam, 256
Internal reflection, 114
Inverse square law, 40
Isoplanasie curves, 542

Jacquinot: apodizing, 206
Jellett-Cornu analyzer, 148
Johnson noise, 471; for thermopile, 478
Jupiter: satellites of, 81

Keplerian, or astronomical, telescope, 338, 680
Kerr cell experiment to determine **c**, 462
Kerr effect, 155

Kingslake, 398
Kirchhoff: differential, 181, 598; diffraction theory, 180; Fraunhofer lines (with Bunsen), 84; Fresnel's theory of diffraction, 4
Knife-edge test, 294
Knife-edge test oscillator (Elby), 362
Kösters, double-image prism, 393 *et seq.*

Lambert surface, 57
Lateral and angular magnification, 680
Lead compound photoconductor cells, 472, 493, 495
Lead sulfide cell: Ektron, 494
Lebedev: microscope, 388
Lens aberration: astigmatism, 362 *et seq.*; coma, 356; curvature, 366; spherical, 314, 315, 355
Lens: achromatic, 320; coma of thin, 356; equations, 306, 311; equivalent refracting surface, 331; field, 349; field of compound, 344; focal power, 287, 308, 309; immersion, 312; optical power of system of lenses, 310; rim, 341; secondary spectrum, 322; split (Billet), 166–167, 375; telephoto, 339; Tessar, 539; testing of, 396–397; thick, 309; thin, 308
Levorotatory quartz, 150
Light: Huygens' ideas on nature of, 1–2; loss of, in fibers, 356; Maxwell's equations on nature of, 46; Michelson's interferometer for, 235, 237; monochromatic, 168; natural, 106; Newton's ideas on nature of, 8; particle nature of, 436; polarized, 105, 112, 119, 120, 144, 146–148, 152, 178, 665; polarizing, Wollaston prism for, 139, 390, 406; source, pinhole, 296; velocity of, 81, *see also* **c**; as wave motion, Hooke early proponent of concept of, 1, 222
Light-chopping experiment: Rupp's, 92
Light waves: differential equation for, 45; velocity of, in glass, 48
Line width: collision, 88; Doppler, 86; due to various line-broadening factors, 91; natural, 85
Lippich analyzer, 148; error of, 665
Lloyd: mirror, 166–167, 375
Localized fringes, 228
L-R-C circuit, 13
Lumen, 55
Luminance, 55
Lummer-Gehrcke interferometer, 268
Lyot, 401, 405

McAlister, 584
Mach: comment on transverse wave motion, 178; description of Huygens' ideas, 2
Magnesium oxide crystals, 594

Magnification, 327, 328, 330, 336; angular and lateral, 680; normal, 348

Magnifiers, 336

Magnitudes: photometric, 54

Maksutov: optical system, 371

Malus: discovery of polarization by reflection, 110; law, 113; polarization detector, calcite rhomb, 139

Mangin mirror, 317

Maraldi: discoverer of Poisson's spot, 186

Maxwell: commutator bridge, 458; equations, 40–42, 46; perfect optical system, 326

Maxwellian view, 294

Meissner: highest resolving power, 263

Mercury (Hg198): radiation "from gold," 168; vapor, scattering by, 103

Metal: films, reflection of, 72; optical constants, at low frequencies, 69; optical constants for visible light, 71

Metallic reflection: Hagen and Rubens formula, 71

Meter expressed in cadmium wavelengths, 241

Mica, 143; cleavages, terraced, Tolansky's measurement of, 255

Michelson: measurement of meter, 238; and Morley experiment, 436; transmission echelon, 265

Michelson interferometer, 381; for microwaves, 515; for monochromatic light, 235; for white light, 237

Microscope: Abbe theory of resolution, 533; defined, 336; diffraction in, 425, 527; equations, 290; interference, 387–389; Lebedev, 388; magnification and depth of focus, 338; numerical aperture for, 335; phase contrast, 379, 530

Microwaves: Brewster's angle for, 523; Faraday effect for, 524; generator and receiver for, 507; interference and diffraction, 511; Michelson interferometer for, 515; polarization of, 522; reflection, 517, 519; standing, 510; transmission, 509

Minimum deviation, 214, 303–304, 678

Minnaert: description of Haidinger's brush, 111

Mirrors: astigmatism of, 365; "elliptical," 362; Fresnel, 375; Lloyd, 166–167, 375; Mangin, 317; parabolic, 291, 357; spherical aberration formula for, 681; telescope, 292; test with auxiliary flat, 294; test of "flat," 364–365

Modulation: periodic, 503; *see also* Chopping

Modulation line broadening, 90

Molecular spectra method for c, 461

Monochromatic light, 168

Monochromaticity of pulse by Fourier transforms, 426

Monochromator, 213; double, resolving power (Sinton and Davis), 220

Morey, 325

Mossotti. *See* Clausius.

Motion, Brownian, 473, 485; simple harmonic, 11

Mouton and Cotton effect, 155

Müller-Geiger counter, 499

Multilayer films, 252

Multilayer filters, 587

Multiple reflections, 242 *et seq.*

Multiple-beam, high-order spectroscopy, 257

Multiple-slit interference-diffraction pattern, 207 *et seq.*

Natural light, 106

Natural line width, 85

Negative and positive crystals, 149

Newton: black spot, 225; ideas about nature of light, 8; rings, 222, 225

Nicol prism, 139

Nitrobenzene for Kerr cell, 155

Nodal points, 547

Noise: background, 471; current, 471; detector, 470; equivalent, 479; in interferometric radiometry, 431; Johnson, for thermopile, 478; shot, 471, 498

Nonreflecting films, 244, 248

Normal magnification, 348

Nörremberg doubler, 116

Nu (ν), reciprocal dispersion, 319

Numerical aperture, 337, 342, 535

Nyquist formula, 471

Object: real and virtual, 274

Object and image space, 327

Obliquity factor, 181

Optical activity, 152; specific rotatory power of quartz, 153; in turpentine liquid and vapor, 153

Optical axis, or center line, $\mathcal{C}$, 127; direction in calcite, 129; velocity in quartz along, 150

Optical constants of metals, 71

Optical defects, 549

Optical efficiency of eye, 488

Optical glass, 319, 323 *et seq.*

Optical path, 225, 333

Optical polishing, Twyman, 294

Optical power: of surface, 287; of system of lenses, 310; of thick lens, 309; of thin lens, 308

Optical surfaces, 278; figuring of, 293; figuring of by thermal evaporation, 677; geometry of, 280; proving of, 229

Optical systems: Bouwers' catadioptric, 318, 370; Czerny-Turner, 371; Ebert, 371; Fastie, 371; field of view of, 344; Maksu-

tov, 371; Maxwell's perfect, 326; power of combination of lenses, 310; telecentric, 348
Order of interference, 163
Ordinary ray, 128
Oscillator: harmonic, equations of motion for, 64; in light field, 63
Oscillator knife-edge test: Elby's, 362
Oscillators of optical medium, 62, 127
Overcoats evaporated on glass, 248 *et seq.*

Parabolic mirror, 291, 357
Parallax, 349; stellar, 81, 82
Paraxial rays, 276
Particle nature of light, 436
Path: difference, 225–226; optical, 225, 333
Pellin-Broca prism, 305
Penetration on total reflection, 124
Period of wave motion, 19
Periscope, 349
Permittivity method, c by, 457, 465
Perot-Fabry: absorption cell, 346; interferometer, 231; resolving power, 260
Perspective, 344
Pfund: selenium polarizer, 123
Phase change: on internal reflection, 118; on external reflection, 25
Phase contrast, 407; in microscopy, 379, 530
Phase plate, 531, 532
Phase retardation plate, 141
Phase reversal zone plate, 189
Phase velocity, 96; method for determination of c, 459
Phases: random, 31
Photoconductive cells: minimum detectable power of, 495; time constant of, 494–495
Photoconductivity, 492
Photoelasticity, 154
Photoelectric effect, 437
Photoelectric surfaces: composite, 497
Photoemissive surfaces: quantum efficiency of, 497
Photographic plate, 491
Photographic sensitizing, 492
Photography: quantum efficiency of, 490
Photometric magnitudes, 54; interrelation of, 57, 59, 60
Photometry: heterochromatic, 56
Photomultiplier tubes: refrigeration of, 498
Photons, 439
Photopic (high light level) vision, 487
Photorefractometer, 574
Phototube, 496
Photovisual triplet: Cooke, 323
Pile-of-plates polarizer, 122
Pinhole light source, 296
Planck: constant, 438; hypothesis, 469; radiation law, 76–79
Plane and angle of incidence, 50, 679–680

Planes: conjugate points and, 326; principal points and, 309
Plate: focal length of zone, 669; geometrical optics of plane parallel glass, 301; phase, 531–532; quarter wave, 142, 385, 386; retardation phase, 141; reversal zone phase, 189; tint, 143, 154; tipping, for measuring small deflections, 301; zone, 188
Points: conjugate, and planes, 326; principal, and planes, 309
Poisson's bright spot, 186
Polariscope: Savart, 401
Polarization: for beam splitting, 384; detector, calcite rhomb as (Malus), 139; electrical, and index of refraction, 64; by interference filter, 386; of microwaves, 522; by reflection, 109, 110; by scattering, 108; of sky light, 107; state of, 107
Polarized light: analysis of, 144; analyzer for, 105, 112, 146–148, 665; azimuth of, 146; circular and elliptical, 119–120; right- and left-circular, 120, 152
Polarized wave motions: superposition of, 27
Polarized waves in string, 16
Polarizer: for infrared radiation, 123, 124; parallel plate, 665; pile-of-plates, 122
Polarizing light: Wollaston prism for, 139, 390, 406
Polaroid, 105
Polaroid glasses, 120
Polish, 278
Polishing: optical (Twyman), 294
Porter, 425
Positive and negative crystals, 149
Power: of combination of lenses, 310; minimum detectable, of photoconductive cells, 495; resolving, *see* Resolving power; of surface, 287; of thick lens, 309; of thin lens, 308
Pressure: radiation, 75
Principal points and planes, 309
Prism: Amici, 305; angular dispersion of, 304; Broca-Pellin, 305; Cornu, 151, 666; deviation of, 304; direct-vision, 678; energy-limited resolution of, 674; Fresnel's, for demonstrating theory of circular light in crystals, 150; Kösters' double image, 393 *et seq.*; Nicol, 139; resolving power of, 214; Rochon, 139, 406; Thallon, 305; transmission of, 662; Wadsworth arrangement, 305; Wernicke, 305; Wollaston, 139, 390, 406; Zenger, 305
Probability concept, 448
Pulfrich refractometer, 664
Pulse, Gaussian, and spectrum, 631
Pupils, 341–342

Quantum efficiency: of photoemissive surfaces, 497; of photography, 490

Quantum mechanics, 442
Quarter wave plate, 142, 385, 386
Quartz: dextrorotatory, 150; fibers, 562; levorotatory, 150; plate interferometer, 420; refractive indices, 149, 150; specific rotation of, 153, 665: velocity in, along optical axis, 150

Radiation: blackbody, 75–76; infrared, polarizers for, 123, 124; laws, intercompared, 78; Planck, Rayleigh-Jeans, Stefan-Boltzmann, Wien laws, 76–79, 476; pressure, 75
Random phases, 31
Ray: chief, 349; extraordinary, 128; graphical ray tracing, 546; ordinary, 128; plotters, 546; tracing, 277, 545, 550
Rayleigh: comments on blazed gratings, 218; dimensional analysis, 104; failure of superposition in sound, 35; interferometer, 375, 377; $\frac{\lambda}{4}$ criterion, proof for, 677; resolving power, 211; superposition with random phases, 33
Rayleigh-Jeans: radiation law, 77–79
Rays: fan of, 303; light, Drude's remarks on, 276; paraxial, 276; skew, 537, 551
Rectangular aperture: diffraction by, 190, 201

Reflectance formula for $\frac{\lambda}{4}$ films, 671

Reflection: coefficients from boundary conditions, Fresnel, 50; double, in infrared, 135; echelon, Williams, 266; external, phase change on, 25; frustrated total, 124, 573; grating, 216, 608 *et seq.*; internal, 114, 118; of metal films, 72; metallic, Hagens and Rubens formula, 71; microwaves, 517, 519; penetration on total, 124; polarization by, 109, 110; total, 117
Reflection-eliminating films, 248
Reflection-enhancing films, 245
Reflection-reducing films, 248
Reflections: multiple, 242 *et seq.*
Reflectors: conical, 556
Refracting surface: equivalent (lens), 331; single, spherical aberration formula for, 681
Refraction: calcite, 130, 135; theory, third-order, 352–353, 540
Refraction index, 65; for artificial dielectrics, 521; complex, 65; of fibers, 562; of gases, 66
Refractions: double, 128 *et seq.*; at single surface, 282
Refractive indices: quartz, 149, 150
Relativity, 436
Replica gratings, 216
Residual-ray crystals, 592–593
Resolving limit, 262

Resolving power: and beam retardation, 675; criterion, Sparrow, 669; diffraction grating, 215; diffraction-limited, 210; double monochromator (Sinton and Davis), 220; energy-limited, 270, 674; Fabry-Perot, 260; with fibers, 571; with Gaussian spectrum lines, 670; highest, Meissner, 263; prism, 214; Rayleigh's, 211; spectroscopic, 212; telescope, 340
Retinene, or visual yellow, 489
Reversibility: principle of, 114, 221
Rhodopsin, or visual purple, 489
Right- and left-circular polarized light, 120, 152
Right- and left-handed crystals, 149
Ripples, 21
Rochon prism, 139, 406
Rods, 487
Roemer: velocity of light by Jupiter's moons, 81
Rotary power of quartz, 665
Rough surfaces, 278
Rowland circle for gratings, 609
Rubens and Hagen: formula for metallic reflection, 71
Ruling of diffraction gratings, 216
Rupp: light-chopping experiment, 92

Saccharimetry, 146; angle uncertainty in, 666
Sagittal and meridional (*i.e.* tangential) focus, 366, 367
Savart, 401; plate, 391; polariscope, 401
Scattering, 100; beam splitting by, 383; coefficient, 104; by ether, liquid and vapor, 101; of light, Wood's experiments, 101–103; by mercury vapor, 103; polarization by, 108; single and multiple, 108; by sky, 104; by sodium vapor, 102; Thomson's formula, 663
Schmidt camera, 369
Scotopic (dark-adapted) vision, 487
Secondary spectrum of lens, 322
Seeing: astronomical, 577
Seidel: third-order theory, 353, 540
Selenium polarizer: Pfund's, 123
Sensitizing: photographic, 492
Shape factor, 316
Shot noise, 471, 498
Shrödinger, 449
Shutter: electro-optical, 92, 156
Sign: conventions of, 52, 286, 288, 308
Silicon filters, 590
Simple harmonic motion, 11
Sine relationship (or condition), 329–330, 534; deduction of, 333–336
Single-slit diffraction, 29, 198, 203, 422
Sinton and Davis: double monochromator resolving power, 220

Sirius and its companion, 205
Skew rays, 537, 551
Sky light: blue color of, 104; polarization of, 107
Slit diffraction: Fraunhofer, 203; Fresnel, 198
Smith, 286, 309, 545, 546, 550, 551, 678
Sodium: D lines of, 82
Sodium vapor: anomalous dispersion in, 98; phase velocity in, 96; scattering by, 102
Soleil compensator, 144
Solid angle, 58
Sparrow: resolving power criterion, 669
Spectral analysis with two-beam interference, 172
Spectral brightness, 54
Spectral lines in absorption and emission, 83; breadth of, 85 *et seq.*
Spectrograph: contrasted with spectrometer, spectroscope, and monochromator, 213
Spectrometer: contrasted with spectrograph, spectroscope, and monochromator, 213; interference, 259
Spectroscope: contrasted with spectrograph, spectrometer, and monochromator, 213
Spectroscopic resolving power, 212
Spectroscopy: high-order multiple-beam, 257
Spectrum: analysis, interference used for, 172; channeled, 225, 232, 234, 673; electromagnetic, from gamma rays to radio, 39; hydrogen line, 172; secondary, of lens, 322; spots, contrasted to lines, 206
Speed of light, *see* c
Sphere: aplanatic points of, 314
Spherical aberration, 314, 315, 355
Spherical aberration formula: for single refracting surface, 681; for spherical mirror, 681
Spiral: Cornu, 193
Spot diagrams, 538, 541
Stabilized retinal images, 487
Standing waves, 24
State of polarization, 107
Stefan-Boltzmann: radiation law, 78–79, 476
Stellar diameters, 177
Stellar interferometer, 173, 395
Stellar parallax, 81, 82
Stigmatic mounting of grating, 612 *et seq.*
Stokes: proof that $\mathbf{r}' = -\mathbf{r}$, 114
Stop, 341, 345
Straight edge: diffraction by, 197
Strain in glass: device for observing, 155
String: polarized waves in, 16
Superconducting bolometer, 482
Superposition: general procedures of, 28; of polarized wave motions, 27; principle,

373; with random phases, Rayleigh, 33; in sound, failure of, Rayleigh, 35; of two cosine functions, 23; of waves of different frequency, 34; of waves with random phase, 31
Surface: liquid, wave motion on, 21; power of, 287; rough, 278; single refracting, spherical aberration formula for, 681; single, refraction of, 282; unit, 309

T-number, 343
Telecentric optical system, 348
Telephoto lens, 339
Telescope, 336–338; Keplerian (or astronomical) and Galilean, compared, 680; mirrors, 292; resolving power, 340
Tellurium films, 594
Tessar lens, 539
Tests: knife-edge, 294; mirrors, 294, 364–365; oscillator knife-edge, Elby, 362
Test plate: Wight, 676
Testing: of gauge blocks, 395; of images, 353; of lens, 396–397; for pure coma and astigmatism, 399
Thallon prism, 305
Thermal evaporation figuring, 677
Thermal evaporation films, 408
Thermistor bolometer, 482
Thermoelectric alloys, Hutchins, 478
Thermopile detector, 475; Johnson noise, 478
Thick lens, 309
Thickness determination by channeled spectrum, 234
Thin lens, 308, 356
Third-order refraction theory, 352–353, 540
Thomson scattering formula, 663
Time constant: amplifiers, 505; bolometers, 482; detectors, 470; eye, 488; photoconductive cells, 494–495; thermopile, 478
Tint plate, 143, 154
Tolansky: measurement of terraced mica cleavages, 255
Total internal reflection of microwaves, 519
Total reflection, 117, 124
Tourmaline, 105
Transients, 34
Transmission: of crystals, 589; echelon, Michelson's, 265; microwaves, 509; of prism, 662; of silver, 72
Triplet: Cooke photovisual, 323
Turner: carpenter's rule method of considering periodic film structures, 252
Turner-Czerny optical arrangement, 371
Twyman: optical polishing, 294
Twyman and Green: interferometer, 375, 382

Uncertainty principle, 441

Uniaxial crystals, 149
Unit surfaces, 309

Vector addition of amplitudes, 25
Vector analysis, 633 *et seq.*
Velocity: group velocity, 92–93; group-velocity experiments, 453; of light in air and water, Foucault, 2, 10; of light waves in glass, 48; of longitudinal waves, 20; phase, in sodium vapor, 96; in quartz, along optical axis, 150; signal, 100; of sound waves, 21; of transverse waves, 20; *see also* c
Vignetting, 346
Visibility: of Fresnel fringes, 668; of fringes, 168; of interference bands, 168, 233; of interference bands for a Gaussian line, 667; theorem, Dyson, 379
Vision: averted, 488; photopic (or high level), 487; scotopic (dark-adapted), 487

Wadsworth grating mounting, 612
Wadsworth prism arrangement, 305
Water vapor absorption lines, 433
Water waves, 21
Wave: guides (cones), 556; mechanics, 449; nature of particles, 439; packet, 447; trainlets, Huygens, 158
Wave motion: constants of, 65; in a drumhead, 23; electromagnetic, history, 40; graphical summation for, 25; Hooke, early proponent of concept of light as, 1, 222; on liquid surface, 21; period of, 19; polarized, superposition of, 27; transverse, Mach's comment on, 178
Wave number, 262
Wave-front division, 161, 223, 374, 381

Wavelength, 19
Wavelengths: cadmium, meter expressed in, 241
Wavelets: spheroidal, in calcite (Huygens), 129
Waves: amplitude division, 223; of different frequency, superposition of, 34; electromagnetic, Hertz discovery of, 47; false back, 181; longitudinal, velocity of, 20; polarized, in string, 16; with random phase, superposition of, 31; sound, velocity of, 21; standing, 24; transverse, velocity of, 20; water, 21; *see also* Light waves
Wernicke prism, 305
White: absorption cell, 347
White-light fringes, 167, 171, 237
Wiedemann-Franz law, 477
Wien: radiation law, 76–78
Wiener's experiment, 51
Wight: test plate, 676
Williams: reflection echelon, 266
Wilson: cloud chamber, 441
Window, 345
Wollaston: polarizing prism, 139, 390, 406
Wood: experiments on scattering of light, 101–103; ultraviolet moon photographs, 72
Wratten: filters, 581–582

Young, 373, 490; double-slit interference experiment, 160; Gibson's thumbnail biography of, 164
Young and Lloyd: interference, colors in, 167

Zenger prism, 305
Zone plate, 188; focal length of, 669